# ESSENTIALS OF HEAT TRANSFER

This introductory textbook is designed to teach students the principles, materials, and applications they need to understand and analyze heat transfer problems that they are likely to encounter in practice. The emphasis is on modern practical problems, clearly evident in the numerous examples, which sets Professor Massoud Kaviany's work apart from the many books available today. Kaviany discusses heat transfer problems and the engineering analysis that motivate the fundamental principles and analytical methods used in problem solving in the search for innovative and optimal solutions. Those familiar with the previous version of this book will notice that this volume is a more manageable length and that the generic problem-solving engine has been replaced with MATLAB software. The materials removed from the book are available on the Web at www.cambridge.org/kaviany. A complete solutions manual for the numerous exercises is also available to qualified instructors.

Massoud Kaviany is Professor of Mechanical Engineering and Applied Physics at the University of Michigan. He is a Fellow of the American Society of Mechanical Engineers (ASME) and the winner of ASME's Heat Transfer Memorial Award (2006) and the prestigious James Harry Potter Gold Medal (2010). He has won the Teaching Award (1999 and 2000) and the Education Excellence Award (2003) from the College of Engineering at Michigan. He is the author of the books *Principles of Heat Transfer in Porous Media*, Second Edition (1999), *Principles of Convective Heat Transfer*, Second Edition (2001), *Principles of Heat Transfer* (2001), and *Heat Transfer Physics* (Cambridge University Press, 2008).

# Essentials of Heat Transfer

## PRINCIPLES, MATERIALS, AND APPLICATIONS

### Massoud Kaviany

*University of Michigan*

CAMBRIDGE
UNIVERSITY PRESS

CAMBRIDGE UNIVERSITY PRESS
Cambridge, New York, Melbourne, Madrid, Cape Town,
Singapore, São Paulo, Delhi, Tokyo, Mexico City

Cambridge University Press
32 Avenue of the Americas, New York, NY 10013-2473, USA

www.cambridge.org
Information on this title: www.cambridge.org/9781107124000

First published 2011

Portions of this work appeared in *Principles of Heat Transfer* by Massoud Kaviany
© 2002 John Wiley and Sons, Inc.

Printed in the United States of America

*A catalog record for this publication is available from the British Library.*

*Library of Congress Cataloging in Publication data*

Kaviany, M. (Massoud)
Essentials of heat transfer / Massoud Kaviany.
    p.   cm.
ISBN 978-1-107-01240-0 (hardback)
1. Heat–Transmission–Textbooks.    I. Title.
QC320.K38   2011
621.402′2–dc23            2011015904

ISBN 978-1-107-01240-0 Hardback

Additional resources for this publication at www.cambridge.org/kaviany

*If the teacher is wise, he does not bid you to enter*
*the house of his wisdom.*
*But leads you to the threshold of your own mind.*

*– K. Gibran*

To my daughters, Saara and Parisa

# Contents

* This section is found on the Web at www.cambridge.org/kaviany.

* This section is found on the Web at www.cambridge.org/kaviany.

* This section is found on the Web at www.cambridge.org/kaviany.

* This section is found on the Web at www.cambridge.org/kaviany.

* This section is found on the Web at www.cambridge.org/kaviany.

* This section is found on the Web at www.cambridge.org/kaviany.

* This section is found on the Web at www.cambridge.org/kaviany.

\* This section is found on the Web at www.cambridge.org/kaviany.

* This section is found on the Web at www.cambridge.org/kaviany.

* This section is found on the Web at www.cambridge.org/kaviany.

# Preface

In the long run men hit only what they aim at. Therefore, though they should fail immediately, they had better aim at something high.

– H. D. Thoreau

Heat transfer is a result of the spatial variation of temperature within a medium, or within adjacent media, in which thermal energy may be stored, converted to or from other forms of energy and work, or exchanged with the surroundings. Heat transfer occurs in many natural and engineered systems. As an engineering discipline, heat transfer deals with the innovative use of the principles of thermal science in solving the relevant technological problems. This introductory textbook aims to provide undergraduate engineering students with the knowledge (principles, materials, and applications) they need to understand and analyze the heat transfer problems they are likely to encounter in practice. The approach of this book is to discuss heat transfer problems (in the search for innovative and optimal solutions) and engineering analyses, along with the introduction of the fundamentals and the analytical methods used in obtaining solutions. Although the treatment is basically analytical, empiricism is acknowledged because it helps in the study of more complex geometries, fluid flow conditions, and other complexities that are most suitably dealt with empirically.

A combination of descriptive and analytical discussions are used to enable students to understand and articulate a broad range of problems. This is achieved by building on a general foundation based on the thermal energy conservation equation (thermodynamics), thermal physics (microscopic and macroscopic thermal transport mechanisms and thermal energy conversion mechanisms), thermal chemistry (change in chemical and physical bonds due to temperature change, e.g., chemical reaction and phase change), mechanics (fluid and solid), and thermal engineering analysis (e.g., thermal nodes and circuit models). These allow the reader to use the textbook and the accompanying solver to arrive at some specific solutions. The reader is also assisted in finding the needed information in the more advanced or specialized cited references (e.g., specialized handbooks). The material is presented in a consistent manner; therefore, the instructor can use classroom time to empha-size some key concepts, to give other examples, and to solve example problems. By addressing engineering heat transfer problems, along with the fundamentals, the book will show students – by the time they have worked through it – how to analyze and solve many practical heat transfer problems.

Transfer of heat occurs by mechanisms of conduction, convection, and radiation. The transfer of heat may cause, or may be the result of, the energy storage or energy conversion processes. Examples of such processes are sensible or phase change heat storage (i.e., a physical- or chemical-bond energy and work done or energy converted, including mechanical and electromagnetic energy). In many heat transfer problems, more than one phase (solid, liquid, gas) is present, leading to single- or multiphase media. In multiphase media, depending on where the source of heat is located, these phases may or may not be in local thermal equilibrium (i.e., have the same local temperature). Because of vast variations in geometry and in the initial and bounding-surface conditions of the medium, no general solution to the energy equation is available and, therefore, specific cases must be considered separately. Here, around a central theme of the heat flux vector and its spatial and temporal variations caused by the volumetric storage and conversion of energy, the intra- and intermedium heat transfers are examined. The heat flux vector, along with the three mechanisms, can be tracked through the heat transfer media. This allows visualiza-tion of heat transfer with the aim of enhanced innovative ideas. This visualization is made progressively more complete by addressing the heat transfer mechanisms, the phases present, the volumetric energy storage and conversion, and the role of geometric, initial, and bounding surface conditions. Thermal circuit models are used and some heat-transfer-based devices are also presented throughout the text.

This book is divided into eight chapters. Chapter 1 gives a general introduction to the analysis of thermal systems, the heat flux vector (and the contributions from various mechanisms), and the integral-volume energy conservation equation. The concept of heat flux vector tracking is emphasized to help with the visualization of heat transfer. This chapter is intended as a descriptive guide and addresses the basic concepts and applications (compared to the following chapters, which are more quantitative). The energy equation is posed with the heat flux vector as the focus (i.e., playing the central role).

Chapter 2 discusses spatial temperature nonuniformity and the need for differential-, integral-, combined differential- and integral-, and the finite-small-length forms of the energy conservation equation, various forms of which are used in succeeding chapters. Energy and work conversion mechanisms and their relations and the bounding-surface thermal conditions are also discussed. After a general, broad, and unified introduction, various heat transfer mechanisms and their respec-tive roles in heat transfer problems are discussed in more detail in the chapters that follow.

Chapter 3 covers steady-state and transient conduction heat transfer. This allows for the examination of the spatial and temporal variation of temperature and heat flow rate within a heat transfer medium or through multimedia composites. The concepts of thermal nodes, thermal conduction resistance (and its relation with the thermal potential and heat flow rate), and thermal circuit models are used. Afterwards, thermoelectric cooling is discussed and analyzed. This chapter also considers multidimensional conduction, lumped-capacitance analysis, finite-small-volume analysis, solid–liquid phase change, and thermal expansion and thermal stress.

Chapter 4 considers radiation heat transfer on the bounding surfaces of the heat transfer medium and examines the concepts related to radiation heat transfer between opaque, diffuse, and gray surfaces. Surface radiative resistances are introduced and added to the circuit analysis of heat transfer, and steady-state and transient substrate heat transfers are addressed. The irradiation of nongray surfaces is also discussed. Radiation heat transfer from lasers, the sun, and flames is treated as energy conversion to allow nongray heat transfer.

Chapter $5^*$ covers simultaneous one-dimensional conduction and convection in unbounded fluid streams. This allows the consideration of simple fluid flow fields (steady, uniform, one-dimensional fluid flow). First, the case of no-energy conversion is considered. Then, examples of physical- and chemical-bond energy and electromagnetic energy conversions, used for cooling or heating of gaseous streams, are considered. These examples allow the demonstration of the simultaneous roles of axial conduction and convection. They also allow the analysis of simultaneous heat and mass transfer.

The surface-convection heat transfer is covered in Chapters 6 and 7. The fluid flow (with nonuniform velocity fields) and heat transfer within the fluid and at the bounding solid (designations of semi-bounded and bounded fluid streams are used to focus on the flow and the thermal behavior of the fluid) are closely examined.

Chapter 6 treats semi-bounded fluid streams and introduces the concepts of viscous and thermal boundary layers and of local and average surface-convection resistance. Fluid-flow characteristics, such as flow transitions, and the solid-surface characteristics, such as geometric effects, are discussed. The liquid-gas phase change is also discussed.

Chapter 7 covers bounded fluid streams; it discusses velocity-area averaged fluid temperature and average convection resistance. This average fluid temperature changes as heat is exchanged with the bounding surface. The heat exchange between two bounded streams separated by solid surfaces (i.e., indirect heat exchangers) is also analyzed. The average convection resistance for the two-stream exchange is introduced and added to circuit analysis.

Chapter $8^*$ discusses the selection of the heat transfer medium. It addresses the selection of heat transfer media and bounding surfaces and discusses the thermal engineering analysis involved. It uses the concepts developed in Chapters 1 through 7 to address thermal problem solving (in some cases, with the aid of solvers).

The appendixes give several thermodynamic relations (Appendix $A^*$), derivation of the differential-volume thermal energy equation (Appendix $B^*$), and tables

---

$^*$ This section is found on the Web at www.cambridge.org/kaviany.

of some thermochemical and thermophysical properties (Appendix C\*). A nomenclature;\* glossary;\* answers to problems; key charts, figures, and tables; and subject index are also included.

The problem (and example) label format follows as *Chapter Number.Problem Number.Purpose.Solver*. Purpose is categorized as *Familiarity* (FAM), which involves the first steps in using the concepts and relations; *Fundamentals* (FUN), which gives further insights into principles and requires connections among two or more concepts and reactions; and *Design* (DES), which searches for an optimum engineering solution using available relations. The *Solver* option (S) indicates that the problem is intended for use with a solver such as MATLAB. Examples of problem labels are EXAMPLE 4.3.FUN and PROBLEM 5.6.DES.S.

Sections and text marked with asterisks (\*) are available online at www.cambridge.org/kaviany. They provide further readings for a more comprehensive presentation of the subject.

Example problems are solved throughout the book to show students how to use what they have learned in solving thermal engineering problems and how to gain further insight. End-of-chapter problems are prepared for homework assignments. A Solutions Manual that gives step-by-step solutions to these end-of-chapter problems is available to qualified instructors.

As outlined in the following Guide to Instructors and Students, parts of this book can be omitted for a semester-long course without loss of continuity. These sections are aimed at second-time readers and those attempting a more thorough reading.\*

It is hoped that, through both induction and deduction, the reader will acquire a fundamental understanding of heat transfer and the required tools for solving thermal engineering problems.

---

\*  This section is found on the Web at www.cambridge.org/kaviany.

# Acknowledgments

> What we must decide is perhaps how we are valuable, rather than how valuable we are.
>
> – F. S. Fitzgerald

I would like to thank all whose contributions to heat transfer have made this book possible. Word processing of the manuscript was done by Greg Aloe, Ashish Deshpande, Gi-Suk Hwang, Jedo Kim, Lance Kincaid, Gayton Liter, Chan-Woo Park, and Xiulin Ruan, and the illustrations were made by Paul Perkins – all students at the University of Michigan – who patiently endured many revisions. Joseph Klamo helped with the solver menu items. Many of my students in the research laboratory (Jae-Dong Chung, Luciana da Silva, Gayton Liter, Alan McGaughey, Minas Mezedur, Amir Oliveira, Chan-Woo Park, and Xiulin Ruan) and in my classrooms have greatly contributed to this project. Gayton Liter, Alan McGaughey, Amir Oliveira, and Xiulin Ruan were constant sources of support and ideas. Using these materials, they also assisted me in teaching the undergraduate heat transfer course. I am very thankful to all of them, without whom this book would not have been completed. I would also like to thank Peter Gordon of Cambridge University Press for his exceptional support, as well as the publishers and authors who have allowed me to use reproductions of their figures and tables in this book.

# Guide to Instructors and Students

La poesia
Non é di
Chi la scrive
Ma di
Chi la usa!
– Il Postino

A typical one-semester course syllabus is given in the Solutions Manual. However, at the discretion of the instructor, different parts of the text can be chosen to emphasize various concepts.

Chapter 1 is, in part, descriptive (as opposed to quantitative) and introduces many concepts. Therefore, this chapter should be read for a qualitative outcome and an overview of the scope of the book. This is needed for a general discussion of heat transfer analysis before specifics are introduced. The drawing of heat flux vector tracking is initially challenging. However, once learned, it facilitates visualization of heat transfer, and then the construction of the thermal circuit diagrams becomes rather easy.

Also, to get a broad and unified coverage of the subject matter, a very general treatment of spatial temperature nonuniformity and the appropriate form of the energy equation is given in Chapter 2. This chapter gives the building blocks and the results to be used in succeeding chapters. Some of the energy conversion mechanisms may be new to students; however, they are described in the text and are easy to follow. Application of Chapter 2 materials unfolds in the chapters that follow. For example, various energy conversion mechanisms are used along with heat transfer mechanisms in interesting and practical problems.

Chapter 3 is the longest chapter (Chapter 6, the second longest). The treatment of transient conduction heat transfer, with the distinction between the distributed (nonuniform temperature) and lumped (uniform temperature) treatments, requires special attention. This is because, although a general numerical treatment that applies to various cases (Section 3.7) is possible, the transient conduction closed-form solutions given in Chapter 3 are applicable only under special conditions. Those special conditions should be reviewed carefully.

In Chapter 4, the physics of volume and surface thermal radiation is given through a short and simple treatment. Surface and intersurface radiation resistances are introduced. The distinction between gray and nongray surfaces should be emphasized, so the limits and possibilities of the analysis of each of these surfaces become clear. The nongray prescribed irradiation is treated as energy conversion to avoid the misuse of the radiation circuit developed for gray opaque surfaces.

Chapter 5* can be omitted without loss of continuity. If time permits, Sections 5.1 and 5.2 introduce combined conduction convection and the Péclet number. Section 5.4 makes an interesting introduction to gaseous premixed combustion and the role of heat transfer. This is a demanding but very rewarding topic because combustion is used in many heat transfer problems, such as energy conversion, process heat transfer, and climate-control applications.

Chapter 6 is the second-longest chapter, but it allows a comparative study of surface-convection heat transfer (for semi-bounded moving fluids) for various flow arrangements with or without phase change. The surface-convection resistances are progressively compared in an attempt to allow design selection. Boiling and condensation are complex topics and, here, enough is given for a basic understanding; therefore, discussions are kept simple.

Chapter 7 treats fluid flow through tubes, as well as other continuous and discontinuous solid surfaces. Here, the concept of thermal resistance is extended to include single- and two-stream heat transfer. This unifies the concept of average-convection resistance and allows its inclusion in the thermal circuit analysis.

Chapter 8* is a review of and showcase for interesting and innovative thermal engineering examples. Both problem formulation and solution are discussed, and the use of software (e.g., MATLAB) is demonstrated. One or two examples can be chosen for in-depth discussions.

Sections and text marked with "*" are available online at www.cambridge.org/kaviany and can be skipped for a semester-long course without negatively affecting the printed book.

Homework is an essential part of the course. Weekly homework should be given and graded. During weekly problem-solving sessions, upcoming homework problems should be discussed and student questions answered. Homework problems are generally not very similar to the example problems, thus allowing the students to extend the applications of the materials they have learned. Problems use clear application of the concepts learned in the chapter. The problems related to Chapters 1 and 2 are more conceptual and fundamental; problems in Chapters 3 to 7 are more familiarity based; and in Chapter 8 (and infrequently in Chapters 3 to 7), design-type problems are given.

* This section is found on the Web at www.cambridge.org/kaviany.

As a general guide, four significant figures are kept in the arithmetic operations and numerical solutions; however, they are not necessarily indicative of the actual uncertainties. For example, some thermophysical properties are available in only two significant figures.

Based on my experience, at the end of the course, students will have learned some of the specifics of thermal engineering design and analysis and will also be able to articulate the innovative use of heat transfer in engineering problems.

Please send corrections and suggestions to kaviany@umich.edu.

# Introduction and Preliminaries

After countless metamorphoses (i.e., conversions) all energy, unless it is
stored (i.e., converted to other than thermal energy), eventually turns into
heat and adds its share to the thermal budget.

– H.C. von Baeyer

In this introductory chapter, we discuss some of the reasons for the study of heat
transfer (applications and history), introduce the units used in heat transfer analysis,
and give definitions for the thermal systems. Then we discuss the heat flux vector
**q**, the heat transfer medium, the equation of conservation of energy (with a refor-
mulation that places **q** as the central focus), and the equations for conservation of
mass, species, and other conserved quantities. Finally we discuss the scope of the
book, i.e., an outline of the principles of heat transfer, and give a description of the
following chapters and their relations. Chart 1.1 gives the outline for this chapter.
This introductory chapter is partly descriptive (as compared to quantitative) in order
to depict the broad scope of heat transfer applications and analyses.

## 1.1 Applications and History

Heat transfer is the transport of thermal energy driven by thermal nonequilibrium
within a medium or among neighboring media. As an academic discipline, it is part of
the more general area of thermal science and engineering. In a broad sense, thermal
science and engineering deals with a combination of thermal science, mechanics,
and thermal engineering analysis and design. This is depicted in Chart 1.2. In turn,
thermal science includes thermal physics, thermal chemistry, and thermal biology.
Thermal physics encompasses the thermodynamics (interplay between energy and
work) and the physics of thermal energy transport and thermal energy conversion
mechanisms and properties (i.e., physics of heat transfer). The thermal energy transport mechanism is by fluid particles, solid crystal quantized lattice vibration (referred
to as phonon transport), mobile electrons (and ions), and emission and absorption
of photons. The physics of thermal energy conversion covers a large number of
phenomena. In an electromagnetic thermal energy conversion, such as microwave
heating, the molecules absorb the electromagnetic energy and convert it into internal

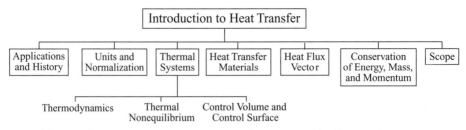

Chart 1.1. A summary of the topics presented in Chapter 1.

energy storage. The storage of thermal energy is by molecular translation, vibration, and rotation, and by change in the electronic or intermolecular bond energy. Thermal chemistry addresses the changes in the chemical and physical bonds caused by, or affecting, a change in the temperature (such as the exothermic, self-sustained chemical reaction of combustion). Thermal biology deals with the temperature control and thermal behavior of biological systems (e.g., plants, animals, or their organs) and subsystems (e.g., cells). In a broad coverage, nearly all aspects of thermal science and engineering are encountered in the heat transfer analyses.

The mechanics addresses force-displacement-motion and includes thermal strain and stress in solids and thermobuoyant motion in fluids.

The thermal engineering analysis and design includes representation of heat transfer on the energy conservation by use of thermal circuit models, use of analytical and numerical methods in solving the energy conservation equation, and optimization.

In this section, a further description of this field, along with its applications and history, is given.

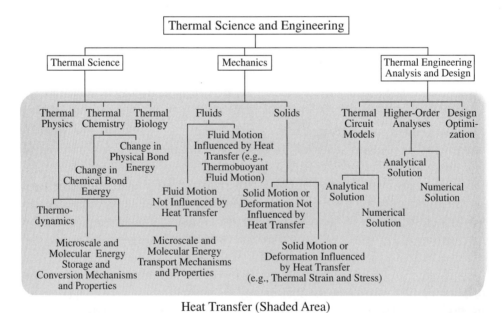

Heat Transfer (Shaded Area)

Chart 1.2. A general definition of thermal science and engineering and the disciplines encountered in the heat transfer analyses.

### 1.1.1 Heat Transfer

#### (A) Thermal Energy Transport, Storage, and Conversion

Heat transfer is the study of thermal energy transport within a medium or among neighboring media by molecular interaction (conduction), fluid motion (convection), and electromagnetic waves (radiation), resulting from a spatial variation in temperature. This variation in temperature is governed by the principle of energy conservation, which when applied to a control volume or a control mass (i.e., a control system), states that the sum of the flow of energy and heat across the system, the work done on the system, and the energy stored and converted within the system, is zero.

#### (B) Heat Transfer Medium: Single- and Multiphase Systems

A heat transfer medium may be a single-phase medium such as gas, liquid, or solid. The medium may be multiphase, such as a brick which has about 20 percent volume void space which is filled by air. So the brick is a solid-gas, two-phase medium.

Another two-phase medium is the water-crushed ice mixture where under gravity, the suspended-ice particles (i.e., the solid phase) move until they reach a mechanical equilibrium. Note that in this water-ice system, a pure-substance, liquid-solid phase change (a change in the physical bond) can occur.

When heat flows into a medium, it may cause (at elevated temperatures) a chemical reaction, such as when a polymeric (e.g., polyurethane) foam, used for insulation, is heated to above 200°C. At this temperature, this heat transfer medium is not chemically inert and in this case a thermochemical degradation (manifested by charring) occurs.

Another example of a multiphase system is the heat exchanger, in which heat transfer within and across the boundaries of two or more media occurs simultaneously (i.e., thermal nonequilibrium exists among them).

#### (C) Thermal Equilibrium Versus Nonequilibrium

In the water-ice example, consider that before adding the ice to water, the water is in a cup and in thermal equilibrium with the ambient air (room temperature, generally assumed 20°C). Then after mixing, heat transfer occurs between the water and ice because of the difference in their temperature. After some elapsed time, exchange of heat, and some melting, the heat transfer between the water and ice becomes less significant than that between the combined cup-water-ice system and the ambient air. Here we may assume that after the first regime, where the water-ice thermal nonequilibrium was dominant, a second regime arrives, where we may assume a thermal equilibrium within the cup-water-ice system and a thermal nonequilibrium between these and the ambient air.

The length scale over which the thermal nonequilibrium is considered can be as small as that between electrons and their lattice nucleus, as is the case in very rapid laser heating of surfaces. Such small-scale thermal nonequilibria are generally transient and short lived. The length scale can also be large, such as in radiation heating of the earth by the sun. These large-scale thermal nonequilibria are generally considered steady state.

As part of the engineering analysis, depending on the regime of interest, justifiable assumptions and approximations need to be made in order to reduce the problem to a level which can be solved with a reasonable effort. In heat transfer analysis, establishment of the dominant thermal nonequilibria is the primary step.

### (D)  *Theory, Empiricism, Semi-Empiricism, and Modeling*

The analysis of heat transfer is based on the principles of thermodynamics (for describing the physical and chemical states and the conservation of energy), the physics of heat conduction, fluid motion (for convection), and electromagnetic fields (for radiation), and the physics of thermal energy conversion. The mathematical analyses of such phenomena are established and continue to be developed. However, for simplicity and ease of use, semi-empirical or empirical relations are also used. This blend of fundamentals and semi-empiricism allows for the engineering analysis of some very complex and yet often encountered systems. Thermal circuit models allow for the reduction of problems involving multiple media and multiple heat transfer mechanisms to readily interpretable and solvable relations.

## 1.1.2  Applications

Heat transfer occurs in natural and engineered systems and over a very large range of temperature $T$, length $L$, time $t$, and mass $m$, scales. These scales and systems are briefly discussed below.

### (A)  *Temperature, Length, Time, and Mass Scales*

In order to allow for a broad introduction to the range of phenomena and scales involved in heat transfer applications, Figure 1.1 gives examples of the temperature $T$ [Figure 1.1(a)], length $L$ [Figure 1.1(a)], time $t$ [Figure 1.1(b)], and mass $M$ [Figure 1.1(b)] scales that are encountered.

At the temperature of absolute zero, the entropy may become zero (as in a perfect crystalline structure), i.e., the highest structural order. The lowest temperature possible is $T = 0$ K (or $T = -273.15°$C), the absolute zero in Kelvin scale. Helium has the lowest boiling temperature ($T = 4.216$ K) among the elements and compounds. The absolute zero is not expected to be reached, but very low temperatures, of the order of $T = 10^{-3}$ K, are achieved by the dilution refrigeration technique [1,25]. Yet lower temperatures, of the order $T = 10^{-5}$ K, are achieved by the adiabatic demagnetization technique. In this method, a paramagnetic salt is first exposed to a magnetic field, thus causing a molecular order. When the magnetic field is removed, this causes a disorder and heat is absorbed (while the paramagnetic salt is at a very low temperature).

(a) Temperature and Length Scales

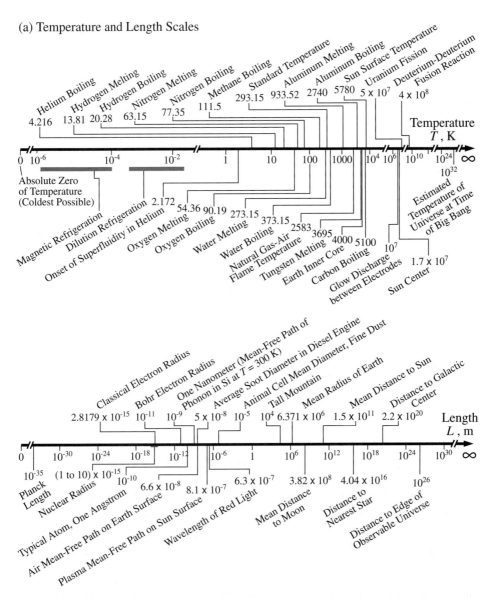

Figure 1.1. (a) Range of temperature and length scales encountered in heat transfer analysis.

For detection of very small temperature differences, the semiconductors are used. The electrical resistivity $\rho_e$ of semiconductors has a rather unusually large dependence on temperature, especially at low temperatures. Then temperature variations as small as one-millionth of one °C can be measured.

The upper bound for temperature is not known. One of the highest temperatures predicted is that based on a theory of formation of the universe. This will be discussed shortly, but the theory predicts a temperature of nearly $T = 10^{32}$ K, at the very beginning of the creation of the universe [8]. In a fusion reactor, temperatures of the order of $T = 10^8$ K are required for a continuous reaction. Thermal plasmas with

(b) Time and Mass Scales

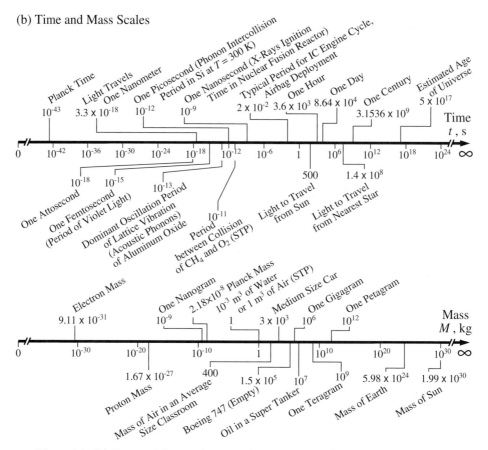

Figure 1.1. (b) Range of time and mass scales encountered in heat transfer analysis.

temperatures in the range of $T = 10^4$ to $10^6$ K are used for materials processing. The prefix cryo- is used in reference to very cold and pyro- in reference to very hot (as in cryogenics and pyrogenics referring to physical phenomena at very low and very high temperatures).

The smallest length with a physical significance is the Planck length, $L_P = 1.616 \times 10^{-35}$ m (obtained by combining the Newton constant of gravity $G_N$, Planck constant $h_P$, and speed of light in vacuum $c_o$), while the radius of the earth is $R = 6.371 \times 10^6$ m, and the mean distance between the earth and the sun is $L = 1.5 \times 10^{11}$ m. The mean-free path of the air (mostly nitrogen and oxygen) at standard temperature ($T = 20°C$) and pressure ($p = 1.013 \times 10^5$ Pa), i.e., atmospheric STP conditions, is about $\lambda_m = 66$ nm. The heat transfer medium may be as small or smaller than the components of a transistor in an integrated circuit (i.e., submicron) or as large as an airplane, if not larger.

The shortest elapsed time for a significant physical change is the Planck time, $t_P = 5.391 \times 10^{-44}$ s (obtained in a similar manner to the Planck length). The shortest time used in the analysis of low-energy interactions is the duration of collision time $\tau_c$, which is the average duration of intramolecular or intermolecular collision (this is of the order of femtoseconds). The next, and larger time scale, is the average time between collisions $\tau_m$ (or mean-free time). The establishment of molecular-level equilibrium requires a yet larger time scale $\tau_e$, called molecular relaxation time. The next larger time required for establishing equilibrium over large length scale is called the hydrodynamic time $\tau_h$. Then $\tau_c < \tau_m < \tau_e < \tau_h$. Pulsed lasers with the duration of pulse being of the order of $10^{-15}$ s (i.e., one femtosecond 1 fs) are used for surface thermal modification of solids. The average period for the cyclic motion of the internal-combustion, gasoline engine is about $t = 2 \times 10^{-3}$ s (20 ms).

The smallest mass (not considering sub-electron particles) is the mass of an electron and is $m_e = 9.109 \times 10^{-31}$ kg. The average classroom contains $M = 400$ kg of air. The mass of the earth is estimated as about $M = 10^{24}$ kg.

The particular application defines the range of the temperature, the length, the time, and the mass (system) of interest. Knowing these ranges would allow for the proper inclusion of the relevant thermal phenomena and the imposition of the various simplifications and approximations needed to make the analysis practical. Generally, scale filtration is used, where, depending on the relative scale of interest, the finer scales and the associated phenomena are averaged and represented at the largest scale of the filter. In many engineering applications, this filter scale is rather large. However, there are also some very short and very small scale thermal problems, such as the picosecond pulsed lasers used in the nanometer-structure manufacturing of integrated circuits and devices. There, the thermal nonequilibrium between the electrons and the inert and/or ionized molecules is important and influences the transient heat transfer and phase change.

The various natural occurrences and engineering applications of heat transfer are summarized in Chart 1.3(a). The major division is natural versus engineered systems.

### (B) Natural Systems

Chart 1.3(a) lists some examples of geological and biological heat transfer. In biological application, in addition to the normal mammalian temperatures, higher and lower temperatures occur or are imposed. Examples are heat therapy, cryo-preservation, and cryo-surgery [17]. Examples of the thermal aspects of natural systems are the universe, the earth, and the human body.

### (C) Engineered Systems

Examples of engineered thermal systems occur in applications ranging across electronics, energy, environment, manufacturing, processing, transportation, sensing, and others. These are listed in Chart 1.3(a).

As an example, Chart 1.3(b) shows the thermal aspects of manufacturing. Manufacturing is making materials into products suitable for use [20]. As expected, this processing of materials would involve heat transfer. In shaping of materials, reduction of mass is achieved by, for example, sintering materials by heating, or the volume is reduced by sintering-compacting while heating. Joining of similar or dissimilar materials, such as welding, soldering, brazing, or thermal fusing, involves

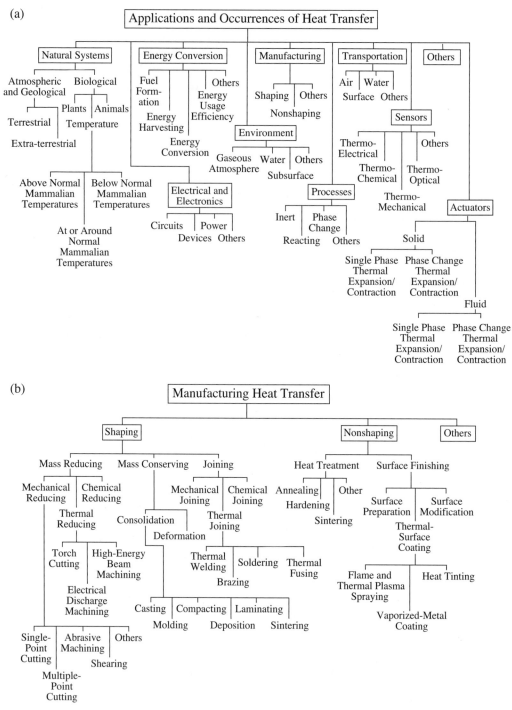

Chart 1.3. (a) Various applications of heat transfer in natural and engineered systems. (b) Various aspects of manufacturing heat transfer.

(c)

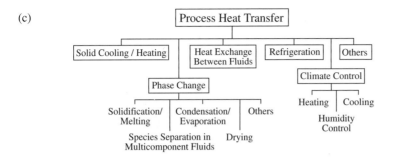

(d)

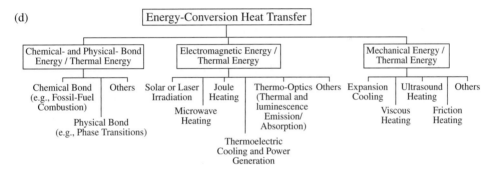

(e)

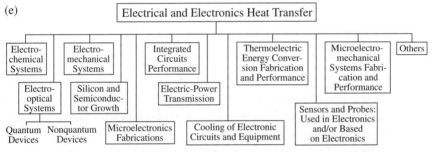

Chart 1.3. (c) Various aspects of process heat transfer. (d) Various aspects of energy-conversion heat transfer. (e) Various aspects of electrical and electronic heat transfer.

heat transfer. In nonshaping manufacturing processes, heat treatment of the metals and ceramics (e.g., glass) and surface coating by deposition of semi-molten particles, can be mentioned. Mechanical energy is converted to thermal energy in machining and grinding (both are among shaping of materials by mass reduction).

As another example of engineering applications, consider the process heat transfer. Chart 1.3(c) gives some aspects of process heat transfer [11]. Process heat transfer is the controlled addition/removal of the heat as a step in a process. In some processes a heat transfer device performs this task (as in heat exchangers, dryers, distillators, molds, etc.). In others, heat transfer is a supplement to another process (as in heat applied for melting in crystal growth, heating of organic-liquid transfer lines for reduction of the liquid viscosity, or in electrical preheating of the automobile catalytic converters for enhancement in performance during the engine start up). It can also be the heating/cooling of solids as in food processing. The heat transfer processes requiring phase change, including the drying of wet porous solids (such as clothes), consume a significant portion of the energy used in the process heat transfer. Another example is the distillation of organic and nonorganic liquid mixtures for purification purposes. Climate control (mobile or immobile indoor climates) is another example of process heat transfer.

Currently, 40% of all energy consumed is from oil, 20% from natural gas, 26% from coal, 6% from nuclear reactions, and 8% from renewable sources. The major areas of consumption of energy are 18% in space heating, 16% in process steam, 14% in automotive, 11% in direct heating, 8% in electric motors, 6% in other transportation, 5% in lighting, 5% in traveling, 4% in water heating, 4% in feedstock, 3% in air conditioning, 2% in refrigeration, 1% in cooking, and 3% others. Energy conversions may involve heat transfer with the thermal energy as the final (or initial) form of energy or as an intermediate form of energy. Direct-energy conversion heat transfer is where energy from chemical- and physical-bond energy (such as fossil-fuel and phase-transition energy), electromagnetic energy (such as solar and laser irradiation and microwave), and mechanical work (such as solid-surface friction work or gaseous expansion/compression pressure work) are converted to (or from) thermal energy [10].

Chart 1.3(d) gives examples of energy-conversion heat transfer. (This topic will be discussed in detail in Section 2.3). In the case of electromagnetic/thermal with indirect energy conversion, electrical power is generated when heat is added to a junction of a semiconductor and a metal (a direct energy conversion). Some of these examples are irreversible (e.g., Joule, microwave, and ultrasound heating) and some are reversible (e.g., phase-change heat storage, and thermoelectric energy conversion). An example of indirect energy conversion would be conversion of kinetic energy of a moving automobile into heat by the intermediate step of brake contact friction. Another example is the electrical power generation involving shaftwork (turbines) as the intermediate form of energy.

Yet another application is electrical and electronics heat transfer. This is shown in Chart 1.3(e). The applications include removal of Joule heating (energy conversion) or heat transfer used in the microelectronic fabrication processes. In some miniaturized, integrated sensors and devices, heat transfer controls the sensor performance or the device actuation. In automobile electrical generators, heat generated by Joule heating must be removed from the induction coils for improved performance (the electrical resistance of the electrically conducting copper wires increases with temperature). In thermal sensor applications, thermal sensing may be used to detect chemicals, determine radiation intensity, etc. [14]. As an example in hot-wire anemometry (discussed in Section 6.8), the rate of heat transfer from a heated wire submerged in a moving fluid is used to measure the velocity of the fluid near the wire.

Throughout the text, in discussions, examples, and end of chapter problems, we will consider various applications of heat transfer in natural and engineered systems. As is expected, in these applications the energy conversion (from and to thermal energy) is an important part of the problems. Therefore, we will address quantitative analysis of the energy conversion, to and from thermal energy, by chemical- and physical-bond electromagnetic, and mechanical energy.

Table 1.2. *Primary and derived units and symbols*

| Primary quantity | Unit |
|---|---|
| $L$, length | m, meter |
| $t$, time | s, second |
| $m$, mass | kg, kilogram |
| $T$, temperature | °C, or K, degree celsius, or kelvin |
| $J_e$, electric current | A, ampere |

| Derived quantity | Unit |
|---|---|
| $c_p$, specific heat capacity | J/kg-K or J/kg-°C |
| $e_c$, electric charge | C, coulomb or A-s |
| $e_e$, electric field intensity | V/m or N/C |
| $E$, energy | J = N-m/s, joule |
| $f$, force per unit volume | N/m$^3$ |
| $F$, force | N = kg-m/s$^2$, newton |
| $j_e$, current density | A/m$^2$ or C/m$^2$-s |
| $k$, thermal conductivity | W/m-K or W/m-°C |
| $p$, pressure | Pa = N/m$^2$, pascal |
| $q$, heat flux | W/m$^2$ |
| $Q$, heat flow rate | W = J/s, watt |
| $R_t$, thermal resistance | °C/W or K/W |
| $u$, velocity | m/s |
| $\alpha$, thermal diffusivity | m$^2$/s |
| $\theta$, plane angle | rad, radian (m/m) |
| $\mu$, viscosity (dynamic) | Pa-s |
| $\rho$, density | kg/m$^3$ |
| $\varphi$, electric potential | V, volt or W/A or J/C |
| $\omega$, solid angle | sr, steradian (m$^2$/m$^2$) |

### 1.1.3 History, Frontiers, and Integration*

This section is found on the Web site www.cambridge.org/kaviany. This section provides a historical perspective on heat transfer as part of the thermal science and engineering and its developed by advances made in thermal science and the integration of these advances into the thermal engineering practice.

There are also four examples, namely Example 1.1 on temperature (cool down) of the universe due to expansion, Example 1.2 on the surface temperature of the Earth and the increase in atmospheric $CO_2$ and global warming, Example 1.3 on human body heat transfer, and Example 1.4 on internal combustion engine heat transfer. The examples introduce the various mechanisms of heat transfer as well as some energy conversions to and from the thermal energy.

## 1.2 Units and Normalization (Scaling)

### 1.2.1 Units

The SI units (Système International d'Unités) are used in this text. The primary quantities that have assigned units are listed in Table 1.2, along with the quantities

Table 1.3. *Unit prefixes*[1]

| Factor | Prefix | Symbol | Factor | Prefix | Symbol |
|--------|--------|--------|--------|--------|--------|
| $10^{24}$ | yotta | Y | $10^{-24}$ | yocto | y |
| $10^{21}$ | zetta | Z | $10^{-21}$ | zepto | z |
| $10^{18}$ | exa | E | $10^{-18}$ | atto | a |
| $10^{15}$ | peta | P | $10^{-15}$ | femto | f |
| $10^{12}$ | tera | T | $10^{-12}$ | pico | p |
| $10^{9}$ | giga | G | $10^{-9}$ | nano | n |
| $10^{6}$ | mega | M | $10^{-6}$ | micro | $\mu$ |
| $10^{3}$ | kilo | k | $10^{-3}$ | milli | m |
| $10^{2}$ | hecto | h | $10^{-2}$ | centi | c |
| $10^{1}$ | deka | da | $10^{-1}$ | deci | d |

[1] One angstrom is 1 Å $= 10^{-10}$ m.

with units derived from these primary units. The symbols used for these quantities are also listed. Some of these quantities such as length, electric current, velocity, force, and heat flux can be assigned directions, i.e., they can be presented as vectors. Others, such as time, mass, density, and temperature, which cannot be presented as vectors are called scalars.

There are prefixes used to designate magnitudes smaller or larger than the units. These are given in Table 1.3. For example, a micrometer $= 1$ $\mu$m $= 10^{-6}$ m and a femtosecond $= 1$ fs $= 10^{-15}$ s.

The Kelvin temperature scale K is a primary scale and its zero value designates the absolute zero for temperature. The Celsius (or centigrade) temperature scale °C is related to the Kelvin scale K through

$$T(\mathrm{K}) = T(\mathrm{°C}) + 273.15 \quad \text{relation between temperature scales K and °C.} \quad (1.1)$$

The single point used to scale the Kelvin scale is the triple point of water, which is 273.16 K or 0.01°C. In the English units, the Fahrenheit °F and Rankine R temperature scales are related to the Celsius temperature scale through

$$T(\mathrm{°F}) = 1.8 T(\mathrm{°C}) + 32 \qquad (1.2)$$

$$T(\mathrm{R}) = T(\mathrm{°F}) + 459.67. \qquad (1.3)$$

The standard atmosphere temperature is $T_o = 15$°C (288.15 K), the pressure is $p_o = 101.325$ kPa, the density is $\rho_o = 1.225$ kg/m$^3$, and the mean molecular weight is $M_o = 28.946$ kg/kmole (referred to as atmospheric STP conditions). For thermodynamic and chemical equilibrium calculations (thermodynamic STP) the reference is taken as 25°C (298.15 K). Some conversion factors between the SI and the English units are given in Table C.1(a), in Appendix C. The universal constants used here are listed in Table C.1(b).

## 1.2.2 Normalization (Scaling)

Once the objective of the heat transfer analysis and the associated temperature, length, time, and mass scales of the system are determined, it is useful to normalize (e.g., scale) the temperature, length, time, mass, and other variables with respect to the appropriate scales of the system. Some ranges of possible scales encountered were discussed in Section 1.1.2.

As an example consider the primary quantity length. Once the length scale $L$ is determined, the space variable $x$ is normalized using $L$. The dimensionless quantities are designated in the text using the superscript *, i.e., $x^* = x/L$.

Other normalizations, some leading to dimensionless numbers (e.g., Reynolds number), are also made with quantities such as force, time, temperature, diffusivity, and thermal resistances. The approach taken here is to scale the energy conservation equation (and if needed the mass and momentum conservation equations), using the known parameters. The dimensionless numbers used in this text are placed in Table C.1(c), in Appendix C, and in the Nomenclature listed after Chapter 8.

Consider as another example, the derived quantity resistance to heat flow $R_t(°C/W)$, where $t$ indicates the mechanism of heat transfer (e.g., $t$ would be $k$ for conduction and $ku$ for surface convection). The normalization of the thermal resistances leads to a noticeable simplification of their representation. For example, we compare the resistance to conduction within a solid $R_{k,s}$ to the conduction outside this solid $R_{k,s\text{-}\infty}$, i.e., $N_k = R_{k,s}/R_{k,s\text{-}\infty}$, and when $N_k \leq 0.1$, we neglect the internal resistance. In another example with surface convection (Chapters 6 and 7), we use the ratio of the fluid conduction resistance $R_{k,f}$ to the surface-convection resistance $R_{ku}$, i.e., $N_{ku,f} = R_{k,f}/R_{ku}$, (i.e., normalized thermal resistance) to show the effect of the fluid motion on the surface heat transfer rate. For an effective surface convection, $N_{ku,f}$ (also called the Nusselt number Nu) should be large. These normalizations also aid in determining the dominant resistances to the heat flow.

## 1.3 Thermal Systems

A system can be defined as the part of the universe chosen for analysis. A thermal system exchanges thermal energy with its thermal surroundings (thermal surroundings being that part of the universe affected by this thermal energy exchange) when it is in thermal nonequilibrium with these surroundings. A thermal system contains matter or substance and this substance may change by transformations and by mass exchange with the surroundings. Throughout the text, we will give examples of many thermal systems, and we will make general reference to thermal systems in Chapter 8 after we examine all mechanisms of heat transfer.

To perform a thermal analysis of a system, we need to use thermodynamics, which allows for a quantitative description of the state of substance. This is done by defining the boundaries of the system, applying the conservation principles, and examining how the system participates in thermal energy exchange and conversion. These topics are addressed in the sections that follow.

### 1.3.1 Thermodynamic Properties

From thermodynamics, the state of a pure, compressible substance is specified by any two independent intensive thermodynamic properties. Generally, these two are selected among temperature $T$($°$C or K), pressure $p$ (Pa), and density $\rho$(kg/m$^3$) (or the specific volume, $v = 1/\rho$ ), as the independent intensive properties. Then the specific internal energy $e$(J/kg), specific enthalpy $h$(J/kg), etc., are determined from these properties. The temperature is defined as $T = (\partial e/\partial s)_v$ and this relation is called the Kelvin temperature. In Appendix A, some of the relations among the thermodynamic properties are reviewed.

The change in the specific internal energy $e$ is related to change in $T$ and $v$ as (the derivation is given in Appendix A)

$$de = c_v dT + \left( T \frac{\partial p}{\partial T} \mid_v - p \right) dv \quad \text{relation among changes in } e, T, \text{ and } v. \qquad (1.4)$$

Similarly the change in the specific enthalpy $h$ is given in terms of change in $T$ and $p$ as

$$dh = c_p dT + \left( v - T \frac{\partial v}{\partial T} \mid_p \right) dp \quad \text{relation among changes in } h, T, \text{ and } p. \qquad (1.5)$$

Note that the specific heat capacity at constant volume $c_v$ and at constant pressure $c_p$ are defined as

$$c_v \equiv \frac{\partial e}{\partial T} \mid_v, \quad c_p \equiv \frac{\partial h}{\partial T} \mid_p \quad \text{two specific heat capacities.} \qquad (1.6)$$

In heat transfer problems, $T$ is the primary variable. Then depending on the process, a constant $v$ or a constant $p$ process is assumed as the idealized process for comparison. In general, for heat transfer involving fluid flow, the constant pressure idealization is used, and therefore, $c_p$ is used more often. For solids, incompressibility is assumed, and therefore, the two heat capacities are equal.

### 1.3.2 Thermal Nonequilibrium

Heat transfer is a result of temperature nonuniformity. The presence of a spatial temperature nonuniformity is stated as the existence of thermal nonequilibrium. When we consider heat transfer between a control volume and its surroundings (i.e., the ambient), we are generally addressing the nonuniformity of temperature within the system (i.e., interior), as well as in the surroundings. The ratio of internal to external resistance is described by the dimensionless numbers called Biot, Bi, and radiation-conduction number $N_r$, depending on the heat transfer mechanisms involved.

The return to equilibrium is designated by the time constant or relaxation time $\tau(s)$. In some cases, we may choose to consider only the heat transfer, i.e., the thermal nonequilibrium, between the surroundings and the surface of the controlled mass or volume. In other cases we may be interested in thermal nonequilibrium between this surface and its interior. This is designated by the dimensionless penetration time or Fourier number Fo. The selection is made based on the dominant role that is played by the exterior or the interior temperature nonuniformity and the information available for the analysis.

At high temperatures where gases ionize, the lighter electrons and the heavier ions and neutrals can be at thermal nonequilibrium. Also fast transients, the electrons and lattice molecules are not in thermal equilibrium in metallic or semi-metallic solids (this also occurs in small regions near interfaces of different materials joined together). In all of these cases, different temperatures are assigned to electrons and other species and through relaxation time (or other models) heat transfer is allowed between them.

### 1.3.3 Control Volume and Control Surface

For thermal analysis, the boundary of the system as well as its contents (what is contained in its volume) need to be identified. These controlled (for the purpose of analysis) boundaries and volume are called the control surface $A(m^2)$ and control volume $V(m^3)$. We consider a control volume (instead of a control mass) to allow for flow of mass across the boundaries. The thermal system analysis will consist of describing/predicting (i) the thermal processes (changes) occurring within the control volume, and (ii) the interaction with surroundings occurring at the control surface.

#### (A)  Control Volume

Selection of the control volume is determined by the information sought. In a thermal system, temperature variation occurs across tiny spatial dimensions containing thousands of molecules (such that a continuum treatment of the medium of heat transfer can be made), and across yet smaller spatial dimensions containing a few molecules. Throughout the text, we cover the range of length from smallest spatial dimension that allows for a continuum description (called the differential or infinitesimal length, designated with $\Delta$), up to the largest dimension influencing the heat transfer aspects of the system (called the integral length). In Section 2.2 we describe the energy equation used for differential, integral, and combined differential-integral lengths.

---

EXAMPLE 1.5. FAM

---

An example of cascading control volume is shown in Figure Ex. 1.5. At the atomic level, designated by the crystalline solid lattice size, conversion of electromagnetic energy to thermal energy by Joule heating [collision of accelerated electrons with the lattice vibrations (phonons)] results in a rise in the temperature of the electric current carrying filament in a lamp. The conversion of thermal energy back to a different electromagnetic energy (i.e., thermal radiation), occurs simultaneously. The lamp is placed inside a cavity (i.e., recessed volume). The cavity and lamp are contained in an environment (surrounding). For the heat transfer from the cavity containing this lamp to its surrounding, we choose a control volume covering the cavity $V_c$. If we are interested in the heat transfer between the lamp and the cavity, we choose a control volume covering the lamp $V_l$. If the heat transfer from the tungsten filament is desired, a control volume is taken enclosing the filament $V_f$. If the temperature distribution within the tungsten filament is needed, a small control volume is chosen within the solid filament $\Delta V_s$. Also, if the temperature distribution and heat transfer in the air inside the cavity is desired, we choose a small control volume within the air $\Delta V_a$.

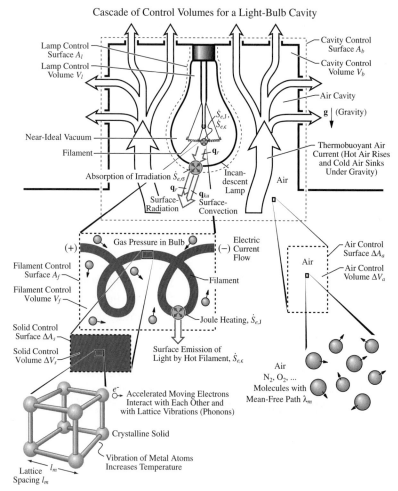

Figure Ex. 1.5. Cascade of control volumes (and the associated control surface) taken for the analysis of a recessed incandescent lamp and its surrounding air.

## COMMENT

Note that for the cavity, lamp, and filament, we have chosen integral volumes $V$, while we have chosen differential volumes $\Delta V$ inside the filament and in the air. This is because for the filament and the air we stated an interest in resolving the variation of temperature within these bodies. Note that the conduction electron interactions (collision with lattice atoms) lead to the Joule heating $\dot{S}_{e,\text{J}}$, and thermal radiation emission is shown by $\dot{S}_{e,\epsilon}$. The absorption of radiation is shown by $\dot{S}_{e,\sigma}$. We will discuss these in Section 2.3.2.

## (B) Control Surface

The control surface is placed on the boundary marking the separation of the system to be studied from its surroundings, i.e., the bounding surface. This boundary can be a phase boundary (e.g., solid-fluid or liquid-gas) where a discontinuity in the thermodynamic and transport properties occurs, or it can be within the same phase or medium. As the first example, consider when a droplet or solid particle is in thermal nonequilibrium with (i.e., has a temperature different from) the surrounding gas. The control surface for the analysis of heat transfer between the (liquid or solid) particle and the gas is taken at the phasic interface. This configuration is rendered in Figure 1.2.

As the second example, consider the control surface wrapped around the phasic interface that gives the proper information about the heat and mass transfer coupling between the particle (i.e., the content of the control volume) and the gas (i.e., the surrounding).

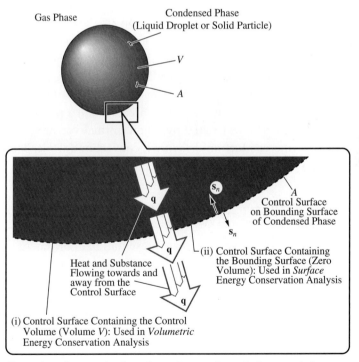

Figure 1.2. Control surface at the interface of a condensed phase and the surrounding gas, showing the two types of control surface. (i) A surface containing a nonzero volume, and (ii) a surface containing no volume.

The first application of the control surface is in a control-volume analysis of the conservation of energy, mass momentum, etc. Then the surface normal unit vector will be the outward normal shown. The second application is in analysis of the control-surface (or interface) energy, mass, momentum, etc., conservation or balances. For this application, the control surface covers both sides of the bounding surface (with no volume enclosed). These two cases are marked as (i) and (ii) in Figure 1.2. As the most general description, we start with very small (i.e., infinitesimal) control volumes within each of these phases. Then we integrate the energy equation over each phase to the bounding surface (e.g., boundary) of each phase (which becomes a common boundary at the phasic interface). The bounding surface thermal conditions will be discussed in Section 2.4. Whenever justifiable, we will use the integral-volume form of the energy equation and assume a uniform temperature within this volume. We use a differential volume when the temperature changes occurring within the volume are much larger than those occurring outside the volume.

## 1.4  Principal Energy Carriers and Heat Flux Vector

### 1.4.1  Macroscopic Heat Transfer Mechanisms

Heat flows in the direction of decreasing temperature. When a moving fluid is present, heat also flows along the direction of this fluid flow. Therefore, the heat flow has a direction and a magnitude, i.e., the heat flow is a vector quantity. The magnitude of the heat transfer rate per unit area and its direction are designated by the heat flux vector $\mathbf{q}(\mathrm{W/m^2})$. Here we use this time rate of heat transfer (per unit area) without using an overlaying dot (such as that used for the time rate of energy conversion $\dot{S}$). This is solely for convenience and has been adapted widely, i.e., $\dot{\mathbf{q}} \equiv \mathbf{q}$.

The heat flux vector gives the thermal energy flow per unit time (unit time around a given time $t$) and per unit area (unit area around a point in space $\mathbf{x}$ on a surface $A$). The surface $A$ is designated by a unit normal vector $\mathbf{s}_n$. The heat flux vector is the sum of the conduction $\mathbf{q}_k(\mathrm{W/m^2})$, convection (also called advection) $\mathbf{q}_u(\mathrm{W/m^2})$, and radiation $\mathbf{q}_r(\mathrm{W/m^2})$ heat flux vectors, i.e.,

$$\mathbf{q} \equiv \mathbf{q}_k + \mathbf{q}_u + \mathbf{q}_r \quad \text{heat flux vector with three mechanisms.} \qquad (1.7)$$

We note that (1.7) allows for a very compact presentation of the energy equation in Chapter 2. To do this, we have combined $\mathbf{q}$ and $\mathbf{q_r}$ in their conventional usage (with their direction dictated by the direction of decrease in temperature) while $\mathbf{q}_u$ has the direction of fluid velocity $\mathbf{u}$. Figure 1.3(a) gives a rendering of the heat flux vector showing the simultaneous flow of heat and fluid in a medium with a spatial variation in temperature. The iconic (i.e., graphical symbol) designation used in this text for $\mathbf{q}$ is also shown. The width of the heat flux vector indicates its relative magnitude.[†]

---

[†]  When applied to the total heat flow rate, this iconic presentation is called the Sankey diagram (arrows and their widths showing various heat flow rates in and out of a control volume).

(a) Heat Flux Vector **q** and Fluid Flow Velocity Vector **u**

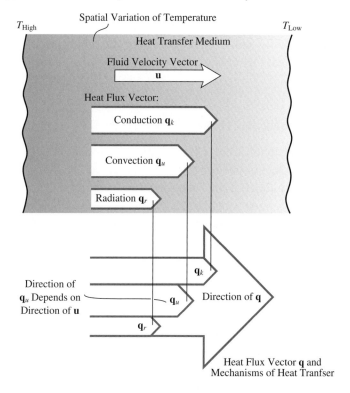

(b) Energy Storage and Energy Conversion $\dot{S}$

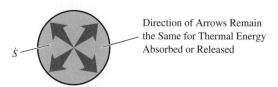

Figure 1.3. (a) Iconic (i.e., graphical) presentation of the heat flux vector and its three contributors: conduction, convection, and radiation. While the conduction and radiation heat flux vectors are in the direction of decreasing temperature, the convection heat flux vector depends on the direction of the velocity vector. The velocity vector is also shown. (b) Iconic presentation of energy storage and energy conversion.

For example, when heat is generated in a medium, this is shown by an increase in the width of the heat flux vector. Figure 1.3(b) gives the iconic presentation of the energy storage and energy conversion. Two crossing double-headed arrows are used. This has other physical interpretations and will be discussed in Section 2.1.1.

In addition to heat flux vectors describing heat transfer within a medium, i.e., $\mathbf{q}_k$, $\mathbf{q}_u$, and $\mathbf{q}_r$, there is a special designation for the fluid conduction at the interface of a fluid stream and a stationary, impermeable solid. This is shown as the surface-convection heat flux $\mathbf{q}_{ku}$, where on the solid surface, due to the fluid stream no-slip condition (due to solid impermeability and fluid viscosity), we have $\mathbf{u} = 0$.

However, this interface heat flux is influenced by the fluid velocity adjacent to the surface interface. Then the heat flowing through a solid by conduction $\mathbf{q}_k$ flows by surface convection $\mathbf{q}_{ku}$ through the interface of this solid and its adjacent moving fluid, and it then flows by convection $\mathbf{q}_u$ through the moving fluid. This solid-fluid surface-convection heat flux is especially designated as $\mathbf{q}_{ku}$ and will be discussed in Chapters 6 and 7.

### 1.4.2 Atomic-Level Heat Carriers (Heat Transfer Physics)

The carrying of heat by molecular contact among molecules in thermal nonequilibrium is called conduction heat transfer. For gases, conduction occurs during the thermal fluctuation motion (i.e., no net motion) and random collision among gaseous molecules (called molecular or fluid particle heat carriers). The average distance traveled between these molecular collisions is called the mean-free path $\lambda_f$.[†] For liquids and solids, conduction occurs between adjacent molecules by propagation of the quantized lattice-vibrational waves (called phonon heat carriers),[‡] or by the movement and collision of free electrons (called electron heat carriers). These are also described by mean-free paths $\lambda_p$ and $\lambda_e$. We will discuss these in Chapter 3. The heat carried by electromagnetic waves (called photon heat carriers) is referred to as the radiation heat transfer. This will be examined in Chapter 4. The heat carried by a net fluid motion (called net-motion fluid particle heat carriers), as compared to the thermal fluctuation motion of otherwise stationary fluids, is referred to as convection heat transfer. Different regimes in convection heat transfer (molecular, viscous-laminar, viscous-transitional, and viscous-turbulent fluid-flow regimes) will be discussed in Chapters 5 to 7. The concept of mean-free path can also apply to the net motion of fluid particles and we can use $\lambda_f$, $\lambda_{f,l}$, and $\lambda_{f,t}$ for molecular, viscous-laminar, and viscous-turbulent fluid-flow regimes.[§]

While the properties of heat transfer medium determine the conduction and radiation heat transfer rates, the convection heat transfer in addition depends on the fluid velocity $\mathbf{u}$. In conduction and radiation, the ability of the heat carriers to move in the medium, before a significant loss of their heat content, makes them effective heat carriers. Since the heat losses occur during collisions resulting in local thermal equilibrium among carriers, the longer the length between collisions, the longer the mean-free path $\lambda_i$ ($i$ stands for any carrier) and the higher the heat transfer rate. In convection, the ability of the fluid particles to move at a high velocity makes them effective heat carriers.

In radiation, as the mean-free path of photons $\lambda_{ph}$ becomes large, the photons initiating from far regions need to be included. This makes the general treatment of

---

[†] Mean-free path is a phenomenological representation of the distance a microscale heat carrier travels before encountering a collision resulting in a significant loss of its energy. Also, see the Glossary for definitions.

[‡] Phonons are also treated using classical molecular dynamics. The wave nature appears more in nanoscale thermal transport.

[§] The turbulent Kolmogoroff mean-free path $\lambda_K$ originates from a requirement of a unity Reynolds number based on this microscale length scale (and a similar microscale velocity scale).

**Gas Molecules**                                    **Liquid Molecules**

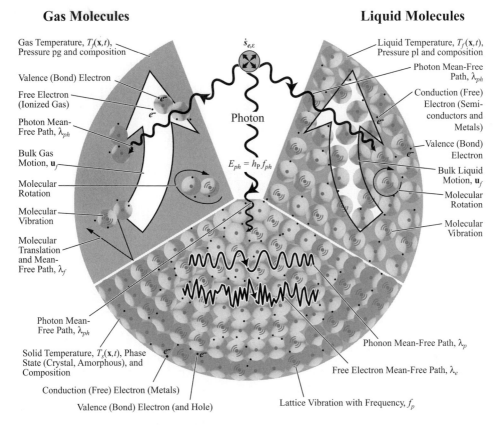

Figure 1.4. Microscale heat carriers, i.e., fluid particle, phonon, electron, and photon in three phases of matter.

the radiation heat transfer more involved, compared to conduction and convection. We will consider some special cases, such as the very large mean-free path treatment of surface-radiation heat transfer of Chapter 4, and the very short mean-free path treatment of volumetric-radiation heat transfer in Chapter 5.

Figure 1.4 gives a rendering of the microscale heat carriers in gas, liquid, and solid phases. The discrete nature of the microscale thermal energy is also shown.

Table 1.4 summarizes some of the characteristics of the microscale heat carriers. The microscale heat carriers are subject to statistical variations associated with their energy levels and fluctuations. For example, the carriers that follow the Bose-Einstein distribution [Section 3.2.2(B)], can have more than one carrier occupying an energy state. However, in the Fermi-Dirac distribution [Section 3.2.2(A)] there is a limit of one carrier per energy state. The Planck distribution for the photons (same as Bose-Eistein) will be discussed in Section 4.1.2. The Maxwell-Boltzmann gas molecular velocity distribution function is mentioned in Section 1.5.1. Therefore, they are represented by statistical distributions of their attributes. However, as with other statistical descriptions, the average description (called the ensemble average) must

Table 1.4. *Statistical attributes of the microscale heat carriers and their concentration $n_i$, mean-free path $\lambda_i$, and speed $u_i$*

| | Conduction | | | Convection | | | Radiation |
|---|---|---|---|---|---|---|---|
| | | | | **Fluid Particle** (In Macroscopic Motion, i.e., with Net Flow) | | | |
| | **Fluid Particle** (In Microscopic Thermal Fluctuation Motion, i.e., Random Motion with no Net Flow) | **Phonon** (Propagation of Quantal Lattice Vibrations, i.e., Slight Deviations from Atomic Equilibrium Position) | **Electron** (Drifting in Solids under Temperature Nonuniformity) | | | | **Photon** (Propagation of Quantal Electromagnetic Wave Emitted as Thermal Radiation) |
| Microscale Heat Carrier | | | | | | | |
| Carrier Regimes | Free Fluid Particles (Dilute Gases), and Interacting Fluid Particles (Liquids and Dense Gases) | Acoustic Phonon (Low Frequency), and Optical Phonon (High Frequency) | Free Electron (Occupying Higher Bond-Energy Levels), and Valence Electron (Lower Bond-Energy Levels) | **Molecular Flow** Molecular Gas Flow (Low Density Flow or Flowing Through Very Small Passages, Called Knudsen Regime) | **Laminar Flow** Fluid Particles in Viscous Laminar Motion (i.e., Orderly Flow without any Turbulent Fluctuations) | **Turbulent Flow** Fluid Particles in Viscous Random, Fluctuation Motion (Carriers Called Turbulent Eddies) | Transparent Media (Distant Radiation Dominated), Semi-Transparent Media (Distant and Local Volumetric Radiation Dominated), and Diffusion (Local Radiation Dominated) |
| Carrier Energy Subject to Statistical Variation | Particle Thermal Fluctuation Velocity and Kinetic Energy State | Phonon Quantum (Number or Modes) Energy State | Electron Quantum (Number or Modes) Energy State, also Called Fermion | Fluid Particle Velocity and Kinetic Energy State | None, other than Thermal Fluctuation Energy State | Fluid Particle Turbulent Fluctuation Velocity or Kinetic Energy | Photon Quantum (Number or Modes) Energy State, also Called Boson |
| Carrier-Attribute Statistics | Classical-Quantum Statistics (Maxwell-Boltzmann Distribution Function) | Quantum Statistics (Bose-Einstein Distribution Function) | Quantum Statistics (Fermi-Dirac Distribution Function) | Classical Statistics (Maxwell-Boltzmann Distribution Function) | None, other than Thermal Fluctuation Motion Statistics | Classical Statistics (Called Turbulent Energy Spectra) | Quantum Statistics (Bose-Einstein Distribution Function) |
| Carrier Collision, Scattering, and Other Interactions | With Other Fluid Particles | With Other Phonons, with Electrons, with Photons, with Lattice Defects and Impurities | With Other Electrons, with Phonons, with Photons, with Lattice Defects and Impurities | Mostly with Interfaces (Fluid-Fluid or Fluid-Solid) | With Other Fluid Particles | With Other Eddies and Decaying by Viscous Dissipation | With Other Photons, with Phonons, with Electrons, with Molecules and Other Scatterers |
| Carrier Mean-Free Path $\lambda_i$ (m) | $0 \leq \lambda_f < L$ Several Times the Interparticle Spacing $l_m$ | $l_m < \lambda_p < L$ Several to Many Times Lattice Dimension $l_m$ | $0 < \lambda_e < L$ Several to Many Times Lattice Dimension $l_m$ | $\lambda_f \leq L$ Linear Dimension of System $L$ | $\lambda_{fl} = L$ Linear Dimension of System $L$ | $\lambda_K \leq \lambda_{ft} < L$ Much Less than Linear Dimension of System $L$ | $0 \leq \lambda_{ph} < L$ In Gases: $\lambda_{ph} > L$ Liquids and Solids: $\lambda_{ph} < L$ |
| Carrier Concentration $n_i$ (1/m³) | Increases with Fluid Density | Increases with Liquid or Solid Density | Increases with Free-Electron Density | Increases with Fluid Density | Increases with Fluid Density | Increases with Fluid Velocity | Increases with Temperature |
| Carrier Speed $u_i$ (m/s) | Thermal Speed (rms of Fluctuation Velocity) $\langle u_f^2 \rangle^{1/2}$ (m/s) | Average of Longitudinal and Transverse Speed of Sound $u_p = a_s$ (m/s) | Electron Drift Speed $u_e$ (m/s) | Average of Fluid Particles Velocity $\langle \mathbf{u}_f \rangle$ (m/s) | Fluid Velocity $\mathbf{u}_f$ (m/s) | rms of Fluid Particle Turbulent Fluctuation Velocity $\overline{(\mathbf{u}_f \cdot \mathbf{u}_f)}^{1/2}$ (m/s) | Speed of Light $u_{ph} = c$ (m/s) |

match with the macroscopic observations (e.g., no net motion and a macroscopic pressure and temperature for the fluctuating gas molecules). The mean-free path of the microscale heat carriers are also compared to the linear dimension $L$ of the thermal system analyzed.

The microscale heat carriers are treated as particles and their distribution, transport and interactions (including scattering) are statistically described (i.e., governed) by the Boltzmann transport equation (which is a general conservation, or rate of change, equation). The scattering of the microscale heat carriers by themselves, other carriers, lattice molecules, lattice vacancies, or impurities, is governed by the classical mechanics if the collision cross section is larger than the thermal de Broglie wavelength $\lambda_{dB}$. This is defined as $\lambda_{dB} = h_P/(2\pi m k_B T)^{1/2}$, where $m$ is the mass of the particle making up the heat transfer medium. Otherwise, wave diffraction occurs and the quantum mechanical scattering theory is used.

In Section 3.2, we will show how the conduction heat flux vector can be expressed as the product of carrier concentration (or density of state) $n_i$, carrier heat capacity $c_{v,i}$, carrier speed $u_i$, and carrier mean-free path $\lambda_i$. In Chapter 4, we will discuss the photon mean-free path, and in Section 6.3.3 we will discuss the relationship between the turbulent eddy mean-free path and the turbulent heat flux vector.

### 1.4.3 Net Heat Transfer Rate $Q|_A$

Here $\mathbf{q}$ is taken to be positive when it is along the surface normal $\mathbf{s}_n$. The net heat transfer $Q|_A(\mathrm{W})$ through (i.e., out of) the boundaries (surface) of a control volume is given by

$$Q|_A = \int_A (\mathbf{q} \cdot \mathbf{s}_n)\, dA = \int_A [(\mathbf{q}_k + \mathbf{q}_u + \mathbf{q}_r) \cdot \mathbf{s}_n] dA \qquad \begin{array}{l}\text{integrated surface heat} \\ \text{flow rate in terms of } \mathbf{q},\end{array} \qquad (1.8)$$

where $\mathbf{s}_n$ is the outward, normal unit vector on surface $A$. With this definition, when the net rate of heat flow is out of the control, then $Q|_A$ is positive. A schematic of the outward, normal unit vector is shown in Figure 1.5(a) on a control volume $V$ having a surface $A$. The dot product of $\mathbf{q}$ and $\mathbf{s}_n$ at a location on a differential surface area $\Delta A$ is shown in Figure 1.5(b). By integration over the entire surface $A$, we will have (1.8). When $\mathbf{q}$ is parallel to the surface, i.e. it is along the surface tangent $\mathbf{s}_t$, the dot product of $\mathbf{q}$ and $\mathbf{s}_n$ is zero and no heat flows across the control surface. When $\mathbf{q}$ is parallel to $\mathbf{s}_n$, then the dot product of $\mathbf{q}$ and $\mathbf{s}_n$ has its maximum magnitude, which is equal to the magnitude of $\mathbf{q}$. When this dot product is a positive quantity, heat flows out of the control surface, and when it is negative, heat flows into the control surface.

The heat flux vector flowing into and out of a control volume is rendered in Figure 1.5(c). A rectangular control volume $V$ and the Cartesian coordinate system $(x, y, z)$ are shown. The total heat flow rate $Q|_A$ is shown as the sum of the contributions from six surfaces, $A_{x,1}$, $A_{x,2}$, $A_{y,1}$, $A_{y,2}$, $A_{z,1}$, and $A_{z,2}$. Note that the surface normal for the surface pairs are in opposite directions. The surface normal for all

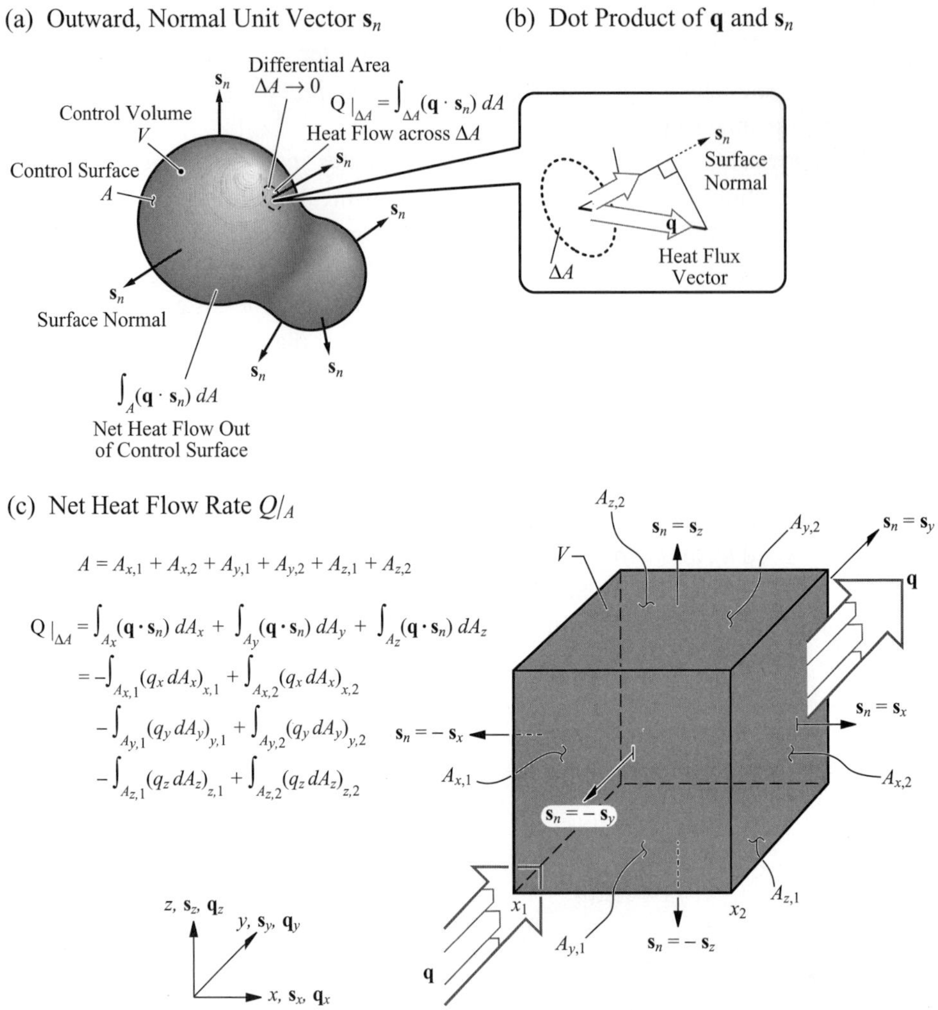

(a) Outward, Normal Unit Vector $\mathbf{s}_n$              (b) Dot Product of $\mathbf{q}$ and $\mathbf{s}_n$

Figure 1.5. (a) The outward, surface-normal unit vector. (b) The dot product of $\mathbf{q}$ and $\mathbf{s}_n$ and its integral over $\Delta A$, $Q|_A$. (c) The net heat flow into and out of a control volume, $Q|_A$.

surfaces is shown. For a control volume, the direction of $\mathbf{s}_n$ changes as the surface orientation changes. This is rendered in Figure 1.4(a). The direction of $\mathbf{q}$ can also change. The dot product of $\mathbf{q}$ and $\mathbf{s}_n$ is also shown in Figure 1.4. The net flow of heat across the control surface is positive if the integral given in (1.8) is positive and this corresponds to a net heat flow out of the control volume. The net inflow of heat corresponds to a negative value for this integral. Then we can write, for a control surface $A$ around a control volume $V$,

$$Q|_A = \int_A (\mathbf{q} \cdot \mathbf{s}_n) dA = \begin{cases} > 0 & \text{heat flows out of the control volume} \\ < 0 & \text{heat flows into the control volume} \\ 0 & \textit{no net heat flow into or out of the control volume.} \end{cases} \quad (1.9)$$

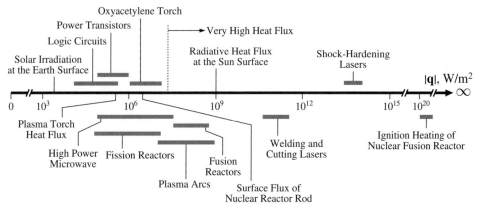

Figure 1.6. Examples of magnitude of the heat flux vector.

To find the component of the heat flux in a direction of a unit vector (for example $\mathbf{s}_x$, in the $x$ direction in the Cartesian coordinate system), we perform a dot product, i.e.,

$$\mathbf{q} \cdot \mathbf{s}_x = q_x = \mathbf{q}_k \cdot \mathbf{s}_x + \mathbf{q}_u \cdot \mathbf{s}_x + \mathbf{q}_r \cdot \mathbf{s}_x$$

$$= q_{k,x} + q_{u,x} + q_{r,x} \quad x\text{-component of heat flux vector}, \qquad (1.10)$$

where $q_x$ is a scalar, compared to $\mathbf{q}$, which is a vector.

Here we did not make a separate designation for $\mathbf{q}_{ku}$, and have included it as a special case of $\mathbf{q}_k$.

### 1.4.4 Magnitude and Representation of $\mathbf{q}$

The magnitude of $\mathbf{q}$, i.e., $|\mathbf{q}| = q$, varies greatly in different applications. The upper bound is unknown, but very high values of $q$ have been engineered (though for only a short time). Figure 1.6 gives examples of $q$ encountered in natural and engineered systems. Generally, $q$ larger than $10^7$ W/m$^2$ is considered very high.

The conduction $\mathbf{q}_k$ (W/m$^2$) and convection $\mathbf{q}_u$ (W/m$^2$) mechanisms (or modes) of the heat flux vectors in (1.7) are defined as

$$\mathbf{q}_k \equiv -k\nabla T \quad \text{Fourier law of conduction heat flux} \qquad (1.11)$$

and

$$\mathbf{q}_u \equiv \rho e \mathbf{u} \quad \text{convection heat flux}, \qquad (1.12)$$

where $\mathbf{q}_u$ is expressed here in terms of the specific internal energy. In some cases, it becomes more convenient to work with the specific enthalpy, as discussed in Appendix B.

Using Equations (1.11) and (1.12), the heat flux vector (1.7) becomes

$$\mathbf{q} = -k\nabla T + \rho e \mathbf{u} + \mathbf{q}_r \quad \text{heat flux vector}. \qquad (1.13)$$

In (1.13), $k$(W/m-K or W/m-°C) is the thermal conductivity (or conductivity) [2], and it is a thermal transport property (a thermophysical property) of the medium representing the contributions from conduction microscale heat carriers, i.e., fluid particles, electrons, and phonons. Also, $\nabla T$(°C/m or K/m) is the gradient of temperature, a vector, $\rho$(kg/m³) is the density, $e$(J/kg) is the specific internal energy, $\mathbf{u}$(m/s) is the velocity vector of the heat transfer medium, and $\mathbf{q}_r$(W/m²) is the thermal radiation heat flux vector. The inverse of $k$ is called the thermal resistivity $1/k$(m-°C/W or m-K/W).

The conduction portion of the heat flux vector $\mathbf{q}_k$ describes the heat flow along the vector $-\nabla T$, i.e., positive heat flow along the direction of decreasing temperature. The conductivity $k$ ($k \geq 0$) will be discussed in Chapter 3. The gradient of temperature for the Cartesian coordinate system is given by its three components in the $x$, $y$, and $z$ directions, i.e.,

$$\nabla T = \frac{\partial T}{\partial x}\mathbf{s}_x + \frac{\partial T}{\partial y}\mathbf{s}_y + \frac{\partial T}{\partial z}\mathbf{s}_z \qquad \begin{array}{l}\text{gradient of temperature in}\\ \text{Cartesian coordinates,}\end{array} \qquad (1.14)$$

where, again, $\mathbf{s}_x$, $\mathbf{s}_y$, and $\mathbf{s}_z$ are the unit vectors.

The convection portion of the heat flux vector $\mathbf{q}_u$, (1.12), describes the heat flow (i.e., fluid particles moving with velocity $\mathbf{u}$ and carrying the internal energy $\rho e$ of the fluid through a unit area at point $\mathbf{x}$ and at time $t$) along the direction of the velocity vector $\mathbf{u}$. The direction of $\mathbf{q}_u$ depends on the direction of $\mathbf{u}$. The velocity vector can be steady or it can vary with time, orderly (i.e., periodic oscillation), or with random fluctuations (turbulent flow). For gases flowing at low pressure or flowing through a very small passage, the probability of interaction between fluid particles becomes comparable with that between the fluid particle and the walls and the flow is called a molecular flow. Otherwise the flow is called a viscous flow. A viscous flow may either be orderly (called a laminar flow) or it may contain velocity fluctuations (called a turbulent flow). The convection heat flux will be discussed in Chapters 5, 6, and 7.

The radiation portion of the heat flux vector $\mathbf{q}_r$ describes the flux of photons, where each photon moves at the speed of light $c$(m/s), has an energy $h_P f$ (J), and has a concentration $n_{ph}$(1/m³). This radiation heat flux is the integral taken over all frequencies $f$ (or wavelengths) and in the direction of the surface normal $\mathbf{s}_n$. This will be discussed in Section 4.1.1, where we examine how the photon flux is influenced by emission, absorption, and scattering of the thermal radiation in the heat transfer medium. Thermal radiation is emitted due to the medium temperature and over the entire electromagnetic radiation spectrum. However, in thermal radiation, the ultraviolet, visible, and infrared portions of the electromagnetic spectrum have the most significant energy content, compared to other wavelength portions. When needed, the remainder of the electromagnetic radiation spectrum, including those portions in the microwave and low frequency electric-power applications, are generally generated by other than thermal radiation and are discussed in relation to the conversion of energy in Section 2.3.2.

The representation of $\mathbf{q}_r$ in terms of the thermal radiation properties is made by the general equation of radiation heat transfer, but the representation can be simplified under some conditions. These conditions will be discussed in Sections 2.3.2(E) and (F), in Chapter 4 and in Section 5.2.5. As will be discussed, in some cases the surface radiation dominates (i.e., the volume between surfaces does not emit, absorb, or scatter the radiation). For example, in most cases, air is treated as a nonemitting, nonabsorbing, and nonscattering medium (and it is therefore called transparent). In some other cases, however, this volume is totally absorbing (like most solids, and it is called opaque). In some cases the absorption and scattering are very significant, such that only locally emitted radiation can penetrate a short distance (as compared to the radiation emanating from distant locations), and this is called diffusion (or local) radiation, and is expressed similar to the conduction heat flux (1.11). The most general treatment includes the distant and local radiation.

### EXAMPLE 1.6. FUN

Heat absorbed from irradiation flows two-dimensionally into the top surface of a solid block, as shown in Figure Ex. 1.6(a). The block is a composite (i.e., made of more than one distinct material). Material A has a thermal conductivity $k$ which is smaller than that of material B. The heat is conducted through the block, but because the lower surface is ideally insulated (i.e., no heat flows through it), the heat flows to the side surfaces and upon reaching the side surfaces, the heat leaves by surface convection and surface radiation.

(a) Track the heat flux vector starting from the top surface and ending on the side surfaces.
(b) What condition should exist for the heat flux across the side walls to be perpendicular to these surfaces?

**SOLUTION**
This is a steady-state (i.e., time-independent) heat transfer problem.

(a) A portion of the irradiation is absorbed at the surface and this is shown as energy conversion $\dot{S}_{e,\alpha}$. Since the heat flowing in from the top surface has to be removed from the side surfaces, we sketch the heat flux vector track as shown in Figure Ex. 1.6(b). The heat flux vector splits and since material A has a lower conductivity, less heat flows toward the left surface. Upon arriving at the side surfaces, heat flows by surface convection $q_{ku}$ and surface radiation $q_r$.
(b) The heat flow along the side surface is by conduction only, i.e.,

$$\mathbf{q}_k = -k\nabla T = -k\frac{\partial T}{\partial x}\mathbf{s}_x - k\frac{\partial T}{\partial y}\mathbf{s}_y.$$

For the heat flux along the $y$ direction to be zero, $\partial T/\partial y$ must be zero. This would mean a uniform temperature along the side walls.

**COMMENT**
The heat flux tracking identifies the media through which the heat flows, and identifies the mechanisms for their transfer. This results in any energy conversion that results in energy release or energy absorption. Ideal insulation indicates no

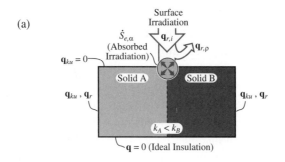

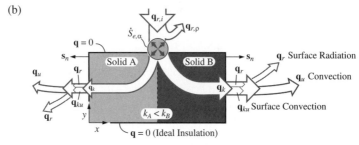

Figure Ex. 1.6. (a) Radiation heating of a composite block. (b) Heat flux vector tracking.

heat transfer by any mechanism. Therefore, the lower surface has no heat flowing across it (i.e., is adiabatic). Also, in some cases the conduction perpendicular to the surface is much larger than that along the surface, i.e.,

$$k\left|\frac{\partial T}{\partial x}\right| \gg k\left|\frac{\partial T}{\partial y}\right| \quad \text{at and near right- and left-side surfaces,}$$

and therefore, a uniform surface temperature is assumed for these surfaces in the heat transfer analysis.

## 1.5 Heat Transfer Materials and Heat Flux Tracking

The heat transfer medium is the medium within which the heat is being transferred and energy storage and conversion may occur. The heat transfer medium may be inert or reactive and may undergo phase change. The medium may be a single phase or multiphase. Here we first discuss the three phases of matter, then composite heat transfer media, and finally the heat transfer media with fluid motion. These topics will give an introduction on how the various properties of the matter enter into the heat transfer analysis.

### 1.5.1 Three Phases of Matter: Intermolecular and Intramolecular Forces*

This section is found on the Web site www.cambridge.org/kaviany. This section discusses how matter consists of localized regions of very high density (i.e., atoms and molecules) separated by regions of nearly zero density. It considers how all properties of matter are determined by the interatomic and intermolecular forces. This section also discusses the principle of the classical equipartition of energy and the

Boltzmann constant and role of potential energy in formation of phases of matter. The kinetic theory of gases and the gas mean-free path are also discussed.

### 1.5.2 Microscale Energies of Matter: Discrete and Continuous Energies*

This section is found on the Web site www.cambridge.org/kaviany. In this section, we consider all the energy of matter divided into sensible heat energy and energy that can be converted to and from heat. This change in energy can be due to conversion of bond (chemical and physical) energy, or electromagnetic energy, or mechanical energy, or a combination of these. We give examples of these sensible and converted heats, in terms of microscopic energy scales such as the Boltzmann constant. Heat transfer is the result of spatial variation of temperature caused by a change in the local energy states. We also discuss these microscopic energy states.

### 1.5.3 Multiphase Heat Transfer Medium: Composites

The medium for heat transfer can be a solid, a liquid, a gas, or a combination of these phases (i.e., a composite). For multiphase media, the phases can be arranged as layered or nonlayered.

Figure 1.8(a) shows some typical layered media. These layers can be made of various solid materials. Examples are composite solids that have thin or thick film coating or added thermal insulation layers. The layers can also be made of solid and fluid (gas or liquid) as in solids partly bounded by a fluid (e.g., combustion chambers), and the fluid can be in motion (e.g., surface cooling by a cold moving fluid). The layered system can also be made of two fluids as in a droplet in a gas or a bubble in a liquid. All three phases may be present as in film condensation on a solid surface (as in condensers).

Figure 1.8(b) renders some typical nonlayered phase distributions starting with solid-solid composites. Particulate flows, such as in fluidized beds, and porous media are examples of nonlayered solid-fluid heat transfer media. The injection of liquid droplets into a gas, as in spray cooling/heating or spray combustion, results in a nonlayered liquid-gas phase distribution. Finally, shown in Figure 1.8(b) is a solid-liquid-gas nonlayered system, namely, nucleate pool boiling on a solid surface.

### 1.5.4 Fluid Motion

The fluid can be stagnant (i.e., quiescent), i.e., $\mathbf{u} = 0$, or it can be in steady, orderly time-varying, or randomly time-varying (i.e., turbulent) motion, i.e., $\mathbf{u} = \mathbf{u}(\mathbf{x}, t) \neq 0$. In Section 1.7.3, we will discuss the fluid motion under four external forces, namely, pressure, viscous stress, gravity force, and electromagnetic force. There are other forces, such as the surface tension at the interface of two immiscible (such as liquid and gas) fluids. When motion is due to an external (or imposed) pressure field, this is called a forced flow.

The fluid flow due to the gravity field acting on a nonuniform density fluid is called a buoyant flow or natural convection (as compared to forced convection).

(a) Layered

(i) Solid-Solid

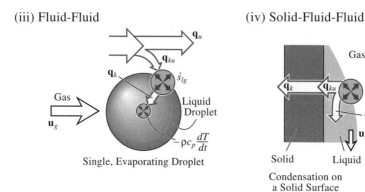

(ii) Solid-Fluid

(iii) Fluid-Fluid

(iv) Solid-Fluid-Fluid

(b) Nonlayered

(i) Solid-Solid

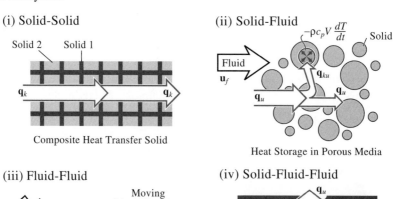

(ii) Solid-Fluid

(iii) Fluid-Fluid

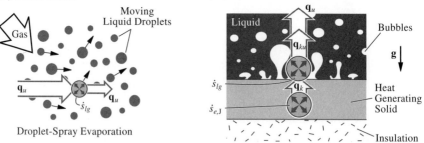

(iv) Solid-Fluid-Fluid

Figure 1.8. Heat transfer in multiphase systems: (a) layered systems and (b) nonlayered systems.

(c) Intramedium Heat Transfer:One-Dimensional,
Premixed Gaseous Combustion

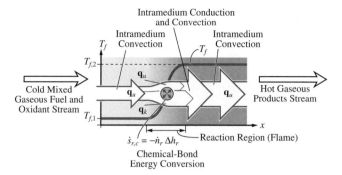

(d) Combined Intramedium and Intermedium Heat Transfer:
Indirect Heat Exchange between Two Streams

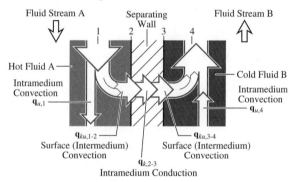

Figure 1.8. (c) An example of intramedium heat transfer. (d) An example of combined intramedium and intermedium heat transfer.

This density variation can be due to variation of density with temperature (called the thermobuoyant flow or thermal convection), or due to the variation of density with species concentration (called mass-diffusiobuoyant flow, or mass-diffusional convection or solutal convection), or due to the simultaneous presence of the gas and liquid phases (called the phase-density buoyant flow) as in bubbly-liquid or droplet-gas or particulate (solid)-gas flows.

The thermobuoyant flows are an example of heat transfer-influenced flows. Another example is the acoustic waves in gases caused by a sudden and drastic heating or cooling. This is called the thermoacoustic flow. In liquid-gas systems, where the surface tension acting on the interface of the gas and liquid is nonuniform (due to the variation of temperature over the interface), a flow is induced which is called the thermocapillary flow.

The fluid stream may be unbounded, or partly or completely bounded by the surface of a solid. The heat transfer in unbounded fluid streams is the simplest to study (Chapter 5). In semi-bounded fluids, the fluid stream is only partly bounded by the surface of a solid, as in fluid stream over a semi-infinite or short flat plate (Chapter 6). In flow through tubes, ducts, porous solids, etc., the fluid stream is bounded by a solid surface (Chapter 7).

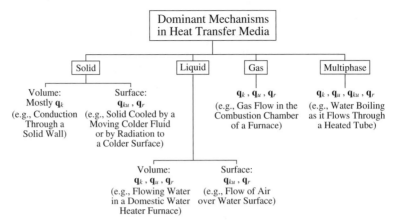

Chart 1.4. Dominant mechanisms of heat transfer in various phases.

### 1.5.5 Intramedium and Intermedium Heat Transfer

Depending on the main (i.e., dominant) heat transfer, we may be addressing the heat transfer among the phases (i.e., across bounding surfaces) making up the heat transfer medium (called the intermedium heat transfer, Chapters 4, 6, and 7) or the heat transfer through a medium (called the intramedium heat transfer, Chapters 3 and 5). These are depicted in Figures 1.8(c) and (d). There are cases where both intra- and intermedium heat transfer need to be addressed simultaneously (Chapters 3, 4, 6, and 7).

For example, consider heating of particles in a hot gas stream for a later deposition on a surface (as in gas or thermal-plasma heated particles for surface coating). We address the heat transfer between the dispersed particles and the gas as an intermedium heat transfer between the solid and the fluid phases making up the medium (i.e., the dominant thermal nonequilibrium is between the phases making up the medium).

In some applications the heat transfer and its effects are important only within a single medium. An example of this intramedium heat transfer is the heat transfer-controlled chemical reaction such as in premixed gaseous combustion (Section 5.2). This is rendered in Figure 1.8(c) where the heat needed to sustain the reaction in a moving gaseous fuel-oxidizer mixture is provided by conduction and opposed by convection (with radiation also playing a role).

In other applications, both intermedium and intramedium heat transfer are of significance. This is the case in most of the heat transfer applications. For example, indirect heat transfer between two fluid streams, is rendered in Figure 1.8(d). The heat is convected in each stream and due to the lack of thermal equilibrium between them, heat is transferred from one stream to the separating wall, through the wall, and then from the wall to the second stream.

### 1.5.6 Heat Flux Vector Tracking in Various Phases

Not all of the three mechanisms of heat transfer are significant in every phase of matter. Chart 1.4 gives an approximate listing of mechanisms significant in each phase (including multiphase). Solids are generally opaque to radiation (exceptions are glasses and other optical solids). On the other hand, gases are generally transparent to radiation. For both opaque and transparent media, volumetric radiation heat transfer is generally neglected. For solids, there is generally no convection heat transfer, unless a moving solid (such as extruding rods) is considered.

Again note that at the interface of a solid and a moving fluid we represent the fluid conduction $\mathbf{q}_k$ at the surface with surface convection $\mathbf{q}_{ku}$. Also for surfaces that have a uniform temperature, the surface-convection heat flux vector is perpendicular to the surface. Then, the heat is transferred through solid-fluid interface by surface convection, where on the solid side the heat transfer is by conduction, and on the fluid side by conduction and convection. The surface radiation may also be significant.

### EXAMPLE 1.7. FUN

In a range-top burner, depicted in Figure Ex. 1.7(a), used for cooking, heat released per unit volume by chemical reaction $\dot{s}_{r,c}$ is transferred from the burner cap and the flame to the pan and surroundings by conduction, convection, and radiation. The heat of combustion causes air currents (thermobuoyant motion), resulting in surface-convection heat transfer to the pan surface.

Using this sketch, track the heat flux vector around the burner-pan area.

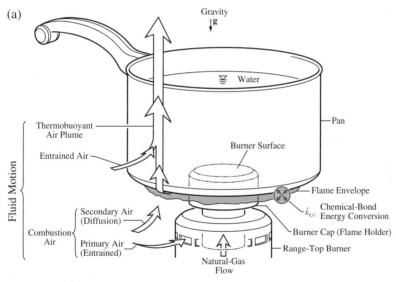

Figure Ex. 1.7. (a) Range-top, natural-gas burner with a pan placed over the flame. The various fluid flows are also shown.

### SOLUTION

The range-top burner is designed to deliver most of heat released $\dot{s}_{r,c}$ from the combustion of natural gas-air to the cooking pan. The combustion air is provided passively (i.e., by entrainment and diffusion). The flame envelope is formed around the permeable burner cap and its shape is determined by the heat losses. As the reaction is completed, the hot combustion products rise due to thermobuoyancy and entrain further air into the thermobuoyant air plume. These are also depicted in Figure Ex. 1.7(b).

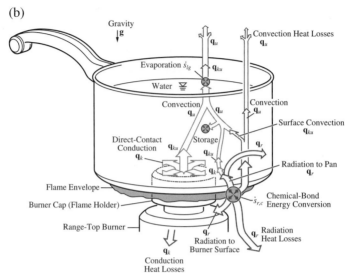

Figure Ex. 1.7. (b) Rendering of heat transfer by various mechanisms around a pan placed on the range-top burner.

The heat is transferred from the burner cap to the pan by direct-contact conduction $(A_k q_k)_c$ and by surface radiation $(A_r q_r)_c$ where $A_k$ is the area over which conduction takes place and $A_r$ is for radiation. The subscript $c$ stands for cap. The heat is transferred from the flame to the pan by surface convection $(A_{ku} q_{ku})_f$ and by radiation $(A_r q_r)_f$. The subscript $f$ stands for fluid. The remainder of heat generated by chemical reaction is considered a heat loss. This heat leaves with the thermobuoyant air plume as convection, to the burner support by conduction, and to the ambient surfaces by surface (burner cap) and flame radiation. The heat flowing into pan water is used for raising the water temperature (as storage) and for water surface evaporation $\dot{s}_{lg}$.

**COMMENT**
The burner size and the flame size can be adjusted to the diameter and height of the pan, for maximum heat-transfer efficiency.

## 1.6 Conservation of Energy

The law of conservation of energy states that the total energy of all the participants in any process must remain unchanged throughout the process. Energy can be transformed and transferred, but the total amount stays the same. The conservation of energy principle (e.g., the first law of thermodynamics) can be applied to any control volume. We consider the general case of matter being able to flow through the boundaries (control surface area $A$) as well as within the volume (control volume

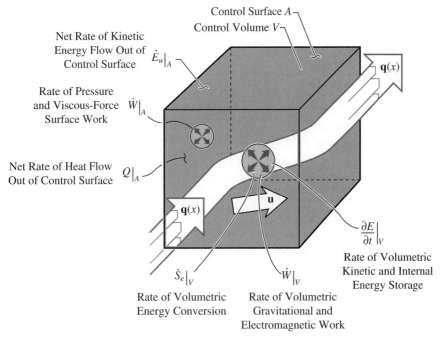

Figure 1.9. Conservation of energy applied to a control volume $V$ having a surface $A$. Heat transfer and various forms of energy and work flowing through the control surface and occurring within the control volume are shown.

$V$) as depicted in Figure 1.9. The case of a stationary matter is then a special case of this general description with the velocity vector $\mathbf{u}$(m/s) set to zero.

The mathematical statement of the energy conservation for a control volume applies the Reynolds transport theorem [21], which addresses the time rate of change within a control volume $V$ and the transport across its control surface $A$, in the conservation statement.

Here we assume that the contribution of the potential energy is negligible compared to the kinetic and internal energy. We begin with the form of the energy conservation equation commonly used in the thermodynamic analysis.[†] This form combines the mechanical (kinetic) and thermal (internal) energy, as designated by $E$, and is [21]

$$\left.\frac{\partial E}{\partial t}\right|_{V} + \dot{E}|_{A} = -Q_{k,r}|_{A} + \dot{W}|_{A,V} \qquad \text{integral-volume energy equation,} \qquad (1.21)$$

①     ②     ③     ④

---

[†] Alternatively, we can derive the energy equation starting from the transport equation for each of the four microscale energy carriers (electron, fluid particle, phonon and photon) and then add them to arrive at the energy equation for all energy carriers. The carrier transport equation is referred to as the Boltzmann transport equation (Section 1.7.4), where each carrier is treated as a particle with a probability distribution function.

where

☐1 rate of energy storage in the control volume
☐2 net rate of energy flow out of the control surface
☐3 negative of net rate of heat flow out of the control surface by conduction and radiation
☐4 rate of total work (surface $\dot{W}|_A$ and volumetric $\dot{W}|_V$) done on the control volume.

Note that we have used the definition of $Q|_A$ given by (1.21) such that when $Q|_A$ is positive, heat flows out of the control volume.

The internal energy depends on the temperature. Then in systems with a nonuniform temperature distribution, the fluid flow through the system combined with the variation of the temperature results in a net flow of thermal energy. This local surface flow of thermal energy is represented as a form of heat called the convection heat flow, $Q_u(W)$. In this text, in order to arrive at a more general definition for the rate of the heat transfer, we include this rate of internal energy flow into the rate of total heat flow and designate this as $Q|_A$. Now $Q|_A$ includes the three contributions discussed in Section 1.4 and given by (1.7). The advantage of this formulation will be made clearer in Section 2.1. As will be discussed in Section 1.7.3, we consider four principal forces, $\sum_i \mathbf{F}_i$. These are the pressure $\mathbf{F}_p$, viscous $\mathbf{F}_\mu$, gravitational $\mathbf{F}_g$, and electromagnetic $\mathbf{F}_e$ forces. We also divide the work term $\dot{W}|_{A,V}$ into the surface work (e.g., pressure work $\dot{W}_p$, viscous work $\dot{W}_\mu$), the volumetric work (e.g., total gravitational and electromagnetic work, $\dot{W}_{g,e}$), and all other work, designated as $\dot{S}_e$. Then (1.21) for the control volume shown in Figure 1.9, becomes

$$
Q_{k,u,r}|_A = Q|_A = -\left.\frac{\partial E}{\partial t}\right|_V -\dot{E}_u|_A +\dot{W}_p|_A +\dot{W}_\mu|_A +\dot{W}_{g,e}|_V +\dot{S}_e|_V \tag{1.22}
$$

☐1 ☐2 ☐3 ☐4 ☐5 ☐6 ☐7

integral-volume energy equation with generalized $Q$,

where

☐1 net rate of heat flow out of the control surface (conduction, convection, and radiation)
☐2 negative of rate of energy storage in the control volume
☐3 negative of net rate of kinetic energy flow out of the control surface
☐4 rate of pressure work done on the control surface
☐5 rate of viscous work done on the control surface
☐6 rate of gravitational and electromagnetic work done on the control volume
☐7 rate of other work done (or conversion of energy) on the control volume; note that the chemical and physical bond energy conversion is included in the internal energy.

Here $Q|_A$(J/s) is the net rate of heat flow through the control surface (positive if the net rate is out of the control surface), $E$(J) designates the internal and kinetic energy, $\dot{E}_u|_A$(J/s) is the net rate of kinetic energy flow, $\dot{W}|_A$(J/s) is the rate of work done on the control surface, and $\dot{W}|_V$(J/s) is the rate of work done on the control volume. $\dot{S}_e$(J/s) designates all other work and energy conversion. These other work modes include the surface-tension work (including that between phases at a liquid-gas interface). The energy conversion term $\dot{S}_e$(J/s) will be discussed in Section 2.3.

Note again that we have stated the energy conservation equation in a manner that includes the net rate of flow of the internal energy (i.e., transport of internal energy), through the control surface, in the heat transfer term. From here on, for brevity, we use $Q$ (and $\mathbf{q}$) to include all three mechanisms of heat transfer.

Note that the work done on the system is divided into a part caused by the surface forces (e.g., pressure and viscous) and a part caused by the volumetric distant body forces (e.g., gravitational and electromagnetic). The work mode and energy conversion modes not directly addressed are included in $\dot{S}_e$.

Each term in (1.22) will now be written in terms of measurable and more elementary variables, e.g., thermodynamic properties, velocity, and forces. This is done using the heat flux vector $\mathbf{q}$(W/m$^2$), volumetric internal energy $\rho e$(J/m$^3$), volumetric kinetic energy $\rho \mathbf{u} \cdot \mathbf{u}/2$(J/m$^3$) [where again $\mathbf{u}$(m/s) is the velocity vector], and volumetric energy conversion $\dot{s}_e$(W/m$^3$). Also included is the work done by the external forces. These are the pressure $p$(N/m$^2$), viscous stress tensor $\mathbf{S}_\mu$(N/m$^2$), volumetric gravitational body force vector $\rho \mathbf{g}$(N/m$^3$), and volumetric electromagnetic force vector $\mathbf{f}_e$(N/m$^3$). We will discuss these forces again in the treatment of the momentum conservation in Section 1.7.3.

The potential energy is assumed negligible compared to the internal and kinetic energy. Using these more elementary variables we rewrite (1.22) as

$$Q|_A \equiv \int_A (\mathbf{q} \cdot \mathbf{s}_n)dA = -\frac{\partial}{\partial t}\int_V \left(\rho e + \frac{\rho}{2}\mathbf{u} \cdot \mathbf{u}\right) dV \quad - \int_A \frac{\rho}{2}(\mathbf{u} \cdot \mathbf{u})(\mathbf{u} \cdot \mathbf{s}_n)dA$$

$$\boxed{1} \qquad\qquad\qquad \boxed{2} \qquad\qquad\qquad\qquad\qquad \boxed{3}$$

$$- \int_A (p\mathbf{u} \cdot \mathbf{s}_n)dA \qquad + \int_A (\mathbf{S}_\mu \cdot \mathbf{u}) \cdot \mathbf{s}_n dA$$

$$\boxed{4} \qquad\qquad\qquad\qquad \boxed{5}$$

$$+ \int_V (\rho \mathbf{g} \cdot \mathbf{u} + \mathbf{f}_e \cdot \mathbf{u})dV \quad + \int_V \dot{s}_e dV$$

$$\boxed{6} \qquad\qquad\qquad\qquad \boxed{7}$$

$$(1.23)$$

integral-volume energy conservation equation.

These various terms are rendered in Figure 1.10. There are four surface integral terms and four (counting separately the graviational and electromagnetic work of term 6) volume integral terms. As will be shown in Chapter 2, some of these terms are removed using the momentum conservation equation. This removal simplifies the energy equation greatly. Also, we will combine presentation of the various work and energy conversion terms to arrive at a yet more compact energy equation.

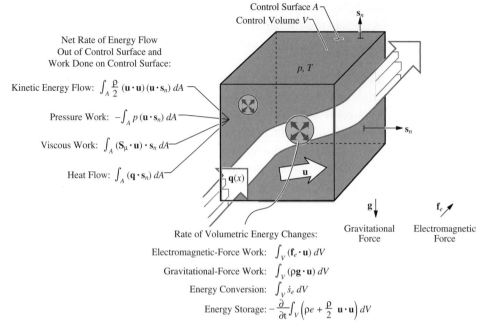

Figure 1.10. Integral form of the energy conservation, showing the various forms of energy, heat, and work flowing through the control surface and occurring within the control volume.

### EXAMPLE 1.8. FUN

Equation (1.23) is the most general form of the energy equation used in heat transfer analysis and includes most forms of energy and work (e.g., excluded is the potential energy, which is assumed negligible).

Comment on each of the terms in (1.23).

### SOLUTION

The seven terms appearing in (1.23) are discussed below.

$\boxed{1}$ Heat Transfer Rate out of the Control Surface : This is significant in all heat transfer applications. This is divided into five parts, i.e.,

$$Q|_A = \int_A (\mathbf{q} \cdot \mathbf{s}_n) dA = Q + Q_k + Q_u + Q_{ku} + Q_r,$$

$Q$ : Heat transfer out of the control surface $A$ (or bounding surface), which is either prescribed or is not represented by conduction, convection, surface convection, and radiation.

$Q_k$ : Heat transfer out of the control surface $A$ by conduction.

$Q_u$ : Heat transfer out of the control surface $A$ by convection. When the surface is a fluid, we use $Q_u$.

$Q_{ku}$ : Heat transfer out of the control surface by surface convection. When part or all of the control surface is an impermeable solid surface and is in thermal nonequilibrium (i.e., having a different temperature) with an adjacent fluid stream, we use $Q_{ku}$ (i.e., surface convection heat transfer) and this heat transfer is due to the motion of the fluid in contact with this impermeable solid surface, while the fluid has a velocity of zero on this surface (i.e., the fluid particle does not slip on the surface, due to fluid viscosity and solid impermeability). The heat is conducted through the solid $Q_k$, is transferred through the solid-fluid interface by surface convection $Q_{ku}$, and is then conducted and convected by the moving fluid $Q_k$ and $Q_u$.

$Q_r$ : Heat transfer out of the control surface $A$ by radiation (surface $A$ having a temperature different than the surrounding surfaces).

In Chapters 3 to 7, we will develop relationships that give $Q_k$, $Q_u$, $Q_{ku}$, and $Q_r$ in terms of thermal potentials and thermal resistances and will use these in thermal circuit analysis of heat transfer problems.

$\boxed{2}$ Time Rate of Change of Energy within the Control Volume : Except in high-speed flows with significant temporal (and spatial) change in the fluid velocity, only the internal energy term is significant in heat transfer applications. The internal energy contains the sensible heat and the bond energy (physical such as phase change and chemical such as reaction).

$\boxed{3}$ Rate of Kinetic Energy Flow out of Control Surface : This is generally insignificant.

$\boxed{4}$ Rate of Pressure Work Done on the Control Surface : This is significant when there is a large fluid expansion or compression (generally gaseous).

$\boxed{5}$ Rate of Viscous Work Done on the Control Surface : This is significant when the fluid is very viscous and experiences a large velocity gradient.

$\boxed{6}$ Rate of Gravitational and Electromagnetic Work Done on the Control Volume : These are generally negligible.

$\boxed{7}$ Rate of Other Energy Conversion in the Control Volume : These can be significant and include microwave heating, Joule heating, Peltier heat absorption/release, ultrasonic heating, etc.

**COMMENT**

In Section 2.2.1 we will present a more compact form of (1.23). In Section 2.3 we will review some key energy conversion mechanisms that are encountered in heat transfer applications.

EXAMPLE 1.9. FUN

A wet adhesive is used to adhere a piece of foam (polyurethane) to a wood (Douglas fir) substrate. The adhesive needs to be heated to cure. The heating is done in a microwave oven, as shown in Figure Ex. 1.9 (a). The heat generation per unit volume (conversion of energy from electromagnetic to thermal $\dot{S}_{e,m}/V = \dot{s}_{e,m}$) in the adhesive layer is shown with the crossing double arrows. The ability for energy conversion is much lower for foam and dry wood (due to smaller dielectric loss factor), as compared to the wet adhesive. The ability to convert the electromagnetic energy to thermal energy depends on a material property called the dielectric loss factor.

(a)

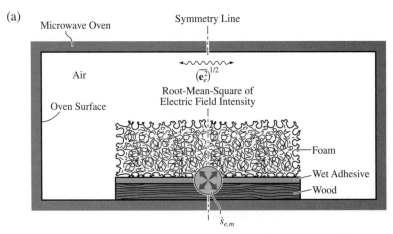

Figure Ex. 1.9. (a) Microwave heating of a foam-wood composite with a wet adhesive connecting the materials.


The generated heat raises the temperature of the adhesive and this heat flows away toward the foam and the wood. As the heat flows through the foam and the wood, it also raises the temperature within them.

Sketch the heat flux vector as it originates from the adhesive and flows into the surrounding layers and ultimately into the oven air and the oven internal surfaces.

## SOLUTION

This is a transient (i.e., time-dependent) heat transfer. Figure Ex. 1.9(b) shows the heat $\mathbf{q}$ flowing away from the source (microwave heated wet adhesive) $\dot{s}_{e,m}$ toward the foam and the wood, while raising the sensible heat (i.e., resulting in a change of temperature; see the Glossary at the end of the book) $-\partial e/\partial t$ of the adhesive.

(b)

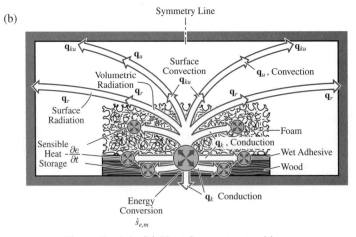

Figure Ex. 1.9. (b) Heat flux vector tracking.

The sensible heat storage per unit volume $-\partial e/\partial t$ is also shown with the crossing double arrows. The heat flowing into the foam and the wood is partly used to raise the local temperature and partly continues to flow toward the surfaces. At the foam surface, heat is transferred by surface radiation $\mathbf{q}_r$ and by surface convection $\mathbf{q}_{ku}$ (the air may be forced to flow or moves due to thermobuoyancy). The radiation heat reaches the oven internal surfaces (assuming that the air is transparent to thermal radiation). The heat flowing into the air will be convected and ultimately transferred to the oven internal surfaces by surface convection.

The heat flow from the wood to the bottom surface of the oven is by conduction $\mathbf{q}_k$.

## COMMENT

As the elapsed time (measured from the time the microwave oven power is turned on) increases, the adhesive temperature and the temperature of the foam and the wood around it increases. For short heating duration, the surfaces of the foam and the wood, away from the adhesive, may not change significantly (i.e., the heat is mostly stored in the adhesive).

### EXAMPLE 1.10. FUN

During friction ignition initiated by heating a piece of wood, heat is generated (mechanical to thermal energy conversion) at the contact surface between a piece of wood and another solid with a large coefficient of surface friction. This is shown in Figure Ex. 1.10(a).

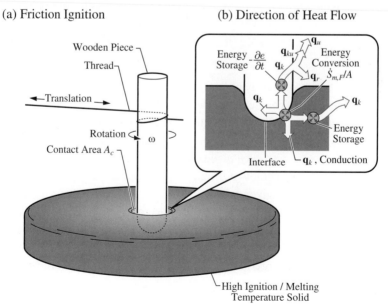

Figure Ex. 1.10. (a) Friction ignition of a piece of wood caused by a rotational contact. (b) The direction of heat flow.

Using the figure, show the location of energy conversion per unit area $\dot{S}_{m,F}/A$ and the transfer of heat through both solids. As the temperature in both solids increases with time, part of the heat conducted is stored in each solid. As the elapsed time increases, the temperature at the contact area increases until the ignition temperature of the pilot (a piece of cotton or wood fiber) is reached. In the sketch show the heat storage and conduction in each piece.

**SOLUTION**

The surface friction and the interfacial motion at the contact area between a low-ignition temperature piece of wood (generally the ignition temperature is around 400°C) and a high ignition/melting temperature (i.e., inert) solid (e.g., a stone) converts the mechanical energy (force and motion) into thermal energy (heat). The heat is generated at the interface between the two solids, $\dot{S}_{m,F}$ and depending on the thermal conductivity and the volumetric heat capacity (i.e., the product of density and mass-specific heat capacity), this heat flows unequally from the interface toward each solid by conduction $\mathbf{q}_k$. This is shown in Figure Ex. 1.10(b). As it flows, a portion of this heat raises the solid temperature $\partial e/\partial t$ (stores heat in the solid, per unit volume) and a portion continues to flow away from the interface by conduction $\mathbf{q}_k$. Some of this heat leaves the surface as surface-convection $\mathbf{q}_{ku}$ and surface radiation $\mathbf{q}_r$ heat transfer.

**COMMENT**

The wood has a relatively small conductivity and most of the heat is stored near the interface. If the wood is moist (the stored water can be assumed to be in the liquid form), part of the heat is used to evaporate the liquid water. This heat generation, storage, and conduction continues until the wood surface temperature (the highest temperature in the wood) reaches the ignition temperature of the pilot.

EXAMPLE 1.11. FUN

---

The human body constantly generates heat (conversion of a food chemical bond to thermal energy). The heat is removed from the body by conduction, convection, and radiation. When the heat is not properly removed, the temperature of the body increases above the normal, a process referred to as hyperthermia. On the other hand, with excessive heat removal from the body, the temperature of the body decreases below normal, called hypothermia.

Using the sketch given in Figure Ex. 1.11(a), consider a person under a condition of hypothermia that is generating a maximum heat under severe shivering of $\dot{S}|_V = 400$ W. However, the heat loss (a combination of conduction $\mathbf{q}_k$, surface-convection $\mathbf{q}_{ku}$, and surface radiation $\mathbf{q}_r$ heat transfer from the surface) is $Q|_A = 800$ W. The stored energy in (1.22) is represented here by the sensible heat storage (neglecting the kinetic energy) as

$$\left.\frac{\partial E}{\partial t}\right|_V = \frac{dE}{dt}\Big|_V = \rho c_v V \frac{dT}{dt} \quad \text{integral-volume energy equation,}$$

(a)

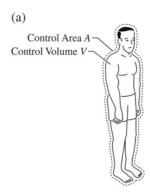

Figure Ex. 1.11. (a) Control surface around a human body.

with $\rho c_v V = 5 \times 10^5$ J/K, where $\rho$ is the density, $c_v$ is the specific heat capacity, and $V$ is the volume.

(a) Sketch the energy generation, energy storage, and surface heat transfer for the human body considered.
(b) Determine how long it will take for the body temperature to drop by 10°C. Neglect any other energy flow, and all other work and energy conversions and use a simplified version of (1.22), i.e.,

$$Q|_A = -\rho c_v V \frac{\partial T}{\partial t} + \dot{S}|_V + 0.$$

**SOLUTION**

(a) Figure Ex. 1.11(b) gives the three terms that are stated to be significant in the energy equation (1.22).
(b) We are asked to determine $\Delta t$ needed for the body to undergo a $\Delta T = -10$°C. Two of these three terms are prescribed, i.e., $\dot{S}|_V$ and $\dot{Q}|_A$. Then we have the simplified form of (1.22), i.e.,

$$Q|_A = -\frac{\partial E}{\partial t} + \dot{S}|_V + 0 \quad \text{energy equation}$$

(b)

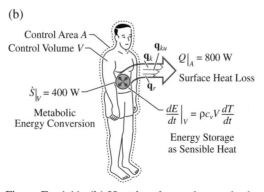

Figure Ex. 1.11. (b) Heat loss from a human body.

or

$$Q|_A = -\rho c_v V \frac{dT}{dt} + \dot{S}|_V .$$

The storage term is the time rate of change of the temperature. When $\dot{S}|_V$ and $Q|_A$ and $\rho c_v V$ do not vary in time, we can write this energy equation as

$$\frac{dT}{dt} = \frac{\dot{S}|_V - Q|_A}{\rho c_v V}$$

or

$$\int dT = \frac{\dot{S}|_V - Q|_A}{\rho c_v V} \int dt.$$

Upon integration, for a temperature change $\Delta T$ over elapsed time interval $\Delta t$ we have

$$\Delta T = \frac{\dot{S}|_V - Q|_A}{\rho c_v V} \Delta t$$

or solving for $\Delta t$ we have

$$\Delta t = \Delta T \frac{\rho c_v V}{\dot{S}|_V - Q|_A}.$$

Now using the numerical values, we have

$$\Delta t = -10(^\circ\text{C}) \frac{5 \times 10^5 (\text{J/K})}{400(\text{W}) - 800(\text{W})} = 12{,}500 \text{ s} = 3.47 \text{ hr.}$$

**COMMENT**

Although more than normal, this is not a severe heat loss rate. Therefore, three and a half hours is a reasonable elapsed time. We will develop relationships for various mechanisms for heat loss $Q|_A$, under various conditions, in Chapters 3 to 7.

EXAMPLE 1.12. FAM

In Example 1.11, the severe surface heat loss rate $Q|_A$ was given as 800 W. The person is wearing cloth and conduction $Q_k$, surface convection $Q_{ku}$ (i.e., no fluid flow through the surface), and surface radiation $Q_r$, respectively, make up 3%, 70%, and 27% of the heat loss. There is no other known surface heat transfer.

Determine the rate of heat transfer by each of the mechanisms.

**SOLUTION**

Since no prescribed surface heat transfer is given, then $Q = 0$. Then we have

$$Q|_A = Q + Q_k + Q_u + Q_{ku} + Q_r$$

$$= 0 + 0.03 Q|_A + 0 + 0.7 Q|_A + 0.27 Q|_A$$

or

$$Q_k = 24 \text{ W}, \qquad Q_{ku} = 560 \text{ W}, \qquad Q_r = 216 \text{ W}.$$

**COMMENT**

This surface-convection heat loss corresponds to a high-speed wind and a low ambient temperature. The energy loss rate due to sweating $\dot{s}_{lg}$, will be introduced through the physical-bond energy conversion mechanism (Section 2.3).

## 1.7 Conservation of Mass, Species, and Momentum*

The following sections are found on the Web site www.cambridge.org/kaviany. We show in a manner similar to that used for the energy equation, the integral-volume conservation equations for the total mass, species, and fluid momentum. We give an example involving heat transfer and phase change, Example 1.13. The Boltzmann transport equation which is also a conservation equation, is briefly mentioned.

### 1.7.1 Conservation of Mass*

### 1.7.2 Conservation of Species*

### 1.7.3 Conservation of Momentum*

### 1.7.4 Other Conserved Quantities*

## 1.8 Scope

Heat transfer is the analysis and innovative use of the (i) the transport of heat (heat flux vector), (ii) the storage of heat, and (iii) the conversion of energy (to and from the thermal energy). The aim of this textbook is to introduce readers to the analysis of heat transfer problems by providing a unified approach that can also lead to the selection of innovative techniques to solve engineering problems.

Figure 1.12 is a rendering of the principles of the heat transfer covered in this text. In the counterclockwise direction from the top, it begins with the heat flux vector and moves to the energy equation, the mechanisms of heat transfer, the energy conversion, the heat transfer medium, the intra- and intermedium heat transfer, and finally ends with the thermal circuit analysis.

The text is divided into Chapters 1 through 8 and Appendices A through C. The organization is summarized in Chart 1.5. Chapters 1 and 2 cover the introduction and preliminaries, and the energy equations and energy conversion mechanisms, respectively. In Chapter 2 the heat flux vector, energy equation, energy conversion, bounding surface thermal conditions, and the methodology for heat transfer analysis are discussed. This chapter lays the foundation for the following five chapters. The following chapters deal with conduction, radiation, and convection (Chapter 5 is on intramedium conduction and convection in unbounded fluid streams, Chapter 6 is on surface convection with semi-bounded fluid streams, and Chapter 7 is on surface convection with bounded fluid streams). The concepts of thermal nodes, thermal resistances, and thermal circuits are used extensively. As each chapter is covered, the appropriate thermal resistances (in analogy with electrical current flow and electric circuits) for that mode of heat transfer are added and used. Chapter 8 addresses heat transfer and thermal systems and contains some thermal design examples and also discusses thermal engineering analysis and problem formulation.

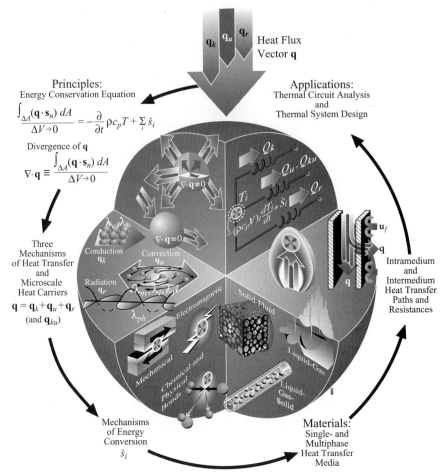

Figure 1.12. Principles of heat transfer analysis covered in the text and the iconic and thermal circuit presentations of heat transfer.

Appendix A gives some thermodynamic relations. Appendix B gives the derivation of the differential-volume energy equation. Appendix C contains tables of thermophysical and thermochemical properties. A nomenclature; glossary; answers to end-of-chapter problems; key charts, figures, and tables; and subject index are found at the end of the book.

## 1.9 Summary

In this chapter the heat flux vector $\mathbf{q}$ and its three mechanisms (conduction $\mathbf{q}_k$, convection $\mathbf{q}_u$, and radiation $\mathbf{q}_r$) were introduced by (1.7). The heat flux vector designates the surface heat flow rate and its direction. Then it was used in the integral control volume energy equation (1.23). The energy equation is cast in a manner that allows for the surface heat transfer $Q|_A$ to lead or follow the energy and work changes occurring within the control volume and across the control surface. The use of vectors has allowed us to write (1.23) in a compact form.

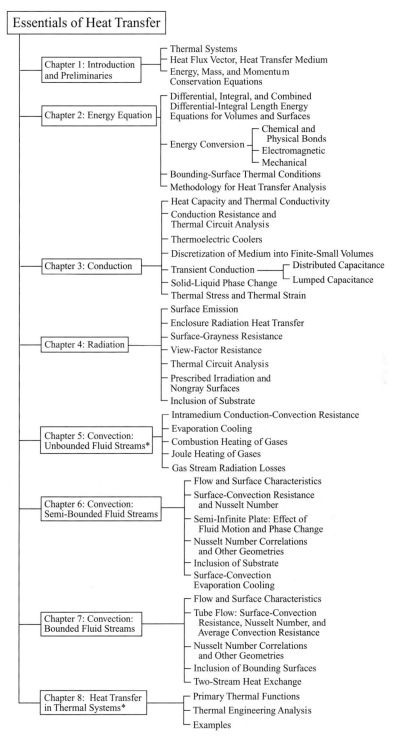

Chart 1.5. Division of the text content into chapters and the major contents of each chapter.
* Indicates online materials at www.cambridge.org/kaviany.

In Chapter 2, we will write (1.23) in a yet more compact form. Then we will apply it to integral as well as differential (as $V \to \Delta V \to 0$) volumes. This will allow us to handle problems in which we can justifiably assume and treat the heat transfer medium as having a uniform temperature, or a continuously varying temperature within it.

## 1.10  References*

This section is found on the Web site www.cambridge.org/kaviany.

## 1.11  Problems

### 1.11.1  Heat Flux Vector Tracking

**PROBLEM 1.1. FAM**[†]
Define the following terms (use words, schematics, and mathematical relations as needed).

(a) Control Volume $V$.
(b) Control Surface $A$.
(c) Heat Flux Vector $\mathbf{q}(\text{W/m}^2)$.
(d) Conduction Heat Flux Vector $\mathbf{q}_k(\text{W/m}^2)$.
(e) Convection Heat Flux Vector $\mathbf{q}_u(\text{W/m}^2)$.
(f) Surface-Convection Heat Flux Vector $\mathbf{q}_{ku}(\text{W/m}^2)$.
(g) Radiation Heat Flux Vector $\mathbf{q}_r(\text{W/m}^2)$.
(h) Net Rate of Surface Heat Transfer $Q|_A(\text{W})$.
(i) Conservation of Energy.

**PROBLEM 1.2. FUN**
An automobile radiator is a cross-flow heat exchanger (which will be discussed in Chapter 7) used to cool the hot water leaving the engine block. In the radiator, the hot water flows through a series of interconnected tubes and loses heat to an air stream flowing over the tubes (i.e., air is in cross flow over the tubes), as shown in Figure Pr. 1.2(a). The air-side heat transfer is augmented using extended surfaces (i.e., fins) attached to the outside surface of the tubes. Figure Pr. 1.2(b) shows a two-dimensional close-up of the tube wall and the fins. The hot water convects heat $\mathbf{q}_u(\text{W/m}^2)$ as it flows through the tube. A portion of this heat is transferred to the internal surface of the tube wall by surface convection $\mathbf{q}_{ku}(\text{W/m}^2)$. This heat flows by conduction $\mathbf{q}_k(\text{W/m}^2)$ through the tube wall, reaching the external tube surface, and through the fins, reaching the external surface of the fins. At this surface, heat is transferred to the air stream by surface convection $\mathbf{q}_{ku}(\text{W/m}^2)$ and to the surroundings (which include all the surfaces that surround the external surface) by surface radiation $\mathbf{q}_r(\text{W/m}^2)$. The heat transferred to the air stream by surface convection is carried away by convection $\mathbf{q}_u(\text{W/m}^2)$.

---

[†] **FAM** stands for FAMILIARITY, **FUN** stands for FUNDAMENTAL, and **DES** stands for DESIGN. Also, the **S** appearing at the end stands for SOLVER. These are explained in the Preface.

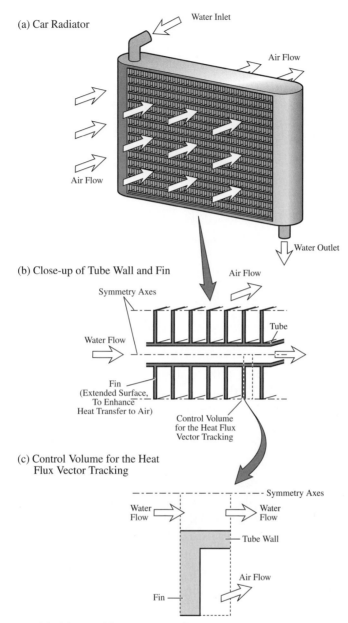

Figure Pr. 1.2. (a), (b), and (c) An automobile radiator shown at various length scales.

On Figure Pr. 1.2(c), track the heat flux vector, identifying various mechanisms, as heat flows from the hot water to the air. Assume that the radiator is operating in steady state.

**PROBLEM 1.3. FUN**
A flat-plate solar collector [Figure Pr. 1.3(a)] is used to convert thermal radiation (solar energy) into sensible heat. It uses the sun as the radiation source and water for storage of energy as sensible heat. Figure Pr. 1.3(b) shows a cross section of a flat-plate solar collector. The space between the tubes and the glass plate is occupied

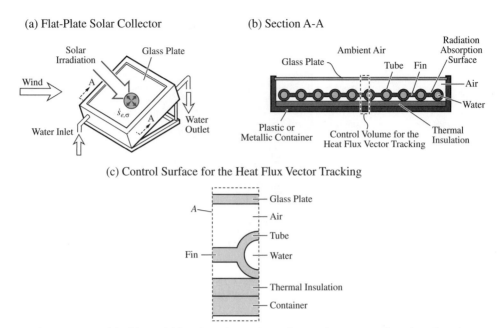

Figure Pr. 1.3. (a), (b), and (c) A flat-plate solar collector shown at various length scales.

by air. Underneath the tubes, a thermal insulation layer is placed. Assume that the glass absorbs a fraction of the irradiation and designate this heat absorbed per unit volume as $\dot{s}_{e,\sigma}(\text{W/m}^3)$. Although this fraction is small when compared to the fraction transmitted, the glass temperature is raised relative to the temperature of the air outside the solar collector. The remaining irradiation reaches the tube and fin surfaces, raising their temperatures. The temperature of the air inside the solar collector is higher than the glass temperature and lower than the tube and fin surface temperatures. Then the thermobuoyant flow (i.e., movement of the air due to density differences caused by temperature differences) causes a heat transfer by surface-convection at the glass and tube surfaces. The net heat transfer at the tube surface is then conducted through the tube wall and transferred to the flowing water by surface convection. Finally, the water flow carries this heat away by convection. Assume that the ambient air can flow underneath the solar collector.

Track the heat flux vector for this thermal system. Note that the tubes are arranged in a periodic structure and assume a two-dimensional heat transfer. Then, it is sufficient to track the heat flux vector for a control volume that includes half of a tube and half of a connecting fin, as shown on Figure Pr. 1.3(c).

**PROBLEM 1.4. FAM**
In printed-circuit field-effect transistors, conversion of electromagnetic energy to thermal energy occurs in the form of Joule heating. The applied electric field is time periodic and the heat generated is stored and transferred within the composite layers. This is shown in Figure Pr. 1.4(a). The dimensions of the various layers are rather small (submicrons). Therefore, large electrical fields and the corresponding

large heat generation can elevate the local temperature beyond the threshold for damage. On Figure Pr. 1.4(b), track the heat flux vector. Note that the electric field is transient.

(a) Physical Description of Field-Effect Transistor  (b) Dimensions

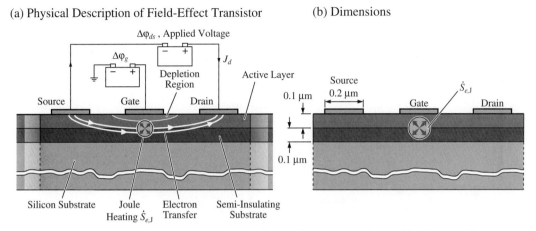

Figure Pr. 1.4. Joule heating energy conversion and heat transfer in a field-effect transistor.

## PROBLEM 1.5. FAM

The attachment of a microprocessor to a printed circuit board uses many designs. In one, solder balls are used for better heat transfer from the heat generating (Joule heating) microprocessor to the printed circuit board. This is shown in Figure Pr. 1.5.

Track the heat flux vector from the microprocessor to the heat sink (i.e., bare or finned surface exposed to moving, cold fluid) and the printed circuit board.

## PROBLEM 1.6. FAM

As part of stem-cell transplantation (in cancer treatment), the donor stem cells (bone marrow, peripheral blood, and umbilical cord blood stem cells) are stored for later transplants. Cryo-preservation is the rapid directional freezing of these cells to temperatures below $-130°$C. Cryo-preservative agents are added to lower freezing point and enhance dehydration. Cooling rates as high as $-dT/dt = 500°$C/s are used (called rapid vitrification). The cells are frozen and kept in special leak-proof vials inside a liquid nitrogen storage system, shown in Figure Pr. 1.6. At one

Physical Model of Microprocessor and
Circuit Board with Solder Balls

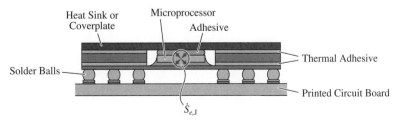

Figure Pr. 1.5. A microprocessor attached to a circuit board.

atmosphere pressure, from Table C.4, $T_{lg}(p = 1 \text{ atm}) = 77.3 \text{ K} = -195.9°\text{C}$. The storage temperature affects the length of time after which a cell can be removed (thawed and able to establish a cell population). The lower the storage temperature, the longer the viable storage period. In one protocol, the liquid nitrogen level in the storage unit is adjusted such that $T = -150°\text{C}$ just above the stored material. Then there is a temperature stratification (i.e., fluid layer formation with heavier fluid at the bottom and lighter fluid on top) with the temperature varying from $T = -196°\text{C}$ at the bottom to $T = -150°\text{C}$ at the top of the unit, as shown in Figure Pr. 1.6.

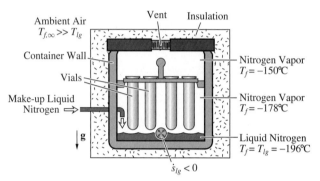

Figure Pr. 1.6. An insulated container used for storage of cyro-preserved stem cells.

Draw the steady-state heat flux vector tracking for the storage container showing how heat transfer by surface convection and then conduction flows through the container wall toward the liquid nitrogen surface. Also show how heat is conducted along the container wall to the liquid nitrogen surface. Note that $\dot{S}_{lg} < 0$ since heat is absorbed during evaporation. In order to maintain a constant pressure, the vapor is vented, and make-up liquid nitrogen is added.

## PROBLEM 1.7. FAM

Induction-coupling (i.e., electrodeless) Joule heating $\dot{S}_{e,J}$ of thermal plasmas, which are high temperature (greater than 10,000 K) ionized-gas streams, is used for particle melting and deposition into substrates. Figure Pr. 1.7 shows a plasma spray-coating system. The powder flow rate strongly influences particle temperature history $T_p(t)$, i.e., the speed in reaching the melting temperature (note that some evaporation of the particles also occurs). This is called in-flight plasma heating of particles. To protect the plasma torch wall, a high-velocity sheath-gas stream is used, along with liquid-coolant carrying tubes embedded in the wall. These are also shown in Figure Pr. 1.7.

(a) Draw the heat flux vector tracking for the (i) plasma-gases-particles, and (ii) sheath-gas streams. Allow for conduction-convection-radiation heat transfer between these two streams. Follow the plasma gas stream to the substrate.
(b) Draw the heat flux vector tracking for (iii) a single particle, as shown in Figure Pr. 1.7. Allow for surface convection and radiation and heat storage as $-\partial E/\partial t$ (this is sensible and phase-change heat storage).

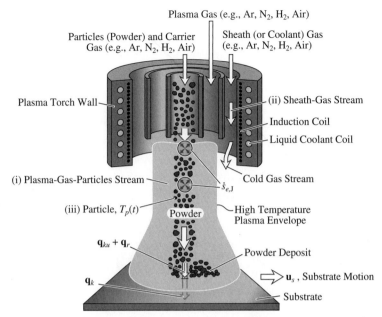

Figure Pr. 1.7. A plasma spray-coating torch showing various streams, Joule heating, and wall cooling.

## PROBLEM 1.8. FUN

A bounded cold air stream is heated, while flowing in a tube, by electric resistance (i.e., Joule heating). This is shown in Figure Pr. 1.8. The heater is a solid cylinder (ceramics with the thin, resistive wire encapsulated in it) placed centrally in the tube. The heat transfer from the heater is by surface convection and by surface-radiation emission (shown as $\dot{S}_{e,\epsilon}$). This emitted radiation is absorbed on the inside surface of the tube (shown as $\dot{S}_{e,\alpha}$) and then leaves this surface by surface convection. The outside of the tube is ideally insulated. Assume that no heat flows through the tube wall.

Draw the steady-state heat flux tracking showing the change in fluid convection heat flux vector $\mathbf{q}_u$, as it flows through the tube.

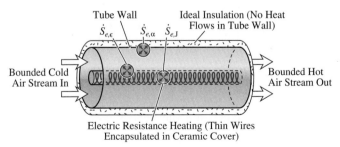

Figure Pr. 1.8. A bounded air stream flowing through a tube is heated by a Joule heater placed at the center of the tube.

## PROBLEM 1.9. FAM

Water is bounded over surfaces by raising the substrate surface temperature $T_s$ above the saturation temperature $T_{lg}(p)$. Consider heat supplied for boiling by electrical resistance heating (called Joule heating) $\dot{S}_{e,J}$ in the substrate. This is shown in Figure Pr. 1.9. This heat will result in evaporation in the form of bubble nucleation, growth, and departure. The evaporation site begins as a bubble nucleation site. Then surface-convection heat transfer $q_{ku}$ is supplied to this growing bubble (i) directly through the vapor (called vapor heating), (ii) through a thin liquid film under the bubble (called micro layer evaporation), and (iii) through the rest of the liquid surrounding the vapor. Surface-convection heat transfer is also supplied (iv) to the liquid (resulting in slightly superheated liquid) and is moved away by liquid motion induced by bubble motion and by thermobuoyancy.

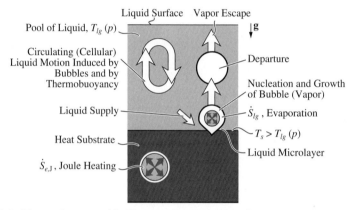

Figure Pr. 1.9. The nucleate pool boiling on a horizontal surface. The Joule heating results in raising the surface temperature above the saturation temperature $T_{lg}$, and bubble nucleation, growth, and departure.

Track the heat flux vector starting from the Joule heating site $\dot{S}_{e,J}$ within the substrate and show the surface-convection heat transfer, (i) to $q_{ku}$ (iv). Also follow the heat-flux vector to the liquid surface. Assume a time-averaged heat transfer in which the bubbles are formed and depart continuously.

## PROBLEM 1.10. FAM

Deep heat mining refers to harvesting of the geothermal energy generated locally by radioactive decay $\dot{S}_{r,\tau}$ and transferred by conduction $Q_k$ from the earth mantle [shown in Figure Ex. 1.2(a)]. Mining is done by the injection of cold water into fractured rocks (geothermal reservoir) followed by the recovery of this water, after it has been heated (and pressurized) by surface-convection $q_{ku}$ in the fractures, through the production wells. These are shown in Figure Pr. 1.10. The heated water passes through a heat exchanger and the heat is used for energy conversion or for process heat transfer. Starting from the energy conversion sources $\dot{S}_{r,\tau}$ and the heat conduction from lower section $Q = Q_k$, draw the steady-state heat flux vector tracking and show the heat transfer to the cold stream by surface convection $q_{ku}$.

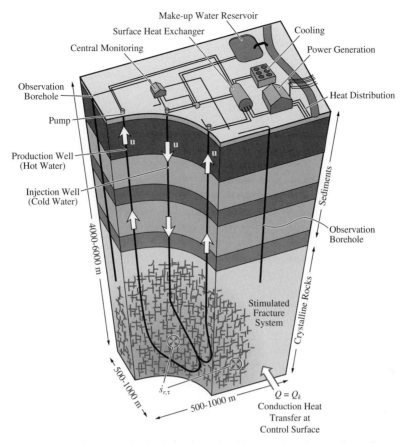

Figure Pr. 1.10. Deep heat mining by injection of cold water into hot rocks and recovery of heated water.

Note that heat is first conducted through the rock before it reaches the water stream. Follow the returning warm water stream to the surface heat exchanger.

## PROBLEM 1.11. FUN

In a seabed hydrothermal vent system, shown in Figure Pr. 1.11, cold seawater flows into the seabed through permeable tissues (fractures) and is heated by the body of magma. The motion is caused by a density difference, which is due to the temperature variations, and is called a thermobuoyant motion (it will be described in Chapter 6). Minerals in the surrounding rock dissolve in the hot water, and the temperature-tolerant bacteria release additional metals and minerals. These chemical reactions are represented by $\dot{S}_{r,c}$ (which can be both endo- and exothermic). Eventually, the superheated water rises through the vent, its plume forming a "black smoker." As the hot water cools, its metal content precipitates, forming concentrated bodies of ore on the seabed.

Draw the heat flux vector tracking for the volume marked as $V$. Note that water flows in the permeable seabed, and therefore, convection should be included (in

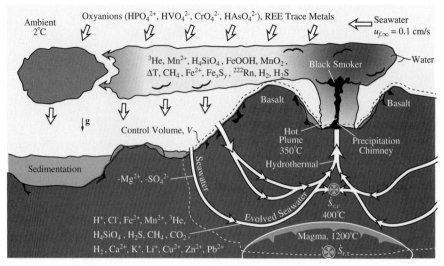

Figure Pr. 1.11. A hydrothermal vent system showing the temperature at several locations and the thermobuoyant flow.

addition to conduction). This is called the intramedium convection (as compared to surface convection) and will be discussed in Chapter 5.

### PROBLEM 1.12. FAM

Electric current-carrying wires are electrically insulated using dielectric material. For low temperature, a polymeric solid (such as Teflon) is used, and for high temperature application (such as in top range electrical oven), an oxide ceramic is used. Figure Pr. 1.12 shows such a wire covered by a layer of Teflon. The Joule heating $\dot{S}_{e,J}$ produced in the wire is removed by a cross flow of air, with air far-field temperature $T_{f,\infty}$ being lower than the wire temperature $T_w$.

Draw the steady-state heat flux vector tracking, starting from the heating source, for this heat transfer problem. Allow for surface radiation (in addition to surface convection).

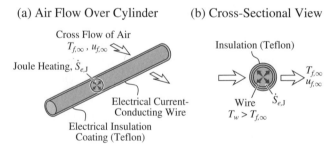

Figure Pr. 1.12. (a) and (b) An electrical current-carrying wire is covered with a layer of electrical insulation, and Joule heating is removed by surface convection and surface-radiation heat transfer.

Oscillating Electric Field Intensity

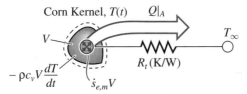

Figure Pr. 1.13. Thermal circuit model for a corn kernel heated by microwave energy conversion.

### 1.11.2 Integral-Volume Energy Equation

**PROBLEM 1.13. FAM**

Popcorn can be prepared in a microwave oven. The corn kernels are heated to make the popcorn by an energy conversion from oscillating electromagnetic waves (in the microwave frequency range) to thermal energy designated as $\dot{s}_{e,m}(\text{W/m}^3)$. With justifiable assumptions for this problem, (1.23) can be simplified to

$$Q\mid_A = -\rho c_v V \frac{dT}{dt} + \dot{s}_{e,m}V, \qquad \text{integral-volume energy equation,}$$

where the corn kernel temperature $T$ is assumed to be uniform, but time dependent. The control volume for a corn kernel and the associated energy equation terms are shown in Figure Pr. 1.13.

The surface heat transfer rate is represented by

$$Q\mid_A = \frac{T(t) - T_\infty}{R_t},$$

where $T_\infty$ is the far-field ambient temperature and $R_t(\text{K/W})$ is the constant heat transfer resistance between the surface of the corn kernel and the far-field ambient temperature.

(a) For the conditions given below, determine the rise in the temperature of the corn kernel for elapsed time $t$ up to 5 min. In Section 3.6.2, we will show that the exact analytical solution is

$$T(t) = T(t = 0) + \dot{s}_{e,m}VR_t(1 - e^{-t/\tau}), \quad \tau = \rho c_p VR_t.$$

(b) At what elapsed time does the temperature reach 100°C?

$\rho = 1{,}000 \text{ kg/m}^3$, $c_v = 1{,}000 \text{ J/kg-K}$, $V = 1.13 \times 10^{-7} \text{ m}^3$, $\dot{s}_{e,m} = 4 \times 10^5 \text{ W/m}^3$, $T(t = 0) = 20°\text{C}$, $T_\infty = 20°\text{C}$, $R_t = 5 \times 10^3 \text{ K/W}$.

**PROBLEM 1.14. FAM**

In severely cold weathers, an automobile engine block is kept warm prior to startup, using a block Joule heater at a rate $\dot{S}_{e,J}$ with the electrical power provided through the household electrical circuit. This is shown in Figure Pr. 1.14. The heat generated conducts through the block of mass $M$ and then is either stored within the volume

$V$ or lost through the surface $A$. The energy equation (1.22) applies to the control surface $A$.

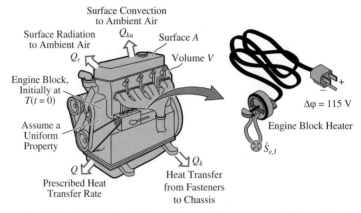

Figure Pr. 1.14. A block Joule heater inserted in an automobile engine block.

Consider that there is no heat transfer by convection across the surface $A$, i.e., $Q_u = 0$. The conduction heat transfer rate (through the fasteners and to the chassis) is $Q_k$, the surface-convection heat transfer rate (to the ambient air) is $Q_{ku}$, and the surface-radiation heat transfer rate (to the surrounding surface) is $Q_r$. In addition, there is a prescribed heat transfer rate $Q$ (not related to any heat transfer mechanism).

(a) Draw the heat flux vector tracking starting from the Joule heating site.
(b) By applying the energy conservation equation to the control volume surface, determine the rate of change of the block temperature $dT/dt$, for the following condition. Use (1.22) and set all terms on the right-hand side except the first and last terms equal to zero. Use $\partial E/\partial t = Mc_v dT/dt$ and the conditions given below. The last term is equal to $\dot{S}_{e,J}$.

$Q = 20$ W, $Q_{ku} = 80$ W, $Q_k = 30$ W, $Q_r = 15$ W, $\dot{S}_{e,J} = 400$ W, $c_v = 900$ J/kg-K, $M = 150$ kg.

## PROBLEM 1.15. FAM

In spark-ignition engines, the electrical discharge produced between the spark plug electrodes by the ignition system produces thermal energy at a rate $\dot{S}_{e,J}$(W). This is called the Joule heating and will be discussed in Section 2.3. This energy conversion results in a rise in the temperature of the electrodes and the gas surrounding the electrodes. This high-temperature gas volume $V$, which is called the plasma kernel, is a mixture of air and fuel vapor. This plasma kernel develops into a self-sustaining and propagating flame front.

About $\int_A Q|_A \, dt = -1$ mJ is needed to ignite a stagnant, stoichiometric fuel-air mixture of a small surface area $A$ and small volume $V$, at normal engine conditions. The conventional ignition system delivers 40 mJ to the spark.

(a) Draw the heat flux vector tracking for the region around the electrodes marked in Figure Pr. 1.15. Start from the energy conversion source $\dot{S}_{e,J}$.

Igniting Gas Kernel by Spark Plug

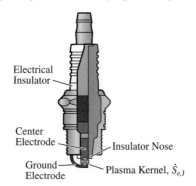

Figure Pr. 1.15. Ignition of a fuel-air mixture by a spark plug in a spark-ignition engine. The plasma kernel is also shown.

(b) Assume a uniform temperature within the gas volume $V$. Assume that all terms on the right-hand side of (1.22) are negligible, except for the first term. Represent this term with

$$\frac{\partial E}{\partial t}\Big|_V = (\rho c_v V)_g \frac{dT_g}{dt}.$$

Then for the conditions given below, determine the final gas temperature $T_g(t_f)$, where the initial gas temperature is $T_g(t = 0)$.
(c) What is the efficiency of this transient heating process?

$\quad (\rho c_v V)_g = 2 \times 10^{-7} \text{ J/}^\circ\text{C}, \ T_g(t = 0) = 200^\circ\text{C}.$

### PROBLEM 1.16. FAM

The temperature distributions for the exhaust gas and the exhaust pipe wall of an automotive exhaust system are shown in Figures Pr. 1.16(a) and (b). The exhaust gas undergoes a temperature difference $\langle T_f \rangle_0 - \langle T_f \rangle_L$ over the upper-pipe region (between the exhaust manifold and the catalytic converter). It can be shown that when the energy equation (1.23) is written for this upper-pipe region, as shown in the figure, and under steady-state conditions, the right-hand side of this equation is zero. Then the energy equation becomes

$$Q\,|_A = 0 \qquad \text{integral-volume energy equation.}$$

The surface heat flows are convection on the left and right surfaces and surface convection and radiation from the other sides, i.e.,

$$Q\,|_A = Q_{u,L} - Q_{u,0} + Q_{ku} + Q_r = 0.$$

The convection heat flow rates are written (as will be shown in Chapter 2 and Appendix B) as

$$Q_{u,L} - Q_{u,0} = \dot{M}_f c_{p,f}(\langle T_f \rangle_L - \langle T_f \rangle_0),$$

where $\dot{M}_f$ (kg/s) is the gas flow rate and $c_{p,f}$ (J/kg-K) is the specific heat capacity at constant pressure.

(a) Temperature Distribution Throughout an Exhaust Pipe

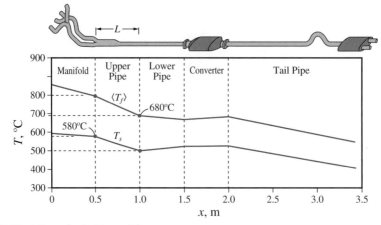

(b) Heat Transfer in Upper Pipe

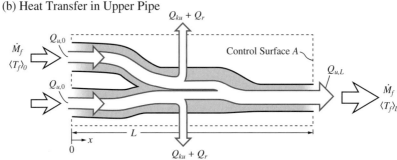

Figure Pr. 1.16. (a) Temperature variation along an automotive exhaust pipe. (b) Control surface and heat transfer for the upper pipe.

For $\dot{M}_f = 0.10$ kg/s and $c_{p,f} = 1,000$ J/kg-K, and using the temperatures given in Figure Pr. 1.16, determine the sum of the surface convection and radiation heat transfer rates.

### PROBLEM 1.17. FUN

On a clear night with a calm wind, the surface of a pond can freeze even when the ambient air temperature is above the water freezing point ($T_{sl} = 0°C$). This occurs due to heat transfer by surface radiation $q_r$(W/m²) between the water surface and the deep night sky. The ice-layer thickness is designated as $\delta_\alpha(t)$. These are shown in Figure Pr. 1.17(a).

In order for freezing to occur and continue, the net heat flow rate from the ice surface must be enough to cool both the liquid water and the ice layer and also allow for the phase change of the water from liquid to solid. Assume that the ambient temperature is $T_\infty = 3°C$, the temperature of the deep sky is $T_{sky} = 0$ K, the earth atmosphere has an average temperature around $T_{atm} = 230$ K, and the temperature of the water at the bottom of the pool is $T_l = 4°C$. Then for this transient heat transfer problem between the deep sky, the ambient air, and the water pool:

(a) Track the heat flux vector for the section shown on Figure Pr. 1.17(b),

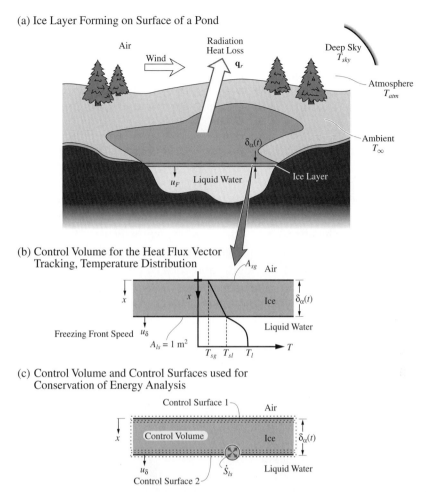

Figure Pr. 1.17. (a), (b), and (c) Ice formation at the surface of a pond and the control volume and control surfaces selected for heat transfer analysis. The temperature distribution is also shown.

(b) For the control volume and control surfaces shown in Figures Pr. 1.17(b) and (c) apply the energy conservation equation (1.22). Note that the control volume and surfaces are for only the ice layer, i.e., the control surface 1 includes only the interface between the ice and the ambient air, and control surface 2 includes only the interface between the ice and the liquid water. For this problem the kinetic energy flux and all the work terms in (1.22) are negligible. For control surface 2 (water/ice interface), due to its zero mass (the control surface is wrapped around the interface), the sensible energy storage is zero but there is a latent heat generated due to the phase change from liquid to solid. (Later in Chapter 2, the latent heat will be separated from the sensible heat and treated as an energy conversion mechanism.) Therefore, for control surface 2, (1.22) becomes

$$Q|_A = \dot{S}_{ls} = -A_{ls}\dot{m}_{ls}\Delta h_{ls}.$$

To evaluate $Q|_A$, use (1.8).

(c) For control surface 2 (water/ice interface), at some elapsed time, the following data applies. The conduction heat flux in the ice is $q_{k,x} = +250$ W/m$^2$, the surface convection heat flux on the water side is $q_{ku,x} = -200$ W/m$^2$, the heat absorbed by the interface solidifying is $\dot{S}_{ls}/A_{ls} = -\dot{m}_{ls}\Delta h_{ls}$ where the heat of solidification $\Delta h_{ls} = -3.34 \times 10^5$ J/kg and $\dot{m}_{ls}$(kg/s-m$^2$) is the rate of solidification and is equal to $\rho_s u_\delta$. For the density of ice use $\rho_s = 913$ kg/m$^3$. Then determine the speed of the ice/water interface movements $u_\delta$(m/s). Assume that the heat flux is one dimensional and $A_{ls} = 1$ m$^2$.

## PROBLEM 1.18. FAM

The temperature of the earth's surface and its atmosphere are determined by various electromagnetic energy conversions and, to a smaller extent, by the radioactive decay (within the earth) $\dot{S}_{r,\tau}$. These are shown in Figure Pr. 1.18 [which is based on the materials presented in Figures Ex. 1.2(a) and (b)]. Starting with solar irradiation $(q_{r,i})_s$, this irradiation is partly absorbed by the atmospheric gases $(\dot{S}_{e,\tau})_s$, partly reflected $(q_{r,i})_\rho$, and the remainder is absorbed by the earth surface $(\dot{S}_{e,\alpha})_s$. The earth's surface also emits radiation $\dot{S}_{e,\epsilon}$ and this mostly infrared radiation is partly absorbed (mostly by the greenhouse gases, such as $CO_2$) in the atmosphere $(\dot{S}_{e,\tau})_i$ and this is in turn re-emitted $(\dot{S}_{e,\tau})_i = (\dot{S}_{e,\epsilon})_i$.

(a) Complete the heat flux vector tracking by drawing the radiation $q_r$ and conduction $q_k$ heat flux vectors arriving and leaving the earth control surface $A$, also shown in Figure Pr. 1.18. Assume a steady-state heat transfer and neglect convection.

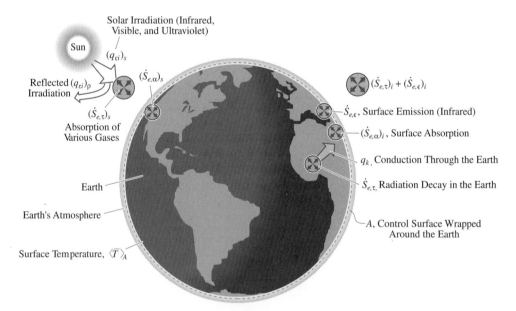

Figure Pr. 1.18. Solar irradiation and internal radiation heating of the earth and its surface and infrared, radiation emission (part of this is absorbed and emitted by the earth atmosphere).

(b) Starting from (1.22) and assuming a steady state with the left-hand side approximated as $Q_k = -Aq_k$, $A = 4\pi R^2$, $q_k = 0.078$ W/m$^2$, and the right-hand side approximated by

$$(\dot{S}_{e,\alpha})_s + \dot{S}_{e,\epsilon} + (\dot{S}_{e,\alpha})_i,$$

where

$$(\dot{S}_{e,\alpha})_s = A \times 172.4(\text{W/m}^2) \quad \text{time and space average solar irradiation}$$
$$\dot{S}_{e,\epsilon} + (\dot{S}_{e,\alpha})_i = -A(1 - \alpha_{r,i})\sigma_{\text{SB}}\langle\overline{T}\rangle_A^4, \quad \sigma_{\text{SB}} = 5.67 \times 10^{-8} \text{ W/m}^2\text{-K}^4,$$

determine the time-space averaged earth surface temperature $\langle\overline{T}\rangle_A$ for $\alpha_{r,i} = 0.55$.

### PROBLEM 1.19. FAM

Sodium acetate (trihydrate) is used as a liquid-solid phase-change heater. It has a heat of melting of $\Delta h_{sl} = 1.86 \times 10^5$ J/kg and melts/freezes at $T_{ls} = 58°$C [Table C.5(a)]. It can be kept in a sealed container (generally a plastic bag) as liquid in a metastable state down to temperatures as low as $-5°$C. Upon flexing a metallic disk within the liquid, nucleation sites are created at the disk surface, crystallization begins, heat is released, and the temperature rises. Consider a bag containing a mass $M = 100$ g of sodium acetate. Assume that the liquid is initially at $T = 58°$C and that during the phase change the transient surface heat transfer rate (i.e., heat loss) is given by

$$Q|_A = Q_o(1 - t/\tau),$$

where $Q_o = 50$ W. This is shown in Figure Pr. 1.19. Start from (1.22) and replace the time rate of change of the internal energy with $\partial E/\partial t|_V = -\dot{S}_{ls}$, where $\dot{S}_{ls} = -\dot{M}_{ls}\Delta h_{ls} = \dot{M}_{ls}\Delta h_{sl}$. This represents isothermal phase change. Then in the absence

Plastic Bag Containing Phase-Change (Liquid-Solid) Material

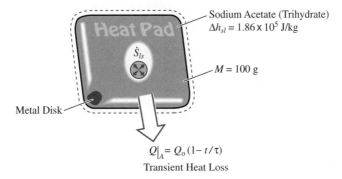

Figure Pr. 1.19. Surface heat transfer from a plastic bag containing phase-change material.

of any other work and energy conversion, this change in internal energy balances with surface heat losses.

Determine $\tau$, the elapsed time during which all the heat released by phase change will be removed by this surface heat transfer.

### PROBLEM 1.20. FAM

Nearly all of the kinetic energy of the automobile is converted into friction heating $\dot{S}_{m,F}$ during braking. The front wheels absorb the majority of this energy. Figure Pr. 1.20 shows a disc brake. This energy conversion raises the rotor temperature $T_r$ and then heat flows from the rotor by conduction (to axle and wheel), by surface radiation to the surroundings and by surface convection to the air. The air flows over the rotor in two parts; one is over the inboard and outboard surfaces, and the other is through the vanes (passages). The air flow is due mostly to rotation of the rotor (similar to a turbomachinary flow).

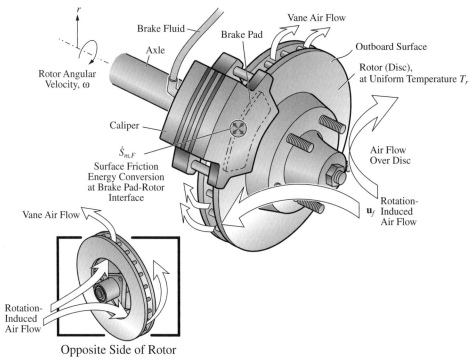

Figure Pr. 1.20. An automobile disc brake showing the air flow over the disc and through the rotor vanes.

Assume that the rotor is at a uniform temperature (this may not be justifiable during rapid braking).

(a) Draw the heat flux vector tracking for the rotor, by allowing for the heat transfer mechanisms mentioned above.

(b) Now consider the heat storage/release mechanism represented by $-\partial E/\partial t \mid_V$, in (1.22). During quick brakes, the magnitude of net rate of heat transfer $Q \mid_A = Q_{A,r} = -\partial E/\partial T \mid_V$ is much smaller than $\partial E/\partial t \mid_V$ and $\dot{S}_{m,F}$. Assume all other terms on the right-hand side of (1.22) are negligible. With no heat transfer, determine the rate of rise in the rotor temperature $dT/dt$, using

$$\frac{\partial E}{\partial t}\bigg|_V = M_r c_v \frac{dT_r}{dt},$$

$M_r = 15$ kg, $\dot{S}_{m,F} = 6 \times 10^4$ W and $c_v = 460$ J/kg-K.

# 2

# Energy Equation

Where the telescope ends, the microscope begins.
Which of the two has the grander view?

– V. Hugo

While the assumption of a uniform temperature within a volume or along a given direction may be valid in some applications, there are applications where the temperature nonuniformity needs to be addressed. To do this, we will use a differential length along directions with nonuniform temperature. Then by analytically integrating over the desired length, we will obtain the continuous variation of the temperature in these directions.

In this chapter we follow our presentation of the energy equation in the form (1.23) and with the heat flux vector $\mathbf{q}$ as the focus in this equation. Then we present, in a compact form, the most general form of the energy equation, which is for a

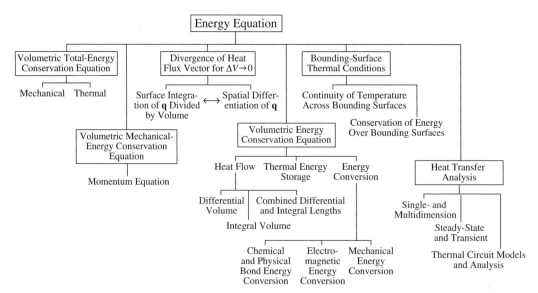

Chart 2.1. A summary of the topics related to the energy equation and those presented in Chapter 2.

differential (infinitesimal) volume $\Delta V \to 0$. Then we rewrite the energy equation for an integral volume and also for a volume made of combined differential and integral lengths. Then we discuss the energy conversion mechanisms. Next the thermal conditions at the bounding surfaces of heat transfer media are discussed. Then, the elements of heat transfer analysis are examined. Chart 2.1 gives a summary of these topics. This chapter lays the foundation, and the needed forms of the energy equation, for the heat transfer analysis presented in the following chapters. As these chapters unfold, and the various forms of the energy equation and bounding surface conditions are used, the central role of this chapter becomes clearer.

### EXAMPLE 2.1. FUN

Since surface-convection heat transfer rate $Q_{ku}$ depends on the surface-convection heat transfer surface area $A_{ku}$, in order to enhance $Q_{ku}$, extra (or extended) surfaces are used (these are called fins). In order to effectively use these extra surfaces, they are made thin (so that a large number of them can be attached) and long (to increase $A_{ku}$). Also, a high thermal conductivity material, such as aluminum, is used. Then for a three-dimensional fin shown in Figure Ex. 2.1(a), the temperature variation along the $x$ direction becomes the most pronounced, as compared to those along the $y$ and $z$ directions. The system is in a steady state (the $\partial/\partial t$ terms are zero, i.e., their arguments do not vary with time).

(a) Show the track of the heat flux vector (for one fin) as it originates from the base, where the energy conversion $\dot{S}$ occurs, and leaves as $\mathbf{q}_u$ with the ambient fluid flow.

(b) Choose a control volume that has an infinitesimal length $\Delta x \to 0$ along $x$, but integral lengths $l$ and $w$ along the $y$ and $z$ directions. Write the expressions for $\Delta V \to 0$, $A_k$, and $\Delta A_{ku} \to 0$.

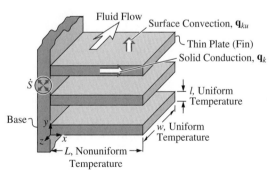

Figure Ex. 2.1. (a) Fins attached to a surface for enhancing surface-convection heat transfer.

### SOLUTION

(a) Figure Ex. 2.1(b) renders the heat flux vector tracking that begins with conduction through the base. This is followed by conduction through the fin and surface convection from the fin surface. Finally, the heat flows by convection along the fluid stream.

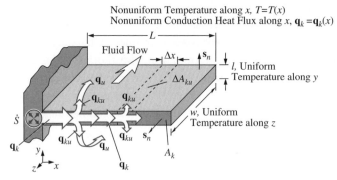

Figure Ex. 2.1. (b) Assumption of nonuniform temperature along the $x$ axis, and uniform temperature along the $y$ and $z$ axes and resulting differential area $\Delta A_{ku}$.

(b) The control volume with differential length $\Delta x$ along the $x$ direction is shown in Figure Ex. 2.1(b). Along the $y$ and $z$ directions, we have the integral lengths $l$ and $w$. The differential volume $\Delta V$ is

$$\Delta V = lw\Delta x, \quad \Delta x \to 0 \text{ differential volume with one-differential length.}$$

The conduction surface area $A_k$ is

$$A_k = lw, \quad \Delta x \to 0 \text{ differential area with one-differential length, for surface convection.}$$

The differential surface-convection surface area is

$$\Delta A_{ku} = 2l\Delta x + 2w\Delta x, \quad \text{finite surface area for conduction,}$$

where we have included the front and back, and the top and bottom surfaces of the differential surface area. Here we have a steady-state, one-dimensional temperature distribution $T = T(x)$ and conduction heat flux vector $\mathbf{q}_k = \mathbf{q}_k(x)$.

**COMMENT**

The choice of directions with significant temperature variations depends on the problem considered and the desired-acceptable (justifiable) extent of the approximation made in the analysis. For example, for $L > 10l$, the assumption of temperature uniformity along the $y$ direction is justifiable for metallic fins. The $z$-direction temperature uniformity is more readily justified for high conductivity base materials, which results in a uniform heat flow through the fin base ($x = 0$). In the limit of $\Delta x \to 0$, we will have a differential spatial limit in the $x$ direction.

## 2.1  Nonuniform Temperature Distribution: Differential (Infinitesimal)-Volume Energy Equation

For a stationary integral control volume $V$, (1.23) gives, on the left-hand side, the net rate of heat flow across the control surface $A$ and, on the right-hand side, the rest of the terms. Equation (1.23) is the integral-volume, total (mechanical and thermal) energy equation.

For the most general treatment, we are interested in the continuous variation of temperature within the heat transfer medium. Therefore, to allow for a pointwise variation of temperature, we need to consider a diminishing volume. This diminishing volume is the limit of an infinitesimal (or differential) volume $\Delta V$ taken as it approaches zero, i.e., $\Delta V \rightarrow 0$. Here we seek a pointwise (or infinitesimally small volume) energy equation.

We consider a volume that is infinitesimal (vanishingly) small, i.e., $\Delta V \rightarrow 0$. However, this infinitesimal volume is large enough such that the local thermodynamic quantities such as density, pressure, and temperature, and local kinematic quantities such as velocity, and local transport properties such as thermal conductivity and viscosity, are all independent of its size.

To this infinitesimal volume, we apply the energy conservation equation (1.23). The details are given in Appendix B and include the derivation of the differential-volume total (mechanical and thermal) and mechanical energy equations. The latter is the dot product of the fluid momentum equation (1.27) and the fluid velocity. It is important to become familiar with the derivation of the differential-volume thermal energy equation and its relationship to the total energy equation. This includes references to the momentum conservation equation (1.27) and the mass conservation equation (1.25). However, to understand and use the materials developed here, we begin with the results of the derivations given in Appendix B and proceed to develop a physical interpretation of the thermal energy equation.

The differential-volume $(\Delta V \rightarrow 0)$ energy equation is (1.23) divided by $\Delta V$ (and allowing $\Delta V \rightarrow 0$, i.e., taking the differential limit). As shown in Appendix B, the differential-volume thermal energy conservation equation, written with the temperature as the variable, is that given by (B.47), for constant $c_p$. This equation is repeated here. Note that $\sum_i \dot{s}_i$ is the sum of the volumetric energy conversion mechanisms, between the thermal energy and the chemical- and physical-bond, electromagnetic, and mechanical energy. Equation (B.47) is

$$
\begin{aligned}
\nabla \cdot \mathbf{q} \quad &\equiv \quad \nabla \cdot (\ \mathbf{q}_k \ + \ \mathbf{q}_u \ + \ \mathbf{q}_r) \\
&= \quad \nabla \cdot (-k \nabla T + \rho c_p \mathbf{u} T + \ \mathbf{q}_r)
\end{aligned}
$$

$$
\underset{\substack{\text{divergence of} \\ \text{heat flux vector}}}{} \qquad \underset{\text{spatial rate of change of heat flux vector}}{}
$$

$$
= \quad \underbrace{-\frac{\partial}{\partial t} \rho c_p T}_{\substack{\text{time rate of sensible-} \\ \text{heat storage/release}}} \quad + \quad \sum_i \dot{s}_i \left\{ \begin{array}{l} \text{chemical- and physical-bond energy conversion} \\ \text{electromagnetic energy conversion} \\ \text{mechanical energy conversion} \end{array} \right.
$$

differential-volume energy equation. \hfill (2.1)

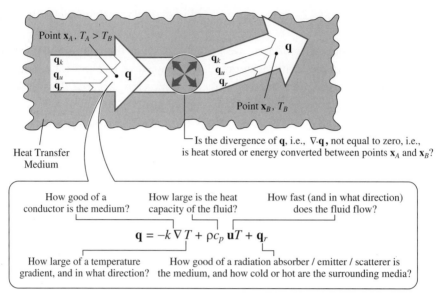

Figure 2.1. Heat flow toward and away from a point in a nonisothermal (i.e., heat transfer) medium. The iconic presentation of the three mechanisms of heat transfer and energy storage-conversion are also shown.

In examining the units we note that $\nabla \cdot \mathbf{q}(1/m)(W/m^2) = (W/m^3)$, $\nabla(-k\nabla T)$ $(1/m)(W/m\text{-}K)(1/m)(^\circ C) = (W/m^3)$, $\nabla \cdot \rho c_p T \mathbf{u}(1/m)(kg/m^3)(J/kg\text{-}^\circ C)(^\circ C)(m/s) = (W/m^3)$, $\nabla \cdot \mathbf{q}_r (1/m)(W/m^2) = (W/m^3)$, $\partial \rho c_p T/\partial t(kg/m^3)(J/kg\text{-}^\circ C)(^\circ C/s) = (W/m^3)$, and $\dot{s}_i(W/m^3)$, so all terms are in $W/m^3$.

In Figure 1.3, we rendered the heat flux vector in a nonisothermal medium. Figure 2.1 shows two points $\mathbf{x}_A$ and $\mathbf{x}_B$ in a heat transfer medium. The center of the heat flux vectors is the location $\mathbf{x}$ for which $\mathbf{q}$ is shown. The mechanisms of heat transfer are also shown. The substance located between $\mathbf{x}_A$ and $\mathbf{x}_B$ may undergo energy storage or energy conversion and this will result in a difference between $\mathbf{q}$ at locations $\mathbf{x}_A$ and $\mathbf{x}_B$.

Figure 2.2 shows the three mechanisms of heat transfer as heat flows about a point in the heat transfer medium. Figure 2.2(a) is for all three mechanisms of heat transfer, i.e., conduction, convection, and radiation. Figure 2.2(b) is for conduction and convection, and Figure 2.2(c) is for conduction only. The associated differential-volume energy equations are also shown.

### 2.1.1 Physical Interpretation of Divergence of $\mathbf{q}$

A physical interpretation of (2.1) can be given in terms of the sign of the divergence of $\mathbf{q}$, i.e., $\nabla \cdot \mathbf{q}$. This is

$$\nabla \cdot \mathbf{q} = -\frac{\partial}{\partial t}\rho c_p T + \sum_i \dot{s}_i = \begin{cases} > 0 & \text{source} \\ < 0 & \text{sink} \\ = 0 & \text{zero divergence.} \end{cases} \tag{2.2}$$

(a) Conduction, Convection, and Radiation

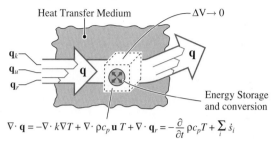

$$\nabla \cdot \mathbf{q} = -\nabla \cdot k\nabla T + \nabla \cdot \rho c_p \mathbf{u} \, T + \nabla \cdot \mathbf{q}_r = -\frac{\partial}{\partial t}\rho c_p T + \sum_i \dot{s}_i$$

(b) Conduction and Convection

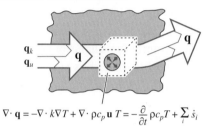

$$\nabla \cdot \mathbf{q} = -\nabla \cdot k\nabla T + \nabla \cdot \rho c_p \mathbf{u} \, T = -\frac{\partial}{\partial t}\rho c_p T + \sum_i \dot{s}_i$$

(c) Conduction

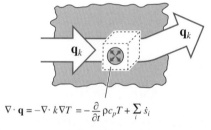

$$\nabla \cdot \mathbf{q} = -\nabla \cdot k\nabla T = -\frac{\partial}{\partial t}\rho c_p T + \sum_i \dot{s}_i$$

Figure 2.2. The heat flux vector and the differential-volume thermal energy equation with energy storage and conversion. (a) All heat transfer mechanisms are significant. (b) Only conduction and convection mechanisms are significant. (c) Only conduction mechanisms are significant.

In Figure 2.2, we made an iconic presentation for $\nabla \cdot \mathbf{q} \neq 0$, i.e., nonzero divergence of the heat flux vector by using two crossing, double-headed arrows, These indicate a "diverging" heat flux field. Divergence is defined as the tendency to move (i.e., branch off) in different directions from the same point. Now, if $\nabla \cdot \mathbf{q}$ is positive (i.e., $> 0$), in an infinitesimal volume around a point $\mathbf{x}$ in the heat transfer medium, this point is said to be a thermal source location. The source is either due to release of stored energy (i.e., $\partial T/\partial t < 0$) or due to conversion of energy to thermal energy ($\dot{s} > 0$). Then it is said that heat is diverging (i.e., moving out) from the location $\mathbf{x}$. If $\nabla \cdot \mathbf{q}$ is negative, then the location $\mathbf{x}$ is a sink. Note that the right-hand side of (2.2) contains two terms, and it is the sum of the two terms that should be positive, negative, or zero. A vector, such as $\mathbf{q}$, with zero divergence is called solenoidal (i.e., $\mathbf{q}$ is called a solenoidal vector field).

Under $\nabla \cdot \mathbf{q} = 0$, we have

$$\nabla \cdot \mathbf{q} \equiv \frac{\displaystyle\int_{\Delta A} (\mathbf{q} \cdot \mathbf{s}_n) dA}{\Delta V \to 0} \equiv \frac{Q \mid_{\Delta A}}{\Delta V \to 0} = 0 \qquad \begin{array}{l} \text{physical significance of} \\ \text{zero divergence for } \mathbf{q} \end{array} \qquad (2.3)$$

or

$$\int_{\Delta A} (\mathbf{q} \cdot \mathbf{s}_n) dA = Q \mid_{\Delta A} = 0 \qquad \text{zero divergence for } \mathbf{q}, \qquad (2.4)$$

which simply states that the area-integral of $\mathbf{q}$ around a (differential) control volume is zero, or no net heat transfer occurs across the (differential) control volume.

Chart 2.2 gives some examples of heat transfer with $\nabla \cdot \mathbf{q} = 0$. Since there is no energy storage or conversion, the local rate of change in $\mathbf{q}$ (change in its direction and in contributions from different mechanisms) is zero. The examples include applications where the direction of the heat flux vector $\mathbf{q}$ does not change, but the dominant mechanisms of heat transfer do change. For example, conduction may be the dominant heat transfer mechanism in medium 1 (for example a gas), but radiation or convection in other medium (for example a solid). In other applications, the direction and the dominant mechanisms both change. For example, heat transfer by conduction in a solid changes direction as it flows by surface convection into a parallel flow, fluid stream. In all of these examples, at any location in any of the media, we have $\nabla \cdot \mathbf{q} = 0$.

### 2.1.2 Relation between Volumetric Differentiation and Surface Integration

As given by (2.3), the area integral (i.e., over control surface $A$ covering the control volume $V$) of $\mathbf{q}$ is related to the volume integral (i.e., over $V$) of $\nabla \cdot \mathbf{q}$. The relation is given by the divergence theorem of Gauss, which states that

$$\int_V (\nabla \cdot \mathbf{q}) \, dV \equiv \int_A (\mathbf{q} \cdot \mathbf{s}_n) \, dA \qquad \begin{array}{l} \text{divergence theorem for volume } V \\ \text{having surface area } A \end{array} \qquad (2.5)$$

or

$$\int_{\Delta V} (\nabla \cdot \mathbf{q}) \, dV \equiv \int_{\Delta A} (\mathbf{q} \cdot \mathbf{s}_n) \, dA = Q \mid_{\Delta A} \qquad \begin{array}{l} \text{divergence theorem for} \\ \text{volume } \Delta V \text{ having} \\ \text{differential surface area } \Delta A. \end{array} \qquad (2.6)$$

This can be rewritten in a form similar to (2.3), i.e.,

$$\nabla \cdot \mathbf{q} \equiv \lim_{\Delta V \to 0} \frac{\displaystyle\int_{\Delta A} (\mathbf{q} \cdot \mathbf{s}_n) \, dA}{\Delta V} = \lim_{\Delta V \to 0} \frac{Q \mid_{\Delta A}}{\Delta V} \qquad \begin{array}{l} \text{relation between divergence} \\ \text{and surface integral of } \mathbf{q}. \end{array} \qquad (2.7)$$

This is the equivalence between the volumetric differentiation and surface integration in the limit of $\Delta V \to 0$ (along with it $\Delta A \to 0$). The steps for the derivation of this relation are given in Appendix B by (B.15) to (B.17). This relation has allowed us

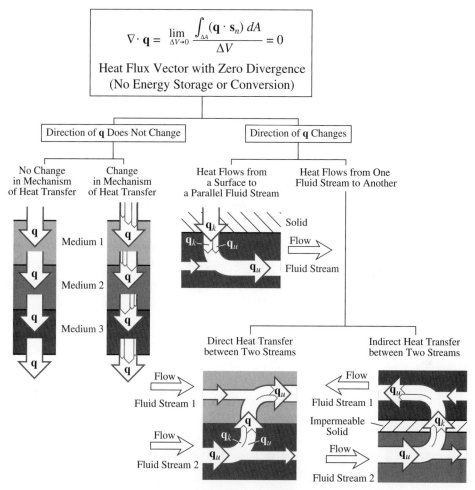

Chart 2.2. Aspects of heat transfer with zero divergence of the heat flux vector.

to move from the integral-volume energy equation (1.23) to the differential-volume energy equation (2.1).

We now combine (2.1) and (2.7) to have

$$\frac{\int_{\Delta A} (\mathbf{q} \cdot \mathbf{s}_n)\, dA}{\Delta V \to 0} = \nabla \cdot \mathbf{q} = -\frac{\partial}{\partial t} \rho c_p T + \sum_i \dot{s}_i \qquad \begin{array}{l}\text{relation between integral- and} \\ \text{differential-volume forms of} \\ \text{the energy equation.}\end{array} \qquad (2.8)$$

This shows that as $\Delta V \to 0$, the area integral of $\mathbf{q}$ over the control surface of $\Delta A$ divided by $\Delta V$ (which is $\nabla \cdot \mathbf{q}$) is equal to the sum of the negative of the volumetric (i.e., per unit volume) storage and the volumetric energy conversion.

EXAMPLE 2.2. FUN

The divergence theorem (2.5) relates the integration of $\nabla \cdot \mathbf{q}$ over the control volume $V$ (or $\Delta V$) to the integration of $(\mathbf{q} \cdot \mathbf{s}_n)$ over the control surface $A$ (or

$\Delta A$) around that control volume. Use the Cartesian coordinates and a volume with finite lengths $L_y$ and $L_z$ in the $y$ and $z$ directions, and differential length $\Delta x$ in the $x$ direction, as shown in Figure Ex. 2.2.

(a) Then, following the steps given by (B.15) to (B.17), prove this theorem for $q_x$ as the only component of $\mathbf{q}$ (i.e., $q_y = q_z = 0$) and $x$ as the only differential spatial variable.
(b) Extend the results to three dimensions, by referring to the results of part (a), and arrive at (2.7).

**SOLUTION**
(a) The mathematical statement of the divergence theorem is given by (2.5) or

$$\lim_{\Delta V \to 0} \int_{\Delta A} (\mathbf{q} \cdot \mathbf{s}_n) dA = Q \mid_{\Delta A} = \int_{\Delta V} (\nabla \cdot \mathbf{q}) dV.$$

Since $\Delta V$ is very small, we assume that $\nabla \cdot \mathbf{q}$ does not change within $\Delta V$. Then on the right-hand side we can move $\nabla \cdot \mathbf{q}$ out of the integral and we have

$$\lim_{\Delta V \to 0} \frac{\int_{\Delta A} (\mathbf{q} \cdot \mathbf{s}_n) dA}{\Delta V} = \frac{Q \mid_{\Delta A}}{\Delta V} = \nabla \cdot \mathbf{q}.$$

Using the steps described in Appendix B, i.e., (B.15) to (B.17), we begin with the left-hand side term and write this for the $x$-direction component of the heat flux, i.e., $q_x$, and assume that $q_x$ has a different magnitude, but is uniform, over areas $(\Delta A_x)_x$ and $(\Delta A_x)_{x+\Delta x}$ (the uniformity assumption is justifiable when $\Delta A_x$ is very small). Then we have

$$
\begin{aligned}
\frac{\int_{\Delta A} (\mathbf{q} \cdot \mathbf{s}_n) \Delta A}{\Delta V} &= \frac{[(\mathbf{q} \cdot \mathbf{s}_n)\Delta A]_x + [(\mathbf{q} \cdot \mathbf{s}_n)\Delta A]_{x+\Delta x}}{L_y L_z \Delta x} \\
&= \frac{-(q_x \Delta A_x)_x + (q_x \Delta A_x)_{x+\Delta x}}{L_y L_z \Delta x} \\
&= \frac{[(q_x)_{x+\Delta x} - (q_x)_x]\Delta A_x}{L_y L_z \Delta x} \\
&= \frac{(q_x)_{x+\Delta x} - (q_x)_x}{\Delta x} \frac{L_y L_z}{L_y L_z}.
\end{aligned}
$$

Note that we have defined $\mathbf{q}_x$ to be pointing along the positive $x$ axis, for both surfaces.

We have used $\Delta V = L_y L_z \Delta x$ and $(\Delta A_x)_x = (\Delta A_x)_{x+\Delta x} = L_y L_z$, as shown in Figure Ex. 2.2. Now taking the limit of $\Delta x \to 0$, we have

$$\Delta V = L_y L_z \Delta x \to 0,$$

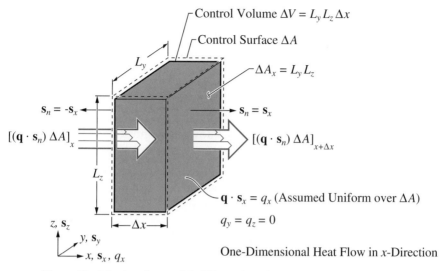

Figure Ex. 2.2. A volume with differential length in the $x$ direction only.

and using the definition of the differentiation as given by (B.16), we have

$$\lim_{\Delta V \to 0} \frac{\int_{\Delta A} (\mathbf{q} \cdot \mathbf{s}_n) dA}{\Delta V} = \lim_{\Delta x \to 0} \frac{(q_x)_{x+\Delta x} - (q_x)_x}{\Delta x} \equiv \frac{\partial q_x}{\partial x}.$$

(b) Now, from the definition (in Cartesian coordinates), we have

$$\nabla \cdot \mathbf{q} \equiv \left( \frac{\partial}{\partial x} \mathbf{s}_x + \frac{\partial}{\partial y} \mathbf{s}_y + \frac{\partial}{\partial z} \mathbf{s}_z \right) \cdot (\mathbf{q}_x + \mathbf{q}_y + \mathbf{q}_z) = \frac{\partial q_x}{\partial x} + \frac{\partial q_y}{\partial y} + \frac{\partial q_z}{\partial z},$$

and since we have assumed that $q_y$ and $q_z$ are zero, we have the final result as

$$\lim_{\Delta V \to 0} \frac{\int_{\Delta A} (\mathbf{q} \cdot \mathbf{s}_n) dA}{\Delta V} = \frac{Q \mid_{\Delta A}}{\Delta V \to 0} = \nabla \cdot \mathbf{q}.$$

**COMMENT**

Note that in order to perform differencing (and differentiation) on $q_x$, we have chosen $q_x$ to be designated as positive when flowing in the positive $x$ direction. Then the unit vector normal to the surface at location $x$ is in the negative $x$ direction, and the product of $\mathbf{q}_x$ and $\mathbf{s}_n \Delta A_x$ at $x$ is negative. At location $x + \Delta x$ this normal unit vector to the surface is in the positive $x$ direction, and therefore, the product of $\mathbf{q}_x$ and $\mathbf{s}_n \Delta A_x$ at $x + \Delta x$ is positive. This has led to the traditional use of the opposite signs for the terms called the heat flow in and out of the control volume.

Here the vectorial forms of $\mathbf{q}$ and $\mathbf{s}_n A$ are retained to aid in the proper usage in various applications.

Also, note that both $Q \mid_{\Delta A}$ and $\nabla \cdot \mathbf{q}$ are scalars (as compared to $\nabla$ and $\mathbf{q}$, which are vectors).

## 2.2 Uniform Temperature in One or More Directions: Energy Equation for Volumes with One or More Finite Lengths

In some applications, with justifications we can assume a uniform temperature $T$ over the entire volume of interest or over the bounding surface of interest. This is called the integral-volume (or length) or lumped-capacitance treatment. In other cases, nonuniform (i.e., distributed) but known (i.e., prescribed) variations of $T$ and $\mathbf{q}$ are available. For these cases, we integrate the energy equation over the medium of interest along the finite lengths. For other applications, depending on the number of spatial directions over which we use an integral length (i.e., one, two, three directions), a combination of finite (i.e., integral) and infinitesimal (i.e., differential) lengths appear in the desired energy equation.

Also, in the numerical solution of the energy equation, the medium is divided into small but finite volumes within which a uniform temperature is assumed. This is called the division of the volume into finite-small volumes.

Here we will discuss the integral volume, the combined differential- and integral-length volume, and the finite-small volume forms of the energy equation. The resulting energy equations will be used in the following chapters.

### 2.2.1 Integral-Volume Energy Equation

In many thermal engineering analyses, the integral-volume form (which assumes a uniform temperature within the medium) of the energy equation is justifiably used. This is referred to as the zeroth-order analysis (indicating a zero spatial direction is selected for the differential length). Recasting the thermal energy equation (2.1) in the integral-volume form using (2.8) and integrating it over the volume and assuming constant density ($c_p$ is already assumed constant) and volume, we have

$$Q\mid_A = \int_A (\mathbf{q} \cdot \mathbf{s}_n) dA$$

$$\equiv Q + Q_k + Q_u + Q_{ku} + Q_r$$

$$= -\rho c_p V \frac{dT}{dt} + \sum_i \dot{S}_i = \begin{cases} > 0 & \text{source} \\ < 0 & \text{sink} \\ = 0 & \text{no net heat flow} \end{cases}$$

integral-volume energy equation, $\qquad\qquad$ (2.9)

where $Q$ is prescribed surface heat transfer rate and others are associated with conduction, convection, surface convection, and radiation.

Note that (2.9) is a special representation of (1.23), where all the terms on the right-hand side of (1.23) are represented by only two terms in (2.9). This simplification in the presentation of the energy equation will readily enable us to perform many heat transfer analyses in Chapters 3 to 8.

(a) Finite Control Volumes Around an Electrical Heater and Its Surrounding Enclosure

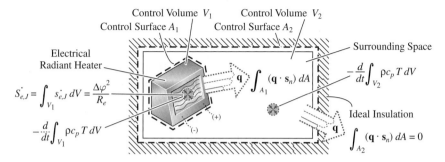

(b) Finite Control Volume Around a Tube with Internal Reactive Flow

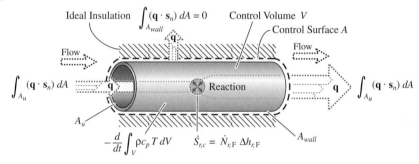

(c) Finite Control Volume Around a Solidification (Crystallization) Cell

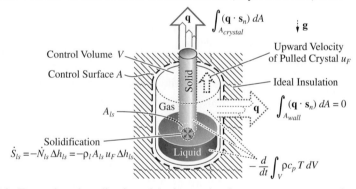

Figure 2.3. Examples of application of the integral-volume energy conservation equation. (a) Joule heating $\dot{S}_{e,J}$. (b) Combustion $\dot{S}_{e,c}$. (c) Phase change $\dot{S}_{ls}$.

Figure 2.3 shows the integral-volume analysis of heat transfer for some applications. Figure 2.3(a) shows an integral volume around an electrical heater and around an ideally insulated enclosure surrounding this heater. The heat generated $\dot{S}_{e,J} = \int_{V_1} \dot{s}_{e,J} dV$ raises the temperature of the enclosed volume $V_2$. Figure 2.3(b) shows an integral volume around a tube with flow and reaction. This can be a combustion tube, where the heat generated $\dot{S}_{r,c}$ is convected along the tube for later usage (i.e., as a hot gas stream). Figure 2.3(c) shows an integral volume around a solidification cell. In the cell shown, a silicon crystal is grown by cooling (by conduction) the melt through a seed crystal, which is being continuously pulled at a velocity $u_F$. The heat released due to solidification $\dot{S}_{ls}$ is removed through the pulled crystal.

EXAMPLE 2.3. FUN

---

In surface coating of a substrate by liquid-droplet impingement and solidification, the particles (polymers, metals, etc.) are first heated as they flow toward the surface. The heating is done by the gas stream that carries the particles toward the surface. Because of the generally short flight time, very hot gases are used. The gas may be heated by chemical reaction or by Joule heating before mixing with the particles (we will discuss these in Chapter 5). Consider a particle with a given $\rho c_p V$. During the particle flight, the high temperature gas (which can be at such a temperature that the gas may become ionized, i.e., a thermal plasma) transfers heat to the particle by surface convection at a constant rate $A q_{ku}$, where $A$ is the particle surface area. The particle loses heat by surface radiation at a constant rate $A q_r$. This is schematically shown in Figure Ex. 2.3(a).

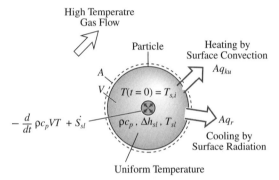

Figure Ex. 2.3. (a) Surface convection and radiative heat transfer to an initially solid particle that undergoes melting upon heating.

The temperature variation within the volume is much smaller than that between its bounding surface $A$ and the surroundings. Therefore we assume a uniform temperature within the particle.

(a) Write the expression for the time it takes to reach the melting temperature $t_{sl}$. This is done by starting from the energy equation describing the transient particle temperature $T = T(t)$ before melting, i.e., for $T(t) < T_{sl}$, where $T_{sl}$ is the solid-liquid phase-change (i.e., melting) temperature.
(b) Write the expression for the time it takes to melt the particle $\Delta t_{sl}$.
(c) Qualitatively show the time variation of the particle temperature as it begins at an initial temperature $T(t = 0) < T_{sl}$ at time $t = 0$ and then reaches the melting temperature $T_{sl}$, and after the complete melting when the liquid is heated and evaporates.

**SOLUTION**
(a) Here we have a given surface heat transfer $A q_{ku} + A q_r$ and a volumetric energy storage and conversion as sensible heat and as phase change. Note that we assume that $q_{ku}$ is positive (i.e., flowing out of the surface). Then for heating

(i) Heat Transfer to and from a Particle

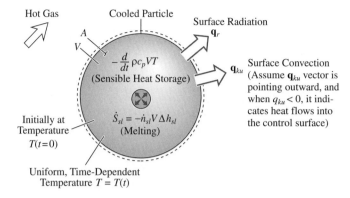

(ii) Temporal Variation of the Particle Temperature

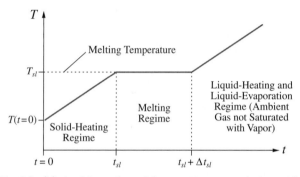

Figure Ex. 2.3. (b) Anticipated particle temperature variation with respect to time.

of the particle, $q_{ku}$ should be negative. We are asked to determine $T = T(t)$. We begin with the integral-volume energy equation (2.9), since we have a uniform temperature $T$. The comparison between the temperature variation within the volume and between its bounding surface and the surroundings is made by comparing the resistances to heat flow occurring inside and outside of the volume. The resistances and the conditions for justification of the assumption of uniform temperature will be discussed in Chapters 4 to 6. When the temperature is assumed uniform, the analysis is called the lumped-capacitance analysis (as compared to allowing for temperature nonuniformity within the volume, which is called the distributed-capacitance analysis). Then for (2.9) we have

$$\int_A (\mathbf{q} \cdot \mathbf{s}_n) dA = Q \mid_A = Aq_{ku} + Aq_r = -\frac{d}{dt}\rho c_p VT$$

$$+ \int_V \dot{s}_{sl} dV \quad \text{integral-volume energy equation.}$$

Before melting, we have $\dot{s}_{sl} = 0$. Then we use this in the above and rearrange the resultant to arrive at

$$A(q_{ku} + q_r) = -\frac{d}{dt}\rho c_p VT \quad T < T_{sl} \quad \text{solid-heating regime.}$$

Assuming constant $\rho c_p V$ and $A(q_{ku} + q_r)$, we integrate this between the initial condition and that at any time and have

$$\int_0^t A(q_{ku} + q_r)dt = -\rho c_p V \int_{T(t=0)}^T dT$$

or

$$T - T(t = 0) = -\frac{A(q_{ku} + q_r)}{\rho c_p V}t \qquad T \leq T_{sl}.$$

This shows a linear rise in the temperature (up to the melting temperature $T_{sl}$), when the surface heat transfer rates are constant. Note that for heating of the particle, it is required that

$$A(q_{ku} + q_r) < 0, \qquad q_{ku} < 0, \quad q_r > 0.$$

Here, this is achieved by a negative $q_{ku}$, i.e., heat flowing into the control surface, while $q_r$ (which is smaller in magnitude than $q_{ku}$) flows out of the control surface. This linear temperature rise is shown in Figure Ex. 2.3(b).

The time required to reach the melting temperature, i.e., $t = t_{sl}$ for $T = T_{sl}$, is

$$t_{sl} = -\frac{[T_{sl} - T(t = 0)]\rho c_p V}{A(q_{ku} + q_r)} \qquad T \leq T_{sl}.$$

(b) During the melting period, i.e., in the melting regime, beginning at $t = t_{sl}$ and lasting for $\Delta t_{sl}$, the heat flowing into the volume is used for phase change. For a pure substance at a constant pressure, the temperature remains constant during the phase change. During the melting, we have $\dot{S}_{sl} = -\dot{n}_{sl}V\Delta h_{sl}$, where $\dot{n}_{sl}$ is the volumetric melting rate, in kg/m³-s, and the temperature does not change while there is any solid left. Then the energy equation (2.9) becomes

$$\int_A (\mathbf{q} \cdot \mathbf{s}_n)dA = A(q_{ku} + q_r) = \dot{S}_{sl} = -\dot{n}_{sl}V\Delta h_{sl} \quad T = T_{sl} \quad \text{melting regime.}$$

We now time integrate this between $t_{sl}$ and $t_{sl} + \Delta t_{sl}$. The result is

$$A \int_{t_{sl}}^{t_{sl}+\Delta t_{sl}} (q_{ku} + q_r)dt = -\int_{t_{sl}}^{t_{sl}+\Delta t_{sl}} \dot{n}_{sl}V\Delta h_{sl}dt.$$

Next we note that at $t_{sl} + \Delta t_{sl}$ all the solid is melted, and we have

$$A(q_{ku} + q_r)\Delta t_{sl} = -\rho V \Delta h_{sl},$$

where $\rho V$ is the mass of the particle $M$. We have used the solid mass conservation equation, which gives $V \int_{t_{sl}}^{t_{sl}+\Delta t_{sl}} \dot{n}_{sl}dt = V\rho = M$.

Now solving for $\Delta t_{sl}$, we have

$$\Delta t_{sl} = -\frac{\rho V \Delta h_{sl}}{A(q_{ku} + q_r)} \qquad T = T_{sl}.$$

(c) After completion of melting, the temperature will rise again, i.e., the liquid-heating and liquid-evaporation regime begins. This is shown, along with the other regimes, in Figure Ex. 2.3(b).

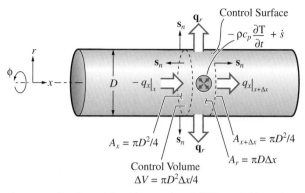

Figure 2.4. An example of combined integral and differential lengths and the associated terms in the combined integral- and differential-length energy equation.

**COMMENT**

Note that it is rather difficult to maintain $A(q_{ku} + q_r)$ constant. As will be shown, $q_r$ and $q_{ku}$ are related to the difference in temperature between the surface and the ambient. Then as $T$ changes, both $q_{ku}$ and $q_r$ also change. We will discuss this in Chapters 4 and 6.

### 2.2.2 Combined Integral- and Differential-Length Energy Equation

In some applications we can justify neglecting the variation of $\mathbf{q}$ or $T$ along a given direction, or we are able to make some assumptions about their variation or uniformity, and therefore, prescribe them. Then we can take the control volume to have a finite (integral) length in that direction. In some other cases we might be able to prescribe the thermal condition on some of the bounding surfaces. Then in these directions, we choose to extend the control volume to these bounding surfaces.

These lead to a combined integral- and differential-length energy equation, which when used in the analysis needs to be integrated in only one or two directions (instead of three). We will use (2.7) to develop these energy equations.

As an example, consider heat transfer in a medium with the geometry of a circular cylinder, as shown in Figure 2.4. This can be a solid circular cylinder (discussed in Section 6.8.2) or the inner portion of a circular tube (discussed in Section 7.2.5). We may be interested in the variation of the axial component of the heat flux vector $\mathbf{q}_x = \mathbf{q}_x(x)$ along the $x$ direction, and therefore, choose a differential length $\Delta x$ along the $x$ axis. In the radial direction $r$, we assume no variation of the radial component of the heat flux vector $\mathbf{q}_r$ along the radial direction and take a finite (integral) length that is the diameter $D$ of the cylinder. In the third direction $\phi$ (i.e., the angular position), we also assume no heat flow, $\mathbf{q}_\phi = 0$, and take the integral angle $\pi$. So, we have $\mathbf{q}_x = \mathbf{q}_x(x)$, $\mathbf{q}_r$ being constant and $\mathbf{q}_\phi$ being zero. This is an example of a heat transfer medium with a volume selected for the analysis that is assumed axisymmetric (or integral along the periphery), radially lumped (or integral along the radius), and axially differential (allowing for variations along $x$). By allowing for the bounding-surface heat transfer $q_r$ to be an integral-length heat transfer, we are then interested in the axial variation of $q_x(x)$.

For this example, the left-hand side of the energy equation (2.8) becomes

$$\lim_{\Delta x \to 0} \frac{\displaystyle\int_{\Delta A} (\mathbf{q} \cdot \mathbf{s}_n) dA}{\Delta V} = \lim_{\Delta x \to 0} \left[ \frac{\displaystyle\int_{\Delta A_r} (\mathbf{q} \cdot \mathbf{s}_n) dA_r}{\pi D^2 \Delta x / 4} + \frac{\displaystyle\int_{A_x} (\mathbf{q} \cdot \mathbf{s}_n) dA_x}{\pi D^2 \Delta x / 4} \right]$$

$$= \lim_{\Delta x \to 0} \left[ \frac{q_r \pi D \Delta x}{\pi D^2 \Delta x / 4} + \frac{[q_x(x + \Delta x) - q_x(x)] \pi D^2 / 4}{\pi D^2 \Delta x / 4} \right]$$

$$= \frac{4 q_r}{D} + \frac{q_x(x + \Delta x) - q_x(x)}{\Delta x \to 0}$$

$$= \frac{4 q_r}{D} + \frac{\partial q_x}{\partial x}. \tag{2.10}$$

Then (2.8) becomes

$$\lim_{\Delta x \to 0} \frac{\displaystyle\int_{\Delta A} (\mathbf{q} \cdot \mathbf{s}_n) dA}{\Delta V} = \frac{4 q_r}{D} + \frac{\partial q_x}{\partial x} = -\rho c_p \frac{\partial T}{\partial t} + \dot{s} \quad \begin{array}{l} \text{combined integral- and} \\ \text{differential-length} \\ \text{energy equation,} \end{array} \tag{2.11}$$

where we have used $A_x = \pi D^2 / 4$, $\Delta A_r = \pi D \Delta x$, and $\Delta V = \pi D^2 \Delta x / 4$.

Later, we will make use of this type of integral-differential volume in Sections 5.6, 6.8.2, 7.2.5, and 7.6.1. We will also have further examples of this throughout the text.

Note that although here we have used a cylindrical geometry as an examples, the treatment for the combined integral-differential volume can be applied to any regular volume geometry (e.g., rectangular, cylindrical, spherical).

### EXAMPLE 2.4. FUN

A thin plate (fin) of thickness $l$, width $w$, and length $L$ is attached to a base planar surface and acts as an extension to this surface for the purpose of surface-convection heat removal enhancement. The heat is removed from the fin by surface convection at a rate $A_{ku} q_{ku}$, where $q_{ku}$ is assumed constant as shown in Figure Ex. 2.4(a). The heat is conducted, $q_k$, within the fin plate and $q_k = q_k(x)$ varies only along the $x$ direction (a justifiable assumption for $l \ll L$).

Derive the combined integral- (for other than the $x$ direction) and differential-length (along the $x$ direction) energy equation for the thin plate, for a steady-state (i.e., $\partial / \partial t = 0$) intramedium conduction and surface convection, with no energy conversion (i.e., $\dot{s} = 0$).

**SOLUTION**

Figure Ex. 2.4(b) shows the mechanisms of heat transfer and the integral-differential control volume and control surface for this problem. We only have two terms in the energy equation; these are the surface convection $A_{ku}q_{ku}$ and the intramedium conduction $A_k q_k$, where $A_{ku}$ is the surface area for surface convection and $A_k$ is the surface area for conduction. The combined-length energy equation is developed from (2.7) and is

$$\lim_{\Delta x \to 0} \frac{\int_{\Delta A}(\mathbf{q} \cdot \mathbf{s}_n)dA}{\Delta V} = \lim_{\Delta x \to 0}\left(\frac{\int_{\Delta A_{ku}}(\mathbf{q}_{ku} \cdot \mathbf{s}_n)\, dA_{ku}}{\Delta V} + \frac{\int_{A_k}(\mathbf{q}_k \cdot \mathbf{s}_n)\, dA_k}{\Delta V}\right) = 0$$

energy equation,

where

$$\Delta A_{ku} = 2l\Delta x + 2w\Delta x$$

$$A_k = lw$$

$$\Delta V = lw\Delta x.$$

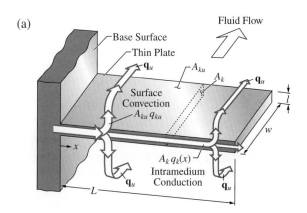

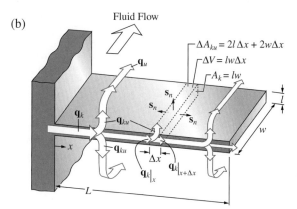

Figure Ex. 2.4. (a) A thin plate attached to a base. (b) The combined integral-differential length control volume taken along the thin plate.

Substituting for these in the above energy equation, we have

$$\lim_{\Delta x \to 0} \left[ \frac{q_{ku}(2l + 2w)\Delta x}{lw\Delta x} + \frac{(q_k |_{x+\Delta x} - q_k |_x)lw}{lw\Delta x} \right] = 0$$

$$\lim_{\Delta x \to 0} \left[ \frac{q_{ku}(2l + 2w)}{lw} + \frac{(q_k |_{x+\Delta x} - q_k |_x)}{\Delta x} \right] = 0$$

or

$$q_{ku}\frac{2l + 2w}{lw} + \frac{dq_k}{dx} = 0 \qquad \begin{array}{l}\text{combined integral- and differential-length} \\ \text{energy equation.}\end{array}$$

Note that at location $x = x$, $\mathbf{q}_k$ and $\mathbf{s}_n$ are in opposite directions, while at $x = x + \Delta x$ they are in the same direction.

### COMMENT

This combined integral-differential volume energy equation shows the spatial rate of change of the intramedium conduction heat flux $q_k$ caused by the surface-convection heat transfer. Since $q_{ku}$ is positive when heat flows away from the control surface, then $dq_k/dx$ is negative for positive $q_{ku}$ (i.e., $q_k$ decreases along the $x$ direction).

Fins (i.e., extended surfaces) are used to enhance surface-convection heat transfer, especially when gas streams (which have low thermal conductivity, such as air) are used to cool surfaces. In Chapter 6 we will show that $q_{ku}$ may also vary along the $x$ direction and we will also discuss the steps for the determination of the temperature distribution along the $x$ direction.

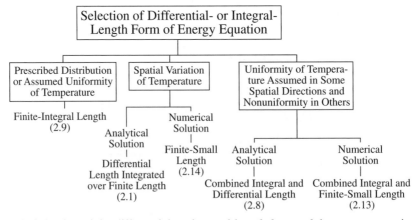

Chart 2.3. Selection of the differential- or integral-length forms of the energy equation, and their corresponding equation numbers.

### 2.2.3 Discrete Temperature Nonuniformity: Finite-Small-Volume Energy Equation*

This section in found on the Web site www.cambridge.org/kaviany. This subsection develops the energy equation for a finite-small volume (i.e., $\Delta V$ is small but not approaching zero). We assume that the temperature is uniform within the finite, but small volume (i.e., a finite-volume averaged $T$) and the energy conversion is also represented by uniform temperature distribution. This allows for a discrete, spatial variation of $T$ and $\mathbf{q}$. This is most useful in obtaining numerical solutions for media with nonuniform temperature distributions.

### 2.2.4 Summary of Selection of Energy Equation Based on Uniformity or Nonuniformity of Temperature

Chart 2.3 summarizes the selection of the finite-integral, finite-small, and differential length scales, or a combination of these, depending on the problem analyzed. The choice depends on the justifiable assumptions regarding the temperature uniformity. The corresponding equation numbers are also listed.

EXAMPLE 2.5. FUN

Consider the case of a negligible convection and radiation heat transfer and a simplified, two-dimensional geometry. Write the finite-small volume energy equation (2.15) for the geometry depicted in Figure Ex. 2.5.

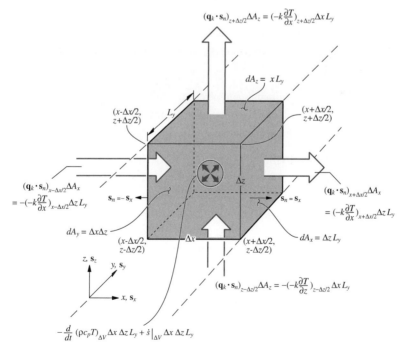

Figure Ex. 2.5. Conduction heat transfer and energy conservation across a finite-small control volume.

**SOLUTION**

The finite-small volume energy equation (2.15) becomes

$$k\frac{\partial T}{\partial x}\mid_{x-\Delta x/2} \Delta zL_y - k\frac{\partial T}{\partial x}\mid_{x+\Delta x/2} \Delta zL_y + k\frac{\partial T}{\partial z}\mid_{z-\Delta z/2} \Delta xL_y - k\frac{\partial T}{\partial z}\mid_{z+\Delta z/2} \Delta xL_y$$

$$= -\frac{d}{dt}(\rho c_p T)\mid_{\Delta V} \Delta x\Delta zL_y + \sum_i \dot{s}_i \mid_{\Delta V} \Delta x\Delta zL_y$$

two-dimensional conduction energy equation for a finite-small volume,

where a prescribed length $L_y$ (in the $y$ direction) is chosen. Note that we have assumed a uniform temperature within $\Delta V$, i.e., uniform $T_{\Delta V}$.

The gradient of the temperature at the four surfaces needs to be related to the average temperature in the finite volume $T_{\Delta V}$, and the temperature of the neighboring finite volumes.

**COMMENT**

In Section 3.7, these gradients will be represented by resistances and potentials. Then a numerical procedure for the simultaneous solution of the temperature in all the finite-small volumes (comprising the physical volume of interest) will be given. This is referred to as distributed capacitance, finite-small volume analysis.

## 2.3 Thermal Energy Conversion Mechanisms

So far we have referred to the chemical-bond (including nuclear), electromagnetic, mechanical, and thermal mechanisms of energy conversion. In (2.1), we have used the temperature form of the energy equation and placed the phase-change energy conversion into the physical-bond energy conversion. The energy conversion term in (2.1) is repeated here:

$$\sum_i \dot{s}_i = \begin{cases} \text{chemical- and physical-bond energy conversion} \\ \text{electromagnetic energy conversion} \\ \text{mechanical energy conversion.} \end{cases} \qquad (2.16)$$

Here we have chosen to include the transport of internal energy (thermal energy) by fluid flow as a mechanism of heat transfer (i.e., convection). We have included the thermal energy and the kinetic energy in the energy storage term. The remainder of the work and energy storage mechanisms have been included in the energy conversion term $\sum_i \dot{s}_i$. In the following, we will discuss the different mechanisms of energy conversion (to and from thermal energy) represented by $\dot{s}_i$.

While $\dot{S}_i/V$ indicates the volumetric energy conversion, there are cases where the energy conversion occurs on the bounding surface; this is designated by $\dot{S}_i/A$. One example would be the friction heating of two solid surfaces in relative motion. We also address their surface energy conversions.

Figure 2.7 shows the volumetric $\dot{S}_i/V$ (W/m$^3$) and bounding-surface $\dot{S}_i/A$(W/m$^2$) energy conversions. Again, we have divided the mechanisms (for energy conversion to or from thermal energy) into the chemical- or physical-bond, electromagnetic, and

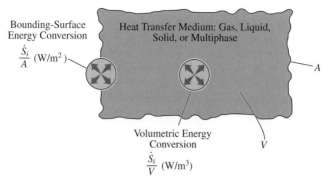

Figure 2.7. Bounding-surface and volumetric energy conversions.

mechanical energy conversions. Chart 2.4 lists these energy conversion mechanisms. Each mechanism (and its subdivisions), and the associated relations, are discussed below.

Some of the energy conversions are the result of interaction between microscale heat carriers [2]. For example, absorption of photon (electromagnetic waves) by electrons in solid (e.g., radiation heating of solids), or between electrons and phonons (lattice) in Joule heating.

### 2.3.1  Chemical- or Physical-Bond Energy Conversion

The heat transfer medium may undergo desired or undesired chemical and physical transformations when its temperature is raised or lowered beyond some threshold values or energy is absorbed or released. As was discussed in Section 1.5.1, the bonding forces are ionic, covalent, or van der Waals. The temperature marking distinct changes (e.g., phase transitions) in these bonding states can be the ignition temperature for a combustible medium or the solidification temperature for a liquid medium. The chemical- or physical-bond energy conversion results in a release or absorption of heat in the heat transfer medium.

These changes in the chemical or physical bonds can be irreversible or reversible, but are mostly irreversible. We begin with chemical and nuclear reactions and then discuss phase change.

### (A)  Chemical Reaction

Chemical thermodynamics is briefly reviewed in Appendix A, Section A.3. We repeat (A.15), which expresses the stoichiometric chemical reaction between fuel F and oxidant O with P as the product as

$$\nu_F F + \nu_O O = \nu_P P \quad \text{stoichiometric chemical reaction,} \qquad (2.17)$$

where $\nu_i$ is the number of moles of species $i$ (i.e., stoichiometry coefficient). The thermal energy converted can be given either in J per kg of the product or one of the reactants (especially if one of the reactants can be called the fuel). Figure 2.8(a) gives a rendering of a chemical reaction. This energy is given per kg of the fuel species, i.e., $\Delta h_{r,F}$(J/kg of fuel) and is called the specific heat of reaction of the fuel.

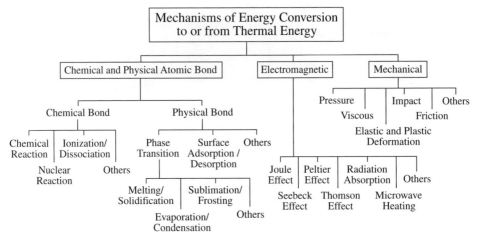

Chart 2.4. A classification of energy conversion mechanisms between thermal energy and other forms of energy.

In general, the volumetric rate of production of a species A $\dot{n}_{r,A}(\text{kg/m}^3\text{-s})$ depends on the concentrations of the reactant species A and B, the product species P, the temperature, and the pressure, i.e., $\dot{n}_{r,F} = \dot{n}_{r,F}(\rho_F, \rho_O, \rho_P, T, p)$. Here $\dot{n}_{r,F}$ is given in kg of species F (i.e., fuel) per unit volume and per unit time. Then the heat released per unit volume $\dot{s}_{r,c}$ is

$$\frac{\dot{S}_{r,c}}{V} = \dot{s}_{r,c} = \dot{n}_{r,F}\Delta h_{r,F} \quad \text{volumetric energy conversion by chemical reaction.} \quad (2.18)$$

(a) Chemical-Bond Energy Conversion

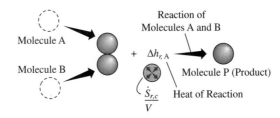

(b) Physical-Bond Energy Conversion

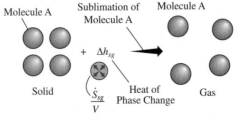

Figure 2.8. (a) Rendering of molecular collision and chemical reaction. (b) Rendering of the change in intermolecular spacing as a result of phase change.

Note that since species F is consumed in the reaction, $\dot{n}_{r,F}$ is negative. For exothermic reactions, $\Delta h_{r,F}$ is negative and for endothermic reactions, $\Delta h_{r,F}$ is positive.

The temperature dependence of the reaction rate is in general rather complex. However, simple power-law relations are used with specified limits on applicability [1,19]. For oxidation of gaseous hydrocarbon fuels (e.g., combustion of natural gas), one of the simple relations (called the chemical kinetic model) for the dependence of the volumetric reaction rate $\dot{n}_{r,F}(\text{kg/m}^3\text{-s})$ on temperature is called the phenomenological Arrhenius rate model and is

$$\dot{n}_{r,F} = -a_r \rho_F^{a_F} \rho_O^{a_O} e^{-\Delta E_a/R_g T} \quad \text{chemical kinetic model.} \tag{2.19}$$

Here $\rho_F(\text{kg/m}^3)$ and $\rho_O(\text{kg/m}^3)$ are the fuel and oxidant densities, $a_r(\text{s}^{-1})$ is the pre-exponential factor, $\Delta E_a(\text{J/kmole})$ is the activation energy, and $R_g(\text{J/kmole-K})$ is the universal gas constant. The pre-exponential factor is related to the collision frequency. The sum of $a_F$ and $a_O$ is called the order of reaction.

Then combining (2.19) and (2.18) we have

$$\frac{\dot{S}_{r,c}}{V} = \dot{s}_{r,c} = -\Delta h_{r,F} a_r \rho_F^{a_F} \rho_O^{a_O} e^{-\Delta E_a/R_g T} \quad \begin{array}{l}\text{volumetric energy conversion} \\ \text{by chemical reaction.}\end{array} \tag{2.20}$$

We note that $a_r$ is a rather large number, because the collision frequency is very large. The collision frequency for similar molecules is approximated by $f_c = (\overline{u_m^2})^{1/2}/\lambda_m$, where $(\overline{u_m^2})^{1/2}$ and $\lambda_m$ are given by (1.16) and (1.19). For example, at $T = 15°\text{C}$ and $p = 1$ atm, in a volume of the methane gas, the molecules collide at a frequency $f = 12.9$ GHz ($1.29 \times 10^{10}$ 1/s) and for an oxygen gas the molecules collide at a frequency $f = 6.4$ GHz.

The activation energy $\Delta E_a$ is a measure of the required association energy of the colliding molecules. Molecules that possess energy greater than $\Delta E_a$ will react. Then chemical reactions requiring high temperatures have a large $\Delta E_a$. The universal gas constant $R_g$ is equal to $8.3145 \times 10^3$ J/kmole-K.

A simpler (and less accurate) chemical kinetic model that is used for the oxidation of hydrocarbons near one atm pressure is the zeroth-order model, which uses $a_F = a_O = 0$. The zeroth-order model can then be written as

$$\dot{n}_{r,F} = -a_r e^{-\Delta E_a/R_g T} \quad \text{zeroth-order chemical-kinetic model.} \tag{2.21}$$

This states that the reaction is independent of the mass fraction of the reactants and products, which is valid only for a certain range of these concentrations.

### EXAMPLE 2.6. FAM

Consider the simplified reaction rate model (2.19) for the oxidation of the gaseous methane (which makes up most of the natural gas) and air. We note that this reaction is rather complicated with many intermediate chemical reactions. For a relatively accurate description, at least four reactions should be included

[13]. However, for a very simplified analysis, such as that intended here, a single-step reaction may be used.

Taking air to be consisting of oxygen and nitrogen with a 1 to 3.76 molar ratio of $O_2$ to $N_2$ (i.e., mass fraction 0.2 and 0.7670), we assume a single-step stoichiometric reaction given by

$$CH_4 + 2O_2 + 2 \times 3.76N_2 \rightarrow 2H_2O + CO_2 + 7.52N_2$$

methane-air stoichiometric reaction

or

$$CH_4 + 2O_2 + 7.52N_2 \rightarrow P + 7.52N_2,$$

where here the fuel species F is $CH_4$ and P stands for all products (water vapor and carbon dioxide).

The empirically determined constants in the reaction rate are: the exponents of densities $a_F = -0.3$, $a_O = 1.3$, the pre-exponential factor $a_r = 5.3 \times 10^8$ s$^{-1}$, and the activation energy $\Delta E_a = 2.0 \times 10^8$ J/kmole [16]. These give for (2.19)

$$\dot{n}_{r,F} = -\rho_F^{-0.3} \rho_O^{1.3} (5.3 \times 10^8) \exp\left(-\frac{2.0 \times 10^8}{R_g T}\right) \quad \begin{array}{l} \text{first-order chemical-kinetic} \\ \text{model for stoichiometric} \\ \text{methane-air reaction,} \end{array}$$

where $\dot{n}_{r,F}$ is in kg/m$^3$-s, and $R_g = 8.3145 \times 10^3$ J/kmole-K.

The heat of combustion is $\Delta h_{r,CH_4} = -5.55 \times 10^7$ J/kg-$CH_4$. Then the heat released per unit volume is found from (2.20) as

$$\dot{s}_{r,c} = -2.942 \times 10^6 \rho_F^{-0.3} \rho_O^{1.3} \exp\left(\frac{-2.0 \times 10^8}{R_g T}\right) \text{ (W/m}^3\text{)} \quad \begin{array}{l} \text{volumetric energy} \\ \text{conversion rate for} \\ \text{stoichiometric methane-} \\ \text{air chemical reaction.} \end{array}$$

The adiabatic flame temperature for this reaction (methane-air) is 2191 K, when the initial nonreacted mixture is at $T = 16°C$. Further examples of $\Delta h_{r,F}$ and adiabatic flame temperature will be given in Section 5.4.

The constants used in the zeroth-order model for the methane-air reaction, i.e., $a_F = a_O = 0$, are $a_r = 1.3 \times 10^8$ kg/m$^3$-s and $\Delta E_a = 2.10 \times 10^8$ J/kmole.

**COMMENT**

The sum of $a_F$ and $a_O$ is called the order of reaction. For $a_F = -0.3$ and $a_O = 1.3$, this sum is equal to one, i.e., the reaction is the first order, but depends on the concentration of the oxidant and the fuel.

### EXAMPLE 2.7. FAM

From the dependence of the chemical reaction rate (i.e., combustion rate) on the temperature, given by chemical kinetic models, e.g., (2.19) and (2.21), we note that the reaction rate $\dot{n}_{r,F}$ increases with increase in temperature $T$.

Using (2.21), determine $\dot{n}_{r,F} = \dot{n}_{r,F}(T)$, and the volumetric energy conversion (volumetric heat generation) rate $\dot{s}_{r,c} = \dot{n}_{r,F}\Delta h_{r,F}$, for the methane-air reaction at (a) $T = 300$ K, (b) 800 K, and (c) 2,000 K.

Use the constants given in Example 2.6.

**SOLUTION**

The exponential dependence of $\dot{n}_{r,F}$ on $T$ suggests a strong temperature dependence.

From (2.21) we have

$$\dot{n}_{r,F}(\text{kg/m}^3\text{-s}) = -a_r(\text{kg/m}^3\text{-s}) \exp\left(-\frac{\Delta E_a}{R_g T}\right).$$

The constants are given in Example 2.5 and are $a_r = 1.3 \times 10^8$ kg/m$^3$ and $\Delta E_a = 2.10 \times 10^8$ J/kmole. Then using the numerical values, we have

(a) $\dot{n}_{r,F}(T = 300\text{ K}) = -1.3 \times 10^8(\text{kg/m}^3\text{-s}) \exp\left[-\dfrac{2.1 \times 10^8(\text{J/kmole})}{8.314 \times 10^3(\text{J/kmole}) \times 300(\text{K})}\right]$

$\qquad = -3.535 \times 10^{-29}$ kg/m$^3$-s  volumetric methane depletion rate.

$\dot{s}_{r,c}(T = 300\text{ K}) \quad = \dot{n}_{r,F}(T = 300\text{ K})\Delta h_{r,A}$

$\qquad = -3.535 \times 10^{-29}(\text{kg/m}^3\text{-s}) \times [-5.55 \times 10^7(\text{J/kg})]$

$\qquad = 1.962 \times 10^{-21}$ J/m$^3$-s

$\qquad = 1.962 \times 10^{-21}$ W/m$^3$  volumetric heat release rate.

(b) $\dot{n}_{r,F}(T = 800\text{ K}) = -1.3 \times 10^8(\text{kg/m}^3\text{-s}) \exp\left[-\dfrac{2.1 \times 10^8(\text{J/kmole})}{8.314 \times 10^3(\text{J/kmole}) \times 800(\text{K})}\right]$

$\qquad = -1.940 \times 10^{-14}$ kg/m$^3$-s

$\dot{s}_{r,c}(T = 800\text{ K}) \quad = (-1.940 \times 10^{-14}) \times (-5.55 \times 10^7) = 1.077 \times 10^{-6}$ W/m$^3$.

(c) $\dot{n}_{r,F}(T = 2,000\text{ K}) = -1.3 \times 10^8(\text{kg/m}^3\text{-s}) \exp\left[-\dfrac{2.1 \times 10^8(\text{J/kmole})}{8.314 \times 10^3(\text{J/kmole}) \times 2,000(\text{K})}\right]$

$\qquad = -4.257 \times 10^2$ kg/m$^3$-s

$\dot{s}_{r,c}(T = 2,000\text{ K}) = (-4.257 \times 10^2) \times (-5.55 \times 10^7) = 2.363 \times 10^{10}$ W/m$^3$.

**COMMENT**

Note that because of the temperature dependency (i.e., kinetic control), the heat generation rate is temperature controlled. Within a flame, to be discussed in Section 5.4, this temperature is in turn controlled by the heat transfer. We note again that the actual chemical reaction between methane and air involves many reaction steps. Also the chemical reaction kinetics have a temperature dependence that can be more complex than the simple exponential relation used here [7].

Note the ratio

$$\frac{\dot{s}_{r,c}(2{,}000\text{ K})}{\dot{s}_{r,c}(800\text{ K})} = \exp\left\{-\frac{\Delta E_a}{R_g}\left[\frac{1}{2{,}000(\text{K})} - \frac{1}{800(\text{K})}\right]\right\} = 2.194 \times 10^{16}.$$

This chemical reaction kinetics is characteristic of the gaseous molecular collision frequency and kinetic energy dependence on the temperature. Both of these increase substantially with temperature, thus allowing for a strong reactivity (i.e., reaction rate) at high temperatures.

### (B) Nuclear Fission Reaction*

This section is found on the Web site www.cambridge.org/kaviany. In Section (B), Nuclear Fission Reaction is discussed and the relations for these energy conversion rates are given along with an example (Example 2.8).

### (C) Nuclear Fusion Reaction*

This section is found on the Web site www.cambridge.org/kaviany. In Section (C), Nuclear Fission Reaction is discussed and the relations for these energy conversion rates are given along with an example (Examples 2.8).

### (D) Nuclear Radioactive Decay

Radioactive materials undergo a nuclear-bond to thermal energy conversion. Radioactive decay transforms a nucleus by emitting different particles. In alpha decay, the nucleus releases a helium nucleus. In beta decay, the nucleus emits an electron or neutrino. In gamma decay, the nucleus lowers its internal energy by emitting a photon.

This energy conversion is present in the earth's subsurface and in the radioactive heat sources used for thermoelectric power generation (also called a radioisotope power source). The radioisotope power sources were originally used in pacemaker cardiac implants, and are currently used in modular power systems for use in space-based systems. Most power systems use plutonium as the fuel with a half life of about 90 years [17]. Radioactive heat generation is also present in nuclear waste. The earth contains the naturally radioactive elements uranium, thorium, and potassium.

The rate of decay is time dependent although the time rate of change may be very small. The radioactive source may or may not be shielded (i.e., may be concentrated or distributed). As an example, consider the model for a low-level radioactive source placed at a location $r_o$(m), which generates volumetric (distributed) heat according to

$$\frac{\dot{S}_{r,\tau}}{V} = \dot{s}_{r,\tau} = \dot{s}_{r,o}e^{-t/\tau - (r-r_o)^2\sigma_{ex}^2/2} \quad \text{volumetric heating by radioactive decay,} \qquad (2.24)$$

where $\dot{s}_{r,o}$(W/m$^3$) is the maximum volumetric heating from the source, $\tau$(s) is the time constant, $r$ is the radial location measured from $r_o$, and $\sigma_{ex}$(1/m) is the extinction coefficient. For nuclear waste deposited in a rock repository, typical values are $\dot{s}_{r,o} = 0.75$ W/m$^3$, $\tau = 63.37$ yr, and $\sigma_{ex} = 0.002$ 1/m.

A high level radioactive source used in thermoelectric energy conversion, such as a high concentration of plutonium, may have $\dot{s}_{r,o} = 5 \times 10^5$ W/m$^3$ and $\tau = 87.7$ yr.

The medical applications of miniature nuclear-powered batteries are divided into energetic devices (power requirement larger than 300 $\mu$W, such as those used in the artificial heart) and cybernetic devices (power requirement smaller than 300 $\mu$W, such as those originally used in the cardiac pacemaker). For cybernetic devices, there are substantial electrical power losses at the electrode-tissue contacts, and therefore, much larger power is actually used (as much as 10 times).

### (E) Phase Change

The physical-bond energy is that bonding similar molecules together. Beginning at the molecular level, the formation of cluster and condensed phases (including physical adsorption) results in energy release. At the thermodynamic phase equilibrium state, when two phases coexist, addition or removal of heat causes phase change. For a pure substance, the phase equilibrium state is designated by the temperature or pressure. The two are related through the Clausius-Clapeyron relation, $T_{ij} = T_{ij}(p)$. This is briefly discussed in Appendix A, Section A.2. For a volumetric rate of phase change from phase $i$ to phase $j$, designated by $\dot{n}_{ij}$ (kg/m$^3$-s), with an energy consumption of $\Delta h_{ij}$ (J/kg), the volumetric term associated with this change in the physical bond between the constituent molecules is given by

$$\frac{\dot{S}_{ij}}{V} = \dot{s}_{ij} = -\dot{n}_{ij}\Delta h_{ij} \text{ at } T_{ij} = T_{ij}(p) \quad \begin{array}{l} \text{volumetric heat absorption/release} \\ \text{by phase change.} \end{array} \quad (2.25)$$

Figure 2.8(b) gives a rendering of a phase change from a solid phase to a gas phase. When the phase change absorbs energy, $\Delta h_{ij} \equiv h_j - h_i$ is positive, and when it releases energy, then $\Delta h_{ij}$ will be negative. For example, for the liquid-gas phase change for water (i.e., evaporation) at $T_{lg} = 100°$C (boiling point at one atmosphere), the heat absorbed is $\Delta h_{lg} = +2.257 \times 10^6$ J/kg. And for the gas-liquid phase change (i.e., condensation), the energy is released and $\Delta h_{gl}$ is negative, i.e., $\Delta h_{gl}(T) = -\Delta h_{lg}(T)$. As the critical state $(T_c, p_c)$ is approached, $\Delta h_{lg}$ vanishes (no distinction is made between the liquid and gas phase), and beyond the critical point a dense gas-like substance exists. Examples of heat of phase change are given in Appendix C, the periodic table, Table C.2, and Tables C.3, C.4, C.5(a), C.6, C.26, C.27, C.28, and C.29.

The heats of adsorption $\Delta h_{ad}$ for some gas-porous solid pairs are listed in Table C.5.(b). Note that $\Delta h_{ad}$(J/kg-gas) is rather substantial, and whenever a porous solid with a large adsorption capability (such as desiccants and wool) is used, the porous solid can have a noticeable rise in temperature during adsorption. During desorption, heat is absorbed by the porous solid.

When phase change at the interface between phase $i$ and $j$ occurs with the mass flux of phase change given by $\dot{m}_{ij}$ (kg/m$^2$-s), then the interface phase change heat flux is given by

$$\frac{\dot{S}_{ij}}{A_{ij}} = -\dot{m}_{ij}\Delta h_{ij} \text{ at } T_{ij} = T_{ij}(p) \quad \begin{array}{l}\text{surface heat absorption/release}\\ \text{by phase change,}\end{array} \qquad (2.26)$$

where $\dot{m}_{ij}$ (kg/m$^2$-s) is the phase-change mass flux.

Traditionally, the enthalpy form of the energy equation is used and the interfacial phase change is not treated as an energy conversion. The results are then the same as the temperature-based energy used here with the phase change treated as an energy conversion. This is discussed in Appendix B, Section B.5.

### (F) Dissociation and Ionization

Similar to phase change, dissociation of gas molecules into neutral atoms occurs at high temperatures. Also at high temperatures atoms become ionized, forming ions and electrons. Then a mixture of ions, electrons, and neutral atoms and molecules is formed. This is significant in thermal plasma (discussed in Section 5.5). We represent the volumetric dissociation-ionization by $\dot{s}_{ij}$ (W/m$^3$).

### EXAMPLE 2.9. FUN

In applications involving liquid purification, e.g., water desalination (removal of mineral impurities), and in humidification, thin liquid films are heated from below and evaporation occurs at the liquid-gas interface. This requires a thin liquid film and a solid surface temperature $T_s$, which is not far from the saturation temperature $T_{lg}$ at one atmosphere (to prevent boiling). This is rendered in Figure Ex. 2.9. The heat is supplied through the solid substrate as $q_k$ (W/m$^2$). This heat flows through the thin liquid film by conduction and then reaches the interface, which is at the liquid-gas phase equilibrium temperature $T_{lg} = T_{lg}(p = 1 \text{ atm})$. The conduction heat flow per unit area (i.e., conduction heat flux) is presented by

$$q_k = k_l \frac{T_s - T_{lg}}{l} \quad \text{conduction heat flux through liquid film,}$$

where $k_l$ is the liquid thermal conductivity and $l$ is the liquid film thickness. Under steady state, this heat flows to the surface and results in evaporation. The energy equation for the interface $A$ is the simple form of (2.9), i.e.,

$$-Aq_k = \dot{S}_{lg}$$

$$= -\dot{M}_{lg}\Delta h_{lg} = -A\dot{m}_{lg}\Delta h_{lg} \quad \text{surface energy equation.}$$

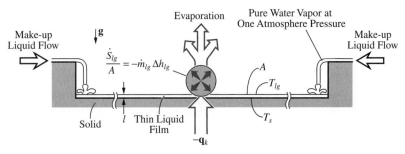

Figure Ex. 2.9. Steady-state heat transfer to, and evaporation from, a thin liquid film.

For a water liquid film with $k_l = 0.68$ W/m-K, film thickness $l = 0.5$ mm, temperatures $T_s = 104°C$ and $T_{lg} = 100°C$ ($p = 1$ atm, or $1.013 \times 10^5$ Pa), and $\Delta h_{lg} = 2.257 \times 10^6$ J/kg, determine the evaporation mass flux $\dot{m}_{lg}$.

## SOLUTION

The rate of heat flow to the interface, per unit area, $q_k$ is given and it results in evaporation of the water liquid film. This would require that the liquid in contact with the solid be slightly superheated [i.e., $T_s > T_{lg}(p)$], while the liquid at the interface is at the saturated state, i.e., $T = T_{lg}(p)$. The superheat should not exceed the bubble incipient superheat at which boiling will occur. We will discuss boiling in Section 6.6.2. We assume that the make-up water (shown in Figure Ex. 2.9) needed for a steady-state evaporation is also at the saturation temperature $T_{lg}(p)$.

From the energy equation given above we solve for $\dot{m}_{lg}$ and then substitute the expression for $q_k$. The result is

$$\dot{m}_{lg} = \frac{q_k}{\Delta h_{lg}} = \frac{k_l(T_s - T_{lg})}{\Delta h_{lg} l}.$$

Note that for a given $T_s - T_{lg}$, $\dot{m}_{lg}$ increases as $l$ decreases.

Using the numerical values we have

$$\dot{m}_{lg} = \frac{0.68(\text{W/m-°C})[104(°C) - 100(°C)]}{2.257 \times 10^6 (\text{J/kg}) \times 10^{-3}(\text{m})} = 1.205 \times 10^{-3} \text{ kg/m}^2\text{-s}$$

$$= 1.205 \text{ kg/m}^2\text{-s}$$

$$= 4.339 \text{ kg/m}^2\text{-hr}.$$

## COMMENT

The evaporation generally takes place in a gaseous ambient that contains species other than the vapor (called the noncondensables or inerts). Then the partial pressure of the vapor and the corresponding saturation temperature at this partial pressure should be used.

## 2.3.2  Electromagnetic Energy Conversion

Interaction of the electromagnetic waves (or fields) with the heat transfer medium results in an energy conversion. We will discuss the classification of electromagnetic waves in Chapter 4, where we discuss thermal radiation. This interaction results in an absorption or release of heat in the medium.

Some of the mechanisms for conversion of electromagnetic energy to or from thermal energy are shown in Figure 2.9. Here we consider the Joule, Seebeck, Peltier, and Thomson effects, volumetric and surface absorption of radiation, the surface emission of radiation, and microwave heating. These are discussed below.

The electric current vector per unit area $\mathbf{j}_e(\mathrm{A/m^2})$, also called the electric current flux vector, has some similarity to the heat flux vector $\mathbf{q}(\mathrm{W/m^2})$. We will discuss and use this similarity in Section 3.3.1, and also throughout the text to construct thermal circuit models.

### (A)  Joule or Ohmic Heating

Joule (or Ohmic) heating (i.e., electrical-resistance heating) converts (irreversibly) the electrical energy to thermal energy. The resistance to an electric current flow is given by the electrical resistance $R_e(\mathrm{ohm})$, $R_e \equiv L/\sigma_e A \equiv \rho_e L/A$, where $\sigma_e(1/\mathrm{ohm\text{-}m})$ is the electrical conductivity and is related to $\rho_e$, as $\sigma_e = 1/\rho_e$. Here $\rho_e(\mathrm{ohm\text{-}m})$ is the electrical resistivity, $L(\mathrm{m})$ is the length of the conductor, and $A(\mathrm{m^2})$ is the cross-sectional area of the conductor. These are shown in Figure 2.9(a). The conduction (free) electrons are accelerated when an electric field is present and they interact with each other, with the lattice vibrations, with defects, and with other heat carriers, and transfer their kinetic energy. This shows that the conduction electrons will not have the same temperature as the lattice atoms. But, since the mass of lattice is much larger, the lattice temperature is used as the medium temperature. The extent of these collisions and energy losses determines the electrical resistivity $\rho_e$. In general, $\rho_e = \rho_{e,p} + \rho_{e,d}$, where $\rho_{e,p}$ is the electron-phonon (or lattice vibration) resistivity, and $\rho_{e,d}$ is the electron-defect resistivity. The electric resistivity can be viewed as the electron scattering by phonons and defects. In Section 3.2.2(A) we will discuss this further. The volumetric rate of energy conversion $\dot{s}_{e,\mathrm{J}}(\mathrm{W/m^3})$ can be given in terms of the change in the electrical potential $\Delta\varphi$ (voltage, V) across the resistor as

$$\dot{s}_{e,\mathrm{J}} = \frac{\Delta\varphi^2}{\rho_e L^2} \quad \text{volumetric Joule heating in terms of electrical resistivity.} \qquad (2.27)$$

This Joule or Ohmic heating occurs in any electric conductor.

The electric field intensity vector $\mathbf{e}_e(\mathrm{V/m})$ is related to the electrical potential $\varphi$ through $\mathbf{e}_e = -\nabla\varphi$. Then for a one-dimensional field with a change in the electrical potential $\Delta\varphi$ occurring over a linear dimension $L$, we write the electric field intensity as $e_e = -\Delta\varphi/L$. For a volume $V = LA$, where $A$ is the cross-sectional area, having a uniform $\dot{s}_{e,\mathrm{J}}$, the rate of energy conversion $\dot{S}_{e,\mathrm{J}}(\mathrm{W}) = \dot{s}_{e,\mathrm{J}}V$ is

$$\dot{S}_{e,\mathrm{J}} = \dot{s}_{e,\mathrm{J}}V = \dot{s}_{e,\mathrm{J}}LA = \frac{\Delta\varphi^2}{R_e} \quad \begin{array}{l}\text{Joule heating in terms of}\\ \text{electrical resistance,}\end{array} \qquad (2.28)$$

(a) Joule or Ohmic Heating

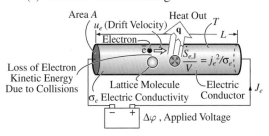

(b) Seebeck Thermoelectric Potential

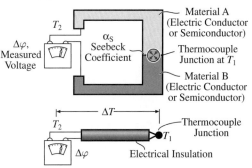

Conventional Representation of Thermocouple Junction

(c) Peltier Cooling/Heating

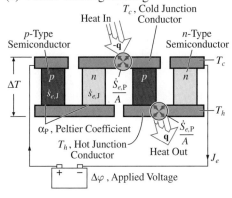

(d) Thomson Electrically-Induced Heat Flow

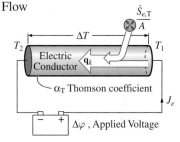

(e) Surface and Volumetric Radiation Attenuation/Absorbtion

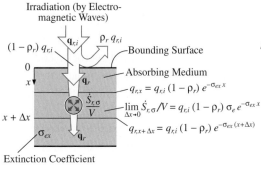

(f) Surface Thermal Radiation Emission

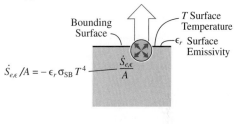

$$\dot{S}_{e,\epsilon}/A = -\epsilon_r \sigma_{SB} T^4$$

(g) Dielectric (Microwave) Heating

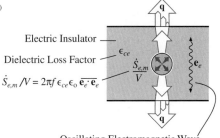

$$\dot{S}_{e,m}/V = 2\pi f \epsilon_{ce} \epsilon_0 \, \overline{\mathbf{e}_e \cdot \mathbf{e}_e}$$

Oscillating Electromagnetic Wave

Figure 2.9. Various energy conversion mechanisms between electromagnetic and thermal energy. (a) Joule or Ohmic heating. (b) Seebeck thermoelectric potential. (c) Peltier thermoelectric cooling (and heating). (d) Thomson electrically-induced heat flow. (e) Surface and volumetric radiation absorption. (f) Surface thermal radiation emission. (g) Microwave heating.

or

$$\dot{S}_{e,\text{J}} = \frac{e_e^2 L^2}{\rho_e L/A} = \frac{e_e^2 LA}{\rho_e} \qquad \text{Joule heating in terms of} \atop \text{electric field intensity.} \qquad (2.29)$$

Using the electrical conductivity $\sigma_e = 1/\rho_e$, the volumetric rate of energy conversion is

$$\dot{s}_{e,\text{J}} = \frac{e_i^2}{\rho_e} = \sigma_e e_e^2 = \sigma_e \frac{\Delta\varphi^2}{L^2} \qquad \text{volumetric Joule heating in} \atop \text{terms of electrical conductivity.} \qquad (2.30)$$

The current flux $j_e(\text{A/m}^2)$, also called the current density, and the current $J_e = j_e A(\text{A})$ are related to the potential $\Delta\varphi$ through the Ohm law as

$$J_e = j_e A = \sigma_e e_e A = \sigma_e \frac{\Delta\varphi}{L} A = \frac{\Delta\varphi}{R_e} \text{ or } j_e = \frac{\Delta\varphi}{R_e A} \qquad \text{electrical current and current flux.}$$
$$(2.31)$$

or

$$R_e = \frac{\Delta\varphi}{j_e A} = \frac{\Delta\varphi}{J_e}, \quad R_e = \frac{\rho_e L}{A} \qquad \text{electrical resistance, Ohm law.} \qquad (2.32)$$

Then using the above definitions for $R_e$, we can write (2.30) as

$$\frac{\dot{S}_{e,\text{J}}}{V} = \dot{s}_{e,\text{J}} = \rho_e j_e^2 = \frac{j_e^2}{\sigma_e} \qquad \text{volumetric Joule heating in} \atop \text{terms of current density.} \qquad (2.33)$$

In an alternating electric field, this volumetric heating becomes $\dot{s}_{e,\text{J}} = \sigma_e \overline{\mathbf{e} \cdot \mathbf{e}}$, where $\overline{\mathbf{e} \cdot \mathbf{e}}$ is the time-averaged, dot product of the electric field vector $\mathbf{e}$. The Joule heating of conducting fluids (e.g., high temperature ionized gases, which are called thermal plasmas) are treated similarly. In magnetic-induction heating, the current is induced in the charged moving fluid or the conducting solid by an external magnetic field. Most Joule heated plasmas are heated by magnetic induction. In electric arc heated plasmas, the current flows between two electrodes with the plasma flow between them. The Joule heating of gases will be discussed in Section 5.3.

Table C.8 in Appendix C lists the electrical resistivity of some pure metals for temperatures ranging from 1 to 900 K. The electrical resistivity increases with an increase in the temperature. The largest value listed is for zirconium at 900 K and is $123.1 \times 10^{-8}$ ohm-m. Metals have the lowest electrical resistivity (at room temperature, $\rho_e$ of the order of $10^{-7}$ ohm-m), the semiconductors have an intermediate resistivity (at room temperature, $10^{-4}$ ohm-m to $10^7$ ohm-m), and electrical insulators have high resistivity (at room temperature, about $10^{12}$ ohm-m). For semiconductors, $\rho_e$ decreases with increasing temperature; however, as noted for metals (as shown in Table C.8), $\rho_e$ increases with an increase in temperature. Also, for semiconductors $\rho_e$ depends on the irradiation by electromagnetic waves, especially in the visible wavelength range.

## EXAMPLE 2.10. FAM

Thin electrical current-carrying wires are used for Joule heating in appliances, climate control heaters, and in process heating. Using copper wire with a diameter $D = 1$ mm and a current $J_e = 4$ A:

(a) Determine the volumetric rate of heat generation $\dot{s}_{e,J}$.
(b) If the diameter is now reduced to 0.1 mm, what will be the volumetric rate of heat generation?

**SOLUTION**
(a) For copper at 20°C the resistivity is found from Table C.8 in Appendix C. This gives $\rho_e = 1.678 \times 10^{-8}$ ohm-m. For a wire of diameter $D = 1$ mm, carrying a direct current $J_e = 4$ A, the rate of energy conversion from electromagnetic energy to thermal energy is found from (2.33) as

$$\frac{\dot{S}_{e,J}}{V} = \dot{s}_{e,J} = \rho_e j_e^2 = \rho_e \frac{J_e^2}{A^2} = \frac{\rho_e J_e^2}{(\pi D^2/4)^2}$$

$$= \frac{1.678 \times 10^{-8}(\text{ohm-m}) \times 4^2(\text{A})}{\left[\dfrac{3.14 \times (10^{-3})^2}{4}\right]^2 (\text{m}^4)} = 4.357 \times 10^5 \text{ W/m}^3.$$

Note that if the current density $j_e$ was specified, the magnitude of the diameter of the wire would not be required for the determination of $\dot{s}_{e,J}$.

(b) In examining the above equation, note that for a given current, reducing the wire diameter by a factor of 10 increases the Joule heating by a factor of $10^4$ or to $\dot{s}_{e,J} = 4.357 \times 10^9$ W/m$^3$ for $D = 0.1$ mm $= 100$ $\mu$m.

**COMMENT**
The wire diameter dependence discussed above is the reason for using very thin wires in the conversion of electrical to thermal energy (e.g., in electric water heaters, electric ovens, and incandescent lamps). The heat release rate is comparable to that of combustion and fission reaction.

### (B) Seebeck Thermoelectric Potential

The Seebeck effect (also called the electrical potential creation) is the electrostatic potential difference between the low and high temperature regions of a conductor. The Seebeck effect is a reversible volumetric or interfacial phenomenon.[†] Here we

---

[†] Under an electric potential gradient $\nabla \varphi_s$ and a temperature gradient $\nabla T$, the current flux is given by

$$j_e \equiv \sigma_e e_e - \alpha_S \sigma_e \nabla T = -\sigma_e \nabla \varphi - \alpha_S \sigma_e \nabla T.$$

Under no current flow, the static electric field vector $e_s$ is found from above as

$$0 = -\sigma_e \nabla \varphi_s - \alpha_S \sigma_e \nabla T$$

or

$$e_s = -\nabla \varphi_s = \alpha_S \nabla T.$$

Integration of this equation gives (2.34).

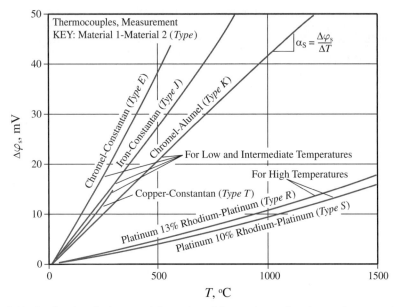

Figure 2.10. Static electrical potential produced by various electric conductor pairs, as a function of junction temperature. The reference temperature is taken equal to 0°C. The slope of the curves is the Seebeck coefficient. (From Benedict, R.P., reproduced by permission ©1977 Wiley.)

consider the interfacial effect for two dissimilar electric conductors A and B joined together with the joint (commonly called the junction) at temperature $T_1$ while the ends are at temperature $T_2$. This is shown in Figure 2.9(b). Then the static electrical potential $\Delta\varphi_s$ is given by

$$\Delta\varphi_s = \varphi_{s,2} - \varphi_{s,1} \equiv \pm\int_{T_1}^{T_2} \alpha_S dT \simeq \pm\alpha_S(T_2 - T_1) \quad \begin{array}{l}\text{Seebeck static electric}\\ \text{potential between two}\\ \text{junctions at different}\\ \text{temperatures,}\end{array} \quad (2.34)$$

where $\alpha_S$(V/°C or V/K, since $\alpha_S$ is defined per increment of 1°C or 1 K change in the temperature) is the Seebeck coefficient or the thermopower. The $\pm$ indicates the choice of materials A and B, which determines the direction of decrease in the electrical potential. The Seebeck coefficient $\alpha_S$ is of the order 10 microvolts per 1°C (or K) temperature change, i.e., $\mu$V/°C or $\mu$V/K. The Seebeck effect is used in the measurement of temperature by thermocouples (i.e., junctions of dissimilar conductors) and in electric power generation. Some of the metallic alloys used are chromel ($Ni_{90}Cr_{10}$), alumel ($Ni_{94}Mn_3Al_2Si_1$), and constantan ($Cu_{57}Ni_{43}$).

Due to the temperature difference across each conductor, the free electrons will have a larger kinetic energy where the temperature is higher. At the junction of two different materials, the more energetic electrons move between the conductors to establish an equilibrium. This causes the charge difference between the two materials, and therefore, the potential difference (with the cold location being the most negative).

To demonstrate the thermocouple (used in thermoelectric thermometry) applications, Figure 2.10 shows the performance, i.e., electrical potential given by (2.34), produced by various electrical conductor pairs, as a function of temperature at junction $T_1$. In addition to pure metals, alloys are also used (e.g., constantan, by volume, is 55% copper and 45% nickel). The reference temperature $T_2$ for which $\Delta\varphi$ is set to zero is taken as $0°C$. For high sensitivity, the type E thermocouples, which have an output, i.e., $\alpha_S$, of about 60 $\mu V/°C$ at $350°C$, are used. As shown in Figure 2.10, the thermocouples that can be used at high temperatures generally have lower sensitivity, especially at lower temperatures.

For detection of very small temperature differences, the semiconductors are used. This is based on the rather large dependence of their electrical resistivity on temperature. Then temperature variations as small as one-millionth of one $°C$ are measured.

### (C) Peltier Thermoelectric Cooling and Heating

The Peltier effect is the evolution or absorption of heat (reversible) when a direct electric current flows through a nonhomogeneous conductor (volumetric) or crosses the metallic conductor connecting two dissimilar semiconductors. The semi-metals and semiconductors have superior thermoelectric properties (including lower thermal conductivity, this will be discussed in Section 3.3.7), and are the materials of choice.

Figures 2.11(a)–(d) give a rendering of the Peltier heat flow at the junction between a $p$- and an $n$-type semiconductor connected through a metallic conductor, Figure 2.11(c). For the $p$-type semiconductors, the electron energy $E_F$ (called the Fermi energy level) is lower than that for the $n$-type semiconductors. The gap energy for the $n$-type material $(\Delta E_{e,g})_n$ is the energy added to a donor (conduction) level electron to move it to the Fermi level which is the same for all three materials after they are in contact. The gap energy for the $p$-type material $(\Delta E_{e,g})_p$ is the energy added to a Fermi level electron to move it to the acceptor (valence) level. These are shown in Figures 2.11(d)(i) and (ii). This energy $\dot{S}_{e,P}/A$ is absorbed in a junction as the electric current density $j_e$ flows from the $n$-type material to the metal and then through the $p$-type material (while the electrons move in the opposite direction). This is an idealized presentation, since these are transition (or depletion) layers around the junction (shown as having thickness $\delta$).

The Peltier heat flow rate is proportional to the current flowing through the junction. This interfacial phenomenon is given by

$$\dot{S}_{e,P} \equiv \pm \, \alpha_P j_e A \quad \text{Peltier interfacial cooling or heating,} \qquad (2.35)$$

where $\alpha_P(V)$ is the Peltier coefficient, $j_e(C/m^2\text{-s or A/m}^2)$ is the current flux, and $A(m^2)$ is the cross-sectional area. Note that the Peltier heat flow is proportional to $j_e$ to the first power, in contrast to the Joule heating, (2.33), which is proportional to

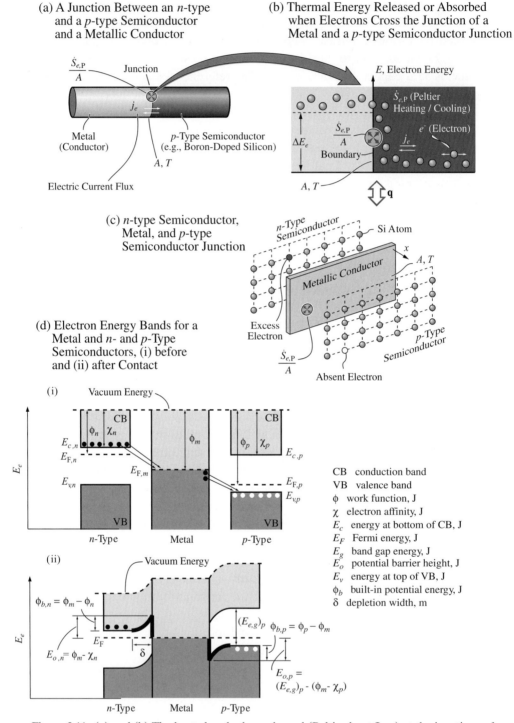

(a) A Junction Between an *n*-type and a *p*-type Semiconductor and a Metallic Conductor

(b) Thermal Energy Released or Absorbed when Electrons Cross the Junction of a Metal and a *p*-type Semiconductor Junction

(c) *n*-type Semiconductor, Metal, and *p*-type Semiconductor Junction

(d) Electron Energy Bands for a Metal and *n*- and *p*-Type Semiconductors, (i) before and (ii) after Contact

CB   conduction band
VB   valence band
$\phi$   work function, J
$\chi$   electron affinity, J
$E_c$   energy at bottom of CB, J
$E_F$   Fermi energy, J
$E_g$   band gap energy, J
$E_o$   potential barrier height, J
$E_v$   energy at top of VB, J
$\phi_b$   built-in potential energy, J
$\delta$   depletion width, m

Figure 2.11. (a) and (b) The heat absorbed or released (Peltier heat flow) at the junctions of the *n*- and *p*-type semiconductor materials, (c) connected through a conductor. The interface (i.e., junction) absorbs (or releases) energy as current flows through the junction. (d)(i) and (ii) The *n*-type-metal-*p*-type junctions, before and after contacts. The *n*-type electrons have higher electron-energy levels than the *p*-type. The Fermi levels ($E_F$), and other electron energies, are also shown.

$j_e$ to the second power. The Peltier coefficient is related to the Seebeck coefficient through

$$\alpha_P \equiv \alpha_S T \quad \text{Peltier coefficient,} \tag{2.36}$$

where $T\,(\text{K})$ is the absolute temperature of the junction. Using this, (2.35) is written as

$$\dot{S}_{e,P} = \pm \alpha_S J_e T \quad \text{thermoelectric (Peltier) heat absorption/release.} \tag{2.37}$$

The negative sign applies when electrons move from the $p$-type material to the $n$-type material (and the current is from the $n$-type material to the $p$-type material).

The semiconductors are divided into the $p$- and $n$-type semiconductors (the thermoelectric semiconductors are discussed in Section 3.3.7). When a $p$-type and an $n$-type material are connected or when there is a metallic conductor that connects a $p$-type and an $n$-type material [as shown in Figure 2.9(c)], then for the junction we have

$$\alpha_S = \alpha_{S,p} - \alpha_{S,n} \quad p\text{-}n \text{ junction Seebeck coefficient.} \tag{2.38}$$

where $\alpha_{S,p} > 0$ and $\alpha_{S,n} < 0$.

For some semiconductors used in thermoelectricity, $(\Delta E_{e,g})_i$ is of the order of 0.1 eV (as shown in Figure 1.7). The product $\alpha_S T$ is energy transport (across $p$-$n$ junction) per unit charge.[†] The junction Seebeck coefficient $\alpha_S$ depends on $T$, $\alpha_S = \alpha_S(T)$, and is independent of the shape or dimension of the junction (except for very small lengths). Examples of $\alpha_{S,p}$ and $\alpha_{S,n}$ are given in Tables C.9(a) and C.9(b).

The Peltier effect is being used for cooling purposes and will be discussed in Section 3.3.7, where it will be shown that parasitic heat transfer tends to deter optimum performance. It is also used for direct electrical power generation (especially in modular electric-power systems). This is similar to heat pumps, except there is no moving part involved. This is depicted in Figure 2.12(a). The electric circuit diagram for thermoelectric power generation is shown in Figure 2.12(b). The electric power is produced by maintaining a $p$-$n$ couple junction at temperature $T_h$, while the electrode connections are at a lower temperature $T_c$. There is an external electrical load $R_{e,o}$ placed in the circuit. The current $J_e$ is determined by noting that across the circuit the sum of the potentials is zero [15], i.e.,

$$\sum_i \Delta\varphi = 0 \quad \text{summation of potentials with no external source.} \tag{2.39}$$

[†] For semiconductors, we have [15]

$$\alpha_{S,i} = \pm \frac{k_B}{e_c}\left[\frac{(\Delta E_{e,g})_i}{k_B T} - \left(s + \frac{5}{2}\right)\right],$$

where, $i = p$ or $n$, and $s$ is a scattering parameter (e.g., $s = -0.5$).

### (a) Thermoelectric Power Generation by Heat Absorption at Hot Junction

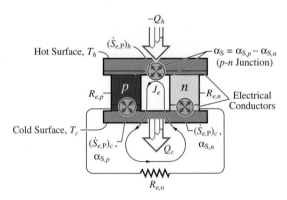

### (b) Electrical Circuit Diagram

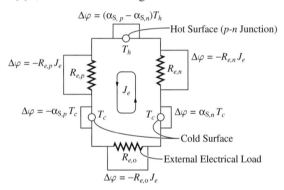

Figure 2.12. (a) The direct electrical power generation in a thermoelectric cell with an external resistance. (b) Electrical circuit diagram for the thermoelectric unit.

From this relation, and by making the proper substitution for the electrical potentials as shown in Figure 2.12(b), the current $J_e$ is found. Next we find that the electrical power generation rate $P_{e,o}$ is given by

$$P_{e,o} = J_e^2 R_{e,o}(\text{W}) = \left[\frac{\alpha_S(T_h - T_c)}{R_{e,o} + R_e}\right]^2 R_{e,o} \quad \text{electrical power generation,} \quad (2.40)$$

where $R_e$ is the combined electrical resistance of the $p$ and $n$ materials (because these resistances are in series, $R_e = R_{e,p} + R_{e,n}$).

By taking the derivate of (2.40) with respect to $R_{e,o}$ and setting the resultant equal to zero, it can be shown that for optimum power output $R_{e,o} = R_e$. In (2.40), it is assumed that $\alpha_{S,p}$ and $\alpha_{S,n}$ are independent of $T$. Because of the large $T_h - T_c$ used in the power generation (as compared to the rather small $T_h - T_c$ used in thermoelectric cooling), their dependence on temperature should be included and generally average values for $\alpha_{S,p}$ and $\alpha_{S,n}$ should be used.

Among materials (alloys) that have large Seebeck coefficients (and low thermal conductivity) are ZnSb, PbTe, Bi$_2$Ti$_3$, PbSe, Bi$_2$Se$_3$, Sb$_2$Te$_3$, MnTe, GeTe, SiGe, and group III-V alloys [14].

---

### EXAMPLE 2.11. FAM

---

In thermoelectric cooling, as depicted in Figure Ex. 2.11, the objective is the Peltier heat absorbed at the cold *p-n* junction. As will be shown in Section 3.3.7, the parasitic Joule and conductive heat also flow to the cold junction, thus deteriorating the cooling potential. For a bismuth telluride cold *p-n* junction [the properties are listed in Table C.9(a)] with $\alpha_{S,p} = 230 \times 10^{-6}$ V/K and $\alpha_{S,n} = -210 \times 10^{-6}$ V/K and a current $J_e = 10$ A, determine the Peltier heat absorbed at the cold junction $(\dot{S}_{e,P})_c$ at (i) $T_c = 120°$C, (ii) $T_c = 20°$C, and (iii) $T_c = -80°$C.

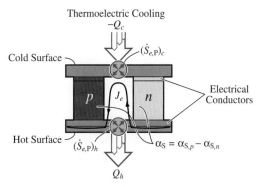

Figure Ex. 2.11. A thermoelectric cooling junction showing the steady-state heat flow into the cold junction and heat flow out of the hot junction.

### SOLUTION

The junction Seebeck coefficient is used in (2.36) and the Peltier heat absorption/release is determined from (2.37) using the junction Seebeck coefficient. So we have for the cold junction

$$(\dot{S}_{e,P})_c = -\alpha_S J_e T_c = -(\alpha_{S,p} - \alpha_{S,n})J_e T_c.$$

Note that $T_c$ is in the absolute temperature scale K.

Now using the numerical values from Table C.9(a), we have

$$\alpha_{S,p} = 230 \times 10^{-6} \text{ V/K}$$

$$\alpha_{S,n} = -210 \times 10^{-6} \text{ V/K}$$

$$(\dot{S}_{e,P})_c = -(230 \times 10^{-6} + 210 \times 10^{-6})(\text{V/K}) \times 10(\text{A}) \times T_c(\text{K}).$$

For the three temperatures we have

(i) $(\dot{S}_{e,P})_c(T_c = 120°C) = -4.40 \times 10^{-3}(\text{V/K}) \times (273.15 + 120)(\text{K})$

$\qquad\qquad\qquad = -1.73 \text{ W}$

(ii) $(\dot{S}_{e,P})_c(T_c = 20°C) = -1.29 \text{ W}$

(iii) $(\dot{S}_{e,P})_c(T_c = -80°C) = -0.850 \text{ W}.$

**COMMENT**

Note that the magnitude of $(\dot{S}_{e,P})_c$ decreases as $T_c$ decreases, and therefore, the colder the cooling junction becomes, the lower will be its heat absorption capability. We have not yet discussed the size of this junction. In Section 3.3.7, this, along with the parasitic Joule and conductive heat flow to the cold junction, will be discussed. In practice, many junctions are used and the heat absorbed at each junction is additive, resulting in the supply of the desired cooling capability.

### (D) Thomson Electrically-Induced Heat Flow

The Thomson effect is related to the heat absorbed (reversible) in a conductor when there is a current passing through it (or potential applied across it), e.g., along the $x$ direction, as shown in Figure 2.9(d). The absorbed heat (a surface phenomenon) flows along the conductor and the temperature gradient $dT/dx$ influences the heat absorbed. The Thomson effect occurs in any electric conductor (i.e., no junction is required). In conductors subject to temperature variations, such as $T_1$ and $T_2$ applied across a conductor of length $L$, the heat absorbed is

$$\dot{S}_{e,T}/A = \int_0^L \alpha_T j_e \frac{dT}{dx} dx = \int_{T_1}^{T_2} \alpha_T j_e dT \simeq \alpha_T j_e (T_2 - T_1) \quad \begin{array}{l}\text{Thomson induced} \\ \text{heat flow,}\end{array}$$

(2.41)

where $\alpha_T(\text{V}/°C)$ is the Thomson coefficient and is of the order of one $\mu\text{V}/°C$. For most metals, the Thomson effect is much smaller than the Peltier effect (but there are metal pairs with large $\alpha_T$). Here we have made the approximation (and used $\simeq$ instead of $=$) that $\alpha_T$ is constant. The Thomson coefficient is positive when a positive current flows toward the warmer end of the conductor (i.e., negative $dT/dx$).

### (E) Volumetric and Surface Radiation Absorption*

This section is found on the Web site www.cambridge.org/kaviany. In Section (E) Volumetric and Surface Radiation Absorption we discuss this energy conversion rate and introduce the radiation extinction or attenuation coefficient and list its values for some common materials. There is also an example (Example 2.12).

### (F) Surface Radiation Emission

Surface radiation emission is radiation emitted by a body due to its temperature (i.e., thermal radiation) and is depicted in Figure 2.9(f). The emitted electromagnetic wave is called photon and its energy is quantized. A surface at temperature $T$ emits radiation in the amount $\dot{S}_{e,\epsilon}/A(\text{W/m}^2)$ (this is also called the total hemispherical emissive power):

$$\frac{\dot{S}_{e,\epsilon}}{A} = -\epsilon_r \sigma_{\text{SB}} T^4 \quad \text{surface radiation emission,} \tag{2.48}$$

where $\epsilon_r (0 \leq \epsilon_r \leq 1)$ is the total, hemispherical surface emissivity (or simply emissivity), $\sigma_{\text{SB}} = 5.670 \times 10^{-8}$ W/m$^2$-K$^4$ is the Stefan-Boltzmann constant [also listed in Table C.1(b) in Appendix C], and $T$ is the absolute temperature (i.e., K). We will discuss this further in Chapter 4. Note that because of the fourth-power temperature dependence, the thermal radiation emission becomes very significant at high temperatures. Table C.18, in Appendix C, lists $\epsilon_r$ for various surfaces and Section 4.1.3 gives examples of magnitude of $\epsilon_r$ for various surfaces.

### (G) Microwave (Dielectric) Heating

Electrically insulating materials (i.e., dielectrics) withstand a large electrical potential, and when subjected to an alternating electric field intensity vector $\mathbf{e}_e = \mathbf{e}_e(t)$, they convert the electromagnetic energy to thermal energy by absorption of electromagnetic energy and conversion to intermolecular vibration. This is an irreversible, volumetric energy conversion and is depicted in Figure 2.9(g).

For an alternating electromagnetic field with an alternating electric field intensity with instantaneous value $e_e(\text{V/m})$, an oscillation frequency $f$ (Hz or 1/s), a dielectric loss factor $\epsilon_{ec}$ (this is the imaginary part of the relative permittivity), and with the permittivity of free space $\epsilon_o (8.8542 \times 10^{-12}$ A$^2$-s$^2$/N-m$^2$), the time-averaged volumetric rate of conversion of electromagnetic energy to thermal energy (also called microwave heating) is given by [11]

$$\frac{\dot{S}_{e,m}}{V} = \dot{s}_{e,m} = 2\pi f \epsilon_{ec} \epsilon_o \overline{e_e^2}, \quad \overline{e_e^2} = \overline{\mathbf{e}_e \cdot \mathbf{e}_e} \quad \text{dielectric (microwave) heating.} \tag{2.49}$$

The overbar indicates time average time averaging is discussed in Section 2.5.3 and $(\overline{e_e^2})^{1/2}$ is the root mean square of the electric field intensity. The angular frequency $\omega$ is related to $f$ through $\omega = 2\pi f$.

The complex, relative permittivity $\epsilon_e$ has a real part, which is called the relative permittivity (or the dielectric constant) and is used in the determination of the electrical capacitance. This is designated as $\epsilon_{er}$ (and has a magnitude between unity and infinity). The imaginary part is called the dielectric loss factor $\epsilon_{ec}$ and is a measure of the interaction of the medium with an oscillating electromagnetic field. For nonmagnetic, dielectric materials (e.g., water and biological tissues), the electromagnetic field interacts with the electric charge or the electric dipole moment. For composite materials (e.g., biological tissue), the effective dielectric loss factor $\langle \epsilon_{ec} \rangle$ is used. In general $\epsilon_{ec}$ and $\langle \epsilon_{ec} \rangle$ depend on the frequency, temperature, and other parameters such as the moisture content.

Both microwave (defined by $3 \times 10^8 < f < 3 \times 10^{11}$ Hz) and radiowave (defined by $3 \times 10^4 < f < 3 \times 10^8$ Hz) electromagnetic radiation are used in communications. Because the microwave electromagnetic radiation is also used in communications, the commercial use of the microwave heating frequencies is mostly restricted to frequency bands, 0.915 GHz (wavelength $\lambda = 327.9$ mm), which is used in industrial ovens for deeper penetration, and 2.450 GHz ($\lambda = 122.4$ mm), which is used in domestic ovens. Microwaves are also used for therapeutic heating of biological tissues. Typical values for $\epsilon_{ec}$ and $\langle \epsilon_{ec} \rangle$ at frequencies around 1 GHz (i.e., $10^9$ Hz) are 39 for beef steak, 0.04 for Douglas fir wood, and 0.364 for distilled water. Some typical values for $\epsilon_{ec}$ and $\langle \epsilon_{ec} \rangle$ are given in Table C.10 of Appendix C.

### EXAMPLE 2.13. FUN

In a microwave oven operating at a frequency of $10^9$ Hz (one gigahertz), water is heated in two paper cups. As shown in Figure Ex. 2.13, in one cup the water used is distilled and in the other it has 0.5% (on molal basis) of ordinary salt (NaCl). The power setting on the microwave oven is chosen for $(\overline{e_e^2})^{1/2} = 10^2$ V/m. For the $\rho c_p$ of water and cup use $4 \times 10^6$ J/m$^3$-°C and $4 \times 10^5$ J/m$^3$-°C. Assume that no heat is transferred between the water and its surroundings and also the cup and its surroundings. Also assume that the electric field is uniform within the cup.

(a) For both cases (i.e., with and without salt), estimate the rate of increase in the temperature of the water and the cup.
(b) For each of the cups, does the temperature of the water rise faster or slower than that of the cup?
    Use the dielectric loss factor at a temperature near 90°C.

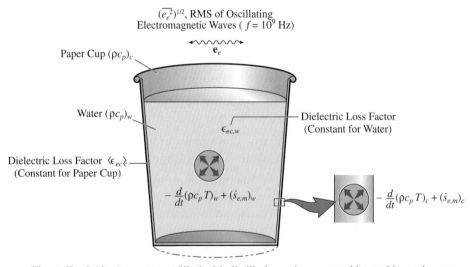

Figure Ex. 2.13. A paper cup filled with distilled or salt water and heated in a microwave oven.

**SOLUTION**

We begin by writing the integral-volume energy equation (2.9) for the water and the cup, assuming a uniform temperature and a uniform energy conversion for each medium, i.e., $T_w$ and $T_c$, and $(\dot{s}_{e,m})_w$ and $(\dot{s}_{e,m})_c$. These energy equations are

$$Q\mid_{A,w} = 0 = -V_w\frac{d}{dt}(\rho c_p)_w T_w + V_w(\dot{s}_{e,m})_w \quad \text{energy equation for water}$$

$$Q\mid_{A,c} = 0 = -V_c\frac{d}{dt}(\rho c_p)_c T_c + V_c(\dot{s}_{e,m})_c \quad \text{energy equation for cup.}$$

The energy conversion terms are the volumetric microwave heating rates given by (2.49). These would become

$$(\dot{s}_{e,m})_w = 2\pi f \epsilon_{ec,w} \epsilon_o \overline{e_e^2} \quad \text{volumetric energy conversion for water}$$

$$(\dot{s}_{e,m})_c = 2\pi f \langle \epsilon_{ec} \rangle_c \epsilon_o \overline{e_e^2} \quad \text{volumetric energy conversion for cup.}$$

Using these in the above energy equations, we solve for the time rate of change in temperature and have (for constant $\rho c_p$)

$$\frac{dT_w}{dt} = \frac{(\dot{s}_{e,m})_w}{(\rho c_p)_w} \quad \text{rate of rise in temperature of water}$$

$$\frac{dT_c}{dt} = \frac{(\dot{s}_{e,m})_c}{(\rho c_p)_c} \quad \text{rate of rise in temperature of cup}$$

or

$$\frac{dT_w}{dt} = \frac{2\pi f \epsilon_{ec,w} \epsilon_o \overline{e_e^2}}{(\rho c_p)_w} \quad \text{and} \quad \frac{dT_c}{dt} = \frac{2\pi f \langle \epsilon_{ec} \rangle_c \epsilon_o \overline{e_e^2}}{(\rho c_p)_c}.$$

(a) For $f = 10^9$ Hz, we will use the values for $\epsilon_{ce,w}$ and $\langle \epsilon_{ec} \rangle_c$ from Appendix C, Table C.10. Note that for the paper we have the dielectric loss factor at $T = 82°$C and for the distilled water at $T = 95°$C and for the distilled water with 0.5% (in molal basis) salt at $T = 25°$C. We use these as the only available properties, noting the lack of consistency in temperatures used for properties.

$$\langle \epsilon_{ec} \rangle_c = 0.216 \quad \text{paper, Table C.10}$$

$$(\epsilon_{ec,w})_{no\ salt} = 0.364 \quad \text{distilled water, Table C.10}$$

$$(\epsilon_{ec,w})_{salt} = 269 \quad \text{distilled water with salt, Table C.10.}$$

Substituting the numerical values in these equations, for the time rate of change in temperatures, we have

$$\frac{dT_c}{dt} = \frac{2\pi \times 10^9 (1/s) \times 0.216 \times 8.8542 \times 10^{-12} (A^2\text{-}s^2/N\text{-}m^2) \times (10^2)^2 (V/m)^2}{4 \times 10^5 (J/m^3\text{-}°C)}$$

$$\frac{dT_w}{dt}\Big|_{no\ salt} = \frac{2\pi \times 10^9 (1/s) \times 0.364 \times 8.8542 \times 10^{-12} (A^2\text{-}s^2/N\text{-}m^2) \times (10^2)^2 (V/m)^2}{4 \times 10^6 (J/m^3\text{-}°C)}$$

$$\frac{dT_w}{dt}\Big|_{salt} = \frac{2\pi \times 10^9 (1/s) \times 269 \times 8.8542 \times 10^{-12} (A^2\text{-}s^2/N\text{-}m^2) \times (10^2)^2 (V/m)^2}{4 \times 10^6 (J/m^3\text{-}°C)}.$$

In reviewing the units, note that

$$\frac{(1/s)(A^2\text{-}s^2/N\text{-}m^2)(V^2/m^2)}{(J/m^3\text{-}°C)} = \frac{(A^2\text{-}V^2\text{-}s^2)(1/N\text{-}m^4\text{-}s)}{(J/m^3\text{-}°C)} =$$

$$\frac{(W^2\text{-}s^2)(1/J\text{-}m^3\text{-}s)}{(J/m^3\text{-}°C)} = \frac{(J^2\text{-}°C)}{(J^2\text{-}s)} = (°C/s),$$

where we have used

$$(V\text{-}A) = (W), \quad (J) = (W\text{-}s), \quad \text{and } (J) = (N\text{-}m).$$

Carrying on the arithmetic, we have

$$\frac{dT_c}{dt} = 3.00 \times 10^{-4}°C/s$$

$$\frac{dT_w}{dt}\Big|_{no\ salt} = 5.06 \times 10^{-5}°C/s$$

$$\frac{dT_w}{dt}\Big|_{salt} = 0.374°C/s.$$

(b) From these numerical results, by comparing the rates of temperature rise for the case of distilled water without salt, we have

$$\frac{dT_w}{dt}\Big|_{no\ salt} < \frac{dT_c}{dt},$$

i.e., the cup heats up faster. For the case of added salt

$$\frac{dT_w}{dt}\Big|_{salt} > \frac{dT_c}{dt},$$

i.e., the salt water heats up faster.

**COMMENT**

Note the drastic effect of an increase in $\epsilon_{ec}$, due to the addition of the salt. Table C.10 shows that food materials also have a relatively high effective dielectric loss factor $\langle \epsilon_{ec} \rangle$.

*(H) Others*

There are other thermo-chemico-electric energy conversions. Some of them are mentioned below.

In alkali metal thermal to electric conversion (i.e., AMTEC), thermal energy is used to drive a current of ions across a selective barrier (beta-alumina solid electrolyte, a ceramic, is used). Heat is supplied to liquid sodium, causing its evaporation at one side of the barrier, and heat is removed from the vapor sodium on the other side. This creates a large vapor pressure difference across the barrier, which in turn causes an electrical potential. The electrodes on each side of the barrier provide sites for electrochemical reactions at the interfaces. On the high activity (high vapor pressure) anode, oxidation of Na occurs, and on the low activity cathode, reduction occurs. An electrical current is generated in a circuit made of these electrodes and an external resistance.

In thermoionic power generation, heat is added to remove the electrons from an electrically conducting surface (hot electrode or emitter). The electrons are collected at the cold electrode (i.e., anode), where heat is removed. Under an external resistance load, a voltage potential (direct current) is formed between the electrodes.

In infrared photovoltaic (also called thermal photovoltaic) energy conversion, a solid or porous surface is heated by gaseous combustion and emits photons to a photocell. This is similar to solar photovoltaic energy conversion, except the emitted radiation has a peak in the infrared range (instead of the visible). The spectrum of the electromagnetic radiation will be discussed in Chapter 4.

In photoelectric power generation, the photons in the irradiation beam impinging on a conducting solid cause emission of electrons from the surface. The photon energy consumed is equal to the energy needed to free the electrons from their bonds (for example, 4.6 eV for silicon). This may require several photons to free a single electron. For irradiation with a wavelength $\lambda$, the energy per photon is $E_{ph} = h_P \lambda / c$, where $h_P$ is the Planck constant, and $c$ is the speed of light.

In laser-induced fluorescent (anti-Stokes) cooling of solids, for example, a glass made of heavy metal-fluoride compounds is doped with ytterbium ions and irradiated by laser and in turn emits photons. The wavelength of the laser is selected (tuned) such that the emitted photons have a larger energy than the absorbed photons. The difference in energy is the heat absorbed from the solid lattice vibration (absorption of phonons). The solid equilibrium temperature is determined by heat flowing into the solid. If this parasitic heat can be reduced, low solid temperatures are possible (e.g., cryocoolers).

### 2.3.3 Mechanical Energy Conversion

The mechanical energy is associated with force and displacement and its energy conversion rate involves the rate of displacement (velocity). The heat transfer medium may undergo motion under an applied force. The medium, which can be a gas, liquid, or solid (or a composite), will interact with this applied force. Depending on the phase of the medium, different mechanical energy conversion mechanisms become significant.

The mechanical energy conversions considered are pressure gradient and frictional (viscous dissipation, surface solid-solid friction, and dissipation of ultrasound waves). The frictional (dissipation) energy conversions are irreversible. As a result of these energy conversions, heat is released or absorbed in the medium.

### (A) Pressure-Compressibility Cooling or Heating*

This section and section B (Viscous Heating) are found on the Web site www.cambridge.org/kaviany. In Sections (*A*) and (*B*), an example of each of these two mechanical-to-thermal energy conversions are given (Examples 2.14 and 2.15).

### (B) Viscous Heating*

### (C) Surface Friction Heating

Surface friction heating is a surface (i.e., interfacial) energy conversion of mechanical energy to thermal energy and is irreversible. The friction (including grinding, cutting, braking, etc.) occurs at the interface (contact surface) of two solid surfaces in contact and in a relative motion. When the solid is not rigid (i.e., deforms), the friction heating becomes volumetric and is concentrated around the volumes with the largest deformation (which is near the contact interface). Here we discuss surface-friction heating. The rate of the thermal energy conversion per unit surface area $q_{m,F}$ is

$$\frac{\dot{S}_{m,F}}{A} = \mu_F p_c \Delta u_i \quad \text{surface friction heating,} \qquad (2.53)$$

where $\mu_F$ is the friction coefficient, $p_c(\text{N/m}^2)$ is the contact or joint (or lining) pressure, and $\Delta u_i(\text{m/s})$ is the interface relative velocity (or speed). For a joint with a relative rotation (as in brakes and clutches) the product of the angular frequency $\omega$ and the radius $R$ is used for the local velocity.

The thermal energy generated flows through the various media surrounding the location of heat generation. In surface grinding, we start with the amount of heat generated over a contact surface $\dot{S}_{m,F}$. A portion of this, $a_1 \dot{S}_{m,F}$, flows into the workpiece, where $a_1$ is a constant between 0 and unity. The rest of the heat flows into the grinding wheel and the coolant fluid (if any). This is rendered in Figure 2.14(a). This example will be discussed in Section 3.3.6 in connection with heat conduction through the workpiece.

Among the mechanical energy conversion mechanisms, friction heating has one of the largest magnitudes.

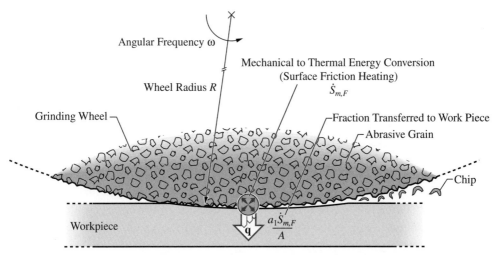

Figure 2.14. (a) Surface mechanical energy conversion in grinding and flow of part of this into the workpiece.

EXAMPLE 2.16. FUN

In automobile brakes, the contact force (and its duration) depends on the speed of the automobile being stopped (among other parameters). At higher speeds, higher forces (or contact pressures) are applied. The braking may also be intermittent (a sequence of small periods of on and off contacts). Consider that the contact pressure $p_c$ is proportional to the relative speed of the brake pad and drum to the first power, i.e., $p_c = a_1 \Delta u_i$, where $a_1$ is a constant and $\Delta u_i$ is the relative speed with the brake pad being stationary.

Show the ratio of the surface-friction heating at two different drum angular frequencies, $\omega_1$ (rad/s) and a lower frequency $\omega_2 = 0.1\omega_1$.

**SOLUTION**

The friction interface relative speed $\Delta u_i$ is the product of the angular frequency $\omega$ (rad/s) and the radius of the contact surface $R$, i.e.,

$$\Delta u_i = \omega R.$$

Now for a contact pressure $p_c$ proportional to the contact velocity $\Delta u_i$, for the surface-friction heating rate given by (2.53), we have

$$\dot{S}_{m,F}/A = \mu_F p_c \Delta u_i$$

$$= \mu_F a_1 \Delta u_i \Delta u_i$$

$$= \mu_F a_1 \Delta u_i^2.$$

Next we substitute for $\Delta u_i$ and have

$$\dot{S}_{m,F}/A = \mu_F a_1 (\omega R)^2.$$

Now the ratio of the two surface-friction heating rates is

$$\frac{(\dot{S}_{m,F}/A)_{\omega_1}}{(\dot{S}_{m,F}/A)_{w_2}} = \frac{\mu_f a_1 R^2 \omega_1^2}{\mu_f a_1 R^2 \omega_2^2} = \frac{\omega_1^2}{\omega_2^2}$$

$$\frac{1}{(0.1)^2} = 10^2.$$

**COMMENT**

At high speeds, and when the brake is applied for a relatively long period (e.g., long downhill travel), excessive surface-friction heating occurs. Without a proper cooling of the pads, this can cause damage to the contact material. The heat generated during the friction contact period $\tau$, over the contact area $A_c$, is

$$\int_0^\tau \dot{S}_{m,F}\,dt = \int_0^\tau (\dot{S}_{m,F}/A_c)A_c\,dt.$$

Note that for a sudden and complete stop, the kinetic energy of the car is converted into the surface-friction heating and this energy is rather substantial at high speeds.

### (D)  Ultrasound Heating

Ultrasound heating is another example of an irreversible volumetric mechanical to thermal energy conversion. The attenuation of longitudinal acoustic (i.e., compressive, mechanical) waves with frequencies above the human hearing frequency (i.e., above $2 \times 10^4$ Hz) are used. This is because the attenuation (or absorption) of these sound waves is proportional to their frequency. In ultrasonic heating applications, frequencies near or larger than 1 MHz ($10^6$ Hz) are used.

Figure 2.14(b) shows the wavelength and frequency of various acoustic waves (infrasound, audible sound, ultrasound, hypersound, Debye [i.e. accoustic], and optical phonon frequency ranges in sound). The Debye or optical phonons are also called thermal phonons since they are due to thermal fluctuations in atomic positions in solids (Section 3.2.2). The movement of the atoms in the medium occurs parallel to the direction of the wave propagation. The imposed and thermal acoustic waves attenuate as they travel the medium and this dissipated energy is converted to heat at a volumetric rate of $\dot{s}_{m,ac}$(W/m$^3$) given by

$$\boxed{\frac{\dot{S}_{m,ac}}{V} = \dot{s}_{m,ac} \equiv -\frac{dI_{ac}(x)}{dx} = 2\sigma_{ac}I_{ac} \qquad \begin{array}{l}\text{volumetric heating by acoustic waves,}\\ \text{and acoustic absorption coefficient,}\end{array}} \quad (2.54)$$

where $\sigma_{ac}$(1/m) is the acoustic absorption (or extinction or attenuation) coefficient[†] and $I_{ac}$(W/m$^2$) is the acoustic intensity. This is similar to (2.43).

---

[†]  The dimensionless acoustic attenuation which is equal to $\sigma_{ae}a_s/\pi f$, is also used. The inverse of the acoustic attenuation is called the quality factor.

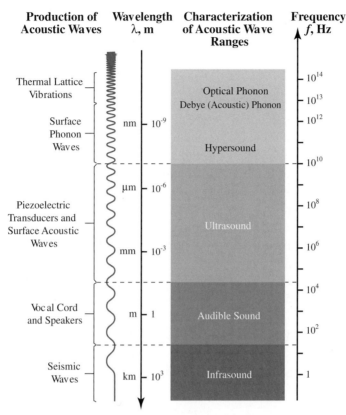

Figure 2.14. (b) Spectra (wavelength and frequency) of various acoustic waves.

For nonpolar, low viscosity and thermal conductivity fluids, the acoustic absorption coefficient $\sigma_{ac}$ is related to the fluid thermophysical properties through [14]

$$\sigma_{ac} = \frac{(2\pi f)^2 \mu}{2\rho a_s^3} \left( \frac{4}{3} + \frac{\gamma - 1}{\text{Pr}} \right), \quad a_s = \lambda f \quad \text{speed of sound}, \qquad (2.55)$$

where $\rho(\text{kg/m}^3)$ is the fluid density, $f(\text{Hz or 1/s})$ is the frequency, $\mu(\text{Pa-s})$ is the dynamic viscosity, $a_s(\text{m/s})$ is the speed of sound in the fluid, $\gamma = c_p/c_v$ is the ratio of specific heat capacities (a measure of compressibility of the fluid), and $\text{Pr} = \mu c_p/k$ is the Prandtl number signifying the ratio of viscous diffusion $\mu/\rho$ to the heat diffusion $k/\rho c_p$. We will discuss the Prandtl number in Chapter 6. Also, in (2.55) the frequency and the wavelength are related through the speed of sound $a_s$. In general, $\sigma_{ac}$ is determined experimentally.

Similar $\sigma_{ac}$ relations are available for solids.[‡] These relations also show a $f^2$ and $f$ dependence for $\sigma_{ac}$. Figure 2.14(c)(i) shows typical values of acoustic extinction coefficient for some fluids, fluid-solid composites, and solids. The speed of sound in some media are also shown in Figure 2.14(c)(ii). As the acoustic waves travel through an attenuating medium, the acoustic intensity decreases. This attenuation is

---

[‡] For dielectric, crystalline solids at low temperatures we have [7]

similar to that discussed for volumetric radiation absorption and is given by (2.43). The acoustic intensity along a direction $x$ is given by [7]

$$\frac{I_{ac}(x)}{I_{ac,o}} = e^{-2\sigma_{ac}x}, \tag{2.56}$$

where $I_{ac,o}$ is the acoustic intensity at location $x = 0$. In some cases a uniform acoustic field distribution is a reasonable assumption (when $2\sigma_{ac}x \ll 1$). Otherwise the drop in the acoustic intensity should be included.

Ultrasonic heating is significant in liquids and in liquid-like composite media (i.e., biological tissues). Ultrasonics is used for welding where solids in contact are made to slide (oscillatory) and create friction heating. Ultrasonics is also used for medical applications at a frequency in the range of 0.5 to 20 MHz (i.e., $5 \times 10^5$ to $20 \times 10^6$ Hz) and generally with an intensity $I_{ac}$ of about $10^3$ to $1.5 \times 10^7$ W/m$^2$ [7, 10]. For example, it is used in the following: diagnostic imaging using pulse-echo (1 to 20 MHz, 17.5 kW/m$^2$) and pulsed Doppler (1 to 20 MHz, 157 kW/m$^2$); in physiotherapy (0.5 to 3 MHz, 250 kW/m$^2$); and in surgery (0.5 to 10 MHz, 15,000 kW/m$^2$). The duration of exposure for surgery is 1 to 16 s. For the high intensities, a focused array of transducers is used. Tissue ultrasonic absorption occurs primarily in tissue proteins. Table C.11 in Appendix C lists some typical values of $\sigma_{ac}$ for some fluids and biological tissues. These typical values for $\sigma_{ac}(1/m)$, at $f = 1$

$$\sigma_{ac} = \gamma_G^2 \frac{c_v T}{2a_s^3} \frac{(2\pi f)^2 \tau_p}{1 + (2\pi f)^2 \tau_p^2},$$

where

$$\gamma_G = \frac{3\beta_s E_p}{\rho c_v},$$

is the Grüneisen constant, $\beta_s$ is linear thermal expansion coefficient, and $E_p$ is the bulk modulus (1.18).

The phonon relaxation time is $\tau_p = \lambda_p/a_s$, where $\lambda_p$ is the phonon mean-free path (3.26), such that

$$\tau_p = \frac{3k^p}{\rho c_v a_s^2},$$

and $k^p$ is the phonon thermal conductivity.

Then for $2\pi f \tau_p < 1$, we have

$$\sigma_{ac} = \gamma_G^2 \frac{(2\pi f)^2 k^p T}{\rho a_s^5}.$$

Also, for $2\pi f \tau_p > 1$, we have

$$\sigma_{ac} = \frac{\pi}{4} \gamma_G^2 \frac{(2\pi f) c_v T}{\rho a_s^5}.$$

(i) Acoustic extinction coefficient for some gases, liquids and solids, at $f = 1$ MHz.

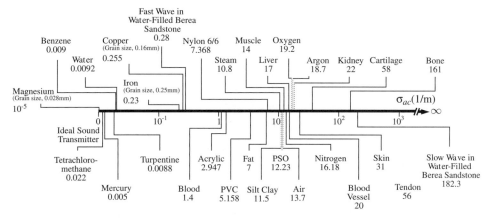

(ii) Sound speed for some gases, liquids and solids, at $f = 1$ MHz.

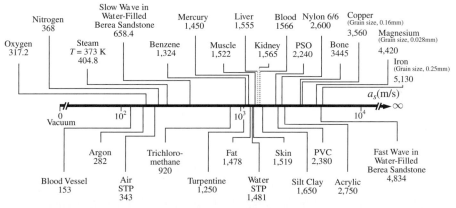

Figure 2.14. (c)(i) Typical values of the acoustic extinction coefficient for some fluids, and fluid-solid composites, and solids. (ii) Typical sound speeds in various materials. Unless specified, $T = 273$ K, and $p = 1$ atm.

MHz, are 16.4 for nitrogen and dry air (larger for humid air), 0.0025 for pure water, 1.4 for blood, 14 for muscle, 31 for skin, and 161 for bone. Then for biological materials (tissues) with large $\sigma_{ac}$, the energy absorbed can be very significant and can cause local damage. Focused ultrasound is also used in surgery for cutting and to stop internal bleeding by cauterizing.

The speed of sound in blood (close to that of water), muscle and skin is about 1,519 m/s and in bone is 3,445 m/s. Then for a frequency $f = 10^6$ Hz, the wavelength of acoustic wave in skin is $\lambda = 1.519$ mm and in bone is $\lambda = 3.445$ mm. The acoustic impedance is defined as the product $\rho a_s$. The large difference in acoustic impedance between air and skin causes nearly complete reflection. By adding coupling gels, the air gap between the generator and skin is removed, causing significant transmission. Equation (2.55) shows that as $f$ increases, the acoustic absorption coefficient increases. This corresponds to a smaller penetration depth for the higher frequency waves.

EXAMPLE 2.17. FUN

Human cells die when exposed to a temperature of 43°C for about 10 minutes. This is used in therapeutic heating of cancerous tumors (called therapeutic hyperthermia) [8]. One method of heating is by local ultrasound energy conversion (another method of heating is the local microwave energy conversion). As heat is generated by ultrasound mechanical waves, part of this heat is removed by the blood flow (this is called blood perfusion cooling). Assume that no heat is lost and that the local ultrasound energy results in raising the local tumor tissue temperature from 37°C to 43°C. An ultrasound applicator is used with the acoustic intensity of $I_{ac} = 10^4$ W/m$^2$ and frequency of $10^6$ Hz (i.e., $f = 1$ MHz). The acoustic absorption coefficient $\sigma_{ac}(1/m)$ is that corresponding to the muscle tissue. Assume a uniform acoustic intensity and a uniform temperature. Determine how long it will take to achieve this rise in temperature. For tissue, the product $\rho c_p$ is taken to be nearly that of water, i.e., $\rho c_p = 4 \times 10^6$ J/m$^3$-°C, Table C.23.

**SOLUTION**
In the absence of any heat losses, we use the integral-volume energy equation (2.9) as

$$Q\mid_A = 0 = V\left(-\frac{d}{dt}\rho c_p T + \dot{s}_{m,ac}\right) \quad \text{integral-volume energy equation}$$

or

$$\frac{d}{dt}\rho c_p T = \dot{s}_{m,ac}.$$

Now assuming that $\rho c_p$ is constant and using (2.54) for $\dot{s}_{m,ac}$, we have

$$\frac{dT}{dt} = \frac{\dot{s}_{m,ac}}{\rho c_p} = \frac{2\sigma_{ac}I_{ac}}{\rho c_p}.$$

Separating the variables, integrating, and using the limits we have

$$\Delta T = \int_{37°C}^{43°C} dT = \frac{2\sigma_{ac}I_{ac}}{\rho c_p}\int_0^t dt.$$

Then solving for $t$, we have

$$t = \frac{\rho c_p \Delta T}{2\sigma_{ac}I_{ac}}.$$

Using Table C.11, we have $\sigma_{ac} = 14$ m$^{-1}$.
    Using the numerical results we have

$$t = \frac{(\rho c_p)(\text{J/m}^3\text{-°C}) \times (43 - 37)(\text{°C})}{2\sigma_{ac}(1/m)I_{ac}(\text{W/m}^3)}$$

$$= \frac{4 \times 10^6(\text{J/m}^3\text{-°C}) \times (43 - 37)(\text{°C})}{2 \times 14(1/m) \times 10^4(\text{W/m}^3)} = 85.72 \text{ s}.$$

**COMMENT**

This is a rather fast response to the ultrasound heating. In practice, blood perfusion (blood being at nearly $37°C$) removes heat from the tissues having temperatures larger than $37°C$. This and other losses (i.e., conduction) result in a slower rise in the local temperature.

Other mechanical energy conversion mechanisms include muscle contraction $\dot{S}_{m,c}$ [about 25% of muscle movement energy consumed, the rest is removed as heat by the blood-flow convection (perfusion)].

### 2.3.4  Summary of Thermal Energy Conversion Mechanisms

Table 2.1 summarizes the direct energy conversion mechanisms and properties described above. The chemical-bond (chemical and nuclear reactions) and the physical-bond (phase change) energy conversions are listed first. They are followed by the electromagnetic energy conversions (dielectric heating, Joule heating, Peltier cooling-heating, Thomson heat flow, surface-radiation absorption heating, volumetric-radiation absorption heating, surface radiation emission). Finally the mechanical energy conversions (pressure cooling-heating, viscous heating, surface friction heating, and volumetric ultrasound heating) are listed. The list is not exhaustive. Throughout the text we will refer to these as we consider various mechanisms of heat transfer along with these energy conversions. Table 2.1 is also repeated as Table C.1(d) in Appendix C for easy reference. Typical magnitudes for these mechanisms will be reviewed and are listed in Section 8.2.

Table 2.1.  *Summary of volumetric $\dot{S}_i/V$ ($W/m^3$) and surface $\dot{S}_i/A$($W/m^2$) thermal energy conversion mechanisms and their properties*

| Mechanism of energy conversion | Relation |
| --- | --- |
| chemical reaction: exo- or endothermic reaction | $\dot{S}_{r,c}/V = -\Delta h_{r,F} a_r e^{-\Delta E_a/R_g T}$ |
| nuclear reaction: fission reaction heating | $\dot{S}_{r,fi}/V = -n_n u_n (A_{nu}/V)\Delta h_{r,fi}$ |
| nuclear reaction: fusion reaction heating | $\dot{S}_{r,fu}/V = -n_1 n_2 (Au)_{12}\Delta h_{r,fu}$ |
| nuclear reaction: radioactive decay heating | $\dot{S}_{r,\tau}/V = \dot{s}_{r,o}\exp[-t/\tau - 1/2(r-r_o)^2\sigma_{ex}^2]$ |
| physical bond: phase-change cooling/heating | $\dot{S}_{ij}/V = -\dot{n}_{ij}\Delta h_{ij}$ or $\dot{S}_{ij}/A = -\dot{m}_{ij}\Delta h_{ij}$ |
| electromagnetic: dielectric (microwave) heating | $\dot{S}_{e,m}/V = 2\pi f \epsilon_{ec}\epsilon_o e_e^2$ |
| electromagnetic: Joule heating | $\dot{S}_{e,J}/V = \rho_e j_e^2 = j_e^2/\sigma_e$ |
| electromagnetic: Peltier cooling/heating | $\dot{S}_{e,P}/A = \pm\, \alpha_P j_e = \pm\alpha_S T j_e$ |
| electromagnetic: Thomson heating/cooling | $\dot{S}_{e,T}/A = \alpha_T j_e (T_2 - T_1)$ |
| electromagnetic: surface absorption heating | $\dot{S}_{e,\alpha}/A = (1-\rho_r)q_{r,i} = \alpha_r q_{r,i}$ |
| electromagnetic: volumetric absorption heating | $\dot{S}_{e,\sigma}/V = (1-\rho_r)q_{r,i}\sigma_{ex}e^{-\sigma_{ex}x}$ |
| electromagnetic: surface emission | $\dot{S}_{e,\epsilon}/A = -\epsilon_r \sigma_{SB} T^4$ |
| mechanical: pressure-compressibility cooling/heating | $\dot{S}_{m,p}/V = \beta_f T (\partial/\partial t + \mathbf{u}\cdot\nabla)p$ |
| mechanical: viscous heating | $\dot{S}_{m,\mu}/V = \mathbf{S}_\mu : \nabla\mathbf{u}$ |
| mechanical: surface-friction heating | $\dot{S}_{m,F}/A = \mu_F p_c \Delta u_i$ |
| mechanical: volumetric ultrasonic heating | $\dot{S}_{m,ac}/V = 2\sigma_{ac}I_{ac}$ |

## 2.4 Bounding-Surface and Far-Field Thermal Conditions

In addition to separating the heat transfer media, the bounding surfaces are selected as locations where the mechanisms of heat transfer are selectively altered. An example will be a gas-solid bounding surface where the surface radiation and surface convection on the gas side is balanced (through the energy equation applied to the surface) by conduction on the solid side. We will discuss the selection of heat transfer medium and bounding surfaces throughout the text, and particularly in Chapter 8.

We have examined the differential-volume energy equation, which applies to any location within the heat transfer medium. We have also used the integral-volume energy equation, which applies to the entire volume marked by the bounding surfaces. We now examine the thermal conditions at the bounding surfaces of the heat transfer medium. A rendering of the bounding-surface thermal conditions is shown in Figure 2.15. These bounding surfaces separate the heat transfer medium from the surroundings. As we discussed in Section 1.3.3 in dealing with control surfaces, the bounding surface may be a location at which there is a distinct change in the material, e.g., solid-fluid interface, liquid-gas interface, solid-solid interface. The bounding surface may also occur within a medium, at a location where a known thermal condition occurs.

The control surface used for the bounding-surface thermal analysis may cover the entire surface or maybe a surface segment (including an infinitesimal surface chosen along a finite bounding surface). Therefore, depending on the application, the entire surface (when no variation occurs along the surface or an average over the surface is used), a finite segment of the surface (when a uniform or average behavior over a finite segment is used, but variations are allowed among the segments), or infinitesimal segment (for allowance in the local variations along the surface) is used.

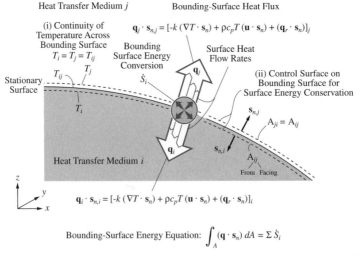

Continuity of Temperature and Surface Energy Conservation at a Bounding Surface

Bounding-Surface Energy Equation: $\int_A (\mathbf{q} \cdot \mathbf{s}_n)\, dA = \Sigma \dot{S}_i$

Figure 2.15. A rendering of the bounding-surface thermal conditions rendering (i) the continuity of temperature, and (ii) the surface energy conservation.

We begin with a general discussion; however, within our scope, we consider uniform thermal conditions over the bounding surface. This uniformity also results in heat transfer vector being perpendicular to the surface. There are two bounding-surface thermal conditions. These are (i) the continuity of temperature across the bounding surface, and (ii) the bounding-surface energy conservation. These are discussed below.

### 2.4.1 Continuity of Temperature across Bounding Surface*

Whether a distinct surface designates the boundary between a medium $i$ and a medium $j$ (for example medium $i$ is a liquid and medium $j$ is a gas), or when the bounding surface is within the same medium, the temperature of the two media separated by the bounding surface at this intermedium surface is set equal. This is assuming the continuity of temperature across the boundary. This condition is written as

$$T_i = T_j \equiv T_{ij} \quad \text{continuity of temperature across bounding surface,} \ A_{ij} = A_{ji}, \quad (2.57)$$

i.e., the continuous temperature condition at the interface designated by the area $A_{ij} = A_{ji}$. This is also shown in Figure 2.15. The indices in $A_{ij}$ indicate that the surface normal unit vector $\mathbf{s}_n$ is pointing from medium $j$ to medium $i$.

This is a local thermal equilibrium condition (i.e., local thermal equilibrium between medium $i$ and medium $j$ at a point on the surface separating the two media) and is a justifiable assumption under most conditions.

### 2.4.2 Bounding-Surface Energy Equation

We now apply the integral-volume energy conservation equation, stated by (2.9), to the bounding surface shown in Figure 2.15. Note that we choose the control surface to cover both sides of (i.e., to wrap around) the interface and to have a zero thickness (i.e., a zero-mass interface). We locate the surface exactly on the intermedium surface. Since there is zero mass, then we neglect all the volumetric terms in (2.9). Traditionally, because the volumetric terms are not present, the bounding-surface energy equation is called the surface energy balance. We also include the surface energy conversion terms $\dot{S} = \Sigma_i \dot{S}_{s,i}$ (e.g., surface-friction heating, Peltier heating-cooling). Then (2.9) for a finite area $A$ or differential $\Delta A$ becomes

$$\int_A (\mathbf{q} \cdot \mathbf{s}_n) \, dA = \dot{S}, \quad \int_{\Delta A} (\mathbf{q} \cdot \mathbf{s}_n) \, dA = \dot{S} \quad \begin{array}{l} \text{integral and differential} \\ \text{surface energy equation} \end{array} \quad (2.58)$$

$$\dot{S} = A(\dot{S}/A) \quad \text{surface energy conversion,} \quad (2.59)$$

where $\dot{S}/A$ is the surface energy conversion per unit area, listed in Table 2.1. For a finite surface area, we now expand the surface area to $A_{ij}$ and $A_{ji}$ and write the above as

$$Q\,|_A = \int_{A_{ij}} (\mathbf{q} \cdot \mathbf{s}_n)_i \, dA_{ji} + \int_{A_{ji}} (\mathbf{q} \cdot \mathbf{s}_n)_j \, dA_{ij}$$

$$= Q\,|_{A_{ji}} + Q\,|_{A_{ij}} = \sum_i \dot{S} \quad \text{on } A_{ij} \text{ and } A_{ji}, \tag{2.60}$$

or by writing each mechanism separately we have

$$\int_{A_{ij}} [-k\,(\nabla T \cdot \mathbf{s}_n) + \rho c_p\, T\,(\mathbf{u} \cdot \mathbf{s}_n) + \mathbf{q}_r \cdot \mathbf{s}_n]_i \, dA_{ji}$$

$$+ \int_{A_{ji}} [-k\,(\nabla T \cdot \mathbf{s}_n) + \rho c_p\, T\,(\mathbf{u} \cdot \mathbf{s}_n) + \mathbf{q}_r \cdot \mathbf{s}_n]_j \, dA_{ij} = \dot{S} \quad \text{on } A_{ij} = A_{ji}. \tag{2.61}$$

This is the most general temperature-based energy balance (or energy conservation equation) for the bounding surface between the heat transfer medium $i$ and the surrounding heat transfer medium $j$. Now if a uniform (no spatial variation) $\mathbf{q}_i$ and $\mathbf{q}_j$ can be assumed over $A_{ij} = A_{ji}$, we have

$$A_{ij}[-k_i\,(\nabla T \cdot \mathbf{s}_n)_i - k_j\,(\nabla T \cdot \mathbf{s}_n)_j + (\rho c_p\, T \mathbf{u} \cdot \mathbf{s}_n)_i + (\rho c_p\, T \mathbf{u} \cdot \mathbf{s}_n)_j$$

$$+ (\mathbf{q}_r \cdot \mathbf{s}_n)_i + (\mathbf{q}_r \cdot \mathbf{s}_n)_j]$$

$$\equiv Q\,|_A = Q\,|_{A_{ij}} + Q\,|_{A_{ji}}$$

$$= \dot{S} \quad \text{integral-surface energy equation with uniform conditions on } A_{ij} = A_{ji}. \tag{2.62}$$

Note that $Q\,|_{A_{ij}}$ and $Q\,|_{A_{ji}}$ designate all the surface heat transfer mechanisms.

For a differential surface area $\Delta A$, a similar result is obtained using differential areas $\Delta A_{ij}$ and $\Delta A_{ji}$

The bounding surface phase change is treated as an energy conversion and derivation of the expression for $\dot{S}_{ij}$ on the bounding surface is given in Appendix B, Section B.7.

In applying this, depending on the two media and the approximations made about the dominant mechanisms, various simplifications are made. For example, when $\dot{S} = 0$ and $\mathbf{u} = 0$ on the bounding surface, or when we can assume that $\mathbf{q}_r = 0$ for the $i$ or $j$ medium, this energy balance becomes very simple. In general the conduction heat transfer, on both sides of the interface, is not neglected. When there is a solid surface (i) in contact with a moving fluid (j), the conduction is represented with surface convection [from (2.57)] (Chapters 6 and 7).

Consider the case of $\mathbf{u} = 0$ with a uniform temperature $T_{sf} = T_{fs} = T_s$ along the surface $A_{sf} = A_{fs}$. Such a case results in the heat flow being perpendicular to the surface (we designate $x$ as the coordinate axis perpendicular to the surface). This corresponds to a solid-fluid interface with a uniform surface temperature. Then we further assume that the radiation in the solid is insignificant and (2.62) becomes

$$A_{sf}(-q_{k,s} + q_{k,f} + q_{r,f}) = A_{sf}\left(k_s\frac{\partial T_s}{\partial x} - k_f\frac{\partial T_f}{\partial x} + q_{r,f}\right) = \dot{S} \qquad (2.63)$$

surface energy equation for $\mathbf{u} = 0$, uniform conditions on $A_{sf} = A_{fs}$.

Note that since we have taken $q_{k,f}$ to be positive, we have chosen $q_{k,s}$ to also be positive when it is along $\mathbf{q}_{k,f}$ (and opposite to $\mathbf{s}_{n,s}$). Again, when this fluid is in motion, the fluid conduction term is represented by surface convection, i.e., $q_{k,f} = q_{ku}$ (Section 6.2.3).

### 2.4.3  Prescribed Bounding-Surface Thermal Conditions*

### 2.4.4  Far-Field Thermal Conditions*

## 2.5  Heat Transfer Analysis*

This section and the sections that follow are found on the Web site www.cambridge.org/kaviany. In this section we discuss aspects of heat transfer analysis that deal with nonuniformity of temperature and require the application of the differential-volume energy equation, multidimensionality of heat flow, and time variation. Then we introduce the thermal circuit modeling. The methodology used for heat transfer analysis in the text is explained by the following general steps. There are also two examples: Examples 2.20 and 2.21.

### 2.5.1  Integration of Differential-Volume Energy Equation*

### 2.5.2  Single- and Multidimensional Heat Flow*

### 2.5.3  Time Dependence and Steady State*

### 2.5.4  Thermal Circuit Models*

### 2.5.5  Summary of Methodology for Heat Transfer Analysis*

### 2.5.6  Solution Format for End-of-Chapter Problems*

## 2.6  Summary

We have developed the energy equation in several forms that allow for the various mechanisms of heat transfer, energy conversion, and sensible heat storage. These various forms would allow for the continuous variation of temperature (differential volume) or for imposing a uniform temperature (integral volume). The various, commonly occurring energy conversions to and from thermal energy were discussed and summarized in Table 2.1. We have distinguished among the surface heat transfer $Q|_{A,i}$, the surface energy conversion, and the volumetric terms (energy conversion and storage). The surface heat transfer for control surface $Q|_{A,i} = Q_i + Q_{r,i} + \sum_j Q_{t,i-j}$, will be represented by the difference in potentials between surfaces (or nodes) $i$ and $j$ and the internodal resistances $R_{t,i-j}$. These resistances are associated with conduction, convection, and radiation, and in some cases represent combined mechanism, such as surface convection and conduction-convection). The relations for $R_t$'s will be developed in the following chapters and summarized in Chapter 8, in Tables 8.1 and 8.2 and Figure 8.2.


## 2.7  References*

This section is found on the Web site www.cambridge.org/kaviany.


## 2.8  Problems

### 2.8.1  Finite- and Differential-Length Energy Equation

**PROBLEM 2.1. FAM**

Consider a steady-state, two-dimensional heat flux vector field given by

$$\mathbf{q} = 3x^2 \, \mathbf{s_x} + 2xy \, \mathbf{s_y}.$$

The control volume is centered at $x = a$ and $y = b$, with sides $2\Delta x$ and $2\Delta y$ (Figure Pr. 2.1).

(a) Using the above expression for $\mathbf{q}$ show that

$$\lim_{\Delta V \to 0} \frac{\int_A \mathbf{q} \cdot \mathbf{s}_n \, dA}{\Delta V} = \nabla \cdot \mathbf{q},$$

where the divergence of the heat flux vector is to be evaluated at $x = a$ and $y = b$. Use a length along $z$ of $w$ (this will not appear in the final answers). (Hint: Show that you can obtain the same final answer starting from both sides.)
(b) If the divergence of the heat flux vector is nonzero, what is the physical cause?
(c) In the energy equation (2.1), for this net heat flow (described by this heat flux vector field), is the sum of the volumetric terms on the right, causing the nonzero divergence of $\mathbf{q}$, a heat source or a heat sink? Also is this a uniform or nonuniform volumetric source or sink? Discuss the behavior of the heat flux field for both positive and negative values of $x$ and $y$.

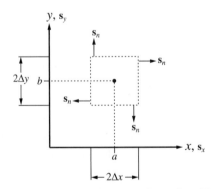

Figure Pr. 2.1. A finite control volume in a two-dimensional heat transfer medium.

## PROBLEM 2.2. FUN

Figure Pr. 2.2(a) shows a flame at the mouth of a cylinder containing a liquid fuel. The heat released within the flame (through chemical reaction) is transferred to the liquid surface by conduction and radiation and used to evaporate the fuel (note that a flame also radiates heat). The flame stabilizes in the gas phase at a location determined by the local temperature and the fuel and oxygen concentrations. The remaining heat at the flame is transferred to the surroundings by convection and by radiation and is transferred to the container wall by conduction and radiation. This heat then conducts through the container wall and is transferred to the ambient, by surface convection and radiation, and to the liquid fuel, by surface convection. The container wall and the liquid fuel also lose some heat through the lower surface by conduction. Figure Pr. 2.2(b) shows a cross section of the container and the temperature profiles within the gas and liquid and within the container wall. In small- and medium-scale pool fires, the heat recirculated through the container wall accounts for most of the heating of the liquid pool. Assume that the liquid pool has a make-up fuel line that keeps the fuel level constant and assume that the system has been operating under steady state (long enough time has elapsed).

(a) On Figure Pr. 2.2(b) track the heat flux vector, identifying the various mechanisms.
(b) For the regions shown in Figure Pr. 2.2(c), apply the integral-volume energy equation. Note that region 1 encloses the flame and it is assumed that the fuel vapor burns completely. Region 2 surrounds the liquid/gas interface, region 3 encloses the liquid, and region 4 is the container wall.
(c) For each of the regions, is the area-integral of the heat flux vector $Q|_A$ equal to zero or not?

(a) Small-Scale Pool Fire

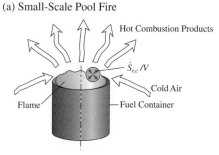

(b) Cutout of Container and Temperature Distributions      (c) Regions of Heat Flow

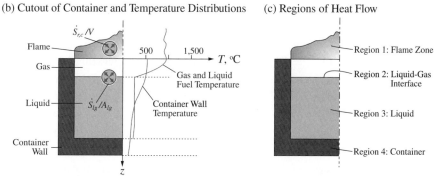

Figure Pr. 2.2. (a), (b), and (c) A small-scale pool fire showing the various regions.

**PROBLEM 2.3. FUN**

The wall of the burning fuel container is made of a metal, its thickness is small compared to its length, and the surface-convection heat fluxes at the inner and outer surfaces of the container wall are designated by $q_{ku,o}$ and $q_{ku,i}$. Under these conditions, the temperature variation across the wall thickness is negligibly small, when compared to the axial temperature variation. Also, assume that the heat transfer from within the container is axisymmetric (no angular variation of temperature). Figure Pr. 2.3 shows the differential control volume (with thickness $\Delta z$), the inner radius $R_i$, and outer radius $R_o$ of the container.

(a) Apply a combined integral- and differential-length analysis for the container wall (integral along the radius and polar angle and differential along the $z$ axis) and derive the corresponding combined integral- and differential-length energy conservation equation.

(b) Sketch the anticipated variations of the conduction heat flux $q_{k,z}$ and the wall temperature $T$, along the container wall (as a function of $z$).

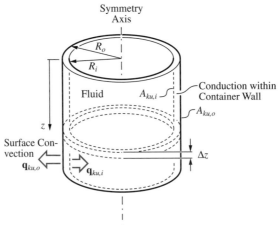

Figure Pr. 2.3. A cylindrical container with a control volume having a differential length along the $z$ direction.

**PROBLEM 2.4. FUN**

A nitrogen meat freezer uses nitrogen gas from a pressurized liquid nitrogen tank to freeze meat patties as they are carried by a conveyor belt. The nitrogen flows inside a chamber in direct contact with the meat patties, which move in the opposite direction. The heat transfer mechanism between the nitrogen gas and the meat patties is surface convection. Meat patties are to be cooled down from their processing (initial) temperature of $T_i = 10°C$ to the storage (final) temperature of $T_o = -15°C$. Each meat patty has a mass $M = 80$ g, diameter $D = 10$ cm, and thickness $l = 1$ cm. Assume for the meat the thermophysical properties of water, i.e., specific heat in the solid state $c_{p,s} = 1,930$ J/kg-K , specific heat in the liquid state $c_{p,l} = 4,200$ J/kg-K, heat of solidification $\Delta h_{ls} = -3.34 \times 10^5$ J/kg, and freezing temperature $T_{ls} = 0°C$. The average surface-convection heat transfer between the nitrogen and the meat patties is estimated as $q_{ku} = 4,000$ W/m$^2$ and the conveyor belt moves with a speed of $u_c = 0.01$ m/s.

(a) Sketch the temperature variation of a meat patty as it moves along the freezing chamber.

(b) Neglecting the heat transfer between the conveyor belt and the meat patties, find the length of the freezing chamber. Use the simplifying assumption that the temperature is uniform within the meat patties. This allows the use of a zeroth-order analysis (lumped-capacitance analysis).

## PROBLEM 2.5. FUN

While the integral-volume energy equation (2.9) assumes a uniform temperature and is applicable to many heat transfer media in which the assumption of negligible internal resistance to heat flow is reasonably justifiable, the differential-volume energy equation (2.1) requires no such assumption and justification. However, (2.1) is a differential equation in space and time and requires an analytical solution. The finite-small volume energy equation (2.13) allows for a middle ground between these two limits and divides the medium into small volumes within each of which a uniform temperature is assumed. For a single such volume (2.9) is recovered and for a very large number of such volumes the results of (2.1) are recovered.

   Consider friction heating of a disk-brake rotor, as shown in Figure Pr. 2.5. The energy conversion rate is $\dot{S}_{m,F}$. The brake friction pad is in contact, while braking, with only a fraction of the rotor surface (marked by $R$). During quick brakes (i.e., over less than $t = 5$ s), the heat losses from the rotor can be neglected. Apply (2.13), with (i) the volume marked as the pad contact region, and (ii) the entire volume in Figure Pr. 2.5, and determine the temperature $T$ after $t = 4$ s for cases (i) and (ii) and the conditions given below.

   Note that the resulting energy equation, which is an ordinary differential equation, can be readily integrated.

   $\dot{S}_{m,F} = 3 \times 10^4$ W, $R_o = 18$ cm, $R_i = 13$ cm, $l = 1.5$ cm, $R = 15$ cm, $\rho c_p = 3.5 \times 10^6$ J/m$^3$-K, $T(t = 0) = 20°$C, $t = 4$ s.

   Note that $\dot{S}_{m,F}$ remains constant, while $\Delta V$ changes.

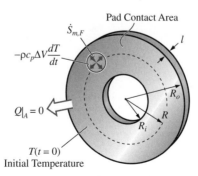

Figure Pr. 2.5. A disc-brake rotor heated by friction heating. The region under the brake pad contact is also shown.

## PROBLEM 2.6. FUN

In laser-induced spark ignition, laser irradiation $q_{r,i}$ is used to cause ionization of the fuel-oxidant mixture at the end of the laser pulse. The ionization is caused by multiphoton ionization. In multiphoton ionization, the ionizing gas molecules absorb a large number of photons.

Consider a pulsed laser, emitting a near-infrared radiation, $\lambda = 1.064\ \mu m$, with a time-dependent, focused irradiation flux given by

$$q_{r,i}(t) = (q_{r,i})_o e^{-t^2/\tau^2},$$

where $-\infty < t < \infty$, $(q_{r,i})_o$ is the peak irradiation, and $\tau(s)$ is time constant. Assume that this irradiation flux is uniform over the focal surface.

(a) Using the maximum photon energy given by

$$h_{\mathrm{P}}f = h_{\mathrm{P}}\frac{\lambda}{c},$$

where $h_{\mathrm{P}}$ is the Planck constant, and $f$ is the frequency, $\lambda$ is the wavelength, and $c$ is the speed of light, determine the photon flux $\dot{m}_{ph}$(photon/m²-s). Use the speed of light in vacuum $c = c_o$.

(b) Using a temperature-dependent extinction coefficient

$$\sigma_{ex}(1/\mathrm{m}) = \frac{a_1 n_e^2}{T^{1/2}}\left(1 - e^{-a_2/T}\right),$$

where $a_1$ and $a_2$ are constants and $T(\mathrm{K})$ is the kernel temperature, determine the energy absorbed in the focal volume, shown in Figure Pr. 2.6, over the time span, $-\infty < t < \infty$, i.e.,

$$\int_{-\infty}^{\infty} \dot{S}_{e,\sigma}dt.$$

(c) Express the results of (b) per kernel volume $V$.

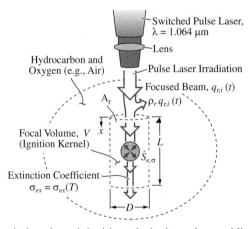

Figure Pr. 2.6. Laser-induced spark ignition of a hydrocarbon-oxidizer gaseous mixture.

$(q_{r,i})_o = 10^{17}$ W/m$^2$, $T = 10^6$ K, $a_1 = 0.1645 \times 10^{-42}$ K$^{1/2}$m$^5$, $n_e = 10^{26}$(1/m$^3$), $a_2 = 1.35 \times 10^4$ K, $D = 16.92$ $\mu m$, $L = 194$ $\mu m$, $\tau = 3.3$ ns, $\rho_r = 0$.

Note that

$$\int_{-\infty}^{\infty} e^{t^2/\tau^2} dt = \pi^{1/2}\tau.$$

## PROBLEM 2.7. FUN

In thermoelectric cooling, a pair of $p$- and $n$-type semiconductors are jointed at a junction. When an electric current, given as current flux (or current density) $j_e$(A/m$^2$), passes through their junction, heat is absorbed.

This current also produces the undesirable (parasitic) Joule heating. This energy conversion (per unit volume) is given by (2.33) as

$$\dot{s}_{e,J} = \rho_e(T)j_e^2,$$

where $\rho_e$(ohm-m) is the electrical resistivity and varies with temperature $\rho_e = \rho_e(T)$. Figure Pr. 2.7 shows a semiconductor slab ($p$- or $n$-type), which is a part of a pair. The energy equation (2.8) would be simplified by assuming that heat flows only in the $x$ direction, that the heat transfer is in a steady state, and that the energy conversion term is given above. A small length $\Delta x$ is taken along the $x$ direction and the conduction heat flux vectors at $x$ and $x + \delta x$ are given as $\mathbf{q}_k|_x$ and $\mathbf{q}_k|_{x+\Delta x}$.

(a) Using (2.8), and assuming that the results for the given small $\Delta x$ are valid for $\Delta x \to 0$, determine the magnitude of $\dot{s}_{e,J}$.

(b) Using the relationship for $\dot{s}_{e,J}$ given above, and the value for $j_e$ given below, determine $\rho_e(T)$ and from Tables C.9(a) and (b) find a material with this electrical resistivity $\rho_e(T)$ (ohm-m).

$\mathbf{q}_k|_x = -1.030 \times 10^4 \mathbf{s}_x$ W/m$^2$, $\mathbf{q}_k|_{x+\Delta x} = 1.546 \times 10^4 \mathbf{s}_x$ W/m$^2$, $j_e = 4 \times 10^6$ A/m$^2$, $\Delta x = 0.1$ mm.

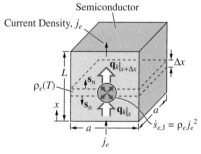

Figure Pr. 2.7. A semiconductor slab with a one-dimensional heat conduction and a volumetric energy conversion (Joule heating). The conduction heat flux vectors are prescribed at locations $x$ and $x + \Delta x$.

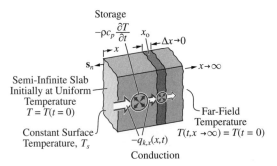

Figure Pr. 2.8. A semi-infinite slab with an initial temperature $T(t = 0)$ has its surface temperature suddenly changed to $T_s$.

## PROBLEM 2.8. FUN

In some transient heat transfer (i.e., temperature and heat flux vector changing with time) applications, that portion of the heat transfer medium experiencing such a transient behavior is only a small portion of the medium. An example is the seasonal changes of the air temperature near the earth's surface, which only penetrates a very short distance, compared to the earth's radius. Then the medium may be approximated as having an infinite extent in the direction perpendicular to the surface and is referred to as a semi-infinite medium. Figure Pr. 2.8 shows such a medium for the special case of a sudden change of the surface temperature from the initial (and uniform throughout the semi-infinite medium) temperature of $T(t = 0)$ to a temperature $T_s$. Under these conditions, the solution for the heat flux is given by (as will be discussed in Section 3.5.1)

$$q_{k,x}(x, t) = \frac{k[T_s - T(t = 0)]}{(\pi \alpha t)^{1/2}} e^{-x^2/4\alpha t},$$

where $\alpha = k/\rho c_p$ is called the thermal diffusivity.

This conduction heat flux changes with time and in space.

(a) Using (2.1), with no energy conversion and conduction as the only heat transfer mechanism, determine the time rate of change of local temperature $\partial T/\partial t$ at location $x_o$ and elapsed time $t_o$.

(b) Determine the location of largest time rate of change (rise) in the temperature and evaluate this for the elapsed time $t_o$.

$k = 0.25$ W/m-K (for nylon), $\alpha = 1.29 \times 10^{-5}$ m²/s (for nylon), $T_s = 105°$C, $T(t = 0) = 15°$C, $x_o = 1.5$ cm, $t_o = 30$ s.

## PROBLEM 2.9. FUN

A device that allows for heat transfer between two fluid streams of different temperatures is called a heat exchanger. In most applications, the fluid streams are bounded by flowing through ducts and tubes and are also kept from mixing with each other by using an impermeable solid wall to separate them. This is shown in Figure Pr. 2.9. Assume that radiation and conduction are not significant in each stream and that there is steady-state heat transfer and no energy conversion. Then there is only bounded fluid stream convection and surface convection at the separating wall,

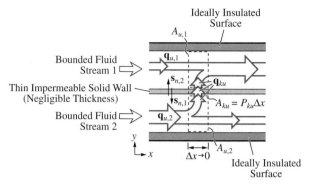

Figure Pr. 2.9. Two streams, one having a temperature higher than the other, exchange heat through a wall separating them.

as shown in Figure Pr. 2.9. Assume that the wall has a zero thickness. Also assume convection heat flux $q_u$ is uniform (an average) across the surface area for convection $A_u$. The surface area for the surface convection over a differential length $\Delta x$ is $\Delta A_{ku}$.

Starting from (2.8), write the energy equations for the control surfaces $\Delta A_1$ and $\Delta A_2$ shown. These control surfaces include both the convection and the surface convection areas. Show that the energy equations become

$$-\frac{P_{ku}}{A_{u,1}}q_{ku} + \frac{d}{dx}q_{u,1} = 0,$$

$$\frac{P_{ku}}{A_{u,2}}q_{ku} + \frac{d}{dx}q_{u,2} = 0.$$

**PROBLEM 2.10. FUN**

A heat transfer medium with a rectangular control volume, shown in Figure Pr. 2.10, has the following uniform heat fluxes at its six surfaces:

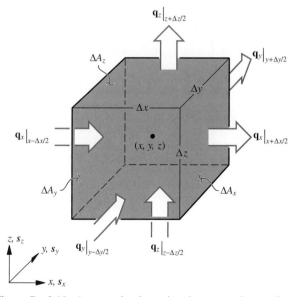

Figure Pr. 2.10. A control volume in a heat transfer medium.

$$\mathbf{q}_x|_{x-\Delta x/2} = -4\mathbf{s}_x \ \text{W/m}^2, \qquad \mathbf{q}_x|_{x+\Delta x/2} = -3\mathbf{s}_x \ \text{W/m}^2,$$
$$\mathbf{q}_y|_{y-\Delta y/2} = \ 6\mathbf{s}_y \ \text{W/m}^2, \qquad \mathbf{q}_y|_{y+\Delta y/2} = \ 8\mathbf{s}_y \ \text{W/m}^2,$$
$$\mathbf{q}_z|_{z-\Delta z/2} = \ 2\mathbf{s}_z \ \text{W/m}^2, \qquad \mathbf{q}_z|_{z+\Delta z/2} = \ 1\mathbf{s}_z \ \text{W/m}^2.$$

The uniformity of heat flux is justifiable due to the small dimensions $\Delta x = \Delta y = \Delta z = 2$ mm.

(a) Assume that $\Delta V \to 0$ is approximately valid for this small, but finite volume and determine the divergence of $\mathbf{q}$ for the center of this control volume, located at $(x, y, z)$.
(b) Is there a sink or a source of heat in this control volume located at $(x, y, z)$?
(c) What could be the mechanisms for this source or sink of heat?

## PROBLEM 2.11. FUN
Although the temperature variation within a heat transfer medium is generally three dimensional, in many cases there is a dominant direction in which the most significant temperature variation occurs. Then, the use of a one-dimensional treatment results in much simplification in the analysis. Consider the steady-state surface temperatures given in Figure Pr. 2.11, for selected locations on a solid, rectangular piece. The heat flows through the solid by conduction and from its surface to the ambient by surface convection.

(a) By examining the gradient of temperature in each direction, determine the dominant conduction heat flow direction. As an approximation, use

$$\frac{\partial T}{\partial x} \simeq \frac{\Delta T_x}{\Delta x}, \quad \frac{\partial T}{\partial y} \simeq \frac{\Delta T_y}{\Delta y}, \quad \frac{\partial T}{\partial z} \simeq \frac{\Delta T_z}{\Delta z},$$

where $\Delta x$ is the length over which the temperature change $\Delta T_x$ occurs.
(b) Select a control volume that has a differential length in the direction of dominant conduction heat flow and an integral length over the other two directions. Schematically show this integral-differential volume.
(c) Write an energy equation for this control volume.

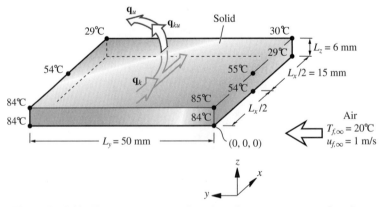

Figure Pr. 2.11. Temperature at various locations on a rectangular plate.

### 2.8.2 Energy Conversion Mechanisms (to and from Thermal Energy)

- Chemical- and Physical-Bond Energy
- Electromagnetic Energy
- Mechanical Energy

**PROBLEM 2.12. FUN**

Many computer disks are read by magnetoresistive transducers. The transducer is located on a thin slider that is situated slightly above the disk, as shown in Figure Pr. 2.12. The transistor is developed using the principle that its resistance varies with the variation of the surrounding magnetic field. Since its resistance is also temperature dependent, any temperature change will result in a noise in the readout. When the slider and disk are at the same temperature, the viscous-dissipation heat generation becomes significant in creating this undesired increase in the temperature. Assume the flow of air at $T = 300$ K between the disk and slider is a Newtonian, one-dimensional, Couette flow, as shown in Figure Pr. 2.12. The distance between the disk and the slider is $L = 20$ nm, and the relative velocity is $\Delta u_i = 19$ m/s.

Determine the magnitude of the volumetric viscous-dissipation heat generation $\dot{s}_{m,\mu}$. Use Table C.22, and the relation $\mu_f = \nu_f / \rho_f$ to determine $\mu_f$ for air at $T = 300$ K.

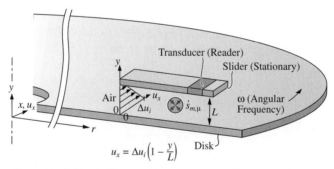

Figure Pr. 2.12. A disk read by a magnetoresistive transducer.

**PROBLEM 2.13. FAM**

Catalytic combustion (and catalytic chemical reaction in general) is an enhancement in the rate of chemical reaction due to the physical-chemical mediation of a solid surface. For example, in the automobile catalytic converter the rate of reaction of exhaust-gas unburned fuel is increased by passing the exhaust gas over the catalytic surface of the converter. The catalytic converter is a solid matrix with a large surface area over which the exhaust gas flows with a relatively small pressure drop. The catalytic effect is produced by a surface impregnated with precious metal particles, such as platinum. For example, consider the following chemical kinetic model for the reaction of methane and oxygen

$$CH_4 + 2O_2 \rightarrow CO_2 + 2H_2O \quad \text{stoichiometric chemical reaction}$$

$$\dot{m}_{r,CH_4} = -a_r \rho_{CH_4} \rho_{O_2}^{1/2} e^{-\Delta E_a / R_g T} \quad \text{chemical, kinetic model,}$$

where $\dot{m}_{r,CH_4}$ (kg/m²-s) is the reaction-rate per unit surface area. The pre-exponential factor $a_r$(cm$^{5/2}$/s-g$^{1/2}$) and the activation energy $\Delta E_a$(J/kmole) are determined empirically. The model is accurate for high oxygen concentrations and for high temperatures. The densities (or concentrations) are in g/cm³. Consider the following catalytic (in the presence of Pt) and noncatalytic (without Pt) chemical kinetic constants:

$$\text{without Pt} : a_r = 1.5 \times 10^{11} \text{ cm}^{5/2}/\text{s-g}^{1/2}, \quad \Delta E_a = 1.80 \times 10^8 \text{ J/kmole},$$

$$\text{with Pt} : a_r = 1.5 \times 10^{12} \text{ cm}^{5/2}/\text{s-g}^{1/2}, \quad \Delta E_a = 1.35 \times 10^8 \text{ J/kmole}.$$

For a mixture of methane and oxygen at a pressure of 1 atm and a temperature of 500°C:

(a) Determine the densities of $CH_4$ and $O_2$ assuming an ideal-gas behavior,
(b) Determine the rate of reaction per unit surface area $\dot{m}_{r,CH_4}$.
(c) Comment on the effect of the catalyst.
    Use the cgs units (cm, g, s).

### PROBLEM 2.14. FUN
In order to produce silicon wafers, single-crystal silicon ingots are formed by the slow solidification of molten silicon at the tip of a cylinder cooled from the base. This was shown in Figure 2.3(c) and is also shown in Figure Pr. 2.14. The heat released by solid to fluid phase change $\dot{S}_{ls}$(W) is removed from the solid-liquid interface $A_k$ by conduction through the ingot $Q_k$. The energy equation for the solid-fluid interface $A_k$ (nonuniform temperature in the liquid), as given by (2.9), is

$$Q_k = \dot{S}_{ls},$$

where the conduction heat flow $Q_k$ is given by

$$Q_k = A_k k \frac{T_{sl} - T_s}{L},$$

where $L$ and $T_s$ are shown in Figure Pr. 2.14 and $T_{ls}$ is the melting temperature. The rate of phase-change energy conversion is

$$\dot{S}_{ls} = -\rho_l A_k u_F \Delta h_{ls} = -\dot{M}_{ls} \Delta h_{ls} = \dot{M}_{ls} \Delta h_{sl},$$

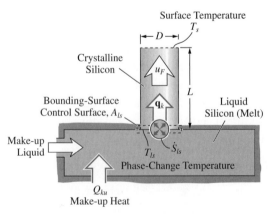

Figure Pr. 2.14. Czochralski method for single-crystal growth of silicon.

where $\Delta h_{sl} > 0$. Using the thermophysical properties given in Tables C.2 (periodic table for $\Delta h_{sl}$) and C.14 (at $T = 1{,}400$ K for $k$), in Appendix C, determine the temperature $T_s$ (at the top of the ingot). Assume that the liquid and solid have the same density.

$$L = 20 \text{ cm}, u_F = 0.4 \text{ mm/min.}$$

### PROBLEM 2.15. FAM

The electrical resistivity of metals increases with temperature. In a surface radiation emission source made of Joule-heating wire, the desired temperature is $2{,}500°$C. The materials of choice are tantalum Ta and tungsten W. The electrical resistivity for some pure metals, up to $T = 900$ K, is given in Table C.8. Assume a linear dependence of the electrical resistivity on the temperature, i.e.,

$$\rho_e(T) = \rho_{e,o}[1 + \alpha_e(T - T_o)].$$

(a) From the data in Table C.8, determine $\alpha_e$ for both metals.
(b) Using the equation above, determine the metals electrical resistivity at $T = 2{,}500°$C.
(c) If the wire has a diameter $D = 0.1$ mm and a length $L = 5$ cm (coiled), determine the electrical resistance $R_e$ for both metals at $T = 25°$C and $T = 2{,}500°$C.
(d) If a Joule heating rate of 100 W is needed, what current should be applied at $T = 2{,}500°$C?
(e) Determine the voltage needed for this power.

### PROBLEM 2.16. FAM

Electrical power is produced from a thermoelectric device. The thermoelectric junctions are heated (heat added) by maintaining the hot junction at $T_h = 400°$C and cooled (heat removed) by maintaining the cold junction at $T_c = 80°$C. This is shown in Figure Pr. 2.16.

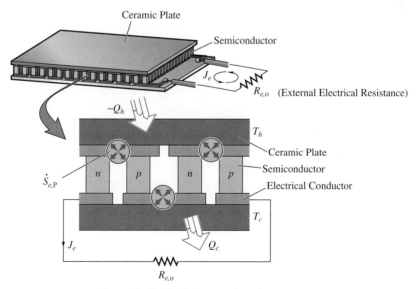

Figure Pr. 2.16. A thermoelectric generator.

There are 120 $p$-$n$ pairs. The pairs are $p$- and $n$-type bismuth telluride ($Bi_2Te_3$) alloy with Seebeck coefficients $\alpha_{S,p} = 2.30 \times 10^{-4}$ V/K and $\alpha_{S,n} = -2.10 \times 10^{-4}$ V/K. The resistance (for all 120 pairs) to the electrical current $J_e$ produced is $R_e = 2.4$ ohm for the thermoelectric path. For an optimum performance, the external resistance $R_{e,o}$ is also equal to 2.4 ohm.

(a) Determine the current produced.
(b) Determine the power produced.

## PROBLEM 2.17. FUN

A premixed mixture of methane $CH_4$ and air burns in a Bunsen-type burner, as shown in Figure Pr. 2.17. Assume that the flame can be modeled as a plane flame. The reactants (methane and air) enter the flame zone at a temperature $T_1 = 289$ K. The concentration of methane in the reactant gas mixture is $\rho_{F,1} = 0.0621$ $kg_{CH_4}/m^3$ and the heat of reaction for the methane/air reaction is $\Delta h_{r,CH_4} = -5.55 \times 10^7$ $J/kg_{CH_4}$ (these will be discussed in Chapter 5). For both reactants and products, assume that the average density is $\rho = 1.13$ kg/m$^3$ and that the average specific heat is $c_p = 1,600$ J/kg-K (these are temperature-averaged values between the temperature of the reactants $T_1$ and the temperature of the products $T_2$).

(a) For the control volume enclosing the flame (Figure Pr. 2.17) apply the integral-volume energy conservation equation. Neglect the heat loss by radiation and assume a steady-state condition.

(b) Obtain an expression for the heat generation inside the flame $\dot{S}_{r,c}$(W) as a function of the cold flow speed $u_f$(m/s), the concentration of methane in the reactant gas mixture $\rho_{F,1}$($kg_{CH_4}/m^3$), the heat of reaction for the methane/air reaction $\Delta h_{r,CH_4}$($J/kg_{CH_4}$), and the area of the control surface $A_u$(m$^2$). Assume a complete combustion of the methane. [Hint: Use the conservation of mass of fuel equation (1.26) to obtain an expression for the volumetric reaction rate $\dot{n}_{r,F}$(kg/m$^3$-s).]

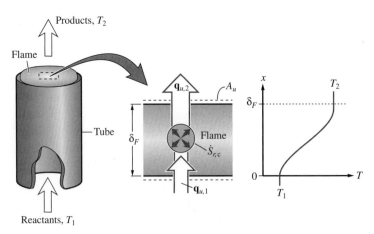

Figure Pr. 2.17. A premixed methane-air flame showing the flame and the temperature distribution across $\delta_F$.

(c) Using the integral-volume energy conservation equation obtained in item (a) and the expression for the heat generation obtained in item (b) calculate the temperature of the reacted gases (i.e., the adiabatic flame temperature, $T_2$).

## PROBLEM 2.18. FUN

A moist-powder tablet (pharmaceutical product) is dried before coating. The tablet has a diameter $D = 8$ mm, and a thickness $l = 3$ mm. This is shown in Figure Pr. 2.18. The powder is compacted and has a porosity, i.e., void fraction, $V_f/V = 0.4$. This void space is filled with liquid water. The tablet is heated in a microwave oven to remove the water content. The rms of the electric field intensity $(\overline{e_e^2})^{1/2} = 10^3$ V/m and the frequency $f = 1$ GHz.

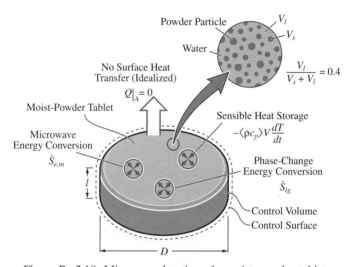

Figure Pr. 2.18. Microwave heating of a moist-powder tablet.

(a) Determine the time it takes to heat the tablet from the initial temperature of $T(t = 0) = 18°C$ to the final temperature of $T = 40°C$, assuming no evaporation.
(b) Determine the time it takes to evaporate the water content, while the tablet is at a constant temperature $T = 40°C$. For the effective (including the liquid and powder) volumetric heat capacity $\langle \rho c_p \rangle$, which includes both water and powder, use $2 \times 10^6$ J/m³-K. For the water density and heat of evaporation, use Table C.27. Assume that the dielectric loss factor for the powder is negligible compared to that for the water.

## PROBLEM 2.19. FUN

The automobile airbag deploys when the pressure within it is suddenly increased. This pressure increase is a result of the inflow of gaseous products of combustion or pressurized air from an inflater connected to the bag. The airbag fabric may be permeable, and this results in expansion of the gas as it flows through the fabric, as rendered in Figure Pr. 2.19. Assume that the pressure gradient is approximated by

$$\frac{\partial p}{\partial x} \simeq \frac{p_o - p_i}{D},$$

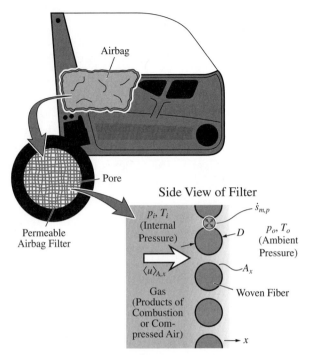

Figure Pr. 2.19.  An automobile airbag system.

where $p_o$ and $p_i$ are the external and internal pressure and $D$ is the woven-fiber diameter. Consider the gas flowing with an average gas velocity $\langle u \rangle_{A,x} = 2$ m/s, and pressures of $p_i = 1.5 \times 10^5$ Pa and $p_o = 1.0 \times 10^5$ Pa. The fabric diameter is $D = 0.5$ mm. Use the expression for the volumetric energy conversion $\dot{s}_{m,p}$ for the one-dimensional flow given in Example 2.14.

(a) Determine the volumetric expansion cooling rate $\dot{s}_{m,p}$.
(b) Write an integral-volume energy equation for the gas flowing through a pore. Allow for expansion cooling and show the control volume, surface convection, and convection heat flows. Designate the flow area for the pore as $A_x$, and use $DA_x$ for the control volume. For the bounding surfaces, choose the two adjacent woven-fabric and imaginary-pore walls.
(c) For the case of no surface convection heat transfer, determine the drop in temperature $T_i - T_o$.

   Use $\rho c_p$ for air at $T = 300$ K (Table C.22).

**PROBLEM 2.20. FUN**
When high viscosity fluids, such as oils, flow very rapidly through a small tube, large strain rates, i.e., $du/dr$ (where $r$ is the radial location shown in Figure Pr. 2.20), are encountered. The high strain rate, combined with large fluid viscosity $\mu_f$, results in noticeable viscous heating. In tube flows, when the Reynolds number $Re_D = \rho_f \langle u \rangle_A D / \mu_f$ is larger than 2,300, transition from laminar to turbulent flow occurs. In general, high cross-section averaged fluid velocity $\langle u \rangle_A$ results in a turbulent flow. The fluid velocity for a laminar flow is shown in Figure Pr. 2.20. For laminar flow, the center-line velocity is twice the average velocity, while for turbulent flow the

coefficient is less than two. Assume that the cross-section averaged viscous heating rate, (2.51), is approximated as

$$\langle \dot{s}_{m,\mu} \rangle_A = \mu_f \frac{a_1^2 \langle u \rangle_A^2}{D^2},$$

where $a_1 = 1$ is a constant.

(a) Determine the volumetric heating rate for engine oil at $T = 310$ K (Table C.23), noting that $\mu_f = \nu_f \rho_f$, $\langle u \rangle_A = 10$ m/s, and $D = 1$ mm.
(b) Apply (2.8) to a differential length along the tube. Allow only for surface convection, convection along $x$, and viscous heating, i.e., similar to (2.11), with added energy conversion.

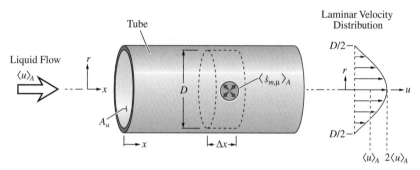

Figure Pr. 2.20. Viscous heating of fluid flow inside a small tube.

### PROBLEM 2.21. FUN
During braking, nearly all of the kinetic energy of the automobile is converted to frictional heating at the brakes. A small fraction is converted in the tires. The braking time, i.e., the elapsed time for a complete stop, is $\tau$. The automobile mass is $M$, the initial velocity is $u_o$, and the stoppage is at a constant deceleration $(du/dt)_o$.

(a) Determine the rate of friction energy conversion for each brake in terms of $M$, $u_o$, and $\tau$. The front brakes convert 65% of the energy and the rear brakes convert the remaining 35%.
(b) Evaluate the peak energy conversion rate for the front brake using $M = 1,500$ kg (typical for a mid-size car), $u_o = 80$ km/hr, and $\tau = 4$ s.

### PROBLEM 2.22. FUN
In therapeutic heating, biological tissues are heated using electromagnetic (i.e., microwave, and in some cases, Joule heating) or mechanical (i.e., ultrasound heating) energy conversion. In the heated tissue, which may be a sore muscle (e.g., an athletic discomfort or injury), some of this heat is removed through the local blood flow and this is called perfusion heating. Under steady state, the local tissue temperature reaches a temperature where the surface heat transfer from the tissue balances with the energy conversion rate. Consider the therapeutic ultrasound heating shown in Figure Pr. 2.22.

(a) Using (2.9) write the integral-volume energy equation that applies to this steady-state heat transfer. Assume no conduction and radiation heat transfer and allow for

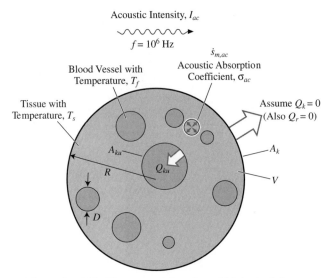

Figure Pr. 2.22. Therapeutic heating of biological tissue.

surface convection $Q_{ku}$ through the blood vessels distributed through the tissue with a surface-convection area $A_{ku}$. Draw a schematic showing the various terms in the energy equation.

(b) In this energy equation, replace the surface-convection heat transfer with

$$Q_{ku} = A_{ku}q_{ku} = A_{ku}\langle Nu\rangle_D \frac{k_f}{D}(T_s - T_f),$$

where $\langle Nu\rangle_D$ is a dimensionless quantity called the dimensionless surface-convection conductance (or Nusselt number), $k_f$ is the blood thermal conductivity, $D$ is the average blood vessel diameter, $T_s$ is the tissue temperature, and $T_f$ is the blood temperature.

(c) Solve the energy equation for $T_s$.

(d) Using the following numerical values, determine $T_s$.

$I_{ac} = 5 \times 10^4$ W/m$^2$, $\sigma_{ac}$ (from Table C.11, for muscle tissue), $V$ (sphere of $R = 3$ cm), $D = 10^{-3}$ m, $A_{ku} = 0.02$ m$^2$, $\langle Nu\rangle_D = 3.66$, $k_f = 0.62$ W/m-K (same as water), $T_f = 37°$C.

### PROBLEM 2.23. FUN

Among the normal paraffins (n-paraffins) are the hydrocarbon fuels, e.g., methane $CH_4$, propane $C_2H_6$, and butane $C_4H_{10}$. Table Pr. 2.23 [19] gives a set of constants for the chemical kinetic model given by

$$\dot{n}_{r,F} = -a_r \rho_F^{a_F} \rho_O^{a_O} e^{-\Delta E_a/R_g T},$$

for the $CH_4$ oxidation represented by a single-step, stoichiometric reaction

$$CH_4 + 2O_2 \rightarrow CO_2 + 2H_2O.$$

This set of parameters is found to give a good agreement between predicted and measured flame speeds as a function of methane/oxygen ratio. Determine the reaction rates $\dot{n}_{r,F}$ at $T = 1,000°$C using the above model (with the constants from Table Pr. 2.23).

(a) Use a reactant-rich condition of $\rho_{O_2} = 0.9307$ kg/m$^3$, $\rho_{CH_4} = 0.2333$ kg/m$^3$, and
(b) a product-rich condition of $\rho_{O_2} = 0.1320$ kg/m$^3$, $\rho_{CH_4} = 0.0328$ kg/m$^3$, to represent two locations within the flame. These are characteristics of CH$_4$ reaction with oxygen (called oxy-fuel reactions as compared to air-fuel reactions) at one atm pressure.
(c) Compare the results with the prediction of the zeroth-order model given in Example 2.6 by (2.21).

Table Pr. 2.23. *Constants in chemical kinetic model for methane oxidation*

| $a_r$, s$^{-1}$ | $\Delta E_a$, J/kmole | $a_F$ | $a_O$ |
|---|---|---|---|
| $5.3 \times 10^8$ | $2.026 \times 10^8$ | $-0.3$ | $1.3$ |

## PROBLEM 2.24. FUN

In addition to being abundant and readily available, air is a fluid whose temperature can be raised well above and below room temperature, for usage as a hot or cold stream, without undergoing any phase change. In the high temperature limit, the main constituents of air, nitrogen (N$_2$) and oxygen (O$_2$), dissociate and ionize at temperatures above $T = 2,000$ K. In the low temperature limit, oxygen condenses at $T = 90.0$ K, while nitrogen condenses at $T = 77.3$ K. Consider creating (a) a cold air stream with $T_2 = 250$ K, (b) a hot air stream with $T_2 = 1,500$ K, and (c) a hot air stream with $T_2 = 15,000$ K.

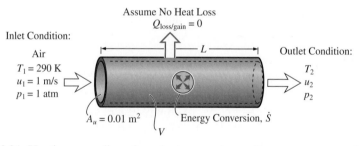

Figure Pr. 2.24. Heating or cooling of an air stream using various energy conversion mechanisms.

The air stream is at atmospheric pressure, has a cross-sectional area $A_u = 0.01$ m$^2$, an inlet temperature $T_1 = 290$ K, and a velocity $u_1 = 1$ m/s, as shown in Figure Pr. 2.24. For each of the cases above, (i) choose an energy conversion mechanism from Table 2.1 that would provide the required energy conversion mechanisms for heating or cooling, (ii) write the integral-volume energy equations (2.9) for a steady-state flow and heat transfer. Give the amount of fluid, electromagnetic energy, etc., that is needed.

## PROBLEM 2.25. FUN

A transparent thin-foil heater is used to keep a liquid crystal display (LCD) warm under cold weather conditions. The thin foil is sandwiched between the liquid crystal and the backlight. A very thin copper foil, with cross section $w = 0.4$ mm and $l = 0.0254$ mm and a total length $L = 70$ cm, is used as the heating element. The foil is embedded in a thin polyester membrane with dimensions $W = 10$ cm and $H = 2$ cm, which also acts as an electrical insulator (Figure Pr. 2.25).

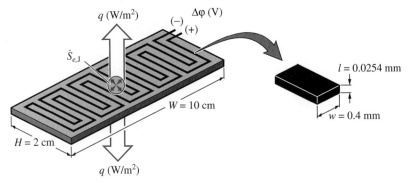

Figure Pr. 2.25. A transparent thin-foil heater.

The thin foil heats the liquid crystal by Joule heating. Assume that the amount of heat flowing to the backlight panel is the same as the amount flowing to the liquid crystal and that the system is operating under a steady-state condition. For the electrical resistivity of copper use $\rho_e = 1.725 \times 10^{-8}$ ohm-m. If the thin foil heater is to provide $q = 1,000$ W/m$^2$ to the liquid crystal, calculate:

(a) The volumetric rate of heat generation in the wire $\dot{s}_{e,J}$(W/m$^3$),
(b) The electrical potential $\Delta\varphi$(V) needed,
(c) The current flowing in the wire $J_e$(A), and
(d) Recalculate items (a) to (c) for twice the length $L$.

### PROBLEM 2.26. FUN

A single-stage Peltier cooler/heater is made of Peltier cells electrically connected in series. Each cell is made of $p$- and $n$-type bismuth telluride (Bi$_2$Te$_3$) alloy with Seebeck coefficients $\alpha_{S,p} = 230 \times 10^{-6}$ V/K and $\alpha_{S,n} = -210 \times 10^{-6}$ V/K. The cells are arranged in an array of 8 by 15 (pairs) cells and they are sandwiched between two square ceramic plates with dimensions $w = L = 3$ cm ( see Figure Pr. 2.26). The current flowing through the elements is $J_e = 3$ A.

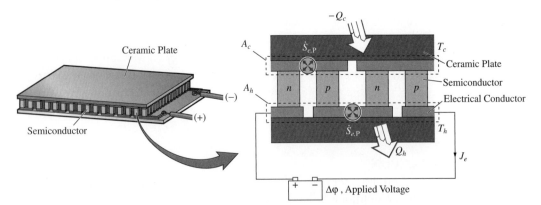

Figure Pr. 2.26. A single-stage Peltier cooler/heater.

(a) If the temperature at the cold junctions is $T_c = 10°C$, calculate the Peltier heat absorbed at the cold junctions $\dot{S}_{e,P}/A(W/m^2)$ (per unit area of the ceramic plate).
(b) If the hot junctions is at $T_h = 50°C$, calculate the Peltier heat released at the hot junctions $\dot{S}_{e,P}/A(W/m^2)$ (per unit area of the ceramic plate).

## PROBLEM 2.27. FAM.S

A pocket combustion heater uses heat released (chemical-bond energy conversion) from the reaction of air with a powder. The powder is a mixture of iron, water, cellulose (a carbohydrate), vermiculite (a clay mineral), activated carbon (made capable of absorbing gases), and salt. Air is introduced by breaking the plastic sealant and exposing the permeable membrane containing the powder to ambient air. Since the air has to diffuse through the powder, and also since the powder is not mixed, the heat release rate is time dependent, decreasing with time. We express this as $\dot{S}_{r,c} = \dot{S}_{r,o}\exp(-t/\tau_r)$, where $\tau_r(s)$ is the reaction rate time constant. The pocket heater has a mass of $M = 20$ g and a heat capacity of $c_p = 900$ J/kg-K. During the usage, heat leaves the pocket heater surface. This heat is expressed as a resistive-type heat transfer and is given by $Q = (T - T_\infty)/R_t$, where $T_\infty$ is the ambient temperature and $R_t(°C/W)$ is the surface heat transfer resistance. Initially the heater is at the ambient temperature, i.e., $T(t = 0) = T_\infty$. This is shown in Figure Pr. 2.27.

Combustion Handwarmer

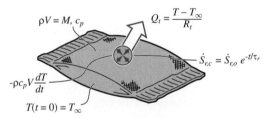

Figure Pr. 2.27. A pocket combustion heater and its heat transfer model.

(a) Write the energy equation for the pocket heater.
(b) Using a software, plot the temperature of the pocket heater $T = T(t)$ versus time, up to $t = \tau_r$.
(c) What is the maximum heater temperature?
(d) Determine the closed-form solution for $T = T(t)$?
$T_\infty = 10°C$, $\dot{S}_{r,o} = 5$ W, $\tau_r = 3$ hr, and $R_t = 7.4°C/W$.

## PROBLEM 2.28. FUN

In electrical power generation using thermoelectric energy conversion, the electrical power can be optimized with respect to the external electrical resistance. Starting from (2.40), show that the maximum power generation occurs for $R_{e,o} = R_e$, i.e., when the external electrical resistance is equal to the thermoelectric electrical resistance.

## PROBLEM 2.29. FUN

The volumetric pressure-compressibility heating/cooling energy conversion $\dot{s}_p$ can be represented in an alternative form using $c_v$ instead of $c_p$ in the energy equation.

Starting from (B.44) in Appendix B, show that for an ideal gas, the volumetric pressure-compressibility energy conversion $\dot{s}_p$ becomes

$$\dot{s}_{m,p} = -p\nabla \cdot \mathbf{u} = \left[\left(\frac{c_p}{c_v} - 1\right)c_v\rho T\right]\nabla \cdot \mathbf{u}.$$

Use the following relation, derived from combining (1.4), (1.5), and (1.6),

$$c_p \equiv c_v + T\left(\frac{\partial p}{\partial T}\Big|_v \frac{\partial v}{\partial T}\Big|_p\right).$$

## PROBLEM 2.30. FAM

A microwave heater is used to dry a batch of wet alumina powder. The microwave source is regulated to operate at $f = 10^9$ Hz and to provide an electrical field with a root-mean-square intensity of $(\overline{e_e^2})^{1/2} = 10^3$ V/m. The effective dielectric loss factor of the alumina powder $\langle\epsilon_{ec}\rangle$ depends on the fluid filling the pores. For a porosity of 0.4, the effective dielectric loss factor of the completely dry alumina powder is $\langle\epsilon_{ec}\rangle = 0.0003$ and the effective dielectric loss factor of the completely wet alumina powder is $\langle\epsilon_{ec}\rangle = 6.0$.

(a) Determine the microwave heating $\dot{s}_{e,m}$(W/m³) for these two cases.
(b) Discuss the efficiency of the use of microwave heating in drying the alumina powder when the moisture content (i.e., amount of water in the pores) is small.
(c) From Table C.10, would a sandy soil dry faster or slower than the alumina powder? Note that although both $\epsilon_{ec}$ and $\langle\epsilon_{ec}\rangle$ are listed, no distinction is made in Table C.10.

## PROBLEM 2.31. FUN

The range-top electrical heater is shown in Figure Pr. 2.31(a). It has electrical elements made of a central electrical conductor (electric current carrying) surrounded by an electrical insulator. The electrical insulator should have a large thermal conductivity to carry the heat generated by Joule heating in the electrical conductor to the surface for surface convection-radiation heat transfer. This is shown in Figure Pr. 2.31(b). During the start-up and turn-off, the transient heat transfer in the heater becomes significant. In order to analyze this transient heating, the temperature distribution in the heater is examined. Since the electric conductor also has a high thermal conductivity, it is treated as having a uniform temperature. However, the electrical insulator (generally an oxide ceramic) has a relatively lower thermal conductivity, and this results in a temperature nonuniformity within it.

(a) Divide the volume of the electrical insulator into three regions, as shown in Figure Pr. 2.31(b).
(b) Select a control volume in the region between $r = R_1$ and $r = R_2$ and render the heat transfer through this control volume.
(c) Show that the energy equation for this control volume allowing for conduction and sensible heat storage in the electrical insulator is given by

$$k\frac{\partial T}{\partial x}\Big|_{R_1} 2R_1 - k\frac{\partial T}{\partial x}\Big|_{R_2} 2R_2 = -(\rho c_p)_2(R_2^2 - R_1^2)\frac{\partial T_2}{\partial t}.$$

Range-Top Electrical Heater

(a) Physical Model

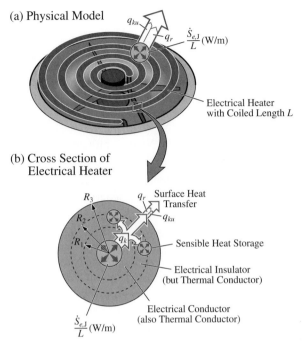

(b) Cross Section of
Electrical Heater

Figure Pr. 2.31. (a) A range-top electrical heater. (b) The various layers within the heating element.

### 2.8.3 Bounding-Surface Thermal Conditions

**PROBLEM 2.32. FUN**

Consider air (fluid) flow parallel to a semi-infinite plate (solid, $0 \leq x \leq \infty$), as shown in Figure Pr. 2.32. The plate surface is at a uniform temperature $T_{sf}$. The flow is along the $x$ axis. The velocity of air $\mathbf{u}_f$ at the solid surface is zero. Starting from (2.61) show that at a location $L$ along the plate, the surface energy equation becomes

$$k_s \left. \frac{\partial T_s}{\partial y} \right|_{y=0^-} - k_f \left. \frac{\partial T_f}{\partial y} \right|_{y=0^+} = 0 \quad \text{on } A_{sf}.$$

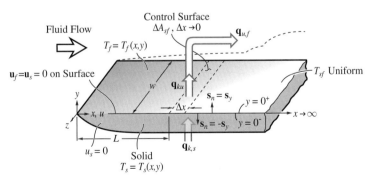

Figure Pr. 2.32. A parallel air flow over a semi-infinite plate ($0 \leq x \leq \infty$), with a uniform surface temperature $T_s$.

Use the definition of surface-convection heat flux $q_{ku}$ (positive when leaving the solid toward the gas) given as

$$-k_s \left.\frac{\partial T_s}{\partial y}\right|_{y=0^-} = -k_f \left.\frac{\partial T_f}{\partial y}\right|_{y=0^+} \equiv q_{ku} \quad \text{on } A_{sf}.$$

Neglect surface radiation heat transfer, and use $\mathbf{u}_s = 0$, and $\mathbf{u}_f = 0$ on $A_{sf}$. There is no surface energy conversion.

### PROBLEM 2.33. FUN

The divergence of the heat flux vector $\nabla \cdot \mathbf{q}$ is indicative of the presence or lack of local heat sources (energy storage/release or conversion). This is stated by (2.2). Consider a gaseous, one-dimensional steady-state fluid flow and heat transfer with a premixed combustion (exothermic chemical reaction) as shown in Figure Pr. 2.33. For this, (2.2) becomes

$$\nabla \cdot \mathbf{q} = \frac{d}{dx}q_x = \dot{s}_{r,c}(x).$$

Here $q_x = q_{k,x} + q_{u,x}$ (assuming no radiation) is idealized with a distribution and the source terms.

$$\dot{s}_{r,c}(x) = -\rho_{F,1}u_{f,1}\Delta h_{r,F}\frac{1}{\sigma(2\pi)^{1/2}}e^{-(x-x_o)^2/2\sigma^2},$$

where $\rho_{F,1}$ is the fluid density far upstream of the reaction (or flame) region, $u_{f,1}$ is the fluid velocity there, and $\Delta h_{r,F}$ is the heat of combustion (per kg of fuel). The exponential expression indicates that the reaction begins to the left of the flame location $x_o$ and ends to its right, with the flame thickness given approximately by $6\sigma$. This is the normal distribution function and represents a chemical reaction that initially increases (as temperature increases) and then decays and vanishes (as products are formed and fuel depletes).

(a) For $\sigma = \delta/6 = 0.1$ mm, plot $q_x/(-\rho_{F,1}u_{f,1}\Delta h_{r,F})$ and $\dot{s}_{r,c}/(-\rho_{F,1}u_{f,1}\Delta h_{r,F})$, with respect to $x$ (use $x_o = 0$ and $x_1 = -\delta < x < x_L = \delta$). Assume $q_x(x = -\delta) = 0$.

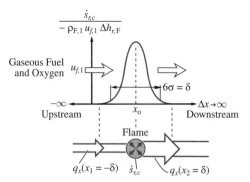

Figure Pr. 2.33. Variable energy conversion (source) term for combustion in a premixed gaseous flow. The source has a normal distribution around a location $x_o$. The flame thickness is approximated as $\delta = 6\sigma$.

(b) Noting that no temperature gradient is expected at $x = x_1 = -\delta$ and at $x = x_2 = \delta$, i.e., $q_{k,x} = 0$ at $x = x_1$ and $x = x_2$, determine $q_{u,x}$ at $x = x_2$, for $\rho_{F,1} = 0.06041$ kg/m$^3$, $u_{f,1} = 0.4109$ m/s, and $\Delta h_{r,F} = -5.553 \times 10^7$ J/kg. These are for a stoichiometric, atmospheric air-methane laminar flame.

## PROBLEM 2.34. FUN

A $p$-$n$ junction is shown Figure Pr. 2.34. The junction (interface) is at temperature $T_j$. The ends of the two materials are at a lower temperature $T_c$ and a higher temperature $T_h$.

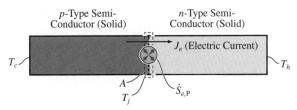

Figure Pr. 2.34. Conduction heat transfer in a slab containing a thermoelectric $p$-$n$ junction.

(a) Starting from (2.62), write the surface energy equation for the interface. Make the appropriate assumptions about the mechanisms of heat transfer expected to be significant.

(b) Express the conduction heat transfer as

$$Q_{k,n} = \frac{T_j - T_h}{R_{k,n}}, \quad Q_{k,p} = \frac{T_j - T_c}{R_{k,p}}.$$

Comment on the signs of $Q_{k,p}$ and $Q_{k,n}$ needed to absorb heat at the junction to produce electrical potential-current.

## PROBLEM 2.35. FUN

Below are described two cases for which there is heat transfer and possibly energy conversion on a bounding surface between two media (Figure Pr. 2.35).

(a) A hot solid surface is cooled by surface-convection heat transfer to a cold air stream and by surface-radiation heat transfer to its surrounding. Note that the air velocity at the surface is zero, $\mathbf{u}_g = 0$ at $x = 0$. Also, assume that the radiation is negligible inside the solid (i.e., the solid is opaque).

(b) Two solid surfaces are in contact with each other and there is a relative velocity $\Delta u_i$ between them. For example, one of the surfaces is a brake pad and the other is a brake drum.

For each of these cases, apply the bounding-surface energy equation (2.62) to the interface separating the two media. Assume that the surfaces are at uniform temperatures. As a consequence, the heat transfer at the interface is one-dimensional and perpendicular to the interface.

(a) Surface-Convection and Surface-Radiation Cooling of a Hot Surface

(b) Friction Heating between Two Sliding Solid Surfaces

Figure Pr. 2.35. Two examples of bounding surface between two media.

## PROBLEM 2.36. FUN

An opaque (i.e., a medium that does not allow for any transmission of radiation across it) solid surface is called a selective radiation surface when its ability to absorb radiation is different than its ability to emit radiation. This is shown in Figure Pr. 2.36. A selective absorber has a higher absorptivity $\alpha_r$ compared to its emissivity $\epsilon_r$.

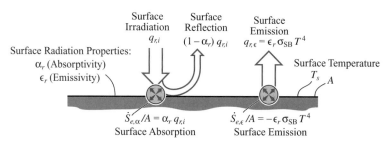

Figure Pr. 2.36. A selective thermal radiation absorber.

(a) From Table C.19, choose four surfaces that are selective absorbers and four that are selective emitters. The data in Table C.19 is for absorption of solar irradiation (a high temperature radiation emission).

(b) Using black-oxidized copper, determine the surface-absorption heat flux for a solar irradiation of 700 W/m$^2$ and surface-emission heat flux at surface temperature of 90°C.

(c) Determine the difference between the heat absorbed and heat emitted.

## PROBLEM 2.37. FUN

A droplet of refrigeration fluid (refrigerant) R-134a, which is used in automobile air-conditioning systems, is evaporating. The initial droplet diameter is $D(t = 0)$ and the diameter decreases as heat is absorbed on the droplet surface from the gaseous ambient by surface convection and radiation.

(a) Starting from (2.62) and by replacing $\mathbf{q}_{k,g}$ by $\mathbf{q}_{ku,g}$, and noting that the difference between the convection terms is represented by $\dot{S}_{lg}$ (Section B.7), i.e., drop convection, write the appropriate surface energy equation. The radiation heat transfer within the droplet can be neglected (i.e., liquid-side heat transfer is zero; $\mathbf{q}_l = 0$). Assume a uniform droplet temperature, i.e., assume the liquid conduction can also be neglected.

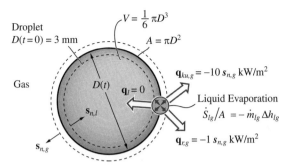

Figure Pr. 2.37. Droplet evaporation by surface convection and radiation.

(b) Using the properties listed in Table C.26 (they are for $p = 1$ atm) and the heat flux rates given in Figure Pr. 2.37, determine the evaporation rate per unit area $\dot{m}_{lg}$.
(c) Starting with (1.25) and setting the outgoing mass equal to the evaporation rate, derive an expression giving the instantaneous droplet diameter $D(t)$, as a function of the various parameters.
(d) Determine the time needed for the droplet diameter to decrease by a factor of 10.

**PROBLEM 2.38. FUN**
A thermoelectric element (TE) is exposed at its cold-junction surface (at temperature $T_c$) partly to an electrical connector ($e$) and partly to the ambient air ($a$). This is shown in Figure Pr. 2.38. Heat is transferred to the surface through the thermoelectric element $\mathbf{q}_{k,TE}$ in addition to a prescribed heat flux, $\mathbf{q}_{TE}$ that combines some parasitic heating. These are over the surface area $A_a + A_e$. Heat is also transferred to the surface from the adjacent air and the connector, over their respective areas $A_a$ and $A_e$. The area in contact with air undergoes heat transfer by surface convection $q_{ku,a}$ and surface radiation $q_{r,a}$. The connector heat transfer is by conduction $\mathbf{q}_{k,e}$. There is a Peltier energy conversion $(\dot{S}_{e,P})_c/A_e$ at the surface (and since it occurs where the current passes, it occurs over $A_e$). The Joule heating $(\dot{S}_{e,J})_c/A$ is also represented as a surface energy conversion (this presentation will be discussed in Section 3.3.6) and is over the entire element area $A_a + A_e$.

(a) Starting from (2.60), write the surface energy equation for the cold junction control surface $A$.

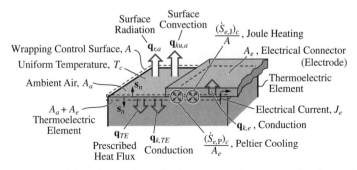

Figure Pr. 2.38. The cold-junction surface of a thermoelectric element showing various surface heat transfer and energy conversions.

(b) Determine $q_{k,e}$ for the conditions given below.

$$A_a = 10^{-6}\,\text{m}^2, \quad A_e = 10^{-6}\,\text{m}^2, \quad q_{k,TE} = -4 \times 10^4\,\text{W/m}^2,$$

$$(\dot{S}_{e,J})_c/A = 2 \times 10^4\,\text{W/m}^2, \qquad (\dot{S}_{e,P})_c/A_e = -2 \times 10^5\,\text{W/m}^2,$$

$$q_{r,a} = 0,\, q_{ku,a} = 0,\, q_{TE} = 0.$$

Assume quantities are uniform over their respective areas.

**PROBLEM 2.39. FUN**

When the ambient temperature is high or when intensive physical activities results in extra metabolic energy conversion, then the body loses heat by sweating (energy conversion $\dot{S}_{lg}$). Figure Pr. 2.39 shows this surface energy exchange, where the heat transfer to the surface from the tissue side is by combined conduction and convection $q_{k,t}$, $q_{u,t}$ and from the ambient air side is by conduction, convection, and surface radiation $q_{k,a}$, $q_{u,a}$, $q_{r,a}$. The surface evaporation is also shown as $\dot{S}_{lg}/A_t$ where $A_t$ is the evaporation surface area. The tissue conduction $q_{k,t}$ is significant for lowering the body temperature or removing extra metabolic heat generation (i.e., when the heat flow is dominantly from the tissue side). Preventing the high ambient temperature from raising the tissue temperature, however, relies only on intercepting the ambient heat transfer on the surface (i.e., when the heat flow is dominated by the ambient air side and the tissue conduction is not significant).

(a) Starting from (2.60), write the surface energy equation for the skin control surface $A$ (wrapped around the surface with $A_t = A_a$) in Figure Pr. 2.39.
(b) For the conditions given below, determine $q_{k,t}$.

$$\dot{S}_{lg}/A_t = -300\,\text{W/m}^2,\, q_{u,a} = q_{u,t} = 0,$$

$$q_{k,a} = q_{ku,a} = -150\text{W/m}^2, \quad q_{r,a} = -10\,\text{W/m}^2.$$

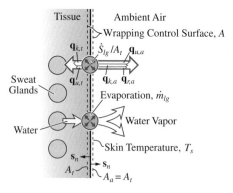

Figure Pr. 2.39. The surface heat transfer, and energy conversion by sweat cooling. across human skin.

Assume that quantities are uniform over their respective surfaces.

## PROBLEM 2.40. FUN

In laser materials processing-manufacturing, high-power, pulsed laser irradiation flux $q_{r,i}$ is used and most of this power is absorbed by the surface. Figure Pr. 2.40 shows the laser irradiation absorbed $\dot{S}_{e,\alpha}/A = \alpha_r q_{r,i}$ (where $\alpha_r$ is the surface absorptivity), the surface radiation emission flux $(\dot{S}_{e,\epsilon})/A = -\epsilon_r \sigma_{SB} T_s^4$, the gas-side surface convection $q_{ku}$, and the solid (substrate or working piece) conduction $q_{k,s}$, over a differential control surface $\Delta A \to 0$. Since the irradiation is time dependent (e.g., pulsed), the heat transfer and energy conversions are all time dependent (and nonuniform over the surface).

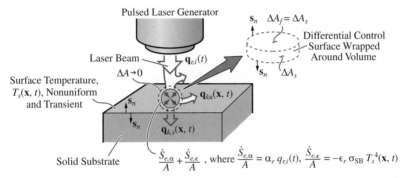

Figure Pr. 2.40. Laser irradiation of a substrate and a differential control surface taken in the laser impingement region.

(a) Starting with (2.58), write the surface energy equation for the differential control surface $\Delta A$, made of $\Delta A = \Delta A_f + \Delta A_s$, with $\Delta A_f = \Delta A_s$.
(b) Determine $q_{k,s}$ for the conditions given below.

$$\epsilon_r = 0.8, \, \alpha_r = 0.9, \, q_{r,i} = 10^{10} \, \text{W/m}^2, \, T_s = 2 \times 10^3 \, \text{K}, \, q_{ku} = 10^7 \, \text{W/m}^2.$$

Note that the entire surface radiation is represented as energy conversions $\dot{S}_{e,\alpha}$ and $\dot{S}_{e,\epsilon}$.

### 2.8.4 General*

This section is found on the Web site www.cambridge.org/kaviany.

**PROBLEM 2.41. FUN***

**PROBLEM 2.42. FUN***

**PROBLEM 2.43. FUN***

**PROBLEM 2.44. FUN***

**PROBLEM 2.45. FUN***

**PROBLEM 2.46. FUN***

# 3

## Conduction

Every solid of finite dimension contains a finite number of atoms, and therefore, a sonic spectral distribution of free vibrations of lattice atoms. These quantized, elastic waves are similar to photons and may be called phonons.

— P. Debye

In this chapter we discuss the steady-state internodal conduction resistance $R_k(°\text{C/W})$ and develop relations for $R_k$. We also consider transient conduction where sensible heat storage, i.e., the time rate of change of temperature, is significant. In transient treatments, whenever justifiable, we assume that the temperature is uniform (but time dependent). Otherwise we allow for both temporal and spatial variation of temperature and heat transfer rate.

Conduction heat transfer examines the magnitude, direction, and spatial-temporal variations of the conductive heat flux vector $\mathbf{q}_k$ in a medium. The conduction heat flux was defined by (1.11), i.e, the Fourier law of conduction. These variations are governed by the volumetric storage/release and conversion of thermal energy and by the bounding-surface and initial conditions of the medium. The most general energy equation is the differential-volume energy equation given by (2.1), which allows for spatial and temporal variation of $\mathbf{q}$ and $T$. In vectorial form, we can write the conduction energy equation as

$$\nabla \cdot \mathbf{q} = \nabla \cdot \mathbf{q}_k \equiv \lim_{\Delta V \to 0} \frac{\displaystyle\int_{\Delta A} (\mathbf{q}_k \cdot \mathbf{s}_n)\,dA}{\Delta V}$$

$$\equiv \underbrace{-\nabla \cdot k\nabla T}_{\substack{\text{spatial-rate} \\ \text{of change of} \\ \text{conductive} \\ \text{heat flux}}} = \underbrace{-\frac{\partial}{\partial t}\rho c_p T}_{\substack{\text{volumetric} \\ \text{storage}}} + \underbrace{\dot{s}}_{\substack{\text{volumetric} \\ \text{energy} \\ \text{conversion}}} \tag{3.1}$$

differential-volume, conduction energy equation.

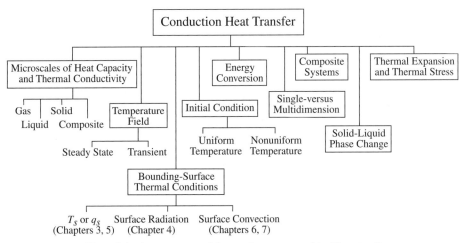

Chart 3.1. A summary of the topics presented in Chapter 3.

Here, other than density $\rho(\text{kg/m}^3)$, we have two thermophysical properties: the conductivity $k$ (W/m-°C or W/m-K, because $k$ is defined per an increment of 1°C, which is identical to 1 K change in the temperature), and the specific heat capacity $c_p$ (J/kg-°C or J/kg-K, again because $c_p$ is defined per an increment of 1°C).

Equation (3.1) is a second-order, partial differential equation in $T$ along with the appropriate bounding-surface and initial thermal conditions, as discussed in Section 2.4. Here we use the simplest forms of the thermal conditions, i.e., the prescribed $T_s$ or $\mathbf{q}_s$ on the bounding surface $A_k$. These boundary conditions are (2.64) or (2.65), i.e.,

$$T = T_s \text{ or } A_k[-k(\nabla T \cdot \mathbf{s}_n) + q_s] = \dot{S} \qquad \begin{array}{l}\text{prescribed bounding-surface} \\ \text{thermal conditions on } A_k,\end{array} \qquad (3.2)$$

where we have assumed that $\mathbf{u} = 0$ and that radiation heat transfer is negligible. Note that $q_s$ is the magnitude normal to the surface. We will consider surfaces over which we can justifiably assume a uniform temperature; then the heat flows perpendicular to the bounding surface.

The initial condition is the condition at the beginning of the time period of interest, with $t$ as the time variable. We choose to set the elapsed time to zero at this initial state, i.e., $t = 0$. The simplest initial thermal condition is that of a prescribed temperature $T(\mathbf{x}, t = 0)$. Then the initial condition is given by

$$T = T(\mathbf{x}, t = 0) \quad \text{initial condition.} \qquad (3.3)$$

Note that $T(\mathbf{x}, t = 0)$ can be uniform throughout the heat transfer medium, or it can be nonuniform.

Chart 3.1 describes aspects of conductive heat transfer considered in this chapter, namely the differential-volume energy equation, thermophysical properties ($c_p$ and $k$), steady-state conduction, transient conduction, solid-liquid phase change, and thermal expansion and thermal stress.

In general, since $\dot{s}$ may vary within $V$, the surface $A$ may be curved, $T_s$ or $q_s$ may be time dependent, and $T(\mathbf{x}, t = 0)$ may not be uniform, a general analytical solution to (3.1) to (3.3) does not exist. However, analytical solutions for many specific conditions do exist. Using numerical techniques, solutions can be found for nearly any condition. This powerful method will be discussed in Section 3.7.

The radiation and convective bounding-surface conditions will be discussed in Chapters 4, 6, and 7.

We begin with the properties ($c_p$ and $k$), then consider steady-state conduction, followed by transient conduction.

The materials encountered in heat transfer applications can be made of pure chemical elements such as nitrogen, mercury, and aluminum. Depending on the temperature (and pressure), these elements can be in the gaseous, liquid, or solid phase or at phase equilibrium conditions where two or three phases may coexist (e.g., melting point, boiling point, or triple point), or at or above the critical state (called the supercritical state) where no distinction is made between the gas and the liquid phase (i.e., characterized as a dense gas).

In many cases, the material is a compound of chemical elements, e.g., carbon dioxide, water, and aluminum oxide. In some cases a composite made of distinctly separated materials is used. For example, in a nonevacuated fiberglass insulation, the air and the glass fibers make up a composite material.

The molecular description (i.e., the fundamental physics) of the thermodynamic properties (such as the specific heat capacities, i.e., $c_v$ and $c_p$) and transport properties (such as the thermal conductivity, $k$) begins by considering a chemical element existing in a phase. Then chemical compounds and composites are addressed. The gas-phase behavior is studied the most, because of the more tractable limit of the dilute gases with negligible intermolecular collisions. In this chapter, we present short summaries of the molecular description of the specific heat capacity and thermal conductivity, beginning with $c_p$ in Section 3.1 and then discussing $k$ in Section 3.2.

## 3.1 Microscale Heat Storage and Specific Heat Capacity $c_p$

In general, the specific heat capacity is the ability of a medium to store heat (J) per unit rise in the temperature (K or °C) of the medium and per unit mass (kg). More specifically, the specific heat capacity at constant volume $c_v$(J/kg-K or J/kg-°C) defined by (1.6), is a measure of the change in the specific internal energy, for a change made in the temperature (°C or K) (at constant volume). The specific heat capacity at constant pressure $c_p$ (J/kg-K or J/kg-°C) defined by (1.6) is similar to $c_v$, except the change is in the specific enthalpy (at constant pressure). Chart 3.2 gives a summary of the different aspects of $c_p$ considered here. These are the mechanisms of heat capacity for different phases of matter (gas, liquid, and solid) and the heat capacity of composites.

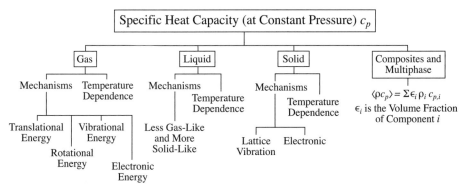

Chart 3.2. The specific heat capacity of various phases and their dependence on the mechanisms of heat storage. The temperature dependence is also indicated.

### 3.1.1 Gases: Thermal Fluctuation Motion

For a monatomic gas, the internal energy $e$(J/kg) resides in the kinetic energy of its atoms [given by (1.15)], i.e.,

$$e = \frac{E_u}{m} = \frac{N_A}{M}\left(\frac{1}{2}m\overline{u_m^2}\right) = \frac{3}{2}\frac{N_A}{M}k_B T \quad \text{for monatomic gas}$$

$$= \frac{3}{2}\frac{R_g}{M}T, \quad R_g \equiv N_A k_B = 8.3145 \times 10^3 \text{ J/kmole-K}, \tag{3.4}$$

where we have used $m = M/N_A$. Now we determine $c_v$ from (1.6) as

$$c_v \equiv \frac{\partial e}{\partial T}\Big|_v = \frac{3}{2}\frac{R_g}{M}, \quad Mc_v = \frac{3}{2}R_g \quad \text{for monatomic gas}, \tag{3.5}$$

where $N_A$(number of molecules per kmole) is the Avogadro number and is equal to $6.0225 \times 10^{23}$, $M$(kg/kmole) is the molecular weight, $k_B$ is the Boltzmann constant and is equal to $1.381 \times 10^{-23}$ J/K, and $R_g$(J/kmole-K) is equal to the product $N_A k_B$ and is called the universal gas constant. The 3/2 factor appearing in (3.5) can be interpreted as the particles having three degrees of freedom (one for each direction) for translation and each having an energy of $k_B T/2$. Now for polyatomic molecules, as rotation, vibration, and other interactions are added, the specific heat capacity is increased, i.e., larger numbers of mechanisms (called internal) are available for storage of energy. Figure 3.1(a) shows the variation of the product $Mc_v$ with respect to $T$ for the diatomic hydrogen $H_2$. At high temperatures, there are six modes of motion (three translational, two rotational, and one vibrational). At low temperatures ($T < T_r = 40$ K), the rotational energy is suppressed, and for $T < T_v = 500$ K, the vibrational energy is suppressed (i.e., the ground-state vibrational energy is excited above 500 K). For $T > 2{,}500$ K, this diatomic gas dissociates (i.e., gradually becomes a thermal plasma) and electronic energy storage occurs. Therefore, for other than the monatomic gases, the specific heat capacity would depend on the temperature (increasing as the temperature increases). Note that $c_v = (Mc_v)/M$, and since the molecular weight of hydrogen is the smallest among elements, it has the

(a) Mechanisms of Energy Storage in Specific Heat Capacity $Mc_v$ of Diatomic $H_2$

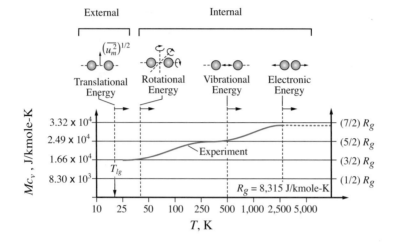

(b) Specific Heat Capacity $Mc_p$ of Some Gases at 1 atm pressure

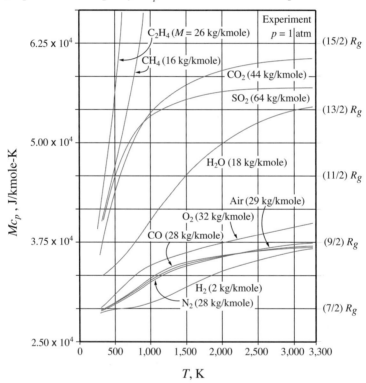

Figure 3.1. Variation of specific heat capacity at constant volume $c_v$ and constant pressure $c_p$ with respect to temperature. (a) $Mc_v$ for gaseous hydrogen molecules (the boiling point at 1 atm pressure is also shown). (b) $Mc_p$ for some other gases, at 1 atm pressure. (From Themelis, N.J., reproduced by permission © 1995 Gordon and Breach.)

largest $c_v$. This can be further examined by reviewing Table C.22, which lists several gases.

For an ideal gas behavior, the relationship between the two specific heat capacities (or specific heats) becomes, using the general relationship between $c_p$ and $c_v$,

$$c_p \equiv c_v + T \left( \frac{\partial p}{\partial T} |_v \frac{\partial v}{\partial T} |_p \right) = c_v + T \left( \frac{\beta^2}{\rho \kappa} \right)$$

$$= c_v + \frac{R_g}{M} \quad \text{for ideal gas,} \tag{3.6}$$

where $\beta(1/K)$ is the volumetric (or cubic) thermal expansion coefficient defined by (A.10), and $\kappa(1/Pa)$ is the isothermal compressibility given by (1.18). Equation (3.6) shows that $c_p \geq c_v$. The heat capacity of gases generally decreases with increase in pressure.

Figure 3.1(b) shows the specific heat capacity at constant pressure for various gases and as a function of temperature, at one atmosphere pressure. The product $Mc_p$ is given and the values for $M$ (molecular weight) are also given in the figure. Note the strong temperature dependence of $c_p$ for methane as compared to air (dry air is a mixture of nitrogen and oxygen, with traces of a few other species), as the available number of modes of motions increases with the number of atoms in the molecule (the total is 3 times the number of atoms in the molecule).

The specific heat capacity of some common gases is given in Table C.22 for one atom pressure and at several temperatures, and for saturated vapors in Tables C.26 to C.29.

For an ideal gas mixture, the mixture specific heat capacity is founded by summing the product of mass fraction $\rho_i/\rho$ and component specific heat capacity $c_{p,i}$. This gives $c_p = \sum \rho_i c_{p,i}/\rho$, where $c_p$ is the mixture heat capacity and $\rho = \sum \rho_i$ is the mixture density.

### 3.1.2 Solids: Lattice Vibration and Phonon

Based on electrical properties, solids are divided into metals (i.e., conductors) and nonmetals (i.e., insulators or dielectrics). There are also semi-metals and semiconductors. These will be discussed in Section 3.2.3. For solids, the thermal (i.e., internal) energy is stored in the lattice vibration[†] (i.e., the concerted harmonic motion of all

---

[†] The elastic, root-mean-square strain for a single lattice is given by

$$\Delta l_m^* = \frac{\Delta l_m}{l_m} = \frac{2k_B T (1 - v_P)}{E_s l_m^2},$$

where $l_m$ is the monatomic, cube unit-cell lattice constant, $E_s$ is the Young modulus of elasticity, and $v_P$ is the Poisson ratio. The effective harmonic oscillator spring constant is equal to $E_s l_m (N/m)$. Also, for isotropic polycrystalline solids, the phonon group speed $u_p$ is given by

$$u_p^2 = \frac{E_s}{3(1 - v_P)\rho}, \quad E_p = \frac{1}{\kappa} = \frac{E_s}{3(1 - 2v_P)},$$

where $\kappa$ is the adiabatic compressibility. For isotropic materials $v_P \to 0$, we have $u_p^2 = (E_s/\rho)^{1/2}$. Note that no distinction is made between transverse and longitudinal waves. The phonon speed is atoms) and in free electrons. The energy of an elastic wave in a crystalline solid is quantized and this quantum of energy is called a phonon, $E_p = h_P f$, where $f$ is the frequency. The dominant phonon frequency is less than the high frequency electromagnetic photons capable of electronic transition and bond breakage. For crystalline solids, the ratio $\beta^2/\rho\kappa$ appearing in (3.6) is nearly equal to zero, and the two specific heat capacities are nearly equal and given by the Debye approximate model (which assumes dominance of long-wavelength modes of low frequencies) as [2]

$$c_p \cong c_v = 9\frac{R_g}{M}\left(\frac{T}{T_D}\right)^3 \int_0^{T_D/T} \frac{x^4 e^x}{(e^x - 1)^2}dx, \quad x \equiv \frac{h_P f}{k_B T} \quad \begin{matrix} \text{Debye heat capacity} \\ \text{model for crystalline} \\ \text{lattice solids.} \end{matrix} \quad (3.7)$$

The Debye temperature $T_D = h_P f_D/k_B$ is a temperature (or a cutoff frequency $f_D$) threshold beyond which the lattice specific heat capacity of the solid does not change with a further increase in temperature. The Debye temperature is related to the speed of sound in the solid $a_s$ (which is the same as the acoustic phonon speed $u_p$) and the number of unit cells per unit volume $n = 1/l_m^3$ ($l_m$ is the average lattice constant), i.e., $2\pi f_D = (6\pi^2 u_p^3 n)^{1/3}$. The magnitude of $T_D$ for some elements is listed in Figure 3.2(a) [12]. For $T > T_D$, all modes (frequencies) of vibration are excited, and for $T < T_D$ some of the modes are frozen (similar to rotational and vibrational temperature for gas molecules discussed in Section 3.1.1). As $a_s = u_p$ and $n$ increase, $c_v$ and $T_D$ increase. As an example of speed of sound in solids, the speed of sound in carbon at STP is 12,288 m/s.

A typical temperature dependence of $c_v$ for crystalline solids is shown for copper in Figure 3.2(b) [32]. The prediction using (3.7) is also shown. For monatomic solid crystals, the maximum value of $c_v$ is equal to $3R_g/M$. At low temperature, $c_v$ is proportional to $T^3$, i.e., $c_v = 12\pi^4 R_g(T/T_D)^3/5M$, and this is called the low-temperature Debye law. Again an increase in $c_v$ is found with an increase in the temperature up to $T_D$. The classical limit of $3R_g$ is for potential and kinetic energies, each having three degrees of freedom. The heat capacity of many metals is expressed as the sum of electronic and lattice-vibration contribution, but the lattice-vibration contribution dominates. For multiatomic solids, the average molecular weight is used in (3.7).

Here for solids, we assume $c_p = c_v$. In general, for solids, $c_p$ is slightly larger than $c_v$, at high temperatures, and for solids (3.6) is written as $c_p = c_v(1 + \beta\gamma_G T)$, where $\gamma_G$ is the Grünesien parameter (or constant) and has a magnitude near one.

The specific heat capacity of some common solids are given in Tables C.16 (metals) and C.17 (nonmetals).

also related to the Debye temperature $T_D$, as stated in text above. Also, for a monatomic linear chain, the maximum frequency for the normal modes of the chain is [6]

$$f_{max} = f_D = \frac{1}{\pi}\left(\frac{N_A E_s l_m}{M}\right)^{1/2},$$

where $M$ is the molecular weight and $N_A$ is the Avogadro number. This maximum frequency is similar to the Debye cutoff frequency $f_D$. Note that since $f_{max}$ is proportional to $u_p$, the peak in $k$ versus $T$ curve moves to the right as the intermolecular forces (i.e., lattice stiffness) increase, resulting in an increase in $u_p$ and $f_{max}$ (i.e., $f_D$).

(a) Debye Temperature, $T_D(K)$, of Some Elements and Compounds

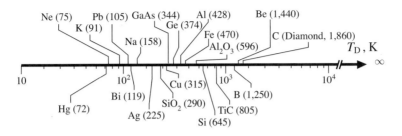

(b) Specific Heat Capacity, $Mc_v$, of Solid Copper with Respect to Temperature

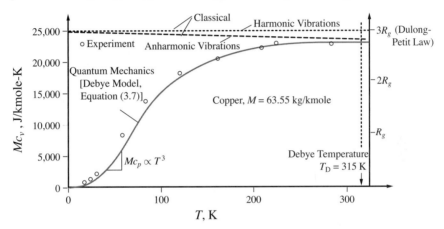

Figure 3.2. (a) Debye temperature $T_D$ of some elements. (b) Variation of the crystalline lattice specific heat capacity, due to lattice vibration, i.e., phonon, for copper with respect to temperature, as predicted by the Debye model. (From Tien, C.-L., and Lienhard, J.M., reproduced by permission © 1971, Holt, Rinehard, and Winston.) Also shown are the experimented values.

### 3.1.3 Liquids: Solid-Like versus Gas-Like

The specific heat capacity of polyatomic liquids can be resolved into contributions from translational, rotational, and vibrational motions. These are all dependent on temperature and pressure. For bipolar liquids there is also polar-polar orientation which is also temperature dependent. The total energy is the sum of potential and kinetic energies.

The specific heat capacity of a liquid has a magnitude closer to its solid state, as compared to its gaseous state heat capacity. This similarity is more evident, closer to the triple point. The exception appears near the critical point, where $c_v$ for the liquid phase is closer to that for the gas phase. Away from the critical point, the liquid may be assumed nearly incompressible and then $c_v \simeq c_p$.

As a comparison between liquid and gaseous $c_p$, the specific heat capacity of water is shown in Figure 3.3 as a function of temperature and for several pressures. Note that $c_{p,l}$ (i.e., for liquid) is twice $c_{p,g}$ (i.e., for gas), near one atm pressure. As the critical state is reached, both $c_{p,l}$ and $c_{p,g}$ increase and at the critical state $c_p \to \infty$.

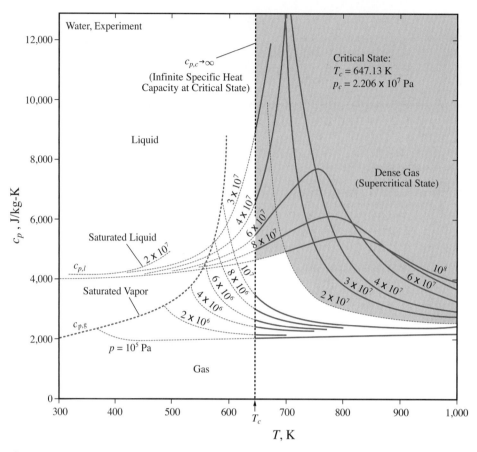

Figure 3.3. Specific heat capacity of water (liquid and vapor) at various temperatures and pressures. (From Liley, P.E., et al., reproduced by permission © 1988 Hemisphere.)

The specific heat capacity of some common liquids is given in Table C.23, for one atm pressure and at several temperatures, and for standard liquids in Tables C.26 to C.29. The difference between $c_p$ and $c_v$ for liquid water at $T = 300$ K, can be determined using (3.6) along with $\beta = 2.57 \times 10^{-4}$ 1/K, $\kappa = 4.52 \times 10^{-10}$ 1/pa and $\rho = 997$ kg/m$^3$. The result is $c_p - c_v = 43.97$ J/kg-K (m$^2$/s$^2$-K), which is small compared to $c_p = 4.179$ J/kg-K (Table C.27).

### 3.1.4 Thermal Materials: From Zero to Infinite Heat Capacity

As a consequence of the third law of thermodynamics, for crystalline materials we have $c_p = 0$ at $T = 0$ K. During a constant temperature phase change, the specific enthalpy $h$ (i.e., the intermolecular bond) changes. Then an apparent specific heat at constant pressure is defined using (1.6) and is infinite during such a phase change, i.e.,

$$c_{p,app} = \frac{\partial h}{\partial T} \Big|_p \to \infty \qquad \begin{array}{l}\text{apparent specific heat capacity during}\\ \text{isothermal } (dT = 0) \text{ phase change.}\end{array} \qquad (3.8)$$

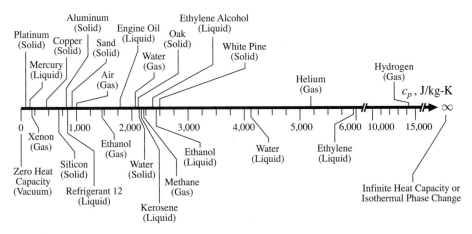

At 300 K and 1 atmosphere pressure, unless stated otherwise.

Figure 3.4. Typical values of the specific heat capacity for various phases.

In general, the molecules with a smaller molecular weight have a higher heat capacity. Some typical values of $c_p$ for various phases are listed in Figure 3.4. Note that hydrogen (gas) has the highest $c_p$ among elements (at room temperature). On the low end of the $c_p$ values, mercury and some of the heavy metals appear.

As shown in the figure, water (liquid) has a relatively high heat capacity (at $T = 300$ K and under saturation pressure, $c_{p,l} = 4{,}179$ J/kg-°C). Helium in gaseous state also has a very large $c_p$ compared to other gases, liquids, and solids. In general, solids do not have very high $c_p$. Note again that pure, metallic elements such as aluminum, copper, and platinum all have a $c_p$ of less than 1,000 J/kg-°C.

We repeat that $c_p$ is defined per unit change in temperature. Then the unit given for change in temperature can be in °C or in K (the changes are identical), i.e., from (1.6), for $\Delta T \to 0$

$$c_p \equiv \frac{\Delta h}{\Delta T}\,|_p \qquad c_p\,(\text{J/kg-°C or J/kg-K}). \qquad (3.9)$$

In the energy equation, the product of density and specific heat capacity $\rho c_p$ (J/m³-°C) appears. The distinct difference in the density between the condensed phase (solid or liquid) and gaseous phase greatly influences $\rho c_p$. The condensed phase substance with a relatively small $c_p$ may have a much larger $\rho c_p$, compared to a gaseous phase substance with a higher $c_p$. Note that $c_p$ increases with an increase in $T$, for all the three phases of matter.

When a mixture of phases is present in a volume, a volume-averaged heat capacity $\langle \rho c_p \rangle$ is used. This is based on the volume fraction $\epsilon_i$, and the individual specific heat capacities $c_{p,i}$, and the densities $\rho_i$. Then

$$\langle \rho c_p \rangle_V = \langle \rho c_p \rangle \equiv \sum_i \rho_i c_{p,i} = \sum_i \epsilon_i \rho_i c_{p,i} \qquad (3.10)$$

or

$$\langle c_p \rangle_V = \langle c_p \rangle \equiv \frac{\sum\limits_i \epsilon_i \rho_i c_{p,i}}{\langle \rho \rangle} \qquad \text{volume-averaged specific heat capacity,} \qquad (3.11)$$

where $\langle \rho \rangle$ is the volume-averaged density defined as

$$\langle \rho \rangle_V = \langle \rho \rangle = \sum_i \epsilon_i \rho_i \qquad \text{composite density.} \qquad (3.12)$$

In Appendix C, the periodic table (Table C.2) and Tables C.6, C.16, C.17, C.22, C.23, C.24, C.26, C.27, C.28, and C.29, give the value of $c_p$ for various substances. In some cases the temperature dependence is also listed.

Since the temperature, and therefore specific heat capacity $c_p$, vary within the heat transfer medium, an average value is generally used for $c_p$. One simple and relatively accurate method of averaging is to use the volume average medium temperature $\langle T \rangle_V$ to evaluate this average $c_p$. This is the constant, average $c_p$ used in the energy equations (2.1) and (2.9). In gaseous combustion, where a large temperature variation occurs over a short distance (i.e., over the flame thickness), the proper use of the average $c_p$ becomes especially important. We will discuss this in Section 5.2.

---

EXAMPLE 3.1. FUN

In sensible heat storage, an added heat flow $Q|_A(\text{W})$ to a control volume $V(\text{m}^3)$ raises the temperature of the content of this volume. This is depicted in Figure Ex. 3.1. Assume a uniform temperature within this volume (i.e., treating the volume as a lumped-capacitance volume) and allow for no heat losses.

Determine the rate of rise in the temperature when the following materials are used (properties evaluated at 300K).

(i) Air (gas)            $\rho = 1.161\,\text{kg/m}^3$, $c_p = 1{,}007\,\text{J/kg-}^\circ\text{C}$ at $T = 300$ K
(ii) Water (liquid)      $\rho = 997\,\text{kg/m}^3$, $c_p = 4{,}179\,\text{J/kg-}^\circ\text{C}$ at $T = 300$ K
(iii) Aluminum (solid)   $\rho = 2{,}702\,\text{kg/m}^3$, $c_p = 903\,\text{J/kg-}^\circ\text{C}$ at $T = 300$ K.

The heat flow rate $Q|_A = -1{,}000$ W is negative, because heat is flowing into the control volume, and $V = 0.1\,\text{m}^3$.

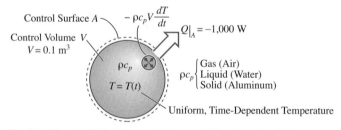

Figure Ex. 3.1. Heat addition to a volume, resulting in a rise in the temperature.

**SOLUTION**
The volumetric energy equation for a uniform temperature within the control volume with no energy conversion is the simplified form of the integral-volume energy equation (2.9), i.e.,

$$Q \mid_A = -\int_V \frac{d}{dt} \rho c_p \, T dV = -\rho c_p V \frac{dT}{dt} \quad \text{integral-volume energy equation,}$$

where we have assumed that the product $\rho c_p$ is constant.
Solving for $dT/dt$, we have

$$\frac{dT}{dt}(^\circ\text{C/s}) = \frac{-Q \mid_A (\text{W})}{V(\text{m}^3)\rho(\text{kg/m}^3)c_p(\text{J/kg-}^\circ\text{C})}.$$

Note that $(\text{J}) = (\text{W-s})$.
Substituting the numerical values of the parameters on the right-hand side, we have

(i) $\dfrac{dT}{dt}(\text{air}) = \dfrac{-(-1,000)(\text{J/s})}{1.161(\text{kg/m}^3) \times 1,007(\text{J/kg-}^\circ\text{C}) \times 0.1(\text{m}^3)} = 8.553^\circ\text{C/s}$

(ii) $\dfrac{dT}{dt}(\text{water}) = \dfrac{-(-1,000)(\text{J/s})}{997(\text{kg/m}^3) \times 4,179(\text{J/kg-}^\circ\text{C}) \times 0.1(\text{m}^3)} = 0.002400^\circ\text{C/s}$

(iii) $\dfrac{dT}{dt}(\text{aluminum}) = \dfrac{-(-1,000)(\text{J/s})}{2,702(\text{kg/m}^3) \times 903(\text{J/kg-}^\circ\text{C}) \times 0.1(\text{m}^3)} = 0.04099^\circ\text{C/s}.$

**COMMENT**
Note that water stores the heat with the lowest rate of rise in the temperature. Since in heat storage the heat loss from the storage volume is important and this heat loss is proportional to the difference in temperature between the storage volume and its ambient, the materials that do store the heat without much rise in the temperature are considered more suitable. Therefore, among these three materials, water would be the choice storage material for heat storage purposes.

## 3.2 Microscale Conduction Heat Carriers and Thermal Conductivity *k*

Thermal conductivity $k$(W/m-K or W/m-$^\circ$C) is the ability of a medium to transfer heat (J) per unit of time (s) and per unit of area (m²), i.e., $q_k$, without any net motion and in the presence of a unit temperature difference $\Delta T$ (K or $^\circ$C) over a unit length $\Delta x$(m) within the medium. This makes the unit for the thermal conductivity as (J-m/s-m²-K) or (W/m-K). Since the temperature difference is used, then (W/m-K) and (W/m-$^\circ$C) are identical.

In contrast to the heat capacity (which is an equilibrium property), the thermal conductivity (which is a transport or nonequilibrium property) depends on the ability of micro heat carriers (referred to in Table 1.4) of heat to travel to exchange this heat. This travel is in the form of fluctuations (or random-motion displacement about an equilibrium location). The ability to transfer heat by conduction is related

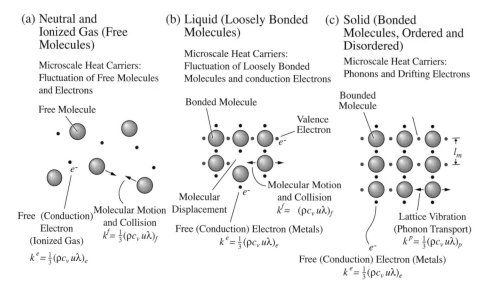

(a) Neutral and Ionized Gas (Free Molecules)

Microscale Heat Carriers: Fluctuation of Free Molecules and Electrons

Free Molecule

Free (Conduction) Electron (Ionized Gas)

$k^e = \frac{1}{3}(\rho c_v u \lambda)_e$

Molecular Motion and Collision

$k^f = \frac{1}{3}(\rho c_v u \lambda)_f$

(b) Liquid (Loosely Bonded Molecules)

Microscale Heat Carriers: Fluctuation of Loosely Bonded Molecules and conduction Electrons

Bonded Molecule

Valence Electron

$e^-$

Molecular Motion and Collision

$k^f = (\rho c_v u \lambda)_f$

Molecular Displacement

Free (Conduction) Electron (Metals)

$k^e = \frac{1}{3}(\rho c_v u \lambda)_e$

(c) Solid (Bonded Molecules, Ordered and Disordered)

Microscale Heat Carriers: Phonons and Drifting Electrons

Bounded Molecule

$l_m$

Lattice Vibration (Phonon Transport)

$k^p = \frac{1}{3}(\rho c_v u \lambda)_p$

$e^-$

Free (Conduction) Electron (Metals)

$k^e = \frac{1}{3}(\rho c_v u \lambda)_e$

Figure 3.5. Microscale mechanisms of heat conduction in various phases. (a) Neutral and ionized gases. (b) Liquids. (c) Solids. The heat carriers shown are molecules ($m$), free (conduction) electrons ($e$), and phonons ($p$). Here $\rho$ is density, $c_v$ is the specific heat capacity, $u$ is the speed, and $\lambda$ is the mean-free path.

to the ability of the molecules (for gases), electrons and holes (for metallic and semi-metallic liquids and solids), and phonons (for all liquids and solids, but dominant in nonmetals) to store/release thermal energy (i.e., heat capacity), and the ability of the electrons (and phonons) and molecules to travel a distance before losing their energy. The phonons represent the lattice and intermediate vibration (for liquids and solids). This frequent collision and randomness is generally described by a mean-free path. Depending on the medium (gas, liquid, solid, or composite), the thermal conductivity $k$ varies greatly. Figure 3.5 shows the dominant mechanisms and heat carriers in heat conduction through various phases. Shown in the figure is $k^m$ for conduction by molecular collision, $k^e$ for free (conduction) electron (or electronic) and hole conduction, and $k^p$ for phonon conduction.

Molecular transport of heat in gases by molecular collision is treated using classical (as compared to quantum) statistical mechanics. The treatment of liquids and solids requires inclusion of quantum-statistical mechanics. As discussed in Sections 1.4.2 and 1.5.1, the prediction of transport properties can be most detailed or may be simplified using more deterministic (for crystalline solids) or more statistical (for gases, liquids, and some solids) phenomenological models such as the intermolecular potential. The statistical description of heat carriers is referred to as the Boltzmann transport, and the related conservation equation for the heat carriers (i.e., molecules, electrons, and phonons) is called the Boltzmann transport equation [6, 12]. This equation governs the conservation of the heat-carrier intensity and is similar to the energy and mass conservation equations. Here we only present a brief discussion of the molecular thermal transport in gases (the so-called kinetic theory of heat conduction). We also present the experimental results for $k$ for the gases,

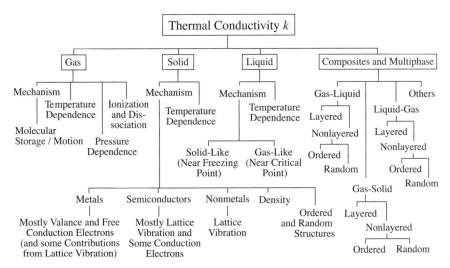

Chart 3.3. Microscale heat carrier of conduction in the three phases and also in composites. The temperature dependence of the thermal conductivity is also indicated.

solids, liquids, and composites. The results will show how the thermal conductivity of gases is related to the microscale properties such as the speed, mean-free path, the molecular density, and the specific heat capacity. Then we also use this relation to describe conduction in liquids and solids involving phonons and electrons.

Chart 3.3 gives a classification of media of heat conduction and the different molecular mechanisms contributing to $k$ for each medium. We note that the temperature dependence of $k$ is not as simple and always monotonic as that of $c_p$ (i.e., $c_p$ increases as $T$ increases). The very-low and very-high temperature behaviors of $k$ are different depending on the conductivity mechanisms and vary greatly among the three phases.

### 3.2.1 Gases: Thermal Fluctuation Motion and Mean-Free Path $\lambda_f$

Assuming an ideal gas behavior, the thermal conductivity can be estimated using the concepts of the most probable thermal speed $\langle u \rangle_f$ (but with no net motion, i.e., $\overline{u}_f = 0$), and the fluid particle mean-free path $\lambda_f$, along with what we know about the specific heat capacity $c_v$. Note that here, since $\overline{u}_f = 0$, we have no net convection heat transfer $\mathbf{q}_u = 0$. We assume a one-dimensional temperature field $T = T(x)$. From the kinetic theory of gases [36], the number of fluid particles (i.e., gaseous molecules) passing through a plane perpendicular to the $x$ axis, per unit area and per unit time, is $n\langle u_f \rangle/4$, where $n$ is the number of molecules per unit volume (or number density).[†] For molecules passing the plane $x = 0$, the molecules can move in different directions. By averaging over the proper angles, the average distance travelled along the $x$ axis becomes $2\lambda_f/3$, where $\lambda_f$ is the mean-free path defined by (1.19). This is rendered in Figure 3.6. The 2/3 factor comes from averaging the

---

[†] The prediction of the thermal conductivity of gases can also be made using the Boltzmann transport equation (Section 1.74 and B.2.1), which is a general method also applicable to electron and phonon (and also photons).

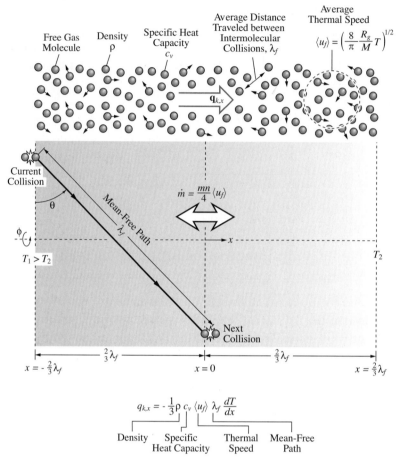

Figure 3.6. Rendering of the mechanism of bulk (i.e., averaged over many molecules) heat conduction in gases by molecular heat capacity, thermal motion, and intermolecular collision.

molecular travel over a hemisphere bounding the surface at $x = 0$. The heat flow per unit time and per unit area (i.e., heat flux) across the $x = 0$ plane is found by examining the molecules coming to the plane from the left and from the right (with no net molecular flow), each carrying with them their internal energy. For any two points separated by a distance $4\lambda_f/3$ along the $x$ axis, where the temperatures are $T_1$ (at $x = -2\lambda_f/3$) and $T_2$ (at $x = 2\lambda_f/3$), the one-dimensional, conduction heat flux $q_{k,x}$ (because $\bar{u}_f = 0$) is given by

$$q_{k,x} = \frac{1}{4}mn\langle u_f\rangle (e_1 - e_2) \qquad \text{molecular description of one-dimensional heat conduction}$$

$$= \frac{1}{4}mn\langle u_f\rangle \frac{de}{dT}\frac{dT}{dx}(-\Delta x)$$

$$= \frac{1}{4}mn\langle u_f\rangle \frac{de}{dT}\frac{dT}{dx}\left(-\frac{4}{3}\lambda_f\right), \qquad (3.13)$$

where $m$ is the mass of the molecules.

Then using the definition of $c_v$ from (1.6), i.e., $c_v \equiv \partial e / \partial T \mid_v$, and since for ideal gases $e = e(T)$, (3.13) becomes

$$q_{k,x} = -\frac{1}{3} \rho c_v \langle u_f \rangle \lambda_f \frac{dT}{dx}, \quad \rho = mn \qquad \text{molecular description of one-dimensional heat conduction.} \qquad (3.14)$$

Now comparing this with the definition of the conduction heat flux (3.1), and noting that here we are considering only the $x$ direction, $\mathbf{q}_k \cdot \mathbf{s}_x = q_{k,x}$, we have

$$q_{k,x} \equiv -k \frac{dT}{dx} \qquad \text{macroscopic description of one-dimensional heat conduction.} \qquad (3.15)$$

Now comparing (3.14) and (3.15) gives the relation for $k$ as

$$k = k^f = \frac{1}{3} m n c_v \langle u_f \rangle \lambda_f$$

$$= \frac{1}{3} \rho c_v \langle u_f \rangle \lambda_f \qquad \text{thermal conductivity due to gaseous molecular fluctuations.} \qquad (3.16)$$

Now using the definition of the most probable thermal speed $\langle u_f \rangle$ from (1.16) and the mean-free path $\lambda_f$ for an ideal gas from (1.19), we have

$$k^f = \frac{1}{3} m n c_v \left( \frac{8 k_B T}{\pi m} \right)^{1/2} \frac{1}{2^{1/2} \pi} \frac{k_B T}{d_m^2 p}. \qquad (3.17)$$

For an ideal gas, we also have

$$p v = p / \rho = \frac{R_g}{M} T \quad \text{or} \quad p = \frac{\rho R_g T}{M} \quad \text{or} \quad p = n k_B T \quad \text{monatomic ideal gas.} \qquad (3.18)$$

Then (3.17) becomes

$$k^f = \frac{2}{3 \pi^{3/2}} c_v \frac{(m k_B T)^{1/2}}{d_m^2} \qquad \text{thermal conductivity for monatomic, ideal gas.} \qquad (3.19)$$

This shows that $k^f$ is independent of pressure. The Boltzmann constant $k_B$, along with other universal constants, are listed in Table C.1(b).

Equation (3.19) shows that the thermal conductivity of an ideal gas is temperature dependent and increases with an increase in temperature [through the variation of $c_v$ with respect to temperature, as shown in Figures 3.1(a) and (b), in addition to the explicit square-root relation appearing in (3.19)].

Reiterating, the heat conductivity in gases is due to their random molecular (because most assumptions made may apply to monatomic gases) motion and the ability of the molecules to store and release heat (translational, rotational, vibrational, and electrical). The most probable thermal speed $\langle u_f \rangle$ used in (3.16) is related to the speed of sound $a_s$. For an ideal gas the speed of sound $a_s$ is found from its general definition listed below and by using (3.18), i.e.,

$$a_s \equiv \left(\frac{\partial p}{\partial \rho}\right)_s^{1/2} \quad \text{definition of speed of sound}$$

$$a_s = \left(\frac{c_p}{c_v}\frac{\partial p}{\partial \rho}\right)_T^{1/2} \quad \text{Maxwell relation between derivates}$$

$$a_s = \left(\frac{c_p}{c_v}\frac{R_g}{M}T\right)^{1/2} \quad \text{speed of sound for ideal gas}$$

$$a_s = \left(\frac{\pi}{8}\frac{c_p}{c_v}\right)^{1/2}\langle u_f \rangle \quad \text{or} \quad \langle u_f \rangle = \left(\frac{8}{\pi}\frac{c_v}{c_p}\right)^{1/2}a_s \quad \begin{array}{l}\text{relation between speed of} \\ \text{sound and thermal speed,}\end{array} \quad (3.20)$$

where again $\langle u_f^2 \rangle^{1/2}$ is given by (1.16). Then we can write (3.16) as

$$\boxed{k^f = \frac{1}{3}\rho c_v a_s \left(\frac{8}{\pi}\frac{c_v}{c_p}\right)^{1/2}\lambda_f = \frac{1}{3}\rho c_v \left(\frac{8}{\pi}\frac{R_g}{M}T\right)^{1/2}\lambda_f \quad \begin{array}{l}\text{thermal conductivity} \\ \text{for monatomic ideal gas.}\end{array}} \quad (3.21)$$

The ratio $c_p/c_v$ is about 1.46 for air, and therefore for most gases, the difference between $a_s$ and $\langle u_f \rangle$ is not very large.

The thermal conductivity of some gases is listed in Table C.12. This includes the temperature dependence.

EXAMPLE 3.2. FUN

As an example of predicting $k$ for gases, consider air at STP, i.e., $T = 15°C$ and $p = 1$ atm. The measured thermal conductivity $k$, density $\rho$, molecular weight, speed of sound $a_s$, and mean-free path $\lambda_f$ are listed in Table C.7. The specific heat capacity at constant volume is related to that at constant pressure through (3.6), i.e., $c_v = c_p - R_g/M$.

Determine the thermal conductivity using

$$k^f = k = \frac{5\pi}{32}\rho\left(c_v + \frac{9}{4}\frac{R_g}{M}\right)a_s\left(\frac{8}{\pi}\frac{c_v}{c_p}\right)^{1/2}\lambda_f \quad \begin{array}{l}\text{thermal conductivity for} \\ \text{polyatomic ideal gas.}\end{array}$$

This is the rigorous extension of (3.21) for polyatomic gases.

**SOLUTION**
From Table C.7, we have

$$\lambda_f = 6.63 \times 10^{-8} \text{ m} \quad \text{Table C.7}$$
$$\rho = 1.225 \text{ kg/m}^3 \quad \text{Table C.7}$$
$$a_s = 340.29 \text{ m/s} \quad \text{Table C.7 [which can also be determined using (3.20)].}$$

Using $c_p$ from Table C.22 ($c_p = 1{,}006$ J/kg-K), this gives

$$c_v = c_p - \frac{R_g}{M} = 1{,}006(\text{J/kg-K}) - 8{,}315(\text{J/kg-K})/28.964(\text{kg/kmole})$$

$$= 1{,}006 - 287.1 = 718.9 \text{ J/kg-K}.$$

Then $k$ is

$$k = \frac{5\pi}{32} \times 1.225(\text{kg/m}^3) \times [718.9 + (9 \times 8{,}315)/(4 \times 28.97)](\text{J/kg-K})$$

$$\times\, 340.29(\text{m/s}) \times \left(\frac{8}{\pi} \times 718.9/1{,}006\right)^{1/2} \times 6.108 \times 10^{-8}(\text{m})$$

$$= \frac{5\pi}{32} \times 1.225 \times 1{,}365 \times 340.29 \times 1.349$$

$$\times\, 6.108 \times 10^{-8} \text{ W/m-K} = 0.02302 \text{ W/m-K}.$$

The measured value, from Table C.22, is $k = 0.02540$ W/m-K.

**COMMENT**

The difference is 9.4%. Further discussion on transport properties from kinetic theory are given in [6, 12, 14, 26, 27, 30, 33, 37].

### 3.2.2 Solids: Electrons and Phonons and Their Mean-Free Paths $\lambda_e$ and $\lambda_p$

Here, we first use an electrical classification of solids (metals, nonmetals, and semiconductors) to point out the roles of the conduction electrons and the thermal lattice vibration. Then in Section 3.2.5 we use a thermal classification (thermal insulators or thermal conductors). Because of the special role that the conduction electrons play in conducting heat, the solid metals that do contain a large number of conduction electrons, and nonmetals that do not contain conduction electrons (i.e., electrical insulators) are examined separately. The semiconductors (and semi-metals) have an intermediate electrical behavior between these two classes of solids, but their thermal conduction is similar to nonmetals, and are discussed last.

### (A) Metals

Most room temperature solid elements are metals. There are also many metallic alloys and compounds. The atomic binding in crystalline metals reduces the energy of the valence electrons (as compared to electrons in free atoms). Therefore, in metals a large number of valence electrons are free to move (called the conduction or free electrons), usually one or two per atom.[†] Therefore, the electrons are categorized as core, valence, and free, depending on the strength of their bonds with the nucleus, with the core electrons being closely orbiting the atoms, valence electrons covering

---

[†] The conduction electrons follow the Fermi-Dirac statistical distribution

$$f_e^\circ(E_e) = \frac{1}{\exp\left(\dfrac{E_e - E_F}{k_B T}\right) + 1},$$

where $E_e$ is the electron energy and $E_F$ is the Fermi energy (the highest energy by conduction electrons).

The electron work function, $\Delta E_{e,w}$ is the energy needed to move the electron from the Fermi level to the vacuum (emission) level [Figure 2.11(d)].

a larger and complex geometrical surface (called electron Fermi surface), and free (conduction) electrons in periodic potentials of crystals (as shown in Figure 2.11). The description of electrons is based on periodic potential in which electrons move.

The high thermal conductivity of metals is closely related to their high electrical conductivity $[\sigma_e(1/\text{ohm-m}) \equiv 1/\rho_e$, where $\rho_e$ is the electrical resistivity]. The high thermal conductivity is due to the acceleration of the conduction electrons in the presence of a temperature gradient $\nabla T$ (for heat flow) or an electrical potential gradient $\nabla \varphi$ (for electric current flow). According to the Drude classical theory (which in principle treats electrons similar to free gas molecules), the acceleration is brought to an end when the electrons collide (or are scattered by) with atoms in the solid crystal, with other electrons, or with other heat carriers. The average distance traveled between collisions is the mean-free path of the electron $\lambda_e$. The electrical conductivity due to free electrons $\sigma_e(1/\text{ohm-m})$ is approximated as

$$\sigma_e = \frac{n_e e_c^2 \tau_e}{m_e} = n_e e_c \mu_e \quad \text{electrical conductivity due to free electrons,} \qquad (3.22)$$

where $n_e(\text{electron/m}^3)$ is the number of free electrons per unit volume, $e_c(\text{C})$ and $m_e(\text{kg})$ are the electron charge and rest mass, and $\mu_e$ is the electron mobility,[†] the Fermi speed, $u_F$. The ratio $\lambda_e/u_F$ is called the electron relaxation time $\tau_e$. From the discussion of Section 2.3.2(A), $\tau_e^{-1} = \tau_{e,p}^{-1} + \tau_{e,d}^{-1}$, where $\tau_{e,p}$ is the electron-phonon and $\tau_{e,d}$ is the electron-defect relaxation time. In terms of the current flux vector, we have $\boldsymbol{j}_e = -n_e e_c \mu_e \boldsymbol{e}_e$.

Assuming that the temperature-gradient accelerated electrons travel the same average distance $\lambda_e$ before transferring their excess thermal energy to the atoms, similar to (3.16), the free-electron (or electronic) thermal (or heat) conductivity $k^e$ is given by

$$k^e = \frac{1}{3} n_e c_{v,e} u_F \lambda_e = \frac{1}{3} n_e c_{v,e} u_F^2 \tau_e \quad \text{electronic thermal conductivity } k^e, \qquad (3.23)$$

where $c_{v,e}(\text{J/electron-K})$ is the specific heat capacity of each electron and $\tau_e$ is the electron relaxation time ($\tau_e = \lambda_e/u_F$). Note that $n_e c_{v,e}(\text{J/m}^3\text{-K or J/m}^3\text{-°C})$ is the volume-specific heat capacity and is equal to $\rho_e c_{v,e}$ when $c_{v,e}$ is given in J/kg-K instead of per electron. Also note that $c_{v,e}$ and $\lambda_e$ are temperature dependent and $\lambda_e$ may be much larger than the intermolecular spacing. The electron mean-free path (or relaxation time) is influenced by the electron-electron, electron-phonon, and electron-lattice defect scattering mechanisms. Different ones of these mechanisms

---

[†]  From (2.31) using $\boldsymbol{j}_e = -n_e e_c \boldsymbol{u}_e$, and $\boldsymbol{u}_e = -e_c \boldsymbol{F}_e \tau_e/m_e$ which is the drift velocity, we have

$$\boldsymbol{j}_e = \left(\frac{n_e e_c^2 \tau_e}{m_e}\right)\boldsymbol{e}_e \equiv \sigma_e \boldsymbol{e}_e$$

or

$$\sigma_e = \frac{n_e e_c^2 \tau_e}{m_e}.$$

Then using $\mu_e = e_c \tau_e/m_e$, (3.22) is obtained.

are dominant in different temperature (absolute) ranges. At room temperature for gold, we have $\lambda_e = 31$ nm, and $u_e = 1.46 \times 10^6$ m/s, and $\rho_e c_{v,e} = 2.1 \times 10^4$ J/m$^3$-K. This gives $k = 315$ W/m-K and $\lambda_e$ is limited by electron-phonon scattering.

Now comparing the two conductivities, from (3.22) and (3.23) we have

$$\frac{k^e}{\sigma_e} = \frac{1}{3} \frac{m_e u_F^2 c_{v,e}}{e_c^2}, \qquad \text{relation between electrical and thermal conductivity for free electrons.} \tag{3.24}$$

From the quantum-statistical mechanics, $c_{v,e}$ is proportional to $T$ [i.e., $c_{v,e} = \pi^2 k_B T/(2T_F)$, where $T_F$ is the Fermi temperature of conduction electron, and is about 50,000 K, thus making $c_{v,e}$ much smaller than $k_B$] and $\langle u_e \rangle$ is assumed independent of $T$. Then using the proper relations, we have for metals [6]

$$\frac{k^e}{\sigma_e} = k^e \rho_e = \frac{\pi^2}{3} \left( \frac{k_B}{e_c} \right)^2 T,$$

or, given in the form of Wiedemann-Franz law as

$$\frac{k^e}{\sigma_e T} = \frac{k^e \rho_e}{T} = \frac{\pi^2}{3} \left( \frac{k_B}{e_c} \right)^2 = 2.442 \times 10^{-8} \text{ W-ohm/K}^2 \quad \begin{array}{l} \text{relation between} \\ k^e \text{ and } \sigma_e T, \end{array} \tag{3.25}$$

where electron charge is $e_c = 1.602 \times 10^{-19}$ C(or A-s), $k_B = 1.381 \times 10^{-23}$ J/kg, and $T$(K) is the absolute temperature. Equation (3.25) shows that the ratio of $k^e$ and $\sigma_e T$ is a constant.

Table C.8 in Appendix C gives the electrical resistivity $\rho_e$ for various metals at different temperatures.

The relationship between the electrical and thermal conductivities given by (3.25) holds well for the pure metals. However, in general, the relationship between $\lambda_e$ and $T$ in (3.23) is rather complicated and varies greatly for low and high temperature regimes. At high temperatures, the lattice vibration (discussed next) effectively scatters the electrons, and therefore, reduces the thermal conductivity. Then, the classical (as compared to quantum) statistical mechanics based electron-gas theory of Drude discussed above does not completely explain the electronic properties. The quantum statistical theory of solids (and the conduction energy band or band gap) explains the observed electrical and thermal conductivity behaviors. Based on this theory, the solid crystal behaves as an electrical insulator when the allowed energy bands are either completely filled or are completely empty. Then no electron can move in an electric field. The solid behaves as metal when one or more energy bands are partly filled. For semi-metals and semiconductors, one or two energy bands are slightly filled or are slightly empty. This theory describes the difference in the electrical resistivity $\rho_e$ of the solids, which can be as large as $10^{30}$ ohm-m between a very poor solid conductor and a very good solid conductor [6].

### (B) Nonmetals (Dielectrics)

Room-temperature solid nonmetals include P, S, Se, Br, and I. The heat conduction in semi-metals and semiconductors is also dominated by nonelectronic heat carriers. We now consider the absence of conduction electrons (i.e., when the electron conduction-gap energy $\Delta E_{e,g}$ is large, $\Delta E_{e,g}$ is the energy required to make a valence electron into a free or conduction electron). Then, the presence of a temperature gradient (i.e., a spatial variation of temperature) in the solid causes a nonuniform, elastic thermal lattice vibration. This nonuniform vibrational energy is transferred along the solid (as an elastic wave).

The transmission is by phonons, which are the quanta of energy in each mode of vibration (in analogy with photons of electromagnetic waves) traveling in the solid phase. The solid lattice is characterized as crystalline (with periodic structure) or amorphous (nonperiodic). Thermally excited phonons are similar to thermally excited photons. The acoustic vibration of the lattice (around the equilibrium location) is described by a three-dimensional harmonic oscillator governed by the intermolecular forces (or potentials) [5]. The phonon internal energy is the sum of energy in all possible phonon numbers (which is the product of the phonon distribution function $f_p^\circ$ and the phonon density of states $D_p$), each having energy $\hbar_P \omega$ and all polarizations (one longitudinal and two transverse). The vibration frequencies (which are associated with the concerted harmonic motion of all atoms and are called the normal modes) are as large as orders of terahertz, $10^{12}$ Hz. The distribution of phonon number (called occupation number) is frequency and temperature dependent. As was referred to in Table 1.4, this distribution (or statistics) is called the Bose-Einstein statistics and is similar to the photon Planck distribution function which will be discussed in Chapter 4.

The relation between the thermal conductivity $k$ and $a_s$, $c_p$, and $\lambda$ remains that given by (3.23). Here we use the lattice specific heat capacity $c_v$, mean phonon velocity (also called lattice heat-carrier group velocity) $u_p$, and the phonon mean-free path (also called heat-carrier, mean-free path) $\lambda_p$. Then the lattice (or phonon) thermal conductivity is [6] found from the Boltzmann transport equation and is given by the Callaway (this is based on the solutions to the Boltzmann transport equation discussed in Section 1.7.4, and the flux of phonons is used to define the thermal conductivity) model as

$$k^p = \frac{1}{3} \int_0^{\omega_D} \tau_p u_p^2 \frac{df_p^\circ}{dT} \frac{h_P}{2\pi} D_p(\omega) d\omega, \quad f_p^\circ = \frac{1}{\exp(-\hbar_P \omega/k_B T) - 1}$$

Bose-Einstein distribution function

$$= (48\pi^2)^{1/3} \frac{k_B^3}{h_P^2} \frac{T^3}{T_D l_m} \int_0^{T_D/T} \tau_p \frac{x^4 e^x}{(e^x - 1)^2} dx, \quad x = \frac{\hbar_P \omega}{k_B T}, \quad \omega = 2\pi f$$

$$\equiv \frac{1}{3} \rho c_v u_p \lambda_p \quad \text{lattice or phonon thermal conductivity } k^p, \tag{3.26}$$

where $f$ is the frequency, $k_B$ is the Boltzmann constant, $h_P$ is the Planck constant, $\hbar_P = h_P/2\pi$, $T_D$ is the Debye temperature, $D_p$ is the phonon density of state, $l_m$ is the cube lattice parameter or the lattice constant (which is equal to $n^{-1/3}$, for an assumed cubic lattice, where $n$ is the number of atoms per unit volume), $\tau_p$ is the total phonon relaxation time (and is related to $\tau_{p,n}$ the normal, and $\tau_{p,r}$ the resistive phonon relaxation times).[†] Note that these are model relaxation times and are generally frequency dependent and different than the integral relation time used in relations such as (3.23). The lattice heat capacity is given by (3.7). An example of use of (3.26) is given as an end of chapter problem.

The low frequency phonons do not undergo any significant scattering and are called ballistic phonons. In general, $u_p$ is the speed of sound in the solid state $a_s$ (thus the name acoustic phonons is used). Generally, the plane longitudinal wave speed $u_{p,l}$ is used, but the average phonon speed is also used and is defined as $3u_p^{-3} = 2u_{p,t}^{-3} + u_{p,l}^{-3}$, where subscript $t$ denotes the two transverse and subscript $l$ denotes the single longitudinal wave speeds. In the transverse waves, the displacement is perpendicular to the direction of sound propagation and in longitudinal waves it is along this direction. In the Debye approximation the velocity of sound is assumed to be the same for all three waves. As an example, for silver at $T = 300$ K, $u_{p,t} = 3,740$ m/s and $u_{p,l} = 7,300$ m/s, while in the Debye approximation $u_p = 6,533$ m/s. For silica, $u_{p,l} = 9,040$ m/s and $u_{p,t} = 5,340$ m/s. The average phonon speed is related to $T_D$ through $u_p = T_D 2\pi k_B/h_P(6\pi^2 n)^{1/3} = T_D k_B l_m(4\pi/3)^{1/3}/h_P$. Note that $n$ is the number of atoms per unit volume and for other than elements the number of atoms per molecule should be included. This gives, for example, $u_p = 2,160$ m/s in lead, 4,100 m/s in silicon dioxide, 5,940 m/s in 1% carbon steel, and 12,890 m/s in beryllium. The specific heat capacity of a phonon is the same as the lattice heat

---

[†] Equation (3.26) may be extended to include various scattering mechanisms which do or do not conserve momentum, then

$$k^P = (48\pi^2)^{1/3}\frac{k_B^3}{h_P^2}\frac{T^3}{T_D l_m}\left[g_1(\omega, T, \tau_p) + \frac{g_2^2(\omega, T, \tau_p, \tau_{p,n})}{g_3(\omega, T, \tau_p, \tau_{p,r})}\right].$$

The integrals $g_1, g_2,$ and $g_3$, and relaxation times $\tau_{p,n}$ (momentum conserving), $\tau_{p,r}$ (not momentum conserving) and $\tau_p$ are defined as

$$g_1 = \int_0^{T_D/T} \tau_p \frac{x^4 e^x}{(e^x - 1)^2} dx$$

$$g_2 = \int_0^{T_D/T} \frac{\tau_p}{\tau_{p,n}} \frac{x^4 e^x}{(e^x - 1)^2} dx$$

$$g_3 = \int_0^{T_D/T} \frac{\tau_p}{\tau_{p,n}\tau_{p,r}} \frac{x^4 e^x}{(e^x - 1)^2} dx$$

$$\frac{1}{\tau_{p,n}} = a_n f T^4 = a_n \frac{k_B}{h_P} T^5 x$$

$$\frac{1}{\tau_{p,r}} = \sum_i \frac{1}{\tau_{p,r,i}} = \frac{1}{\tau_{p,r,u}} + \frac{1}{\tau_{p,r,b}} + \frac{1}{\tau_{p,r,v}} + \frac{1}{\tau_{p,r,p}} + \frac{1}{\tau_{p,r,e}}$$

$$\frac{1}{\tau_p} = \frac{1}{\tau_{p,n}} + \frac{1}{\tau_{p,r}},$$

where $a_n$ is a material constant and the resistive mechanisms included in the summation $\tau_{p,r,i}$ are the three-phonon umklapp processes $\tau_{p,r,u}$, boundary scattering $\tau_{p,r,b}$, lattice-vacancy scattering $\tau_{p,r,v}$, point-defect scattering $\tau_{p,r,p}$, and phonon-electron scattering $\tau_{p,r,e}$. These resistive relaxation times are, in general, frequency, temperature, size, and impurity dependent and are referred to as modal relaxation times (as compared to integral relation times)[6].

capacity and is temperature dependent, as given by (3.7) and shown in Figure 3.2(b). It is proportional to the third power of temperature at low temperatures and becomes a constant at high temperatures. The mean-free path of the phonons $\lambda_p$ is also temperature and defect dependent. Around the room temperature, the interphonon collisions are significant. At low temperatures the interphonon interactions become less significant and the electron-lattice-defect and boundary scattering (elastic and inelastic) become important. The high temperature interactions result in $\lambda_p$ becoming proportional to $1/T$.

The Debye model for thermal conductivity is based on acoustic phonons, however, in polyatomic solids the optical phonons also make contributions (with significant vibration bands in the infrared frequency range).

The predicted temperature dependencies of $k$ for both pure metallic (i.e., $k^e$) and nonmetallic (i.e., $k^p$) solids are rendered in Figures 3.7(a) and (b). For nonmetal, both crystalline and amorphous phases are shown. Note that the temperature dependence of the two phases are very different. We use the index $i$ to designate $e$ or $p$. Note that since $c_{v,i}$ decreases with a decrease in $T$, while $\lambda_i$ increases, then there is a maximum in both curves $k^e$ and $k^p$. The maxima are generally occurring at temperatures $T(k_{max})$ much lower than the room temperature. Note that, in general, metals do not have a thermal conductivity that is greatly different than nonmetals (as compared to their electrical conductivity). However, for many solids, the thermal conductivity around the room temperature is many orders of magnitude lower than its maximum value. Figure 3.7(c) gives the measured $k$ for some metals and nonmetals, as a function of temperature. The measured values for two pure metals and two pure nonmetals are shown. Note that the peaks for nonmetals occur at higher temperatures, compared to metals. Also, note that gallium is a very good conductor at low temperatures. Figure 3.7(d) shows the range of various electron and phonon interaction (scattering) relaxation times. The electron-electron interaction $\tau_{e-e}$, electron-phonon $\tau_{e-p}$, and phonon-phonon $\tau_{p-p}$, have increasingly larger relaxation times (from less than femtosecond to nanosecond). Typical phonon mean-free paths for some of the nonmetallic solids, at room temperature, are listed in Figure 3.7(e). Diamond has a crystalline structure that allows for a much larger phonon mean-free path, compared to amorphous silicon dioxide (silica).

### (C) Semiconductors

Metals (e.g., K, Na, Cu, and most of the elements in Table C.2) are characterized by having conduction electron concentrations $n_e$ larger than $10^{22}$ conduction electrons/cm$^3$. Those elements (and alloys) that have an electronic behavior between metals and nonmetals are called the metalloids or intermediates or semi-metal-semiconductors. Semi-metals (e.g., B, Sb, Te, As) have concentrations between $10^{17}$ and $10^{22}$ conduction electrons/cm$^3$. Semiconductors (e.g., Ge, Si) have concentrations less than $10^{17}$ conduction electrons/cm$^3$. This electron concentration is temperature dependent. Semiconductors have an electrical resistivity $\rho_e$ of $10^{-4}$ to $10^7$ ohm-m at room temperature, with a strong temperature dependence. An ideal dielectric

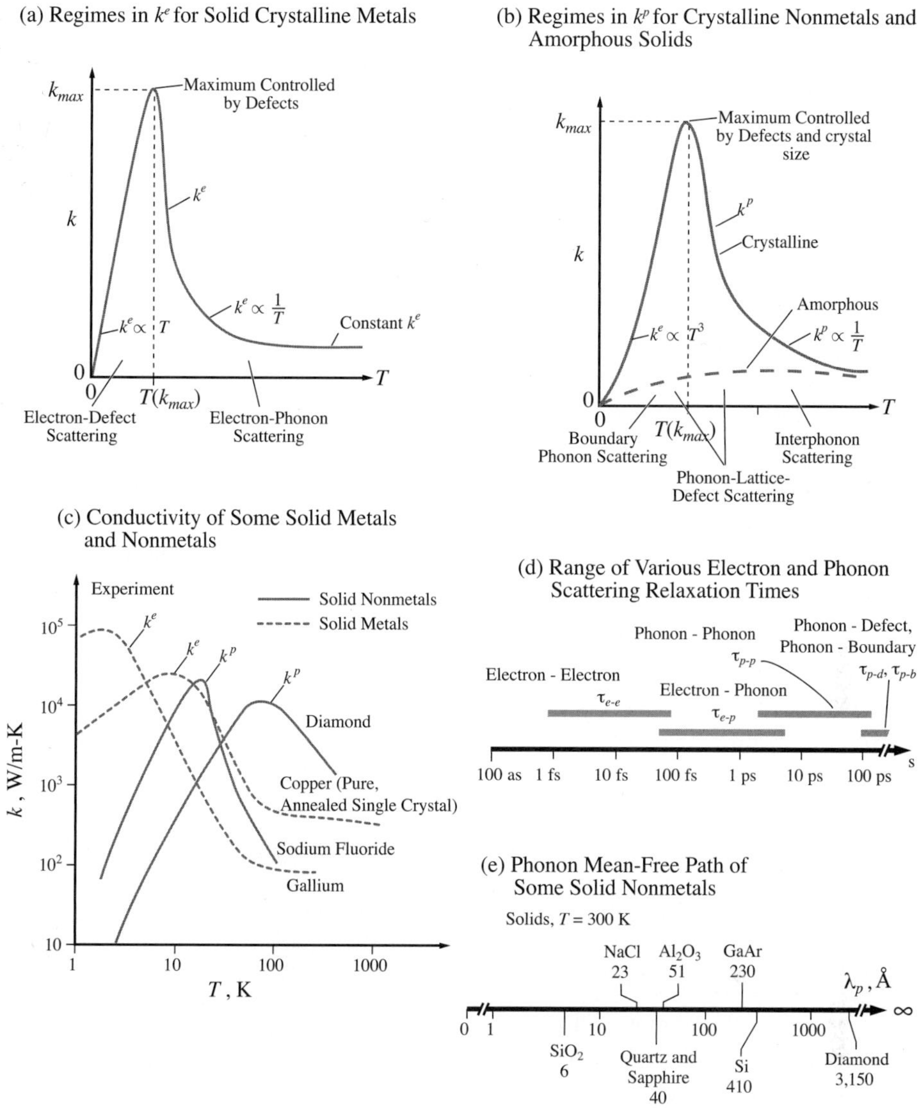

(a) Regimes in $k^e$ for Solid Crystalline Metals

(b) Regimes in $k^p$ for Crystalline Nonmetals and Amorphous Solids

(c) Conductivity of Some Solid Metals and Nonmetals

(d) Range of Various Electron and Phonon Scattering Relaxation Times

(e) Phonon Mean-Free Path of Some Solid Nonmetals

Figure 3.7. (a) and (b) Predicted variation of the thermal conductivity with temperature for solids: (a) crystalline metals; (b) crystalline and amorphous nonmetals. (c) Measured thermal conductivity of some solid crystalline metals and nonmetals. Note that $T(k_{max})$ is below the room temperature. (d) Range of various electron and phonon scattering relaxation times. (e) Phonon mean-free path of some dielectric solids, at room temperature.

material has an infinite electrical resistivity (i.e., zero electrical conductivity). In practice, materials having $\rho_e$ as low as $10^4$ ohm-m are considered dielectrics. The behavior of semiconductors is described by the conduction band energy theory.

Semiconductors have contributions to their thermal conductivity from the electrons $k^e$ (i.e., electronic thermal conductivity), as well as the lattice vibration (i.e., phonons) $k^p$ (lattice thermal conductivity). These contributions are considered additive, i.e., $k = k^e + k^p$. For Si at $T = 300$ K, from Table C.14, we have

$k = k^e + k^p = 149$ W/m-K. Using the Wiedemann-Franz relation (3.25) even with a high electrical conductivity given in Table C.9 for doped Si we find that electric conductivity $k^e = 0.18$ W/m-K. Therefore, for Si the phonon conductivity dominates. While for semiconductors, $k$ has a space lattice contribution, the electrical conductivity $\sigma_e$ only has an electronic contribution.

Materials that have no imperfections are called intrinsic materials (a semiconductor is intrinsic when its electronic properties are dominated by electrons thermally excited from valence to conduction band). Semiconductors such as silicon are made impure by adding other elements (e.g., P, As, and Sb); this is called doping and the resulting materials are called a extrinsic materials. When these impurities can provide free electrons (such as added arsenic or phosphorus), they are called donor elements. The resulting material is called the *n*-type because the electrical conduction is by electrons. When these impurities are deficient in electrons (such as added gallium), they are called acceptor elements (such as boron B) and the resulting material is called the *p*-type, because the electrical conduction is by holes.

These impurities and imperfections drastically affect the electrical properties of semiconductors. The addition of boron to silicon, in proportion of 1 boron atom to $10^5$ silicon atoms, decreases the electrical resistivity $\rho_e = 1/\sigma_e$ drastically, by a factor of $10^3$ at room temperature. The semiconductor thermal conductivity is generally dominated by phonon transport.[†] The presence of dopants can increase or decrease the thermal conductivity depending on the extent of the extra scattering caused by the impurities [6].

Also note that unlike metals, in semiconductors $\rho_e$ decreases with an increase in $T$. Semiconductors also interact with electromagnetic irradiation (a phenomenon called photoelectricity).

### (D) Comparison among Various Solid-State Microscale Heat Carriers

We now consider solid metals, nonmetals, and semiconductors, with the thermal conductivity expressed by (3.23) and (3.26). Some typical values for $l_m$ (lattice spacing, assuming cubic lattice), $T_D$ (Debye temperature), $k^i$, $(\rho c_v)_i$, $u_i$, and $\lambda_i$ ($i = e$ or $p$) are listed in Table 3.1 [8,12]. The simple model $k^i = (\rho c_v)_i u_i \lambda_i / 3$ is used with single (average) speed. This simplification (not allowing for variation of speed with frequency), is not generally justified. However, it gives a comparative measure. In Table 3.1, for gold the electronic heat capacity and the Fermi speed is used (because the electronic conduction dominates), while for other materials the bulk heat capacity is used.

Also note the high electron (Fermi) speed for gold (for most metals the Fermi speed is about $10^6$ m/s). The lattice constant $l_m$, i.e., the interatomic distance, is also listed for comparison with $\lambda_i$. In the case of amorphous silicon oxide, the phonon mean-free path is nearly equal to the lattice spacing, and therefore, the lattice vibration is rather localized.

---

[†] Thermal conductivity in semiconductor superlattices (a structure made of many alternating thin film layers resulting in lower thermal conductivity) is controlled by phonons and electrons.

Table 3.1. *Some typical solid-state microscale heat carrier properties at $T = 293$ K.*

| Material | $l_m$, Å | $T_D$, K | Carrier | $k^i$, W/m-K | $(\rho c_v)_i$, MJ/m³-K | $u_i$, m/s | $\lambda_i$, nm |
|---|---|---|---|---|---|---|---|
| aluminum oxide (crystalline) | 4.759 | 596 | phonon | 36 | 3.04 | 7,009 | 5.08 |
| diamond | 3.567 | 1,860 | phonon | 2,300 | 1.78 | 12,288 | 315 |
| gallium arsenide | 5.650 | 344 | phonon | 44 | 1.71 | 3,324 | 23 |
| germanium | 5.658 | 374 | phonon | 60 | 1.67 | 3,900 | 27.5 |
| gold | 4.078 | 165 | electron | 315 | 0.021 | 1,460,000 | 31 |
| manganese | 5.210 | 410 | electron | 7.8 | 0.021 | 1,960,000 | 0.57 |
| silicon | 5.430 | 645 | phonon | 148 | 1.66 | 6,533 | 41 |
| silicon dioxide (amorphous) | 4.910 | 290 | phonon | 1.4 | 1.79 | 4,100 | 0.6 |

(Here *i* stands for electron or phonon heat carriers. For phonons, a single (average) velocity is used.)

There is a very large range for the mean-free path for the different heat carriers and they differ greatly among the crystalline, polycrystalline, and amorphous (i.e., noncrystalline, such as amorphous silicon dioxide $SiO_2$ or aluminum oxide $Al_2O_3$) solids. The longest mean-free path listed in Table 3.1 is for phonon transport in diamond (a form of carbon) and the shortest mean-free path is again for phonon, but in silicon dioxide. Note that for silicon dioxide the phonon mean-free path $\lambda_p$ is nearly the same as the average lattice constant (average intermolecular spacing) $l_m$. This is due to the local distortions of the lattice structure. Note that the actual lattice is not cubic and is represented by two or more dimensions and for oblique lattices, also by an angle. The values listed are fitted with the Debye model. Sapphire is a crystalline, clear, hard bright-blue variety of corundum (a mineral based on alumina).

As an example of the frequency of lattice vibration, polycrystalline aluminum oxide has a Debye frequency (cutoff frequency) of $f_D = 1.24 \times 10^{13}$ Hz (far into ultrasound range of sound waves).[†] Considering the entire range of temperature for solids (from $T = 0$ K to the melting temperature), gallium is a better heat conductor than copper at or near $T = 1$ K, while diamond is a better conductor than copper and gold around room temperature, as was shown in Figure 3.7(c).

Figure 3.8 gives a list of high melting temperature solids. Listings of the thermal conductivity of some solids [35] are also given in Tables C.14, C.15, C.16 and C.17.

Figure 3.9 shows the variation of thermal conductivity of some typical solids with respect to temperature.

---

[†] Using the Lennard-Jones intermolecular potential parameters $\Delta E_{LJ}$ and $r_o$ (Section 1.5.1), a time scale can be defined as

$$\tau_{LJ} = r_o \left( \frac{m}{\Delta E_{LJ}} \right)^{1/2},$$

where $m$ is the atomic mass. From this, typical time scales are of the order of $10^{-12}$ s (picosecond), for $\Delta E_{LJ}$, $r_o$ and $m$ of the order of $10^{-21}$ J, $10^{-10}$ m, and $10^{-26}$ kg, respectively.

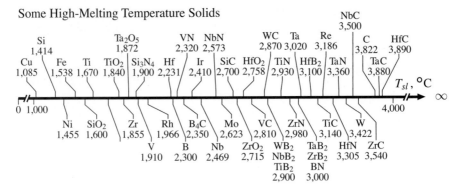

Figure 3.8. Examples of high melting temperature solids.

Some compounds may form many different crystal structures and are called polymorphs. For example, silica ($SiO_2$) may have densities as low as $\rho = 2{,}318 \ kg/m^3$ (for $\alpha$-cristobalite structure) to as high as $\rho = 4{,}283 \ kg/m^3$ (for stishovite structure). Crystalline solids may have an anisotropic (i.e., directional dependent) conductivity. The anisotropy may also be related to their grain structure. In small grains, the grain

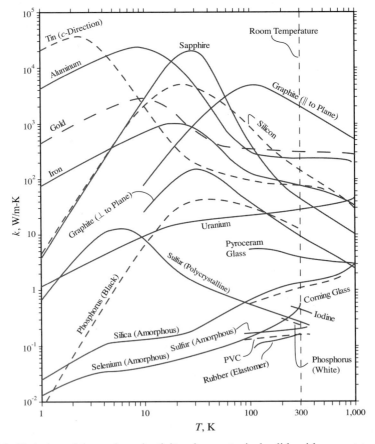

Figure 3.9. Variation of thermal conductivity of some typical solids with respect to temperature. Note that $T(k_{max})$ occurs below the room temperature. (Adapted from [34]).

boundary scattering becomes significant and the thermal conductivity decreases with size (e.g., film thickness in coatings or grain size in small-grain alloys). For example, the thermal conductivity also depends on the method of fabrication (e.g., extrusion, sintering, and vapor-phase deposition).

Glass is an amorphous (i.e., noncrystalline) solid having no repeating order in its structure. The main constituent of glass is silica, but since silica melts at over 1,700°C, impurities (i.e., fluxing agents) are added to reduce the melting temperature for the low temperature forming. Borosilicate and solid-lime glasses are among these (Table C.17).

---

### EXAMPLE 3.3. FAM

To evaluate the accuracy of the $k$-$\sigma_e$ relation (3.25), consider the following example. Using the measured electrical resistivity $\rho_e$(ohm-m) of metals given in Table C.8, determine the predicted thermal conductivity of (a) aluminum, (b) copper, and (c) tungsten at $T = 300$ K. Then compare these with the measured values listed in Tables C.2. and C.14, i.e.,

$$(i)\ k\ \text{(aluminum)} = 237\ \text{W/m-K} \quad \text{at } T = 300\ \text{K}$$
$$(ii)\ k\ \text{(copper)} \quad\ = 401\ \text{W/m-K} \quad \text{at } T = 300\ \text{K}$$
$$(iii)\ k\ \text{(tungsten)} = 173\ \text{W/m-K} \quad \text{at } T = 300\ \text{K}.$$

Comment on the accuracy of the predictions.

### SOLUTION
From Table C.8, we have the electrical resistivities for these pure metals as

$$(i)\ \rho_e\ \text{(aluminum)} = 2.733 \times 10^{-8}\ \text{ohm-m} \quad \text{at } T = 300\ \text{K}$$
$$(ii)\ \rho_e\ \text{(copper)} \quad\ = 1.725 \times 10^{-8}\ \text{ohm-m} \quad \text{at } T = 300\ \text{K}$$
$$(iii)\ \rho_e\ \text{(tungsten)} = 5.440 \times 10^{-8}\ \text{ohm-m} \quad \text{at } T = 300\ \text{K}.$$

Now using the predicted relationship between $k$ and $\sigma_e(= 1/\rho_e)$, i.e., (3.25), we have

$$\frac{k\rho_e}{T} = 2.442 \times 10^{-8}\ \text{W-ohm/K}^2, \quad \text{or}$$

$$k(\text{W/m-K}) = \frac{2.442 \times 10^{-8}(\text{W-ohm/K}^2)}{\rho_e(\text{ohm-m})} T(\text{K}).$$

With $T = 300$ K and using the values for $\rho_e$ listed above, we have

$$(i)\ k(\text{aluminum}) = \frac{2.442 \times 10^{-8}(\text{W-ohm/K}^2)}{2.733 \times 10^{-8}(\text{ohm-m})} \times 300(\text{K}) = 268.1\ \text{W/m-K}$$

$$(ii)\ k(\text{copper}) \quad\ = \frac{2.442 \times 10^{-8}}{1.725 \times 10^{-8}} \times 300 = 424.7\ \text{W/m-K}$$

$$(iii)\ k(\text{tungsten}) = \frac{2.442 \times 10^{-8}}{5.440 \times 10^{-8}} \times 300 = 134.7\ \text{W/m-K}.$$

**COMMENT**

These predicted results are higher than the experimental results for aluminum and copper, but lower for tungsten. The largest difference between the prediction $k_{pr}$ and experiment $k_{ex}$ is for tungsten and is (based on the measured $k_{ex}$)

$$\frac{k_{pr} - k_{ex}}{k_{ex}} \times 100\% = \frac{134.7 - 173}{173} = -22.1\%.$$

As was indicated, better predictions are made by including various scattering (i.e., interaction) mechanisms occurring among phonons, electrons, and lattice atoms.

### 3.2.3 Liquids

The speed of sound $a_s$ and the specific heat capacity $c_v$ in liquids are larger compared to gases, while the mean-free path $\lambda$ is smaller. At STP, the speed of sound in distilled water is about 1,497 m/s and in acetone it is 1,174 m/s. As expressed in (3.16), depending on the magnitude of these quantities $a_s$, $c_p$, and $\lambda_f$, the thermal conductivity $k$ of the liquids can be very different than that for the gases. The thermal conductivity of water is shown in Figure 3.10, as a function of temperature and for several pressures. Note that $k_l \gg k_g$ for near one atm pressure. At the critical state (i.e., where no distinction is made between the gas and liquid states) $k$ is finite and its magnitude is 0.238 W/m-K, which is intermediate between the values of $k_l$ and $k_g$ at lower temperatures. This is also shown in Figure 3.10. This behavior is unlike $c_p$ which tends to infinity at the critical state. The thermal conductivity of liquids first increases with increase in $T$, and then decreases. This is unlike gases and similar to solids.[†] Table C.23 lists the thermal conductivity of some liquids.

For liquid metals, the contribution due to conduction electrons needs to be included and can substantially increase the conductivity. Table C.24 lists thermal conductivity of some liquid metals. Tables C.23, C.26, and C.29 list the thermal conductivity of mercury and sodium in liquid and vapor states.

### 3.2.4 Composites: Local Thermal Equilibrium and Effective Conductivity

A composite material, for example nonevacuated fiberglass insulation, is a solid-gas composite (i.e., a porous solid) with the glass solid phase distributed as fibers and the air filling the remaining space. This is shown in Figure 3.11. A resin-filled fiberglass is a solid-solid composite with the space between the fibers filled with a solid thermoplastic polymer. If the temperature variation across each fiber is

---

[†]   The thermal conductivity of the liquids is predicted from the kinetic theory as

$$k = \frac{3R_g}{N_A^{1/3}} \frac{\rho^{2/3}}{M^{2/3}} \left(\frac{c_p}{c_v} \frac{1}{\rho \kappa}\right)^{1/2}.$$

This gives for water at $T = 300$ K, a $k = 0.6437$ W/m-K, for a $\kappa = 4.46 \times 10^{-10}$ 1/Pa, which is close to the measured value of $k = 0.613$ W/m-K (Table C.27).

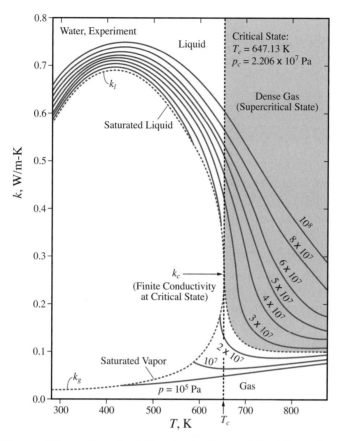

Figure 3.10. Measured thermal conductivity of water (gas and liquid) as a function of temperature for various pressures. (From Liley, P.E., et al., reproduced by permission © 1988 Hemisphere.)

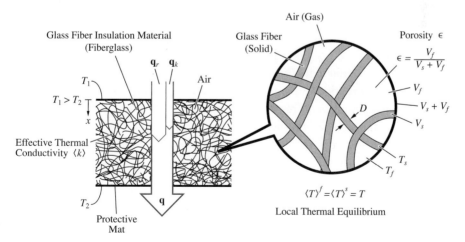

Figure 3.11. Effective thermal conductivity of randomly arranged fiberglass-air composite showing the local temperature for gas and solid.

negligibly small, compared to that across many fibers (for example, across a thick layer of insulation), then we can assume that the fiber and the adjacent filling air (or thermoplastic polymers) are in local thermal equilibrium, i.e., locally having the same temperature. This will allow us to define an effective conductivity $\langle k \rangle$ for this composite medium that can then be used to describe the heat transfer across the medium made of many fibers. This is called the assumption of local thermal equilibrium. Under this assumption, locally the gas-phase, volume-averaged temperature $\langle T \rangle^g$ and the solid-phase volume-averaged temperature $\langle T \rangle^s$ are equal and are simply represented by $T$. This is also shown in Figure 3.11. The qualifier local indicates that at the larger length scales we can have thermal nonequilibrium, i.e., the temperature variation and heat flow across a thick insulation layer. In porous composites, the heat is also transferred by radiation. We will discuss porous media radiation in Section 5.2.

Since the conductivity of air is less than that of glass, the more air space, the lower the effective conductivity of the fiberglass-air composite. Also, if the fibers are randomly arranged, the effective conductivity can be the same in all directions (i.e., isotropic). For fibers regularly arranged, the effective conductivity will be direction dependent (i.e., anisotropic). Wood is an example of an anisotropic composite. It is radially layered.

Liquid emulsions are examples of liquid-liquid composites with the immiscible liquid droplets being dispersed in another liquid (as in composite liquid fuels).

As an example of effective conductivity, consider packed particles with particles being in contact with each other (each particle having several contact points with the surrounding particles, the number of contact points is called the coordination number). For example, we may have packed spherical particles and nearly monosize particles with their interstitial space filled with a fluid. In solid-fluid composites, the porosity $\epsilon$ defined as

$$\epsilon = \frac{V_f}{V_s + V_f} \quad \text{porosity } \epsilon, \tag{3.27}$$

where $V$ stands for volume and subscripts $s$ and $f$ stand for solid and fluid.

The effective conductivity $\langle k \rangle$ can be correlated using various related parameters and properties. For randomly arranged solid-fluid phases, an empirical (i.e., obtained by curve fitting experimental results) correlation for $\langle k \rangle$ as a function of $\epsilon$ and the conductivity of the solid $k_s$ and the fluid $k_f$ is [11]

$$\mathbf{q}_k = -\langle k \rangle \nabla T, \quad \frac{\langle k \rangle}{k_f} = \left( \frac{k_s}{k_f} \right)^{0.280 - 0.757 \log(\epsilon) - 0.057 \log(k_s/k_f)} \quad 0.2 < \epsilon \le 0.6$$

correlation for random porous solids (continuous solid and fluid phases).     (3.28)

For gases, this relation is valid for pore linear dimensions $L$ larger that the mean-free path of the gas in otherwise free space $\lambda_m$, given by (1.19). Random packing of monosize spheres results in a porosity of about 0.40. Note that (3.28) is also valid when the interstitial space between the particles is filled with another solid (as in solid-solid composites).

Further correlations for porous media will be given in Section 3.3.2(B). The effective thermal conductivity of some composites are listed in the Appendix C, Tables C.15 and C.17. In Table C.15, some insulation materials are also listed.

---

EXAMPLE 3.4. FAM

---

The effective conductivity $\langle k \rangle$ of fluid-solid, particle packed-bed composite depends on the shape of the particles and the overall structure of the solid phase (e.g., nonconsolidated or consolidated particles, the extent of the particle-particle contact, etc.), in addition to porosity $\epsilon$ and the conductivity of fluid and solid ($k_f$ and $k_s$). A rendering of a packed bed of spherical particles is given in Figure Ex. 3.4. The particle diameter does not directly influence $\langle k \rangle$, but has an effect on the contact area per unit volume for consolidated particles. Assuming that the relation (3.28) is applicable, determine the effective conductivity of a packed bed of aluminum oxide $Al_2O_3$ (an oxide ceramic) particles filled with (a) air, and (b) water at $T = 300$ K. The properties of material are $k_s = 36$ W/m-K (from Table C.17 at $T = 293$ K), with porosity $\epsilon = 0.4$, for air $k_f = 0.0267$ W/m-K (Table C.22 at $T = 300$ K), and for water $k_f = 0.6065$ W/m-K (from Table C.23, at $T = 300$ K).

Packed Beds (Porous) of Monosize Spherical Particles

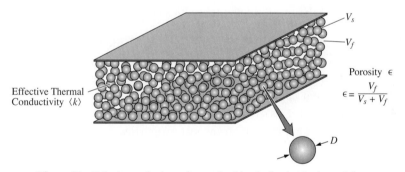

Figure Ex. 3.4. A rendering of a packed bed of spherical particles.

**SOLUTION**
From (3.28) we have

$$\langle k \rangle = k_f \left( \frac{k_s}{k_f} \right)^{0.280 - 0.757 \log \epsilon - 0.057 \log k_s/k_f}.$$

Here we have

$$\frac{k_s}{k_f} = \frac{36(\text{W/m-K})}{k_f(\text{W/m-K})}$$

and $\epsilon = 0.4$.

(a) For air

$$\frac{k_s}{k_f} = \frac{36}{0.0261} = 1,348$$

$$\langle k \rangle = 0.0267(\text{W/m-K}) \times (1,334)^{0.280-0.757\log 0.4-0.057\log 1,348}$$

$$= 0.0267(\text{W/m-K}) \times (1,334)^{0.280+0.3012-0.1784}$$

$$= 0.0267(\text{W/m-K}) \times (1,334)^{0.4028} = 0.4866 \text{ W/m-K}.$$

(b) For water

$$\frac{k_s}{k_f} = \frac{36}{0.606} = 59.41$$

$$\langle k \rangle = 0.606(\text{W/m-K}) \times (59.41)^{0.280-0.757\log 0.4-0.057\log 58.73}$$

$$= 0.606(\text{W/m-K}) \times (59.41)^{0.280+0.3012-0.1008}$$

$$= 0.606(\text{W/m-K}) \times (59.41)^{0.4801} = 4.317 \text{ W/m-K}.$$

**COMMENT**

Note that the nonlinear relation between $\langle k \rangle$ and $k_f$ results in a ratio of $\langle k \rangle$ for the two filling fluids, which is

$$\frac{\langle k \rangle \,|_{\text{with water}}}{\langle k \rangle \,|_{\text{with air}}} = \frac{4.317(\text{W/m-K})}{0.4860(\text{W/m-K})} = 8.871,$$

while the rate of the two fluid conductivities is

$$\frac{k_f \,|_{\text{water}}}{k_f \,|_{\text{air}}} = \frac{0.606}{0.0267} = 22.70.$$

Also note that if we had used a simple volume averaging (as was correctly applied to the heat capacity $\rho c_p$), i.e., $\langle k \rangle = \epsilon k_f + (1 - \epsilon)k_s$, an erroneous $\langle k \rangle$ would have resulted. This shows that for $k_s/k_f$ far from unity, this simple volume averaging does not apply to the transport properties.

### 3.2.5 Thermal Materials: From Ideal Insulators to Ideal Conductors

Materials are chosen for heat transfer applications to perform as good insulators, such as in fiberglass insulation used for thermal insulation of buildings, or as good conductors, such as aluminum nitride used as a high thermal conductivity (but very low electrical conductivity) substrate for silicon-based integrated circuits. In addition to the ideal insulator (having no matter, i.e., vacuum is the ideal insulator) and the superconductors (certain solids, at low temperatures), there are many intermediate-conductivity applications. On the low conductivity end of the commonly used materials, there are wood and bricks and other porous solid materials and on the high conductivity end there are stainless steel, carbon steel, brass, aluminum, and other dense solid materials. Some of these materials are described below.

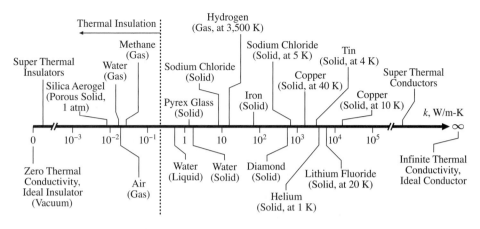

At $T = 300$ K and one atmosphere pressure, unless stated otherwise.

Figure 3.12. Typical values of thermal conductivity $k$ or $\langle k \rangle$, from ideal insulators to ideal conductors.

Figure 3.12 gives the range of thermal conductivity, from ideal insulators to ideal conductor, and in between. Appendix C contains several listings for $k$ (Tables C.12 to C.17) and in some cases their temperature dependence (Tables C.12 to C.14). The thermal conductivities of some substances are also shown in Tables C.22 to C.24, and C.26 to C.29.

Among the ceramics, the nonoxide ceramics (e.g., aluminum nitride) are good thermal conductors while they are also electrical insulators (unlike metals, which are good thermal and electrical conductors).

Since temperature, and therefore, conductivity $k$ varies within the heat transfer medium, an average value is generally used. One simple and relatively accurate method of averaging is to use the average medium temperature $\langle T \rangle_V$ (where $V$ is the volume of the medium) to evaluate the average $k$.

### (A) Insulators

Insulators are roughly classified as materials with $k \lesssim 0.2$ W/m-K. The ideal insulator is absolute vacuum and has $k = 0$. The materials with the lowest conductivity are gases (i.e., air) with the limit of zero conductivity at zero pressure (absolute). Maintaining a steady vacuum condition requires continuous pumping due to leakage, generally adds weight, and is costly. Although air is a poor conductor, in a nonuniform temperature field and under nonzero gravity, it experiences a thermobuoyant motion. Under thermobuoyant motion, the convective component of the heat flux vector becomes significant and the heat transfer rate increases. The presence of a solid matrix (e.g., the fibers in the fiberglass insulation) will suppress this motion.

When the temperature gradient is perpendicular to the gravity vector (i.e., in the horizontal direction), the fluid undergoes a thermobuoyant motion. This is due to the dependence of the density on the temperatures, which, under this condition, will result in a variation of density perpendicular to the gravity vector, i.e., a density variation in the horizontal direction. This thermobuoyant motion is the rise of the lighter fluid and fall of the heavier fluid, and will be discussed in Section 6.5.

When a vertical temperature gradient exists, as long as the heavier fluid is at the bottom and the lighter fluid on the top, there will not be a thermobuoyant flow. Also, when the heavier fluid is on the top, unless a threshold of the gradient of the density (or temperature) is exceeded, the fluid will not move. This is called the threshold of instability. Once this threshold is exceeded, there will be thermobuoyant motion (in the form of cellular fluid currents).

Fiber-based or foam insulations are designed to have a large void fraction (i.e., small solid fraction) to maximize this suppression of the thermobuoyant motion and minimize the effect of the higher conductivity of the solid. Therefore, porous solids (i.e., porous media) are used as nearly ideal insulators (i.e., $\langle k \rangle$ will be small, but not zero). The solid material can be a polymer (as in polyurethane foam) or ceramic (e.g., silicon dioxide $SiO_2$, or zirconium dioxide $ZrO_2$). A list of thermal insulators is given in Table C.15.

When the pores of the solid are smaller than the mean-free path of the gas molecules [i.e., large Knudsen number (1.20)], and when the solid fraction is small, an effective conductivity lower than the conductivity of the gas $k_f$ can be obtained. This is discussed in one of the end of chapter problems. As an example, the effective conductivity $\langle k \rangle$ of a carbon-doped silica (i.e., silicon dioxide $SiO_2$) aerogel [0.06 solid fraction with pore size of 15 to 20 nm and the effective density, from (3.12), $\langle \rho \rangle \equiv \epsilon \rho_f + (1 - \epsilon)\rho_s = 110 \text{ kg/m}^3$] is shown as a function of pressure (at $T = 300$ K) in Figure 3.13(a). The gas in the pores is air. The convective heat transfer is suppressed due to the small pores, but the radiation heat transfer becomes significant and contributes to the measured conductivity. The added carbon black (e.g., soot) acts as the scatterer of phonons and photons. Less than 10% (by weight) of carbon black is used.

Note in Figure 3.13(a) that the effective conductivity $\langle k \rangle$, at STP (standard temperature and pressure, 25°C and one atm pressure), is less than that for the air alone.

One disadvantage of high porosity materials is their limitations to weight bearing (i.e., structural instabilities). Thus depending on the structural load, an optimization of the structural and conductivity performance is required.

As another example, the measured effective conductivity $\langle k \rangle$ for packed fibers (3 to 6 $\mu$m in diameter and 1.6 mm in length) of zirconia (cubic zirconium oxide $ZrO_2$, melting temperatures about 2,600°C) is shown in Figure 3.13(b), as a function of temperature. This bulk fiber is chemically inert and can be used at temperatures higher than alumina and silica insulators. Many commercial powders have a chemical composition with mostly $ZrO_2$ and some $HfO_2$ and $Y_2O_3$. The color is white and a powder form or slabs made of the fiber are used in furnaces. The measured $\langle k \rangle$ for several powder effective densities $\langle \rho \rangle$, from 240 to 1,440 kg/m$^3$, are given in Figure 3.13(b) [39]. Again, at high temperatures, the effect of radiation becomes dominant, especially for low solid fractions (i.e., low effective density $\langle \rho \rangle$). Note the small value for $\langle k \rangle$, when small $\langle \rho \rangle$ is used. The bulk value of $k$ for $ZrO_2$ at $T = 100$°C is 1.68 W/m-K.

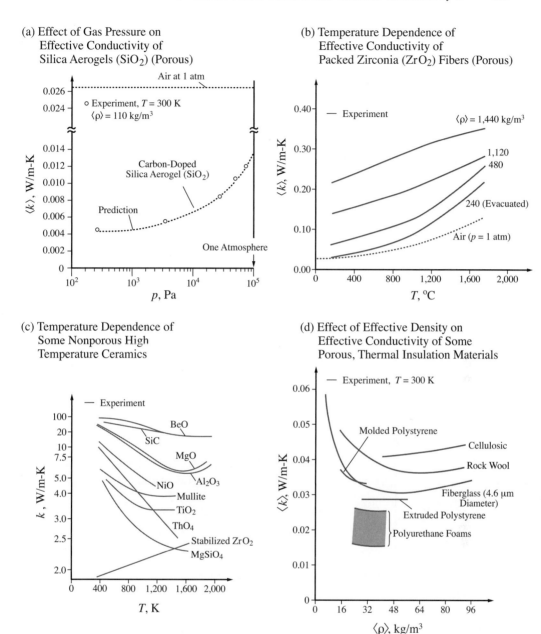

Figure 3.13. (a) Variation of the effective thermal conductivity $\langle k \rangle$ of carbon-doped silica aerogels, with respect to the interstitial air pressure $p$ (Pa). (From Zeng, S.Q., et al., reproduced by permission © 1995 ASME.) (b) Variation of the effective thermal conductivity $\langle k \rangle$ for zirconia powder, with respect to temperature and for various packing densities. (c) Variation of the thermal conductivity of some high temperature ceramics, with respect to temperature. (From Themelis, N.J., reproduced by permission © 1995 Gordon and Breach.) (d) Effective thermal conductivity for some insulation building material, given as a function of density. (From ASHRAE Handbook, reproduced by permission © 1997 ASHRAE.)

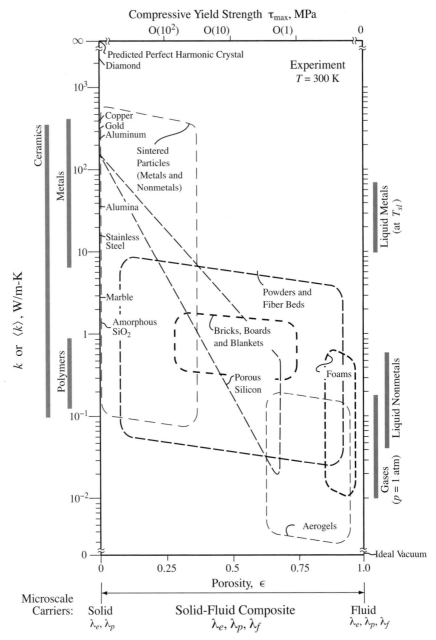

Figure 3.13. (e) Thermal conductivity of solids as a function of porosity, at $T = 300$ K. From very high conductivity (low porosity, or solid) to very low conductivity (high porosity). The compressive yield strength $\tau_{max}$ is also shown at the top axis.

The resin-filled fibers offer high structural stability (including machinability) and relatively low conductivity. The density of these composites is generally around that of water, i.e., slightly less than 1,000 kg/m$^3$.

The conductivity of some high-temperature ceramics is given in Figure 3.13(c), as a function of temperature. Note the rather low conductivity of stabilized ZrO$_2$

(especially when used in porous form). Mullite is a silicate of aluminum, $Al_6Si_2O_{13}$, has a low $k$, and is corrosion resistant at very high temperatures. Silicon carbide SiC has a high $k$ and is suitable for furnace applications. The conductivity depends on the method used for the material production. For example, chemical vapor deposition (CVD) produces a different crystalline structure than the sintered-powder fabrication. Therefore, the method of production should also be listed along with the value for $k$ (e.g., Table C.17 for SiC).

Finally the effective conductivity $\langle k \rangle$ of building insulation materials is given in Figure 3.13(d), as a function of the effective density $\langle \rho \rangle$. The polyurethane foams offer the lowest $\langle k \rangle$, but do not have high structural strength. The fiber glass insulation is also shown. The conductivity of insulation is also listed in Table C.15.

### (B) Conductors

A host of nonmetals $k^p$ and metals $k^e$ exhibit high conductivity. At room temperature, diamond, graphite, silver, and copper provide very high conductivity. Thin layers of diamond (a particular, crystalline structure of carbon) can be deposited on surfaces by a gas-phase reaction synthesis.

In superconductors, i.e., certain solids at or below their superconductivity temperature (called the transition temperature), the resistance to current (or heat flow) is decreased substantially resulting in significant electronic and lattice thermal conductivities. Below the transition temperature, the effective number of conduction electrons decreases, thus decreasing $k^e$. However, the lattice thermal conductivity $k^p$ increases. In general, disordered alloys (mixtures of metals and nonmetals) undergo a large increase in their $k$ at low temperatures.

Among the readily available metals that are not very reactive to most liquid and gas ambients, at near room temperature, is aluminum.

Figure 3.13(e) summarizes the conductivity of solids with respect to porosity, for $T = 300$ K. The very-high conductivity, nonporous ($\epsilon = 0$) solids (with phonon or electron carries), as well as the very-low conductivity, highly porous ($\epsilon \rightarrow 1$) evacuated solids are shown. The approximate compressive yield strength $\tau_{max}$ is also shown for some of these solids on the top axis. Note, the low weight- (or load-) bearing capability (low $\tau_{max}$) of the highly porous solids. In general, $\tau_{max}(t) = \tau_{max}(t = 0)\epsilon^a$, where $a$ is between 2.5 and 4.0. Porous silica is formed by the electrochemical etching of the doped crystalline silicon. In the measurement of the effective thermal conductivity of porous media, the interstitial radiation effect is also induced in $\langle k \rangle$. Therefore, in addition to the fluid ($\lambda_f$ or $\lambda_p$) and solid ($\lambda_e$ and/or $\lambda_p$) microscale conduction heat carriers, the microscale radiation heat carrier, photon ($\lambda_{ph}$), is also included. For $\epsilon = 1$, only the fluid microscale conduction heat carries ($\lambda_m$ and $\lambda_p$, and for charged fluids $\lambda_e$) are present.

For solids ($\epsilon = 0$) the predictions for an ideal harmonic crystal and for disordered solid $SiO_2$ are also shown. The amorphous solids (including ceramics), and the polymers and elastomers, have lower conductivities. Amorphous selenium has

$k = 0.245$ W/m-K at $T = 300$ K. Crystalline or polycrystalline (or quasi-crystalline) solids have higher conductivities.

In solid electrolytes, ions cause mobile ionic conductivity and this also contributes to the thermal conductivity. These mobile ions also scatter phonons, so the net effect depends on the ionic mobility and the phonon-ion scattering cross section.

<hr/>

### EXAMPLE 3.5. FAM

<hr/>

Insulation materials are chosen in part for their mechanical properties. For example, a pipe with free space on its outside is insulated using an insulation wrapping, where this insulation must sustain its weight and in many cases have an outside cover that is impermeable to keep out moisture and reactive gases. In contrast, insulation fillings are (fill-in powders and fill-in foams) added in double-wall pipes or in other confined spaces. These are shown in Figure Ex. 3.5.

Using Table C.15, find (a) a wrapping pipe insulation for use at $T = 300$ K, (b) a foam formed in place for use at $T = 300$ K, (c) an insulating cement for use at 450 K, (d) a blanket insulation for use at 300 K, and (e) a mineral powder insulation for use at 1,100 K.

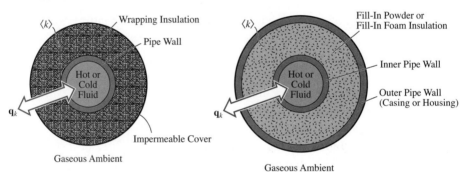

Figure Ex. 3.5. (a) Wrapping, and (b) fill-in type insulations used with a pipe.

### SOLUTION

(a) For a wrapping pipe insulation, from Table C.15, we choose

pipe insulation (slag or glass):

$$\langle \rho \rangle = 48 \text{ to } 64 \text{ kg/m}^3, \quad T = 297 \text{ K}, \quad \langle k \rangle = 0.033 \text{ W/m-K}.$$

The denser wrapping $\langle \rho \rangle = 160$ to 240 kg/m$^3$, adds strength without much increase in $\langle k \rangle$ (i.e., $\langle k \rangle = 0.048$ W/m-K).

(b) For a foam formed in place, from Table C.15, we choose

polyurethane foam (formed in place):

$$\langle \rho \rangle = 70 \text{ kg/m}^3, \quad T = 300 \text{ K}, \quad \langle k \rangle = 0.026 \text{ W/m-K}.$$

There are a variety of foams formed in place and depending on the strength needed, different $\langle \rho \rangle$ are selected.

(c) For an insulating cement that can be applied as a coating to the outside of a pipe, from Table C.15, we choose

cement (insulating):

$$\langle \rho \rangle = 380 \text{ to } 480 \text{ kg/m}^3, \quad T = 297 \text{ K}, \quad \langle k \rangle = 0.071 \text{ W/m-K}.$$

(d) For a blanket insulation, from Table C.15, we choose

blanket and felt (aluminosilicate fibers):

$$\langle \rho \rangle = 96 \text{ to } 128 \text{ kg/m}^3, \quad T = 297 \text{ K}, \quad \langle k \rangle = 0.036 \text{ W/m-K}.$$

(e) For a mineral powder insulation, from Table C.15, we choose

diatomaceous-earth powder (coarse):

$$\langle \rho \rangle = 320 \text{ kg/m}^3, \quad T = 1{,}144 \text{ K}, \quad \langle k \rangle = 0.062 \text{ W/m-K}.$$

Note that choosing a higher density will generally increase $\langle k \rangle$ noticeably.

**COMMENT**
Among the insulations listed above, the foam, formed in place, has the lowest $\langle k \rangle$. In practice, the temperature of the operation and the mechanical constraints determine the insulation selected.

## 3.3 Steady-State Conduction

When it is justifiable, we assume that the sensible heat storage/release term in (3.1) vanishes, i.e., the time variation of the temperature is negligible compared to the spatial variation and the volumetric energy conversion rate. We can then focus on solving the steady-state conduction equation. This simplified form of (3.1) [along with the repetition of (3.2)] is

$$\nabla \cdot \mathbf{q}_k \equiv \lim_{V \to \Delta V \to 0} \frac{\int_{\Delta A} (\mathbf{q}_k \cdot \mathbf{s}_n) dA}{\Delta V} = \frac{Q_k \mid_{\Delta A}}{\Delta V \to 0}$$

$$= -\nabla \cdot k \nabla T = \dot{s} \text{ in } V \qquad \begin{array}{l}\text{energy equation for} \\ \text{steady-state conduction}\end{array}$$

$$T = T_s \text{ or } A_k[-k(\nabla T \cdot \mathbf{s}_n) + q_s] = \dot{S} \qquad \begin{array}{l}\text{thermal condition specified} \\ \text{on bounding surface } A_k,\end{array} \qquad (3.29)$$

where $A_k$ is the conduction heat transfer area.

Chart 3.4 gives the various aspects of the steady-state heat conduction analysis. These are medium, geometry, thermal conduction resistance and circuit analysis, composite media, energy conversion, and the number of space dimensions considered.

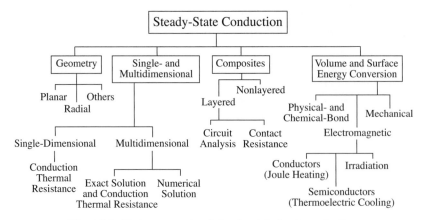

Chart 3.4. Various aspects of steady-state heat conduction.

In examining conduction heat transfer through heat transfer media, we consider media with planar or curved bounding surfaces (i.e., geometries). For the planar surfaces, we use the Cartesian coordinate system $(x, y, z)$, where $x$, $y$, and $z$ are the principal axes. For cylinders (curved with respect to one angle), we use the cylindrical coordinate system $(r, \phi, z)$, where $r$ and $z$ are the principal axes and $\phi$ is the principal angle. For spheres (same radius of curvature with respect to both angles), we use the spherical coordinate system $(r, \phi, \theta)$, where $r$ is the principal axis and $\phi$ and $\theta$ are the principal angles. These are shown in Figure 3.14. The surface area $A$ and volume $V$ for each geometry are also listed. These are also listed in Table C.1(e), along with other geometries.

If the bounding surface $A_k$ is planar, then we choose the Cartesian coordinate system, and upon using (B.17), (3.29) becomes

$$\nabla \cdot \mathbf{q}_k \equiv \lim_{V \to \Delta V \to 0} \frac{\int_{\Delta A} (\mathbf{q}_k \cdot \mathbf{s}_n) dA}{\Delta V} = \frac{\partial q_{k,x}}{\partial x} + \frac{\partial q_{k,y}}{\partial y} + \frac{\partial q_{k,z}}{\partial z}$$

$$\equiv -\frac{\partial}{\partial x} k \frac{\partial T}{\partial x} - \frac{\partial}{\partial y} k \frac{\partial T}{\partial y} - \frac{\partial}{\partial z} k \frac{\partial T}{\partial z} = \dot{s}. \tag{3.30}$$

Under the conditions we have, the spatial variation of the $x$ component of the conduction heat flux vector $\mathbf{q}_k$, i.e., $\partial q_{k,x}/\partial x$, is much larger than the spatial variation in other directions, i.e., $\partial q_{k,y}/\partial y \ll \partial q_{k,x}/\partial x$, $\partial q_{k,z}/\partial z \ll \partial q_{k,x}/\partial x$. Then the one-dimensional version of (3.30) is

$$-\frac{d}{dx} k \frac{dT}{dx} = \dot{s}$$

$$\begin{array}{cc} \text{spatial-rate of} & \text{volumetric} \\ \text{change in } \mathbf{q}_k & \text{energy} \\ \text{in } x \text{ direction} & \text{conversion.} \end{array} \tag{3.31}$$

This is said to apply to steady-state, planar and heat conduction.

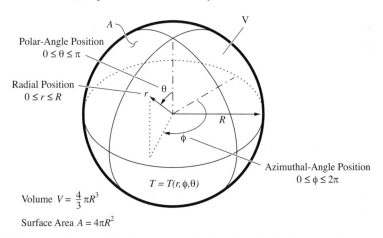

(a) Cartesian Coordinate System

(b) Cylindrical Coordinate System

(c) Spherical Coordinate System

Figure 3.14. The coordinates of the Cartesian, cylindrical, and spherical coordinate systems.

If the bounding surface $A_k$ is curved in one direction with a constant radius of curvature, then we can use a cylindrical coordinate system and (3.29) becomes

$$\nabla \cdot \mathbf{q}_k \equiv \lim_{V \to \Delta V \to 0} \frac{\int_{\Delta A} (\mathbf{q}_k \cdot \mathbf{s}_n) dA}{\Delta V}$$

$$\equiv -\left( \frac{1}{r} \frac{\partial}{\partial r} kr \frac{\partial T}{\partial r} + \frac{\partial}{r \partial \phi} k \frac{\partial T}{r \partial \phi} + \frac{\partial}{\partial z} k \frac{\partial T}{\partial z} \right) = \dot{s}, \qquad (3.32)$$

where $r$ is the radial position, $\phi$ is the angular position (or latitude angle), and $z$ is along the cylinder axis, as shown in Figure 3.14(b).

When the variation of the radial component of the heat flux vector is much larger than the variation of the other components, then (3.32) is written as

$$-\frac{1}{r}\frac{d}{dr}kr\frac{dT}{dr} = \dot{s}. \tag{3.33}$$

This is said to apply to a steady-state, radial-cylindrical, heat (axisymmetric) conduction.

When $A_k$ is the bounding surface of a sphere (i.e., curved with respect to two principal angles with equal principal radii of curvature), then we convert to the spherical coordinate system and (3.29) becomes

$$\nabla \cdot \mathbf{q}_k = \lim_{V \to \Delta V \to 0} \frac{\displaystyle\int_{\Delta A} (\mathbf{q}_k \cdot \mathbf{s}_n) dA}{\Delta V}$$

$$\equiv -\left( \frac{1}{r^2}\frac{\partial}{\partial r}kr^2\frac{\partial T}{\partial r} + \frac{\partial}{r\sin\theta\partial\phi}k\frac{\partial T}{r\sin\theta\partial\phi} + \frac{\partial}{r\sin\theta\partial\theta}k\frac{\sin\theta\partial T}{r\partial\theta} \right) = \dot{s}, \tag{3.34}$$

where $\phi$ is the latitude and $\theta$ is the azimuthal angle [Figure 3.14(c)].

When the radial variation of the radial component of the heat flux vector dominates over the other variations, we write (3.34) as

$$-\frac{1}{r^2}\frac{d}{dr}kr^2\frac{dT}{dr} = \dot{s}. \tag{3.35}$$

This is said to apply to steady-state, radial-spherical heat conduction. The difference between (3.33) and (3.35) is due to the cylindrical versus spherical surface area, where the first is proportional to $r$ and the second to $r^2$. We now define the thermal resistance $R_k(^\circ\mathrm{C/W})$ for the one-dimensional planar and radial layers.

### 3.3.1 One-Dimensional, Intramedium Conduction: Electrical Circuit Analogy and Thermal Resistance $R_k(^\circ\mathrm{C/W})$

#### (A) One-Dimensional Planar Geometry

For the geometry with a dominant flow of heat in only one direction, e.g., $x$ direction, and with a uniform conduction cross-sectional area $A_k$, we can write (3.31) as

$$\frac{d}{dx}q_{k,x} = \dot{s} \quad \text{differential-volume conduction energy equation.} \tag{3.36}$$

For the case of no energy conversion, this becomes

$$\frac{d}{dx}q_{k,x} = 0. \tag{3.37}$$

This states that $q_{k,x}$ is constant (or uniform along $x$). Here we have the simplest form of heat flow, i.e., unidirectional and uniform. This is shown in Figure 3.15(a). Now assuming that $k$ does not vary with $x$ (i.e., uniform $k$), and by substituting for

(a) Physical Model

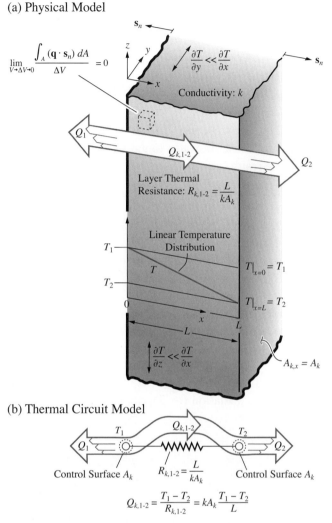

(b) Thermal Circuit Model

$$Q_{k,1\text{-}2} = \frac{T_1 - T_2}{R_{k,1\text{-}2}} = kA_k \frac{T_1 - T_2}{L}$$

Figure 3.15. (a) A schematic of one-dimensional steady-state conduction through a thin slab. (b) Thermal circuit model for this conduction heat transfer.

$q_{k,x}$, we have

$$-k\frac{d^2 T}{dx^2} = 0. \tag{3.38}$$

Equation (3.38) is integrated to give

$$\frac{dT}{dx} = a_1, \tag{3.39}$$

and when integrated again we have

$$T(x) = a_1 x + a_2, \tag{3.40}$$

where $a_1$ and $a_2$ are the integration constants to be determined from the boundary conditions. We now consider the prescribed temperature on the bounding surface, as stated by (3.29). For the case of two prescribed temperatures at locations $x = 0$ and $x = L$, we write the boundary conditions as

$$T(x = 0) = T_1, \quad T(x = L) = T_2. \tag{3.41}$$

Using the first of these in (3.40), we have the second constant of integration as

$$T(x = 0) = T_1 = a_2. \tag{3.42}$$

The relation for the first constant is also found from using the second equation in (3.41) in (3.40), and the result is

$$T(x = L) = T_2 = a_1 L + a_2. \tag{3.43}$$

Substituting (3.42) into (3.43) and solving for $a_1$, we have, for the first constant of the integration,

$$a_1 = \frac{T_2 - T_1}{L}. \tag{3.44}$$

Now substituting for $a_1$ and $a_2$ in (3.40), the temperature distribution is given in terms of the prescribed temperatures $T_1$, $T_2$ and the length $L$ as

$$T(x) = T_1 + \frac{T_2 - T_1}{L}x. \tag{3.45}$$

This describes a linear variation (or distribution) of $T$ along the $x$ axis, and is depicted in Figure 3.15. The heat flux at any location $x$ is determined by differentiating (3.45), i.e.,

$$Q_k = A_k q_{k,x}, \quad q_{k,x} = -k\frac{dT}{dx} = -k\frac{T_2 - T_1}{L}. \tag{3.46}$$

Note that $q_{k,x}$ is proportional to $T_2 - T_1$ (the temperature difference in the $x$ direction, i.e., $\Delta T$). This linear relationship between $q_k$ and $\Delta T$ is a characteristic of the steady-state conduction heat transfer.

Equation (3.46) shows that $q_{k,x}$ is uniform throughout the planar layer, which was expected, because here we have a uniform cross-sectional area and $\nabla \cdot \mathbf{q}_k = dq_{k,x}/dx = 0$. Note that if $T_2 > T_1$, then $q_{k,x}$ is negative, i.e., heat flows opposite to the $x$ direction. Because $q_{k,x}$ is uniform we can write for any locations $x$ and $x + \Delta x$

$$q_{k,x} = -k\frac{dT}{dx} = -k\frac{T(x + \Delta x) - T(x)}{\Delta x} \quad \text{linear temperature distribution.} \tag{3.47}$$

The heat flow rate $Q_{k,x}$(W) is the product of $q_{k,x}$ and the area $A_{k,x}$ or

$$Q_{k,1\text{-}2} = Q_{k,x} = A_{k,x}q_{k,x} = -A_k k\frac{T_2 - T_1}{L} = \frac{T_1 - T_2}{L/(A_k k)} = \frac{\Delta T}{L/(A_k k)} \quad \text{Fourier law,} \tag{3.48}$$

where $A_k$ is the area through which heat is conducted and since we have a one-dimensional heat flow, we can drop the subscript $x$.

Table 3.2. *Conduction resistance for one-dimensional, steady conduction in planar and radial systems.*

| Current | Potential | Resistance |
|---------|-----------|------------|
| $J_{e,1\text{-}2} = A j_{e,1\text{-}2}$ $= (\varphi_1 - \varphi_2)/R_{e,1\text{-}2}$, C/s, or A | $\varphi_1 - \varphi_2$, V | $R_{e,1\text{-}2} = L/(\sigma_e A)$, V/A, or ohm, slab |
| $Q_{k,1\text{-}2} = A_k q_k$ $= (T_1 - T_2)/R_{k,1\text{-}2}$, W | $T_1 - T_2$, °C | $R_{k,1\text{-}2} = L/(A_k k)$, °C/W, slab |
| | | $R_{k,1\text{-}2} = \ln(R_2/R_1)/(2\pi L k)$, cylindrical shell |
| | | $R_{k,1\text{-}2} = (1/R_1 - 1/R_2)/(4\pi k)$, spherical shell |
| | | $R_{k,1\text{-}2} = 1/(4\pi R_1 k)$, spherical shell, $R_2 \to \infty$ |

Now that we have written the Fourier law (3.48) in terms of a linear, one-dimensional temperature distribution, we compare this to the Ohm law (2.32) regarding the current flow $J_e$. The evident similarity is used to assist in the analysis of the conduction heat transfer using the concepts used in the electrical circuit analysis.

We now consider an electric current flow $J_e$ when a potential $\varphi_1 - \varphi_2$ is applied across a conductor of thickness $L$ with an electrical conductivity $\sigma_e$, and a cross section $A$. Then the electrical resistance is $R_{e,1\text{-}2} = (\rho_e L)/A$ or $R_{e,1\text{-}2} = L/(\sigma_e A)$. This is written similar to (2.32) as

$$J_{e,1\text{-}2} = j_{e,1\text{-}2}A = \frac{\Delta\varphi}{R_{e,1\text{-}2}} = -\frac{\varphi_2 - \varphi_1}{L/(\sigma_e A)} \quad \text{Ohm law.} \tag{3.49}$$

By comparing (3.48) and (3.49), and based on this analogy[†] and the relation between the electrical resistance and electrical conductivity, the thermal resistance is designated by $R_{k,1\text{-}2}$(°C/W or K/W) and is defined as

$$R_{k,1\text{-}2}(°C/W) \equiv \frac{T_1 - T_2}{Q_{k,1\text{-}2}} = \frac{L}{A_k k} \quad \text{conduction resistance for slab.} \tag{3.50}$$

This conduction resistance $R_{k,1\text{-}2}$ is the measure of the temperature difference needed (which can be written as °C or K because only a difference in temperature is needed) for the flow of one Watt of thermal energy through a layer of conductivity $k$, thickness $L$, and area $A_k$. This is also shown in Figure 3.15(b). The inverse of this resistance is the thermal conductance $(A_k k)/L$(W/°C). This analogy between the current density and the heat flux is summarized in Table 3.2.

Note that the magnitude of $R_k$ is between zero (no resistance, ideal conductor) and infinity (infinite resistance, ideal insulator) and the sign of $R_k$ is always positive, i.e., $0 \leq R_k \leq \infty$. The case of variable conduction cross-sectional area $A_k$ is left as an end of chapter problem.

---

[†] The analogy between electric current and heat flow rate is not complete and in some cases does not exist. This is because the general form of the Maxwell equations of electromagnetism are vastly different than the energy equation and its constitutive relations. However, the concept and visualization of the thermal circuit models and diagrams facilitate the presentation of the heat flow paths and is used throughout the text.

In commercial applications, English units are used. The thermal resistance of a conductor with $k$ (Btu/hr-ft-°F) and length $L$(ft) and with an area $A_k = 1$ ft$^2$, is given as the $R_k$-value [$R_k$-value $= L/(A_k=1$ ft$^2)k$]. The higher the $R_k$-value, the better the conductor's insulation potential. Typical materials for building insulations are fiberglass, rockwood, cellulose, vermiculite, and foam. Table C.15 lists the thermal conductivity of some building and insulation materials. For examples in building applications, the magnitudes of $R_k$-value of 30, 19, and 13°F/(Btu/hr) are typically used for attics, floor spaces, and walls, respectively. Table C.1(a) lists the SI-English unit conversion constants. Figures 3.13(a), (b), and (c) and Table C.15 give $\langle k \rangle$ for some insulation materials.

The inverse of $R_k$ is $G_k = 1/R_k$ conductance $G$(W/K), and is measure of ease of heat flow, i.e., $Q_{k,1\text{-}2} = G_{k,1\text{-}2}(T_1 - T_2)$.

### (B) One-Dimensional Radial Geometry

(i) *Cylindrical Shell*

For a cylindrical shell, with a constant conductivity $k$, no volumetric energy conversion, and with a dominant radial temperature gradient, we write (3.33) as

$$-\frac{1}{r}\frac{d}{dr}r\frac{dT}{dr} = 0. \tag{3.51}$$

For the boundary conditions corresponding to the prescribed temperatures at $r = R_1$ and $r = R_2 > R_1$, we have

$$T(r = R_1) = T_1, \quad T(r = R_2) = T_2. \tag{3.52}$$

This is shown in Figure 3.16(a). Integrating (3.51) once we have

$$\frac{dT}{dr} = \frac{a_1}{r}. \tag{3.53}$$

Integrating this once again, we have

$$T(r) = a_1 \ln r + a_2. \tag{3.54}$$

The two constants of the integration are determined by applying the boundary conditions (3.52) to (3.54). Then we have

$$T_1 = a_1 \ln R_1 + a_2 \tag{3.55}$$

$$T_2 = a_1 \ln R_2 + a_2. \tag{3.56}$$

Solving these for $a_1$ and $a_2$, we have

$$a_1 = \frac{T_2 - T_1}{\ln(R_2/R_1)} \tag{3.57}$$

$$a_2 = T_1 - \frac{T_2 - T_1}{\ln(R_2/R_1)} \ln R_1. \tag{3.58}$$

(a) Physical Model

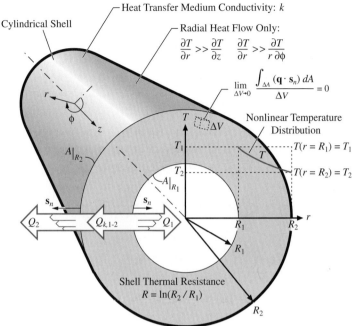

(b) Thermal Circuit Model

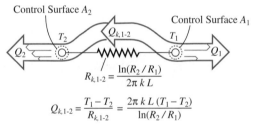

Figure 3.16. (a) A schematic of radial heat conduction through a long cylindrical shell. (b) Thermal circuit model for this conduction heat transfer.

Now substituting these in (3.54), the temperature distribution $T(r)$ is nonlinear and given by

$$T(r) = T_1 + \frac{T_2 - T_1}{\ln(R_2/R_1)} \ln \frac{r}{R_1}. \qquad (3.59)$$

The nonlinear ln relation should be compared with the linear relation in (3.45) for the slab. Now the heat flow rate at any location $r$ is determined by differentiating (3.59) and is

$$Q_{k,1\text{-}2} = (A_k q_k)_r = -2\pi r L k \frac{\partial T}{\partial r}\Big|_r = -\frac{2\pi L k (T_2 - T_1)}{\ln(R_2/R_1)}, \qquad (3.60)$$

where $L$ is the length of the cylinder.

Next using the analogy with the electrical circuit, we have the cylindrical-shell thermal resistance $R_k(°C/W)$ as

$$R_{k,1\text{-}2}(°C/W) \equiv \frac{T_1 - T_2}{Q_{k,1\text{-}2}} = \frac{\ln{(R_2/R_1)}}{2\pi L k} \quad \begin{array}{l} \text{conduction resistance for} \\ \text{cylindrical shell.} \end{array} \tag{3.61}$$

The thermal circuit model is shown in Figure 3.16(b).

(ii) *Spherical Shell*

For spherical shells, we follow the same steps as above and start from (3.35); the expressions for $T(r)$, $Q_{k,1\text{-}2}$, and $R_{k,1\text{-}2}(°C/W)$ are

$$T(r) = T_1 + \frac{T_2 - T_1}{1/R_1 - 1/R_2}\left(\frac{1}{R_1} - \frac{1}{r}\right) \tag{3.62}$$

$$Q_{k,1\text{-}2} = -\frac{4\pi k (T_2 - T_1)}{1/R_1 - 1/R_2}, \tag{3.63}$$

and

$$R_{k,1\text{-}2}(°C/W) = \frac{T_1 - T_2}{Q_{k,1\text{-}2}} = \frac{1/R_1 - 1/R_2}{4\pi k} \quad \begin{array}{l} \text{conduction resistance} \\ \text{for spherical shell.} \end{array} \tag{3.64}$$

(iii) *Spherical Shell with Very Large Outer Radius*

For $R_2 \to \infty$, (3.62) to (3.64) become

$$T(r) = T_1 + (T_2 - T_1)\left(1 - \frac{R_1}{r}\right) = T_2 - (T_2 - T_1)\frac{R_1}{r} \tag{3.65}$$

$$Q_{k,1\text{-}2} = -4\pi R_1 k(T_2 - T_1) \quad \text{or} \quad q_{k,1\text{-}2}(r = R) = \frac{Q_{k,1\text{-}2}}{4\pi R_1^2} = \frac{k(T_1 - T_2)}{R_1}, \tag{3.66}$$

and

$$R_{k,1\text{-}2}(°C/W) = \frac{T_1 - T_2}{Q_{k,1\text{-}2}} = \frac{1}{4\pi R_1 k} = \frac{1}{2\pi D_1 k} \quad \begin{array}{l} \text{conduction resistance for} \\ \text{a very thick spherical shell,} \end{array} \tag{3.67}$$

where $D_1$ is the diameter, $D_1 = 2R_1$.

This shows that under steady-state condition, a sphere of diameter $D_1$ maintained at temperature $T_1$, which is different than the far-field ambient temperature $T_2$, will transfer heat to this large ambient of conductivity $k$ at a finite and steady rate according to (3.66). The semi-infinite planar and one-dimensional, radial-cylindrical geometries (very long cylinder) do not allow for such a steady-state heat flow by conduction. This is because in (3.48) and in (3.60), as $L \to 0$ and $R_2/R_1 \to \infty$, $Q_{k,1\text{-}2}$ would tend to zero, i.e., no heat flows, while, in (3.63) for $R_2/R_1 \to \infty$, we obtain (3.66). Cubes and short cylinders (of any cross-sectional geometry) conducting heat to an infinitely large ambient also have steady-state behavior (similar to spheres), but the constant in (3.67) has a different value. This will be further discussed when considering multidimensional conduction in Section 3.3.8.

A summary of conduction resistances for radial shells is also listed in Table 3.2.

---

### EXAMPLE 3.6. FAM

A 1 m$^2$ surface area designated as $A_1$ is to be thermally insulated by placing an insulation of thickness $l$ and conductivity $k$ over it. The surface is at temperature $T_1$ and the insulation surface is at temperature $T_2$. This 1 m$^2$ surface area can be in the form of (a) a planar surface with area $A_1 = L_y L_z$, (b) a cylindrical surface with area $A_1 = 2\pi R_1 L_y$, or (c) a spherical surface with area $A_1 = 4\pi R_1^2$. These surfaces are shown in Figure Ex. 3.6, along with the geometry of their respective insulations. For the planar surface, we have $L_y = L_z = 1$ m, and for the cylindrical surface we have $L_y = 1$ m. The thickness of the insulation $l = 5$ cm and its conductivity $k = 0.1$ W/m-K. The temperatures are $T_1 = 90°$C and $T_2 = 40°$C. Determine the rate of heat loss from each surface and comment on any differences among them.

**SOLUTION**
The three surfaces have the same surface area $A_1 = 1$ m$^2$; however the insulation surface away from $A_1$, i.e., $A_2$, is not the same for all the three geometries. The heat flow rate at the surface $A_1$ with the insulation added is given by the appropriate resistance listed in Table 3.2. For the three cases, these are

$$\text{(a) } Q_{k,1\text{-}2}(\text{planar}) = \frac{T_1 - T_2}{R_{k,1\text{-}2}(\text{planar})} = \frac{T_1 - T_2}{\dfrac{l}{A_1 k}} = \frac{T_1 - T_2}{\dfrac{l}{L_y L_z k}} \qquad \text{Table 3.2}$$

$$\text{(b) } Q_{k,1\text{-}2}(\text{cylinder}) = \frac{T_1 - T_2}{R_{k,1\text{-}2}(\text{cylinder})} = \frac{T_1 - T_2}{\dfrac{\ln(R_2/R_1)}{2\pi L k}} = \frac{T_1 - T_2}{\dfrac{\ln\dfrac{R_1 + l}{R_1}}{2\pi L_y k}} \qquad \text{Table 3.2}$$

$$\text{(c) } Q_{k,1\text{-}2}(\text{sphere}) = \frac{T_1 - T_2}{R_{k,1\text{-}2}(\text{sphere})} = \frac{T_1 - T_2}{\dfrac{\dfrac{1}{R_1} - \dfrac{1}{R_2}}{4\pi k}} = \frac{T_1 - T_2}{\dfrac{\dfrac{1}{R_1} - \dfrac{1}{R_1 + l}}{4\pi k}} \qquad \text{Table 3.2.}$$

The inner radii, $R_1$, for the cylindrical and spherical geometries are determined from

$$A_1 = 1 \text{ m}^2 = 2\pi R_1 L_z = 2\pi R_1(\text{m}) \times 1(\text{m}) \quad \text{or } R_1(\text{cylinder}) = 0.1590 \text{ m}$$

$$A_1 = 1 \text{ m}^2 = 4\pi R_1^2 \quad \text{or } R_1(\text{sphere}) = 0.2822 \text{ m}.$$

(a) Planar Surface

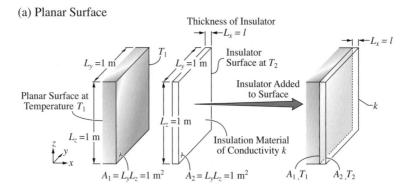

(b) Cylindrical Surface

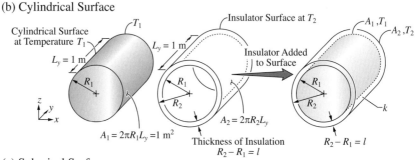

(c) Spherical Surface

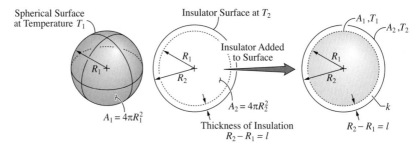

Figure Ex. 3.6. (a), (b), and (c) Insulation layer of thickness $l$ placed over planar, cylindrical, and spherical surfaces of surface area equal to 1 m$^2$.

Using the numerical values for $k$ and $l$, along with $L_y$, $L_z$, and $R_1$, we have

(a) $\quad Q_{k,1\text{-}2}(\text{planar}) = \dfrac{50(^\circ\text{C})}{\dfrac{0.05(\text{m})}{0.1(\text{W/m-}^\circ\text{C}) \times 1(\text{m}) \times 1(\text{m})}} = 100.0 \text{ W}$

(b) $\quad Q_{k,1\text{-}2}(\text{cylinder}) = \dfrac{50(^\circ\text{C})}{\dfrac{\ln\left[\dfrac{0.159(\text{m}) + 0.05(\text{m})}{0.159(\text{m})}\right]}{2\pi \times 0.1(\text{W/m-}^\circ\text{C}) \times 1(\text{m})}} = 114.8 \text{ W}$

(c) $\quad Q_{k,1\text{-}2}(\text{sphere}) = \dfrac{50(^\circ\text{C})}{\dfrac{\dfrac{1}{0.2822(\text{m})} - \dfrac{1}{0.2822(\text{m}) + 0.05(\text{m})}}{4\pi \times 0.1(\text{W/m-}^\circ\text{C})}} = 117.7 \text{ W}.$

**COMMENT**

For the same inner surface area, note that heat flow rate increases as the surface is curved. This is due to the increase in surface area occurring for the curved geometries and away from the inner surface. At the outer surface, the surface areas are

(a)  $A_2(\text{planar}) = L_y L_z = 1 \text{ m}^2$
(b)  $A_2(\text{cylinder}) = 2\pi(R_1 + l)L_z = 1.313 \text{ m}^2$
(c)  $A_2(\text{sphere}) = 4\pi(R_1 + l)^2 = 1.386 \text{ m}^2$.

Note the radial expansion at the outer surface. Radial contraction of surface area occurs away from the outer surface. Then opposite results are obtained (comparative reduction in $Q_{k,1\text{-}2}$ for radial insulations) when the outer surface area is kept the same. This is posed in Problem 3.2.

### 3.3.2 One-Dimensional Treatment of Composites

Composites are made of two or more distinct materials or phases. We have already discussed the fiberglass-air composite insulation in Section 3.2.4. There we used an empirical (i.e., experimentally determined) relation for effective conductivity $\langle k \rangle$ for complex-geometry composites. Here we analytically determine $\langle k \rangle$ and $R_{k,\Sigma}$ for some simple-geometry composites. This is done by the inclusion of the conductance (or resistance) for each element of the composite. This will result in an overall conductance for the composite that can be used, along with the temperature difference, to determine the heat conducted through the composite. Chart 3.5 gives a division of the composites into layered and nonlayered structures. The layers can be perpendicular to the heat flow, resulting in the series arrangement of the resistances. The layers can be parallel to the heat flow and this is called parallel arrangement of resistances. Figure 3.17(a) gives a rendering of these arrangements for planar layers. The insulation wrapped around a tube will be a radial layer and the heat flowing across the tube wall and this insulation will be flowing across the composite of the two radial materials with their resistances placed in series.

For nonlayered composites, one of the materials is continuous and the others can be continuous or discontinuous in one or more directions. Figure 3.17(b) shows

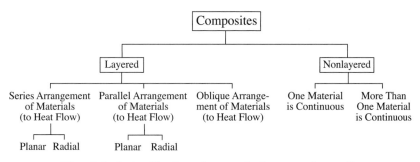

Chart 3.5. A classification of composite heat transfer media.

(a)  Layered Composite

   (i)  Layers Perpendicular                    (ii)  Layers Parallel
        to Heat Flux Vector                           to Heat Flux Vector

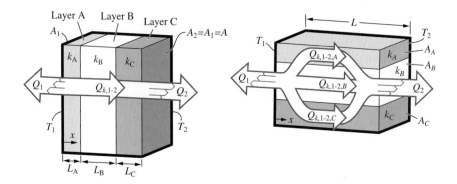

(b)  Nonlayered Composite

   (i)  One Material Continuous            (ii)  Both Materials Continuous
                                                 in Direction of Heat Flux Vector

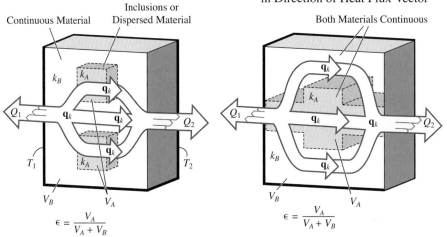

Figure 3.17.  (a) and (b) Rendering of layered and nonlayered composites.

examples of nonlayered composites. The first is for a continuous material $B$ and a discontinuous material $A$. This is called an inclusion (material $A$) composite. In the second example both materials are continuous in one direction. We begin by examining the heat transfer in layered composites and then move to the nonlayered composites.

*(A) Layered Composites: Series Arrangement of Resistances*

Consider a composite made of three materials arranged in layers as shown in Figure 3.18(a)(i). Here we consider the heat flowing perpendicular to the interfaces, i.e., when the layers are offering a series arrangement of resistances to the heat flow.

(a) (i) Physical Model and (ii) Temperature Distribution

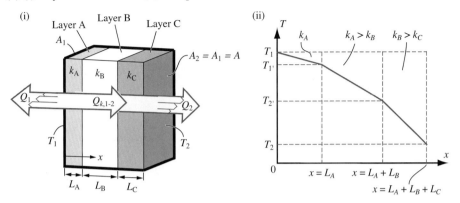

(b) Thermal Circuit Model

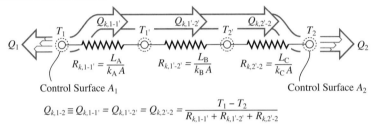

$$Q_{k,1\text{-}2} \equiv Q_{k,1\text{-}1'} = Q_{k,1'\text{-}2'} = Q_{k,2'\text{-}2} = \frac{T_1 - T_2}{R_{k,1\text{-}1'} + R_{k,1'\text{-}2'} + R_{k,2'\text{-}2}}$$

Figure 3.18. (a) Physical model and temperature distribution, and (b) circuit analogy of series layered-composite conduction.

For the assumed case of $k_A > k_B > k_C$, the temperature distribution for the case $T_1 > T_2$ is shown in Figure 3.18(a)(ii). We now consider the energy equation for the surface with no surface energy conversion. The heat flows out of the surface to the left [Figure 3.18(a)] as designated by $Q_1$ and to the right as designated by $Q_{k,1\text{-}1'}$. These are to be compared with $Q|_{A_{ij}}$ and $Q|_{A_{ji}}$ in (2.62). Then (2.62) is written for surface 1 as

$$Q_1 + Q_{k,1\text{-}1'} = 0 \quad \text{or} \quad Q_1 = -Q_{k,1\text{-}1'} \quad \text{energy equation for surface node } T_1. \quad (3.68)$$

This can be done for all the boundary surface. Then the energy equation for each of the surfaces shown in Figure 3.17(a) are

$$Q_1 + Q_{k,1\text{-}1'} = 0 \quad \text{for surface node 1}$$

$$-Q_{k,1\text{-}1'} + Q_{k,1'\text{-}2'} = 0 \quad \text{for surface node } 1'$$

$$-Q_{k,1'\text{-}2'} + Q_{k,2'\text{-}2} = 0 \quad \text{for surface node } 2'$$

$$-Q_{k,2'\text{-}2} + Q_2 = 0 \quad \text{for surface node 2.} \quad (3.69)$$

Then from these, we have

$$-Q_1 = Q_{k,1\text{-}1'} = Q_{k,1'\text{-}2'} = Q_{k,2'\text{-}2} \equiv Q_{k,1\text{-}2} = Q_2, \quad (3.70)$$

where we note that same heat flow rate $Q_{k,1\text{-}2}$ flows through each resistance.

For each resistance we have

$$Q_{k,1\text{-}2} = -\frac{Ak_A}{L_A}(T_{1'} - T_1) = \frac{T_1 - T_{1'}}{L_A/Ak_A} = \frac{T_1 - T_{1'}}{R_{k,1\text{-}1'}} \tag{3.71}$$

$$Q_{k,1\text{-}2} = -\frac{Ak_B}{L_B}(T_{2'} - T_{1'}) = \frac{T_{1'} - T_{2'}}{L_B/Ak_B} = \frac{T_{1'} - T_{2'}}{R_{k,1'\text{-}2'}} \tag{3.72}$$

$$Q_{k,1\text{-}2} = -\frac{Ak_C}{L_C}(T_2 - T_{2'}) = \frac{T_{2'} - T_2}{L_C/Ak_C} = \frac{T_{2'} - T_2}{R_{k,2'\text{-}2}}. \tag{3.73}$$

Figure 3.18 shows the physical and circuit models. We can solve for $T_{1'}$ from (3.71), and the result is

$$T_{1'} = T_1 - \frac{Q_{k,1\text{-}2}L_A}{Ak_A}. \tag{3.74}$$

The result for $T_{2'}$ from (3.72) and when using (3.70) is

$$T_{2'} = T_{1'} - \frac{Q_{k,1\text{-}2}L_B}{Ak_B} = T_1 - \frac{Q_{k,1\text{-}2}L_A}{Ak_A} - \frac{Q_{k,1\text{-}2}L_B}{Ak_B}$$

$$= T_1 - Q_{k,1\text{-}2}\left(\frac{L_A}{Ak_A} + \frac{L_B}{Ak_B}\right). \tag{3.75}$$

The result for $T_2$ from (3.73) and when using (3.70) is

$$T_2 = T_{2'} - \frac{Q_{k,1\text{-}2}L_3}{Ak_C} = T_1 - Q_{k,1\text{-}2}\left(\frac{L_A}{Ak_A} + \frac{L_B}{Ak_B} + \frac{L_C}{Ak_C}\right). \tag{3.76}$$

Now using this equation with the temperature difference across the three layers as $T_1 - T_2$ and the definition of thermal resistance given by (3.50), we have for the heat flow through the three layers $Q_{k,1\text{-}2}$

$$Q_{k,1\text{-}2} = \frac{T_1 - T_2}{L_A/k_AA + L_B/k_BA + L_C/k_CA} = \frac{T_1 - T_2}{R_{k,1\text{-}1'} + R_{k,1'\text{-}2'} + R_{k,2'\text{-}2}}. \tag{3.77}$$

This shows that for this layered arrangement of layers perpendicular to the heat flow, the thermal resistances are added as series resistances with the temperature difference across the composite as the potential. Figure 3.18 shows this series arrangement of resistances.

We now generalize the results to $n$ layers placed perpendicular to the heat flow $Q$ between surfaces 1 and 2. For these $n$ layers in the considered arrangement, we have

$$Q_{k,1\text{-}2} = \frac{T_1 - T_2}{\displaystyle\sum_{i=1}^{n} R_{k,i}} \equiv \frac{T_1 - T_2}{R_{k,\Sigma}}, \quad R_{k,\Sigma} \equiv \sum_{i=1}^{n} R_{k,i}, \quad G_{k,\Sigma} \equiv \frac{1}{\displaystyle\sum_{i=1}^{n} R_{k,i}} \quad \begin{array}{l}\text{series} \\ \text{resistances.}\end{array} \tag{3.78}$$

where $R_{k,\Sigma}$ is the overall conduction resistance for the composite. We could have also used $\langle R_k \rangle$ in place of $R_{k,\Sigma}$; however, in $R_{k,\Sigma}$ the summation symbol has a more

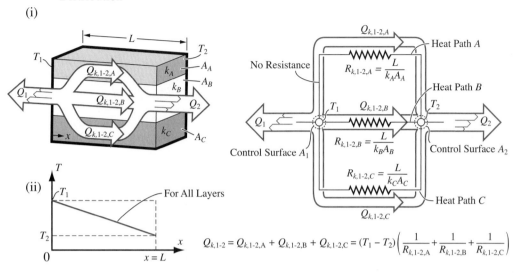

Figure 3.19. (a) Physical model and temperature distribution, and (b) circuit analogy of parallel, layered-composite conduction.

general use (this will also be used in the later chapters). The conductance $G_{k,\Sigma}$ is the inverse of resistance.

Although we have used planar layers as an example, $R_{k,i}$ can be any of those listed in Table 3.2.

Note that in the series arrangement, as one of the resistances becomes negligible, the other resistances dominate and provide the resistance to the heat flow.

### (B) Layered Composites: Parallel Arrangement of Resistances

Now consider the parallel, layered arrangement of Figure 3.17(a)(ii). This is also shown in Figure 3.19(a)(i). Again, we use (2.62) with no surface energy conversion. The temperature distribution in each layer is shown in Figure 3.19(a)(ii). Here to the right of surface 1 we have three surface heat flow rates. For surface nodes 1 and 2, we have the bounding-surface energy equation (2.62) as

$$Q_1 + Q_{k,1\text{-}2,A} + Q_{k,1\text{-}2,B} + Q_{k,1\text{-}2,C} = 0 \quad \text{energy equation for surface 1}$$

$$Q_2 - Q_{k,1\text{-}2,A} - Q_{k,1\text{-}2,B} - Q_{k,1\text{-}2,C} = 0 \quad \text{energy equation for surface 2.} \quad (3.79)$$

Note that when the heat is flowing toward the surface (i.e., opposite to surface normal), we have used a minus sign.

Then from (3.79) we have

$$-Q_1 = Q_2 = Q_{k,1\text{-}2,A} + Q_{k,1\text{-}2,B} + Q_{k,1\text{-}2,C} \equiv Q_{k,1\text{-}2}. \quad (3.80)$$

Now using the conduction resistances given by (3.50) we have

$$Q_{k,1\text{-}2} = Q_{k,1\text{-}2,A} + Q_{k,1\text{-}2,B} + Q_{k,1\text{-}2,C}$$

$$= \frac{T_1 - T_2}{L/A_A k_A} + \frac{T_1 - T_2}{L/A_B k_B} + \frac{T_1 - T_2}{L/A_3 k_C}$$

$$= (T_1 - T_2)\left(\frac{1}{L/k_A A_A} + \frac{1}{L/k_B A_B} + \frac{1}{L/k_C A_C}\right). \qquad (3.81)$$

Figure 3.19(b) shows this parallel arrangement of resistances.

We now generalize the results to $n$ layers placed parallel to the heat flux $Q$. For these $n$ layers, in parallel arrangement, we have

$$\boxed{Q_{k,1\text{-}2} = (T_1 - T_2)\sum_{i=1}^{n}\frac{1}{R_{k,i}} \equiv \frac{T_1 - T_2}{R_{k,\Sigma}}, \qquad \frac{1}{R_{k,\Sigma}} \equiv G_{k,\Sigma} \equiv \sum_{i=1}^{n}\frac{1}{R_{k,i}} \quad \begin{array}{l}\text{parallel}\\\text{resistances,}\end{array}}$$

$$(3.82)$$

where $R_{k,\Sigma}$ is the overall conduction resistance for the composite. Note that overall conductance $G_{k,\Sigma}$ is clearly different than (3.78).

Again, although we used a planar layer example, $R_{k,i}$ can be any of those listed in Table 3.2.

Note that in the parallel arrangement, as one of the resistances becomes negligible, this negligible resistance dominates and shorts the circuit, i.e., there is no resistance to the heat flow (this is unlike the series arrangement).

Also note that in the series arrangement we used nodal temperatures $T_{1'}$ and $T_{2'}$ in addition to $T_1$ and $T_2$, If needed, these temperatures can be determined using (3.74) and (3.75).

We will consider an example of two conduction resistances in parallel arrangement during the discussion on the thermoelectric coolers in Section 3.3.7.

### (C) Nonlayered Composites: Ordered Inclusions, Modeled with One-Dimensional Heat Flow

There are many geometric variations where a material $A$ is in a nonlayered arrangement with another material $B$. Here we consider two-material structures with ordered periodic arrangement (also called ordered lattice) with one of the materials being discontinuous. We assume that there are many of these unit cells arranged in a periodic arrangement. We will not address the temperature drop across each unit cell. However, we assume that over a unit cell, local thermal equilibrium exists. This is similar to the treatment of the effective thermal conductivity given in Section 3.2.4.

As an example, consider the regular arrangement of spherical inclusions of diameter $D$ in a square array arrangement with the distance between centers of the adjacent spheres being $l$. This is shown in Figure 3.20(a). One of the materials is the dispersed (i.e., discontinuous) material, designated as material $A$, in a second material (i.e., parent or continuous material), designated as material $B$. Assuming

Nonlayered Composites: One Material Discontinuous

(a) Square Array of Spherical Inclusions

(i) Physical Model

(ii) Thermal Circuit Model
for Each Unit Cell

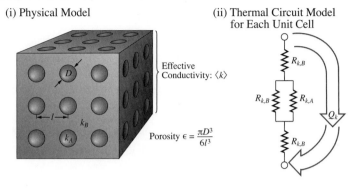

(b) Square Array of Cubic Inclusions

(i) Physical Model

(ii) Thermal Circuit Model
for Each Unit Cell

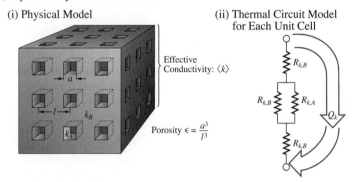

Figure 3.20. Nonlayered, ordered inclusion composites, where one material is discontinuous. (a) Square array of spherical inclusions. (b) Square array of cubic inclusions. The corresponding one-dimensional heat flow thermal circuit models are also shown.

a one-dimensional heat flow with parallel-series [as shown in Figure 3.20(a)(ii)] arrangement of the resistances in each unit cell, we can find the effective conductivity $\langle k \rangle$ in terms of $k_A$ (inclusions), $k_B$ (continuous), and porosity $\epsilon$. Note that this is an approximation, because the heat flow is three-dimensional. However, the results are good approximations. The results are [25]

$$\text{for } \frac{k_B}{k_A} < 1, \quad \frac{\langle k \rangle}{k_B} = \frac{1}{1 - (6\epsilon/\pi)^{1/3}\left[1 - \dfrac{\gamma}{2(\gamma+1)^{1/2}}\ln\dfrac{(\gamma+1)^{1/2}+1}{(\gamma+1)^{1/2}-1}\right]} \quad \begin{array}{l}\text{spherical}\\\text{inclusions}\end{array}$$

(3.83)

$$\text{for } \frac{k_B}{k_A} > 1, \quad \frac{\langle k \rangle}{k_B} = \frac{1}{1 - (6\epsilon/\pi)^{1/3}\left[1 + \dfrac{\gamma}{(-\gamma-1)^{1/2}}\tan^{-1}\dfrac{1}{(-\gamma-1)^{1/2}}\right]} \quad \begin{array}{l}\text{spherical}\\\text{inclusions,}\end{array}$$

(3.84)

where

$$\epsilon = \frac{\pi D^3}{6l^3}, \quad \gamma = \left(\frac{16}{9\pi\epsilon^2}\right)^{1/3}\frac{k_B/k_A}{1 - k_B/k_A} \quad \text{for } \epsilon < 0.3.$$

Note that for $\epsilon = 0$, we have $\langle k \rangle = k_B$, as expected.

The conduction heat transfer through such a composite of length and cross-sectional area $A_k$ is given by (3.48) using $\langle k_r \rangle$, i.e.,

$$Q_{k,i\text{-}j} = \frac{T_i - T - j}{R_{k,i\text{-}j}} = \frac{T_i - T_j}{\dfrac{L}{A_k \langle k \rangle}}. \tag{3.85}$$

Similarly, consider cubic inclusions (of side length $a$) in a square array arrangement with the distance between centers of the adjacent cubes being $l$. This is shown in Figure 3.20(b). Then for an imposed (i.e., idealized) parallel-series arrangement of resistances [as shown in Figure 3.20(b)(ii)] in each unit cell, we have $\langle k \rangle$ as a function of $k_A$, $k_B$, and $\epsilon$ as [25]

$$\frac{\langle k \rangle}{k_B} = \frac{1 + \epsilon^{2/3}(k_A/k_B - 1)}{1 + \epsilon^{2/3}(1 - \epsilon^{1/3})(k_A/k_B - 1)}, \quad \epsilon = \frac{a^3}{l^3} \quad \text{cubic inclusions.} \tag{3.86}$$

Note again that for $\epsilon = 0$, we have $\langle k \rangle = k_B$, as expected. Also for $k_A = k_B$, (3.86) gives $\langle k \rangle = k_B$. Also for $k_A \ll k_B$, we have $\langle k \rangle = k_B(1 - \epsilon^{2/3})/(1 - \epsilon^{2/3} + \epsilon)$.

Depending on the continuity of the materials and their distributions throughout the composite, two- or three-dimensional heat flow will occur. This multidimensional heat flow in complex geometries does not lead to simple analytical expressions for the effective conductivity $\langle k \rangle$. For complex structures, generally empiricism is used and the results are given in the form of correlations. The correlation (3.28) is an example where the packed particles are allowed to touch and become a continuous medium, and therefore, two continuous materials are present. Then, similar to the above, the volume fraction $\epsilon$ and the conductivity of each material, $k_A$ and $k_B$, are used in the correlation.

### EXAMPLE 3.7. FAM

To reduce the heat loss from the upper body, layered clothing is conventionally recommended. In the layered clothing, the air between the layers provides additional and substantial resistance to the heat flow. Consider a single, planar layer of fabric of thickness $L_1 = 4.5$ mm and conductivity of $k_f = 0.1$ W/m-K. This is shown in Figure Ex. 3.7(a). Then consider three fabric layers each having a thickness $L_2$ such that the sum of the three thicknesses is equal to $L_1$, i.e., $3L_2 = L_1$. For the three-layer clothing allow for an air gap of thickness $L_a = 1$ mm between the adjacent fabric layers, with $k_a = 0.025$ W/m-K, as shown in Figure Ex. 3.7(b). The temperatures at either side of the single and multilayer (i.e., fabric and air layers) composites are prescribed as $T_1$ and $T_2$, with $T_1 = 28°C$ and $T_2 = 15°C$.

(a) Draw the thermal circuit diagram.

(b) Determine the heat loss per unit area $q_k (\text{W/m}^2)$ for the single and multilayer composites.

(c) Write the expression for the composite resistance in terms of the porosity defined by (3.27).

(a) A Single, Thick Fabric Layer

(b) Three Fabric Layers and Two Air Layers

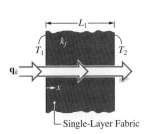

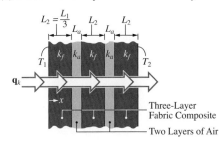

(c) Thermal Circuit Models for both Cases

Single-Layer Fabric

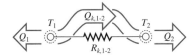

Multilayer Composite

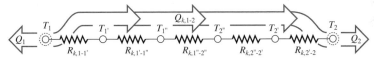

Figure Ex. 3.7. (a) Single-layered fabric clothing. (b) Multilayered fabric clothing. Also shown are the air gaps (layers) between layered fabrics. (c) Thermal circuit model for the two conditions.

## SOLUTION

(a) Figure Ex. 3.7(c) shows the thermal circuit model for the two conditions.

(b) The conduction heat flow per unit area through the single and multilayer composites is one dimensional and given by (3.78), i.e.,

$$Q_{k,1\text{-}2} = A_k q_k = \frac{T_1 - T_2}{R_{k,\Sigma}}.$$

For the single-layer fabric we have the only resistance as

$$R_{k,\Sigma} = R_{k,1\text{-}2} = \frac{L_1}{A_k k_f}.$$

For the multilayer composite we have a series arrangement of five resistances, i.e., (3.78), as shown in Figure Ex. 3.7(b), and these are

$$R_{k,\Sigma} = \sum_{i=1}^{5} R_{k,i} = R_{k,1\text{-}1'} + R_{k,1'\text{-}1''} + R_{k,1''\text{-}2''} + R_{k,2''\text{-}2'} + R_{k,2'\text{-}2}$$

$$= \frac{L_2}{A_k k_f} + \frac{L_a}{A_k k_a} + \frac{L_2}{A_k k_f} + \frac{L_a}{A_k k_a} + \frac{L_2}{A_k k_f}$$

$$= \frac{3L_2}{A_k k_f} + \frac{2L_a}{A_k k_a}.$$

So by substituting for these resistances, for the single-layer fabric we have

$$Q_{k,1\text{-}2} = A_k q_k = \frac{T_1 - T_2}{\dfrac{L_1}{A_k k_f}}$$

or

$$q_k = \frac{T_1 - T_2}{\dfrac{L_1}{k_f}} = \frac{(28 - 15)(^\circ\text{C})}{\dfrac{0.0045(\text{m})}{0.1(\text{W/m-K})}} = \frac{13(^\circ\text{C})}{0.045[^\circ\text{C}/(\text{W/m}^2)]}$$

$$= 288.9 \ \text{W/m}^2 \ \text{for single-layer fabric.}$$

For the multilayer composite, we have

$$q_k = \frac{T_1 - T_2}{\dfrac{3L_2}{k_f} + \dfrac{2L_a}{k_a}} = \frac{(28 - 15)(^\circ\text{C})}{\dfrac{3 \times 0.0015(\text{m})}{0.1(\text{W/m-}^\circ\text{C})} + \dfrac{2 \times 0.001(\text{m})}{0.025(\text{W/m-}^\circ\text{C})}}$$

$$= \frac{13(^\circ\text{C})}{0.045[^\circ\text{C}/(\text{W/m}^2)] + 0.080[^\circ\text{C}/(\text{W/m}^2)]}$$

$$= 104.0 \ \text{W/m}^2 \ \text{for multilayer composite.}$$

(c) The porosity is given by (3.27) and using $V_a$ and $V_f$ for air and fiber volumes, we have

$$\epsilon = \frac{V_a}{V_f + V_a} = \frac{A_k 2 L_a}{A_k 3 L_2 + A_k 2 L_a} = \frac{2L_a}{3L_2 + 2L_a}.$$

Then we write for $R_{k,\Sigma}$

$$R_{k,\Sigma} = \frac{1}{A_k} \left( \frac{3L_1}{k_f} + \frac{2L_a}{k_a} \right)$$

$$= \frac{3L_f + 2L_a}{A_k} \left( \frac{1 - \epsilon}{k_f} + \frac{\epsilon}{k_a} \right),$$

where we have used

$$\frac{3L_2}{3L_2 + 2L_a} = 1 - \frac{2L_a}{3L_f + 2L_a} = 1 - \epsilon.$$

**COMMENT**

Note that the two air layers have provided more resistance, i.e.,

$$A_k R_k(\text{air}) = 2L_a/k_a = 0.080^\circ\text{C}/(\text{W/m}^2).$$

This is larger than that of the single-layer fabric;

$$A_k R_k(\text{fabric}) = L_1/k_f = 0.045^\circ\text{C}/(\text{W/m}^2).$$

The heat loss per unit area $q_k$ is reduced by

$$\frac{q_k(\text{with air layers}) - q_k(\text{without air layers})}{q_k(\text{without air layers})} \times 100\%$$

$$= \frac{(104.0 - 288.9)(\text{ W/m}^2)}{(288.9)(\text{ W/m}^2)} \times 100\% = -64\%.$$

Also note that we could readily determine the temperature at all the four nodes between $T_1$ and $T_2$ (i.e., $T_{1'}$, $T_{1''}$, $T_{2''}$ and $T_{2'}$), by using relations similar to (3.74) and (3.75).

### EXAMPLE 3.8. FAM

A hot water pipe is insulated using an outside casing with formed-in-place, polyurethane foam filling the spacing. This is shown in Figure Ex. 3.8(a). The pipe is made of copper with an inner radius $R = 3.7$ cm and a wall thickness $l_c = 2$ mm. The Teflon casing has an inner radius $R_2 = 7.2$ cm and a wall thickness $l_t = 2$ mm. The pipe has length $L = 1$ m. The inner surface of the pipe is at temperature $T_1 = 45°$C and the other surface of the casing is at $T_2 = 25°$C.

(a) Draw the thermal circuit diagram for steady-state conduction through the pipe-foam-casing composite. Label the outside surface temperature of the pipe as $T_{1'}$ and the inside surface temperature of the casing as $T_{2'}$.

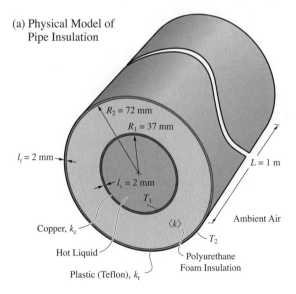

(a) Physical Model of Pipe Insulation

$R_2 = 72$ mm
$R_1 = 37$ mm
$l_t = 2$ mm
$l_c = 2$ mm
$T_1$
$L = 1$ m
Copper, $k_c$
Ambient Air
$\langle k \rangle$
$T_2$
Hot Liquid
Polyurethane Foam Insulation
Plastic (Teflon), $k_t$

(b) Thermal Circuit Diagram

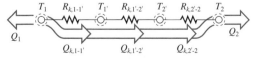

$T_1$  $R_{k,1-1'}$  $T_{1'}$  $R_{k,1'-2'}$  $T_{2'}$  $R_{k,2'-2}$  $T_2$
$Q_1$
$Q_2$
$Q_{k,1-1'}$  $Q_{k,1'-2'}$  $Q_{k,2'-2}$

Figure Ex. 3.8. (a) Physical model for insulation. (b) Thermal circuit diagram.

(b) Determine $-Q_1$ flowing out of surface 1.

(c) Determine $T_{1'}$ and $T_{2'}$. Determine the thermal conductivities at or near 300 K.

**SOLUTION**

(a) The thermal circuit diagram is shown in Figure Ex. 3.8(b). Note that $Q_{k,1\text{-}1'}$ flows through all the resistances and at surface 2 is labeled as $Q_2$. Also note that $T_1'$ and $T_2'$ are not initially given. In other words, by using the overall resistance $R_{k,\Sigma}$, we have from (3.68)

$$Q_1 + Q_{k,1\text{-}1'} = 0 \qquad\qquad \text{energy equation for node } T_1$$

$$Q_1 = -Q_{k,1\text{-}1'} = -\frac{T_1 - T_{1'}}{R_{k,1\text{-}1'}} = -\frac{T_1 - T_2}{R_{k,\Sigma}}$$

$$R_{k,\Sigma} = R_{k,1\text{-}1'} + R_{k,1'\text{-}2'} + R_{k,2'\text{-}2} \qquad \text{overall resistance between nodes } T_1 \text{ and } T_2.$$

So, we can solve for $Q_1$ without evaluation $T_{1'}$.

(b) We now evaluate $Q_1$ using the above energy equation. The resistances are all for radial geometries. From Table 3.2, for cylindrical shells we have

$$R_{k,1\text{-}1'} = \frac{\ln \dfrac{R_1 + l_c}{R_1}}{2\pi L k_c}$$

$$R_{k,1'\text{-}2'} = \frac{\ln \dfrac{R_2}{R_1 + l_c}}{2\pi L \langle k \rangle}$$

$$R_{k,2'\text{-}2} = \frac{\ln \dfrac{R_2 + l_t}{R_2}}{2\pi L k_t}.$$

The thermal conductivities for copper (Table C.16), polyurethane foam formed in place (Table C.17), and Teflon (Table C.17), are

$$\begin{aligned}
&\text{copper:} \quad k_c = 385 \qquad &&\text{W/m-K at } T = 300 \text{ K} \qquad &&\text{Table C.16}\\
&\text{foam:} \quad \langle k \rangle = 0.026 \qquad &&\text{W/m-K at } T = 298 \text{ K} \qquad &&\text{Table C.17}\\
&\text{Teflon:} \quad k_t = 0.26 \qquad &&\text{W/m-K at } T = 293 \text{ K} \qquad &&\text{Table C.17.}
\end{aligned}$$

Using the numerical values, we have

$$R_{k,1\text{-}1'} = \frac{\ln \dfrac{0.039(\text{m})}{0.037(\text{m})}}{2\pi \times 1(\text{m}) \times 385(\text{W/m-K})} = 2.177 \times 10^{-5}\,{}^\circ\text{C/W}$$

$$R_{k,1'\text{-}2'} = \frac{\ln \dfrac{0.074(\text{m})}{0.039(\text{m})}}{2\pi \times 1(\text{m}) \times 0.026(\text{W/m-K})} = 3.755\,{}^\circ\text{C/W}$$

$$R_{k,2'\text{-}2} = \frac{\ln \dfrac{0.074(\text{m})}{0.072(\text{m})}}{2\pi \times 1(\text{m}) \times 0.26(\text{W/m-K})} = 1.678 \times 10^{-2}\,{}^\circ\text{C/W}.$$

Then

$$Q_1 = -\frac{(45 - 25)(°C)}{(2.177 \times 10^{-5} + 3.755 + 1.678 \times 10^{-2})(°C/W)} = -5.303 \text{ W}.$$

This is relatively low heat loss from the water and this insulation is considered a good insulation. Note that the foam resistance is dominant (lowest $k$ and largest thickness among the three layers).

(c) The temperatures $T_{1'}$ and $T_{2'}$ are found by either writing the nodal energy equation or using Figure Ex. 3.8(b) and noting that

$$Q_1 = -\frac{T_1 - T_{1'}}{R_{k,1\text{-}1'}} \quad \text{or} \quad T_{1'} = T_1 + (Q_1 R_{k,1\text{-}1'})$$

$$Q_1 = -\frac{T_{1'} - T_{2'}}{R_{k,1'\text{-}2'}} \quad \text{or} \quad T_{2'} = T_{1'} + (Q_1 R_{k,1'\text{-}2'}).$$

Using the numerical results, we have

$$T_{1'} = 45(°C) - [-5.303(W) \times 2.177 \times 10^{-5}(°C/W)]$$

$$= 45(°C) \quad \text{(no noticeable change in temperature across the copper tube)}$$

$$T_{2'} = 45(°C) - [-5.303(W) \times 3.755(°C/W)] = 25.08(°C) \quad \text{(close to } T_2\text{)}.$$

**COMMENT**

In many applications, we are interested in the overall resistance $R_\Sigma$ and the overall temperature drop and heat flow. Therefore, although we include the intermediate nodes, such as $T_{1'}$ and $T_{2'}$, we do not always need to evaluate them.

### 3.3.3 Thermal Circuit Analysis

A thermal node may be a bounding-surface node or a volumetric node. Under steady-state condition, the difference between the bounding-surface energy equation (2.78) and volumetric energy equation (2.79) is in the surface versus volumetric energy conversion. Here we designate the thermal node $T_i$ as either a surface or a volumetric node and $S_i$ represents either surface or volumetric energy conversion. Then the energy equations (2.72) and (2.73), in absence of radiation and convection, become

$$Q|_{A,i} = Q_i + \sum_{j=1}^{n} Q_{k,i\text{-}j} = \dot{S}_i \quad \text{steady-state energy equation for thermal node } i, \quad (3.87)$$

where we have divided the surface heat transfer into a prescribed $Q_i$ and a conduction heat transfer $Q_{k,i\text{-}j}$. Then for each node $i$ in the thermal system, an energy equation, i.e., (3.87), is written.

To have a closure, we prescribe $T_i$ or $Q_i$, i.e., treat them as independent quantities that are either known or need to be determined. Consider the thermal circuit shown in Figure 3.21.

The nodal energy equation for each node and the internodal heat flow rates are given as follows.

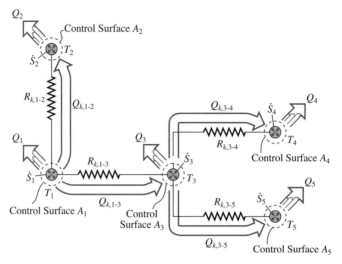

Figure 3.21. Thermal circuit model for steady-state heat transfer.

The energy equations are

$$
\begin{aligned}
\text{node 1:} \quad & Q_1 + Q_{k,1\text{-}2} + Q_{k,1\text{-}3} = \dot{S}_1 \\
\text{node 2:} \quad & Q_2 + Q_{k,2\text{-}1} = \dot{S}_2 \\
\text{node 3:} \quad & Q_3 + Q_{k,3\text{-}1} + Q_{k,3\text{-}4} + Q_{k,3\text{-}5} = \dot{S}_3 \\
\text{node 4:} \quad & Q_4 + Q_{k,4\text{-}3} = \dot{S}_4 \\
\text{node 5:} \quad & Q_5 + Q_{k,5\text{-}3} = \dot{S}_5.
\end{aligned}
\tag{3.88}
$$

The internodal heat flow rates are

$$
\begin{aligned}
\text{element resistance 1-2:} \quad & Q_{k,1\text{-}2} = (T_1 - T_2)/(R_{k,1\text{-}2}) \\
\text{element resistance 1-3:} \quad & Q_{k,1\text{-}3} = (T_1 - T_3)/(R_{k,1\text{-}3}) \\
\text{element resistance 3-4:} \quad & Q_{k,3\text{-}4} = (T_3 - T_4)/(R_{k,3\text{-}4}) \\
\text{element resistance 3-5:} \quad & Q_{k,3\text{-}5} = (T_3 - T_5)/(R_{k,3\text{-}5}).
\end{aligned}
\tag{3.89}
$$

The quantities of interest are

$$
\begin{aligned}
\text{five } Q_i\text{'s:} \quad & Q_1,\, Q_2,\, Q_3,\, Q_4,\, Q_5 \\
\text{five } \dot{S}_i\text{'s:} \quad & \dot{S}_1,\, \dot{S}_2,\, \dot{S}_3,\, \dot{S}_4,\, \dot{S}_5 \\
\text{five } T_i\text{'s:} \quad & T_1,\, T_2,\, T_3,\, T_4,\, T_5 \\
\text{four } Q_{k,i\text{-}j}\text{'s:} \quad & Q_{k,1\text{-}2},\, Q_{k,1\text{-}3},\, Q_{k,3\text{-}4},\, Q_{k,3\text{-}5} \\
\text{four } R_{k,i\text{-}j}\text{'s:} \quad & R_{k,1\text{-}2},\, R_{k,1\text{-}3},\, R_{k,3\text{-}4},\, R_{k,3\text{-}5}.
\end{aligned}
\tag{3.90}
$$

Here in (3.88) and (3.89) we have 9 equations that can be solved for any 9 unknowns, among five $Q_i$'s, five $\dot{S}_i$'s, five $T_i$'s, four $Q_{k,i\text{-}j}$'s, and four $R_{k,i\text{-}j}$'s.

Using the reciprocity of resistances, i.e., $R_{k,i\text{-}j} = R_{k,j\text{-}i}$, the extra resistances are also found. Note that there is also reciprocity of heat flow rates, i.e., $Q_{k,i\text{-}j} = -Q_{k,j\text{-}i}$. This is called the internodal heat transfer reciprocity. Then, we need to specify as many of the variables as needed to have just as many unknowns as we have relations. The solution for these unknowns are found by simultaneously solving the above algebraic equations. In cases with only a few unknowns, simple successive substitutions reduce the number of equations that need to be solved for the unknowns.

Otherwise, numerical solutions are sought. These are linear, algebraic equations in $T_i$'s and $Q_i$'s and nonlinear in $R_{k,i-j}$'s (because the resistances are in the denominator and in general cannot be removed from the denominators).

### EXAMPLE 3.9. FAM

Consider heat losses from a house heated with a furnace, $\dot{S}_1 = \dot{S}_{r,c}$, to its surroundings through the attic side wall, and floor. This is shown in Figure Ex. 3.9(a).

(a) Draw the thermal circuit diagram starting from the furnace.

(b) Determine the temperature $T_1$ that results in a given fraction $a_1\dot{S}_{r,c}$ of the energy converted from chemical bond to thermal energy (combustion) $\dot{S}_{r,c}$ reaching the groundwater, which is at temperature $T_5$, i.e., $Q_{k,4-5} = a_1\dot{S}_{r,c}$. Note that in order to maintain the temperature $T_5$, this heat has to be removed from the groundwater, i.e., $Q_5 = a_1\dot{S}_{r,c}$. All the resistances and temperatures (except $T_1$) are known.

(c) For $\dot{S}_{r,c} = 2{,}000$ W, $a_1 = 0.15$, $T_2 = T_3 = 5°C$, $R_{k,1-2} = R_{k,1-3} = 0.01°C/W$, determine $T_1$ and $Q_5$.

### SOLUTION

(a) Figures Ex. 3.9(a) and (b) show the circuit diagram.

(b) We begin by writing the nodal energy equations (3.87) for nodes 1 and 5. We note that $Q_1 = 0$ (no prescribed heat transfer rate for node 1). Then we

(a) Physical Model and Thermal Circuit Model         (b) Thermal Circuit Model

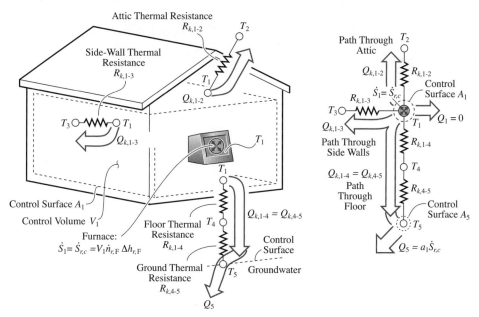

Figure Ex. 3.9. An example of thermal circuit analysis. (a) Heat generation and losses from a house to the water table beneath the house. (b) Thermal circuit model for the problem.

substitute for all the internodal heat flow rates using the known $T$'s and $R_k$'s. The results are

$$Q_1 + \sum_j^n Q_{k,1\text{-}i} = \frac{T_1 - T_2}{R_{k,1\text{-}2}} + \frac{T_1 - T_3}{R_{k,1\text{-}3}} + \frac{T_1 - T_5}{R_{k,1\text{-}4} + R_{k,4\text{-}5}}$$

$$= \dot{S}_1 = \dot{S}_{r,c} \quad \text{energy equation for node 1}$$

$$Q_5 + \sum_j^n Q_{k,5\text{-}i} = Q_5 - \frac{T_1 - T_5}{R_{k,1\text{-}4} + R_{k,4\text{-}5}} = 0.$$

Since $Q_5 = a_1 \dot{S}_{r,c}$, we have for the second energy equation

$$a_1 \dot{S}_{r,c} - \frac{T_1 - T_5}{R_{k,1\text{-}4} + R_{k,4\text{-}5}} = 0 \quad \text{energy equation for node 5.}$$

Now adding the second energy equation to the first, we have

$$\frac{T_1 - T_2}{R_{k,1\text{-}2}} + \frac{T_1 - T_3}{R_{k,1\text{-}3}} + a_1 \dot{S}_{r,c} = \dot{S}_{r,c}.$$

Solving for $T_1$, we have

$$T_1 = \frac{\dot{S}_{r,c}(1\text{-}a_1) + \dfrac{T_2}{R_{k,1\text{-}2}} + \dfrac{T_3}{R_{k,1\text{-}3}}}{\dfrac{1}{R_{k,1\text{-}2}} + \dfrac{1}{R_{k,1\text{-}3}}}.$$

(c) Using the numerical results we have

$$T_1 = \frac{2{,}000 \,(\text{W}) \times (1 - 0.15) + 2 \times \dfrac{5(^\circ\text{C})}{0.01(^\circ\text{C/W})}}{\dfrac{2}{0.01(^\circ\text{C/W})}} = 13.5^\circ\text{C}$$

$$Q_5 = a_1 \dot{S}_{r,c} = 0.15 \times 2{,}000(\text{W}) = 300 \text{ W}.$$

**COMMENT**

Note that since $Q_5$ was prescribed, it was required that $R_{k,1\text{-}4}$ and $R_{k,4\text{-}5}$ be given. In a typical house, during winter, 25% of the heat loss is through the roof, 35% through side walls, 25% through windows and doors, and 15% through the floor.

EXAMPLE 3.10. DES.S

A combustion system consists of a cylindrical annulus combustion chamber, as shown in Figure Ex. 3.10(a). The heat from the combustion is to flow mostly to the outside surface designated by $T_2$. Therefore, a high conductivity solid material is used for the outer cylindrical shell. Here we use silicon carbide SiC, a nonoxide ceramic, with $k$ given in Figure 3.13(c) as a function of temperature. Silicon carbide has a relatively large $k$ and remains inert at high temperatures.

The inner cylindrical shell is made of a lower conductivity material and here we have selected zirconium oxide $ZrO_2$, an oxide ceramic, $k$ also given in Figure 3.13(c). The inner radius for the $ZrO_2$ shell is selected as $R_1 = 25$ mm. The wall thickness for the $ZrO_2$ and SiC shells are selected as $l_1 = l_2 = 3$ mm.

(a) Draw the thermal circuit diagram.
(b) Determine the inner radius for the SiC shell $R_2$ such that the heat flowing through the SiC shell is 10 times that which leaves through the $ZrO_2$ shell.

Use $\langle T_g \rangle = 2,000$ K, $T_2 = 1,000$ K, $T_1 = 800$ K, and evaluate $k$'s at $T = 1,200$ K.

**SOLUTION**

(a) The thermal circuit diagram for this energy conversion and heat transfer is shown in Figure Ex. 3.10(b). We have used nodes $g$, 1, and 2.

(a) A Combustion Chamber ⠀⠀⠀⠀⠀⠀⠀⠀⠀⠀⠀ (b) Thermal Circuit Diagram

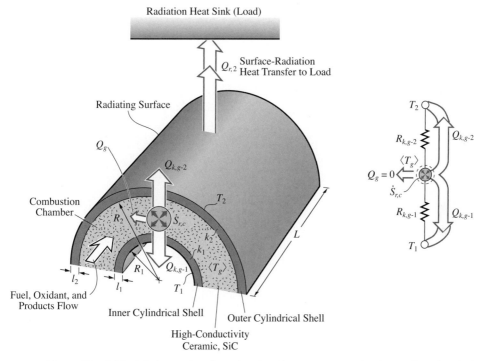

Figure Ex. 3.10. (a) A combustion chamber between a conducting and an insulating cylindrical shell.

(b) The energy equation (3.87), with $Q_g = 0$ (i.e., no prescribed heat transfer is given), gives

$$Q_g + \sum_{j=1}^{2} Q_{k,g-i} = \frac{\langle T_g \rangle - T_1}{R_{k,g-1}} + \frac{\langle T_g \rangle - T_2}{R_{k,g-2}} = \dot{S}_{r,c}.$$

The resistances are those given in Table 3.2, i.e.,

$$R_{k,g\text{-}1} = \frac{\ln \dfrac{R_1 + l_1}{R_1}}{2\pi L k_1}$$

$$R_{k,g\text{-}2} = \frac{\ln \dfrac{R_2 + l_2}{R_2}}{2\pi L k_2}.$$

The energy conversion rate is given as

$$\dot{S}_{r,c} = \dot{s}_{r,c} A_u L = \dot{s}_{r,c} \pi [(R_2^2 - (R_1 + l_1)^2)] L.$$

The design requires that

$$\frac{Q_{k,g\text{-}2}}{Q_{k,g\text{-}1}} = \frac{\langle T_g \rangle - T_2}{\langle T_g \rangle - T_1} \frac{R_{k,g\text{-}1}}{R_{k,g\text{-}2}} = 10$$

or

$$\frac{\langle T_g \rangle - T_2}{\langle T_g \rangle - T_1} \frac{\ln \dfrac{R_1 + l_1}{R_1}}{\ln \dfrac{R_2 + l_2}{R_2}} \frac{k_2}{k_1} = 10$$

or

$$\ln \frac{R_2 + l_2}{R_2} = \left( \ln \frac{R_1 + l_1}{R_1} \right) \left( \frac{\langle T_g \rangle - T_2}{\langle T_g \rangle - T_1} \right) \frac{k_2}{k_1} \times \frac{1}{10}.$$

From Figure 3.13(c), we have for $T = 1{,}200$ K

$$ZrO_2: \quad k_1 \simeq 2.2 \text{ W/m-K} \quad \text{Figure 3.13(c)}$$

$$SiC: \quad k_2 \simeq 20 \text{ W/m-K} \quad \text{Figure 3.13(c)}.$$

Then using the numerical values, we have

$$\ln \left[ \frac{R_2 + 0.003(\text{m})}{R_2} \right] = \ln \left[ \frac{(0.025 + 0.003)(\text{m})}{0.025(\text{m})} \right]$$

$$\times \frac{(2{,}000 - 1{,}000)(\text{K})}{(2{,}000 - 800)(\text{K})} \times \frac{20(\text{W/m-K})}{2.2(\text{W/m-K})} \times \frac{1}{10}$$

The solution is found using a solver, and is

$$R_2 = 0.03346 \text{ m} = 33.46 \text{ mm}.$$

**COMMENT**

Note that among the two heat-flow branches, although $Q_{k,g\text{-}2}$ flows through a smaller temperature difference, it is larger. This is due to $k_2/k_1 = 9.091$ and the larger area of the outer shell. Also note that we only need to have a gap of $R_2 - (R_1 + l_1) = 33.46 - (25 + 3) = 5.46$ mm for the combustion chamber.

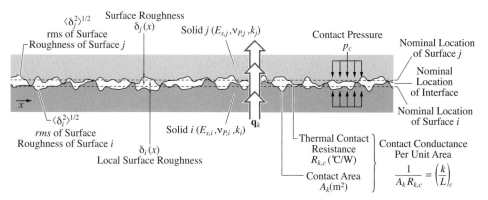

Figure 3.24. A rendering of contact between two solid surfaces showing the surface roughness.

### 3.3.4 Conduction Thermometry*

This sectino in gound on the Web site www.cambridge.org/kaviany. In this section we consider two heat transfer devices, one for the measurement of the surface heat flux and one for the measurement of the conductivity, namely, (A) Heat Flux Meter, and (B) Thermal-Conductivity Meter.

### 3.3.5 Contact Resistance

Thermal contact resistance $R_{k,c}$ is the resistance to heat flow across the interface of two similar or dissimilar materials (i.e., two heat transfer media). In general, the contact resistance between two conducting solids is of significance. This is because the resistance of each media (i.e., intramedia resistance) is generally low compared to the contact (i.e., intermedia) resistance. The two materials may be held in contact by an adhesive bond or by pressure (e.g., gravity force, external force). When an adhesive is used, the gap between the two surfaces may be filled with a high conductivity fluid with small, dispersed high conductivity dielectric solid particles (i.e., high conductivity grease or paste with $k$ of about 2 to 3 W/m-K) to reduce the contact resistance.[†] When electrical conductivity is not a constraint, thermal conductivity epoxies with $k$ over 20 W/m-K, are available.

For two solids (node of materials $i$ and $j$) in contact, the contact resistance $R_{k,c}$ depends on the following parameters: the surface roughness of each material, the hardness (or softness) of the materials, the pressure applied (i.e., joint pressure), the gap material, the temperature, and the aging effects. The surface roughness $\delta$ (i.e., the deviation of surface location from the nominal surface location) varies along the surface and the root mean square (rms) of the roughness $\langle \delta^2 \rangle^{1/2}$ is used to characterize the average surface roughness. These are shown in Figure 3.24. The surface roughnesses are anisotropic (i.e., the geometry is direction dependent). The unit roughness element is called an asperity. The surface finish classification uses

---

[†] An example is Aremco Heat-Away 638 Grease, an aluminum nitride filled silicone system with $k = 2.23$ W/m-K and $\rho_e = 10.14$ ohm-m. These are called Thermal Interface Materials (TIM).

the name grinded surface for a $\langle\delta^2\rangle^{1/2}$ between 1 to 5 $\mu$m. The superpolished surface has a $\langle\delta^2\rangle^{1/2}$ between 0.05 to 0.5 $\mu$m. The electropolished surface has a $\langle\delta^2\rangle^{1/2}$ between 0.01 and 0.1 $\mu$m.

The contact resistance $R_{k,c}$ is measured and is generally presented through the resistance over a unit contact surface area ($A_k = 1$ m$^2$). If we assume that this resistance over a unit surface area is designated as $(L/k)_c$, we have

$$R_{k,c} \equiv \frac{1}{A_k}\left(\frac{L}{k}\right)_c \quad \text{or} \quad A_k R_{k,c}[^\circ\text{C}/(\text{W/m}^2)] \equiv \left(\frac{L}{k}\right)_c \qquad \text{contact resistance.} \quad (3.94)$$

Figure 3.25 shows the measured contact conductance per unit area $(k/L)_c$ for some metallic pairs [25]. For contact of the same materials, the results are given as a function of the joint or contact pressure $p_c$. The theory of elastic contact predicts that $(k/L)_c$ is nearly proportional to $[(1 - v_P^2)p_c/E_s]^{1/3}$, where $v_P$ is the Poisson ratio and $E_s$ is the Young modulus of elasticity. This relation is also shown in the figure. At higher pressures, plastic deformation of the roughnesses occurs. As the joint pressure is increased, the contact conductance increases. For plastic deformations, some correlations for $A_k R_{k,c}$ are available using the material hardness $E_h$.[†]

Note that in general the softer the material is (i.e., the smaller the Young modulus of elasticity), the higher is the contact conductance. For example, aluminum pairs have a much larger contact conductance as compared to stainless-steel pairs. Also, in general, the smaller the roughness (i.e., $\langle\delta^2\rangle^{1/2}$), the higher is the contact conductance. Physical bounding (by sintering) or adding high-conductivity bounding materials reduces the thermal conductance resistance.

The contact conductivity $k_c$, for two completely-joining materials $i$ and $j$, takes on an intermediate value between $k_i$ and $k_j$ and depends on the bonding.

Since the measured quantity is $(L/k)_c$ and $L$ is very small, it is generally assumed that $L \to 0$, and therefore, the contact resistance is envisioned as a jump in temperature $\Delta T_c$ (i.e., change in temperature over unspecified, but small distance). This change in temperature across the contact gap depends on $q_k$ and $R_{k,c}$ and is defined by

$$\frac{\Delta T_c}{A_k R_{k,c}} \equiv q_k \qquad (3.95)$$

or

$$\Delta T_c \equiv Q_k R_{k,c} = q_k A_k R_{k,c} = q_k \left(\frac{L}{k}\right)_c \qquad \text{contact-resistance temperature drop.} \quad (3.96)$$

[†] One of these correlations is

$$\frac{1}{A_k R_{k,c}} = \frac{5.25 \times 10^{-4}\, k_m}{\langle\delta^2\rangle^{1/2}}\left(\frac{p_c}{E_h}\right)^{0.51}, \qquad \text{where} \qquad \frac{2}{k_m} = \frac{1}{k_i} + \frac{1}{k_j}.$$

Here, $k_m$ is the harmonic mean conductivity, and $E_h$ is the Vickers microhardness of the softer material. An example of eleastic-plastic contact resistance is given as an end of chapter problem.

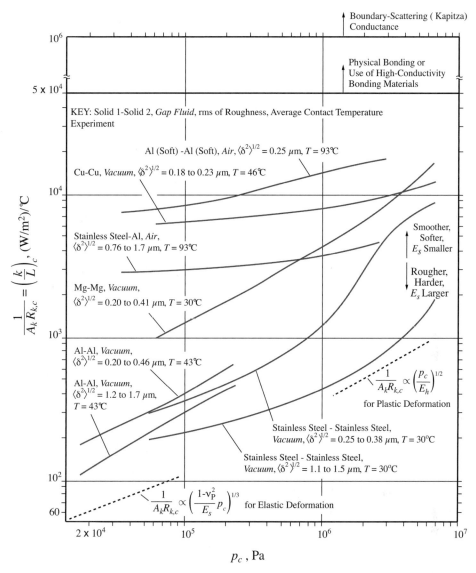

Figure 3.25. Thermal contact conductance for pairs of metallic solids (material $i$ the same as $j$). The trends in increase of smoothness and softness are also shown. (From Schneider, P.J., reproduced by permission © 1985 McGraw-Hill.) In physical bonding (two materials well-fused together), the conductance increases substantially.

As the contact resistance $R_{k,c}$ increases, this jump in temperature across the contact gap increases.

We note that due to boundary (bounding-surface) scattering, there is a phonon boundary resistance $A_k R_{k,b}$ that is of the order of $10^{-7}$ to $10^{-6}$ K/(W/m$^2$). This resistance (also called the Kapitza resistance) is due to the acoustic mismatch between two joining materials and is considered internal to each material [27]. This boundary-scattering conductance is also shown in Figure 3.25.

## EXAMPLE 3.11. FAM

When two metallic pieces are in contact, because of the intrinsic high conductivity of metals, the resistance to heat flow through them is dominated by the contact resistance at their interface. Use aluminum surface pairs in contact with three different surface roughnesses (i) $\langle \delta^2 \rangle^{1/2} = 1.2$ to $1.7\ \mu$m, and (ii) $\langle \delta^2 \rangle^{1/2} = 0.20$ to $0.46\ \mu$m, for a harder aluminum (2024-T3), and (iii) $\langle \delta^2 \rangle^{1/2} = 0.25\ \mu$m for a softer aluminum (75S-T6). The contact conductances for these are shown in Figure 3.25. Assume that the effect of the temperature of the gap fluid is negligible (this is justifiable when the temperature is far from the softening temperature and for low conductivity gap fluid, i.e., other than liquid metals).

(a) Then using a contact pressure $p_c = 10^5$ Pa, compare the contact resistance $A_k R_{k,c}$ among these three contact interfaces.

(b) For a heat flux of $q_k = 10^4$ W/m², determine the temperature jump across the contact interface $\Delta T_c$, for the three surface roughnesses.

### SOLUTION
Since we have assumed a negligible temperature and gap-fluid dependence, we can use the results of Figure 3.25 which are for various temperatures.

(a) From Figure 3.25, at $p_c = 10^5$ Pa, we read the contact conductance per unit area, $1/A_k R_{k,c}$, for Al-Al for the three different RMS roughness as

$$\text{(i)} \quad \frac{1}{A_k R_{k,c}} \equiv \left( \frac{k}{L} \right)_c = 3 \times 10^2 \ (\text{W/m}^2)/^\circ\text{C}$$

$$\text{for } \langle \delta^2 \rangle^{1/2} = 1.2 \text{ to } 1.7\ \mu\text{m}, \qquad\qquad \text{harder Al}$$

$$\text{(ii)} \quad \frac{1}{A_k R_{k,c}} \equiv \left( \frac{k}{L} \right)_c = 4 \times 10^2 \ (\text{W/m}^2)/^\circ\text{C}$$

$$\text{for } \langle \delta^2 \rangle^{1/2} = 0.20 \text{ to } 0.46\ \mu\text{m}, \qquad\qquad \text{harder Al}$$

$$\text{(iii)} \quad \frac{1}{A_k R_{k,c}} \equiv \left( \frac{k}{L} \right)_c = 8 \times 10^3 \ (\text{W/m}^2)/^\circ\text{C}$$

$$\text{for } \langle \delta^2 \rangle^{1/2} = 0.25\ \mu\text{m}, \qquad\qquad \text{softer Al.}$$

Note that the softer aluminum with the smallest roughness has the smallest contact resistance (i.e., highest contact conductance).

(b) From (3.96) we have

$$\Delta T_c \equiv q_k \left( \frac{L}{k} \right)_c = q_k \frac{1}{(k/L)_c}.$$

Using this, for the three contact resistances we have the temperature jumps as

(i) $\Delta T_c = 10^4 (\text{W/m}^2) \dfrac{1}{3 \times 10^2 (\text{W/m}^2\text{-}^\circ\text{C})} = 33.3^\circ\text{C or K}$

for $\langle \delta^2 \rangle^{1/2} = 1.2$ to $1.7\ \mu\text{m}$ \hfill harder Al

(ii) $\Delta T_c = 10^4 (\text{W/m}^2) \dfrac{1}{4 \times 10^2 (\text{W/m}^2\text{-}^\circ\text{C})} = 25.0^\circ\text{C or K}$

for $\langle \delta^2 \rangle^{1/2} = 0.20$ to $0.46\ \mu\text{m}$ \hfill harder Al

(iii) $\Delta T_c = 10^4 (\text{W/m}^2) \dfrac{1}{8 \times 10^3 (\text{W/m}^2\text{-}^\circ\text{C})} = 1.25^\circ\text{C or K}$

for $\langle \delta^2 \rangle^{1/2} = 0.25\ \mu\text{m}$ \hfill softer Al.

**COMMENT**

This heat flux of $q_k = 10\ \text{kW/m}^2$ is considered relatively high (but not too high). It is shown above that when a large roughness, or a harder metal (or a small contact pressure) is used, a large temperature jump occurs across the contact interface.

### 3.3.6 Conduction and Energy Conversion

As discussed in Section 2.3, conversion of chemical- and physical-bond, electromagnetic, and mechanical energy to or from thermal energy may occur at the bounding surfaces or within the volume. Examples of bounding-surface energy conversion are friction heating (mechanical energy conversion) surface irradiation upon an opaque, absorbing surface, or Peltier cooling at a junction of dissimilar electrical conductors and semiconductors (electromagnetic energy conversion). The volumetric energy conversion occurs in epoxy curing (chemical-bond energy conversion), nuclear fission (also chemical-bond energy conversion), or Joule heating (electromagnetic conversion). Here we discuss two simple energy conversion examples, one surface and one volumetric, and discuss their roles in heat conduction.

### (A) Bounding Surface

As an example, consider the irreversible surface-friction heating energy conversion, i.e., conversion of mechanical to thermal energy on the interface of two solids in contact and in relative motion. Here we consider grinding heating where the workpiece undergoes material removal by a grinding wheel. Figure 3.26 gives a schematic of the problem considered. A fraction $a_1$ of this surface thermal energy conversion $\dot{S}_{m,F}$ flows into the workpiece. This heat flux is $a_1 \dot{S}_{m,F}/A_k$ and under steady-state condition is conducted through the workpiece to the other surfaces, which are cooled. For a one-dimensional (planar) heat flow, we have the rendering given in Figure 3.26. If

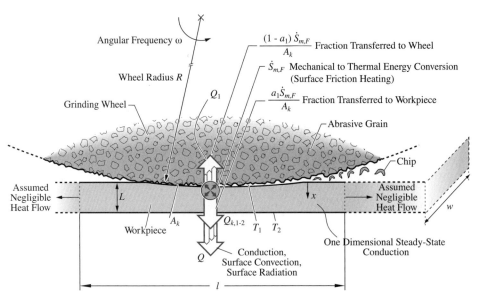

Figure 3.26. A schematic of the conversion of mechanical energy to thermal energy at a grinding surface. The heat conduction through the workpiece substrate is also shown.

the workpiece has a thickness $L$ and the cooled surface is at a temperature $T_2$, the friction surface temperature $T_1$ needs to be kept lower than a critical temperature for the workpiece (e.g., its melting temperature) $T_c$, or the critical temperature for the grinder (e.g., abrasive-belt binding damage temperature).

Applying the bounding surface energy equation (2.62) and treating the surface-friction heating as

$$Q|_{A,1} = \dot{S} = Q_1 + Q_{k,1\text{-}2} = \dot{S}_{m,F} \quad \text{surface energy equation,} \tag{3.97}$$

this energy equation, with $Q_1 = (1 - a_1)\dot{S}_{m,F}$, becomes

$$Q_{k,1\text{-}2} = a_1\dot{S}_{m,F} \quad \text{energy equation for grinding surface.} \tag{3.98}$$

Substituting for $Q_{k,1\text{-}2}$ from (3.50) into (3.97), we have

$$\frac{T_1 - T_2}{R_{k,1\text{-}2}} = \frac{T_1 - T_2}{L/(A_k k)} = a_1\dot{S}_{m,F}, \quad A_k = lw, \tag{3.99}$$

where $l$ is the contact-affected length and $w$ is the width of the workpiece.

Now solving for $T_1$ we have

$$T_1 = T_2 + \frac{a_1\dot{S}_{m,F}L}{lwk} \quad \text{with } T_1 < T_c \quad \text{for no damage.} \tag{3.100}$$

Then for a surface-friction heating rate $a_1\dot{S}_{m,F}$, and a workpiece width $w$, length $l$, and thickness $L$, and conductivity $k$, the temperature at the surface away from grinding wheel, i.e., $T_2$, should be cooled (actively or passively) to prevent thermal damage at the grinding surface 1.

The thicker the workpiece, the smaller its conductivity, and the larger $a_1\dot{S}_{m,F}/wl$, the higher is the temperature at the wheel-workpiece interface.

EXAMPLE 3.12. FAM

In grind polishing of a stainless-steel plate (316) of width and length $w = l = 20$ cm, and thickness $L = 0.05$ cm, heat is generated by friction heating at an amount $\dot{S}_{m,F} = 4 \times 10^3$ W. From this energy, the fraction flowing into the workpiece (i.e., plate) is $a_1 = 0.7$.

(a) Draw the thermal circuit diagram.
(b) If the grinding surface of the stainless-steel (316) plate should have a temperature $T_1 = 350°$C, what should be the temperature at the opposite side of the plate $T_2$?

The thermal conductivity can be evaluated at $T = 350°$C and the heat transfer may be assumed one dimensional.

**SOLUTION**
(a) The physical model of the problem is that shown in Figure Ex. 3.12(a) and the thermal circuit model is that shown in Figure Ex. 3.12(b).
(b) The temperature at surface 1 is found by (3.100), which is repeated here as

$$T_1 = T_2 + \frac{a_1 \dot{S}_{m,F} L}{lwk},$$

and this gives

$$T_2 = T_1 - \frac{a_1 \dot{S}_{m,F} L}{lwk}.$$

The thermal conductivity of the stainless-steel (316) at $T = 300$ K is found from Table C.16 and is

$$k = 13 \text{ W/m-K} \quad \text{Table C.16.}$$

(a) Physical Model                                    (b) Thermal Circuit Model

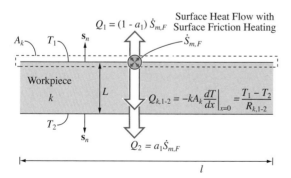
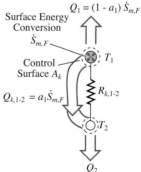

Figure Ex. 3.12. (a) Physical model of grinding, surface-friction heating and workpiece, and the conduction through the substrate. (b) Thermal circuit diagram.

Now substituting into the expression for $T_2$, we have

$$T_2 = 350(°C) - \frac{0.7 \times 4 \times 10^3 (W/m^2) \times 0.05(m)}{0.2(m) \times 0.2(m) \times 13(W/m\text{-}K)} = 80.77°C.$$

**COMMENT**

Note that $T_1 - T_2 = 269.23°C$, which is substantial. The rate of energy conversion depends on the speed of the grinding wheel rotation and the applied contact pressure. Also note that the heat transferred through the workpiece $Q_{k,1\text{-}2}$ has to be removed from surface 2 as $Q_2$, for a steady-state process.

### (B) Volumetric

In addition to the bounding-surface energy conversion, consider an example of volumetric energy conversion by Joule heating of an electric conductor. This is depicted in Figure 3.27(a). The temperature distribution is assumed one dimensional. The medium has an electrical resistivity $\rho_e$ and thermal conductivity $k$. The thickness is $L$. Then we write the differential-volume energy equation (3.29) as

$$\nabla \cdot \mathbf{q} = -\nabla \cdot k\nabla T = \dot{s}_{e,J}. \tag{3.101}$$

For a one-dimensional heat flow, and by using (2.30) for $\dot{s}_{e,J}$, we have

$$\underbrace{\frac{d}{dx}\mathbf{q}_k \cdot \mathbf{s}_x}_{\substack{\text{spatial-rate} \\ \text{change of } \mathbf{q}_k \\ \text{in } x\text{-direction}}} = \underbrace{-k\frac{d^2 T}{dx^2}}_{} = \underbrace{\rho_e j_e^2}_{\substack{\text{volumetric} \\ \text{Joule} \\ \text{heating}}} \quad \begin{array}{l}\text{differential-volume \ energy} \\ \text{equation.}\end{array} \tag{3.102}$$

(a) Physical Model     (b) Temperature Distribution     (c) Thermal Circuit Model

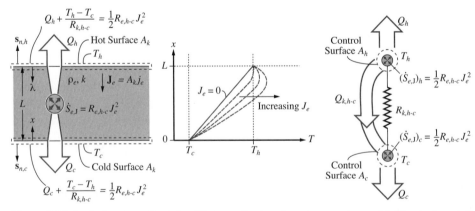

Figure 3.27. Joule heating of a slab with prescribed surface temperatures. (a) Physical model. (b) Temperature distribution. (c) Thermal circuit diagram.

Now using prescribed temperatures at $x = 0$ and $x = L$ as $T_h$ and $T_c$, the boundary conditions in (3.29) become

$$T(x = 0) = T_c, \quad T(x = L) = T_h. \tag{3.103}$$

Solving (3.102) for $T$ by integrating it twice and substituting the boundary conditions (3.103) and determining the constants, we have the result for $T(x)$ as

$$T = T_c + \frac{T_h - T_c}{L}x + \frac{\rho_e j_e^2}{2k}(L - x)x. \tag{3.104}$$

This temperature distribution is depicted in Figure 3.27(b) for $j_e = 0$ and for $j_e > 0$.

Now, the rate of heat flow at $x = 0$ and $x = L$, $Q_c$ and $Q_h$, are determined. Using (2.62) and evaluating the derivatives, these are

$$Q_c = (\mathbf{q}_k \cdot \mathbf{s}_n)_{x=0}A_k = Q_{x=0} \equiv A_k k \frac{dT}{dx}\Big|_{x=0}$$

$$= \underbrace{\frac{A_k k}{L}(T_h - T_c)}_{\substack{\text{due to} \\ \text{conduction}}} + \underbrace{A_k L \frac{1}{2}\rho_e j_e^2}_{\substack{\text{due to Joule} \\ \text{heating}}} \tag{3.105}$$

$$Q_c = \frac{T_h - T_c}{R_{k,h\text{-}c}} + \frac{1}{2}R_{e,h\text{-}c}J_e^2, \quad R_{k,h\text{-}c} = \frac{L}{A_k k}, \quad R_{e,h\text{-}c} = \frac{\rho_e L}{A_k} \tag{3.106}$$

$$Q_h = (\mathbf{q}_k \cdot \mathbf{s}_n)_{x=L}A_k = Q_{x=L} \equiv -A_k k \frac{dT}{dx}\Big|_{x=L}$$

$$= -\frac{A_k k}{L}(T_h - T_c) + A_k L \frac{1}{2}\rho_e j_e^2 \tag{3.107}$$

$$Q_h = -\frac{T_h - T_c}{R_{k,h\text{-}c}} + \frac{1}{2}R_{e,h\text{-}c}J_e^2. \tag{3.108}$$

Note that $J_e = A_k j_e$. For the case of $T_h > T_c$, as depicted in Figure 3.27(b), and for $J_e = 0$, we have $Q_h = -(A_k k/L)(T_h - T_c) < 0$, i.e., heat flows into the hot surface. As $J_e$ increases, a value of $J_e$ is reached for which $Q_h = 0$ and as $J_e$ is further increased, $Q_h > 0$, i.e., heat also flows out of the hot surface.

At the cold surface, for $T_h > T_c$, $Q_c > 0$ for all values of $J_e$, i.e., heat flows out of the cold surface.

To confirm that the volumetric Joule heat generation is removed from the surfaces, we apply the integral-surface energy equation (2.70) to the surface that encloses the hot and cold surfaces, i.e.,

$$Q_h + Q_c = \dot{S}_{e,J} = A_k L \rho_e j_e^2 = R_{e,h\text{-}c}J_e^2 \quad \text{integral-surface energy equation.} \tag{3.109}$$

The energy converted by Joule heating leaves the heat transfer medium in an equal split between the top and bottom bounding surfaces. To show this, we examine (3.106) and (3.108). We note that the difference between $Q_c$ and $Q_h$ is the conduction heat transfer rate which flows from the hot surface toward the cold surface. We have used this result to move the Joule heating from the central volumetric node to the two boundary nodes, as shown in Figure 3.27(c). Then we write the two surface energy equations (3.106) and (3.108) as

$$Q_h + Q_{k,h\text{-}c} = Q_h + \frac{T_h - T_c}{R_{k,h\text{-}c}} = (\dot{S}_{e,\text{J}})_h = \frac{1}{2} R_{e,h\text{-}c} J_e^2 \quad \begin{array}{l} \text{surface energy} \\ \text{equation for } T_h \text{ node} \end{array} \quad (3.110)$$

$$Q_c + Q_{k,c\text{-}h} = Q_c + \frac{T_c - T_h}{R_{k,h\text{-}c}} = (\dot{S}_{e,\text{J}})_c = \frac{1}{2} R_{e,h\text{-}c} J_e^2 \quad \begin{array}{l} \text{surface energy} \\ \text{equation for } T_c \text{ node.} \end{array} \quad (3.111)$$

### 3.3.7 Thermoelectric Coolers

The thermoelectric cooler is a collection of *p*-*n* units (with *p*-*n* junctions), where each unit has the electrical and heat flow aspects that are rendered in Figures 3.28(a) and (b). Heat is removed from the surroundings of the cold surface (or plates). This is called thermoelectric cooling. The temperature distribution within the unit is shown in Figure 3.28(c). Heat is absorbed at $T_c$ and released at $T_h$.

For the semiconductor and the current flow arrangement shown in the Figure 3.28(b), the rate of Peltier heat absorption (or energy conversion) from the cold *p*-*n* junction is given by (2.37), i.e.,

$$(\dot{S}_{e,\text{P}})_c = -\alpha_S J_e T_c, \quad \alpha_S = \alpha_{S,p} - \alpha_{S,n} \quad \text{Peltier cooling at cold junction,} \quad (3.112)$$

where $T_c$ is the absolute temperature of the cold junction. Note that the junction $\alpha_S$ is the sum of $\alpha_{S,p}$ and $-\alpha_{S,n}$.

The product of the Seebeck coefficient and the absolute temperature is the Peltier coefficient (2.36).

Using the results of Section 3.3.6(B) above, i.e., (3.111) for the energy, and adding the Peltier cooling effect given by (3.112), we have for the cold surface

$$Q\mid_{A,c} = \sum_i \dot{S}_{c,i} \quad \text{surface energy equation for node } T_c \quad (3.113)$$

or

$$Q_c + Q_{k,c\text{-}h} = (\dot{S}_{e,\text{J}})_c + (\dot{S}_{e,\text{P}})_c. \quad (3.114)$$

This is shown, as a thermal circuit diagram for node $T_c$, in Figure 3.28(d). We now substitute for $Q_{k,c\text{-}h}$ and the energy conversions, and (3.114) becomes

$$Q_c = \underset{\substack{\text{due to Peltier} \\ \text{heat absorption}}}{-\ \alpha_S J_e T_c} \quad + \underset{\substack{\text{due to} \\ \text{conduction}}}{R_{k,h\text{-}c}^{-1}(T_h - T_c)} \quad + \underset{\substack{\text{due to Joule} \\ \text{heating}}}{\frac{1}{2} R_{e,h\text{-}c} J_e^2} \qquad (3.115)$$

$$\text{energy equation for node } T_c,$$

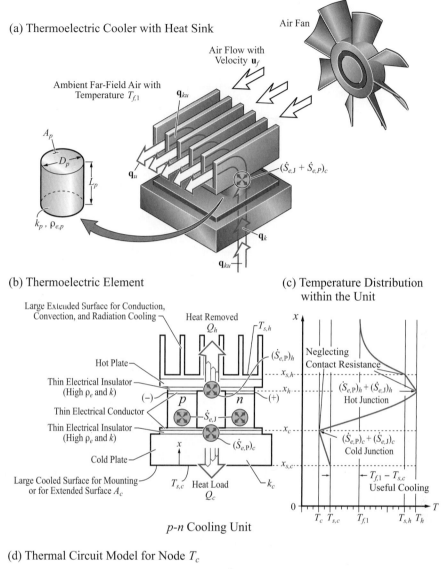

(a) Thermoelectric Cooler with Heat Sink

(b) Thermoelectric Element

(c) Temperature Distribution within the Unit

(d) Thermal Circuit Model for Node $T_c$

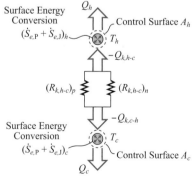

Figure 3.28. (a) Thermoelectric cooling, showing the semiconductor $p$-type and $n$-type materials and the direction of heat flow. (b) Thermoelectric element. (c) Temperature distribution. (d) Thermal circuit model.

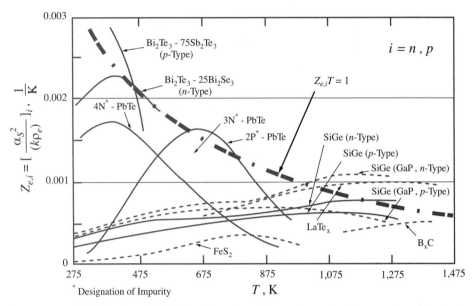

Figure 3.28. (e) Variation of individual junction element ($n$ and $p$) figure of merit, $Z_{e,i}$ for various high performance (low and high temperature ranges) thermoelectric materials, with respect to temperature [24]. The line of $Z_{e,i}T = 1$ is also shown.

where in terms of the area $A_k$ and length $L$, we have

$$\frac{1}{R_{k,h\text{-}c}} = \frac{1}{(R_{k,h\text{-}c})_p} + \frac{1}{(R_{k,h\text{-}c})_n} = \left(\frac{A_k k}{L}\right)_p + \left(\frac{A_k k}{L}\right)_n,$$

$$R_{e,h\text{-}c} = \left(\frac{\rho_e L}{A_k}\right)_p + \left(\frac{\rho_e L}{A_k}\right)_n. \qquad (3.116)$$

Note that the electrical resistances are arranged in series [from (3.78)], while the thermal resistances are arranged in parallel [from (3.82)].

In (3.112) and (3.116), $\alpha_{S,p}$, $\alpha_{S,n}$, $\rho_{e,p}$, $\rho_{e,n}$, $k_p$ and $k_n$ are the Seebeck coefficients, electrical resistivity, and thermal conductivity of the $n$-type and $p$-type materials. $A_k$ is the conductor cross-sectional area and $L$ is the conductor length.

In order to cool a surface, many couples (pairs) can be attached to the surface being cooled, with each pair having the cooling power given above. This would allow for a substantial cooling capability $Q_c(\text{W})$.

Note that both the conduction heat transfer and the Joule-heating contribution to the heat flow toward the cold surface. Only the Peltier effect removes heat from this surface, and when it is larger than the negative of the conductive and the Joule components, it can result in net heat removal from the cold surface. This depends on the current used and the magnitude of $T_h - T_c$. Also note again that the Peltier cooling is proportional to $J_e$, while Joule heating is proportional to $J_e^2$. These are discussed next.

### (A) Maximum Cooling Rate and Maximum Temperature Difference

The current for the maximum in $|Q_c|$, i.e., $Q_{c,max}$, is found by the differentiation of (3.115) with respect to $J_e$ and by setting the resultant to zero. When this resultant equation is solved for $J_e$, we have

$$J_e(Q_{c,max}) = \frac{\alpha_S T_c}{R_{e,h-c}} \qquad \text{optimal current for } Q_{c,max}. \qquad (3.117)$$

The maximum cooling rate $Q_{c,max}$ is found using this current. When (3.117) is substituted in (3.115), we have

$$Q_{c,max} = -\frac{\alpha_S^2 T_c^2}{2R_{e,h-c}} + R_{k,h-c}^{-1}(T_h - T_c) \qquad \text{optimal cooling rate.} \qquad (3.118)$$

For $T_h \to T_c$, this would yet be further maximized.

The maximum temperature difference between the hot and cold plates $T_h - T_c$ is found by setting $Q_c = 0$ in (3.118). The result is

$$(T_h - T_c)(Q_{c,max} = 0) = \frac{\alpha_S^2 T_c^2}{2R_{e,h-c}/R_{k,h-c}} \qquad \begin{array}{l}\text{temperature difference for}\\ \text{optimal current and } Q_{c,max} = 0.\end{array} \qquad (3.119)$$

### (B) Figure of Merit

The thermoelectric-cooling figure of merit $Z_e(1/K \text{ or } 1/°C)$ is defined as

$$Z_e = \frac{\alpha_S^2}{R_{e,h-c}/R_{k,h-c}} \qquad \text{or} \quad (T_h - T_c)_{max} = \frac{Z_e T_c^2}{2} \qquad \begin{array}{l}\text{thermoelectric-cooling}\\ \text{figure of merit.}\end{array} \qquad (3.120)$$

Figure 3.28(e) shows the variation of the individual figure of merit for the $p$- and $n$-type materials, $Z_{e,p}$ and $Z_{e,n}$, having high values for figure of merit, as a function of temperature. The line $Z_{e,i}T = 1$ is also shown. Some materials have values of figure of merit over this line and are considered superior.

Table C.9(a) gives some typical values for the thermoelectric cooling materials. Table C.9(b) lists the thermoelectric properties of a silicon-germanium alloy (mixed crystals) as a function of temperature. This alloy is suited for the high temperature applications. Note that the thermal conductivity of the alloy is much less than that of the component semiconductor (for Si, $k$ is 150 W/m-K while for Ge it is 60 W/m-K, at $T = 300$ from Table C.14). This is due to phonon scattering at the grain boundaries (conduction in semiconductors and semi-metals is dominated by phonons). As discussed in Section 3.2.2(B), the high-frequency phonons are scattered in proportion to $f_p^4$ (i.e., short $\tau_{p,r,p}$ and $\tau_{p,r,v}$), so conduction is mostly by low-frequency phonons having long mean-free paths. The boundary scattering (due to small crystals, i.e., grain size) is also significant. In general, a couple composed of two elements each having nearly the same magnitude in $\alpha_S/\rho_e$, but opposite in the sign in their Seebeck coefficients, is used. A comprehensive listing of suitable thermoelectric materials is given in [24].

### (C) Maximizing the Figure of Merit

By choosing $L_n = L_p$, the quantity $R_{e,h-c}R_{k,h-c}^{-1}$ in (3.120) is minimized with respect to $A_{k,p}/A_{k,n}$ and the results are (left as an end of chapter problem).

$$\frac{A_{k,p}}{A_{k,n}} = \left(\frac{\rho_{e,p}k_n}{\rho_{e,n}k_p}\right)^{1/2} \qquad \text{for maximum figure of merit.} \qquad (3.121)$$

Using this, the thermoelectric-cooling figure of merit (3.120) would not include the geometric parameters and becomes

$$Z_e = \frac{\alpha_S^2}{[(k\rho_e)_p^{1/2} + (k\rho_e)_n^{1/2}]^2} \qquad \text{optimized figure of merit.} \qquad (3.122)$$

The thermal conductivity of semiconductors $k$ can be reduced by adding ultra-fine particles, without significantly affecting the electrical property $\rho_e$ [24]. Otherwise, decrease in $k$ follows an increase in $\rho_e$.

### (D) Coefficient of Performance

As was done for the Joule heating alone in (3.109), for the control surface containing the cold and hot surfaces, as shown in Figures 3.29(a) and (b), the integral-surface energy equation, (2.70), becomes

$$\int_A (\mathbf{q} \cdot \mathbf{s}_n)dA = \int_V \dot{s}_e dV \qquad (3.123)$$

or

$$Q|_A = Q_h + Q_c = \sum \dot{S}_i = \dot{S}_{e,J} + \dot{S}_{e,P} = R_{e,h-c}J_e^2 + \alpha_S J_e(T_h - T_c), \quad (3.124)$$

where the various energy conversion terms are defined in Figure 3.29(b).

This is similar to (3.109) except for the added Peltier thermoelectric energy conversion. The above energy equation can be written as

$$Q_h = -Q_c + R_{e,h-c}J_e^2 + \alpha_S J_e(T_h - T_c), \qquad (3.125)$$

indicating that the hot surface will reject the heat flowing into the cold plate $Q_c$ plus the Joule heating and the net Peltier energy conversion $\alpha_S J_e(T_h - T_c)$. The optimum coefficient of performance $\eta_{cop}$ (ratio of cooling load to electric power required) is

(a) Physical Model for Thermoelectric Cooling Unit    (b) Thermal Circuit Model

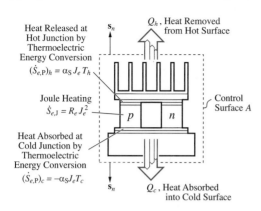
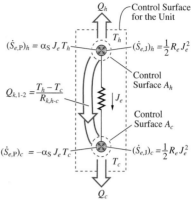

Figure 3.29. Overall energy conservation of a thermoelectric unit. (a) Physical model. (b) Circuit model.

then given by

$$\eta_{cop} \equiv \frac{Q_c}{P_e} = \frac{-Q_c}{R_{e,h\text{-}c}J_e^2 + \alpha_S J_e(T_h - T_c)}, \qquad \text{coefficient of performance } \eta_{cop}$$

$$= \frac{\alpha_S J_e T_c - R_{k,h\text{-}c}^{-1}(T_h - T_c) - R_{e,h\text{-}c}J_e^2/2}{R_{e,h\text{-}c}J_e^2 + \alpha_S J_e(T_h - T_c)} = \frac{T_c}{T_h - T_c} \frac{(1 + Z_e T_o)^{1/2} - \dfrac{T_h}{T_c}}{(1 + Z_e T_o)^{1/2} + 1}$$

$$T_o = \frac{T_h + T_c}{2}, \quad J_e = \frac{\alpha_S(T_h - T_c)}{R_{e,h\text{-}c}[(1 + Z_e T_o)^{1/2} - 1]}, \qquad (3.126)$$

where the optimum current used is that corresponding to maximize $\eta_{cop}$, i.e., $\partial \eta_{cop} / \partial J_e = 0$.

### (E) Conduction through Cold Junction Substrate

Note that under steady state, the heat transfer to the cold junction $Q_c$ flows through the cold plate of the thickness $x_c - x_{s,c}$, shown in Figures 3.28(b) and (c), by conduction and we have

$$Q_c = \frac{T_c - T_{s,c}}{R_{k,c\text{-}sc}} = k_c A_{k,c} \frac{T_c - T_{s,c}}{x_c - x_{s,c}}, \qquad (3.127)$$

where $k_c$ is the conductivity of the cold plate. Here $T_c < T_{s,c}$, thus making $Q_c < 0$, as expected for heat flowing into the control surface. Some dielectric, oxide ceramics such as alumina offer high thermal conductivity (Table C.14). Note that we have assumed that the thin electrical insulator has a negligible heat-flow resistance. The contact resistances have been neglected (but can be significant). Note that the extent of cooling below the ambient, $T_{f,\infty} - T_{s,c}$, is reduced when allowing for the resistances in the load and the heat sink.

### (F) Operating above or below Optimal Current

When operating at a current other than (3.117), the cooling rate is not maximized. Below, we show by a numerical example that the cooling rate drops when a current larger than or smaller than (3.117) is used.

---

EXAMPLE 3.13. FUN.S

---

Consider a bismuth telluride ($Bi_2Te_3$) $p$- and $n$-type thermoelectric unit. The conductors have a circular cross section with $D_n = D_p = 3$ mm and have a length $L_n = L_p = 6$ mm. Plot the variation of cooling power $-Q_c$ with respect to the hot-cold junction temperature difference $T_h - T_c$, for different currents $J_e$. Use $T_h = 308.60$ K.

**SOLUTION**
Using (3.112), and from the properties in Table C.9(a), we have

$$\alpha_S = \alpha_{S,p} - \alpha_{S,n} = (2.30 \times 10^{-4} + 2.10 \times 10^{-4})(\text{V/K}) = 4.4 \times 10^{-4} \text{V/K}.$$

Using (3.116), and from the properties in Table C.9(a), we have

$$R_{e,h\text{-}c} = \left(\frac{\rho_e L}{A}\right)_p + \left(\frac{\rho_e L}{A}\right)_n = \frac{\rho_{e,p} L}{\pi D^2/4} + \frac{\rho_{e,n} L}{\pi D^2/4}$$

$$= \frac{4 \times 2 \times 10^{-5}(\text{ohm-m}) \times 6 \times 10^{-3}(\text{m})}{\pi \times (3 \times 10^{-3})^2(\text{m}^2)} = 1.698 \times 10^{-2} \text{ ohm}$$

$$R_{k,h\text{-}c}^{-1} = \left(k\frac{A}{L}\right)_p + \left(k\frac{A}{L}\right)_n = \frac{k_p \pi D^2/4}{L} + \frac{k_n \pi D^2/4}{L}$$

$$= \frac{(1.7 + 1.45)(\text{W/m-K}) \times \pi \times (3 \times 10^{-3})^2(\text{m}^2)}{4 \times 6 \times 10^{-3}(\text{m})}$$

$$R_{k,h\text{-}c} = 269.5 \text{ K/W}.$$

The thermoelectric figure of merit (3.122) for this system is $Z_e = 0.003078$ $K^{-1}$. For $T_h = 35.45°C = 308.6$ K, using (3.115), the variation of the heat removal rate from the cold junction $Q_c$ is plotted in Figure Ex. 3.13, as a function of $T_h - T_c$, for various currents $J_e = A_k j_e$.

The results show that as the current increases, first $-Q_c$ increases and when the Joule heating contributions become significant, then $-Q_c$ decreases. For the system considered, maximum in $-Q_c$ occurs at $J_e = A_k j_e = 7.997$ A, and the maximum in $T_h$-$T_c$ occurs at 5.92 A. In practice, a smaller current is used.

The maximum $-Q_c$ occurs for $T_h = T_c$ (i.e., $T_h = T_c = 0$) and here it is $Q_{c,max}(T_h = T_c) = -0.5430$ W. Also, for a given current, there is a maximum value for $T_h - T_c$ corresponding to $Q_c = 0$, i.e., no heat added at the cold junction. For this system and for $J_e = 7.997$ A, this gives $(T_h - T_c)_{max} = 146.3°C$.

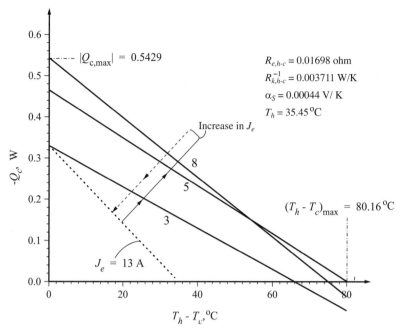

Figure Ex. 3.13. Variation of heat removal rate at the cold junction of a thermoelectric cooler as a function of the current and temperature difference.

**COMMENT**

Note that the required potential drop across the element is found from (2.32) as $\Delta\varphi = J_e R_{r,h\text{-}c}$.

---

EXAMPLE 3.14. FAM

---

A thermoelectric unit is made of bismuth telluride $p$-$n$ pairs. Each pair is made of $p$- and $n$-type elements having a square cross section with each side $a = 1$ mm and each element having a length $L = 1.5$ mm. The elements are placed between two ceramic plates as shown in Figure Ex. 3.14. Each side of the ceramic plates is $a_c = 30$ mm which allows for 14 elements along each side. This results in $N = 7 \times 14$ pairs over the cross section $a_c \times a_c$.

(a) Determine the maximum cooling power $Q_{c,max}$ for $T_h - T_c = 0$.
(b) Determine the maximum cooling power $Q_{c,max}$ for $T_h - T_c = 50°C$.
(c) Determine the corresponding current $J_e(Q_{c,max})$.

The cold plate is at $T_c = 290$ K.

**SOLUTION**

The maximum cooling power is given by (3.118) and for $N$ pairs we have

$$|Q_c|_{max} = N\left[\frac{\alpha_S^2 T_c^2}{2R_{e,h\text{-}c}} - R_{k,h\text{-}c}^{-1}(T_h - T_c)\right].$$

We need to determine $\alpha_S$, $R_{e,h\text{-}c}$, and $R_{k,h\text{-}c}$. From (3.112) for $\alpha_S$ we have

$$\alpha_S = \alpha_{S,p} - \alpha_{S,n}.$$

For $R_{e,h\text{-}c}$, from (3.116) we have, based on the variables defined above,

$$R_{e,h\text{-}c} = \left(\frac{\rho_e L}{A}\right)_p + \left(\frac{\rho_e L}{A}\right)_n = \frac{\rho_{e,p} L}{a^2} + \frac{\rho_{e,n} L}{a^2}.$$

For $R_{k,h\text{-}c}$, from (3.116) we have

$$R_{k,h\text{-}c}^{-1} = \left(k\frac{A}{L}\right)_p + \left(k\frac{A}{L}\right)_n = \frac{k_p a^2}{L} + \frac{k_n a^2}{L}.$$

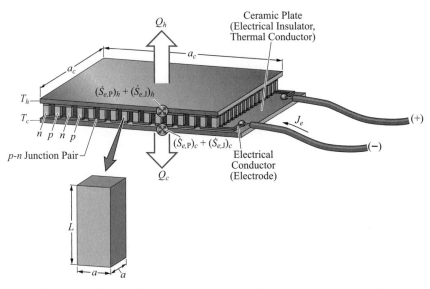

Figure Ex. 3.14. A thermoelectric cooler with many $p$-$n$ cooling units.

Now using the numerical values, we have

$$\alpha_S = 2.3 \times 10^{-4}\ (\text{V/}^\circ\text{C}) + 2.1 \times 10^{-4}\ (\text{V/K}) \qquad \text{Table C.9(a)}$$

$$= 4.4 \times 10^{-4}\ \text{V/}^\circ\text{C or V/K}$$

$$R_{e,h\text{-}c} = 2 \times \frac{1.0 \times 10^{-5}(\text{ohm-m}) \times 1.5 \times 10^{-3}(\text{m})}{10^{-3}(\text{m}) \times 10^{-3}(\text{m})} \qquad \text{Table C.9(a)}$$

$$= 0.030\ \text{ohm}$$

$$R_{k,h\text{-}c}^{-1} = \frac{1.7(\text{W/m-K}) \times 10^{-3}(\text{m}) \times 10^{-3}(\text{m})}{1.5 \times 10^{-3}(\text{m})} \qquad \text{Table C.9(a)}$$

$$+ \frac{1.45(\text{W/m-K}) \times 10^{-3}(\text{m}) \times 10^{-3}(\text{m})}{1.5 \times 10^{-3}(\text{m})}$$

$$= 2.100 \times 10^{-3}\ \text{W/K}.$$

Substituting these numerical values in the above expressions for $|Q_{c,max}|$, we have the following.

(a) The maximum heat flow rate, for $T_h - T_c = 0$ K, is

$|Q_c|_{max}$

$$= 98 \left[ \frac{(4.4 \times 10^{-4})^2 (V/K)^2 \times (290)^2 (K)^2}{2 \times 0.030(ohm)} - 2.100 \times 10^{-3}(W/K) \times (0)(K) \right]$$

$$= 26.60 \text{ W} \quad \text{for} \quad T_h = T_c.$$

(b) The maximum heat flow rate, for $T_h - T_c = 50$ K, is

$|Q_c|_{max}$

$$= 98 \left[ \frac{(4.4 \times 10^{-4})^2 (V/K)^2 \times (290)^2 (K)^2}{2 \times 0.030(ohm)} - 2.100 \times 10^{-3}(W/K) \times (50)(K) \right]$$

$$= 16.30 \text{ W}.$$

(c) The current corresponding to the maximum cooling power is given by (3.117), i.e.,

$$J_e(Q_{c,max}) = \frac{\alpha_S T_c}{R_{e,h-c}}.$$

Using the numerical values, we have

$$J_e(Q_{c,max}) = \frac{4.4 \times 10^{-4}(V/K) \times (290)(K)}{0.03(ohm)}$$

$$= 4.253 \text{ A}.$$

**COMMENT**

Note that using a $T_h - T_c = 50°C$ did not deteriorate the cooling power significantly. However, as shown in Figure Ex. 3.13, at higher values of $T_h - T_c$, a significant drop occurs in the cooling power.

### 3.3.8 Multidimensional Conduction from Buried and Surface Objects*

This section is found on the Web site www.cambridge.org/kaviany. In this section we list the conduction resistance relation for some simple multidimensional media. When no single and dominant direction for the heat transfer exists, the conduction equation needs to be solved for more than one dimension. Some closed-form solutions exist for the steady-state, two- and three-dimensional heat conduction in media with prescribed bounding-surface temperatures. The geometries are categorized as buried objects and surface objects. The resistances are listed in Tables 3.3(a) and (b), and there is an example (Example 3.15).

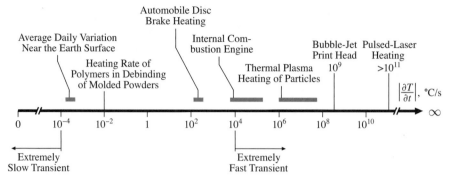

Figure 3.30. Time rate of change of temperature, $|\partial T/\partial t|$, occurring in various applications.

## 3.4 Transient Conduction

Transient conduction addresses time variation of the temperature within a conducting medium. The time rate of change in temperature $\partial T/\partial t$ (i.e., the heating rate) can be small (as in temporal change in the average temperature of the earth atmosphere, i.e., global warming). It can also be very large. For example, in heating of the ink-delivery substrate in thermal inkjet printers (bubble-jet printing), the heated surface is about 50 $\mu m^2$ and the heating rate is about $10^{9}$ °C/s [4,10,21]. Figure 3.30 gives examples of very small, intermediate, and very large values for $|\partial T/\partial t|$(°C/s). Note that under extremely fast heating of the solids and liquids, molecular level nonequilibrium may become significant. For example, in thermal plasmas the lighter electrons heat up much faster than the heavier neutral or ionized atoms and molecules. Also, in high-speed transistors and in fast pulsed laser heating, the electrons heat up to a much higher temperature compared to the lattice atoms. In this section on transient conduction, we first address the simultaneous changes in $T$ with time and with location, i.e., $T = T(\mathbf{x}, t)$. Then we address when the temperature is assumed uniform within the medium, but changes with time.

For example, when an annealed spherical piece of metal is cooled on its surface, through the idealized condition of suddenly maintaining its surface at $T_s$, then the temperature within its interior begins to change. We can determine (i.e., predict) these variations, i.e., $T = T(\mathbf{x}, t)$, using (3.1) to (3.3). This temperature history for the case of a sphere of radius $R$ with a uniform initial temperature $T(t = 0)$, suddenly exposed and maintained at the surface temperature $T_s$, is shown in Figure 3.31. Note that the effect of this change, which begins at the surface, penetrates though the solid and given sufficient elapsed time $t$, the entire sphere will reach $T_s$.

The fluids also undergo transient conduction heat transfer, but because of thermobuoyancy, and due to thermo-surface-tension effect when a liquid-gas interface is present, the fluid may not remain stationary, and therefore, the convection heat transfer should be included.

### 3.4.1 Heat Conductivity Versus Capacity: Thermal Diffusivity $\alpha$

We now examine the intramedium heat transfer with the local change in the temperature with respect to time, $\partial T/\partial t$, not being negligible. The differential-volume

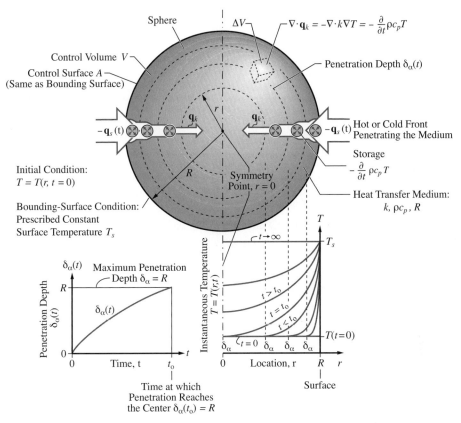

Figure 3.31. Transient temperature distribution in a sphere of initially uniform temperature $T_i$ suddenly subjected to a constant surface temperature $T_s$.

$(\Delta V \rightarrow 0)$ energy equation including the energy conversion term is that given by (3.1), and for a constant $k$ and $\rho$ ($c_p$ was already assumed constant) we have

$$\frac{\int_{\Delta A}(\mathbf{q}_k \cdot \mathbf{s}_a)dA}{\Delta V \rightarrow 0} = \underbrace{-k\nabla^2 T}_{\substack{\text{spatial-rate} \\ \text{of change} \\ \text{of } \mathbf{q}_k}} = \underbrace{-\rho c_p \frac{\partial T}{\partial t}}_{\substack{\text{volumetric} \\ \text{storage}}} + \underbrace{\dot{s}}_{\substack{\text{volumetric} \\ \text{energy} \\ \text{conversion}}} \quad \substack{\text{transient, conduction} \\ \text{energy equation.}}$$

(3.129)

This can be written as

$$-\alpha\nabla^2 T = -\frac{\partial T}{\partial t} + \frac{\dot{s}}{\rho c_p}, \quad \alpha \equiv \frac{k}{\rho c_p} \quad \text{thermal diffusivity,} \quad (3.130)$$

where $\alpha$(m$^2$/s) is the thermal diffusivity and is the ratio of the conductivity to the volumetric (or volume-specific) heat capacity (i.e., the product $\rho c_p$). If $\alpha$ is very large, then the heat transfer medium conducts the heat through it much better than it stores (thus passing the heat through much faster). In contrast, the materials with smaller $\alpha$ do not pass the heat through as much as they store or keep it. Figure 3.32 gives some typical values for the thermal diffusivity for materials in different phases.

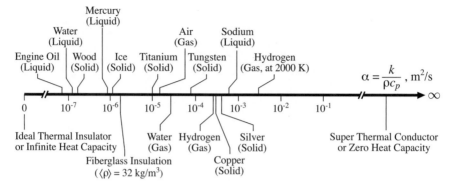

At 300 K and 1 atmosphere pressure, unless stated otherwise.

Figure 3.32. Typical values of thermal diffusivity $\alpha$ for various materials in different phases.

The ideal insulators and/or those materials with large heat capacity will have a very small $\alpha$. The reverse is true for ideal conductors and/or ideal noncapacitors.

Note that a fluid such as the liquid water is a better conductor than air,[†] but has a smaller $\alpha$, because of the larger $\rho c_p$ of the liquid water. Liquid metals have large $\alpha$, because their $k$ is very large.

### 3.4.2 Short and Long Elapsed-Time Behavior: Fourier Number Fo

When the bounding surface of a heat transfer medium is suddenly exposed to a change in the surface temperature $T_s$ or in the heat flux $q_s$, this effect penetrates into the medium. As the elapsed time $t$ increases, this effect penetrates further into the medium. The extent of the penetration into the medium (the distance measured from the surface) is called the thermal penetration distance or the thermal boundary-layer thickness. This distance is generally designated by $\delta_\alpha = \delta_\alpha(t)$. Figure 3.31 shows this thermal boundary-layer thickness for several elapsed times for a sphere subjected to a sudden change in the surface temperature. The initial temperature is assumed to be uniform, i.e., $T(r, t = 0) = T(t = 0)$. The surface temperature is also assumed to be uniform such that the temperature varies only in the radial $r$ direction. Note that $\delta_\alpha$ increases until the effect of the change in the surface temperature reaches the center, i.e., until $\delta_\alpha = R$. The time at which $\delta_\alpha = R$ is designated as the penetration time $t_o$. As will be shown, this time would depend on $R$ and $\alpha$. The

---

[†] For ideal gases, using (3.16), we have

$$\alpha = \frac{k}{\rho c_p} = \frac{1}{3}\frac{c_v}{c_p}\langle u_f\rangle\lambda_f,$$

now using (3.21), we have

$$\alpha = \frac{1}{3}\frac{c_v}{c_p}(\frac{8}{\pi}\frac{R_g}{M}T)^{1/2}\lambda_f$$

This shows that the thermal diffusivity is proportional to the fluid particle mean-free path. The same applies to dielectric solids, where the speed of sound and the phonon mean-free path are used.

ratio of the elapsed time to the diffusion time, i.e., the dimensionless elapsed time, is $t\alpha/R^2$, which is also called the Fourier number $\mathrm{Fo}_R$,

$$\mathrm{Fo}_R \equiv \frac{t\alpha}{R^2} \qquad \text{Fourier number.} \qquad (3.131)$$

Then the dimensionless time for the penetration is given in terms of a constant assigned to the Fourier number. This constant value for the Fourier number is designated as $\mathrm{Fo}_{R,o} \equiv t_o\alpha/R^2$, where $t_o$ is the time required for the penetration. This will be discussed in Section 3.5.2.

### 3.4.3 Distributed versus Lumped Capacitance: Internal-External Conduction Number $N_k$

When the temperature variation within a volume (i.e., a heat transfer medium) is comparable or larger than the difference in temperature between the surface of this volume and its surroundings, then we need to address both the interior and exterior thermal nonequilibrium. Noting that when the thermal resistance (such as conduction) becomes negligibly small we will then have thermal equilibrium between them, then when the thermal resistance interior to a volume is significant compared to the external resistance, we will include this interior temperature nonuniformity. The inclusion of the internal thermal nonequilibrium is called the distributed-capacitance treatment.

On the other hand, when the interior temperature nonuniformity is insignificant compared to the exterior temperature nonuniformity, we assume a uniform interior temperature and this is called the lumped-capacitance treatment. As is expected, high conductivity materials or those with small conduction resistance length tend to have nearly uniform temperatures. In Section 3.6 we will define a dimensionless number, the ratio of internal to external conduction resistances $N_k$, called the internal-external conduction number. For small $N_k$ a uniform temperature can be assumed. Similarly, in Sections 4.6, 5.2, and 6.8, we will define $N_r$, $N_u$, and $N_{ku}$, which are similar ratios but for surface radiation, convection, and surface convection external resistances. However, when considering very short elapsed times (i.e., very small Fourier number $\mathrm{Fo}_{R\,\text{or}\,L}$), most media should be treated as distributed.

As we consider various external resistances to the surface heat transfer, we will continuously compare the interior (conduction) and exterior thermal nonequilibria and when justifiable, we will assume a uniform interior temperature. This assumption greatly simplifies the solutions.

### 3.5 Distributed-Capacitance (Nonuniform Temperature) Transient: $T = T(\mathbf{x}, t)$

We consider temperature nonuniformity caused by a change in the surface thermal conditions. This change then penetrates through the medium.

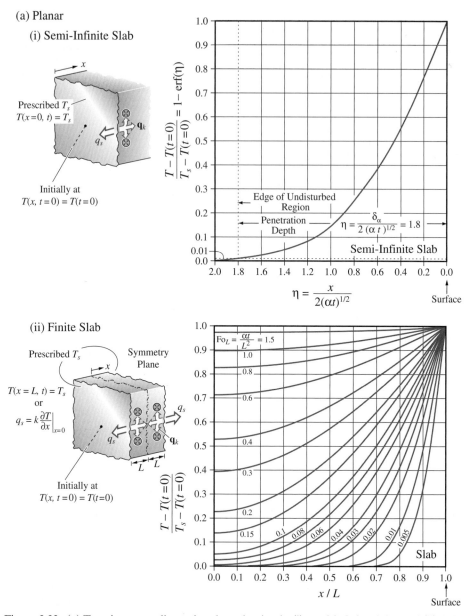

(a) Planar

(i) Semi-Infinite Slab

Prescribed $T_s$
$T(x=0, t) = T_s$

Initially at
$T(x, t=0) = T(t=0)$

$\frac{T - T(t=0)}{T_s - T(t=0)} = 1 - \text{erf}(\eta)$

Edge of Undisturbed Region

Penetration Depth

$\eta = \frac{\delta_\alpha}{2(\alpha t)^{1/2}} = 1.8$

Semi-Infinite Slab

$\eta = \frac{x}{2(\alpha t)^{1/2}}$

Surface

(ii) Finite Slab

Prescribed $T_s$

Symmetry Plane

$T(x = L, t) = T_s$
or
$q_s = k \frac{\partial T}{\partial x}\Big|_{x=0}$

Initially at
$T(x, t=0) = T(t=0)$

$\frac{T - T(t=0)}{T_s - T(t=0)}$

$Fo_L = \frac{\alpha t}{L^2} = 1.5$

Slab

$x / L$

Surface

Figure 3.33. (a) Transient, one-dimensional conduction in (i) semi-infinite slabs, and (ii) finite slabs. (From Carslaw, H.S., and Jaeger, J.C., reproduced by permission © 1986 Clarendon.)

We will examine examples of transient heat transfer that have a sudden change in $T_s$ or $q_s$ specified on the surface. For a one-dimensional heat transfer, planar or radial geometries can be considered. The planar geometry may be a finite slab (a very long and wide rectangular body) or a semi-infinite slab (a very long, wide, and deep body with only one surface being the bounding surface of heat transfer interest). The radial geometry can be a long cylinder or a sphere. Figures 3.33(a) and (b) show these four geometries [7]. For the finite slab, long cylinder, and sphere we

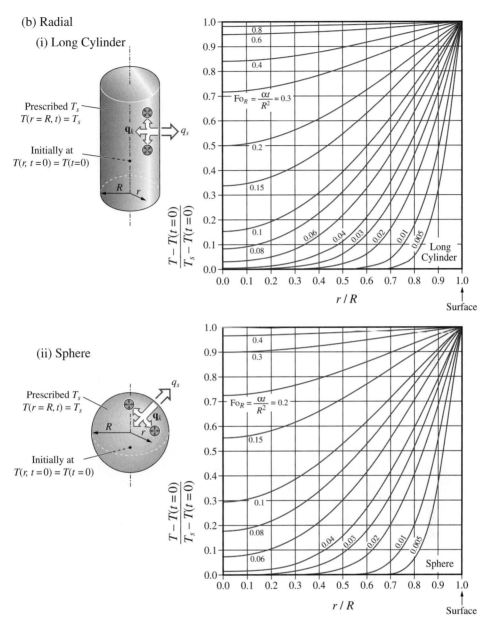

Figure 3.33. (b) Transient, one-dimensional conduction in (i) long cylinders, and (ii) spheres. (From Carslaw, H.S., and Jaeger, J.C., reproduced by permission © 1986 Clarendon.)

assume symmetry around the center plane, line, or point, and place the origin of the coordinate at this location.

The initial condition is assumed to be a uniform temperature within the heat transfer medium. The energy equation has been solved for the temperature for these and many other conditions and reference [7] has a compilation of the solutions.

Table 3.4 gives the solutions as $T = T(\mathbf{x}, t)$, i.e., the temperature distribution at any location and time, for a uniform initial temperature distribution $T(t = 0)$

(c) Average Temperature $\langle T \rangle_V$ (Finite Geometry)

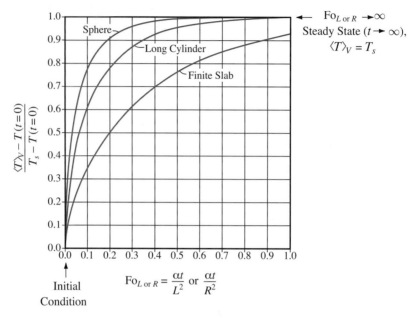

Figure 3.33. (c) Average temperature for a finite slab (thickness $2L$), long cylinder (radius $R$), and sphere (radius $R$), as a function of the Fourier number. (From Carslaw, H.S., and Jaeger, J.C., reproduced by permission © 1986 Clarendon.)

with a sudden change in the surface temperature to $T_s \neq T(t=0)$. For the case of finite slabs, i.e., long cylinders, and spheres, the solution is in the forms of series. The time dependence is expressed with exponential terms, showing that for $t \to \infty$ the steady-steady temperature is reached, i.e., $T(x, t \to \infty) = T_s$. The spatial variations are in terms of the sinusoidal functions or other special functions called the Bessel functions. In general, for the short-time (i.e., small $t$) solutions, many terms are needed in the series solution and graphical presentations are preferable. For the case of a semi-infinite slab, the solution to an imposition of a prescribed surface heat flux $q_s$ is also given in Table 3.4. Graphical presentations of the results for the temperature distributions are given in Figures 3.33(a) and (b). The results for the semi-infinite slab are in terms of the error function, erf. This is a special integral function and will be discussed in Section 3.5.1, where a tabulated presentation of the error function is also made (Table 3.5). Note that for all finite geometries, the effect of change in the surface temperature penetrates until it completely covers the entire body, resulting in a uniform temperature. For the semi-infinite slab, the change continues.

For the case of a semi-infinite medium with a sudden imposition of a surface heat flux $-q_s$, the solution is also given in Table 3.4. For this case the surface temperature changes with time, i.e., $T(x=0, t) = T_s(t)$. The change in surface temperature is found by setting $x = 0$ in the expression given in Table 3.4. The temperature at any location $x$ is found similarly. Note that following our convention, $q_s$ is taken as

Table 3.4. *Solution to transient, one-dimensional temperature distribution for semi-infinite slabs, finite slabs, long cylinders, and spheres.*

| Geometry | Prescribed Boundary Condition on $A_s$ | Transient Temperature Distribution $T = T(\mathbf{x}, t)$ |
|---|---|---|
| semi-infinite slab | (i) $T_s$ constant [Figure 3.33 (a)(i)] | $T(x, t) = T(t = 0) + [T_s - T(t = 0)]\{1 - \text{erf}[\frac{x}{(4\alpha t)^{1/2}}]\}$ <br><br> from bounding surface energy balance, <br><br> $q_s = k\frac{\partial T}{\partial x}\mid_{x=0} = -\frac{k[T_s - T(t = 0)]}{(\pi\alpha t)^{1/2}}$ <br><br> $= -\frac{(k\rho c_p)^{1/2}[T_s - T(t = 0)]}{(\pi t)^{1/2}},$ <br><br> this heat flux is not constant <br> $\text{erf}(\eta), \eta = x/(4\alpha t)^{1/2}$, is the *error function* |
| | (ii) $q_s$ constant | $T(x, t) = T(t = 0) - \frac{q_s(4\alpha t)^{1/2}}{\pi^{1/2}k}e^{-\frac{x^2}{4\alpha t}}$ <br><br> $+ \frac{q_s x}{k}\{1 - \text{erf}[\frac{x}{(4\alpha t)^{1/2}}]\}$ <br><br> surface temperature $T(x = 0, t)$ as well as $T(x, t)$ change depending on $q_s$, $q_s$ is negative when flowing into the slab |
| finite slab | $T_s$ constant [Figure 3.33 (a)(ii)] | $T(x, t) = T_s - \frac{4[T_s - T(t = 0)]}{\pi}$ <br><br> $\times \sum\limits_{i=0}^{\infty}\frac{(-1)^i}{2i + 1}e^{-[\alpha(2i+1)^2\pi^2 t]/(4L^2)}\cos\frac{(2i + 1)\pi x}{2L}$ |
| long cylinder | $T_s$ constant [Figure 3.33 (b)(i)] | $T(r, t) = T_s - \frac{2[T_s - T_i]}{R}\sum\limits_{i=1}^{\infty}e^{-\alpha a_i^2 t}\frac{J_o(r\,a_i)}{a_i J_1(R\,a_i)}$ <br><br> $a_i$'s are *roots* found from $J_o(R\,a_i) = 0$, and $J_o$ and $J_1$ are the *Bessel functions*; $a_i$'s are listed in [6] |
| sphere | $T_s$ constant [Figure 3.33 (b)(ii)] | $T(r, t) = T_s + \frac{2[T(t = 0) - T_s]}{r}\sum\limits_{i=1}^{\infty}\frac{\sin(a_i r)}{a_i}(-1)^{i+1}e^{-\alpha a_i^2 t}$ <br><br> $a_i$'s are roots found from $a_i R = \pi i$ |

(From Carslaw, H.S., and Jaeger, J.C., reproduced by permission © 1986 Clarendon.)

positive when flowing away from the surface, as shown in Figures 3.33(a) and (b). Therefore, for $q_s < 0$ heat flows into the surface and for $q_s > 0$ heat is removed from the surface.

For the finite geometries, for large elapsed times ($t \to \infty$), a steady-state uniform temperature, i.e., $T(\mathbf{x}, t \to \infty) = T_s$, is reached. Figure 3.33(c) gives the average temperature $\langle T\rangle_V = \int_V T dV$ for the three finite geometries, as a function of the Fourier number. Note that the average temperature for the cylinder and the sphere reach $T_s$ at a smaller Fourier number, compared to that for the finite slab. A uniform temperature, i.e., $\langle T\rangle_V = T_s$, is nearly reached for $\text{Fo}_R > 0.5$ for the sphere, for $\text{Fo}_R > 0.9$ for the long cylinder, and for $\text{Fo}_L > 1.7$ for the finite slab. The average temperature can be used to determine the instantaneous, relative heat content $\Delta\int_0^t Qdt = \rho c_p V[\langle T\rangle_V(t) - T(t = 0)]$.

## EXAMPLE 3.16. FAM

In a thick film, thermal barrier coating process, spherical zirconia ($ZrO_2$) particles are heated in a thermal plasma, melted, and deposited on a substrate. This
is shown in Figure Ex. 3.16. The average particle diameter is $D$. The particle
arrives in the plasma with an initial uniform temperature $T(t = 0)$. Assume that
the surface-convection resistance is negligible such that the particle surface temperature suddenly changes to $T_s = T_{f,\infty}$.

(a) Determine the temperature at the center of the particle after an elapsed time
$t$ equal to the time of particle flight $L/u_p$ (elapsed time for impaction).
(b) Neglect the effect of melting on the transient conduction within the particle.
Has the center of the particle melted?
(c) If the distance to the substrate is reduced by $1/2$, what would be the temperature at the center of the particle at the time of impaction?
(d) What is the average temperature $\langle T \rangle_V$ for (c) above? What fraction of the
maximum stored energy has the particle reached in (c)?

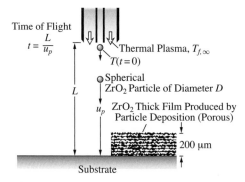

Figure Ex. 3.16. Zirconia particle heated in a thermal plasma to be melted and then
deposited on a substrate.

$$T_{f,\infty} = 2{,}800°C, \quad L = 2.5 \text{ cm}, \quad D = 30 \ \mu\text{m}, \quad u_p = 100 \text{ m/s}, \quad T_{sl} = 2{,}764°C,$$
$$T(t = 0) = 100°C.$$

Use properties from Table C.17.

**SOLUTION**
(a) To determine the temperature at the center of the sphere at time $t$, we use
Figure 3.33(b)(ii). The center is marked by $r = 0$ and the temperature is given
in the dimensionless form $T - T(t = 0)/[T_s - T(t = 0)]$. From Table C.17, we
have for zirconia

$$\rho = 5{,}680 \text{ kg/m}^3 \qquad\qquad \text{Table C.17}$$

$$c_p = 610 \text{ J/kg-K} \qquad\qquad \text{Table C.17}$$

$$k = 1.675 \text{ W/m-K} \qquad\qquad \text{Table C.17}$$

$$\alpha = \frac{k}{\rho c_p} = 4.834 \times 10^{-7} \text{ m}^2/\text{s}.$$

The Fourier number is

$$\mathrm{Fo}_R = \frac{\alpha t}{R^2} = \frac{4\alpha t}{D^2} = \frac{4\alpha L}{u_p D^2}$$

$$= \frac{4 \times (4.834 \times 10^{-7})(\mathrm{m}^2/\mathrm{s}) \times 2.5 \times 10^{-2}(\mathrm{m})}{100(\mathrm{m/s}) \times (3 \times 10^{-5})^2(\mathrm{m}^2)}$$

$$= 0.5371.$$

By examining Figure 3.33(b)(ii), we extrapolate to higher $\mathrm{Fo}_R$ and find that

$$\frac{T(r = 0, \mathrm{Fo}_R = 0.5371) - \tau(t = 0)}{T_s - T(t = 0)} = 1,$$

i.e.,

$$T(r = 0, \mathrm{Fo}_R = 0.5371) = T_s = 2{,}800°\mathrm{C}.$$

(b) As discussed in Section 3.8, the melting of solid reduces the penetration speed. Here we neglect this effect and from the result of (a), we have

$$T(r = 0, \mathrm{Fo}_R = 0.5371) = 2{,}800°\mathrm{C} > T_{sl} = 2{,}764°\mathrm{C} \quad \text{center of particle melts.}$$

(c) For $t = L/2u_p$, i.e., half of the time of flight used in (a), we have

$$\mathrm{Fo}_R = \frac{4\alpha L}{2u_p D^2} = 0.2686.$$

From Figure 3.33(b)(ii), we interpolate that the temperature at the center is

$$\frac{T(r = 0, \mathrm{Fo}_R = 0.2686) - T(t = 0)}{T_s - T(t = 0)} \simeq 0.81$$

or

$$T(r = 0, \mathrm{Fo}_R = 0.2686) = T(t = 0) + 0.81[T_s - T(t = 0)]$$

$$= 100(°\mathrm{C}) + 0.81(2{,}800 - 100)(°\mathrm{C})$$

$$= 2{,}287°\mathrm{C}.$$

Here we note that this is less than the melting temperature $T_{sl} = 2{,}760°\mathrm{C}$.
(d) From Figure 3.33(c), for spheres, we find that for $\mathrm{Fo}_R = 0.2686$ we have

$$\frac{\langle T \rangle_V - T(t = 0)}{T_s - T(t = 0)} \simeq 0.95.$$

Note that although the dimensionless temperature was 0.81 for the center of the sphere, most of the sphere is at a much higher temperature and closer to $T_s$.

The maximum energy that is stored in the sphere occurs where the entire sphere is at $T_s$, i.e., $\langle T \rangle_r = T_s$. Then

$$\frac{\Delta \int_0^t Q dt}{\rho c_p V [T_s - T(t = 0)]} = \frac{\langle T \rangle_V - T(t = 0)}{T_s - T(t = 0)} = 0.95.$$

**COMMENT**

For dimensionless elapsed time (Fourier number) larger than 0.5, the entire sphere has reached $T_s$, as expected and evident also in Figure 3.33(c). Also note that for $Fo_R = 0.03$, the change in surface temperature just reaches the center of the sphere.

EXAMPLE 3.17. FAM

Electrical heaters are used in contact heating (i.e., conduction heating) for various thermal processes. A common example is the fabric iron, rendered in Figures Ex. 3.17(a) and (b). A large fraction of the electrical power flowing into the iron, flows into the fabric. This is designated as $Q_s$. The contact area is $A_k$. Then

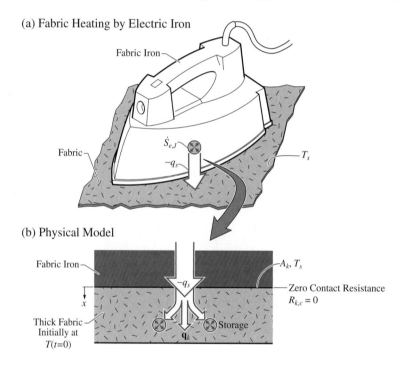

(a) Fabric Heating by Electric Iron

(b) Physical Model

(c) Transient Temperature Distribution

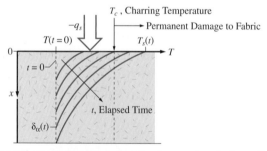

Figure Ex. 3.17. (a) Rendering of fabric heating by an electric iron. (b) Physical model. (c) The transient temperature distribution.

the heat flux is $q_s = Q_s/A_k$. Consider a relatively thick fabric that is initially at $T_i$. Here thick indicates that we are interested in elapsed times such that the effect of heating at the surface (designated by $x = 0$) has not yet penetrated to the other side of the fabric.

The fabric has thermal diffusivity $\langle \alpha \rangle = \langle k \rangle / \langle \rho c_p \rangle = 10^{-7}$ m²/s, and $\langle k \rangle = 0.2$ W/m-K. The heat flux is $q_s = -2 \times 10^4$ W/m², and the initial temperature is $T(t = 0) = 20°$C. The elapsed time is $t = 30$ s. The surface temperature may exceed the charring (or scorching) temperature of fabric $T_c = 180°$C, and then permanent damage is caused to the fabric.

(a) Determine the surface temperature $T_s(t) = T(x = 0, t = 30 \text{ s})$.
(b) Determine the temperature at location $x = 3$ mm from the surface, for this elapsed time, i.e., $T(x = 3 \text{ mm}, t = 30 \text{ s})$.

**SOLUTION**
Because of the thick-layer assumption, the semi-infinite slab geometry can be used. The solution for the transient temperature distribution is given in Table 3.4 and the magnitude of the error function is listed in Table 3.5. The transient temperature distribution is given in Table 3.4 as

$$T(x, t) = T(t = 0) - \frac{q_s(4\langle\alpha\rangle t/\pi)^{1/2}}{\langle k \rangle} e^{-x^2/(4\langle\alpha\rangle t)} + \frac{q_s x}{\langle k \rangle} \left\{ 1 - \text{erf}\left[ \frac{x}{2(\langle\alpha\rangle t)^{1/2}} \right] \right\}.$$

(a) For $x = 0$ and $t = 30$ s, we have

$$T(x = 0, t) = T_s(t) = T(t = 0) - \frac{2q_s(\langle\alpha\rangle t/\pi)^{1/2}}{\langle k \rangle}.$$

This shows that the surface temperature changes with $q_s$ to the first power and $t$ to the 1/2 power.

Using the numerical values, we have

$T(x = 0, t = 30 \text{ s})$

$$= 20(°\text{C}) - \frac{2 \times [-(2 \times 10^4)(\text{W/m}^2)][10^{-7}(\text{m}^2/\text{s}) \times 30(\text{s})/\pi]^{1/2}}{0.2(\text{W/m-}°\text{C})}$$

$$= 20(°\text{C}) + 195.5(°\text{C}) = 215.5°\text{C}.$$

This shows that the surface temperature exceeds the charring temperature at an elapsed time of $t = 30$ s.

(b) For $x = 3$ mm and at $t = 30$ s, we have

$$\frac{x^2}{4\langle\alpha\rangle t} = \frac{(3 \times 10^{-3})^2(\text{m}^2)}{4 \times 10^{-7}(\text{m}^2/\text{s}) \times 30(\text{s})} = 0.75$$

$$\text{erf}[(0.75)^{1/2}] = \text{erf}(0.8660) = 0.7850 \quad \begin{array}{l}\text{from Figure 3.33(a)(i) or for more} \\ \text{accurate results from interpolation} \\ \text{in Table 3.5}\end{array}$$

$T(x = 3 \text{ mm}, t = 30 \text{ s})$

$$= 20(^\circ\text{C}) - \frac{2 \times [-2 \times 10^4(\text{W/m}^2)][10^{-7}(\text{m}^2/\text{s}) \times 30(\text{s})/\pi]^{1/2}}{0.2(\text{W/m-}^\circ\text{C})} e^{-0.75}$$

$$+ \frac{[-2 \times 10^4(\text{W/m}^2)] \times 3 \times 10^{-3}(\text{m})}{0.2(\text{W/m-}^\circ\text{C})} [1 - \text{erf}(0.8660)]$$

$$= 20(^\circ\text{C}) + 92.34(^\circ\text{C}) - 64.49(^\circ\text{C}) = 47.85^\circ\text{C}.$$

**COMMENT**

Note the rather sharp drop in the temperature occurring between the two locations $x = 0$ and $x = 3$ mm. This is due to the rather low effective diffusivity $\langle\alpha\rangle$ of the fabric. Figure Ex. 3.17(c) gives the quantitative, transient temperature distribution within the fabric.

### 3.5.1 Derivation of Temperature Distribution for Semi-Infinite Slab: Thermal Effusivity $(\rho c_p k)^{1/2}$

As an example of derivation of the transient temperature distributions given in Table 3.4, consider the transient conduction in a semi-infinite slab, with no energy conversion. The energy equation (3.129), for the coordinate system shown in Figure 3.33(a)(i) with $x = 0$ at the surface of the semi-infinite slab, becomes

$$\underbrace{\frac{\partial}{\partial x}\mathbf{q}_k \cdot \mathbf{s}_x}_{\substack{\text{spatial-rate of} \\ \text{change of } \mathbf{q}_k \\ \text{in } x \text{ direction}}} = \underbrace{-\rho c_p \frac{\partial T}{\partial t}}_{\substack{\text{volumetric} \\ \text{storage}}} \quad \text{differential-volume energy equation} \tag{3.132}$$

or

$$-\alpha\frac{\partial^2 T}{\partial x^2} = -\frac{\partial T}{\partial t} \tag{3.133}$$

$$\alpha\frac{\partial^2 T}{\partial x^2} - \frac{\partial T}{\partial t} = 0 \quad \text{a partial differential equation.} \tag{3.134}$$

We consider the case of a prescribed surface temperature $T_s$, given by (3.2) for the surface and by (2.66) for the far field. The uniform initial condition is given by (3.3) as $T = T(x, t = 0) = T(t = 0)$. Then the bounding-surface and initial conditions are

$$T(x = 0, t) = T_s$$

$$T(x \to \infty, t) = T(t = 0)$$

$$T(x, t = 0) = T(t = 0). \tag{3.135}$$

We note that both $x$ and $t$ have a semi-infinite range, i.e., $0 \leq x < \infty$, and $0 \leq t < \infty$. This and the characteristics of the second-order, partial differential energy equation (3.134) allow for the introduction of a similarity variable $\eta$ that combines $x$ and $t$ and reduces this partial differential equation to an ordinary differential equation. The dimensionless similarity variable $\eta$ is defined as

$$\eta = \frac{x}{2(\alpha t)^{1/2}} \quad \text{dimensionless similarity variable.} \tag{3.136}$$

The power of $x$ is unity and that of $t$ is 1/2. This is due to the order of the derivatives for $x$ and $t$ in (3.134). We also make the temperature dimensionless by the proper scaling. The temperature scale we have is $T_s - T(t = 0)$, so we define the dimensionless temperature $T^*$ (the asterisk indicates nondimensional quantity) as

$$T^*(\eta) = \frac{T - T(t = 0)}{T_s - T(t = 0)}, \quad 0 \leq T^* \leq 1 \quad \text{dimensionless temperature.} \tag{3.137}$$

Using $\eta$, the boundary and initial conditions given in (3.135) are reduced from three to two, i.e., are now

$$T^*(\eta = 0) = 1,$$

$$T^*(\eta \to \infty) = 0 \quad \text{surface and far-field thermal conditions.} \tag{3.138}$$

Substituting (3.136) and (3.137) in (3.134), the result is

$$\frac{d^2 T^*}{d\eta^2} + 2\eta \frac{dT^*}{d\eta} = 0 \quad \text{dimensionless energy equation, an ordinary differential equation.} \tag{3.139}$$

The solution to (3.139) subject to the conditions of (3.138) is

$$T^* = \frac{T - T(t = 0)}{T_s - T(t = 0)} = 1 - \text{erf}(\eta) \quad \text{dimensionless temperature distribution,} \tag{3.140}$$

where the (Gaussian) error function is defined as

$$\text{erf}(\eta) = \frac{2}{\pi^{1/2}} \int_0^\eta e^{-z^2} dz \quad \text{error function.} \tag{3.141}$$

Substituting for $\eta$ in (3.140), the temperature distribution is given in terms of $x$ and $t$ as

$$T(x, t) = T(t = 0) + [T_s - T(t = 0)] \left\{ 1 - \text{erf} \left[ \frac{x}{2(\alpha t)^{1/2}} \right] \right\} \quad \begin{array}{l} \text{dimensional} \\ \text{temperature} \\ \text{distribution,} \end{array} \tag{3.142}$$

which is the form given in Table 3.4.

Table 3.5 lists the variation of $\text{erf}(\eta)$ with respect to $\eta$. Note that the variation of $\text{erf}(\eta) = \text{erf}[x/2(\alpha t)^{1/2}]$, with respect to $\eta = x/2(\alpha t)^{1/2}$, is also shown graphically in Figure 3.33(a). For $t > 0$ and $x = 0$, $x/2(\alpha t)^{1/2}$ is also zero and $\text{erf}(0) = 0$; then

Table 3.5. *Error function erf($\eta$) as a function of $\eta$*

| $\eta$ | erf($\eta$) | $\eta$ | erf($\eta$) | $\eta$ | erf($\eta$) |
|---|---|---|---|---|---|
| 0.000 | 0.00000 | 0.600 | 0.60386 | 1.600 | 0.97635 |
| 0.040 | 0.04511 | 0.700 | 0.67761 | 1.800 | 0.98909 |
| 0.080 | 0.09008 | 0.800 | 0.74210 | 1.829 | 0.99000 |
| 0.100 | 0.11246 | 0.900 | 0.79673 | 2.000 | 0.99532 |
| 0.200 | 0.22270 | 1.000 | 0.84270 | 3.000 | 0.99998 |
| 0.300 | 0.32863 | 1.100 | 0.88006 | 5.000 | 1.00000 |
| 0.400 | 0.42839 | 1.200 | 0.91031 | $\infty$ | 1.00000 |
| 0.500 | 0.52033 | 1.400 | 0.95228 | | |

(3.142) gives $T(x = 0, t) = T_s$, as expected. For $t > 0$, as $x \to \infty$, i.e., $x/2(\alpha t)^{1/2} \to \infty$, we have erf($\infty$) = 1; then (3.142) gives $T(x \to \infty, t) = T(t = 0)$, again as expected. Note that for $\eta > 2$, erf($\eta$) does reach its asymptotic value of unity and does no longer change as $\eta$ is further increased. This is also evident in Figure 3.33(a).

The rate of heat flow at $x = 0$, i.e., the heat transfer rate to or from any of these surfaces, is found from the bounding-surface energy equation (2.63) with no radiation and $Q_{\rho ck}$ flowing away from the surface. Here for the case of the semi-infinite slab (with $\mathbf{s}_n = -\mathbf{s}_x$), we have

$$(\mathbf{q} \cdot \mathbf{s}_n)A_k = -Q_{\rho ck}(t) = -Q_s(t) = -A_k q_{\rho ck}$$

$$= -A_k q(x = 0, t) = -A_k k \frac{\partial T}{\partial x}\Big|_{x=0} \quad \text{surface energy equation,} \quad (3.143)$$

where we have taken $Q_{\rho ck}$ as positive when it is flowing away from the surface.

By differentiating (3.142), using the chain rule, and evaluating the resultant at $x = 0$, we have

$$q_{\rho ck}(t) = q_s(t) = -k[T_s - T(t = 0)]\frac{\partial T^*}{\partial x}\Big|_{x=0}$$

$$= -k[T_s - T(t = 0)] = -\frac{k}{(\pi \alpha t)^{1/2}}[T_s - T(t = 0)]\frac{\partial \eta}{\partial x}\frac{dT^*}{d\eta}$$

$$= -\frac{(\rho c_p k)^{1/2}}{\pi^{1/2}t^{1/2}}[T_s - T(t = 0)] \quad \begin{matrix}\text{transient surface heat flux,}\\ (\rho c_p k)^{1/2} \text{ is thermal effusivity,}\end{matrix} \quad (3.144)$$

where from (3.136) $\partial \eta/\partial x = 1/(2\alpha t)^{1/2}$ and from (3.141) $d[\text{erf}(\eta)]/d\eta = (2/\pi^{1/2})e^{-\eta^2}$.

This shows that the magnitude of surface heat flux is initially (i.e., $t = 0$) infinite, because the derivative in (3.144) at $t = 0$ is infinite (i.e., the change in the surface temperature has not yet penetrated into the semi-infinite slab). As the elapsed time $t$ increases, the magnitude of the surface heat flux decreases. As $t \to \infty$, the temperature distribution adjacent to the surface becomes nearly uniform, and then $q_{\rho c_p k}$ tends to zero. The quantity $(\rho c_p k)^{1/2}$ is called the thermal effusivity and has a unit of (W-s$^{1/2}$/m$^2$-K). Materials with large effusivity transport a large amount of heat away from their surface, with a small $T_s - T(t = 0)$.

Table 3.6. *Transient conduction resistance*

| Heat Flow Rate | Potential | Resistance |
|---|---|---|
| $Q_{\rho ck}(t) = A_k q_{\rho ck},$ | $T(t=0) - T_s, {}^\circ C$ | $R_{\rho ck}(t) = \left(\dfrac{\pi t}{\rho c_p k}\right)^{1/2} / A_k, {}^\circ C/W$, semi-infinite |
| $W = [T(t=0) - T_s]/R_{\rho ck}(t)$ | | slab |

The results (3.142) and (3.144) can be extended to the case where $T_s = T_s(t)$. This is discussed in Section 3.5.3.

Similar to the steady-state conduction resistance given by (3.50), we can write (3.140) as

$$Q_{\rho ck}(t) = \frac{T(t=0) - T_s}{R_{\rho ck}(t)} = -\frac{T_s - T(t=0)}{R_{\rho ck}(t)}$$

$$= A_k q_{\rho c_p k} \tag{3.145}$$

$$= A_k \frac{(\rho c_p k)^{1/2}}{\pi^{1/2} t^{1/2}} [T(t-0) - T_s]$$

or

$$R_{\rho ck}(t) = \frac{T(t=0) - T_s}{Q_{\rho ck}(t)} = \frac{\pi^{1/2} t^{1/2}}{A_k (\rho c_p k)^{1/2}} \quad \begin{array}{l} \text{transient conduction resistance} \\ \text{for semi-infinite slab,} \end{array} \tag{3.146}$$

where $R_{\rho ck}(t)$ is the transient conduction resistance. The temperature difference $T(t=0) - T_s$ used is based on $Q_{\rho ck}$ being positive when flow is out of the surface. Similar expressions can be found for $R_{\rho ck}(t)$ for slabs, long cylinders, and spheres, but the expressions are lengthy and are not given here. We note that the transient conduction resistance does not have an electrical counterpart, but allows for a simple thermal circuit presentation of this semi-bounded medium.

Table 3.6 lists the transient conduction resistance.

## EXAMPLE 3.18. FUN

Those who demonstrate barefoot stepping on hot (glowing) rocks (or charcoal), rely on high porosity materials (e.g., volcanic rocks and charred wood) with a very small effective conductivity $\langle k \rangle$ and a very small effective volume-specific heat capacity $\langle \rho c_p \rangle$ [i.e., a small thermal effusivity $(\rho c_p k)^{1/2}$]. This is rendered in Figure Ex. 3.18(a).

When two solids initially at two different temperatures $T_1(t=0)$ and $T_2(t=0)$, are brought together with a zero contact resistance, the interface temperature will remain constant at $T_s$. Upon using the continuity of temperature (2.57), and applying the interfacial energy equation (2.62), with no convection, radiation, and energy conversion, and using (3.144) for the interfacial conduction

(a) Physical Model

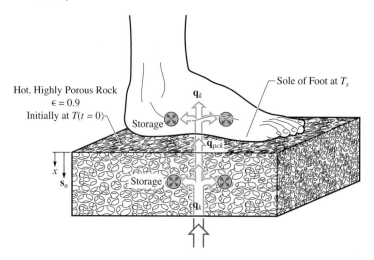

(b) Temperature Distribution

(c) Transient Surface Heat Flux

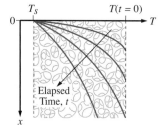

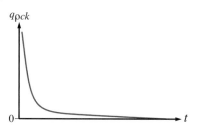

Figure Ex. 3.18. Sole of a foot in contact with a hot high porosity rock. (a) Physical model. (b) Temperature distribution. (c) Transient surface heat flux.

heat flux for each medium, it can be shown (left as an end of chapter problem) that

$$T_{12} = \frac{(\rho c_p k)_1^{1/2} T_1(t=0) + (\rho c_p k)_2^{1/2} T_2(t=0)}{(\rho c_p k)_1^{1/2} + (\rho c_p k)_2^{1/2}}$$

$$= \frac{T_1(t=0) + \dfrac{(\rho c_p k)_2^{1/2}}{(\rho c_p k)_1^{1/2}} T_2(t=0)}{1 + \dfrac{(\rho c_p k)_2^{1/2}}{(\rho c_p k)_1^{1/2}}} \qquad \begin{array}{l}\text{interfacial temperature for two} \\ \text{semi-infinite solids of different} \\ \text{temperatures suddenly brought} \\ \text{into contact.}\end{array}$$

From the above, when a solid, say material 1 with high values for $k$ and $\rho c_p$ and uniform temperature $T_1(t=0) = T_s$, suddenly comes in contact with another solid, say material 2 with a much lower $k$ and $\rho c_p$ and at a uniform temperature $T_2(t=0) = T(t=0)$, then the contact surface will be nearly at $T_s$. This is the result of setting the thermal effusivity ratio $(\rho c_p k)_2^{1/2}/(\rho c_p k)_1^{1/2}$ equal to zero in the above equation. Physically, this is because the material with a much higher $\rho c_p$ will not have much change in its temperature. The solid with the lower values for $k$ and $\rho c_p$ will have a transient temperature distribution given by (3.142) and the surface heat flux given by (3.144).

The human body has much higher $k$ and $\rho c_p$ than the high porosity rock. Therefore, the rock surface in contact with the sole of a foot will be nearly at $T_s$. To avoid damage to the living cells under the dead skin cells, the heat flow rate to the body (i.e., the sole of foot) should be small enough such that it can be removed by the interstitial blood flow.

Consider a rock with a porosity $\epsilon = 0.9$, with the solid properties being that of silicon dioxide. Use (3.28) for $\langle k \rangle$ and (3.10) for $\langle \rho c_p \rangle$ and air as the fluid filling the pores. Using $T_s = 32°C$ and $T(t = 0) = 500°C$, determine the heat flow rate to the surface after (i) 1, (ii) 10, and (iii) 100 s elapsed times.

$\rho_s = 2{,}200$ kg/m$^3$, $c_{p,s} = 745$ J/kg-°C, $k_s = 1.38$ W/m-K, $\rho_f = 1.161$ kg/m$^3$, $c_{p,f} = 1{,}007$ J/kg-K, and $k_f = 0.0263$ W/m-K.

**SOLUTION**

We assume that the elapsed time of interest is sufficiently short to allow for the treatment of the porous rock as a semi-infinite slab with the bounding surface maintained at the constant surface temperature $T(x = 0)$. The heat flux out of the rock at $x = 0$ is given by (3.144), i.e.,

$$q_{\rho ck} = -\frac{\langle k \rangle}{(\pi \langle \alpha \rangle t)^{1/2}}[T_s - T(t = 0)] = -\frac{\langle k \rangle^{1/2}\langle \rho c_p \rangle^{1/2}}{\pi^{1/2}t^{1/2}}[T_s - T(t = 0)]$$

$$q_{\rho ck} = -\frac{1}{\pi^{1/2}}(\langle k \rangle \langle \rho c_p \rangle)^{1/2}\frac{1}{t^{1/2}}[T_s - T(t = 0)].$$

The effective conductivity $\langle k \rangle$ is determined using the correlation (3.28), i.e.,

$$\frac{\langle k \rangle}{k_f} = \left(\frac{k_s}{k_f}\right)^{0.280-0.757\log(\epsilon)-0.057\log(k_s/k_f)}.$$

The effective heat capacity is given by (3.10) as

$$\langle \rho c_p \rangle = \epsilon(\rho c_p)_f + (1 - \epsilon)(\rho c_p)_s.$$

The numerical values for $\langle k \rangle$, $\langle \rho c_p \rangle$, and $\langle \alpha \rangle$ are

$$k_s = 1.38 \text{ W/m-K}$$

$$k_f = 0.0263 \text{ W/m-K}$$

$$(\rho c_p)_s = 1.654 \times 10^6 \text{ J/m}^3\text{-K}$$

$$(\rho c_p)_f = 1.169 \times 10^3 \text{ J/m}^3\text{-K}$$

$$\epsilon = 0.9$$

$$k_s/k_f = 52.47$$

$$\langle k \rangle = 0.03266 \text{ W/m-K}$$

$$\langle \rho c_p \rangle = 1.665 \times 10^5 \text{ J/m}^3\text{-K}$$

$$\langle \alpha \rangle = \frac{\langle k \rangle}{\langle \rho c_p \rangle} = 1.959 \times 10^{-7} \text{ m}^2\text{/s}.$$

Now we use $\langle \alpha \rangle$, $T_s - T(t = 0)$, and $t$ in the relation for $q_s$ given above, and we have

$$q_{\rho c k} = -\frac{1}{\pi^{1/2}}\frac{[0.03266(\text{W/m-}^\circ\text{C}) \times 1.665 \times 10^5 (\text{J/m}^3\text{-}^\circ\text{C})]^{1/2}}{t^{1/2}}[30(^\circ\text{C}) - 500(^\circ\text{C})]$$

(i) $t = 1\,\text{s}$:     $q_{\rho c k} = 1.955 \times 10^4$ W/m$^2$
(ii) $t = 10\,\text{s}$:    $q_{\rho c k} = 6.183 \times 10^3$ W/m$^2$
(iii) $t = 100\,\text{s}$:   $q_{\rho c k} = 1.955 \times 10^3$ W/m$^2$.

**COMMENT**

Note that $\langle k \rangle$ is close to that of air alone, i.e., a very small conductivity. Also note that the $\langle \alpha \rangle$ is also very small, i.e., smaller than air diffusivity of $2.250 \times 10^{-5}$ m$^2$/s.

Note that $q_{\rho c_p k}$ at $t = 1$ s is not small. Also note that because of the $t^{-1/2}$ relation, this heat flux decreases by a factor 1/10 when $T$ increases by a hundredfold. Therefore, only the heat flow during the small elapsed times can have a potentially damaging effect. By increasing the thickness of the skin (i.e., layer of dead cells), this surface heat flux will be partly absorbed in the skin, and therefore, will be prevented from reaching the living cells. This would further avoid the damage. Evaporation of any moisture content of the skin would also tend to decrease $q_{\rho c k}$. The temperature distribution and the time variation of $q_{\rho c k}$ are quantitatively shown in Figures Ex. 3.18(b) and (c).

### 3.5.2 Penetration Depth $\delta_\alpha$, Penetration Fourier Number Fo$_\delta$, and Penetration Speed $u_\delta$

We now examine the history of the penetration of the change initialized at the surface (i.e., at $x = 0$) as it travels into the heat transfer medium.

### (A) Semi-Infinite Slab

First note from Figure 3.33(a) and Table 3.6, that for $\eta = x/2(\alpha t)^{1/2} = 1.8$, we approximately have erf$(\eta) = 0.99$. Then the dimensionless temperature $T^*$ from (3.142) is 0.01, i.e.,

$$T^*\left[\frac{x}{2(\alpha t)^{1/2}} = 1.8\right] \equiv \frac{T\left[\frac{x}{2(\alpha t)^{1/2}}\right] - T(t = 0)}{T_s - T(t = 0)} = 0.01 \qquad \text{criterion for penetration depth for semi-infinite slab.}$$

$$(3.147)$$

This indicates that the dimensionless temperature is changed by only 0.01 of the maximum change $T_s - T(t = 0)$. This is small enough such that we can choose to designate $x/2(\alpha t)^{1/2} = 1.8$ as the edge of the undisturbed region. This is marked in Figure 3.33(a)(i). Now if we designate the length of the disturbed region by $\delta_\alpha$, we have

$$\frac{\delta_\alpha}{2(\alpha t)^{1/2}} = 1.8 \quad \text{or} \quad \delta_\alpha = 3.6(\alpha t)^{1/2} = 3.6\alpha^{1/2}t^{1/2} \quad \begin{array}{l}\text{penetration depth}\\ \text{for semi-infinite slab.}\end{array} \qquad (3.148)$$

This length $\delta_\alpha$ is called the penetration depth. Then at any time $t$ we can determine the penetration length (or depth) $\delta_\alpha$ using (3.142). Using (3.131), we can write (3.147) in terms of the penetration Fourier number $\mathrm{Fo}_\delta$ as

$$\mathrm{Fo}_\delta = \frac{t\alpha}{\delta_\alpha^2} = \frac{1}{3.6^2} = 0.077 \simeq 0.08 \quad \begin{array}{l}\text{penetration Fourier number}\\ \text{for semi-infinite slab.}\end{array} \qquad (3.149)$$

### (B) Finite Slab, Long Cylinder, and Sphere

The solution for the temperature distributions in finite slabs, long cylinders, and spheres, with uniform initial temperatures and a sudden change (at time $t = 0^+$) in the surface temperature to $T_s$ were also given in Table 3.4 and the results were shown graphically in Figures 3.33(a) and (b).

The dimensionless time for the penetration to travel the distance $L$ is found in Figure 3.33(a)(ii) and is nearly 0.07. Since we have used $T^* = 0.01$, the semi-infinite and finite slab have slightly different penetrations due to the different conditions used at $x = L$. For the finite slab we have

$$\mathrm{Fo}_{L,o} = \frac{t\alpha}{L^2} \simeq 0.07 \quad \text{penetration Fourier number for finite slab.} \qquad (3.150)$$

For $\mathrm{Fo}_L > 0.07$ the temperature at location $x = 0$ (center-plane of the slab) begins to change noticeably.

For long cylinders, Figure 3.33(b)(i) shows that $\mathrm{Fo}_R$ corresponding to noticeable change in the temperature at the centerline is

$$\mathrm{Fo}_{R,o} = \frac{t\alpha}{R^2} \simeq 0.04 \quad \text{penetration Fourier number for long cylinder.} \qquad (3.151)$$

For spheres, Figure 3.33(b)(ii) shows that $\mathrm{Fo}_R$ corresponding to noticeable change in the temperature at the center is

$$\mathrm{Fo}_{R,o} = \frac{t\alpha}{R^2} \simeq 0.03 \quad \text{penetration Fourier number for sphere.} \qquad (3.152)$$

### (C) Penetration Speed for Semi-Infinite Slab $u_\delta$

For the case of a sudden change in the surface temperature, the effect of this change penetrates into the medium at a speed that decreases as the elapsed time $t$ increases. The penetration speed (or front speed) $u_\delta$ is defined as

$$u_\delta \equiv \frac{d\delta_\alpha}{dt} \qquad \text{penetration speed.} \tag{3.153}$$

For example, for the case of a semi-infinite slab, $u_F$ is found by differentiating (3.148) and the result is

$$u_\delta = \frac{1.8\alpha^{1/2}}{t^{1/2}} \qquad \text{penetration speed for semi-infinite slab.} \tag{3.154}$$

As was mentioned above, this shows that the front speed decreases with an increase in the elapsed time, as $t^{-1/2}$.

### (D) Approaching Steady State

Figures 3.33(a)(i), (b)(i), and (b)(ii), show that a steady state, i.e., $T(x, t) = T_s$ is nearly reached (asymptote), for finite slabs when $Fo_L \simeq 1.7$, for long cylinders when $Fo_R \simeq 0.9$, and for spheres when $Fo_R \simeq 0.5$. For the semi-infinite slab, Figure 3.33(b)(i), there is no steady-state asymptote.

---

### EXAMPLE 3.19. FUN

---

The effective thermal diffusivity $\langle \alpha \rangle$ of soil is rather low, under both dry and wet conditions. Because of this, the seasonal variations of the earth's surface temperature do not penetrate very far from the surface. The seasonal variation of surface temperature for a location is given in Figure Ex. 3.19(a). The data is the low and high monthly averaged surface temperature. The results are for 24 months and show a nearly periodic time variation. Assume that the difference in extremes in the average earth surface temperature, $T_h - T_c$, occurs instantly. This is rendered in Figure Ex. 3.19(b).

(a) Estimate how far this change penetrates after six months.
(b) Estimate the speed of the penetration $u_F$ after three months, i.e., $t = \tau/2$, where $\tau$ is the period.

Use the correlation (3.28) for the effective thermal conductivity $\langle k \rangle$ of a packed bed of particles. For $k_s$ use sand that has a high concentration of silicon dioxide (i.e., silica). The properties of silica are listed in Table C.17 and we have $\rho_s = 2{,}200$ kg/m$^3$, $c_{p,s} = 745$ J/kg-°C, and $k_s = 1.38$ W/m-°C. For porosity use $\epsilon = 0.4$ corresponding to random packing of spheres. The properties of air are listed in Table C.22. At $T = 300$ K, we have $\rho_f = 1.177$ kg/m$^3$, $c_{p,f} = 1{,}005$ J/kg-°C, and $k_f = 0.0263$ W/m-°C. The properties of water are listed in Table C.23. At $T = 293$ K, we have $\rho_f = 1{,}000$ kg/m$^3$, $c_{p,f} = 4{,}181$ J/kg-°C, and

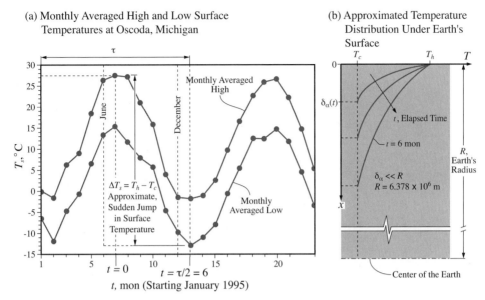

Figure Ex. 3.19. (a) Measured monthly averaged low and high surface temperatures at a location. (b) Assumed sudden change in the earth surface temperature and its penetration into the substrate.

$k_f = 0.597$ W/m-°C. For the effective volume-specific heat capacity $\langle \rho c_p \rangle$, use the simple volume averaging given by (3.10).

**SOLUTION**

We assume that the penetration depth is much smaller than the radius of the earth, and therefore, use the semi-infinite slab geometry. The penetration depth and the front speed $u_\delta$, for this time-periodical surface temperature variation, are estimated using (3.148) and (3.154), i.e.,

$$\delta_\alpha = 3.6(\langle \alpha \rangle t)^{1/2}$$

$$u_\delta = 1.8 \langle \alpha \rangle^{1/2} t^{-1/2}.$$

This is a rough approximation, because to obtain these results we have assumed that $T_s$ is maintained constant. The effective conductivity of a packed bed of particles is given by (3.28) as

$$\frac{\langle k \rangle}{k_f} = \frac{k_s}{k_f}^{0.280 - 0.757 \log \epsilon - 0.057 \log(k_s/k_f)}.$$

The effective volume-specific heat capacity is given by (3.10) as

$$\langle \rho c_p \rangle = \epsilon (\rho c_p)_f + (1 - \epsilon)(\rho c_p)_s.$$

Using the numerical values, the effective properties are

$$k_s = 1.38 \text{ W/m-K, silica}$$

$$k_f = 0.0267 \text{ W/m-K, air}$$

$$k_f = 0.597 \text{ W/m-K, water}$$

$$(\rho c_p)_s = 1.639 \times 10^6 \text{ J/m}^3\text{-K, silica}$$

$$(\rho c_p)_f = 1.183 \times 10^3 \text{ J/m}^3\text{-K, air}$$

$$(\rho c_p)_f = 4.181 \times 10^6 \text{ J/m}^3\text{-K, water}$$

$$\epsilon = 0.4$$

$$k_s/k_f = 51.69, \quad \text{silica-air}$$

$$\langle k \rangle = 0.1797 \text{ W/m-K, silica-air}$$

$$k_s/k_f = 2.312, \quad \text{silica-water}$$

$$\langle k \rangle = 1.074 \text{ W/m-K, silica-water}$$

$$\langle \rho c_p \rangle = 9.839 \times 10^5 \text{ J/m}^3\text{-K, silica-air}$$

$$\langle \rho c_p \rangle = 2.656 \times 10^6 \text{ J/m}^3\text{-K, silica-water}$$

From these, the effective thermal diffusivities, (3.130), are

$$\langle \alpha \rangle = \frac{\langle k \rangle}{\langle \rho c_p \rangle} = 1.826 \times 10^{-7} \text{ m}^2/\text{s, silica-air}$$

$$\langle \alpha \rangle = \frac{\langle k \rangle}{\langle \rho c_p \rangle} = 4.044 \times 10^{-7} \text{ m}^2/\text{s, silica-water.}$$

(a) Penetration Depth

$$t = 6(\text{mon}) \times 30(\text{day/mon}) \times 24(\text{hr/day}) \times 60(\text{min/hr})$$

$$\times 60(\text{s/min}) = 1.555 \times 10^7 \text{ s}$$

$$\delta_\alpha = 3.6 \times (1.826 \times 10^{-7})^{1/2}(\text{m}^2/\text{s})^{1/2}$$

$$\times (1.555 \times 10^7)^{1/2}(\text{s})^{1/2} = 6.066 \text{ m, silica-air}$$

$$\delta_\alpha = 3.6 \times (4.044 \times 10^{-7})^{1/2}(\text{m}^2/\text{s})^{1/2}$$

$$\times (1.555 \times 10^7)^{1/2}(\text{s})^{1/2} = 9.039 \text{ m, silica-water.}$$

(b) Penetration Speed

$$t = 3 \times 30 \times 24 \times 60 \times 60 = 7.776 \times 10^6 \text{ s}$$

$$u_\delta = 1.8 \times (1.826 \times 10^{-7})^{1/2}(\text{m}^2/\text{s})^{1/2}/(7.776 \times 10^6)^{1/2}(1/\text{s})^{1/2}$$

$$= 2.758 \times 10^{-7} \text{ m/s, air-silica}$$

$$u_\delta = 1.8 \times (4.054 \times 10^{-7})^{1/2}(\text{m}^2/\text{s})^{1/2}/(7.776 \times 10^6)^{1/2}(1/\text{s})^{1/2}$$

$$= 4.110 \times 10^{-7} \text{ m/s, water-silica.}$$

**COMMENT**

Note that the seasonal variations of the surface temperature do not penetrate very far beneath the surface. This leaves the soil temperature for depths larger than about 10 m nearly unchanged. Also, note that after three months the penetration speed is very small.

Allowing for a sinusoidal variation of the surface temperature results in a solution with a similar penetration depth [17], but the constant is 0.8620 instead of 3.6. This is left as an end of chapter problem. Therefore, the actual penetration distance is smaller by a factor of nearly four from the estimation made above.

The measured thermal conductivity of dry sand and soil at $T = 300$ K are given in Tables C.15 and C.17 as $\langle k \rangle = 0.33$ and 0.40-0.55 W/m-K, respectively. The soil is a mixture of very fine clay and larger sand particles. There are many chemical elements present in soil.

### 3.5.3 Time-Dependent Surface Temperature: Semi-Infinite Slab*

This section is found on the Web site www.cambridge.org/kaviany. In this section we give the relation for the case of transient conduction in a semi-infinite slab when the surface temperature is a prescribed function of time.

### 3.5.4 Thermal Diffusivity Meter*

This section is found on the Web site www.cambridge.org/kaviany. In this section we discuss how the thermal diffusivity of materials is measured from transient conduction behavior of these materials when they are subjected to sudden (or gradual) changes in their thermal boundary conditions. In particular we discuss the flash method of measurement of thermal diffusivity.

## 3.6 Lumped-Capacitance (Uniform Temperature) Transient: Internal-External Conduction Number $N_{k,i} < 0.1$, $T = T(t)$

We now expand the thermal circuit analysis of Section 3.3.3. For the case of a medium undergoing transient heat transfer, here we assume that the temperature within this control volume, i.e., volumetric node, is uniform.

This is called the lumped-capacitance treatment. This treatment is justified when the nonuniformity of temperature within the volume is negligible compared to its time variation or compared with the temperature difference between that within the volume and its surroundings. If the resistance to heat flow from the surface of this medium (medium $i$) to the surrounding (medium $j$) is designated by $R_{t,i-j}$ (here $t$ stands for any mechanism of heat transfer) and the internal conduction resistance of medium $i$ is designated by $R_{k,i}$, then when the ratio $R_{k,i}/R_{k,i-j} \equiv N_{k,i}$ (internal-external conduction number) is small ($N_{k,i} < 0.1$), the internal resistance is assumed negligible and a uniform temperature is assumed within medium $i$. In

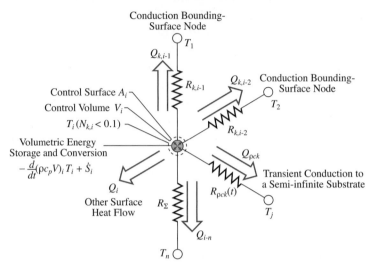

Figure 3.36. A schematic of transient thermal node showing the various terms in the energy equation within for the control volumes around the node represented by lumped-capacitance approximation, $N_{k,i} < 0.1$.

Chapter 4 we consider surface-radiation external resistance $R_{r,i\text{-}j}$ and discuss the dimensionless ratio $R_{k,i}/R_{r,i\text{-}j} = N_r$ (radiation-conduction ratio) and in Chapter 6 we consider surface-convection external resistance $R_{ku,i\text{-}j}$ and discuss the dimensionless ratio $R_{k,i}/R_{ku,i\text{-}j} = N_{ku,i} \equiv \text{Bi}$ (surface-convection conduction number conventionally called the Biot number).

Consider an $n$-node thermal circuit model. Each control surface $i$ is a node and is designated with a $T_i$ ($i = 1, 2...n$) and a surface heat flow shown by $Q|_{A,i}$ and volumetric energy storage-conversion $-d/dt(\rho c_p V)_i T_i + \dot{S}_i$. Between two adjacent surfaces $i$ and $j$ there is an elemental conduction resistance $R_{k,i-j}$. Figure 3.36 gives a schematic of node $T_i$.

For each node (or surface) $i$, we can write a node energy equation as an integral-volume energy equation, i.e., (2.9), as

$$Q\,|_{A,i} = Q_i + \sum_{j=1}^{n} Q_{k,i\text{-}j} = -\frac{d}{dt}(\rho c_p V)_i T_i + \dot{S}_i \qquad \begin{array}{l} \text{energy equation for} \\ \text{thermal node } T_i, \\ \text{lumped capacitance} \\ \text{(uniform temperature)} \end{array} \qquad (3.161)$$

$$N_{k,i} \equiv \frac{R_{k,i}}{R_{k,i\text{-}j}} < 0.1 \qquad \begin{array}{l} \text{internal-external} \\ \text{conduction number} \end{array}$$

Note that $Q_{k,i\text{-}i} = 0$, but is included in the summation. This is similar to (3.87), but includes the energy storage term.

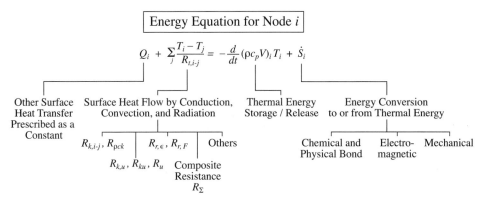

Chart 3.6. Elements of energy equation for lumped capacitance analysis of thermal node $i$.

For each resistance we write the relation among $Q_{k,i-j}$, $R_{k,i-j}$, and $T_i - T_j$ as given by (3.50), i.e.,

$$Q_{k,i-j} = \frac{T_i - T_j}{R_{k,i-j}} \quad \text{elemental surface heat transfer resistance } i\text{-}j. \tag{3.162}$$

Note that $R_{k,i-j} = R_{k,j-i}$ and $R_{k,i-i} = 0$ (also $T_i - T_i = 0$). Also note that $Q_{k,i-j} = -Q_{k,j-i}$.

Then for an $n$-node thermal circuit, we have $n$ energy equations (i.e., $n$ equations) and for any thermal resistance we have a relation. This gives the needed equations that we can solve for the unknowns selected from the heat flow rates $Q_i$ ($i = 1, 2 \ldots n$), surface temperatures $T_i$ ($i = 1, 2 \ldots n$), thermal resistances $R_{k,i-j}(i = 1, 2 \ldots n, j = 1, 2 \ldots n)$, and internodal heat flow rates $Q_{k,i-j}$ ($i = 1, 2 \ldots n$, $j = 1, 2 \ldots n$).

Chart 3.6 gives a summary of the various terms in the energy equation for the volumetric node $i$ subject to the lumped-capacitance approximation.

For transient, lumped capacitance analysis, we need to solve (3.161). For node $i$ we have assumed a uniform, but time varying temperature. For the conduction resistance, we assume that $R_{k,i-j}$ is time independent. This would require that the transient temperature effect on the conduction resistance be negligible compared to the temporal changes in $\rho c_p V T$. When transient conduction heat transfer from a control surface to a semi-infinite medium (or large substrate) occurs, then we use the transient conduction resistance $R_{\rho ck}(t)$. Under this assumption, we can time integrate (3.161). The initial condition for each node is prescribed as $T_i(t = 0)$. Below, we first consider a single, volumetric node. For this case we can integrate (3.161) by assuming that $T_j$ is constant. We first consider the case of surface heat transfer being prescribed $Q_i$. Then we consider a resistant-type surface heat transfer. For multinodes with all nodes having time variations, the energy equations for all nodes are solved simultaneously using a numerical method. This requires solution to a system of ordinary differential equations.

(a) Physical Rendering

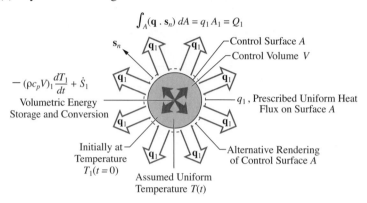

(b) Thermal Circuit Model

Figure 3.37. Prescribed surface heating and volumetric energy conversion with an assumed uniform (but time varying) temperature in a control volume $V_1$. (a) Physical rendering. (b) Thermal circuit diagram.

### 3.6.1 Single Node with Constant, Prescribed Surface Heat Transfer $Q_1$

Figure 3.37 gives a rendering of the constant surface heat transfer rate problem considered. Under this assumption, the integral-volume energy equation (3.161) for the single node considered for a constant, prescribed surface heat transfer rate $Q_1$ (with no surface heat transfer) becomes

$$Q\mid_{A,1} = q_1 A_1 = Q_1 = -(\rho c_p V)_1 \frac{dT}{dt} + \dot{S}_1, \tag{3.163}$$

$$T = T(t=0) \quad \text{initial temperature,} \tag{3.164}$$

where we have already assumed that $c_p$ is constant and here we have added constant $\rho$ and $V$.

Upon rearranging, we have

$$(\rho c_p V)_1 \frac{dT_1}{dt} = -Q_1 + \dot{S}_1 \tag{3.165}$$

or

$$\frac{dT_1}{dt} = \frac{\dot{S}_1 - Q_1}{(\rho V c_p)_1}. \tag{3.166}$$

For the case of $\dot{S}_1 - Q_1$ being constant, this is integrated with respect to time by separating the variables, i.e.,

$$\int dT_1 = \int \frac{\dot{S}_1 - Q_1}{(\rho c_p V)_1} dt. \tag{3.167}$$

For the initial condition of $T_1 = T_1(t = 0)$, we now define the intervals of integration as

$$\int_{T_1(t=0)}^{T_1} dT_1 = T_1 - T_1(t = 0) = \int_0^t \frac{\dot{S}_1 - Q_1}{(\rho c_p V)_1} dt. \tag{3.168}$$

After integration, we have

$$T_1 = T_1(t = 0) + \frac{\dot{S}_1 - Q_1}{(\rho c_p V)_1} t = T_1(t = 0) + \frac{\dot{S}_1 - q_1 A_1}{(\rho c_p V)_1} t$$

$$= T_1(t = 0) + a_1 t \qquad \text{constant surface heat flow rate.} \tag{3.169}$$

As expected, this shows that the assumed uniform temperature changes linearly with respect to time. The decrease or increase in $T$ [compared to $T_1(t = 0)$] depends on the sign of $a_1 \equiv \dot{S}_1 - Q_1$. So when a net heat is added, i.e., $\dot{S}_1 - Q_1 > 0$, the temperature rises. When $\dot{S}_1 - Q_1$ is time-dependent, but prescribed, the integration can be made analytically (using software).

### EXAMPLE 3.20. FAM

A rubber ring (i.e., rubber O-ring) with cross-sectional diameter $D = 3$ mm and ring radius $R_r = 3$ cm, as shown in Figure Ex. 3.20, is softened by heating in a microwave oven. The density of rubber is $\rho = 1,000$ kg/m$^3$ and its specific heat capacity is $c_p = 2,010$ J/kg-K. A surface heat loss per unit area $q_1 = 10$ W/m$^2$ (positive for heat flow out, i.e., along the surface normal) and a uniform temperature $T_1(t)$ is assumed. The RMS of the oscillating electric field $(\overline{e_e^2})^{1/2}$ is $8 \times 10^3$ V/m and the frequency is 3 GHz.

Determine the elapsed time needed to raise the temperature of the rubber by 50°C.

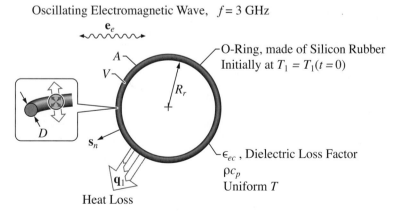

Oscillating Electromagnetic Wave, $f = 3$ GHz

$\mathbf{e}_e$

$A$

$V$

$R_r$

$D$

$\mathbf{s}_n$

$\mathbf{q}_1$

Heat Loss

O-Ring, made of Silicon Rubber
Initially at $T_1 = T_1(t = 0)$

$\epsilon_{ec}$ , Dielectric Loss Factor
$\rho c_p$
Uniform $T$

Figure Ex. 3.20. An O-ring placed in a microwave oven for transient heating.

**SOLUTION**

The transient, uniform temperature of a body of volume $V_1$, i.e., node 1, subject to constant surface heat transfer and volumetric energy conversion is given by (3.169), i.e.,

$$T_1 - T_1(t = 0) = \frac{\dot{s}_1 V_1 - q_1 A_1}{(\rho c_p V)_1} t, \quad \dot{s}_1 = 2\pi f \epsilon_{ec} \epsilon_o \overline{e_e^2},$$

$$A_1 = 2\pi R_r (\pi D) = 2\pi^2 R_r D, \quad V_1 = \left(\pi \frac{D^2}{4}\right) 2\pi R_r = \frac{\pi^2 D^2 R_r}{2}.$$

Now, solving for $t$ from above we have

$$t = \frac{[T_1 - T_1(t = 0)](\rho c_p V)_1}{\dot{s}_1 V_1 - q_1 A_1}.$$

Using the numerical values, we have

$$\epsilon_{ec} = 0.48 \quad \text{from Table C.10, using } f = 3 \text{ GHz}$$

$$\dot{s}_1 = 2\pi \times 3 \times 10^9 (1/\text{s}) \times 0.48 \times 8.8542 \times 10^{-12} (\text{A}^2\text{-s}^2/\,\text{N-m}^2)$$

$$\times \ (8 \times 10^3)^2 (\text{V}^2/\text{m}^2)$$

$$= 5.124 \times 10^6 \text{ W/m}^3$$

$$A_1 = 2\pi^2 \times 3 \times 10^{-2} (\text{m}) \times 3 \times 10^{-3} (\text{m}) = 1.775 \times 10^{-3} \text{ m}^2$$

$$V_1 = \pi^2 (3 \times 10^{-3})^2 (\text{m}^2) \times 3 \times 10^{-2} (\text{m})/2 = 1.331 \times 10^{-6} \text{ m}^2.$$

Then, inserting the numerical values in the relation for $t$, we have

$$t = \frac{50°\text{C} \times 1{,}100(\text{kg/m}^3) \times 2{,}010(\text{J/kg-}°\text{C}) \times 1.331 \times 10^{-6}(\text{m}^3)}{5.124 \times 10^6(\text{W/m}^3) \times 1.331 \times 10^{-6}(\text{m}^3) - 10(\text{W/m}^2) \times 1.775 \times 10^{-3}(\text{m}^2)}$$

$$= 21.63 \text{ s.}$$

**COMMENT**

Note that this volumetric heating rate is rather high and the heat loss rate is rather small.

### 3.6.2 Single Node with Resistive Surface Heat Transfer $Q_{k,1\text{-}2}(t)$

Consider the case of a single volumetric node with uniform temperature $T_1$, i.e., node 1 with a resistance-type surface heat transfer, e.g., conduction heat transfer, to a thermal node with a prescribed temperature $T_2$. This is shown in Figure 3.38(a) and the energy equation (3.161) becomes

$$Q_1 + Q_{k,1\text{-}2}(t) = -\frac{d}{dt}(\rho c_p V)_1 T_1 + \dot{S}_1 \tag{3.170}$$

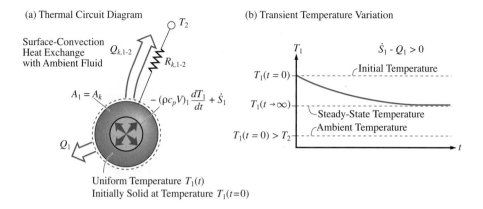

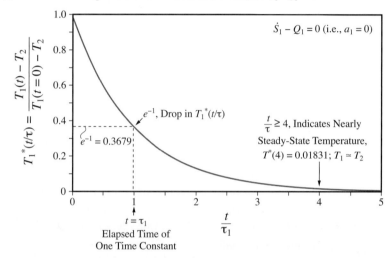

Figure 3.38. (a) Thermal circuit diagram. (b) The anticipated temperature history for a single node with surface heat transfer and volumetric energy conversion. (c) Variation of dimensionless temperature with respect to dimensionless time, for $\dot{S}_1 - Q_1 = 0$.

or

$$Q_1 + \frac{T_1 - T_2}{R_{k,1-2}} = -\frac{d}{dt}(\rho c_p V)_1 T_1 + \dot{S}_1. \qquad (3.171)$$

Here $R_{k,1-2}$ is assumed constant (i.e., no transient effect in $R_{k,1-2}$).

Now assuming a constant $\rho c_p$, we can integrate (3.171) with a prescribed initial temperature $T_1(t = 0)$. The solution is[†]

| $T_1(t) = T_2 + [T_1(t = 0) - T_2]e^{-t/\tau_1} + a_1\tau_1(1 - e^{-t/\tau_1});$ | transient nodal temperature for time-dependent surface heat transfer rate, (3.172) |

[†] The explicit solution for $t$ is

$$t = -\tau_1 \ln \frac{T_1(t) - T_2 - a_1\tau_1}{T_1(t = 0) - T_2 - a_1\tau_1}.$$

where

$$\tau_1 = (\rho c_p V)_1 R_{k,1\text{-}2}, \quad a_1 = \frac{\dot{S}_1 - Q_1}{(\rho c_p V)_1}. \tag{3.173}$$

Here $\tau_1(s)$ is the time constant[†] or the relaxation time for the time response of the solid to the conduction heat transfer. The time constant $\tau_1$ is the ratio of heat capacitance $(\rho c_p V)_1$ to the surface heat transfer conductance $R_{k,1\text{-}2}^{-1}$. For $R_{k,1\text{-}2}^{-1} = 0$, (3.172) gives a solution similar to (3.168), but a Taylor series of the exponential term is needed.

With $\dot{S}_1 - Q_1 \neq 0$, the steady-state temperature $T_1(t \to \infty)$ will be different than $T_2$ and is found by setting $t \to \infty$ in (3.168). Then (3.172) becomes

$$T_1(t \to \infty) = T_2 + a_1\tau_1 = T_2 + (\dot{S}_1 - Q_1)R_{k,1\text{-}2} \quad \text{steady-state temperature.} \tag{3.174}$$

Figure 3.38(b) also shows the anticipated variation of $T_1(t)$ for the case of $\dot{S}_1 - Q_1 > 0$ and $T_2 < T_1(t = 0)$.

For $\dot{S}_1 - Q_1 = 0$, the dimensionless temperature $T^*(t/\tau_1)$ is plotted in Figure 3.38(c) versus dimensionless time $t/\tau_1$. Note that for $t/\tau_1 = 1$, $T^*(t/\tau_1 = 1)$ drops to $e^{-1}$ or $T^*(t/\tau_1 = 1) = 0.3679$. For $t/\tau_1 \geq 4$, $T_1 \simeq T_2$, i.e., $T_1^*(t/\tau_1 = 4) = 0.01839 \simeq 0$.

---

### EXAMPLE 3.21. FAM

---

To detect the presence of water vapor in an otherwise pure air, a heater-condensate detector is used. The presence of the condensate on the heater results in a slower rise in the heater temperature, as compared to when there is no condensate present. The heater-detector, which is based on the Joule heating of a silicon layer, is shown in Figure Ex. 3.21. A thin layer of water of thickness $L_w$ is present on each side of the heater and the silicon-water contact area is $A_k$. The initial temperature of the silicon is $T_1(t = 0)$ and suddenly a current $J_e$ is passed through the silicon, which has an electrical resistance $R_e$. This resistance $R_e$ is achieved by allowing for a matrix structure within the silicon layer (but is not shown). The heater is suspended such that the heat loss to the ambient, other than the condensate, is negligible.

(a) For a given water-layer surface temperature $T_2 = T_{lg}$ (at local vapor pressure $p_g$), and for the parameters specified below, determine the elapsed time required for the heater temperature (assumed to be uniform, i.e., $N_{k,1} < 0.1$) to reach $T_1 = T_{lg} + 50°C$.

(b) Show that $N_k < 0.1$, by using $L_1$ for the silicon layer conduction length.

$L_w = 0.1\,\text{mm}$, $k_w(\text{water}) = 0.65\,\text{W/m-K}$, $A_k = 10^{-4}\,\text{cm}^2$, $L_1 = 1\,\text{mm}$, $R_e = 10\,\text{ohm}$, $J_e = 2\,\text{A}$, $(\rho c_p)_1 = 1.6 \times 10^6\,\text{J/m}^3\text{-K}$, $T_1(t = 0) = 5°C$, $T_2 = T_{lg} = 15°C$.

---

[†]  Note that using (3.173), then (3.171) becomes

$$\frac{dT_1}{dt} + \frac{T_1 - T_2}{\tau_1} = a_1.$$

With $a_1 = 0$, this shows how $T_1$ returns to thermal equilibrium value $T_2$ during the relaxation period $\tau_1$. Under steady-state, $dT_1/dt = 0$, and we have $T_1(t \to \infty) = T_1 + a_1\tau_1$.

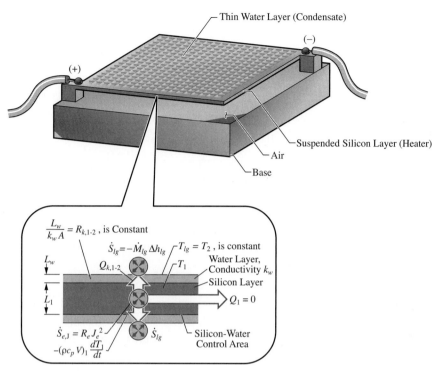

Figure Ex. 3.21. A heater-condensate detector, showing the heated, suspended membrane and the condensate film.

## SOLUTION

The thermal circuit diagram for the heater node $T_1(t)$ is that shown in Figure 3.38(a). Here the Joule heating $\dot{S}_{e,\mathrm{J}}$ is partly stored in the heater and partly conducted away (through the adjacent water layers).

(a) The temperature history of the suspended silicon heater, node 1, is given by (3.172), i.e.,

$$\frac{T_1(t) - T_2}{T_1(t=0) - T_2} = e^{-t/\tau_1} + \frac{a_1\tau_1}{T_1(t=0) - T_2}(1 - e^{-t/\tau_1}).$$

Here

$$R_{k,1\text{-}2} = \frac{L_w}{k_w A_k}, \quad \tau_1 = (\rho c_p V)_1 R_{k,1\text{-}2} = (\rho c_p)_1 A_k L_1 \frac{L_w}{k_w A} = (\rho c_p)_1 \frac{L_1 L_w}{k_w},$$

$$a_1 = \frac{\dot{S}_1 - Q_1}{(\rho c_p V)_1} = \frac{R_e J_e^2}{(\rho c_p)_1 A_k L_1}.$$

From the above equation for $T_1(t)$, we solve for $t$ as the unknown (by first solving for $e^{-t/\tau_1}$ and then taking the ln of this expression) and the result is

$$t = -\tau_1 \ln \frac{\dfrac{T_1(t) - T_2}{T_1(t=0) - T_2} - \dfrac{a_1\tau_1}{T_1(t=0) - T_2}}{1 - \dfrac{a_1\tau_1}{T_1(t=0) - T_2}}$$

or

$$t = -\tau_1 \ln \frac{T_1(t) - T_2 - a_1\tau_1}{T_1(t=0) - T_2 - a_1\tau_1}.$$

Using the numerical values, we have

$$\tau_1 = \frac{1.6 \times 10^6 (\text{J/m}^3\text{-}^\circ\text{C}) \times 10^{-4}(\text{m}) \times 10^{-3}(\text{m})}{0.65(\text{W/m-}^\circ\text{C})} = 0.246 \text{ s}$$

$$a_1 = \frac{10(\text{ohm}) \times (2)^2(\text{A})^2}{1.6 \times 10^6(\text{J/m}^3\text{-}^\circ\text{C}) \times 10^{-4}(\text{m}^2) \times 10^{-3}(\text{m})} = 250.0 \,^\circ\text{C/s},$$

and

$$t = -0.246(\text{s}) \ln \frac{15(^\circ\text{C}) + 50(^\circ\text{C}) - 15(^\circ\text{C}) - 250.0(^\circ\text{C/s}) \times 0.246(\text{s})}{5(^\circ\text{C}) - 15(^\circ\text{C}) - 250.0(^\circ\text{C/s}) \times 0.246(\text{s})}$$

$$= -0.246(\text{s}) \ln(0.1608) = 0.4495 \text{ s} = 449.5 \text{ ms}.$$

(b) The silicon surface for condensation can be micro-machined for selective condensation on this surface only. From Table C.14, the thermal conductivity of silicon is $k_1 = 149$ W/m-K at $T = 300$ K. Then using $L_1$ for the internal conduction resistance length, we have from (3.161), for the internal-external conduction number $N_{k,i}$

$$N_{k,1} = \frac{R_{k,1}}{R_{k,1\text{-}2}} = \frac{L_1/k_1 A_k}{L_w/k_w A_k}$$

$$= \frac{L_1}{L_w} \frac{k_w}{k_1}$$

$$= \frac{10^{-3}(\text{m})}{10^{-4}(\text{m})} \times \frac{0.65(\text{W/m-K})}{149(\text{W/m-K})}$$

$$= 0.04363 < 0.1.$$

Therefore, the assumption of uniform temperature is justifiable for the silicon layer.

**COMMENT**

Note the relatively rapid response of this detector (i.e., sensor). The time constant $\tau_1$ can be reduced by decreasing the volume of the heater.

### 3.6.3 Multiple Nodes*

This section is found on the Web site www.cambridge.org/kaviany. In this section we consider transient heat transfer among multiple media, with each medium represented by a transient, lumped node. Then, the coupled (i.e., dependent) energy equations must be solved. These are coupled ordinary differential equations that need to be solved simultaneously. We also give an example (Example 3.22).

## 3.7 Discretization of Medium into Finite-Small Volumes*

This section is found on the Web site www.cambridge.org/kaviany. In this section we discuss the general numerical solution to transient conduction using the finite-small volume formulation. For the more general bounding-surface conditions, including those with surface heat transfer represented by resistances, the coupled finite-small volume energy equations are solved numerically. We discuss the one- and two-dimensional formulation and give two examples (Examples 3.23 and 3.24).

### 3.7.1 One-Dimensional Transient Conduction*

### 3.7.2 Two-Dimensional Transient Conduction*

### 3.7.3 Nonuniform Discretization*

## 3.8 Conduction and Solid-Liquid Phase Change: Stefan Number $Ste_l$*

This section is found on the Web site www.cambridge.org/kaviany. As an example of a heat transfer-controlled solid-liquid phase change, in this section we consider melting of a thick slab of solid. For a pure substance, the solid and liquid phases will have the same constituents and a distinct interface will appear during melting. We derive the analytical solution for the melt (freezing front propagation speed). There is also an example (Example 3.25).

## 3.9 Thermal Expansion and Thermal Stress*

This section is found on the Web site www.cambridge.org/kaviany. We consider the thermal strain which is an elastic strain that results from expansion of solids with increase in their temperature. When the solid displacement is constrained, this thermal strain results in thermal stress and the study of such mechanical behavior is called the thermoelasticity. We discuss thermal strain and stress and give an example (Example 3.26).

## 3.10 Summary

After discussing the physical aspects of the specific heat capacity and the thermal conductivity, including their temperature dependence, we considered steady-state condition. The internodal conduction resistance $R_{k,i\text{-}j}$ was defined for various geometries and the results were summarized in Table 3.2. Then we considered series and parallel arrangement of resistances and the overall resistance $R_{k,\Sigma}$. In the thermal circuit analysis we used these resistances in the energy equation. In dealing with the transient conduction, we have considered the case of uniform initial temperature

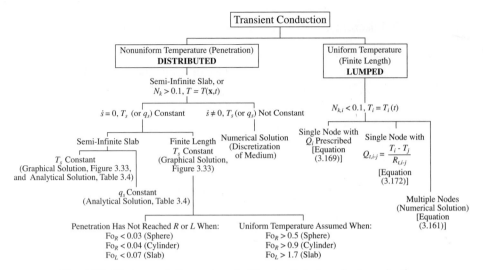

Chart 3.7. Various solutions, given in Chapter 3, to transient conduction.

$T(x, t = 0) = T(t = 0)$. Then for the case of no volumetric energy conversion and where the surface temperature is suddenly changed to a prescribed and a constant value $T_s$ (or a constant $q_s$), we listed the one-dimensional transient temperature distributions in Table 3.4.

In the case of a semi-infinite slab, the temperature distribution evolves with no steady-state solution. For the case of finite lengths (slab, long cylinder, or sphere) the penetration time for the change to reach the center of the body ($x = R$ or $L$) is given by the appropriate value for the Fourier number. These values are listed in Chart 3.7.

When the surface temperature is not constant and/or when there is a volumetric energy conversion, no general solution for the transient temperature distribution exists. For $N_{t,i} < 0.1$ [where $t$ stands for $k$, $r$, or $ku$, and $N_{t,i}$ is defined in (3.161)], i.e., when the external resistance $R_{t,i-j}$ to the heat flow is larger than that of the internal conduction resistance $R_{k,i}$, the temperature is assumed uniform within the body. Under this circumstance, we apply the lumped-capacitance analysis and can allow for a volumetric energy conversion and/or surface heat transfer and using (3.161). For a single node, and when the surface heat transfer can be prescribed and is constant, we have the solution given by (3.169). When the surface heat transfer is a resistive-type and is given by a constant surface resistance $R_{k,i-j}$, then the solution is given by (3.172).

For the transient temperature distribution (with or without an energy conversion) with a time-dependent surface temperature or heat transfer rate, numerical solutions are found. These numerical solutions are discussed in Section 3.7.

The various solutions to the transient conduction heat transfer are also listed in Chart 3.7, along with references to the appropriate equations, figures, and tables.

We also discussed melting under a simple condition and thermal expansion of solids that, under spatial constraints, can lead to thermal stress.

## 3.11 References*

This section is found on the Web site www.cambridge.org/kaviany.

## 3.12 Problems

### 3.12.1 Microscales of Specific Heat Capacity and Thermal Conductivity

**PROBLEM 3.1. FUN**

Equation (3.25) relates the thermal conductivity to the electrical resistivity of pure solid metals. Values for the electrical resistivity as a function of temperature are listed in Table C.8 for different pure metals.

(a) Using (3.25), calculate the predicted thermal conductivity of copper $k_{pr}$ for $T = 200, 300, 500$, and $1,000$ K. For $T = 1,000$ K, extrapolate from the values in the table.

(b) Compare the results obtained in (a), for $k_{pr}$, with the values given in Table C.14, $k_{ex}$. Calculate the percentage difference from using

$$\Delta k(\%) = \left( \frac{k_{pr} - k_{ex}}{k_{ex}} \right) \times 100.$$

(c) Diamond is an electrical nonconductor ($\sigma_e \simeq 0$). However, Figure 3.7(c) shows that the thermal conductivity of diamond is greater than the thermal conductivity of copper for $T > 40$ K. How can this be explained?

**PROBLEM 3.2. FAM**

An airplane flies at an altitude of about 10 km (32,808 ft). Use the relation for the polyatomic ideal-gas thermal conductivity given in Example 3.2.

(a) Determine the air thermal conductivity at this altitude. Use the thermophysical properties given in Table C.7, and assume that $c_v$ and $c_p$ are constant.

(b) Compare the predicted $k$ with the measured value given in Table C.7.

(c) Comment on why $k$ does not change substantially with altitude.

**PROBLEM 3.3. FUN**

Due to their molecular properties, the elemental, diatomic gases have different thermodynamic properties, e.g., $\rho$ and $c_p$, and transport properties, e.g., $k$ properties. Consider (i) air, (ii) helium, (iii) hydrogen, and (iv) argon gases at $T = 300$ K and one atmosphere pressure.

(a) List them in order of the increasing thermal conductivity. Comment on how a gas gap used for insulation may be charged (i.e., filled) with different gases to allow none or less heat transfer.

(b) List them in the order of the increasing thermal diffusivity $\alpha = k/\rho c_p$. Comment on how the penetration speed $u_F$ can be varied by choosing various gases.

(c) List them in order of increasing thermal effusivity $(\rho c_p k)^{1/2}$. Comment on how the transient heat flux $q_{pck}(t)$ can be varied by choosing various gases.

## PROBLEM 3.4. FUN.S

The bulk (or intrinsic) conductivity refers to the medium property not affected by the size of the medium. In gases, this would indicate that the mean-free path of the gas molecules in thermal motion $\lambda_f$ is much smaller than the linear dimension of gas volume $L$. When the linear dimension of the gas volume is nearly the same as or smaller than the mean-free path, then the gas molecules collide with the bounding surface of the gas with a probability comparable to that of the intermolecular collisions. This so-called size effect will occur either at low pressure or for very small $L$. There are simple, approximation expressions describing this size (or low-dimensionality) effect. These expressions include parameters modeling the gas molecule-bounding surface collision and energy exchange. One of these models that is used to predict the size dependence occurring at low gas pressures is

$$k_f(p, T) = \frac{k_f(p = 1 \text{ atm}, T)}{1 + \dfrac{4a_1(2 - \gamma)}{\gamma(c_p/c_v + 1)}\text{Kn}_L},$$

where $\text{Kn}_L$ is the Knudsen number defined in (1.20), i.e.,

$$\text{Kn}_L = \frac{\lambda_f}{L},$$

and $\lambda_f$ is given by (1.19). Here $0 \leq \gamma \leq 1$ is the accommodation factor and $a_1$ is another semi-empirical constant. For example, for nitrogen in contact with ceramic surfaces, $a_1 = 1.944$, $c_p/c_v = 1.401$, and $\gamma = 0.8$.

For nitrogen gas with $L = 10 \ \mu\text{m}$, use $T = 300$ K, and $d_m = 3 \times 10^{-10}$ m and plot $k_f/k_f(\lambda_f \ll L)$ versus the pressure and the Knudsen number.

Use Table C.22 for $k_f$ ($p = 1$ atm, $T = 300$ K).

## PROBLEM 3.5. FUN

The lattice (phonon) specific heat capacity is related to the internal energy $e$, which in turn is given by the energy of an ensemble of harmonic oscillators as

$$e = \frac{N_A}{M} \sum_i E_i$$

$$E_i = h_P f_i n_{p,i}, \qquad n_{p,i} = \frac{1}{e^{x_i} - 1}, \qquad x_i = \frac{h_P f_i}{k_B T},$$

where $h_P$ is the Planck constant, $k_B$ is the Boltzmann constant, $N_A$ is the Avogadro number, $M$ is the molecular weight, and $E_i$ is the average energy per vibrational mode $i$ of each oscillator.

This represents the solid as a collection of harmonic oscillators, vibrating over a range of frequencies $f$, with the number of phonons having a frequency $f_i$ given by $n_{p,i}$.

Starting from (1.6), and using the above, show that the lattice specific heat capacity is

$$c_v = \frac{R_g}{M} \sum_i \frac{x_i^2 e^{x_i}}{(e^{x_i} - 1)^2}.$$

Note that from (3.4), $R_g \equiv k_B N_A$.

## PROBLEM 3.6. FUN

In the Debye approximation model for the lattice (phonon) specific heat capacity given by (3.7), the number of vibrational modes or density of states (per unit frequency around a frequency $f$) is given by the distribution function

$$D_p(f) = \frac{3(2\pi)^2 f^2 V}{2\pi^2 u_p^3}, \quad V = l_m^3 = n^{-1},$$

where $V$ is the volume, $l_m$ is the cubic lattice constant, $u_p$ is the speed of sound (phonon speed), $f$ is the frequency, and $n$ is the number of oscillators (or atoms) per unit volume. The actual lattice may not be cubic and would then be represented by two or more lattice parameters and, if the lattice is tilted, also by a lattice angle. Using this expression, the lattice specific heat capacity is approximated (as an integral approximation of the numerically exact summation) as

$$c_v = \frac{R_g}{M} \sum_i \frac{x_i^2 e^{x_i}}{(e^{x_i} - 1)^2} = \frac{R_g}{M} \int_0^{f_D} \frac{x^2 e^x}{(e^x - 1)^2} D_p(f) 2\pi df, \quad x = \frac{h_p f}{k_B T}.$$

The Debye distribution function (or density of state), when integrated over the frequencies, gives the total number of vibrational modes (three per each oscillator)

$$3n = \frac{1}{V} \int_0^{f_D} D(f) 2\pi df.$$

(a) Show that

$$2\pi f_D = (6n\pi^2 u_p^3)^{1/3}.$$

(b) Using this, derive (3.7), i.e., show that

$$c_v = 9 \frac{R_g}{M} \left(\frac{T}{T_D}\right)^3 \int_0^{T_D/T} \frac{x^4 e^x}{(e^x - 1)^2} dx, \quad T_D = \frac{h_p f_D}{k_B}.$$

## PROBLEM 3.7. FUN

A simple approximate expression is found for the lattice thermal conductivity by only considering the normal (i.e., momentum conserving) phonon scattering mechanisms. This is done using the expression for $c_v$, given by (3.7) in the first part of the expression for $k^p$ given by (3.26), i.e.,

$$k^p = \frac{1}{3} \rho c_v u_p \lambda_p,$$

and noting that

$$\lambda_p = u_p \tau_p,$$

where $\tau_p$ is the phonon relaxation time.

As is done in the Debye approximation, use

$$c_v \lambda_p = \int_0^{T_D/T} c_v(x)\lambda_p(x)dx, \quad x = \frac{h_P f}{k_B T} \quad \text{and} \quad x_D = \frac{h_P f_D}{k_B T} = \frac{T_D}{T},$$

and

$$T_D = \frac{h_P f_D}{k_B} = \frac{h_P}{2\pi k_B}(6n\pi^2 u_p^3)^{1/3},$$

to derive an expression for $k^p$ as a function of $l_m$ as

$$k^p = (48\pi^2)^{1/3} \frac{1}{l_m} \frac{k_B^3}{h_P^2} \frac{T^3}{T_D} \int_0^{T_D/T} \tau_p \frac{x^4 e^x}{(e^x - 1)^2} dx,$$

where, for a cubic crystal lattice, $l_m$ is a lattice constant related to the number of unit cells per unit volume by $l_m^{-1} = n^{1/3}$. From (1.19), use $\rho R_g / M = n k_B$.

## PROBLEM 3.8. FUN.S

The crystal size influences the phonon thermal conductivity due to phonon scattering caused by variation of phonon propagation properties across the crystal surface (similar to light scattering at the interface of two media of different light propagation properties). This boundary scattering is one of the resistive scattering mechanisms included in (3.26). Consider aluminum oxide ($Al_2O_3$, also called alumina) single crystals at $T = 300$ K. The effect of crystal size $L$ can be described by a simple relation for the boundary scattering relaxation time constant $\tau_b$ as

$$\tau_b = \frac{L}{u_p},$$

where $u_p$ is the average phonon velocity. Using the material constants for alumina and at $T = 300$ K, the lattice conductivity given by (3.26) becomes

$$k^p = b_k T^3 \left[ g_1(x, L) + \frac{g_2^2(x, L)}{g_3(x, L)} \right] = b_k \left[ h_1(x, L) + \frac{h_2^2(x, L)}{h_3(x, L)} \right],$$

where $b_k = 2.240 \times 10^5$ W/m-K$^4$ and the $g_i$'s and $h_i$'s represent integrals as defined below.

Some numerical solvers (e.g., MATLAB) have limitations to the size of the numbers which they may use. To avoid this limitation, the $T^3$ may be taken into the integral by defining $\theta_i = \tau_i T^3$ and then rewriting the integrals in (3.26) as

$$h_1 = g_1 T^3 = \int_0^{T_D/T} \theta_p \frac{x^4 e^x}{(e^x - 1)^2} dx, \quad h_2 = g_2 = \int_0^{T_D/T} \frac{\theta_p}{\theta_{p,n}} \frac{x^4 e^x}{(1 - e^x)^2} dx,$$

$$h_3 = \frac{g_3}{T^3} = \int_0^{T_D/T} \frac{\theta_p}{\theta_{p,n}\theta_{p,r}} \frac{x^4 e^x}{(1 - e^x)^2} dx,$$

where

$$\frac{1}{\theta_p} = \frac{1}{\theta_{p,n}} + \frac{1}{\theta_{p,r}} = \frac{1}{\theta_{p,n}} + \left( \sum_i \frac{1}{\theta_{p,r,i}} \right)$$

$$= b_n x + \left( 2 \times b_p x^4 + b_u x^2 + \frac{b_b}{L} \right),$$

where $T_D = 596$ K, $b_n = 3.181 \times 10^3$ $1/K^3$-s, $b_p = 3.596 \times 10^1$ $1/K^3$-s, $b_u = 1.079 \times 10^4$ $1/K^3$-s, $b_b = 2.596 \times 10^{-4}$ m/$K^3$-s.

Use a solver to plot $k_p$ versus grainsize, $L$, for $10^{-9} \le L \le 10^{-4}$ m.

## PROBLEM 3.9. FUN.S

For thin film deposited on surfaces, the thermal conductivity of the film becomes film-thickness dependent, if the film thickness $L$ is near or smaller than the heat-carrier, mean-free path. Consider a ceramic, amorphous silicon dioxide ($SiO_2$, also called silica) where the heat carriers are phonons. This film-thickness dependence (or size effect) of the thermal conductivity may be approximated as

$$k = \frac{k(L \gg \lambda_p)}{1 + \frac{4}{3} \frac{\lambda_p}{L}},$$

where $k(L \gg \lambda_p)$ is the bulk (or size-independent) thermal conductivity, and $\lambda_p$ is the phonon mean-free path.

The reduction in the thermal conductivity (as $\lambda_p/L$ increases) is due to the scattering of the phonons at the boundaries of the thin film.

Using Tables 3.1 and C.17, plot the variation of $k$ for amorphous silica for $0.6 \le L \le 6$ nm, for $T = 293$ K.

## PROBLEM 3.10. FUN.S

The effective thermal conductivity $\langle k \rangle$ is used to describe the conductivity of porous solids (a fluid-solid composite). In many applications requiring a large surface area for surface convection $A_{ku}$, such as in heat storage in solids, packed beds of particles are used. For example, spherical particles are packed randomly or in an ordered arrangement (e.g., simple, body-centered or face-centered, cubic arrangement). Figure Pr. 3.10 shows a simple (also called square-array) cubic arrangement of particles (porosity $\epsilon = 0.476$). Due to their weight or by a contact pressure $p_c$, these elastic particles deform and their contact area changes, resulting in a change in the effective thermal conductivity $\langle k \rangle$.

For spheres having a uniform radius $R$, a Young modulus of elasticity $E_s$, a Poisson ratio $\nu_P$, and a conductivity $k_s$, the effective conductivity for the case negligible fluid conductivity ($k_f = 0$) and subject to contact pressure $p_c$ is predicted as

$$\frac{\langle k \rangle}{k_s} = 1.36 \left[ \frac{(1 - \nu_P^2)p_c}{E_s} \right]^{1/3},$$

which is independent of $R$.

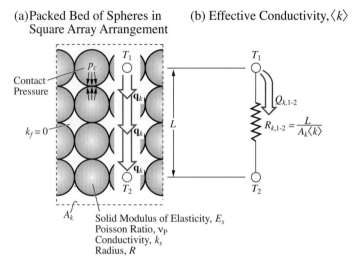

Figure Pr. 3.10. (a) Packed bed of spherical particles with a simple cubic arrangement and a contact pressure $p_c$. (b) The effective thermal conductivity $\langle k \rangle$.

Plot $\langle k \rangle$ versus $p_c$ for $10^5 \leq p_c \leq 10^9$ Pa for packed beds of (i) aluminum, (ii) copper, and (iii) magnesium spherical particles.

For aluminum, $E_s = 68$ GPa, $\nu_P = 0.25$, $k_s = 237$ W/m-K.

For copper (annealed), $E_s = 110$ GPa, $\nu_P = 0.343$, $k_s = 385$ W/m-K.

For magnesium (annealed sheet), $E_s = 44$ GPa, $\nu_P = 0.35$, $k_s = 156$ W/m-K.

## PROBLEM 3.11. FUN.S

The crystalline lattice thermal conductivity $k_p$ is given by (3.26) as

$$k^p = (48\pi^2)^{1/3} \frac{1}{l_m} \frac{k_B^3}{h_P^2} \frac{T^3}{T_D} \left[ g_1(f, T, \tau_p) + \frac{g_2^2(f, T, \tau_p, \tau_{p,n})}{g_3(f, T, \tau_p, \tau_{p,n}, \tau_{p,r})} \right],$$

The integrals $g_n$, $g_{r,1}$, and $g_{r,2}$ and the relaxation times $\tau_{p,n}$, $\tau_{p,r}$, and $\tau_p$ are defined as

$$g_1 = \int_0^{T_D/T} \tau_p \frac{x^4 e^x}{(e^x - 1)^2} dx, \quad g_2 = \int_0^{T_D/T} \frac{\tau_p}{\tau_{p,n}} \frac{x^4 e^x}{(e^x - 1)^2} dx,$$

$$g_3 = \int_0^{T_D/T} \frac{\tau_p}{\tau_{p,n}\tau_{p,r}} \frac{x^4 e^x}{(e^x - 1)^2} dx,$$

$$\frac{1}{\tau_p} = \frac{1}{\tau_{p,n}} + \frac{1}{\tau_{p,r}}, \quad \frac{1}{\tau_{p,n}} = a_n \frac{2\pi k_B}{h_P} T^5 x, \quad \frac{1}{\tau_{p,r}} = \sum_i \frac{1}{\tau_{p,r,i}},$$

where $x = h_P f / k_B T$, $a_n$ is a material constant, and the resistive mechanisms in the summation for $\tau_{p,r}^{-1}$ include the three-phonon umklapp processes, $\tau_{p,r,u}$, boundary scattering, $\tau_{p,r,b}$, point defect scattering, $\tau_{p,r,p}$, lattice vacancy scattering, $\tau_{p,r,v}$, and phonon-electron scattering, $\tau_{p,r,p\text{-}e}$, among others [6]. For alumina ($Al_2O_3$), the phonon-electron scattering is negligible compared to the other resistive mechanisms, and for simplicity is not considered here. The overall resistive time constant, due to these resistive mechanisms, is then given by

$$\frac{1}{\tau_{p,r}} = \frac{1}{\tau_{p,r,p}} + \frac{1}{\tau_{p,r,v}} + \frac{1}{\tau_{p,r,u}} + \frac{1}{\tau_{p,r,b}}$$

$$= A\left(\frac{2\pi k_B}{h_P}\right)^4 T^4 x^4 + A\left(\frac{2\pi k_B}{h_P}\right)^4 T^4 x^4 + a_u\left(\frac{2\pi k_B}{h_P}\right)^2 T^3 e^{-T_D/(\alpha T)} x^2 + \frac{u_p}{L},$$

where $A$, $a_u$, and $\alpha$ are also material constants, $u_p$ is the mean phonon velocity, and $L$ is a characteristic length scale of the crystal or grain boundaries. Note that vacancies and point defects behave identically as resistance mechanisms.

Consider a single alumina ($Al_2O_3$) crystal with linear dimension $L = 4.12$ mm, and the empirically determined material constants, $a_n = 2.7 \times 10^{-13}$ K$^{-4}$, $A = 4.08 \times 10^{-46}$ s$^3$, $a_u = 1.7 \times 10^{-18}$ K$^{-1}$, and $\alpha = 2$. Also from Table 3.1, we have $T_D = 596$ K, $l_m = 0.35$ nm, and $u_p = 7,009$ m/s.

Substituting these values into the expression for the total phonon relaxation time constant, we have

$$\frac{1}{\tau_p} = \frac{1}{\tau_{p,n}} + \left(\frac{1}{\tau_{p,r}}\right)$$

$$= b_n T^5 x + (2 \times b_p T^4 x^4 + b_u T^3 e^{-298/T} x^2 + b_b),$$

where these new $b_i$ constants combine the above $a_i$ constants with the other coefficients and are $b_n = 3.535 \times 10^{-2}$ 1/K$^5$-s, $b_p = 1.199 \times 10^{-1}$ 1/K$^4$-s, $b_u = 2.914 \times 10^4$ 1/K$^3$-s, and $b_b = 1.701 \times 10^6$ 1/s.

Then the expression for the lattice thermal conductivity becomes

$$k^p = b_k T^3 \left[ g_1(x, T) + \frac{g_2^2(x, T)}{g_3(x, T)} \right],$$

where $b_k = 2.240 \times 10^5$ W/m-K$^4$.

(a) The integrals in the expression for the crystalline lattice thermal conductivity must be evaluated for a given temperature. For various temperatures, between $T = 1$ and 400 K, use a solver and determine $k^p$, and then plot $k^p$ versus $T$.

(b) Compare the result with typical $k^p$ vs. $T$ curves for crystalline nonmetals, as shown in Figure 3.7(c).

Hint: To avoid overflow errors that might occur depending on the solver, factor the $T^3$ into the brackets containing the $g_i$ integrals (i.e., into the $b_i$ constants) before solving.

### 3.12.2 Thermal Conduction Resistance and Thermal Circuit Analysis

**PROBLEM 3.12. FUN**

Similar to Example 3.6, consider the internal surface of the three surfaces to be covered with an insulation layer of thickness $l = 5$ cm and thermal conductivity

(a) Plane Surface            (b) Cylindrical Surface            (c) Spherical Surface

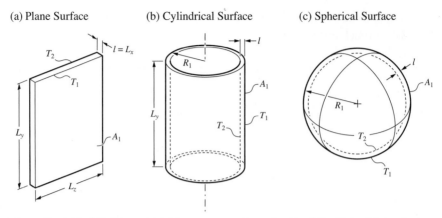

Figure Pr. 3.12. (a), (b), and (c) Three geometries to be lined (inside) with insulation.

$k = 0.1$ W/m-K. The outside surface is at temperature $T_1 = 90°$C and the temperature at the inside surface of the insulation is $T_2 = 40°$C. The surfaces have an outside area $A_1 = 1$ m$^2$ and are of the geometries shown in Figure Pr. 3.12, i.e., (a) a planar surface with area $A_1 = L_y L_z$, (b) a cylinder with area $A_1 = 2\pi R_1 L_y$ and length $L_y = 1$ m, and (c) a sphere with surface area $A_1 = 4\pi R_1^2$.

For each of these geometries, calculate the rate of heat loss through the vessel surface $Q_{k,2-1}$(W). Compare your results with the results of Example 3.6 and comment on the differences among the answers. Neglect the heat transfer through the ends (i.e., assume a one-dimensional heat transfer).

## PROBLEM 3.13. FUN
Consider an infinite plane wall (called a slab) with thickness $L = 1$ cm, as shown in Figure Pr. 3.13.

(a) Determine the conduction thermal resistance $A_k R_{k,1-2}[°C/(W/m^2)]$, if the wall is made of copper.
(b) Determine the conduction thermal resistance $A_k R_{k,1-2}[°C/(W/m^2)]$, if the wall is made of silica aerogel.

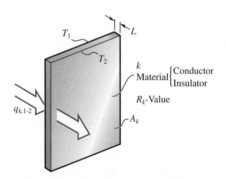

Figure Pr. 3.13. One-dimensional conduction across a slab.

(c) If the heat flux through the wall is $q_{k,1-2} = 1,000$ W/m$^2$ and the internal wall temperature is $T_2 = 60°$C, calculate the external wall temperature $T_1$ for the two materials above.

(d) Express the results for items (a) and (b) in terms of the $R_k$-value.

The thermophysical properties of copper and silica aerogel are to be evaluated at 25°C and 1 atm [Table C.14 and Figure 3.13(a)].

## PROBLEM 3.14. FAM
A furnace wall (slab) is made of asbestos ($\rho = 697$ kg/m$^3$) and has a thickness $L = 5$ cm [Figure Pr. 3.14(i)]. Heat flows through the slab with given inside and outside surface temperatures. In order to reduce the heat transfer (a heat loss), the same thickness of asbestos $L$ is split into two with an air gap of length $L_a = 1$ cm placed between them [Figure Pr. 3.14(ii)].

Determine how much the heat flow out of the wall would decrease (show this as a percentage of the heat flow without the air gap).

Use Tables C.12 and C.17 to evaluate the conductivity at $T = 273$ K or $T = 300$ K.

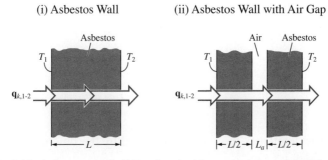

Figure Pr. 3.14. A furnace wall. (i) Insulated without an air gap. (ii) With an air gap.

## PROBLEM 3.15. FUN
A low thermal-conductivity composite (solid-air) material is to be designed using alumina as the solid and having the voids occupied by air. There are three geometric arrangements considered for the solid and the fluid. These are shown in Figure Pr. 3.15. For all three arrangements, the fraction of volume occupied by the fluid (i.e., porosity) $\epsilon$ is the same. In the parallel arrangement, sheets of solid are separated by fluid gaps and are placed parallel to the heat flow direction. In the series arrangement, they are placed perpendicular to the heat flow. In the random arrangement, a nonlayered arrangement is assumed with both solid and fluid phase continuous and the effective conductivity is given by (3.28).

(a) Show that the effective thermal conductivity for the parallel arrangement is given by

$$\langle k \rangle = k_f \epsilon + k_s (1 - \epsilon)$$

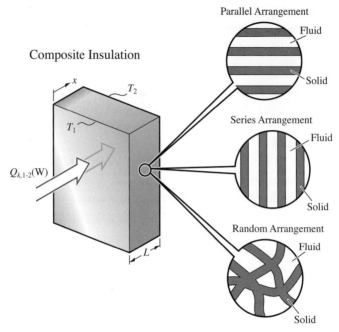

Figure Pr. 3.15. Three solid-fluid arrangements for obtaining a low thermal conductivity composite.

and that, for the series arrangement, the effective thermal conductivity is given by

$$\frac{1}{\langle k \rangle} = \frac{\epsilon}{k_f} + \frac{(1-\epsilon)}{k_s}.$$

The porosity $\epsilon$ is defined as the volume occupied by the fluid divided by the total volume of the medium, i.e.,

$$\epsilon = \frac{V_f}{V_f + V_s}.$$

Also note that $1 - \epsilon = V_s/(V_f + V_s)$.

(b) Compare the effective conductivity for the three arrangements for $\epsilon = 0.6$ using the conductivity of alumina (Table C.14) and air (Table C.22) at $T = 300$ K.
(c) Comment regarding the design of low-conductivity composites.

### PROBLEM 3.16. FAM

During hibernation of warm-blooded animals (homoisotherms), the heart beat and the body temperature are lowered and in some animals the body waste is recycled to reduce energy consumption. Up to 40% of the total weight may be lost during the hibernation period. The nesting chamber of the hibernating animals is at some distance from the ground surface, as shown in Figure Pr. 3.16(i). The heat transfer from the body is reduced by the reduction in the body temperature $T_1$ and by the insulating effects of the body fur and the surrounding air (assumed stagnant). A

(i) Diagram of Woodchuck Home              (ii) Simple Thermal Model

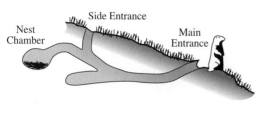
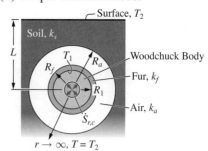

Figure Pr. 3.16. Conduction heat transfer from a warm-blooded animal during hibernation. (i) Diagram of woodchuck home. (ii) Thermal model.

simple thermal model for the steady-state, one-dimensional heat transfer is given in Figure Pr. 3.16(ii). The thermal resistance of the soil can be determined from Table 3.3(a). An average temperature $T_2$ is used for the ground surrounding the nest. The air gap size $R_a - R_f$ is an average taken around the animal body.

(a) Draw the thermal circuit diagram.
(b) Determine $Q_{1\text{-}2}$ for (i) $L = 2.5R_a$, and (ii) $L = 10R_a$.
  $R_1 = 10$ cm, $R_f = 11$ cm, $R_a = 11.5$ cm, $T_1 = 20°C$, $T_2 = 0°C$.

Evaluate air properties at $T = 300$ K, use soil properties from Table C.15, and for fur use Table C.15 for hair.

**PROBLEM 3.17. FAM**
A spherical aluminum tank, inside radius $R_1 = 3$ m, and wall thickness $l_1 = 4$ mm, contains liquid-vapor oxygen at 1 atm pressure (Table C.26 for $T_1 = T_{lg}$). The ambient is at a temperature higher than the liquid-gas mixture. Under steady-state, at the liquid-gas surface, the heat flowing into the tank causes boil off at a rate $\dot{M}_{lg} = \dot{M}_g$. In order to prevent the pressure of the tank from rising, the gas resulting from boil off is vented through a safety valve. This is shown in Figure Pr. 3.17. Then, to reduce the amount of boil-off vent $\dot{M}_g$(kg/s), insulation is added to the tank. First a low pressure (i.e., evacuated) air gap, extending to location $r = R_2 = 3.1$ m, is placed where the combined conduction-radiation effect for this gap is represented by a conductivity $k_a = 0.004$ W/m-K. Then a layer of low-weight pipe insulation (slag or glass, Table C.15) of thickness $l_2 = 10$ cm is added. The external surface temperature is kept constant at $T_2 = 10°C$.

(a) Draw the thermal circuit diagram.
(b) Determine the rate of heat leak $Q_{k,2-1}$.
(c) Determine the amount of boil off $\dot{M}_g$.
(d) Determine the temperature at the inner-surface ($r = R_2$) of the insulation $T_2$.

Evaluate the thermal conductivity of aluminum at $T = 200$ K using Table C.14.

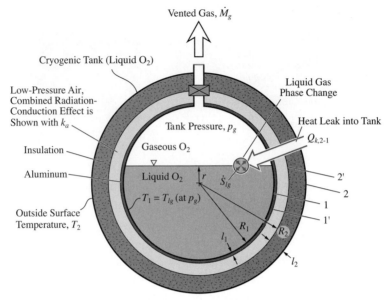

Figure Pr. 3.17. A tank containing a cryogenic liquid and having heat leak to it from a higher temperature ambient.

## PROBLEM 3.18. FAM

A teacup is filled with water having temperature $T_w = 90°C$. The cup is made of (i) porcelain (Table C.15), or (ii) stainless steel 316. The cup-wall inside diameter is $R$ and its thickness is $L$. These are shown in Figure Pr. 3.18. The water is assumed to be well mixed and at a uniform temperature. The ambient air is otherwise quiescent with a far-field temperature of $T_{f,\infty}$, and adjacent to the cup the air undergoes a thermobuoyant motion resulting in a surface-convection resistance $R_{ku}$.

(a) Draw the thermal circuit diagram.
(b) Determine the cup outside surface temperature $T_s$ for cases (i) and (ii).
  $T_{f,\infty} = 20°C$, $L = 3$ mm, $A_{ku}R_{ku} = 10^{-3}$ K/(W/m²).
  Use $L \ll R$ to approximate the wall as a slab and use $A_{ku} = A_k$.

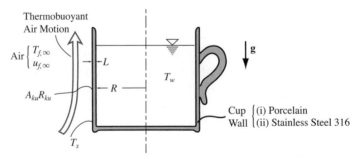

Figure Pr. 3.18. A cup filled with hot water. The cup is made of (i) porcelain, or (ii) stainless steel 316. The ambient air is otherwise quiescent with a thermobuoyant motion adjacent to the cup wall.

## PROBLEM 3.19. FAM

Gaseous combustion occurs between two plates, as shown in Figure Pr. 3.19. The energy converted by combustion $\dot{S}_{r,c}$ in the gas flows through the upper and lower bounding plates. The upper plate is used for surface radiation heat transfer and is made of solid alumina (Table C.14). The lower plate is porous and is made of silica (Table C.17, and include the effect of porosity). The porosity $\epsilon = 0.3$ and the randomly distributed pores are filled with air (Table C.22, use $T = T_{s,2}$). Each plate has a length $L$, a width $w$, and a thickness $l$. The outsides of the two plates are at temperatures $T_{s,1}$ and $T_{s,2}$.

(a) Draw the steady-state thermal circuit diagram.
(b) Determine the effective conductivity of the lower plate.
(c) Determine the uniform gas temperature $T_g$.
(d) Determine the fraction of heat flow through each plate.

$\dot{S}_{r,c} = 10^4$ W, $T_{s,1} = 1,050°C$, $T_{s,2} = 500°C$, $L = 0.3$ m, $w = 0.3$ m, $l = 0.02$ m.

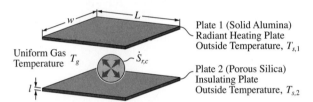

Figure Pr. 3.19. Combustion between two plates; one plate is a conductor while the other is an insulator.

## PROBLEM 3.20. FAM

In IC engines, during injection of liquid fuel into the cylinder, it is possible for the injected fuel droplets to form a thin liquid film over the piston. The heat transferred from the gas above the film and from the piston beneath the film causes surface evaporation. This is shown in Figure Pr. 3.20. The liquid-gas interface is at the boiling temperature, $T_{lg}$, corresponding to the vapor pressure. The heat transfer from the piston side is by one-dimensional conduction through the piston and then

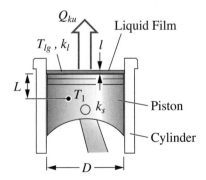

Figure Pr. 3.20. An IC engine, showing liquid film formation on top of the piston.

by one-dimensional conduction through the thin liquid film. The surface-convection heat transfer from the gas side to the surface of the thin liquid film is prescribed as $Q_{ku}$

(a) Draw the thermal circuit diagram and write the corresponding energy equation for the liquid-gas interface.
(b) For the conditions given, determine the rate of evaporation of the liquid film, $\dot{M}_{lg}$ (kg/s).
(c) Assuming that this evaporation rate remains constant, determine how long it will take for the liquid film to totally evaporate.

$Q_{ku} = -13{,}500$ W, $\Delta h_{lg} = 3.027 \times 10^5$ J/kg (octane at one atm pressure, Table C.4), $k_l = 0.083$ W/m-K (octane at 360 K, Table C.13), $T_{lg} = 398.9$ K (octane at 1 atm pressure, Table C.4), $\rho_l = 900$ kg/m$^3$, $k_s = 236$ W/m-K (aluminum at 500 K, Table C.14), $T_1 = 500$ K, $L = 3$ mm, $l = 0.05$ mm, $D = 12$ cm.

**PROBLEM 3.21. FUN**
A two-dimensional, periodic porous structure has the solid distribution shown in Figure Pr. 3.21. This is also called a regular lattice. The steady-state two-dimensional conduction can be shown with a one-dimensional, isotropic resistance for the case of $k_A \ll k_B$.

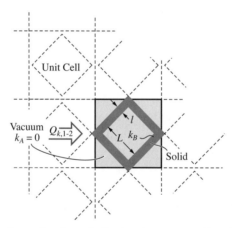

Figure Pr. 3.21. A two-dimensional, periodic structure composite with material $A$ having a conductivity much smaller than $B$.

(a) Draw the thermal circuit model.
(b) Show that for $k_A/k_B \ll 1$, the effective thermal conductivity $\langle k \rangle$ is

$$\frac{\langle k \rangle}{k_B} = 1 - \epsilon^{1/2} \qquad \frac{k_A}{k_B} \ll 1,$$

where $\epsilon$ is the porosity (void fraction) defined by (3.27).
    Use a depth $w$ (length perpendicular to the page).

## PROBLEM 3.22. FUN

The effective thermal conductivity of two-dimensional, periodic-structure (i.e., regular lattice) composites can be estimated using one-dimensional resistance models. Figure Pr. 3.22 shows a simple, two-dimensional unit cell with material $B$ being continuous and material $A$ being the inclusion [similar to the three-dimensional, periodic structure of Section 3.3.2(C)].

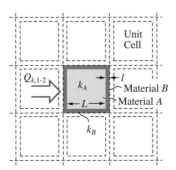

Figure Pr. 3.22. A two-dimensional, periodic structure with material $B$ being continuous.

(a) Derive an expression for the effective conductivity $\langle k \rangle$ of this composite using a series-parallel arrangement of resistances.
(b) Derive an expression for the effective conductivity $\langle k \rangle$ of this composite using a parallel-series arrangement of resistances.
(c) Show that for the case of $k_A/k_B \ll 1$, the result for the parallel-series arrangement is

$$\frac{\langle k \rangle}{k_B} = 1 - \epsilon^{1/2},$$

where $\epsilon$ is the porosity (this result is also obtained in Problem 3.34).

Use only the porosity (void fraction) $\epsilon$ and the conductivities $k_A$ and $k_B$. Use a depth $w$ (length perpendicular to the page).

## PROBLEM 3.23. FUN

In the one-dimensional, steady-state conduction treatment of Section 3.3.1, for planar geometries, we assumed a constant cross-sectional area $A_k$. In some applications, although the conduction is one-dimensional and cross section is planar, the cross-sectional area is not uniform. Figure Pr. 3.23 shows a rubber leg used for the vibration isolation and thermal insulation of a cryogenic liquid container. The rubber stand is in the form of a truncated cone [also called a frustum of right cone, a geometry considered in Table C.1(e)].

(a) Starting from (3.29), with $\dot{s} = 0$, use a variable circular conduction area $A_k(x) = \pi R^2(x)$, while $R(x)$ varies linearly along the $x$ axis, i.e.,

$$R(x) = \frac{R_1}{L_1}x,$$

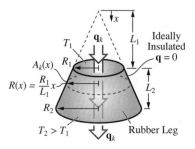

Figure Pr. 3.23. One-dimensional, steady-state heat conduction in a variable area rubber leg.

as shown in Figure Pr. 3.23. Then derive the expression for the temperature distribution $T = T(x)$.

(b) Using this temperature distribution, determine $Q_{k,1\text{-}2}$ and $R_{k,1\text{-}2}$, by using (3.46) and noting that there are no lateral heat losses.

(c) Evaluate $Q_{k,1\text{-}2}$, for the conditions given below.

(d) Use a constant surface area with $\langle R \rangle = (R_1 + R_2)/2$, and the conduction resistance for a slab, and compare $Q_{k,1\text{-}2}$ with that from part (c).

Note that $\Delta V = \pi R^2(x)\Delta x$, as $\Delta x \to 0$.

$T_2 = 20°\text{C}$, $T_1 = 0°\text{C}$, $L_2 = 4$ cm, $L_1 = 10$ cm, $R_1 = 1.5$ cm, $k = 0.15$ W/m-K.

### 3.12.3 Conduction Contact Resistance

### PROBLEM 3.24. FAM

A pair of aluminum slabs with a surface roughness of $\langle \delta^2 \rangle^{1/2} = 0.25$ $\mu$m are placed in contact (with air as the interstitial fluid). A heat flux of $q_k = 4 \times 10^4$ W/m$^2$ flows across the interface of the two slabs.

Determine the temperature drop across the interface for contact pressures of $10^5$ and $10^6$ Pa.

### PROBLEM 3.25. FAM

Thin, flat foil heaters are formed by etching a thin sheet of an electrical conductor such as copper and then electrically insulating it by coating with a nonconductive material. When the maximum heater temperature is not expected to be high, a polymer is used as coating. When high temperatures are expected, thin sheets of mica are used. Mica is a mineral silicate that can be cleaved into very thin layers. However, a disadvantage of the use of mica is that the mica surface offers a much higher thermal-contact resistance, thus requiring a larger joint pressure $p_c$.

Figure Pr. 3.25 shows a thin circular heater used to deliver heat to a surface (surface 1). The solid between surface 1 and the heater is aluminum (Table C.14) and has a thickness $L_1 = 5$ mm. In order to direct the heat to this surface, the other side of the heater is thermally well insulated by using a very low conductivity fiber insulating board (Table C.15) with thickness $L_2 = 10$ mm. The temperature of

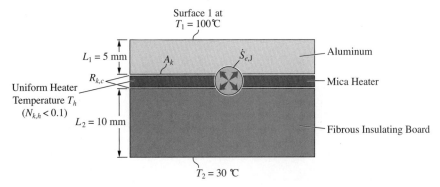

Figure Pr. 3.25. A thin-foil heater encased in mica and placed between an aluminum and an insulation layer.

the aluminum surface is maintained at $T_1 = 100°C$, while the outer surface of the thermal insulation is at $T_2 = 30°C$. The heater generates heat by Joule heating at a rate of $\dot{S}_{e,J}/A_k = 4 \times 10^4$ W/m$^2$ and is operating under a steady-state condition.

(a) Draw the thermal circuit diagram.
(b) Determine the heater temperature $T_h$ for the case of contact resistances of (i) $A_k R_{k,c} = 10^{-4}$ [K/(W/m$^2$)], and (ii) $A_k R_{k,c} = 4 \times 10^{-2}$ [K/(W/m$^2$)].
(c) Comment on the answers obtained above if the heater is expected to fail at $T_{max} = 600°C$.

Use the thermal conductivities at the temperatures given in the tables or at 300 K.

## PROBLEM 3.26. FAM
The automobile exhaust catalytic converter (for treatment of gaseous pollutants) is generally a large surface area ceramic or metallic monolith that is placed in a stainless steel housing (also called can). Figure Pr. 3.26 shows a ceramic (cordierite, a mineral consisting of silicate of aluminum, iron, and magnesium) cylindrical monolith that is placed inside the housing with (i) direct ceramic-stainless contact, and (ii) with a blanket of soft ceramic (vermiculite, a micacious mineral, mat) of conductivity $k_b$ placed between them.

The blanket is placed under pressure and prevents the gas from flowing through the gap. The direct contact results in a contact resistance similar to that of stainless steel-stainless steel with $\langle \delta^2 \rangle^{1/2} = 1.1$ to $1.5$ $\mu$m and $p_c = 10^5$ Pa. The soft blanket (vermiculite mat) has a thickness $l_1 = 3$ mm and $k_b = 0.4$ W/m-K.

(a) Draw the thermal circuit diagrams.
(b) Determine the heat flow between surface at temperature $T_1$ and surface at temperature $T_2$ (i) without and (ii) with the soft ceramic blanket, i.e., $Q_{k,1-2}$.

Use $T_1 = 500°C$, $T_2 = 450°C$, and stainless steel AISI 316 for thermal conductivity.

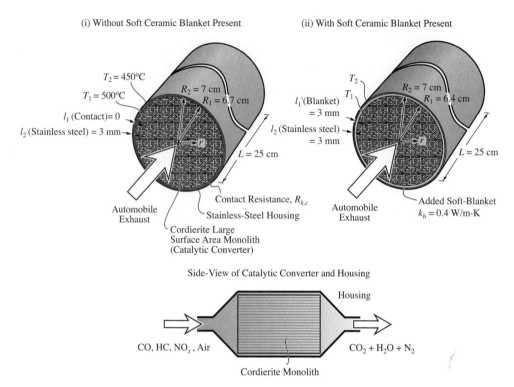

(i) Without Soft Ceramic Blanket Present      (ii) With Soft Ceramic Blanket Present

Side-View of Catalytic Converter and Housing

Figure Pr. 3.26. An automobile catalytic converter (i) without and (ii) with a soft ceramic blanket.

## PROBLEM 3.27. FAM

A thermoelectric power generator uses the heat released by gaseous combustion to produce electricity. Since the low temperature thermoelectric materials undergo irreversible damage (such as doping migration) at temperatures above a critical temperature $T_{cr}$, a relatively low conductivity material (that withstands the high flame temperature; this is referred to as a refractory material) is placed between the flame and the hot junction, as shown in Figure Pr. 3.27. Additionally, a copper thermal spreader is placed between the refractory material and the hot junction to ensure even distribution of the heat flux into the thermoelectric device (assume a uniform temperature for the copper slab). It is desired to generate 20 W of electricity from the thermoelectric module, where this power is 5% of the heat supplied $(-Q_h)$ at the hot junction $T_h$. The refractory material is amorphous silica with conductivity $k_s$. In addition, there is a contact resistance $R_{k,c}$ between the copper thermal spreader and the hot junction. The surface area of the hot junction is $a \times a$.

(a) Draw the thermal circuit diagram for node $T_h$ using the combustion flue gas temperature $T_g$.

(b) Determine the thickness of the refractory material $L$, such that $T_h = T_{cr}$.

$T_g = 750°C, T_{cr} = 250°C, k_s = 1.36 \text{ W/m-K}, a = 6 \text{ cm}, A_k R_{k,c} = 10^{-4} \text{ K/(W/m}^2).$

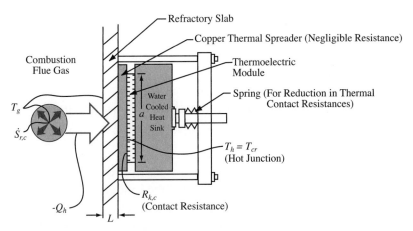

Figure Pr. 3.27. A thermoelectric module, used for power generation, receives heat from a combustion flue gas stream. To reduce the temperature of the hot junction, a refractory slab is used.

### PROBLEM 3.28. FUN.S

In order to reduce the contact resistance, the contacting solids are physically bonded by using high temperatures (as in fusion or sintering) or by using material deposition (as in physical vapor deposition or solidification of melts). This creates a contact layer which has a contact thickness $L_c$ (that is nearly twice the rms roughness) and a contact conductivity $k_c$. This contact conductivity is intermediate between the conductivity of the joining materials A and B (with $k_A < k_B$), i.e.,

$$k_A \leq k_c \leq k_B \quad \text{for} \quad k_A < k_B.$$

Contact Layer in Physical Bonding of Contacting
Materials and Contact Conductivity, $k_c$

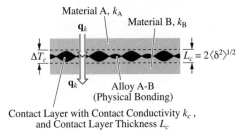

Contact Layer with Contact Conductivity $k_c$,
and Contact Layer Thickness $L_c$

Figure Pr. 3.28. A contact layer in a physical bounding of contacting materials; also shown is the contact conductivity $k_c$.

Consider a contact resistance between a bismuth telluride slab (material A) and copper (material B) slab. This pair is used in thermoelectric coolers, where the semiconductor, doped bismuth telluride is the thermoelectric material and copper is the electrical connector. Then use a general relationship $k_c = k_A + a_1(k_B - k_A)$, $0 \leq a_1 \leq 1$, and plot (semilog scales) the temperature drop across the junction $\Delta T_c$, for the following conditions, as a function of $a_1$. Here $a_1$ depends on the fabrication method used.

$$k_A = 1.6 \text{ W/m-K}, \; k_B = 385 \text{ W/m-K}, \; L_c = 2\langle\delta^2\rangle^{1/2} = 0.5 \; \mu\text{m}, \; q_k = 10^5 \text{ W/m}^2.$$

### 3.12.4 Conduction and Energy Conversion and Thermoelectric Cooling

**PROBLEM 3.29. FUN**

In a thermoelectric cell, the Joule heating that results from the passage of the electric current is removed from the hot and cold ends of the conductor. When the electrical resistivity $\rho_e$ or the current density $j_e$ are large enough, the maximum temperature can be larger than the temperature at the hot-end surface $T_h$. This is shown in Figure 3.27(b).

(a) From the temperature distribution given by (3.104), determine the expression for the location of the maximum temperature in the conductor.
(b) Both the $p$- and the $n$-type legs have the same length $L = 2$ cm. The cross-sectional area of the $n$-type leg is $A_n = 2.8 \times 10^{-5}$ m². Calculate the cross-sectional area for the $p$-type leg $A_p$ if the figure of merit is to be maximized. Use the electrical and thermal properties of the $p$-type and $n$-type bismuth telluride given in Table C.9(a).
(c) Determine the magnitude and the location of the maximum temperature for the $p$-type leg, when $T_h = 40°$C, $T_c = -2°$C, and $J_e = 6$ A.
(d) Determine the rate of heat removal from the hot and cold ends per unit cross-sectional area $(A_n + A_p)$.

**PROBLEM 3.30. FUN**

A thermoelectric cooler has bismuth telluride elements (i.e., $p$- and $n$-type pairs) that have a circular cross section of diameter $d = d_n = d_p = 3$ mm and a length $L = L_n = L_p = 2$ cm, as shown in Figure Pr. 3.30. The temperatures of the hot and

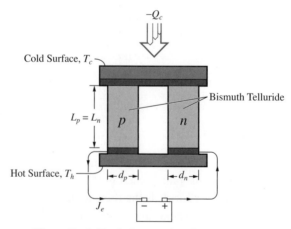

Figure Pr. 3.30. A thermoelectric cooler unit.

cold ends are $T_h = 40°C$ and $T_c = -2°C$. Determine the cooling power $Q_c$ for each junction, if the current corresponds to

(a) the current that maximizes the cooling power $Q_c$, (b) the current that is half of this optimum current, and (c) the current that is twice the optimum current.

## PROBLEM 3.31. FUN

A thermoelectric device is used for cooling a surface to a temperature $T_c$. For the conditions given below, determine (a) the minimum $T_c$, (b) the current for this condition, and (c) the minimum $T_c$ for a current $J_e = 1$ A.

For each bismuth telluride thermoelectric, circular cylinder conductor, use $d_n = d_p = 1.5$ mm, and $L_n = L_p = 4$ mm. The hot junction is at $T_h = 40°C$.

## PROBLEM 3.32. FUN

A thin-film thermoelectric cooler is integrated into a device as shown in Figure Pr. 3.32. In addition to heat conduction through the $p$- and $n$-type conductors, heat flows by conduction through the substrate. Assume a one-dimensional parallel conduction through the $p$- and $n$-type conductors and the substrate.

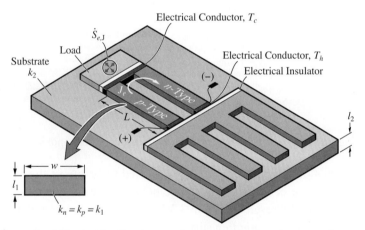

Figure Pr. 3.32. A thin-film thermoelectric cooler placed over a substrate.

(a) Draw the thermal circuit diagram for heat flow between the $T_h$ and $T_c$ nodes.
(b) Show that the maximum temperature difference is

$$(T_h - T_c)_{max}(Q_c = 0) = \frac{Z_e T_c^2}{2\left(1 + \dfrac{l_2 k_2}{l_1 k_1}\right)}.$$

Model the conduction through the substrate as two conduction paths (one underneath each of the $p$- and $n$-type legs). Begin with (3.115) and use the optimum current.

## PROBLEM 3.33. DES.S

A miniature vapor sensor is cooled, for enhanced performance, by thermoelectric coolers. The sensor and its thermoelectric coolers are shown in Figure Pr. 3.33. There are four bismuth-telluride thermoelectric modules and each module is made of four *p*-*n* layers (forming four *p*-*n* junctions) each *p*- and *n*-layer having a thickness *l*, length *L*, and width *w*. The sensor and its substrate are assumed to have the $\rho c_p$ of silicon and a cold junction temperature $T_c(t)$.

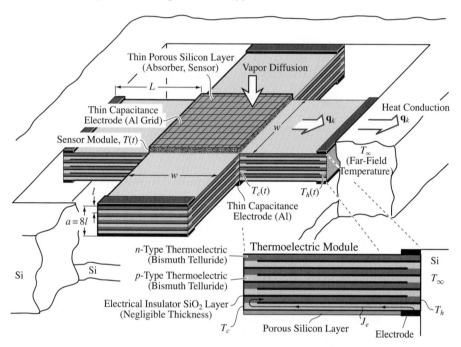

Figure Pr. 3.33. A thermoelectrically cooled, miniature vapor sensor.

The hot junction temperature $T_h(t)$ is expected to be above the far-field solid temperature $T_\infty$. The conduction resistance between the hot junctions and $T_\infty$ is approximated using the results of Table 3.3(b), for steady-state resistance between an ambient placed on the bounding surface of a semi-infinite slab ($T_h$) and the rest of the slab ($T_\infty$) [shown in Table 3.3(b), first entry]. This is

$$R_{k,h\text{-}\infty} = \frac{\ln\dfrac{4w}{2a}}{\pi kw} = \frac{\ln\dfrac{2w}{a}}{\pi kw}, \qquad Q_{k,h\text{-}\infty} = \frac{T_h(t) - T_\infty}{R_{k,h\text{-}\infty}}.$$

Initially there is a uniform sensor temperature, $T_c(t=0) = T_h(t=0) = T_\infty$. For heat storage of the thermoelectric modules, divide each volume into two with each portion having temperature $T_c$ or $T_h$.

(a) Draw the thermal circuit diagram.
(b) Determine and plot the sensor temperature $T_c$.
(c) Determine the steady-state sensor temperature $T_c(t \to \infty)$.

    $l = 3\ \mu m$, $w = 100\ \mu m$, $L = 300\ \mu m$, $a = 24\ \mu m$, $T_\infty = 20°C$, $J_e = 0.010$ A.

**PROBLEM 3.34. FUN**

A highly localized Joule heating applied to myocardium via a transvenous catheter can destroy (ablate) the endocardial tissue region that mediates life-threatening arrhythmias. Alternating current, with radio-frequency range of wavelength, is used. This is shown in Figure Pr. 3.34(i). The current flowing out of the spherical tip of the catheter flows into the surrounding tissue, as shown in Figures Pr. 3.34(ii) and (iii). Due to the rapid decay of the current flux, the Joule heating region is confined to a small region $R_e \leq r \leq R_1$ adjacent to the electrode tip. The total current $J_e$ leaving the spherical tip results in a current density

$$j_e = \frac{J_e}{4\pi r^2}.$$

The tissue having a resistivity $\rho_e$ will have a local energy conversion rate $\dot{s}_{e,J} = \rho_e j_e^2$.

Assuming that all the energy conversion occurs in the region $R_e \leq r \leq R_1$, then the heat is conducted from this region toward the remaining tissue. The far-field temperature is $T_2$, i.e., as $r \to \infty$, $T \to T_2$.

(i) Radio-Frequency Catheter Ablation of Myocardium

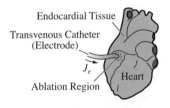

(ii) Joule-Heating Region

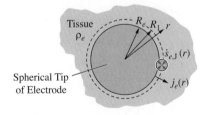

(iii) Temperature Decay Region

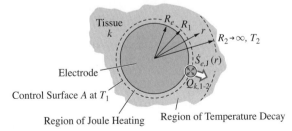

Figure Pr. 3.34. (i) The Joule heating of myocardium region by a spherical electrode tip. (ii) The small heated region. (iii) Heat flow out of the heated region by conduction.

(a) Derive the expression for the local $\dot{s}_{e,J}(r)$ and comment on its distribution.

(b) Draw the thermal circuit diagram and write the surface energy equation for its surface located at $r = R_1$. Use the conduction resistance for a spherical shell (Table 3.2).

(c) Determine $T_1(R_1)$ for the following conditions.

$T_2 = 37°C$, $R_e = 1.0$ mm, $R_1 = 1.3$ mm, $\rho_e = 2.24$ ohm-m, $J_e = 0.07476$ A.

Use Table C.17 for $k$ of muscle. Assume steady-state heat transfer.

### PROBLEM 3.35. FUN

Refractive surgical lasers are used to correct the corneal refractive power of patients who are near-sighted, far-sighted, or astigmatic in their vision. These corneal reshaping procedures can be performed via several mechanisms, including ablation of the corneal surface to change the focal length, removal of a section of the cornea causing reformation, and reshaping of the corneal tissue by thermal shrinkage effects. In order to achieve minimal tissue thermal damage due to the laser ablation, we need to investigate the thermal behavior of the corneal tissue to understand the local thermal effects of laser heating, and to predict the potential for an unintentional injury during laser surgery.

Heat transfer in corneal tissue is modeled as a sphere of radius $R_2 = 5$ mm, with the energy source positioned at the center of the sphere (Figure Pr. 3.35). A small diameter laser beam with $\dot{S}_{e,\alpha} = A_r \alpha_r q_{r,i} = 220$ mW is used. Assume a steady-state conduction heat transfer.

(a) Draw the thermal circuit diagram.

(b) Determine $R_{k,1\text{-}2}$.

(c) Determine $T_1$ of the laser beam at $R_1 = 1$ mm, using the energy equation.

(d) Plot the variation of $T$ with respect to $r$.

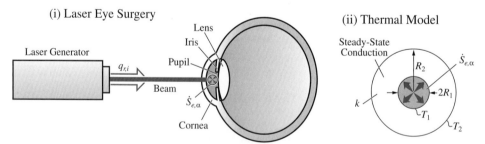

(i) Laser Eye Surgery

(ii) Thermal Model

Figure Pr. 3.35. (i) Laser eye surgery showing the absorbed irradiation. (ii) The thermal model.

### PROBLEM 3.36. FUN

Cardiac ablation refers to the technique of destroying heart tissue that is responsible for causing alternations of the normal heart rhythm. A well-localized region of the endocardial tissue is destroyed by microwave heating. The electric field is provided through a catheter that is inserted into the heart through a vein. One example of this cardiac ablation and the microwave catheter design is shown in Figure Pr. 3.36. The region $R_e < r < R_o$ is heated with $R_e = 1$ mm and frequency $f = 3.00 \times 10^{11}$ Hz. A long catheter is assumed, so the heat transfer is dominant in the cylindrical cross-sectional plane (shown in Figure Pr. 3.36).

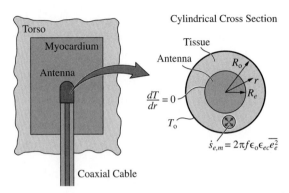

Figure Pr. 3.36. A microwave catheter used for ablation of cardiac tissue.

The dielectric loss factor for heart tissue is $\epsilon_{ec} = 17.6$ and its thermal conductivity is $k = 0.45$ W/m-K.

(a) Assuming a zero temperature gradient $r = R_e$, start with the one-dimensional (radial direction) differential energy equation with $\dot{s}_{e,m} = \dot{s}_{e,m}(r)$, then integrate this differential equation to obtain the radial distribution of the temperature $T = T(r)$. It is known that $\bar{e}_e^2 = e_{e,o}^2 (R_e/r)^2$. In addition to thermal conditions $dT/dr = 0$ at $r = R_e$ use $T = T_o$ at $r = R_o$.

(b) In order to perform a successful ablation, the microwave antenna needs to produce a temperature $T_e = 353$ K (80°C) at $r = R_e$. Given that temperature decreases to the normal tissue temperature $T_o = 310.65$ K (37.5°C) at $r = 1$ cm, determine the required electric field intensity $e_{e,o}$ to produce this temperature.

(c) Plot the variation of $T(r)$ with respect to $r$ for several values of $e_{e,o}$.

**PROBLEM 3.37. FUN**

The thermoelectric figure of merit $Z_e = \alpha_S^2/(R_e/R_k)$ is maximized by minimizing $R_e/R_k$. Begin with the relations for $R_e$ and $R_k$, and the electrical and thermal properties, i.e., $\rho_{e,n}, \rho_{e,p}$. Then calculate $k_n, k_p$, and the geometrical parameters $L_n, A_{k,n}, L_p$, and $A_{k,p}$, as given by (3.116).

(a) Then assume that $L_n = L_p$ and minimize $R_e/R_k$ with respect to $A_{k,p}/A_{k,n}$. Then show that

$$\frac{L_n A_{k,p}}{L_p A_{k,n}} = \frac{A_{k,p}}{A_{k,n}} = \left( \frac{\rho_{e,p} k_n}{\rho_{e,n} k_p} \right)^{1/2} \quad \text{for optimum } Z_e,$$

which is (3.121).

(b) Use this in (3.120), and show that

$$Z_e = \frac{\alpha_S^2}{[(k\rho_e)_p^{1/2} + (k\rho_e)_n^{1/2}]}, \qquad \text{optimized figure of merit.}$$

## PROBLEM 3.38. FUN

Begin with the differential-length energy equation (3.102), and use the prescribed thermal boundary conditions (3.103) for a finite length slab of thickness $L$.

(a) Derive the temperature distribution given by (3.104).
(b) Show that the location of the maximum temperature is

$$x(T_{max}) = \frac{L}{2} + \frac{k(T_h - T_c)}{\rho_e j_e^2 L}.$$

(c) Comment on this location for the case of (i) $j_e \to 0$, and (ii) $j_e \to \infty$.

## PROBLEM 3.39. FUN

The optimum coefficient of performance for the thermoelectric cooler $\eta_{cop}$, given by (3.126), is based on a current that optimizes it. This current is also given in (3.126). Derive the expression for the optimum current, i.e.,

$$J_e[\partial\eta_{cop}/\partial J_e = 0] = \frac{\alpha_S(T_h - T_c)}{R_{e,h\text{-}c}[(1 + Z_e T_o)^{1/2} - 1]},$$

where $Z_e$ is given by (3.120) and

$$T_o = \frac{T_h + T_c}{2}.$$

## PROBLEM 3.40. FUN.S

With thermoelectric coolers, the lowest temperature $T_c$ at the cold junction is achieved when $Q_c = 0$. To further lower this temperature, a technique called staging is used, whereby thermoelectric units are stacked on top of one another. However, staging $p$-$n$ pairs in series is inefficient, since the cooling effect at the common junction is cancelled out by the joule heating in the adjacent pair. Therefore a pyramid structure as shown in Figure Pr. 3.40, is often used rather than a series or parallel configuration. Assume no heat loss at the intermediate junctions, and negligible temperature drops across the electrical insulators and conductors.

(a) Draw thermal circuit diagrams for each of the following configurations of three staging configurations: (i) series, (ii) parallel, and (iii) pyramid.
(b) Determine a general expression for $T_c$ for the pyramid configuration.
(c) Make a plot of $T_c$ versus $J_e$ for each of the three cases in (a), and show them on the same graph, using the given parameters for $0 < J_e \le 5$ A. The expression from (b) can simply be plotted, as can a similar expression for the parallel case, but it may be helpful to generate data for the series configuration using MATLAB. From the

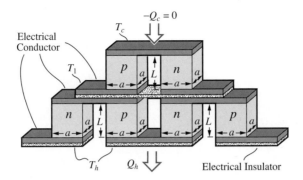

Figure Pr. 3.40. Staging a thermoelectric cooler in a pyramid arrangement.

graph, what is the lowest $T_c$ that can be obtained, and what is the necessary current? Explain why the series and parallel configurations actually give better performance over certain ranges of $J_e$.

(Hint: Think about how $J_e$ affects heating and cooling in the cooler.)

(d) For the pyramid structure, what voltage $\Delta\varphi$ must be applied to the three pairs to obtain a $J_e$ of 4 A.

$T_h = 313.15$ K, and for each bismuth telluride $p$-$n$ pair, $\alpha_s = 4.4 \times 10^{-4}$ V/K, $R_e = 0.030$ ohm, $R_k = 4.762 \times 10^2$ K/W.

## PROBLEM 3.41. DES.S

An in-plane thermoelectric device is used to cool a microchip. It has bismuth telluride elements with dimensions given below. A contact resistance $R_{k,c} = l_c/A_k k_c$, given by (3.94), is present between the elements and the connector. These are shown in Figure Pr. 3.41. The contact conductivities $k_c$ is empirically determined for two different connector materials and are

$$k_c = 10k_{TE} \qquad \text{copper connector}$$

$$k_c = k_{TE} \qquad \text{solder connector,}$$

where $k_{TE}$ is the average of $p$- and $n$-type materials.

(a) Draw the thermal circuit diagram showing the contact resistance at each end of the elements.
(b) Determine the optimum current for cold junction temperature $T_c = 275$ K.
(c) Determine the minimum $T_c$ [i.e., (3.119)] for these conditions.
(d) For $T_c = 275$ K, determine $Q_{c,max}$ from (3.118).
(e) Using this $Q_{c,max}$ and (3.96), determine $\Delta T_c = T_{s,c} - T_c$ and plot $T_{s,c}$ versus $l_c$ for $0 < l_c < 10$ $\mu$m, for both the copper and solder connectors.
    $L = 150$ $\mu$m, $w = 75$ $\mu$m, $a = 4$ $\mu$m, $T_h = 300$ K.

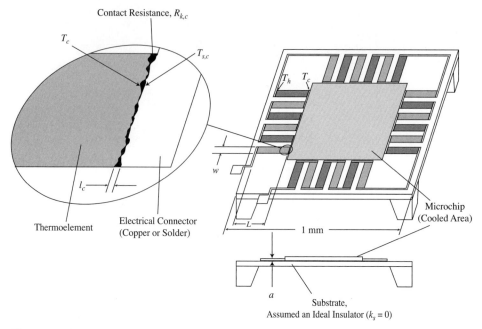

Figure Pr. 3.41. A miniaturized thermoelectric device with a thermal contact resistance between the thermoelectric elements and the connectors.

### 3.12.5 Multidimensional Conduction

## PROBLEM 3.42. FAM

To melt the ice forming on a road pavement (or similarly to prevent surface freezing), pipes are buried under the pavement surface, as shown in Figure Pr. 3.42. The pipe surface is at temperature $T_1$, while the surface is at temperature $T_2$. The magnitude of the geometrical parameters for the buried pipes are given below. Assume that the conduction resistances given in Table 3.3(a) are applicable.

Melting of Surface Ice by Buried Pipes Carrying Hot Water

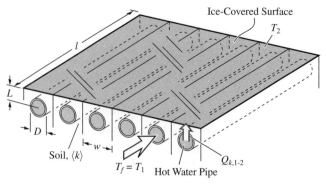

Figure Pr. 3.42. Hot-water carrying buried pipes used for melting of an ice layer on a pavement surface.

(a) Draw the thermal circuit diagram for each pipe.

(b) From Table 3.3(a), determine the conduction resistance for (i) a single pipe (i.e., cylinder) independent of the adjacent pipes, and (ii) a pipe in a row of cylinders with equal depth and an axial center-to-center spacing $w$.

(c) Determine $Q_{k,1\text{-}2}$ per pipe for both cases (i) and (ii), and then compare.

$L = 5$ cm, $D = 1$ cm, $l = 2$ m, $w = 10$ cm, $T_1 = 10°$C, $T_2 = 0°$C.

Use the thermophysical properties of soil (Table C.17).

## PROBLEM 3.43. FUN

The steady-state conduction in a two-dimensional, rectangular medium, as shown in Figure Pr. 3.43, is given by the differential volume energy equation (B.55), i.e.,

$$\nabla \cdot \boldsymbol{q}_k = -k\frac{\partial^2 T}{\partial x^2} - k\frac{\partial^2 T}{\partial y^2} = 0.$$

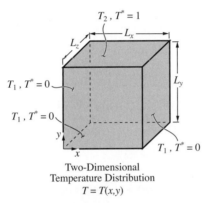

Two-Dimensional
Temperature Distribution
$T = T(x,y)$

Figure Pr. 3.43. A rectangular, two-dimensional geometry with prescribed surface temperatures.

This is called a homogeneous, linear, partial differential equation and a general solution that separates the variable is possible, if the boundary (bounding-surface) conditions can also be homogeneous. This would require that the temperatures on all surfaces be prescribed, or the bounding-surface energy equation be a linear resistive type. When the four surface temperatures are prescribed, such that three surfaces have a temperature and different from the fourth, the final solution would have a simple form.

(a) Using the dimensionless temperature distribution

$$T^* = \frac{T - T_1}{T_2 - T_1},$$

show that the energy equation and boundary conditions become

$$\frac{\partial^2 T^*}{\partial x^2} + \frac{\partial^2 T^*}{\partial y^2} = 0, \ T^* = T^*(x, y)$$

$$T^*(x = 0, y) = 0, \ T^*(x = L_x, y) = 0, \ T^*(x, y = 0) = 0, \ T^*(x, y = L_y) = 1.$$

(b) Use the method of the separation of variables (which is applicable to this homogeneous differential equation with all but one boundary condition being also homogeneous), i.e.,

$$T^*(x, y) = X(x)Y(y)$$

to show that the energy equation becomes

$$-\frac{1}{X}\frac{d^2 X}{dx^2} = \frac{1}{Y}\frac{d^2 Y}{dy^2}.$$

(c) Since the left-hand side is only a function of $x$ and the right-hand side is a function of $y$, then both sides should be equal to a constant. This is called the separation constant. Showing this constant as $b^2$, show also that

$$\frac{d^2 X}{dx^2} + b^2 X = 0$$

$$\frac{d^2 Y}{dy^2} - b^2 Y = 0.$$

Then show that the solutions for $X$ and $Y$ are

$$X = a_1 \cos(bx) + a_2 \sin(bx)$$

$$Y = a_3 e^{-by} + a_4 e^{by}$$

or

$$T^* = [a_1 \cos(bx) + a_2 \sin(bx)](a_3 e^{-by} + a_4 e^{by}).$$

(d) Apply the homogeneous boundary conditions to show that

$$a_1 = 0$$

$$a_3 = -a_4$$

$$a_2 a_4 \sin(bL_x)(e^{by} - e^{-by}) = 0.$$

Note that the last one would require that

$$\sin(bL_x) = 0$$

or

$$bL_x = n\pi, \quad n = 0, 1, 2, 3, \cdots$$

(e) Using these, show that

$$T^*(x, y) = a_2 a_4 \sin\left(\frac{n\pi x}{L_x}\right)(e^{n\pi y/L_x} - e^{-n\pi y/L_x})$$

$$\equiv a_n \sin\left(\frac{n\pi x}{L_x}\right)\sinh\left(\frac{n\pi y}{L_x}\right).$$

Since $n = 1, 2, 3, \cdots$ and differential equation is linear, show that

$$T^* = \sum_{n=0}^{\infty} a_n \sin\left(\frac{n\pi x}{L_x}\right) \sinh\left(\frac{n\pi y}{L_x}\right).$$

(f) Using the last (i.e., nonhomogeneous) boundary condition and the orthogonality condition of the special function $\sin(z)$, it can be shown that

$$a_n = \frac{2[1 + (-1)^{n+1}]}{n\pi \sinh(n\pi L_y/L_x)}, \quad n = 0, 1, 2, 3, \cdots.$$

Then express the final solution in terms of this and verify that all boundary conditions are satisfied.

### PROBLEM 3.44. FAM

In order to maintain a permanent frozen state (permafrost) and a firm ground, heat pipes are used to cool and freeze wet soil in the arctic regions. Figure Pr. 3.44 shows a heat pipe, which is assumed to have a uniform temperature $T_1$, placed between the warmer soil temperature $T_2$, and the colder ambient air temperature $T_{f,\infty}$. The heat transfer between the heat pipe surface and the ambient air is by surface convection and this resistance is given by $A_{ku}R_{ku,1-\infty}$. The heat transfer between the pipe and soil is by conduction $R_{k,1-2}$. Assume a steady-state heat transfer.

(a) Draw the thermal circuit diagram.
(b) Determine the heat pipe temperature $T_1$ and the amount of heat flow rate $Q_{k,2-1}$.
(c) If this heat is used entirely in phase change (solidification of liquid water), determine the rate of ice formation around the buried pipe.

$D = 1$ m, $L_{ku} = 5$ m, $L_k = 2$ m, $T_{f,\infty} = -20°C$, $T_2 = 0°C$, $A_{ku}R_{ku,1-\infty} = 10^{-1}$ K/ (W/m²).

Use Table 3.3(b) to determine the resistance $R_{k,1-2}$.

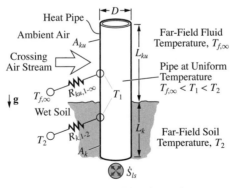

Figure Pr. 3.44. A rendering of a heat pipe used for the maintenance of a permafrost layer in the arctic regions.

**PROBLEM 3.45. FAM.S**

In scribing of disks by pulsed laser irradiation (also called laser zone texturing) a small region, diameter $D_1$, is melted and upon solidification a protuberance (bump) is formed in this location. The surface of the liquid pool formed through heating is not uniform and depends on the laser energy and its duration, which in turn also influences the depth of the pool $L_1(t)$. These are shown in Figure Pr. 3.45(i). Assume that the irradiated region is already at the melting temperature $T_1 = T_{sl}$ and the absorbed irradiation energy $(\dot{S}_{e,\alpha})_1$ is used to either melt the substrate $\dot{S}_{sl}$, or is lost through conduction $Q_{k,1\text{-}2}$ to the substrate. This simple, steady-state thermal model is also shown in Figure Pr. 3.45(ii). The irradiation is for an elapsed time of $\Delta t$.

## (a) Laser Zone Texturing

## (b) Simple, Steady-State Thermal Model

Figure Pr. 3.45. (a) Laser zone texturing of a disk. (b) The associated simple, steady-state heat transfer model.

(a) Draw the thermal circuit diagram.

(b) Determine the depth of the melt $L_1$, after an elapsed time $\Delta t$.

$D_1 = 10 \ \mu\text{m}$, $(\dot{S}_{e,\alpha})_1 = 48$ W, $\Delta t = 1.3 \times 10^{-7}$ s, $T_2 = 50°$C.

Use the temperature, density, and heat of melting of nickel in Table C.2 and, the thermal conductivity of nickel at $T = 1,400$ K in Table C.14.

The energy conversion rate $\dot{S}_{sl}$ is given in Table 2.1 and note that $\dot{M}_{sl} = M_l/\Delta t$, where $M_l = \rho V_l(t)$. Use Table 3.3(b) for the conduction resistance and use $L_1(t = \Delta t)$ for the depth.

### 3.12.6 Distributed-Capacitance Transient Conduction and Penetration Depth

**PROBLEM 3.46. FUN**

To estimate the elapsed time for the penetration of a change in the surface temperature of the brake rotor, the results of Table 3.4 and Figure 3.33(a) can be used. Consider the brake rotor idealized in Figure Pr. 3.46, with symmetric heating.

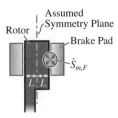

Figure Pr. 3.46. Surface friction heating and its penetration into the brake rotor, during the braking period.

(a) For the conditions given below, determine the penetration time.
(b) If the brake is on for $t = 4$ s, is the assumption of a uniform rotor temperature valid during the braking period?
(c) If the surface-convection cooling occurs after braking and over a time period of 400 s, is the assumption of a uniform rotor temperature valid during the cooling period?

Use carbon steel AISI 1010 for the rotor at $T = 20°C$, and $2L = 3$ cm.

**PROBLEM 3.47. FUN.S**

The time-periodic variation of the surface temperature of a semi-infinite slab (such as that shown in Figure Ex. 3.17) can be represented by an oscillating variation

$$T_s = T(t = 0) + \Delta T_{max}\cos(\omega t),$$

where $\omega = 2\pi f$ is the angular frequency, $f$ (1/s) is the linear frequency, and $\Delta T_{max}$ is the amplitude of the surface temperature change. The solution to the energy equation (3.134), with the above used for the first of the thermal conditions in (3.135), is [17]

$$\frac{T(x,t) - T(x,t=0)}{\Delta T_{max}} = \exp\left[-x\left(\frac{\omega}{2\alpha}\right)^{1/2}\right]\cos\left[\omega t + x\left(\frac{\omega}{2\alpha}\right)^{1/2}\right].$$

Show that the penetration depth $\delta_\alpha$, defined by

$$\frac{T(x,t) - T(x,t=0)}{\Delta T_{max}} = 0.01,$$

is given by

$$\frac{\delta_\alpha}{(2\alpha t)^{1/2}} = 0.4310,$$

which is similar to the penetration depth given by (3.148).

Evaluate the penetration depth after an elapsed time equal to a period, i.e., $t = \tau = 1/f$, where $\tau(s)$ is the period of oscillation.

## PROBLEM 3.48. FAM

Pulsed lasers provide a large power $q_{r,i}$ for a short time $\Delta t$. In surface treatment of materials (e.g., laser-shock hardening), the surface is heated by laser irradiation using very small pulse durations. During this heating, the transient conduction through the irradiated material can be determined as that of a semi-infinite solid subject to constant surface heating $-q_s = q_{r,i}$; this is shown in Figure Pr. 3.48, with the material being a metallic alloy (stainless steel AISI 316, Table C.16). The heated semi-infinite slab is initially at $T(t = 0)$. In a particular application, two laser powers (assume all the laser irradiation power is absorbed by the surface), with different pulse lengths $\Delta t$, are used. These are (i) $-q_s = 10^{12}$ W/m², $\Delta t = 10^{-6}$ s, and (ii) $-q_s = 10^{10}$ W/m², $\Delta t = 10^{-4}$ s ($T(t = 0) = 20°$ C).

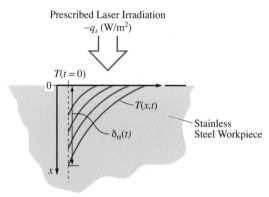

Figure Pr. 3.48. Pulsed laser irradiation of a stainless steel workpiece and the anticipated transient temperature distribution within the workpiece.

(a) Determine the surface temperature $T(x = 0, \ t = \Delta t)$ after elapsed time $t = \Delta t$, for cases (i) and (ii).
(b) As an approximation, use the same expression for penetration depth $\delta_\alpha(t)$ as that for the semi-infinite slabs with a prescribed surface temperature, and determine the penetration depth after the elapsed time $t = \Delta t$, for cases (i) and (ii).
(c) Comment on these surface temperatures and penetration depths.

## PROBLEM 3.49. FUN

In a solidification process, a molten acrylic at temperature $T(t = 0)$ is poured into a cold mold, as shown in Figure Pr. 3.49, to form a clear sheet. Assume that the heat of solidification can be neglected.

(a) Determine the elapsed time for the cooling front to reach the central plane of the melt.
(b) Determine the elapsed time for the temperature of the central plane of the melt to reach the glass transition temperature $T_{ls}$.

$\quad L = 2.5$ mm, $T_{ls} = 90°$C, $T(t = 0) = 200°$C, $T_s = 40°$C.

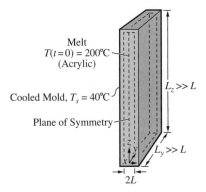

Figure Pr. 3.49. Solidification of an acrylic melt in a mold having a constant temperature $T_s$.

**PROBLEM 3.50. FAM**

During solidification, as in casting, the melt may locally drop to temperatures below the solidification temperature $T_{ls}$, before the phase change occurs. Then the melt is in a metastable state (called supercooled liquid) and the nucleation (start) of the solidification resulting in formation of crystals (and their growth) begins after a threshold liquid supercool is reached. Consider solidification of liquid paraffin (Table C.5) in three different molds. These molds are in the form of (a) a finite slab, (b) a long cylinder, and (c) a sphere, and are shown in Figure Pr. 3.50. The melt in the molds is initially at its melting temperature $T(t = 0) = T_{sl}$. Then at $t = 0$ the mold surface is lowered and maintained at temperature $T_s$.

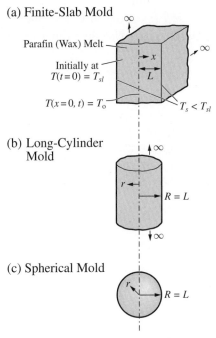

Figure Pr. 3.50. Paraffin (wax) melt is cooled in a mold (three different geometries) and is gradually solidified.

Determine the elapsed time needed for the temperature at the center of the mold, i.e., $T(x = 0, t) = T(r = 0, t)$ to reach a threshold value $T_o$, for molds (i), (ii) and (iii). Assume that solidification will not occur prior to this elapsed time.

$T_s = 25°C$, $L = R = 2$ cm, $T(x = 0, t) = T(r = 0, t) = T_o = 302$ K.

Use the properties of polystyrene (Table C.17).

### PROBLEM 3.51. FAM

An apple (modeled as a sphere of radius $R = 4$ cm), initially at $T(t = 0) = 23°C$ is placed in a refrigerator at time $t = 0$, and thereafter, it is assumed that its surface temperature is maintained at $T_s = 4°C$.

(a) Determine the elapsed time it takes for the thermal penetration depth to reach the center of the apple.

(b) Determine the elapsed time for the center temperature to reach $T = 10°C$.

Use the thermophysical properties of water at $T = 293$ K from Table C.23. Use the graphical results given in Figure 3.33(b).

### PROBLEM 3.52. FAM

In a summer day, the solar irradiation on the surface of a parking lot results in an absorbed irradiation flux $q_s = -500$ W/m$^2$, as shown in Figure Pr. 3.52. The parking lot surface is covered with an asphalt coating that has a softening temperature of 55°C.

Determine the elapsed time it takes for the surface temperature of the parking lot to rise to the softening temperature.

The initial temperature is $T(t = 0) = 20°C$. Assume that all the absorbed heat flows into the very thick asphalt layer. Use the properties of asphalt in Table C.17.

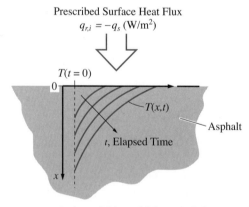

Figure Pr. 3.52. Temperature variation within a thick asphalt layer, suddenly heated by solar irradiation.

### PROBLEM 3.53. FAM

In a shaping process, a sheet of Teflon (Table C.17) of thickness $2L$, where $L = 0.3$ cm, is placed between two constant-temperature flat plates and is heated. The initial temperature of the sheet is $T(t = 0) = 20°C$ and the plates are at $T_s = 180°C$. This is shown in Figure Pr. 3.53.

Determine the time it takes for the center of the sheet to reach 20°C below $T_s$.

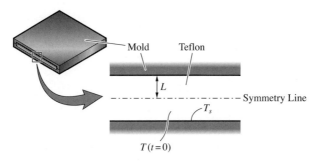

Figure Pr. 3.53. Thermal forming of a Teflon sheet.

## PROBLEM 3.54. FUN.S

When human skin is brought in contact with a hot surface, it burns. The degree of burn is characterized by the temperature of the contact material $T_s$ and the contact time $t$. A first-degree burn displays no blisters and produces reversible damage. A second-degree burn is moist, red, blistered, and produces partial skin loss. A third-degree burn is dry, white, leathery, blisterless, and produces whole skin loss. A pure copper pipe with constant temperature $T_s = 80°C$ is brought in contact with human skin having $\rho c_p = 3.7 \times 10^6$ J/m$^3$-°C, $k = 0.293$ W/m-°C, and $T(t = 0) = 37°C$ for a total elapsed time of $t = 300$ s.

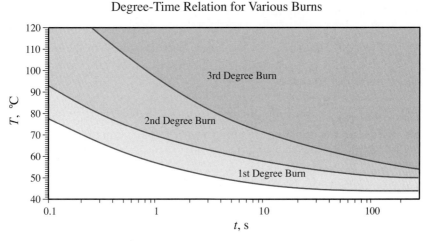

Figure Pr. 3.54. Thermal damage (burn) to a human skin and the regimes of various degrees of burn.

Use the solution for the transient conduction through a semi-infinite slab with a prescribed surface temperature to answer the following questions.

(a) Plot the temperature distribution $T(x, t)(°C)$ as a function of position $x$(mm) at elapsed times of $t = 1, 10, 20, 40, 50, 100, 150, 200, 220, 240, 280,$ and 300 s.
(b) Use this plot, along with the plot shown in Figure Pr. 3.54, to estimate the maximum depths for the first-, second-, and third-degree burns, after an elapsed time $t = 300$ s.

## PROBLEM 3.55. FAM

A hole is to be drilled through a rubber bottle stopper. Starting the hole in a soft room-temperature rubber often results in tears or cracks on the surface around the hole. It has been empirically determined that the rubber material at the surface can be hardened sufficiently for crack- and tear-free drilling by reducing the surface temperature at $x = 1$ mm beneath the surface to below $T = 220$ K. This reduction in temperature can be achieved by submerging the rubber surface into a liquid nitrogen bath for a period of time. This is shown in Figure Pr. 3.55.

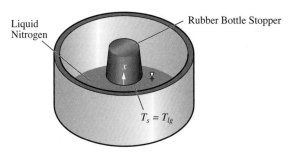

Figure Pr. 3.55. A rubber bottle stopper is temporarily submerged in a liquid nitrogen bath.

Assume that by submerging, the surface temperature drops from the initial uniform temperature $T(x, t = 0) = 20°C$ to the boiling temperature of the nitrogen $T_{lg}$, and then remains constant.

Determine after an elapsed time of $t = 10$ s, (a) the temperature 1 mm from the surface $T(x = 1 \text{ mm}, t = 10 \text{ s})$, (b) the temperature 3 mm from the surface $T(x = 3 \text{ mm}, t = 10 \text{ s})$, and (c) the rate of heat flowing per unit area out of the rubber surface $q_s(x = 0, t = 10 \text{ s})(\text{W/m}^2)$.

(d) Is $t = 10$ s enough cooling time to enable crack- and tear-free drilling of the hole?

Use the saturation temperature of nitrogen $T_{lg}$ at one atm pressure (Table C.26) and the properties of soft rubber (Table C.17).

## PROBLEM 3.56. FAM

The friction heat generation $\dot{S}_{m,F}$ (energy conversion) occurring in grinding flows *into* a workpiece (stainless steel AISI 316 at $T = 300$ K) and the grinder (use properties of brick in Table C.17 at $T = 293$ K). This is shown in Figure Pr. 3.56. For a thick (i.e., assumed semi-infinite) grinder and a thick workpiece, the fraction of the heat flowing into the workpiece $a_1$ can be shown to be

$$a_1 = \frac{(\rho c_p k)_w^{1/2}}{(\rho c_p k)_w^{1/2} + (\rho c_p k)_g^{1/2}},$$

where the properties of the workpiece are designated by $w$ and that of the grinder by $g$.

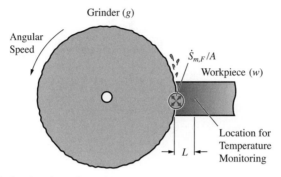

Figure Pr. 3.56. Friction heating of a stainless steel workpiece and the division of the generated heat.

(a) Draw the thermal circuit diagram.

(b) Use this relation for $a_1$ and the transient temperature distribution resulting from the sudden heating of a semi-infinite slab at a constant heat flux, given in Table 3.4, to determine the temperature of the workpiece at location $L$ from the interface and after an elapsed time of $t$.

$t = 15$ s, $L = 1.5$ mm, $\dot{S}_{m,F}/A = 10^5$ W/m$^2$, $T(t = 0) = 300$ K.

**PROBLEM 3.57. FUN**

Two semi-infinite slabs having properties $(\rho, c_p, k)_1$ and $(\rho, c_p, k)_2$ and uniform, initial temperatures $T_1(t = 0)$ and $T_2(t = 0)$, are brought in contact at time $t = 0$.

(a) Show that their contact (interfacial) temperature is constant and equal to

$$T_{12} = \frac{(\rho c_p k)_1^{1/2} T_1(t = 0) + (\rho c_p k)_2^{1/2} T_2(t = 0)}{(\rho c_p k)_1^{1/2} + (\rho c_p k)_2^{1/2}}.$$

(b) Under what conditions does $T_{12} = T_2(t = 0)$? Give an example of material pairs that would result in this limit.

**PROBLEM 3.58. FAM.S**

For thermal treatment, the surface of a thin-film coated substrate [initially at $T(t = 0)$] is heated by a prescribed heat flux $q_s = -10^9$ W/m$^2$ (this heat flux can be provided for example by absorbed laser irradiation) for a short period. This heating period $t_o$ (i.e., elapsed time) is chosen such that only the temperature of the titanium alloy (Ti-2 Al-2 Mn, mass fraction composition) thin film is elevated significantly (i.e., the penetration distance is only slightly larger than the thin-film thickness). The thin film is depicted in Figure Pr. 3.58. Determine the required elapsed time $t_o$ for the temperature of the interface between the thin film and the substrate, located at distance $l = 5\ \mu$m from the surface, to raise by $\Delta T(l, t_o) = T(l, t_o) - T(t = 0) = 300$ K. Note that this would require determination of t from an implicit relation and would require iteration or use of a software.

Evaluate the thermophysical properties at 300 K, as given in Table C.16.

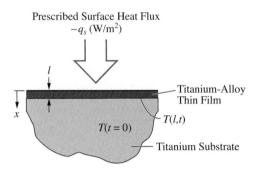

Figure Pr. 3.58. A thin-film coated, semi-infinite substrate heated by irradiation.

## PROBLEM 3.59. FAM

An automobile tire rolling over a paved road is heated by surface friction, as shown in Figure Pr. 3.59. The energy conversion rate for each time (assuming a simple average) divided by the tire surface area is $\dot{S}_{m,F}/A_t$, and this is related to the vehicle mass $M$ and speed $u_o$ through

$$\frac{\dot{S}_{m,F}}{A_t} = \frac{1}{4}\frac{Mg\mu_F u_o}{A_t}.$$

A fraction of this, $a_1\dot{S}_{m,F}/A_t$, is conducted through the tire. The tire has a cover-tread layer with a layer thickness $L$, which is assumed to be much smaller than the tire

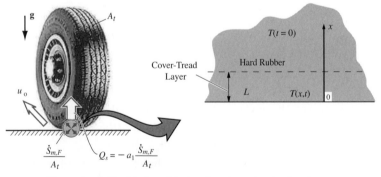

Figure Pr. 3.59. Surface friction heating of a tire laminate.

thickness, but is made of the same hard rubber material as the rest of the tire. The deep unperturbed temperature is $T(t = 0)$.

Determine the temperature at this location $L$, after an elapsed time $t_o$, using (a) $M = 1,500$ kg, and (b) $M = 3,000$ kg.

$T(t = 0) = 20°C$, g = 9.807 m/s$^2$, $u_o$ = 60 km/hr, $L$ = 4 mm, $A_t$ = 0.4 m$^2$, $t_o$ = 10 min, $\mu_F$ = 0.015, $a_1$ = 0.4.

The properties for hard rubber are listed in Table C.17.

## PROBLEM 3.60. FAM

In ultrasonic welding (also called ultrasonic joining), two thick slabs of polymeric solids to be joined are placed in an ultrasonic field that causes a relative motion at their joining surfaces. This relative motion combined with a joint pressure causes a surface friction heating at a rate of $\dot{S}_{m,F}/A$. This heat flows and penetrates equally into these two similar polymeric solids. The two pieces are assumed to be very thick and initially at a uniform temperature $T(t = 0)$.

How long would it take for the contacting surfaces of the two polymers in contact to reach their melting temperature $T_{sl}$?

$\dot{S}_{m,F}/A = 10^4$ W/m$^2$, $T_{sl} = 300°C$, $T(t = 0) = 25°C$, and use the properties of Teflon (Table C.17).

### 3.12.7 Lumped-Capacitance Transient Conduction

## PROBLEM 3.61. FAM.S

A thin film is heated with irradiation from a laser source as shown in Figure Pr. 3.61. Assume that all the radiation is absorbed (i.e., $\alpha_{r,1} = 1$). The heat losses from the film are by substrate conduction only. The film can be treated as having a uniform temperature $T_1(t)$, i.e., $N_{k,1} < 0.1$, and the conduction resistance $R_{k,1-2}$ through the substrate can be treated as constant.

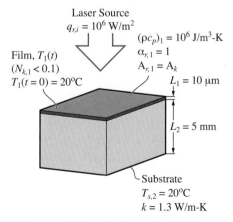

Figure Pr. 3.61. Laser-radiation heating of a thin film over a substrate.

(a) Draw the thermal circuit diagram.

(b) Determine the time needed to raise the temperature of the film $T_1(t)$ to 500°C.

## PROBLEM 3.62. FUN

Carbon steel AISI 4130 spheres with radius $R_1 = 4$ mm are to be annealed. The initial temperature is $T_1(t = 0) = 25$°C and the annealing temperature is $T_a = 950$°C. The heating is done by an acetylene torch. The spheres are placed on a conveyor belt and passed under the flame of the acetylene torch. The surface-convection heat flux delivered by the torch (and moving into the spheres) is given as a function of the position within the flame by (see Figure Pr. 3.62)

$$q_s(t) = q_{ku} = q_o \sin\left(\pi \frac{x}{L}\right),$$

where $L = 2$ cm is the lateral length of the flame and $q_o = -3 \times 10^6$ W/m² is the heat flux at the center of the flame.

Using a lumped-capacitance analysis ($N_{ku,1} < 0.1$) and starting from (3.160), find the speed of the conveyor belt $u_b$ needed for heating the spheres from $T_1(t = 0)$ to $T_a$.

Assume that the surface of the spheres is uniformly heated and neglect the heat losses. Use the properties at 300 K, as given in Table C.16.

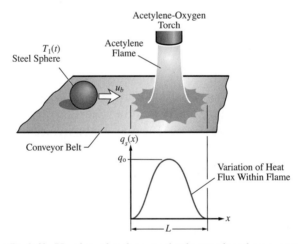

Figure Pr. 3.62. Heating of carbon steel spheres placed on a conveyor.

## PROBLEM 3.63. FAM.S

An electrical resistance regulator is encapsulated in a rectangular casing (and assumed to have a uniform temperature $T_1$, i.e., $N_{k,1} < 0.1$) and attached to an aluminum slab with thickness $L = 3$ mm. The slab is in turn cooled by maintaining its opposite surface at the constant ambient temperature $T_2 = 30$°C. This slab is called a heat sink and is shown in Figure Pr. 3.63. Most of the time, the regulator provides no resistance to the current flow, and therefore, its temperature is equal to $T_2$. Intermittently, the regulator control is activated to provide for an ohmic resistance

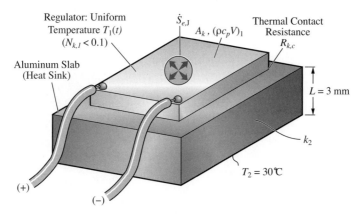

Figure Pr. 3.63. An electrical resistance regulator attached to a substrate with a thermal contact resistance.

and then energy conversion from electromagnetic to thermal energy occurs. A joint pressure is exerted to reduce the contact thermal resistance at the interface between the regulator and the heat sink. However, as the regulator temperature reaches a threshold value $T_{1,o} = 45°C$, the thermal stresses warp the regulator surface and the contact resistance changes from $A_k R_{k,c} = 10^{-3}$ K/(W/m$^2$) to a larger value of $A_k R_{k,c} = 10^{-2}$ K/(W/m$^2$). The regulator has $(\rho c_p V)_1/A_k = 1.3 \times 10^5$ J/K-m$^2$ and the amount of heat generated by Joule heating is $\dot{S}_{e,J}/A_k = 2 \times 10^4$ W/m$^2$. Neglect the heat losses from the regulator to the ambient. Assume that the conduction resistance in the aluminum slab is steady state (i.e., constant resistance) and the energy storage in the slab is also negligible.

(a) Draw the thermal circuit diagram.
(b) Use the lumped-capacitance analysis to determine the required time to reach the threshold temperature $T_c = 45°C$.
(c) Starting with $T_{1,o}$ as the initial temperature and using the new contact resistance $A_k R_{k,c}$, determine the time required to reach $T_1 = T_2 + 2\,T_c$.
(d) Determine the steady-state temperature.
(e) Make a qualitative plot of the regulator temperature versus time, showing (i) the transition in the contact resistance, and (ii) the steady-state temperature.

Use the thermal conductivity of aluminum at $T = 300$ K.

**PROBLEM 3.64. FAM**
In laser back-scribing, a substrate is heated and melted by radiation absorption. Upon solidification, a volume change marks the region and this is used for recording. An example is given in Figure Pr. 3.64, where irradiation is provided through a thick glass layer and arrives from the backside to a thin layer of alumina. The alumina layer absorbs the radiation with an extinction coefficient $\sigma_{ex,1}$ that is much larger than that of glass. Assume that the alumina layer is at a uniform, but time-varying temperature (because of the high thermal conductivity of alumina compared to

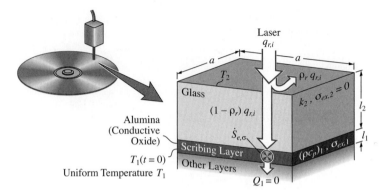

Figure Pr. 3.64. Laser back-scribing on a compact disk storage device.

the glass, i.e., $N_k < 0.1$). Also assume that the conduction resistance in the glass is constant.

(a) Draw the thermal circuit diagram.
(b) Determine the time it takes to reach the melting temperature of the alumina $T_{sl,1}$.

$a = 100\ \mu m$, $l_1 = 0.6\ \mu m$, $l_2 = 3\ \mu m$, $q_{r,i} = 3 \times 10^9$ W/m$^2$, $\sigma_{ex,1} = 10^7$ 1/m, $\rho_r = 0.1$, $T_1(t = 0) = 20°C$, $T_2 = 20°C$, $T_{sl,1} = 1{,}900°C$.

**PROBLEM 3.65. FUN**
In applications such as surface friction heat generation during automobile braking, the energy conversion rate $\dot{S}_{m,F}$ decreases with time. For the automobile brake, this is modeled as

$$\dot{S}_{m,F}(t) = (\dot{S}_{m,F})_o \left(1 - \frac{t}{t_o}\right) \qquad t \le t_o.$$

For a semi-infinite solid initially at $T(t = 0)$, when its surface at $x = 0$ experiences such time-dependent surface energy conversion, the surface temperature is given by the solution to (3.134). The solution, for these initial and bounding-surface conditions, is

$$T(x = 0, t) = T(t = 0) + \left(\frac{5}{4}\right)^{1/2} \frac{(\dot{S}_{m,F})_o}{A_k k} (\alpha t)^{1/2} \left(1 - \frac{2t}{3t_o}\right),$$

where $\dot{S}_{m,F}/A_k$ is the peak surface heat flux $q_s$.

Consider a disc-brake rotor made of carbon steel AISI 1010.

(a) Plot the surface temperature $T(x = 0, t)$ for the conditions given below and $0 \le t \le t_o$.
(b) By differentiating the above expression for $T(x = 0, t)$ with respect to $t$, determine the time at which $T(x = 0, t)$ is a maximum.

$(\dot{S}_{m,f})_o/A_k = 10^5$ W/m$^2$, $t_o = 4$ s, $T(t = 0) = 20°C$.

## PROBLEM 3.66. FUN

Derive the solution (3.172) for $T = T_1(t)$ starting from the energy equation (3.171), which applies to a lumped-capacitance system with a resistive-type surface heat transfer. The initial condition is $T_1 = T_1(t = 0)$ at $t = 0$.

## PROBLEM 3.67. FAM

Water is heated (and assumed to have a uniform temperature, due to thermobuoyant motion mixing) from $T_1(t = 0)$ to $T_1(t = t_f) = T_f$ by Joule heating in a cylindrical, portable water heater with inside radius $R_1$ and height $l$, as shown in Figure Pr. 3.67. The ambient air temperature $T_2$ is rather low. Here we assume that the outside surface temperature (located at outer radius $R_2$) is the same as the ambient temperature (i.e., we neglect the resistance to heat transfer between the outside surface and the ambient). Two different heater wall designs, with different $R_2$ and $k_s$, are considered.

(a) Draw the thermal circuit diagram.
(b) Determine the water temperature $T_f$ after an elapsed time $t_f$ using, (i) a thin AISI 302 stainless-steel wall (Table C.16) with outer wall radius $R_2 = 7.1$ cm, and (ii) a thicker nylon wall (Table C.17) with $R_2 = 7.2$ cm.
(c) Compare the results of the two designs.

$R_1 = 7$ cm, $l = 15$ cm, $T_1(t = 0) = T_2 = 2°C$, $t_f = 2,700$ s, $\dot{S}_{e,J} = 600$ W.

Evaluate the water properties at $T = 310$ K (Table C.23). Neglect the heat transfer through the top and bottom surfaces of the water heater, and treat the wall resistance as constant.

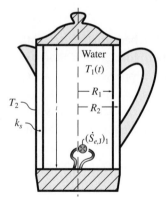

Figure Pr. 3.67. A portable water heater.

## PROBLEM 3.68. FAM

A hot sphere having an initial temperature $T_1(t = 0)$, a diameter $D_1$, and made of lead, is cooled by rolling over a cold surface $T_2 < T_1(t = 0)$ at a constant velocity $u_p$. This is shown in Figure Pr. 3.68. There is contact resistance $(R_{k,c})_{1\text{-}2}$ at the contact between the sphere and the surface. This resistance can be estimated using Figure 3.25 for $p_c = 10^5$ Pa and a pair of soft aluminum surfaces with rms roughness $\langle \delta^2 \rangle^{1/2} = 0.25$ $\mu$m. For simplicity, the surface-convection and surface-radiation heat transfers are assumed to be constant and represented by a prescribed heat transfer rate $Q_1$.

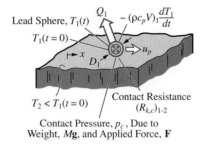

Figure Pr. 3.68. A hot lead sphere is cooled by rolling over a cold surface.

(a) Draw the thermal circuit diagram.
(b) Assume a uniform sphere temperature $T_1(t)$ and determine the elapsed time and the length required for the sphere temperature to reach 50°C above $T_2$.
(c) Use this elapsed time and evaluate the Fourier number $Fo_R$ for the sphere. From this magnitude and by using Figure 3.33(b)(ii) for estimation, is the assumption of a uniform temperature valid from the transient penetration view-point?
(d) By approximating the internal, steady-state resistance as $R_{k,1} = (D_1/2)/\pi D_1^2 k_1$, evaluate $N_{k,1}$ and comment on the validity of a uniform sphere temperature assumption, from the relative temperature variations inside and outside the object.

$D_1 = 2$ mm, $T_1(t = 0) = 300°$C, $T_2 = 30°$C, $u_p = 0.5$ m/s, $A_{k,c} = 0.1$ mm$^2$, $Q_1 = 0.1$ W.

## PROBLEM 3.69. FAM.S

A thin, flexible thermofoil (etched foil) heater with a mica (a cleavable mineral) casing is used to heat a copper block. This is shown in Figure Pr. 3.69. At the surface between the heater and the copper block, there is a contact resistance $(R_{k,c})_{1-2}$. Assume that the heater and the copper block both have small internal thermal resistances ($N_{k,1} < 0.1$), so they can be treated as having uniform temperatures $T_1(t)$ and $T_2(t)$, with the initial thermal equilibrium conditions $T_1(t = 0) = T_2(t = 0)$. The other heat transfer rates, from the heater and the copper block are prescribed (and constant) and are given by $Q_1$ and $Q_2$. If the heater temperature $T_1(t)$ exceeds a

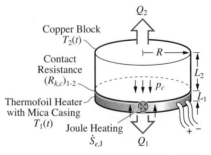

Figure Pr. 3.69. A thermofoil heater is used to heat a copper block. There is a contact resistance between the two.

threshold value of $T_c = 600°C$, the heater is permanently damaged. To avoid this, the thermal contact resistance is decreased by the application of an external pressure (i.e., large contact pressure $p_c$).

(a) Draw the thermal circuit diagram.
(b) Plot $T_1(t)$ and $T_2(t)$ with respect to time, for $0 \le t \le 100$ s, for (i) $A_k(R_{k,c})_{1-2} = 10^{-3}$ K/(W/m²), and (ii) $A_k(R_{k,c})_{1-2} = 10^{-2}$ K/(W/m²).

For mica, use the density and specific heat capacity for glass plate in Table C.17.
$R = 3$ cm, $L_1 = 0.1$ cm, $L_2 = 3$ cm, $\dot{S}_{e,J} = 300$ W, $Q_1 = 5$ W, $Q_2 = 50$ W, $T_1(t = 0) = T_2(t = 0) = 20°C$.

## PROBLEM 3.70. FAM

An initially cold $T_1(t = 0)$, pure aluminum spherical particle is rolling over a hot surface of temperature $T_s$ at a constant speed $u_b$ and is heated through a contact conduction resistance $R_{k,c}$. This is shown in Figure Pr. 3.70.

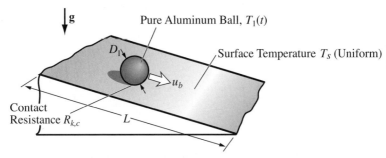

Figure Pr. 3.70. A pure aluminum ball rolls over a hot surface and is heated by contact conduction.

(a) Draw the thermal circuit diagram.
(b) Is the assumption of a uniform temperature valid? Use $R_{k,1} = 1/(4\pi D_1 k_1)$ for the internal conduction resistance.
(c) Determine the length $L$ the ball has to travel before its temperature reaches $T_f$. Assume that the contact conduction is the only surface heat transfer (surface-convection and radiation heat transfer are assumed negligible).

$T_1(t = 0) = 20°C$, $T_s = 300°C$, $T_f = 200°C$, $D_1 = 4$ mm, $R_{k,c} = 1,000$ K/W, $u_b = 0.1$ m/s.

Determine the pure aluminum properties at $T = 300$ K.

## PROBLEM 3.71. FUN

In printed-circuit field-effect transistors, shown in Figure Pr. 3.71, the electrons are periodically accelerated in the active layer and these electrons are scattered by collision with the lattice molecules (which is represented as collision with the lattice phonons), as well as collision with the other electrons and with impurities. These collisions result in the loss of the kinetic energy (momentum) of the electrons (represented by the Joule heating) and this energy is transferred to the lattice molecules due to local thermal nonequilibrium between the electrons having temperature $T_e$ and the lattice (i.e., phonon) having temperature $T_l = T_p$.

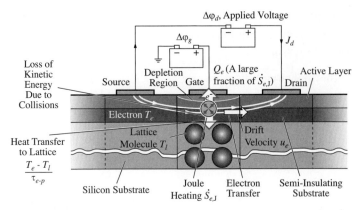

Figure Pr. 3.71. The printed-circuit field-effect transistor and heat conversion $\dot{S}_{e,J}$ by electron kinetic energy loss to lattice attoms, due to collisions. The electron heat transfer as a unit cell is also shown.

An estimate of the electron temperature $T_e$ can be made using the concept of relaxation time. As discussed in the footnote of Section 3.6.2, the energy equation for the electron (transport) in a lattice unit cell can be written as

$$\frac{T_e(t) - T_p}{\tau_{e-p}} + \frac{dT_e(t)}{dt} = a_e = 0.1\frac{m_{e,e}u_e^2}{3k_B}\left(\frac{2}{\tau_{e,m}} - \frac{1}{\tau_{e-p}}\right)$$

$$u_e = -\mu_e e_e,$$

where the terms on the left is the heat transfer from electron to the lattice, and the storage, and the right-hand side term is the energy conversion (Joule heating). Here the coefficient 0.1 in $a_e$ represents that 0.9 of the heat generated is conducted to surroundings. The electron drift velocity $u_e$ is related to the electric field $e_e$ and the electron mobility $\mu_e$ and the mass used $m_{e,e}$ is the effective electron mass and is $0.066m_e$ for GaAs [12]. The effective (as compared to free $m_e$) electron mass represents the electron in a periodic potential (lattice). The two relaxation times are the momentum relaxation time $\tau_{e,m}$, and the energy relation time $\tau_{e-p}$ for electron-lattice (phonon) scattering. Assume that the lattice temperature is constant (due to the much larger mass of the lattice molecules, compared to the electrons).

For the conditions given below, plot the electron temperature, using the solution (3.172), with respect to time, up to an elapsed time of $t = 100$ ps.

$m_{e,e} = 0.066m_e$,   $m_e = 9.109 \times 10^{-31}$ kg,   $k_B = 1.3807 \times 10^{-23}$ J/K,   $\mu_e = 0.85$ m²/V-s, $e_e = 5 \times 10^5$ V/m, $\tau_{e,m} = 0.3$ ps, $\tau_{e-p} = 8$ ps, $T_p = 300$ K, $T_e(t=0) = T_p$.

### 3.12.8 Multinode Systems and Finite-Small-Volume Analysis*

This section and the Problems that follow are found on the Web site www.cambridge.org/kaviany.

**PROBLEM 3.72. FAM***

**PROBLEM 3.73. FUN***

**PROBLEM 3.74. FUN***

**PROBLEM 3.75. FUN***

**PROBLEM 3.76. FUN.S***

### 3.12.9  Solid-Liquid Phase Change*

**PROBLEM 3.77. FAM***

**PROBLEM 3.78. FUN***

**PROBLEM 3.79. FAM**

**PROBLEM 3.80. DES***

### 3.12.10  Thermal Expansion and Thermal Stress*

**PROBLEM 3.81. FAM***

**PROBLEM 3.82. FUN***

**PROBLEM 3.83. FUN***

### 3.12.11  General*

**PROBLEM 3.84. DES.S***

**PROBLEM 3.85. FUN.S***

**PROBLEM 3.86. FUN.S***

**PROBLEM 3.87. FUM***

**PROBLEM 3.88. FUN.S***

**PROBLEM 3.89. FAM***

**PROBLEM 3.90. FAM***

**PROBLEM 3.91. FAM***

**PROBLEM 3.92. FAM***

# 4

# Radiation

The internal energy of oscillators of frequency $f$ is made of entirely determined number of finite equal parts with each oscillator (called photon)
having energy $h_P f$ with $h_P$ being a natural constant (now called the Planck
constant).

– M. Planck

Radiative heat flux is the result of emission of thermal radiation by objects due to
their temperature. This emitted radiation (photons) covers the entire electromagnetic spectrum. The radiation heat flux vector $\mathbf{q}_r$ is the average of the photon flux
(moving at speed of light $c$) taken over all the frequencies $f$ (or wavelengths $\lambda$)
and in the desired direction designated by the surface normal $\mathbf{s}_n$. This photon flux
is altered as it passes through a medium (by emission, absorption, and scattering in
this medium).

   The general differential-volume thermal radiation analysis is beyond our scope
here. The extinction coefficient $\sigma_{ex}$, (which is more accurately decomposed into
absorption and scattering contributions) which is the inverse of the photon mean-free
path $\lambda_{ph} \equiv 1/\sigma_{ex}$, is a measure of the radiation penetration. The general treatment for
arbitrary $\sigma_{ex}(1/\text{m})$ must address both distant radiation (i.e., integral) and local (i.e.,
differential) radiation heat transfer. This is done using the equation for conservation
of radiation energy (called the equation of radiation transport). Here we assume
that the medium occupying the space between the radiating surfaces is transparent,
i.e., $\sigma_{ex}^* \ll 1$. The optical thickness was defined by (2.46), i.e., $\sigma_{ex}^* = \sigma_{ex}L = L/\lambda_{ph}$,
with $L$ being the average radiation path for the system (e.g., the average distance
between surfaces). Chart 4.1 gives a classification of the medium occupying the
volume between radiating surface. These are thin, intermediate, and large optical
thicknesses. Typical values for $\sigma_{ex}$, for various phases, was given in Figure 2.13. The
case of optically thick media, defined by (2.46), i.e., $\sigma_{ex}^* > 10$, will be discussed in
Chapter 5, Section 5.4.6.

   In this chapter we discuss mostly the surface-radiation heat transfer. Then for
surface-radiation heat transfer, we will only need to use the energy equation. We
will consider surfaces over which we can justifiably assume a uniform temperature,

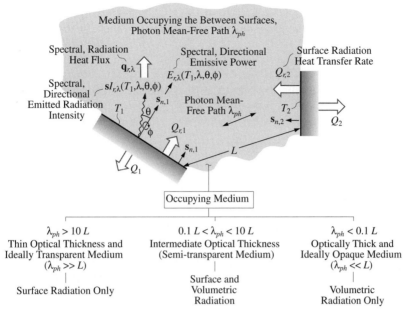

Chart 4.1. A classification of the medium occupying the volume between radiating surfaces in terms of the photon mean-free path $\lambda_{ph} = 1/\sigma_{ex}$.

where the heat flows only perpendicular to these surfaces. In practice, any nonuniform-temperature surface can be divided into small surfaces (or segments) until the uniform surface temperature assumption becomes justified. Here we will examine the surface-radiation heat flux vector $\mathbf{q}_r$ and the parameters influencing its magnitude and direction. The integral-surface and integral-volume energy equations with surface-radiation heat transfer are (2.72) and (2.9), i.e.,

$$Q|_A = Q + Q_r = Q + \int_A (\mathbf{q}_r \cdot \mathbf{s}_n)dA = \dot{S} \quad \text{for integral surface}$$

$$Q|_A = Q + Q_r = -\frac{d}{dt}\rho c_p VT + \dot{S} \quad \text{for integral volume}$$

energy equation with surface-radiation heat transfer. (4.1)

Whenever the surfaces exchanging radiation are gray, diffuse, and opaque, then we combine the surface emission, absorption, and reflection, and then use the concepts of radiosity and irradiation to arrive at the surface-radiation heat transfer rate $Q_r$. Otherwise the surface-radiation and the volumetric radiation are treated as an energy conversion term $\dot{S}_e$. This alternative presentation is due to the photon emission, absorption, and scattering expressed by $\dot{S}_e$ and in turn expressed by the radiation heat flux and its spatial variations. This alternative was discussed in Section 2.3.2(E) and, depending on the problem analyzed, the one that is more physically relevant to the problem is used. This is further discussed in Sections 4.5.4 and 5.6.

For gray, diffuse, and opaque surfaces making up an enclosure, we will define the surface-radiation resistances $R_{r,\epsilon}(1/m^2)$ and $R_{r,F}(1/m^2)$ for heat exchange among these surfaces and use them in thermal circuit analyses. Next we consider pre-scribed irradiation and nongray surfaces. Then we will include the substrate heat transfer.

## 4.1 Microscale Radiation Heat Carrier: Photon and Surface Thermal Radiation Emission

### 4.1.1 Thermal Radiation

Thermal radiation heat transfer is the electromagnetic radiation that is emitted by solids, liquids, and gases as a result of their temperature and is generally detected as heat or light. These electromagnetic emissions are due to the molecular electronic, rotational, and vibrational energy transitions of the matter. Figure 4.1 gives the classification of the electromagnetic waves (to the right) and the source of their emission (to the left). The frequency $f$ (Hz = 1/s) and wavelength $\lambda$ (m, but generally the unit $\mu$m is used) are related through the speed of light $c$(m/s). Here we assume a nonparticipating medium occupies the space between the bounding surfaces (e.g., vacuum, resulting in no refraction or attenuation of radiation) and use the speed of light in vacuum $c_o$. Then the speed of light is related to wavelength and frequency through $c_o = \lambda f$, where $c_o = 2.9979 \times 10^8$ m/s, and $\omega$(rad/s) $= 2\pi f$ is the angular (or circular) frequency. Thermal radiation has a significant energy in the wavelength range of $2 \times 10^{-7}$ to $10^{-3}$ m. This wavelength range is covered by the ultraviolet range (but only a small portion of this wavelength range), visible range, and infrared range of the electromagnetic waves. However, for low to moderate temperature applications, thermal radiation is dominated by the infrared wavelengths and is dominated by molecular rotational and vibrational transitions.

The source of electromagnetic radiation emission can be other than thermal emission, e.g., fluorescent radiation is emitted by absorbing X-rays or ultraviolet rays and X-rays can be emitted by electron bombardment of a surface. These are examples of electromagnetic-radiation wavelengths with a significant energy, but outside the thermal radiation range and include X-ray, microwave, radiowave, and the large-wavelength (i.e., low-frequency) power range. Therefore, we included these in the energy conversion term $\dot{s}_e$.

### 4.1.2 Ideal Photon Emission: Emissive Power $E_{b,\lambda}$, $E_b$

In Chapter 3, we dealt with some microscale heat carriers [i.e., free molecules (gases), electrons, and phonons] and described the macroscopic thermodynamic and trans-port properties, $c_p$ and $k$, for such heat carriers. We now discuss microscale radiation heat carrier, the photon. A photon is a quantum of a single mode (i.e., single wave-length $\lambda$, direction $\theta$, and $\phi$, and polarization) of an electromagnetic field. The photon energy of frequency $f$ is $E_{r,\lambda} = E_{ph,\lambda} = n_{ph}h_\text{P}f = n_{ph}h_\text{P}\lambda/c$, where $n_{ph}$ is an integer, $h_\text{P}$(J-s) is the Planck constant ($h_\text{P} = 6.6261 \times 10^{-34}$ J-s). For example, for $\lambda = 0.5\,\mu$m, green light, and for $n_{ph} = 1$, i.e., a single photon, $E_{r,\lambda} = 3.975 \times 10^{-19}$ J = 2.481 eV

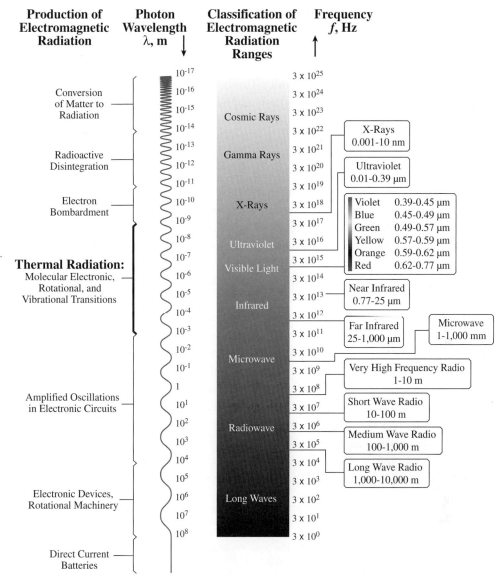

Figure 4.1. Classification of electromagnetic waves (i.e., radiation or photon), based on the production, wavelength range, common names, and frequency.

and for $\lambda = 1.0$ μm this is 1.240 eV. This follows that energies of the elementary quantum oscillator making up the emitted radiation take on discrete values. The distribution $n_{ph} = n_{ph}(\lambda, t)$ for an ideal oscillator-emitter (called the blackbody oscillator-emitter), will be given below. In Figure 4.1, the gamma and cosmic rays have the largest electromagnetic energy and the long waves have the lowest energy. The earth's electromagnetic field prevents these powerful radiations from reaching and harming life on the earth.

Radiation energy (and its conservation equation, called the equation of radiation transfer) is expressed in terms of the radiation intensity and its direction **s**. The radiation intensity is flux of photon and is the product of the concentration $n_{ph}(1/m^3)$, energy content $h_{P}f$ (J), velocity of photons $c$(m/s), along the desired direction $\mathbf{s}$. This would be integrated over the frequencies (or wavelengths). In dealing with the emitted surface-radiation (i.e., emitted photons), we use the emissive power, which is related to the emitted radiation intensity. The spectral, directional emission intensity $I_{r,\lambda}$ is radiation energy emitted per unit time, per unit small wavelength interval around the wavelength $\lambda$, per unit surface area normal to $\mathbf{s}$, and per unit solid angle. The spectral, directional emissive power $E_{r,\lambda}(T, \lambda, \theta, \phi)$, where $\theta$ and $\phi$ show the directions, is related to the emission intensity $I_{r,\lambda}(T, \lambda, \theta, \phi)$ through a simple cosine relation.[†]

The spectral and total radiation heat flux vectors $\mathbf{q}_{r,\lambda}$ and $\mathbf{q}_r$ are related to the spectral radiation intensity.[‡] The intensity and emission power and the angles are shown in Chart 4.1. To obtain the total radiation heat flux vector, integration is needed over the entire range of wavelengths $0 \leq \lambda < \infty$ and the entire range of solid angle (a hemisphere covering the surface).

A surface maintained at temperature $T$ continuously emits electromagnetic radiation. The amount of energy emitted $E_{\lambda,i}$ varies with the wavelength $\lambda$ (this is called the spectral emission, because of the wavelength dependence). A surface that emits the maximum amount possible (ideal emitter) at all wavelengths is called the blackbody emitter. A blackbody emits in all wavelengths (i.e., has a continuous spectrum). A blackbody also emits equally in all directions (this is called diffuse emission, because of the direction independence), and therefore, its emission is independent of $\phi$ and $\theta$ (these are the angles used in the spherical coordinate system defined in Figure 3.14).

The blackbody, wavelength- (i.e., spectral), and temperature-dependent emission per unit area $E_{b,\lambda} = E_{b,\lambda}(T, \lambda)$(W/m²-$\mu$m) is called the spectral, hemispherical blackbody emissive power. The spectral, hemispherical blackbody emissive power

---

[†] For diffuse radiation, with $\theta$ being the angle between the surface and its normal, we have $E_{r,\lambda} = I_{r,\lambda} \cos\theta$. This is called the Lambert cosine law. Also as will be described in Section 4.1.3, the hemispherical, spectral emissive power is related to intensity through $E_{r,\lambda}(T, \lambda) = \int_0^{2\pi} \int_0^{\pi/2} I_{r,\lambda}(T, \lambda, \theta, \phi) \cos\theta \sin\theta d\theta d\phi$. The equation describing the conservation of radiation (photon) intensity is called the equation of radiative transfer and is discussed in detail in [2, 7, 8, 10]. This is similar to other conservation equation discussed in Sections 1.6 and 1.7.

[‡] The spectral radiation heat flux vector $\mathbf{q}_{r,\lambda}$ is related to the spectral radiation intensity $I_{r,\lambda}$ (photon flux) through

$$\mathbf{q}_{r,\lambda} = \int_0^{2\pi} \int_{-\pi/2}^{\pi/2} \mathbf{s} I_{r,\lambda} \cos\theta \sin\theta d\theta d\phi,$$

where this integration is over complete sphere (i.e., a solid angle of $4\pi$) and $\mathbf{s}$ is unit vector along which $\mathbf{q}_{r,i}$ is evaluated. This intensity is not only due to emission and is influenced by scattering and absorption. The conservation of intensity is described by the equation of radiative transfer [2, 7, 8, 10]. Then the total radiation heat flux vector $\mathbf{q}_r$ is determined from

$$\mathbf{q}_r = \int_0^\infty \mathbf{q}_{r,\lambda} d\lambda.$$

Then $\mathbf{q}_r$ is the net radiation passing through a surface having a normal unit vector $\mathbf{s}$. Then $\mathbf{q}_r$ will not be zero, whenever due to scattering, absorption, or emission the integral of $I_{r,\lambda}$ over the complete solid angle and wavelengths is not equal.

$E_{b,\lambda}$ is the summation over all possible energy modes $E_{\lambda,i}$ (quantum states of photon gas) and is given by the Planck for ideally emitted photons (or Planck law of electromagnetic energy spectrum) as [11]

$$E_{b,\lambda}(T,\lambda)(\text{W/m}^2\text{-}\mu\text{m}) = \pi I_{r,\lambda} = \frac{c_o}{4} E_{ph} f_{ph}^o D_{ph} = \frac{2\pi h_P c_o^2}{\lambda^5 (e^{h_P f/k_B T} - 1)}$$

$$= \frac{a_1}{\lambda^5 (e^{h_P c_o/\lambda k_B T} - 1)}, \quad a_1 = 2\pi h_P c_o^2$$

$$= \frac{a_1}{\lambda^5 (e^{a_2/\lambda T} - 1)} \quad \begin{array}{l} \text{Planck distribution function} \\ \text{(spectral, hemispherical blackbody} \\ \text{emissive power),} \end{array} \quad (4.2)$$

where $f_{ph}^o$ is the Bose-Einstein distribution function (Table 1.4), $D_{ph}$ is the photon density of states [same as in (3.26)], $T$ (absolute temperature) is in K, $\lambda$ is in $\mu$m, and the constants are $a_1 = 2\pi h_P c_o^2 = 3.742 \times 10^8$ W-$\mu$m$^4$/m$^2$ and $a_2 = h_P c_o/k_B = 1.439 \times 10^4$ $\mu$m-K ($k_B$ is the Boltzmann constant). This is for emission into a nondiffracting (i.e., index of refraction $n$ is unity) medium (i.e., vacuum or a gas that does not influence the pathway of the electromagnetic waves) considered here. The 4 in (4.2) is from the isotropic radiation over the total solid angle of $4\lambda$.

The unit for $E_{b,\lambda}$ is W/m$^2$-$\mu$m, and it will be integrated over the entire wavelength range $0 < \lambda(\mu\text{m}) < \infty$ to result in the heat flow rate per unit area. This is discussed below.

Figure 4.2(a) shows the variation of the spectral, hemispherical blackbody emissive power with respect to the wavelength. The results are for four different temperatures corresponding to the sun surface temperature, the temperature at which a heated object will first appear as dull (i.e., long wavelength) red (called the Draper point), i.e., $T = 798$ K, the average human body temperature, and the boiling temperature of nitrogen at one atm pressure.

The temperature color relation for human eyes are 500–600°C (dark red), 600–800°C (dull red), 800–1,000°C (bright cherry red), 1,000–1,200°C (orange), 1,200–1,400°C (bright yellow), and 1,400–1,600°C (white). The spectral, blackbody emissive power $E_{b,\lambda}$ corresponding to $\lambda = 0.77$ $\mu$m (the longest wavelength in the visible region and corresponding to the color red) and $T = 798$ K is also marked for reference.

The peak of $E_{b,\lambda}$ is at a wavelength given by $\lambda_{max} T = 2,898$ $\mu$m-K. This is called the Wien displacement law [wavelength of $(E_{b,\lambda})_{max}$ is inversely proportional to temperature]. For $T = 798$ K, this gives $\lambda_{max} = 3.632$ $\mu$m, i.e., in the near infrared range. From Figure 4.2(a), we can observe that for the spectral, blackbody emissive power to peak in the ultraviolet range, temperatures above 6,000 K are needed.

Note that for a given $\lambda$, $E_{b,\lambda}$ increases as $T$ increases. The human eyes sense the red light ($0.62 \leq \lambda \leq 0.77$ $\mu$m) when $E_{b,\lambda}$ reaches a threshold value. This threshold value is reached when the temperature is $T \simeq 798$ K (505°C) and the dull red emission appears (in a dark background).

We can rewrite (4.2) such that the right-hand side would have the product $\lambda T (\mu\text{m-K})$ as the only variable, and the result is

$$\frac{E_{b,\lambda}}{T^5}(\text{W/m}^2\text{-}\mu\text{m-K}^5) = \frac{a_1}{(\lambda T)^5(e^{a_2/\lambda T} - 1)} \qquad \text{Planck law with } \lambda T \text{ as variable.} \qquad (4.3)$$

The total, blackbody emissive power $E_b(\text{W/m}^2)$ is found by integrating the spectral, blackbody emissive power over the wavelength range ($0 \le \lambda < \infty$), and this gives

$$E_b(T) \equiv \int_0^\infty E_{b,\lambda}(T, \lambda)d\lambda \qquad (4.4)$$

$$E_b(T) = a_1 T^4 \int_0^\infty \frac{d(\lambda T)}{(\lambda T)^5(e^{a_2/\lambda T} - 1)} \qquad (4.5)$$

$$E_b(T) = \frac{\pi^4 a_1}{15 a_2^4} T^4 = \frac{2\pi^5 k_{\text{B}}^4}{15 h_{\text{P}}^3 c_o^2} \equiv \sigma_{\text{SB}} T^4 (\text{W/m}^2) \qquad \begin{array}{l}\text{Stefan-Boltzmann relation for}\\ \text{total, blackbody emissive power,}\end{array} \qquad (4.6)$$

where $\sigma_{\text{SB}} = 5.670 \times 10^{-8}$ W/m$^2$-K$^4$ is the Stefan-Boltzmann constant. This integration (i.e., the area under the curve) is also shown in Figure 4.2(a).

As stated by (4.6), the total emissive power of a blackbody surface $E_b$ is proportional to its temperature to the fourth power.

Based on the relation between the emission power and the blackbody radiation intensity discussed in the first footnote of Section 4.1.2, we have $I_{r,b} = \sigma_{\text{SB}} T^4/\pi$.

### 4.1.3 Band Fraction $F_{0\text{-}\lambda T}$

Figure 4.2(b) shows the variation of $E_{b,\lambda}/T^5$ with respect to $\lambda T (\mu\text{m-K})$. This is the expression given by (4.3). Note the peak in this quantity, appearing at $\lambda T = 2,898$ $\mu$m-K.

The band fraction of the total blackbody emissive power $F_{0\text{-}\lambda T} = F_{0\text{-}\lambda T}(\lambda T)$ is defined as the ratio of the blackbody emissive power within a wavelength band 0 to $\lambda$ (and at temperature $T$) to the total, blackbody emissive power (for the same temperature $T$), i.e.,

$$F_{0\text{-}\lambda T}(\lambda T) \equiv \frac{\displaystyle\int_0^\lambda E_{b,\lambda}(T, \lambda)d\lambda}{\displaystyle\int_0^\infty E_{b,\lambda}(T, \lambda)d\lambda} = \frac{\displaystyle\int_0^\lambda E_{b,\lambda}(T, \lambda)d\lambda}{\sigma_{\text{SB}} T^4} \qquad \begin{array}{l}\text{band fraction of}\\ \text{total blackbody}\\ \text{emissive power.}\end{array} \qquad (4.7)$$

$F_{0\text{-}\lambda T}(\lambda T)$ is tabulated in Table 4.1 and is also shown graphically in Figure 4.2(b).

We now examine Table 4.1. Reading the $F_{0\text{-}\lambda T}$ column, for $F_{0\text{-}\lambda T} = 0.01$, i.e., 1% of the total, blackbody emissive power, we have (by interpolation) $\lambda T = 1,444$ $\mu$m-K. This indicates that for a given $T$, the wavelength between $\lambda = 0$ and

(a) Spectral, Hemispherical Blackbody Emissive Power $E_{b,\lambda} = E_{b,\lambda}(T,\lambda)$ versus $\lambda$

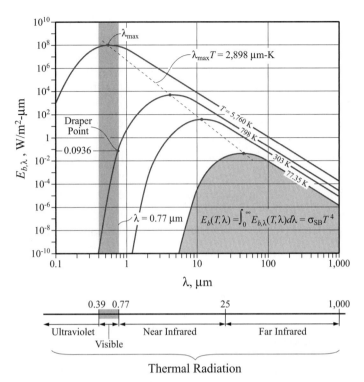

(b) Spectral, Hemispherical Blackbody Emissive Power / $T^5$ versus $\lambda T$

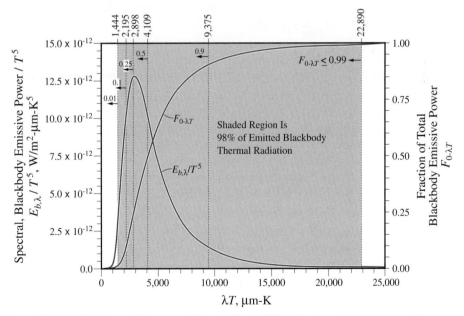

Figure 4.2.  (a) Variation of the spectral, hemispherical blackbody emissive power with respect to wavelength. (b) Variations of the spectral, hemispherical blackbody emissive power divided by $T^5$, and the band fraction of the emissive power $F_{0\text{-}\lambda T}$, with respective to the product $\lambda T$.

$\lambda = 1,444$ ($\mu$m-K)/$T$(K) contains only 1% of the total, blackbody emissive power. Also note that $E_{b,\lambda}/T^5$ peaks at $\lambda T = 2,898$ $\mu$m-K and this corresponds to $F_{0-\lambda T} = 0.25$ [i.e., 25% of the blackbody emissive power is contained between $\lambda = 0$ and $\lambda = 2,898(\mu$m-K)/$T$(K)]. We also note that $F_{0-\lambda T} = 0.99$ for $\lambda T = 22,890$ $\mu$m-K.

As shown in Figure 4.2(b), nearly all (i.e., 98%) of the radiation energy emitted from a blackbody surface is in the range of the product of the wavelength and temperature, $\lambda T$, between 1,444 and 22,890 $\mu$m-K. Therefore, for the surface of the sun, which is at nearly 5,780 K, the wavelength range is 0.25 to 3.96 $\mu$m (which includes the visible). For the human body with a bare surface temperature of 303 K, this range is 4.78 to 75.54 $\mu$m. In cryogenic applications, for a surface cooled to the boiling temperature of nitrogen 77.35 K ($-195.8°$C), this range is 18.7 to 296 $\mu$m.

These examples show that for a given $T$, the wavelengths with significant thermal energy are within less than four orders of magnitude (as compared to the total possible range of $0 \le \lambda \le \infty$).

Also note that from (4.7) we can write for the band fraction $F_{0-\lambda T}(\lambda T)$

$$F_{\lambda_1 T - \lambda_2 T} = \frac{\int_{\lambda_1}^{\lambda_2} E_{b,\lambda}(T, \lambda)d\lambda}{\sigma_{\text{SB}} T^4} = F_{0-\lambda_2 T} - F_{0-\lambda_1 T}. \qquad (4.8)$$

## EXAMPLE 4.1. FAM

Determine (a) the band fraction of emitted radiation energy, and (b) the blackbody radiation energy emitted, which are in the visible range ($0.39 \le \lambda \le 0.77\mu$m), when a thin tungsten wire (i.e., a filament) is heated to (i) $T = 500$ K, (ii) $T = 1,500$ K, (iii) $T = 2,500$ K, and (iv) $T = 3,500$ K. (The melting temperature of tungsten is $T_{sl} = 3,698$ K.)

**SOLUTION**
(a) The fraction of emitted radiation energy in a band $\lambda_1 T$ to $\lambda_2 T$ is found from the fraction in band $0$-$\lambda_1 T$ and $0$-$\lambda_2 T$ from (4.8), i.e.,

$$F_{\lambda_1 T - \lambda_2 T} = F_{0-\lambda_2 T} - F_{0-\lambda_1 T}.$$

(b) The energy emitted is then found from (4.8), i.e.,

$$\int_{\lambda_1 T}^{\lambda_2 T} E_{b,\lambda}(T, \lambda)d\lambda = (F_{\lambda_1 T - \lambda_2 T})\sigma_{\text{SB}} T^4.$$

(i) For $T = 500$ K, we have

$$\lambda_1 T = 0.39(\mu\text{m}) \times 500(\text{K}) = 195 \ \mu\text{m-K}$$

$$\lambda_2 T = 0.77(\mu\text{m}) \times 500(\text{K}) = 385 \ \mu\text{m-K}$$

$$F_{0-\lambda_1 T} \simeq 0, \quad F_{0-\lambda_2 T} \simeq 0 \quad \text{Table 4.1}$$

$$F_{\lambda_1 T - \lambda_2 T} = F_{0-\lambda_2 T} - F_{0-\lambda_1 T} = 0 - 0 = 0,$$

$$\int_{\lambda_1 T}^{\lambda_2 T} E_{b,\lambda}d\lambda = (F_{\lambda_1 T - \lambda_2 T})\sigma_{\text{SB}} T^4 = 0 \times 5.67 \times 10^{-8}(\text{W/m}^2\text{-K}^4) \times 500^4(\text{K}^4) = 0 \ \text{W/m}^2.$$

Table 4.1. *Variation of $E_{b,\lambda}/T^5\,(W/m^2\text{-}\mu m\text{-}K^5)$ and the band fraction of blackbody emissive power $F_{0\text{-}\lambda T}$,[a] with respect to product $\lambda T\,(\mu m\text{-}K)$*

| $\lambda T$, $\mu$m-K | $E_{b,\lambda}/T^5$, W/m²-$\mu$m-K⁵ | $F_{0\text{-}\lambda T}$ | $\lambda T$, $\mu$m-K | $E_{b,\lambda}/T^5$, W/m²-$\mu$m-K⁵ | $F_{0\text{-}\lambda T}$ |
|---|---|---|---|---|---|
| 1000 | $0.02110 \times 10^{-11}$ | 0.00032 | 8200 | $0.21101 \times 10^{-11}$ | 0.86396 |
| 1400 | $0.23932 \times 10^{-11}$ | 0.00779 | 8600 | $0.18370 \times 10^{-11}$ | 0.87786 |
| 1800 | $0.66872 \times 10^{-11}$ | 0.03934 | 9000 | $0.16051 \times 10^{-11}$ | 0.88999 |
| 2000 | $0.87858 \times 10^{-11}$ | 0.06672 | 9400 | $0.14075 \times 10^{-11}$ | 0.90060 |
| 2200 | $1.04990 \times 10^{-11}$ | 0.10088 | 9800 | $0.12384 \times 10^{-11}$ | 0.90992 |
| 2400 | $1.17314 \times 10^{-11}$ | 0.14025 | 10,400 | $0.10287 \times 10^{-11}$ | 0.92188 |
| 2600 | $1.24868 \times 10^{-11}$ | 0.18311 | 11,200 | $0.08121 \times 10^{-11}$ | 0.93479 |
| 2800 | $1.28242 \times 10^{-11}$ | 0.22788 | 12,000 | $0.06488 \times 10^{-11}$ | 0.94505 |
| 3000 | $1.28245 \times 10^{-11}$ | 0.27322 | 12,800 | $0.05240 \times 10^{-11}$ | 0.95329 |
| 3200 | $1.25702 \times 10^{-11}$ | 0.31809 | 13,600 | $0.04275 \times 10^{-11}$ | 0.95998 |
| 3400 | $1.21352 \times 10^{-11}$ | 0.36172 | 14,400 | $0.03520 \times 10^{-11}$ | 0.96546 |
| 3600 | $1.15806 \times 10^{-11}$ | 0.40359 | 15,200 | $0.02923 \times 10^{-11}$ | 0.96999 |
| 3800 | $1.09544 \times 10^{-11}$ | 0.44336 | 16,000 | $0.02447 \times 10^{-11}$ | 0.97377 |
| 4000 | $1.02927 \times 10^{-11}$ | 0.48085 | 10,000 | $0.11632 \times 10^{-11}$ | 0.91415 |
| 4200 | $0.96220 \times 10^{-11}$ | 0.51599 | 10,800 | $0.09126 \times 10^{-11}$ | 0.92872 |
| 4400 | $0.89607 \times 10^{-11}$ | 0.54877 | 11,600 | $0.07249 \times 10^{-11}$ | 0.94021 |
| 4600 | $0.83212 \times 10^{-11}$ | 0.57925 | 12,400 | $0.05823 \times 10^{-11}$ | 0.94939 |
| 4800 | $0.77117 \times 10^{-11}$ | 0.60753 | 13,200 | $0.04728 \times 10^{-11}$ | 0.95680 |
| 5000 | $0.71366 \times 10^{-11}$ | 0.63372 | 14,000 | $0.03875 \times 10^{-11}$ | 0.96285 |
| 5200 | $0.65983 \times 10^{-11}$ | 0.65794 | 14,800 | $0.03205 \times 10^{-11}$ | 0.96783 |
| 5400 | $0.60974 \times 10^{-11}$ | 0.68033 | 15,600 | $0.02672 \times 10^{-11}$ | 0.97196 |
| 5600 | $0.56332 \times 10^{-11}$ | 0.70101 | 16,400 | $0.02245 \times 10^{-11}$ | 0.97542 |
| 5800 | $0.52046 \times 10^{-11}$ | 0.72012 | 17,200 | $0.01899 \times 10^{-11}$ | 0.97834 |
| 6000 | $0.48096 \times 10^{-11}$ | 0.73778 | 18,000 | $0.01617 \times 10^{-11}$ | 0.98081 |
| 6200 | $0.44464 \times 10^{-11}$ | 0.75410 | 18,800 | $0.01385 \times 10^{-11}$ | 0.98293 |
| 6400 | $0.41128 \times 10^{-11}$ | 0.76920 | 19,600 | $0.01193 \times 10^{-11}$ | 0.98474 |
| 6600 | $0.38066 \times 10^{-11}$ | 0.78316 | 22,000 | $0.00786 \times 10^{-11}$ | 0.98886 |
| 6800 | $0.35256 \times 10^{-11}$ | 0.79609 | 26,000 | $0.00426 \times 10^{-11}$ | 0.99297 |
| 7000 | $0.32679 \times 10^{-11}$ | 0.80807 | 30,000 | $0.00250 \times 10^{-11}$ | 0.99529 |
| 7200 | $0.30315 \times 10^{-11}$ | 0.81918 | 34,000 | $0.00156 \times 10^{-11}$ | 0.99669 |
| 7400 | $0.28246 \times 10^{-11}$ | 0.82949 | 38,000 | $0.00103 \times 10^{-11}$ | 0.99759 |
| 7600 | $0.26155 \times 10^{-11}$ | 0.83906 | 42,000 | $0.00070 \times 10^{-11}$ | 0.99819 |
| 7800 | $0.24326 \times 10^{-11}$ | 0.84796 | 46,000 | $0.00049 \times 10^{-11}$ | 0.99861 |
| 8000 | $0.22646 \times 10^{-11}$ | 0.85625 | 50,000 | $0.00036 \times 10^{-11}$ | 0.99890 |

[a]We note that $F_{0\text{-}\lambda T} = F_{0\text{-}\lambda T}(\lambda T)$ is accurately curve-fitted by [10]

$$F_{0\text{-}\lambda T}(\lambda T) = \frac{15}{\pi^4} \sum_{i=1}^{4} \frac{e^{-ix}}{i}\left(x^3 + \frac{3x^2}{i} + \frac{6x}{i^2} + \frac{6}{i^3}\right) \qquad \begin{array}{l}\text{curve fit for band fraction}\\ \text{of total blackbody}\\ \text{emissive power,}\end{array}$$

where $x = 14{,}338\,(\mu\text{m-K})/\lambda\text{T}$.

(ii) For $T = 1{,}500$ K, we have

$$\lambda_1 T = 0.39(\mu\text{m}) \times 1{,}500(\text{K}) = 585\ \mu\text{m-K}$$

$$\lambda_2 T = 0.77(\mu\text{m}) \times 1{,}500(\text{K}) = 1{,}155\ \mu\text{m-K}$$

$$F_{0\text{-}\lambda_1 T} \simeq 0, \quad F_{0\text{-}\lambda_2 T} = 0.00321 \quad \text{Table 4.1}$$

$$F_{\lambda_1 T\text{-}\lambda_2 T} = F_{0\text{-}\lambda_2 T} - F_{0\text{-}\lambda_1 T} = 0.00321 - 0 = 0.00321$$

$$\int_{\lambda_1 T}^{\lambda_2 T} E_{b,\lambda} d\lambda = 0.00321 \times 5.67 \times 10^{-8} (\text{W/m}^2\text{-K}^4) \times 1,500^4 (\text{K}^4) = 9.214 \times 10^2 \text{ W/m}^2.$$

(iii) For $T = 2{,}500$ K, we have

$$\lambda_1 T = 0.39(\mu m) \times 2{,}500(\text{K}) = 975 \ \mu m\text{-K}$$

$$\lambda_2 T = 0.77(\mu m) \times 2{,}500(\text{K}) = 1{,}925 \ \mu m\text{-K}$$

$$F_{0\text{-}\lambda_1 T} = 0.000312, \quad F_{0\text{-}\lambda_2 T} = 0.0565 \quad \text{Table 4.1}$$

$$F_{\lambda_1 T\text{-}\lambda_2 T} = F_{0\text{-}\lambda_2 T} - F_{0\text{-}\lambda_1 T} = 0.0562$$

$$\int_{\lambda_1 T}^{\lambda_2 T} E_{b,\lambda} d\lambda = 0.0561 \times 5.67 \times 10^{-8} (\text{W/m}^2\text{-K}^4) \times 2{,}500^4 (\text{K}^4) = 1.251 \times 10^5 \text{ W/m}^2.$$

(iv) For $T = 3{,}500$ K, we have

$$\lambda_1 T = 0.39(\mu m) \times 3{,}500(\text{K}) = 1{,}365 \ \mu m\text{-K}$$

$$\lambda_2 T = 0.77(\mu m) \times 3{,}500(\text{K}) = 2{,}695 \ \mu m\text{-K}$$

$$F_{0\text{-}\lambda_1 T} = 0.00714, \quad F_{0\text{-}\lambda_2 T} = 0.204 \quad \text{Table 4.1}$$

$$F_{\lambda_1 T\text{-}\lambda_2 T} = F_{0\text{-}\lambda_2 T} - F_{0\text{-}\lambda_1 T} = 0.197$$

$$\int_{\lambda_1 T}^{\lambda_2 T} E_{b,\lambda} d\lambda = 0.0197 \times 5.67 \times 10^{-8} (\text{W/m}^2\text{-K}^4) \times 3{,}500^4 (\text{K}^4) = 1.676 \times 10^8 \text{ W/m}^2.$$

**COMMENT**

Note that as the temperature increases, a larger fraction of the emitted energy is in the visible range. If higher temperatures were possible, then this fraction would reach a maximum and then begin to decrease after that. Also note that when such filaments are used in the incandescent lamps, the efficiency (i.e., the energy emitted in the visible range divided by the total energy provided to the surface) is rather low.

### 4.1.4 Deviation from Ideal Emission: Emissivity $\epsilon_r$

A realbody emits less than the blackbody. The emission from realbody surfaces depends on the direction, as well as $T$ and $\lambda$, and is given by the spectral, directional emissive power $E_{r,\lambda} = E_{r,\lambda}(T, \lambda, \theta, \phi)$. The total, emissive power is designated by $E_r(T)$ and is found by the integration of $E_{r,\lambda}(T, \lambda, \theta, \phi)$ over all the wavelengths $(0 \leq \lambda \leq \infty)$ and the angles completing a hemisphere on the surface $(0 \leq \phi \leq 2\pi, 0 \leq \theta \leq \pi/2)$. These angles are shown in Figure 4.3(a). As shown, the direction is defined with respect to the surface normal using two angles $\theta$ and $\phi$.

(a)  Spectral, Directional Emissive Power $E_{r,\lambda}(T,\lambda,\theta,\phi)$, Emissivity $\epsilon_{r,\lambda}(T,\lambda,\theta,\phi)$

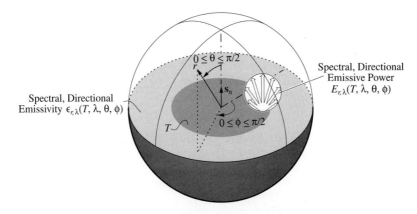

(b)  Behaviors of a Gray (No Wavelength Dependence), Diffuse (No Directional Dependence), and Opaque (No Transmission) Surface

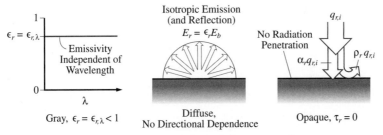

Figure 4.3. (a) Spectral, directional emissive power, and the designated direction. (b) Surface-radiation properties of a gray, diffuse, and opaque surface.

The deviation from the blackbody emission is shown by the emissivity. The spectral, directional emissivity $\epsilon_{r,\lambda}(T, \lambda, \theta, \phi)$ is defined as

$$\epsilon_{r,\lambda}(T, \lambda, \theta, \phi) \equiv \frac{E_{r,\lambda}(T, \lambda, \theta, \phi)}{E_{b,\lambda}(T, \lambda)} \quad \text{spectral, directional emissivity,} \qquad (4.9)$$

where the appearance of $\phi$ and $\theta$ indicates directional dependence.

In (4.9), the spectral, directional emissive power $E_{r,\lambda}(T, \lambda, \theta, \phi)$ is

$$E_{r,\lambda}(T, \lambda, \theta, \phi) \equiv \epsilon_{r,\lambda}(T, \lambda, \theta, \phi)E_{b,\lambda}(T, \lambda) \quad \begin{array}{l}\text{spectral, directional}\\ \text{emissive power.}\end{array} \qquad (4.10)$$

An emitter that has an emissivity that is independent of the wavelength is called a graybody emitter. An emitter that emits equally in all directions is called a diffusebody emitter. A surface that allows no radiation penetration is called an opaque surface. Figure 4.3(b) shows the behaviors of gray, diffuse, and opaque surfaces. These are surfaces that emit with no wavelength or directional dependence and allow no radiation penetration.

Figures 4.4(a)–(c) show the variation of spectral, directional emissivity $\epsilon_{r,\lambda}$ for a realbody surface with respect to temperature [Figure 4.4(a)], wavelength

(a) Temperature Dependence of Spectral Emissivity $\epsilon_{r,\lambda}$ , Given: $\lambda, \phi, \theta$

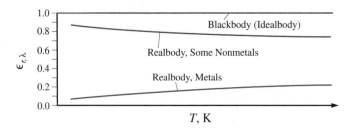

(b) Wavelength Dependence of Spectral Emissivity $\epsilon_{r,\lambda}$ , Given: $T, \phi, \theta$

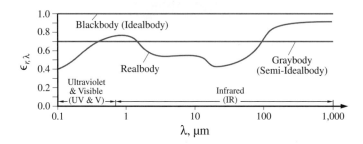

(c) Directional Dependence of Spectral Emissivity $\epsilon_{r,\lambda}$ , Given: $T, \lambda, \phi$

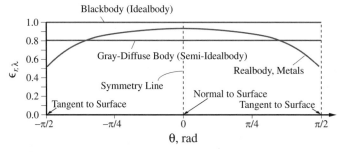

Figure 4.4. Depicted variations of spectral, directional emissivity $\epsilon_{r,\lambda}$ with respect to (a) temperature $T$, (b) wavelength $\lambda$, and (c) polar angle $\theta$.

[Figure 4.4(b)], and polar angle $\theta$ [Figure 4.4(c)]. The ideal (blackbody), semi-ideal (graybody), and realbody behaviors are rendered. It is very convenient to assume a graybody behavior for realbodies. The analysis is then greatly simplified.

The total, hemispherical emissivity $\epsilon_r$ is found by integrating $\epsilon_{r,\lambda} E_{b,\lambda}$ over all the wavelengths and directions (over a unit hemisphere).

For metals, $\epsilon_{r,\lambda,\theta,\phi}$ increases with an increase in the temperature. For nonmetals, $\epsilon_{r,\lambda}$ generally decreases with temperature. This is shown in Figure 4.4(a), where the ideal emitter (blackbody) is also shown ($\epsilon_r = \epsilon_{r,\lambda} = 1$, independent of the wavelength) for blackbody emitters. Figure 4.4(b) shows a realbody ($\epsilon_{r,\lambda} < 1$, and wavelength dependent) is approximated (for simplicity) to be a graybody ($\epsilon_r = \epsilon_{r,\lambda} < 1$, but independent of wavelength). This spectral-average or total, directional emissivity is found by averaging $E_{r,\lambda}$ over all wavelengths, i.e.,

$$\epsilon_r(T, \theta, \phi) \equiv \frac{\int_0^\infty E_{r,\lambda}(T, \lambda, \theta, \phi)d\lambda}{\int_0^\infty E_{b,\lambda}(T, \lambda)d\lambda}$$

$$= \frac{\int_0^\infty \epsilon_{r,\lambda}(T, \lambda, \theta, \phi)E_{b,\lambda}(T, \lambda)d\lambda}{\sigma_{\text{SB}}T^4} \quad \begin{array}{l} \text{total, directional} \\ \text{emissivity.} \end{array} \quad (4.11)$$

Figure 4.4(c) shows that $\epsilon_{r,\lambda}$ varies with respect to the angle relative to the normal to the surface $\mathbf{s}_n$ (i.e, polar angle). In general, the normal emissivity is the largest. The angular average is taken over the entire solid angle covering one side of a plane (called the hemispherical solid angle). This is $0 \leq \phi \leq 2\pi$ and $0 \leq \theta \leq \pi/2$ shown in Figure 4.3(a). Then the wavelength- and direction-averaged emissivity is called the total, hemispherical emissivity $\epsilon_r$ (this is conventionally called the emissivity) and is given by

$$\epsilon_r \equiv \frac{\int_0^{2\pi} \int_0^{\pi/2} \epsilon_r(T, \theta, \phi)E_b \sin\theta \cos\theta d\theta d\phi}{\int_0^{2\pi} \int_0^{\pi/2} E_b \sin\theta \cos\theta d\theta d\phi} \quad \begin{array}{l} \text{total, hemispherical} \\ \text{emissivity (or emissivity).} \end{array} \quad (4.12)$$

Here we have used the concept of solid angle to perform the spatial integration.

Using (4.12), we have the total, hemispherical emissive power $E_r(T) \equiv \epsilon_r E_b(T)$ of a realbody. We have already used the total emissivity in (2.52) where we presented the surface-emission heat flux (same as total emissive power) as

$$q_{r,\epsilon} \equiv q_{e,\epsilon} \equiv E_r(T) \equiv \epsilon_r E_b(T) \equiv \epsilon_r \sigma_{\text{SB}} T^4 \quad \begin{array}{l} \text{total, hemispherical emissivity} \\ \epsilon_r \text{ and total, hemispherical} \\ \text{emissive power } E_r(T). \end{array} \quad (4.13)$$

For convenience, we have used the subscript $r$ for radiation (instead of $e$ for electromagnetic). Figure 4.5(a) shows the temperature dependence of $\epsilon_r$ for some metals. Most data are for total, normal emissivity, but due to lack of other data, they are also used as the total, hemispherical emissivity. Polished metal surfaces have small $\epsilon_r$, while organics and oxidized metals have high $\epsilon_r$. For metals, $\epsilon_r$ increases with temperature. A more comprehensive listing is given in Tables C.18 and C.19 of Appendix C. Typical values for $\epsilon_r$, for different materials, are shown in Figure 4.5(b). In general, the dielectrics (i.e., $\sigma_e \to 0$), such as oxide and nonoxide ceramics, have a higher $\epsilon_r$, compared to metals.

## 4.2 Interaction of Irradiation and Surface

In the last section, we discussed radiation emission from a surface $E_r$. Here we discuss interaction of the irradiation (i.e., impinging radiation) $q_{r,i}$ and a surface. Later, we combine these to analyze radiation heat transfer between surfaces.

(a) Total, Hemispherical Emissivity $\epsilon_r$ of Some Oxidized and Polished Metals, as a Function of Temperature

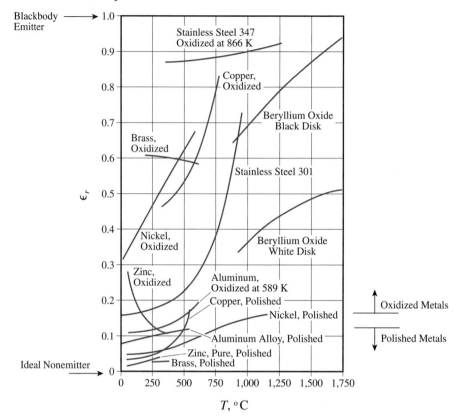

(b) Typical Total, Hemispherical Emissivity $\epsilon_r$ of Some Materials

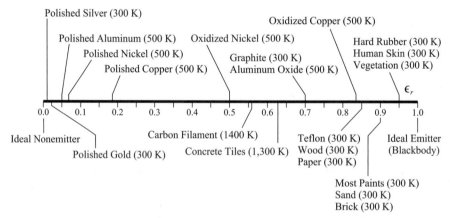

Figure 4.5. (a) Variation of total, hemispherical emissivity $\epsilon_r$, for some metallic solid surfaces, with respect to temperatures. (From Ozisik, M.N., reproduced by permission ©1973, Wiley.) (b) Typical values of the total, hemispherical emissivity $\epsilon_r$, for some solid surfaces.

### 4.2.1 Surface-Radiation Properties: Absorptivity $\alpha_r$, Reflectivity $\rho_r$, and Transmissivity $\tau_r$

Interaction of radiation with a surface is phenomenologically (i.e., based on observables) represented by absorption, reflection, and emission. Similarly, in volumetric radiation, surfaces, particles, or molecules that influence the radiation are distributed within the volume. Then the interaction of radiation with the medium filling a volume is phenomenologically represented by absorption, scattering, and emission. In solar irradiation upon the earth, the scattering gives rise to the blue color of the sky in daylight and the redness of the sun at dusk.

We are interested in interaction between radiation arriving from a source and impinging on a bounding surface. This is depicted in Figure 4.6. In the classical, continuum treatment, the surface is marked by a change occurring in the electromagnetic wave as it crosses the interface separating medium $i$ from medium $j$. This change is due to the change in the electromagnetic properties across the bounding surface. The electromagnetic fields are governed by a collection of conservation equations and relations called the Maxwell electromagnetic equations.

The quantum (quasi-particle) description treats the direct interaction of photons of energy $E_{r,\lambda}$ with the electronic $E_e$, lattice vibration (phonon) $E_p$, and free carriers energy states.

The classical electromagnetism is described by the Maxwell equations of electromagnetic fields and waves, with $\boldsymbol{e}_e$ and $\boldsymbol{h}_e$ being the electric field and the magnetic field intensity vectors. There are three fundamental, classical-continuum electromagnetic properties. These are the complex electrical permittivity (ability to store electrical potential, also called electric inductive capacity) $\epsilon_o\epsilon_e$, the complex, spectral magnetic permeability (ability to modify magnetic flux, also called magnetic inductive capacity) $\mu_o\mu_{e,\lambda}$, and the spectral electrical conductivity $\sigma_{e,\lambda}$. These quantities are also listed in Figure 4.6. The complex, spectral relative electrical permeability is $\epsilon_{e,\lambda} = \epsilon_{er,\lambda} - i\epsilon_{ec,\lambda}$, where here $i = (-1)^{1/2}$. The real part $\epsilon_{er,\lambda}$ is called the spectral dielectric constant and the imaginary part $\epsilon_{ec,\lambda}(0 \leq \epsilon_{ec,\lambda} \leq \infty)$ is called the dielectric loss factor (used in the microwave heating expression in Section 2.3.2).

The speed of propagation of electromagnetic waves, i.e., the spectral speed of light, is defined as $c_\lambda \equiv (\mu_o\mu_{e,\lambda}\epsilon_o\epsilon_{e,\lambda})^{-1/2}$, where for vacuum the relative spectral electrical permittivity $\epsilon_{e,\lambda}$ and the relative magnetic permeability $\mu_{e,\lambda}$, are both unity. Then we have $c_o = (\mu_o\epsilon_o)^{-1/2}$.

All the optical properties (also called propagation properties) and surface-radiation properties of materials and interfaces are determined from these three electromagnetic properties, plus the frequency $f$ (or angular frequency $\omega$) [2, 7, 8, 10].

The optical properties of a medium signify the ability of the medium to refract (i.e., bend) an oblique incident wave, and its ability to absorb. The optical properties are defined through the relationship between the wave vector and the frequency (this relationship is called the dispersion relation). The optical properties are combined in the spectral complex refractive index $m_{r,\lambda} = n_\lambda - i\kappa_\lambda$. Here $n_\lambda$ is the spectral

Interaction of Electromagnetic Radiation with an Intermedium Interface

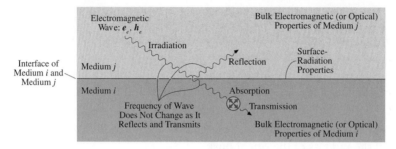

Electromagnetic Properties (Fundamental Properties) from Maxwell EM Equations:

$$\epsilon_0 \epsilon_{e,\lambda}$$

Free-Space (Vacuum), Electrical Permittivity $\epsilon_o = 8.8542 \times 10^{-12}$ A²-s²/N-m² 　　Complex, Relative Electrical Permittivity

$$\epsilon_{e,\lambda} = \epsilon_{er,\lambda} - i\,\epsilon_{ec,\lambda}$$

$$\mu_0 \mu_{e,\lambda}$$

Free-Space Magnetic Permeability $\mu_o = 1.2566 \times 10^{-6}$ N/A²　　Complex, Relative Magnetic Permeability

$$\mu_{e,\lambda} = \mu_{er,\lambda} - i\,\mu_{ec,\lambda}$$

$\sigma_{e,\lambda}$ —— Electrical Conductivity (1/ohm-m)
$$0 \le \sigma_{e,\lambda} < \infty$$

$\epsilon_{e,\lambda}$, $\mu_{e,\lambda}$, and $\sigma_{e,\lambda}$ Are Generally Measured [Also, $c_\lambda = (\mu_o \mu_{e,\lambda} \epsilon_o \epsilon_{e,\lambda})^{-1/2}$]

Optical Properties (First-Level Derived Properties)
Derived through Wave Dispersion Relations :

$$m_{r,\lambda} = n_\lambda - i\,\kappa_\lambda = [\mu_o \mu_{e,\lambda} c_o^2 (\epsilon_o \epsilon_{e,\lambda} - \frac{i\,\sigma_{e,\lambda}}{\omega})]^{1/2}, \quad \omega = 2\pi f$$

Complex Refraction Index　　Extinction Index

Refraction index

$n_\lambda$ and $\kappa_\lambda$ depend on $\epsilon_{e,\lambda}$, $\mu_{e,\lambda}$, and $\sigma_{e,\lambda}$

Surface-Radiation Properties (Second-Level Derived Properties)
Derived through Interfacial Relations:

$\rho_{r\lambda}$ —— Spectral Reflectivity, $0 \le \rho_{r\lambda} \le 1$

$\alpha_{r\lambda}$ —— Spectral Absorptivity, $0 \le \alpha_{r\lambda} \le 1$

$\rho_{r\lambda}$ and $\alpha_{r\lambda}$ Depend on $n_\lambda$ and $\kappa_\lambda$ of
Two Media Making Up the Interface

Figure 4.6. The classical, continuum interaction of electromagnetic radiation with an interface marking a change in radiation properties. The electromagnetic (fundamental), optical (derived), and surface-radiation (derived) properties are also shown.

refraction index ($n_\lambda = c_o/c_x = \lambda_o/\lambda$, since the frequency does not change as the waves travel through different media). Also, $\kappa_\lambda$ is the spectral extinction index. Commonly, $n_{lambda}$ and $\kappa_{lambda}$ are also called the optical constants. So, the optical properties are related to the electromagnetic properties. In practice, in some cases the optical properties are measured directly.

In surface-radiation heat transfer analysis, the surface-radiation properties are used. These are related directly to the electromagnetic properties or to the optical properties.

(a) Rendering of Spectral Reflection, Absorption, and Transmission

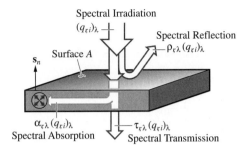

(b) Transparency of Some Solids, Based on Photon Absorption

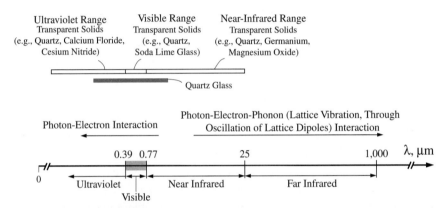

Figure 4.7. (a) Spectral irradiation on a surface and the resulting spectral reflection, absorption and transmission. (b) Transparency of some solids, based on a photon absorption [interaction of a photon with electron] (some through oscillation of crystal dipoles).

The surface-radiation properties are also shown in Figure 4.6. The radiation impinging on a solid surface is called the irradiation and this irradiation is absorbed, reflected, or transmitted as it interacts with the solid. Similar to our discussion of emissivity, the interaction of the electromagnetic waves with matter is strongly dependent on the wavelength. This wavelength-dependent behavior is emphasized and is referred to as spectral behavior.

For ideal dielectrics (i.e., ideal insulators), $\sigma_{e,\lambda} = \kappa_\lambda = 0$ (i.e., nonattenuating), and radiation properties are related to $n$ only. For metals, both $n_\lambda$ and $\kappa_\lambda$ (which are related to $c_\lambda$, $\sigma_{e,\lambda}$, and $\mu_{e,\lambda}$) are much larger and are used to predict the radiation properties.[†]

---

[†] For metals, the photon mean-free path (also called skin depth) is

$$\lambda_{ph} = \left(\frac{2\epsilon_\circ c_\lambda^2}{4\pi\mu_{e,\lambda}f\sigma_{e,\lambda}}\right)^{1/2} = \left(\frac{2}{2\pi f\mu_o\mu_{e,\lambda}\sigma_{e,\lambda}}\right)^{1/2}.$$

This penetration length is generally very small (less than one $\mu$m). Note that metal surfaces are also highly reflecting.

Figure 4.7(a) depicts these phenomenological mechanisms of radiation-surface interactions. Shown in the figure is the spectral (wavelength dependent) irradiation flux $(q_{r,i})_\lambda$ (W/m$^2$), the spectral absorptivity $\alpha_{r,\lambda}$, the spectral reflectivity $\rho_{r,\lambda}$, and the spectral transmissivity $\tau_{r,\lambda}$. Note that $(q_{r,i})_\lambda$ is also related to electromagnetic field and medium properties.[†] From these definitions we have

$$(q_{r,i})_\lambda = (q_{r,\alpha})_\lambda + (q_{r,\sigma})_\lambda + (q_{r,\tau})_\lambda$$

$$= \alpha_{r,\lambda}(q_{r,i})_\lambda + \rho_{r,\lambda}(q_{r,i})_\lambda + \tau_{r,\lambda}(q_{r,i})_\lambda \quad \begin{array}{l}\text{division of spectral irradiation flux} \\ \text{into spectral absorbed, reflected,} \\ \text{and transmitted radiation fluxes.}\end{array} \quad (4.14)$$

Then,

$$\alpha_{r,\lambda} + \rho_{r,\lambda} + \tau_{r,\lambda} = 1 \quad \begin{array}{l}\text{summation law for spectral absorptivity,} \\ \text{reflectivity, and transmissivity.}\end{array} \quad (4.15)$$

The surface-radiation properties can be related to the electromagnetic or the optical properties. However, in most cases in practice, the surface-radiation properties are also directly measured.

The absorption of photon by matter is by match between the single photon energy $E_{r,\lambda} = h_P f$ and a single electronic, phonon (through oscillation of crystal dipoles, resulting from nucleus electron charge distribution), or free carrier energy in the matter. In some cases, more than one of each of these carriers are involved (this is called the multi-state interaction). The low frequency, i.e., infrared photons, are absorbed by phonons and the high frequency, i.e., visible and ultraviolet photons, are absorbed by electronic transitions (band-gap energies). Figure 4.7(b) shows the ultraviolet, visible, and infrared absorption and transmission in solids. When transparency is needed in a wavelength range, materials with no photon absorption energy transitions in this range are used. Quartz is transparent over the ultraviolet, visible, and a portion of near infrared range.

### 4.2.2 Opaque Surface $\tau_r = 0$

An opaque medium is that which does not allow for the irradiation to transmit, while traveling through the medium ($\tau_{r,\lambda} = 0$). An opaque surface is an idealization of the bounding surfaces of media that do not allow for any transmission beyond a very thin surface layer. This would correspond to having a very large optical thickness, i.e., $\sigma_{ex}^* \to \infty$, or the photon mean-free path $\lambda_{ph} = 1/\sigma_{ex}$ tending to zero. The optical thickness is defined by (2.44). Many solids can be considered as

---

[†] For example, the spectral irradiation flux $(q_{r,i})_\lambda$ is related to incoming electric field intensity vector $\boldsymbol{e}_e$, for a planar wave, through

$$(q_{r,i})_\lambda = \frac{1}{2}|m_{r,\lambda}|(\frac{\epsilon_\circ}{\mu_\circ})^{1/2}|\boldsymbol{e}_e|^2.$$

This is found by cross product of the electric and magnetic field intensity vectors ($\boldsymbol{e}_e \times \boldsymbol{h}_e$).

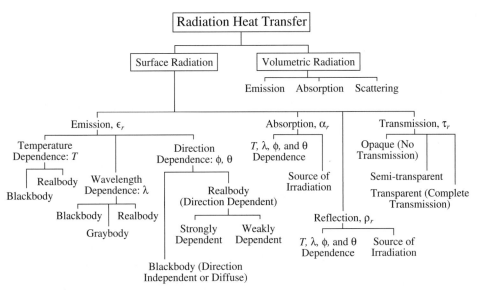

Chart 4.2. Surface-radiation properties and aspects of volumetric radiation.

having an opaque surface, but others, like glass, do allow for a significant transmission of some wavelengths (i.e., are nearly transparent to some wavelengths), and most thin film coatings (i.e., those with thickness of the order of one micron) are considered semi-transparent. Here we limit our discussion to opaque surfaces and consider the radiation heat transfer among such surfaces. The opaque surface is shown in Figure 4.3(b).

We assume that the enclosure volume between these surfaces is completely transparent, i.e., that this medium between these surfaces has a zero optical thickness, i.e., $\sigma_{ex}^* \rightarrow 0$, and that the photon mean-free path $\lambda_{ph}$ tends to infinity in this medium. Thus, we will not discuss the semi-transparent media, i.e., the media participating in the radiation. The most general treatment of the volumetric radiation is made using the conservation of radiation energy represented by the radiation intensity (or more specifically, the photon flux). This conservation equation is called the equation of radiative transfer and includes the emission, absorption, and scattering of radiation as it travels through a medium. Fundamentally, this conservation equation is similar to the Boltzmann transport equation mentioned in Section 3.2 in connection with conduction heat carriers (e.g., gaseous molecules, free electrons, and phonons). In Section 4.5.3 we briefly cover flame radiation to its bounding surface, without directly addressing the equation of radiation transfer. In Section 5.4.6, we briefly discuss the case of an optically thick volume and radiation heat losses from flames; otherwise we defer the discussion of the volumetric radiative heat transfer to advanced courses.

With $\tau_{r,\lambda} = 0$, in general $\alpha_{r,\lambda}$ and $\rho_{r,\lambda}$ depend on the surface material, the source and surface temperatures, the wavelength, and the direction. Chart 4.2 gives some of the characteristics of these surface-radiation properties.

### 4.2.3 Relation among Absorptivity, Emissivity, and Reflectivity
### for Opaque Surfaces

For simplicity, we assume that the surface-radiation properties are independent of direction, i.e., $\alpha_{r,\lambda} = \alpha_{r,\lambda}(\lambda, T)$, etc. From thermodynamic equilibrium considerations for diffuse, opaque surfaces, it can be shown that the spectral, hemispherical absorptivity and emissivity are equal, i.e.,

$$\alpha_{r,\lambda}(\lambda, T) = \epsilon_{r,\lambda}(\lambda, T) \quad \text{diffuse, opaque surface.} \qquad (4.16)$$

This is called the spectral Kirchhoff law.

Under this condition, for an opaque surface, $\tau_{r,\lambda} = 0$, (4.15) becomes

$$\alpha_{r,\lambda} + \rho_{r,\lambda} = \epsilon_{r,\lambda} + \rho_{r,\lambda} = 1 \quad \text{or} \quad \rho_{r,\lambda} = 1 - \alpha_{r,\lambda}$$

$$= 1 - \epsilon_{r,\lambda} \quad \text{diffuse, opaque surface.} \qquad (4.17)$$

For simplicity and ease of analysis (but not generally justifiable), it is assumed that the irradiation is diffuse and gray (direction and wavelength independent) and this leads to a special case of the Kirchhoff law for the total, hemispherical absorptivity $\alpha_r$ and emissivity $\epsilon_r$. This is

$$\boxed{\alpha_r(T) = \epsilon_r(T) \quad \text{total absorptivity for gray, diffuse, and opaque surface.} \qquad (4.18)}$$

Then the values given in Figure 4.5 and Tables C.18 and C.19 for $\epsilon_r$ can also be used for the total hemispherical absorptivity $\alpha_r$.

Then similar to (4.17), we have for the total, hemispherical reflectivity (or simply the total reflectivity) $\rho_r$

$$\boxed{\alpha_r + \rho_r = \epsilon_r + \rho_r = 1 \quad \text{or} \quad \rho_r = 1 - \alpha_r = 1 - \epsilon_r \quad \begin{array}{l} \text{total reflectivity for} \\ \text{gray, diffuse, and} \\ \text{opaque surface.} \end{array} \qquad (4.19)}$$

The blackbody surfaces have $\epsilon_r = 1$. Then using (4.19), $\alpha_r = 1$ and $\rho_r = 0$ for the blackbody surfaces.

In the enclosure radiation heat transfer treatment (i.e., many surfaces making up an enclosure with surface-radiation heat transfer occurring among these surfaces) of Section 4.4, we will assume gray, diffuse, and opaque surfaces and thus we can use (4.19). This will allow for a significant simplification and the use of radiosity and irradiative concepts to express surface radiation heat transfer $Q_r$. In Section 4.5 we consider nongray (i.e., $\epsilon_r \neq \alpha_r$) surfaces. There we allow for the difference between $\alpha_r$ and $\epsilon_r$ and will also address prescribed surface irradiation. This will be done by treating the surface and volumetric radiation as an energy conversion $\dot{S}_r$. Chart 4.3 gives a summary of these subjects, as they are covered in this chapter. Before further analysis, in the next section we demonstrate examples of how radiation heat transfer is measured and used.

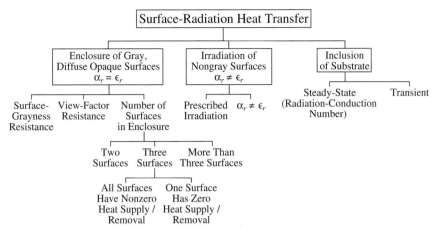

Chart 4.3. Summary of topics related to surface-radiation covered in Chapter 4.

EXAMPLE 4.2. FAM

Consider nonideal emission occurring from real surfaces.

(a) Determine the surface-radiation emission per unit area for a surface at $T = 275°C$ made of (i) a highly polished aluminum, (ii) an oxidized aluminum, and (iii) a Pyrex glass. The surface is depicted in Figure Ex. 4.2. Use the average, total emissivity listed in Table C.18.

(b) Assuming that these surfaces are gray and opaque, determine their surface reflectivity.

(c) Comment on which one of these surfaces is closest to a blackbody surface.

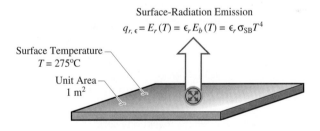

Surface: (i) Highly Polished Aluminum
(ii) Oxidized Aluminum
(iii) Pyrex Glass

Figure Ex. 4.2. Emission from a surface.

**SOLUTION**

(a) The surface-radiation emitted per unit area is given by (4.13), i.e.,

$$q_{r,\epsilon} = E_r(T) = \epsilon_r E_b(T) = \epsilon_r \sigma_{SB} T^4,$$

where $E_r$ is total emissive power.

From Table C.18, we have (using linear interpolation)

(i)  $\epsilon_r$(highly polished aluminum) $= 0.04157$     Table C.18
(ii)  $\epsilon_r$(aluminum oxide) $= 0.63$                           Table C.18
(iii)  $\epsilon_r$(Pyrex glass) $= 0.9446$                           Table C.18.

Then the surface-radiation emitted per unit area is

(i) highly polished aluminum:

$$q_{r,\epsilon} = 0.04157 \times 5.67 \times 10^{-8}(\text{W/m}^2\text{-K}^4)(275 + 273.15)^4(\text{K})^4$$
$$= 2.128 \times 10^2 \text{ W/m}^2$$

(ii) aluminum oxide:

$$q_{r,\epsilon} = 0.63 \times 5.67 \times 10^{-8}(\text{W/m}^2\text{-K}^4)(250 + 273.15)^4(\text{K})^4$$
$$= 3.225 \times 10^3 \text{ W/m}^2$$

(iii) Pyrex glass:

$$q_{r,\epsilon} = 0.9446 \times 5.67 \times 10^{-8}(\text{W/m}^2\text{-K}^4)(250 + 273.15)^4(\text{K})^4$$
$$= 4.835 \times 10^3 \text{ W/m}^2.$$

(b) For gray, opaque surfaces, the total reflectivity is given by (4.19) as

$$\rho_r = 1 - \epsilon_r \quad \text{gray, opaque surface.}$$

Then

(i)   highly polished aluminum:     $\rho_r = 1 - 0.04157 = 0.9584$
(ii)  oxidized aluminum:              $\rho_r = 1 - 0.63 = 0.37$
(iii) Pyrex glass:                        $\rho_r = 1 - 0.9446 = 0.05540.$

(c) For a blackbody surface, $\epsilon_r = 1$. Then from (4.19)

$$\alpha_r = \epsilon_r = 1 \quad \text{blackbody surface}$$
$$\rho_r = 1 - \epsilon_r = 0 \quad \text{blackbody surface.}$$

Thus the Pyrex glass is the closest to a blackbody surface, among the three surfaces.

**COMMENT**
From Table C.18, note that $\epsilon_r$ varies greatly with the temperature and with the physical and chemical state of the surface. The values used here suggest the large difference in $\epsilon_r$ among the very low emissivity surfaces, such as the polished aluminum, and the very high emissivity surfaces, such as the Pyrex glass.

### 4.3 Thermal Radiometry*

This section and the sections that follow are found on the Web site www. cambridge.org/kaviany. In this section we discuss the radiation measurements (i.e., radiometry) and address thermal radiometry which is mostly in the visible and infrared range. Here we examine the pyranometer and the infrared surface-temperature sensor.

#### 4.3.1 Pyranometer*

#### 4.3.2 Infrared Surface-Temperature Sensor*

### 4.4 Enclosure Surface-Radiation Heat Transfer $Q_{r,i}$ among Gray, Diffuse, and Opaque Surfaces

To derive the relations for enclosure radiation, we begin by addressing the radiation heat transfer between a surface (gray, diffuse, and opaque) and its surrounding surfaces (also gray, diffuse, and opaque). The space between these surfaces (i.e., the enclosed volume) does not absorb, scatter, or emit radiation. This space can be evacuated or be occupied by air. In other words, air is assumed nonparticipating in this thermal radiation. For all these surfaces we repeat (4.18) and (4.19) as

$$\epsilon_r = \alpha_r \quad \text{gray, diffuse, opaque surface} \qquad (4.20)$$

$$\rho_r = 1 - \alpha_r = 1 - \epsilon_r \quad \text{gray, diffuse, opaque surface.} \qquad (4.21)$$

The diffuseness allows for treating the irradiation, reflection, and emission all as diffuse (i.e., without a need to specify the direction of incident, reflected, and emitted radiation).

The surrounding surfaces are those surfaces that, along with the surface of our interest, make a complete enclosure. For example, the walls, ceiling, and floor of a room are the surroundings of an electrical radiation heater placed on the floor, on one of the walls of this room, or suspended. For a steel rod laid on the ground in an open area to cool down, the ground and the sky are the surroundings. The sky is a special surface, which is generally assumed to be a blackbody surface, $\epsilon_r = 1$, and has a relatively low temperature. The average temperature used for the universe was discussed in Example 1.1 and is the background thermal radiation emitted from a source with $T \sim 3$ K. The radiation temperature used for the earth's atmosphere depends on the water-vapor content (fog or cloud) and altitude.

Therefore, for any surface of interest we can construct a physical model of the surroundings that completely encloses the surfaces of interest (and when appropriate, we use sky or other far-field media as a surface). We can divide this enclosure into $n$ surfaces. This choice may be based on the geometry of the enclosure or may be based on the accuracy expected of our analysis (e.g., subdivision of a planar surface into smaller planar surfaces). Figure 4.9(a) shows the surfaces in an $n$-surface enclosure.

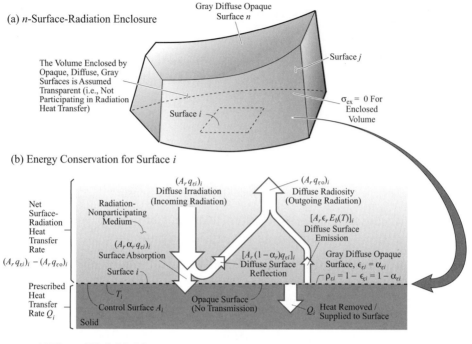

(a) *n*-Surface-Radiation Enclosure

(b) Energy Conservation for Surface *i*

(c) Thermal Node Model

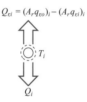

Figure 4.9. (a) An enclosure made of *n*, gray, diffuse, and opaque surfaces. (b) The energy conservation on surface *i*. (c) Thermal node model.

Each surface is irradiated upon and for $A_r$ being the surface area for one of the surfaces, the irradiation for this surface is $A_r q_{r,i}$ ($q_{r,i}$ is the irradiation flux). From each surface, radiation leaves in the amount $A_r q_{r,o}$ and this is called the radiosity ($q_{r,o}$ is the radiosity flux).

### 4.4.1 Surface-Grayness Resistance $R_{r,\epsilon}(1/m^2)$

As shown in Figure 4.9(b), the radiosity flux $q_{r,o}$ is defined as the sum of the surface emission flux $\epsilon_r E_b(T)$ and the reflected irradiation flux $(1 - \alpha_r)q_{r,i}$. Then for any radiation surface *i* in the enclosure (radiation surface $A_{r,i}$), we have

$$(A_r q_{r,o})_i \equiv (A_r q_{r,\epsilon})_i + (A_r \rho_r q_{r,i})_i = [A_r \epsilon_r E_b(T)]_i + [A_r(1 - \alpha_r)q_{r,i}]_i$$

$$= [A_r \epsilon_r E_b(T)]_i + [A_r(1 - \epsilon_r)q_{r,i}]_i \quad \begin{array}{l}\text{radiosity of surface } i \ (A_r q_{r,o})_i \\ \text{is the sum of emission } (A_r \epsilon_r E_b)_i \\ \text{and reflection } (A_r \rho_r q_{r,i})_i.\end{array} \quad (4.22)$$

or

$$(q_{r,i})_i = \frac{(A_r q_{r,o})_i - [A_r \epsilon_r E_b(T)]_i}{[A_r(1 - \epsilon_r)]_i}. \tag{4.23}$$

The surface energy equation is given as the special case of (2.62). This surface energy equation for the control surface $A_i$ (wrapped around the surface) will have surface-radiation heat transfer $Q_{r,i}$ on the surface $A_{r,i}$. All other surface heat transfers are combined into $[(\mathbf{q} \cdot \mathbf{s}_n)A_r]_i = Q_i$. Then using the notations defined in Figure 4.9(c), we have the integral-surface energy equation (4.1) as

$$Q_i + Q_{r,i} = 0, \quad Q_i + (A_r q_{r,o})_i - (A_r q_{r,i})_i = 0 \quad \text{surface energy equation.} \tag{4.24}$$

Here for simplicity, we have not included the energy conversion term, but as we will discuss in Section 4.4.3, this can be readily added. Note that the outgoing heat flow is taken as positive, consistent with the definition given in Section 1.4. We rewrite (4.24), using (4.22), as

$$-Q_i = Q_{r,i} = (A_r q_{r,o})_i - (A_r q_{r,i})_i \tag{4.25}$$

$$-Q_i = Q_{r,i} = [A_r \epsilon_r E_b(T)]_i + [A_r(1 - \epsilon_r)q_{r,i}]_i - (A_r q_{r,i})_i, \tag{4.26}$$

which simplifies to

$$-Q_i = Q_{r,i} = [A_r \epsilon_r E_b(T)]_i - (A_r \epsilon_r q_{r,i})_i, \tag{4.27}$$

and finally to

$$-Q_i = Q_{r,i} = (A_r \epsilon_r)_i [E_b(T) - q_{r,i}]_i. \tag{4.28}$$

We have used the gray surface assumption (4.20), i.e., $\epsilon_r = \alpha_r$.

Now we rewrite $Q_{r,i}$ in terms of the $E_b$ and $q_{r,o}$ by using (4.23) to replace for $q_{r,i}$ in (4.28). Then we have

$$-Q_i = Q_{r,i} = (A_r \epsilon_r)_i \left[ E_b(T) - \frac{A_r q_{r,o} - A_r \epsilon_r E_b(T)}{A_r(1 - \epsilon_r)} \right]_i \tag{4.29}$$

or

$$-Q_i = Q_{r,i} = \frac{[E_b(T) - q_{r,o}]_i}{\left( \dfrac{1 - \epsilon_r}{A_r \epsilon_r} \right)_i} \equiv \frac{[E_b(T) - q_{r,o}]_i}{(R_{r,\epsilon})_i} \quad \begin{array}{l} \text{surface-radiation} \\ \text{heat flow rate with} \\ E_b(T) - q_{r,o} \text{ as potentials,} \end{array} \tag{4.30}$$

Definition of View Factor, $F_{i\text{-}j} \equiv \dfrac{1}{A_{r\,i}} \displaystyle\int_{A_{r\,i}} \int_{A_{r\,j}} \dfrac{\cos\theta_j \cos\theta_i}{\pi R^2}\, dA_{r\,j}\, dA_{r\,i}$

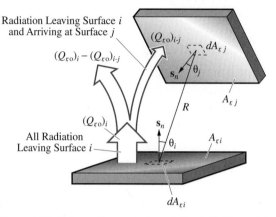

Figure 4.10. Schematic rendering of the definition of the view factor.

where

$$(R_{r,\epsilon})_i \equiv \left(\dfrac{1-\epsilon_r}{A_r \epsilon_r}\right)_i \qquad \text{radiation surface-grayness resistance.} \qquad (4.31)$$

Here $R_{r,\epsilon}(1/m^2)$ is the radiative surface-grayness resistance. It is zero for blackbody surfaces ($\epsilon_r = 1$) and is the measure of the surface grayness.

For blackbody surfaces ($\epsilon_r = 1$) from (4.29), we have $E_b(T) = q_{r,o}$, i.e., radiosity flux and emissive power are equal for blackbody surfaces.

We now need to relate $A_r q_{r,o}$ to the radiation heat transfer of the other surfaces. For this, we need to address the geometry of the enclosure and the concept of the view factor.

### 4.4.2 View-Factor Resistance $R_{r,F}(1/m^2)$

We are considering diffuse surfaces and we are interested in the fraction of radiosity $(A_r q_{r,o})_i$ leaving a surface $i$, that arrives at surface $j$, $(Q_{r,o})_{i\text{-}j}$. This fraction is called the view factor $F_{i\text{-}j}$ (also called shape factor). Figure 4.10 schematically gives the definition of the view factor, which in terms of the surface integrals over surfaces $i$ and $j$, is given by

$$F_{i\text{-}j} \equiv \dfrac{(Q_{r,o})_{i\text{-}j}}{(Q_{r,o})_i} = \dfrac{(Q_{r,o})_{i\text{-}j}}{(A_r q_{r,o})_i} = \dfrac{1}{A_{r,i}} \int_{A_{r,i}} \int_{A_{r,j}} \dfrac{\cos\theta_i \cos\theta_j}{\pi R^2}\, dA_{r,j}\, dA_{r,i}$$

$$(Q_{r,o})_{i\text{-}j} = F_{i\text{-}j}(Q_{r,o})_i \qquad \begin{array}{l}\text{view factor}\\ 0 \le F_{i\text{-}j} \le 1.\end{array} \qquad (4.32)$$

Since for any surface $i$ the sum of all fractions $F_{i\text{-}j}$, for $j = 1, \ldots, n$, is unity, we can write the summation rule for the view factors as

$$\sum_{j=1}^{n} F_{i\text{-}j} = 1 \quad \text{summation rule for any surface } j. \tag{4.33}$$

We can also show that $F_{i\text{-}j}$ and $F_{j\text{-}i}$ are related through the reciprocity rule of the view factors. This is

$$A_{r,i}F_{i\text{-}j} = A_{r,j}F_{j\text{-}i} \quad \text{reciprocity rule between surfaces } i \text{ and } j. \tag{4.34}$$

The view factors for various surface pairs have been compiled [10] and some of these are given in Figures 4.11(a)–(e) and Table 4.2.

Using the view factor, we can write the irradiation on surface $i$, $(Q_{r,i})_i$ as the sum of the fraction of radiosities of all surrounding surfaces $(Q_{r,o})_j$ that arrive at surface $i$, $(Q_{r,o})_j$, $F_{j-i}$, i.e.,

$$(A_r q_{r,i})_i = \sum_{j=1}^{n} (A_{r,j} q_{r,o})_j F_{j\text{-}i} = \sum_{j=1}^{n} A_{r,i} F_{i\text{-}j}(q_{r,o})_j \quad \begin{array}{l} \text{irradiation of surface } i, \\ (A_r q_{r,o})_i, \text{ in terms of radiosity} \\ \text{of surfaces } (q_{r,o})_j, \end{array} \tag{4.35}$$

where we have used the reciprocity rule (4.34).

We can now rearrange this so that we can write $Q_i$ in terms of $A_{r,i}[(q_{r,o})_i - (q_{r,o})_j]$. We begin by substituting (4.35) in (4.24), i.e.,

$$-Q_i = Q_{r,i} = A_{r,i}[(q_{r,o})_i - (q_{r,i})_i]$$

$$= Q_{r,i} = A_{r,i}[(q_{r,o})_i - \sum_{j=1}^{n} F_{i\text{-}j}(q_{r,o})_j]. \tag{4.36}$$

Now using the summation rule (4.33) we write

$$(q_{r,o})_i = 1 \times (q_{r,o})_i = \sum_{j=1}^{n} F_{i\text{-}j}(q_{r,o})_i. \tag{4.37}$$

Substituting this into (4.36), we have the relation among the radiosity fluxes as

$$-Q_i = Q_{r,i} = A_{r,i}\left[\sum_{j=1}^{n} F_{i\text{-}j}(q_{r,o})_i - \sum_{j=1}^{n} F_{i\text{-}j}(q_{r,o})_j\right]$$

$$= Q_{r,i} = \sum_{j=1}^{n} \frac{(q_{r,o})_i - (q_{r,o})_j}{\dfrac{1}{A_{r,i}F_{i\text{-}j}}} \equiv \sum_{j=1}^{n} \frac{(q_{r,o})_i - (q_{r,o})_j}{(R_{r,F})_{i\text{-}j}} \quad \begin{array}{l} \text{energy equation} \\ \text{for node } i. \end{array} \tag{4.38}$$

(a) Bottom Surface of Circular Disk to Top Surface of Coaxial Circular Disk

$$F_{1\text{-}2} = \frac{1}{2}\left\{\gamma - \left[\gamma^2 - 4\left(\frac{R_2^*}{R_1^*}\right)^2\right]^{1/2}\right\} \qquad \gamma = 1 + \frac{1 + R_2^{*2}}{R_1^{*2}}$$

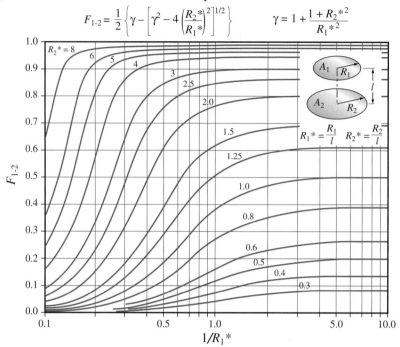

(b) Top Surface of Rectangular Plate to Bottom Surface of Parallel Rectangular Plate

$$F_{1\text{-}2} = \frac{2}{\pi w^* a^*}\left\{\ln\left[\frac{(1 + w^{*2})(1 + a^{*2})}{1 + w^{*2} + a^{*2}}\right]^{1/2} + w^*(1 + a^{*2})^{1/2}\tan^{-1}\frac{w^*}{(1 + a^{*2})^{1/2}} + \right.$$
$$\left. a^*(1 + w^{*2})^{1/2}\tan^{-1}\frac{a^*}{(1 + w^{*2})^{1/2}} - w^*\tan^{-1}w^* - a^*\tan^{-1}a^*\right\}$$

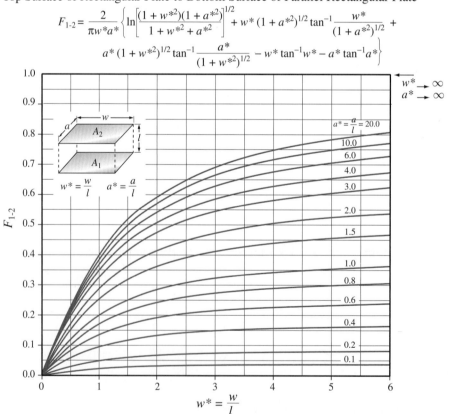

Figure 4.11. (a) View factor for surface pairs, parallel disks. (b) View factor for surface pairs, parallel plates. (From Ozisik, M.N., reproduced by permission. ©1973 Wiley.)

**(c) Top Surface of Rectangular Plate to Top Surface of Perpendicular, Adjacent Plate**

$$F_{1\text{-}2} = \frac{1}{\pi w^*}\left( w^*\tan^{-1}\frac{1}{w^*} + l^*\tan^{-1}\frac{1}{l^*} - (l^{*2}+w^{*2})^{1/2}\tan^{-1}\frac{1}{(l^{*2}+w^{*2})^{1/2}} + \right.$$

$$\left. \frac{1}{4}\ln\left\{\frac{(1+w^{*2})(1+l^{*2})}{1+w^{*2}+l^{*2}}\left[\frac{w^{*2}(1+w^{*2}+l^{*2})}{(1+w^{*2})(w^{*2}+l^{*2})}\right]^{w^{*2}}\left[\frac{l^{*2}(1+l^{*2}+w^{*2})}{(1+l^{*2})(l^{*2}+w^{*2})}\right]^{l^{*2}}\right\}\right)$$

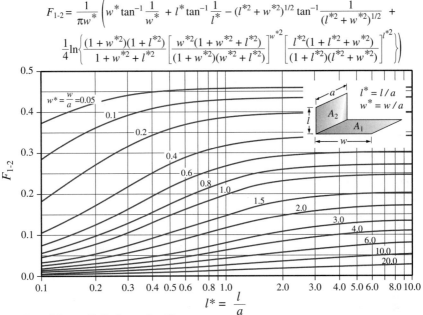

**(d) Interior of Outer Cylinder to Itself with Inner Coaxial Cylinder Present**

$$F_{2\text{-}2} = 1 - \frac{1}{R^*} + \frac{2}{\pi R^*}\tan^{-1}\frac{2(R^{*2}-1)^{1/2}}{l^*} -$$

$$\frac{l^*}{2\pi R^*}\left\{\frac{(4R^{*2}+l^{*2})^{1/2}}{l^*}\sin^{-1}\frac{4(R^{*2}-1)+(l^{*2}/R^{*2})(R^{*2}-2)}{l^{*2}+4(R^{*2}-1)} - \right.$$

$$\left. \sin^{-1}\frac{R^{*2}-2}{R^{*2}} + \frac{\pi}{2}\left[\frac{(4R^{*2}+l^{*2})^{1/2}}{l^*}-1\right]\right\}$$

**(e) Interior of Outer Cylinder to Inner Coaxial Cylinder**

$$F_{2\text{-}1} = \frac{1}{R^*} - \frac{1}{\pi R^*}\left\{\cos^{-1}\frac{b}{a} - \frac{1}{2l^*}\left\{[(a+2)^2 - \right.\right.$$

$$\left.\left. (2R^*)^2]^{1/2}\cos^{-1}\frac{b}{aR^*} + b\sin^{-1}\frac{1}{R^*} - \frac{\pi a}{2}\right\}\right\}$$

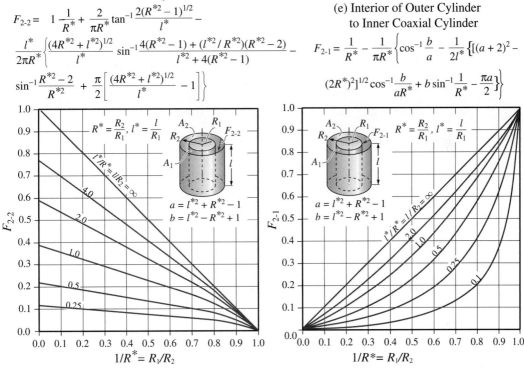

Figure 4.11. (c) View factor for surface pairs, plates at right angles. (d) View factor for surface pairs, coaxial, finite cylinders, interior of outer cylinder to itself. (e) View factor for surface pairs, coaxial, finite cylinders, interior of outer cylinder to inner cylinder. (From Ozisik, M.N., reproduced by permission. ©1973 Wiley.)

Table 4.2. *View factor for some typical surface pairs. (From Siegel, R., and Howell, J.R., reproduced by permission. © 1992 McGraw-Hill.)*

Sphere to Rectangle, $R < l$

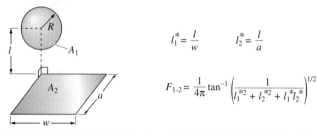

$$l_1^* = \frac{l}{w} \qquad l_2^* = \frac{l}{a}$$

$$F_{1\text{-}2} = \frac{1}{4\pi} \tan^{-1} \left( \frac{1}{l_1^{*2} + l_2^{*2} + l_1^* l_2^*} \right)^{1/2}$$

Sphere to Coaxial Disk

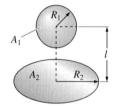

$$R_2^* = \frac{R_2}{l}$$

$$F_{1\text{-}2} = \frac{1}{2} \left[ 1 - \frac{1}{(1 + R_2^{*2})^{1/2}} \right]$$

Outer Surface of Cylinder to Annular Disk at End of Cylinder

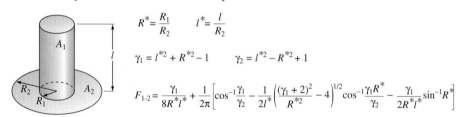

$$R^* = \frac{R_1}{R_2} \qquad l^* = \frac{l}{R_2}$$

$$\gamma_1 = l^{*2} + R^{*2} - 1 \qquad \gamma_2 = l^{*2} - R^{*2} + 1$$

$$F_{1\text{-}2} = \frac{\gamma_1}{8R^* l^*} + \frac{1}{2\pi} \left[ \cos^{-1} \frac{\gamma_1}{\gamma_2} - \frac{1}{2l^*} \left( \frac{(\gamma_1 + 2)^2}{R^{*2}} - 4 \right)^{1/2} \cos^{-1} \frac{\gamma_1 R^*}{\gamma_2} - \frac{\gamma_1}{2R^* l^*} \sin^{-1} R^* \right]$$

Inner Surface of a Ring to Inner Surface of Another Ring on the Same Cylinder

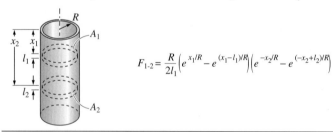

$$F_{1\text{-}2} = \frac{R}{2l_1} \left( e^{x_1/R} - e^{(x_1 - l_1)/R} \right) \left( e^{-x_2/R} - e^{(-x_2 + l_2)/R} \right)$$

In (4.38), the radiation view-factor resistance $(R_{r,F})_{i\text{-}j}$ is defined as

$$(R_{r,F})_{i\text{-}j} \equiv \frac{1}{A_{r,i} F_{i\text{-}j}} = \frac{1}{A_{r,j} F_{j\text{-}i}} = (R_{r,F})_{j\text{-}i} \quad \text{radiation view-factor resistance.} \quad (4.39)$$

Between (4.30) and (4.38) we can determine the enclosure radiation heat transfer. To facilitate the analysis, we use the resistances $R_{r,\epsilon}$ and $R_{r,F}$ in the thermal circuit analysis.

EXAMPLE 4.3. FUN

Determine the view factor from a sphere of radius $R_1$ to (a) a square placed centrally and a distance $l$ away from the sphere with each side being $2R_2$, and to (b) a disk of radius $R_2$. These are shown in Figure Ex. 4.3. The numerical values are $l = 20$ cm, and (i) $R_2 = 20$ cm and (ii) $R_2 \to \infty$. Compare $F_{1-2}$ for the two geometries, for the finite and infinite $R_2$.

(a) Sphere to Coaxial Square Plate        (b) Sphere to Coaxial Disk

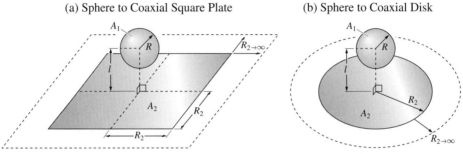

Figure Ex. 4.3. View factor from a sphere. (a) To a coaxial square. (b) To a coaxial disk.

**SOLUTION**

From Table 4.2, we use the relevant view factors.

(a) Sphere to Coaxial Square Plate
    Figure Ex. 4.3 shows that

$$F_{1\text{-}2}\,|_{to\ complete\ square} = 4F_{1\text{-}2}\,|_{to\ \frac{1}{4}\ of\ a\ sphere}$$

$$= 4 \times \frac{1}{4\pi} \tan^{-1}\left(\frac{1}{l_1^{*2} + l_2^{*2} + l_1^* l_2^*}\right)^{1/2} \quad \text{Table 4.2,}$$

where $l_1^* = l/R_2$, $l_2^* = l/R_2$.

(b) Sphere to Coaxial Disk

$$F_{1\text{-}2}\,|_{disk} = \frac{1}{2}\left[1 - \frac{1}{(1 + R_2^{*2})^{1/2}}\right] \quad \text{Table 4.2,}$$

where

$$R_2^* = R_2/l.$$

(i) For finite $R_2$:
Now using the numerical values, we have for the finite $R_2$, $l_1^* = 1$, $l_2^* = 1$, and $R_2^* = 1$, and the results are

$$\text{(a) } F_{1\text{-}2}\,|_{square\ plate} = 4 \times \frac{1}{4\pi} \tan^{-1}\left(\frac{1}{1 + 1 + 1}\right)^{1/2} = 0.1668$$

$$\text{(b) } F_{1\text{-}2}\,|_{disk} = \frac{1}{2}\left[1 - \frac{1}{(1 + 1)^{1/2}}\right] = 0.1464.$$

As expected, the view factor for the square plate is slightly larger.

Table 4.3. *Surface-radiation resistances*

| Heat flow rate | Potential | Resistance |
|---|---|---|
| $Q_{r,i} = A_{r,i} q_{r,i},$ $W = [E_{b,i} - (q_{r,o})_i]/(R_{r,\epsilon})_i,$ | $E_{b,i} - (q_{r,o})_i$, W/m$^2$ | $(R_{r,\epsilon})_i = (1 - \epsilon_r)_i/(A_r \epsilon_r)_i$, 1/m$^2$ |
| $Q_{r,i-j} = A_{r,i} q_{r,i-j},$ W $W = (q_{r,o})_i - (q_{r,o})_j/(R_{r,F})_{i,j}$ | $(q_{r,o})_i - (q_{r,o})_j$, W/m$^2$ | $(R_{r,F})_{i-j} = 1/(A_{r,i} F_{i-j})$, 1/m$^2$ |

(ii) For $R_2 \to \infty$:

Here we have $l_1^* = l_2^* = 0$, $R_2^* \to \infty$, and noting that $\tan^{-1}(\infty) = \pi/2$, we have

$$\text{(a)} \ F_{1\text{-}2} \ |_{square \ plate} = 4 \times \frac{1}{4\pi} \tan^{-1} \left( \frac{1}{0+0+0} \right)^{1/2} = \frac{1}{2}$$

$$\text{(b)} \ F_{1\text{-}2} \ |_{disk} = \frac{1}{2} \left[ 1 - \frac{1}{(1+\infty)^{1/2}} \right] = \frac{1}{2}.$$

**COMMENT**

For $R_2 \to \infty$, the result is expected, because half of the radiation from a diffuse sphere is emitted to its lower plane and half to its upper plane.

### 4.4.3 Thermal Circuit Analysis of Enclosures

As with the steady-state conduction heat transfer between surfaces $i$ and $j$, discussed in Section 3.3.1, the enclosure surface-radiation heat transfer between gray, diffuse, and opaque surfaces $i$ and $j$ is represented by resistances, potentials, and heat flow rates. These electrical analogies are summarized in Table 4.3. Note that the unit for $R_{r,\epsilon}$ and $R_{r,F}$ is 1/m$^2$ (and not K/W). Also, note that the potential is the blackbody emission power and radiosity flux (and not temperature).

We include surface energy conversion $\dot{S}_i$, and for simplicity, we continue to combine all other surface heat transfer into $Q_i$. For each surface in the enclosure, we define the radiation surface-grayness and view-factor resistances as shown in Figure 4.12. Then for each $(q_{r,o})_i$ node we make a nodal energy equation using (4.38), i.e.,

$$-Q_i + \dot{S}_i = Q_{r,i} = \sum_{j=1}^{n} \frac{(q_{r,o})_i - (q_{r,o})_j}{\dfrac{1}{A_{r,i} F_{i-j}}}$$

$$= Q_{r,i} = \sum_{j=1}^{n} \frac{(q_{r,o})_i - (q_{r,o})_j}{(R_{r,F})_{i-j}} \qquad \begin{array}{l} \text{energy equation in terms of} \\ (q_{r,o})_i \text{ for node } i. \end{array} \qquad (4.40)$$

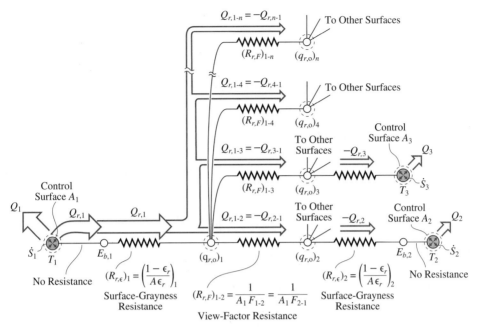

Figure 4.12. Network diagram of radiation heat transfer among $n$ surfaces. The surface-grayness and view-factor resistances are shown for surfaces 1 and 2.

We also use (4.30) for the $E_{b,i}$ node, i.e.,

$$-Q_i + \dot{S}_i = Q_{r,i} = \frac{E_{b,i}(T_i) - (q_{r,o})_i}{\left(\dfrac{1-\epsilon_r}{A_r\epsilon_r}\right)_i}, \quad E_{b,i}(T_i) = \sigma_{\mathrm{SB}} T_i^4 \quad \begin{array}{l}\text{energy equation}\\ \text{in terms of } E_{b,i}, \\ \text{for node } i.\end{array} \quad (4.41)$$

Figure 4.12 gives a thermal circuit diagram for node $i = 1$. There are $n$ radiation surfaces making up the radiation enclosure. There is a $T_1$ node, an $E_{b,1}$ node, and a $(q_{r,o})_1$ node associated with surface 1. This $T_1$ node is connected to the $E_{b,1}$ node, which in turn is connected to the $(q_{r,o})_1$ node. There are no resistances between the $T_1$ and $E_{b,1}$ nodes. The $(q_{r,o})_1$ node is connected to all other $(q_{r,o})_j$ nodes.

In general, all $(q_{r,o})_i$'s are unknown, and for any surface, we either know $Q_i$ and $\dot{S}_i$, or $T_i$. Then, for any surface, we solve for $q_{r,o}$ and either $T$ or $Q$. If all $Q_i$'s and $\dot{S}_i$'s are known, then equation (4.40) is used to solve for the $(q_{r,o})_i$'s. If some $Q_i$'s, $\dot{S}_i$'s, and $T_i$'s are known, then for surfaces with known $T$, we combine (4.40) and (4.41) to have

$$\frac{E_{b,i}(T_i) - (q_{r,o})_i}{\left(\dfrac{1-\epsilon_r}{A_r\epsilon_r}\right)_i} = \sum_{i=1}^{n} \frac{(q_{r,o})_i - (q_{r,o})_j}{\dfrac{1}{A_{r,i}F_{i-j}}} = Q_{r,i}. \quad (4.42)$$

Using (4.40) or (4.41) and (4.42), we write $2n$ equations for $n$ surfaces (and again for each surface either $Q_i - \dot{S}_i = -Q_{r,i}$ or $T_i$ is unknown, and $q_{r,o}$ also is unknown). These can be solved for the $2n$ unknowns made of $(q_{r,o})_i$'s and for each surface either of $T_i$ or $q_i$.

Note that the overall energy conservation is satisfied, i.e.,

$$-\sum_{i=1}^{n}(Q - \dot{S})_i = \sum_{i=1}^{n} Q_{r,i} = 0 \quad \begin{array}{l}\text{overall energy conservation} \\ \text{among surfaces of an enclosure,}\end{array} \qquad (4.43)$$

because all circuits are connected (radiation energy flows only between the nodes included in the circuit).

We now consider the special cases of two- and three-surface enclosures.

---

### EXAMPLE 4.4. FUN

---

Consider two gray, diffuse, and opaque surfaces exchanging radiation heat and being surrounded by a far-away surface, as depicted in Figure Ex. 4.4. The two surfaces are at $T_1$ and $T_2$, while the surrounding surface is at $T_\infty$.

(a) Describe an imaginary surface 3 that can be used to complete the enclosure.
(b) Write the energy equations (4.41) and (4.42) for the resulting three-surface enclosure.

There are no surface energy conversions ($\dot{S}_i = 0$).

**SOLUTION**
(a) The far-away surface or unbounded volume can be represented by an imaginary surface. This surface is shown by the dotted lines in Figure Ex. 4.4. These imaginary surfaces complete the enclosure, i.e., a three-surface enclosure is formed. These surfaces are considered as blackbody surfaces, because they absorb all the radiation that comes from the enclosure surfaces. Therefore, the emissivity of the surface, i.e., surface 3 in Figure Ex. 4.4, is $\epsilon_{r,3} = 1$. The temperature of this surface is that of the surrounding, i.e., $T_3 = T_\infty$. Note that surface 3 may be made of several segments (here two segments are shown). However, it is shown as one collective surface.

(b) Since $\epsilon_{r,3} = 1$, then (4.41) gives

$$E_{b,3} - (q_{r,o})_3 = -Q_3 \left(\frac{1-\epsilon}{A_r \epsilon_r}\right)_3 = 0$$

or

$$E_{b,3} = (q_{r,o})_3 \quad \begin{array}{l}\text{for a blackbody surface with equal} \\ \text{emissive power and radiosity.}\end{array}$$

Here we assume planar surfaces such that $F_{i\text{-}i} = 0$.

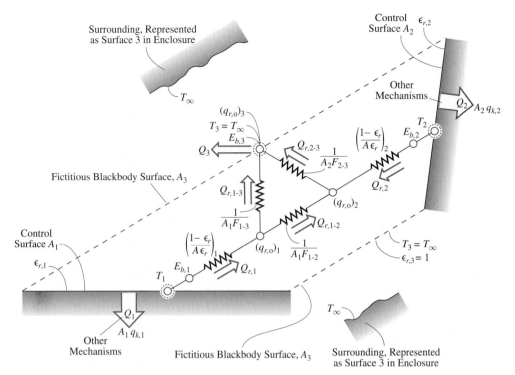

Figure Ex. 4.4. Radiation heat transfer between two surfaces. The two gray, diffuse surfaces do not make a complete enclosure. A fictitious surface 3 is used to complete the enclosure.

Using this, we will replace $(q_{r,o})_3$ by $E_{b,3}$ in the energy equations (4.40) and (4.41) and we have

$$-Q_1 = Q_{r,1} = \frac{(q_{r,o})_1 - (q_{r,o})_2}{\dfrac{1}{A_1 F_{1\text{-}2}}} + \frac{(q_{r,o})_1 - E_{b,3}}{\dfrac{1}{A_1 F_{1\text{-}3}}}$$

$$-Q_2 = Q_{r,2} = \frac{(q_{r,o})_2 - (q_{r,o})_1}{\dfrac{1}{A_2 F_{2\text{-}1}}} + \frac{(q_{r,o})_2 - E_{b,3}}{\dfrac{1}{A_2 F_{2\text{-}3}}}$$

$$-Q_3 = Q_{r,3} = \frac{E_{b,3} - (q_{r,o})_1}{\dfrac{1}{A_3 F_{3\text{-}1}}} + \frac{E_{b,3} - (q_{r,o})_2}{\dfrac{1}{A_3 F_{3\text{-}2}}}$$

$$-Q_1 = \frac{E_{b,1} - (q_{r,o})_1}{\left(\dfrac{1 - \epsilon_r}{A \epsilon_r}\right)_1}$$

$$-Q_2 = \frac{E_{b,2} - (q_{r,o})_2}{\left(\dfrac{1 - \epsilon_r}{A \epsilon_r}\right)_2},$$

where $A_r$'s, $\epsilon_r$'s, and $F_{i\text{-}j}$'s are known and the geometry and materials are specified.

**COMMENT**

The above five equations are solved for five unknowns. Two of three unknowns are $(q_{r,o})_1$, and $(q_{r,o})_2$. The other three are among three $Q_i$'s and three $E_{b,i}$'s (i.e., $T_i$'s). Note that by adding the first three above equations, we confirm what is expected, i.e., the sum of the $Q_{r,i}$'s is zero.

### 4.4.4 Two-Surface Enclosures

When there are only two surfaces making up the enclosure, the circuit simplifies to that shown in Figure 4.13(a). The energy equations for these two nodes are

$$Q_1 + Q_{r,1} = \dot{S}_1 \quad \text{energy equation for node 1} \tag{4.44}$$

$$Q_2 + Q_{r,2} = \dot{S}_2 \quad \text{energy equation for node 2,} \tag{4.45}$$

where in Figure 4.13(a) we have taken $\dot{S}_1 = \dot{S}_2 = 0$. We also note that the same heat flows through all the resistances, i.e.,

$$Q_{r,1} = Q_{r,1\text{-}2} = -Q_{r,2\text{-}1} = -Q_{r,2}, \quad \text{continuity of heat flow in circuit.} \tag{4.46}$$

(a) Enclosure Radiation: Two-Surface Enclosure, Circuit Diagram

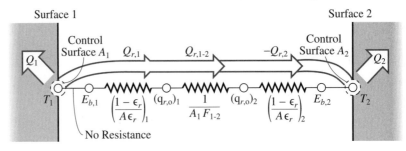

(b) Enclosure Radiation: Single Radiation Shield Placed Between Surfaces 1 and 2, Circuit Diagram

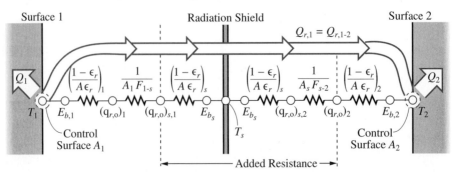

Figure 4.13. Circuit diagram for two-surface enclosures. (a) Without a shield. (b) With one shield.

From Figure 4.13(a), we write for $Q_{r,1\text{-}2}$

$$Q_{r,1\text{-}2} \equiv \frac{E_{b,1}(T_1) - E_{b,2}(T_2)}{R_{r,\Sigma}} \qquad \text{radiation heat transfer in two-surface enclosure,}$$

$$= \frac{E_{b,1}(T_1) - E_{b,2}(T_2)}{(R_{r,\epsilon})_1 + (R_{r,F})_{1\text{-}2} + (R_{r,\epsilon})_2} = \frac{E_{b,1}(T_1) - E_{b,2}(T_2)}{\left(\dfrac{1-\epsilon_r}{A_r\epsilon_r}\right)_1 + \dfrac{1}{A_{r,1}F_{1\text{-}2}} + \left(\dfrac{1-\epsilon_r}{A_r\epsilon_r}\right)_2} \qquad (4.47)$$

i.e., the three radiative resistances are added in series.

In some special cases, expression (4.47) is further simplified, as some of the surface-grayness resistances vanish (as $\epsilon_r$ becomes equal to unity), or the view-factor resistance either becomes infinite (i.e., $F_{1\text{-}2} = 0$) or reaches its minimum value (i.e., $F_{1\text{-}2} = 1$). For example, $F_{1\text{-}2} = 1$, when a small sphere $A_{r,1}$ is placed centrally inside a larger sphere $A_{r,2}$, or when a finite (or infinite) planar surface $A_{r,1}$ radiates to a hemisphere $A_{r,2}$, completely covering it. Then (4.47) becomes

$$Q_{r,1\text{-}2} = \frac{E_{b,1}(T_1) - E_{b,2}(T_2)}{\left(\dfrac{1-\epsilon_r}{A_r\epsilon_r}\right)_1 + \dfrac{1}{A_{r,1}} + \left(\dfrac{1-\epsilon_r}{A_r\epsilon_r}\right)_2} \qquad \text{for } F_{1\text{-}2} = 1. \qquad (4.48)$$

Now if in addition, $A_{r,1}$ is much smaller than $A_{r,2}$, as in the above example, with a very large outside sphere or a very large hemisphere, then the last term $[(1 - \epsilon_r)/A_r\epsilon_r]_2$ vanishes. Also if $\epsilon_{r,2} = 1$, the last resistance in (4.48) also vanishes (surface 2 is a blackbody, i.e., $\epsilon_{r,2} = 1$). Then noting that $[(1 - \epsilon_r)/A_r\epsilon_r]_1 + 1/A_{r,1}$ is equal to $1/A_{r,1}\epsilon_{r,1}$, (4.48) becomes

$$Q_{r,1\text{-}2} = A_{r,1}\epsilon_{r,1}[E_{b,1}(T_1) - E_{b,2}(T_2)] \quad \text{for } F_{1\text{-}2} = 1 \text{ with } \frac{A_{r,1}}{A_{r,2}} \simeq 0 \text{ or } \epsilon_{r,2} = 1. \quad (4.49)$$

The above expression is applicable to surfaces exposed to large surrounding. In these situations, the assumption $A_{r,1}/A_{r,2} \simeq 0$ holds.

### EXAMPLE 4.5. FAM

In a rapid thermal processing, solid objects are heated or cooled for a very short time such that the effect of heating or cooling does not penetrate far into the substrate and/or high temperatures are sustained for a short period to avoid material-property degradation (e.g., segregation of dopants in semiconductors). In a continuous process (as compared to a batch process) thermal processing oven, a large electrically heated plate irradiates upon the integrated-circuit chips facing it as they move through the oven. This is depicted in Figure Ex. 4.5(a).

Assume that the conveyer-chips and the heater make up a two-surface enclosure. Then the planar chip surface has a view factor to the heater that is unity. The chip surface has areas where the emissivity is large, i.e., $\epsilon_{r,1} = 1$, and areas where the surface emissivity is small, $\epsilon_{r,1} = 0.2$. Assume that the emissivity nonuniformity would still allow for a two-surface treatment.

(a) Draw the thermal circuit diagram.
(b) For $T_1 = 250°C$ and $T_2 = 900°C$, determine the surface radiative heat flux $q_{r,1\text{-}2}$ to the high and low emissivity portions of the surface.

### SOLUTION
(a) The thermal circuit model is shown in Figure Ex. 4.5(b).
(b) Since $A_{r,2} \gg A_{r,1}$ and $F_{1\text{-}2} = 1$, we use (4.49) and for $q_{r,1\text{-}2} = Q_{r,1\text{-}2}/A_{r,1}$, we have

$$q_{r,1\text{-}2} = \epsilon_{r,1} \left[ E_{b,1}(T_1) - E_{b,2}(T_2) \right]$$

$$= \epsilon_{r,1} \left( \sigma_{\text{SB}} T_1^4 - \sigma_{\text{SB}} T_2^4 \right)$$

$$= \epsilon_{r,1} \sigma_{\text{SB}} \left( T_1^4 - T_2^4 \right).$$

We now use the numerical values, and for areas with $\epsilon_{r,1} = 1$ we have

$$q_{r,1\text{-}2} = 1 \times 5.670 \times 10^{-8} (\text{W/m-K}^4) [(523.15)^4 (\text{K})^4 - (1{,}173.15)^4 (\text{K})^4]$$

$$q_{r,1\text{-}2} = -1.032 \times 10^5 \text{ W/m}^2 \quad \text{for } \epsilon_{r,1} = 1.$$

For $\epsilon_{r,1} = 0.2$, we have

$$q_{r,1\text{-}2} = 0.2 \times 5.670 \times 10^{-8} (\text{W/m-K}^4) [(523.15)^4 (\text{K})^4 - (1{,}173.15)^4 (\text{K})^4]$$

$$= -2.063 \times 10^4 \text{ W/m}^2.$$

### COMMENT
Note that $q_{r,1-2} < 0$, because this heat is added to the surface. The nonuniformity of the emissivity causes a significant variation in the radiative heating rate, $q_{r,1-2}$. Through substrate conduction, heat is redistributed in the solid, thus avoiding large temperature nonuniformities. However, this transient substrate conduction is not instantaneous and requires an elapsed time which may be larger than the oven transit time for the chips.

### 4.4.5 Radiation Insulators (Shields)

Radiation shields (or radiation insulators) are placed between surfaces to reduce the radiation heat transfer between them. The shields are considered opaque and have low surface emissivity $\epsilon_{r,s}$ (i.e., high reflectivity). The presence of one shield adds two radiative surface-grayness resistances and one view-factor resistance to the heat flow path, as shown in Figure 4.13(b). For $n$ shields we have $2n$ of $R_{r,\epsilon}$ and $n$ of $R_{r,F}$ resistances added. These added resistances, placed in the path of the radiation heat flow, reduce the heat flow significantly. The rate of surface-radiation heat transfer from surface 1 to 2, for the thermal circuit diagram shown in Figure 4.13(b), but with added $n$ identical shields, becomes

(a) Physical Model

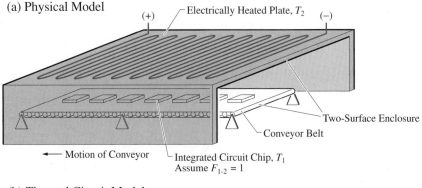

(b) Thermal Circuit Model

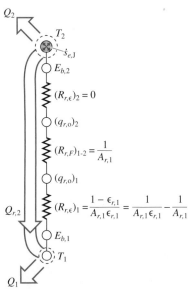

Figure Ex. 4.5. Rapid thermal processing by surface-radiation heat transfer. (a) Physical model. (b) Thermal circuit diagram.

$$Q_{r,1\text{-}2} = \frac{E_{b,1}(T_1) - E_{b,2}(T_2)}{(R_{r,\epsilon})_1 + (R_{r,F})_{1-2} + (R_{r,\epsilon})_2 + n[2(R_{r,\epsilon})_s + (R_{r,F})_s]} \quad \begin{array}{l} n \text{ shields added} \\ \text{between surfaces} \\ 1 \text{ and } 2. \end{array}$$

$$(4.50)$$

This is similar to (4.47), except for the added shield resistances.

For large parallel surfaces separated by radiation shields, we have $A_{r,1} = A_{r,2} = A_{r,s}$, and $F_{1-s} = F_{s-2} = 1$. Then when $\epsilon_{r,1} = \epsilon_{r,2} = \epsilon_{r,s}$, we use (4.47) for $Q_{r,1\text{-}2}$ (without shield) and (4.50) for $Q_{r,1\text{-}2}$ (with $n$ shields), and the ratio of $Q_{r,1\text{-}2}$ with and without shields is

$$\frac{Q_{r,1\text{-}2}(\text{with } n \text{ shields})}{Q_{r,1\text{-}2}(\text{without any shield})} = \frac{1}{n+1}, \quad \begin{array}{l} \text{for } \epsilon_{r,1} = \epsilon_{r,2} = \epsilon_{r,s} \\ \text{and } A_{r,1} = A_{r,2} = A_{r,s}, \end{array} \quad (4.51)$$

i.e., for the case where all of the surface areas and emissivities are equal. From (4.51), for $n$ shields the radiation heat flow is reduced by a factor of $1/(n+1)$.

**EXAMPLE 4.6. FAM**

A double-wall, evacuated glass thermos is used for storing cryogenic liquids (e.g., liquid nitrogen). The evacuation diminishes the conduction heat transfer between the walls, as depicted in Figure Ex. 4.6(a). However, heat transfer by radiation occurs between the glass wall surfaces. The glass walls can be reasonably assumed as very long parallel plates, i.e., $F_{1\text{-}2} = 1$. The glass surface emissivity is $\epsilon_{r,1} = \epsilon_{r,2} = 0.5$. The surface-radiation heat transfer $q_{r,1\text{-}2}$ is reduced when a high reflectivity radiation shield, of emissivity $\epsilon_{r,s} = 0.05$ (e.g., aluminum foil, which is similar to highly polished aluminum, Table C.18), is placed between the surfaces. For liquid nitrogen storage, at one atm pressure, from Table C.3, $T_{lg} = -195.9°\text{C}$. Here, surface 1, which is the cold surface, is at $T_1 = -190°\text{C}$, while the hot surface is at $T_2 = 2°\text{C}$.

Using the radiation shield model depicted in Figure 4.13(b), determine the radiation heat flux $q_{r,1\text{-}2}$, for the case of (a) no shield, (b) one shield, and (c) two shields present.

The shields have the same surface areas (on each side) as $A_{r,1}$ and $A_{r,2}$.

**SOLUTION**

The thermal circuit model of the radiation heat transfer is shown in Figure Ex. 4.6(b). For $F_{1\text{-}2} = 1$, (4.50) becomes

$$q_{r,1\text{-}2} = \frac{E_{b,1}(T_1) - E_{b,2}(T_2)}{A_{r,1}[(R_{r,\epsilon})_1 + (R_{r,F})_1 + (R_{r,\epsilon})_2] + A_{r,1}n[2(R_{r,\epsilon})_s + (R_{r,F})_s]},$$

where

$$A_{r,1}(R_{r,\epsilon})_1 = \frac{1 - \epsilon_{r,1}}{\epsilon_{r,1}} = \frac{1 - 0.5}{0.5} = 1$$

$$A_{r,1}(R_{r,F})_1 = 1 \quad \text{since } F_{1-2} = 1$$

$$A_{r,1}(R_{r,\epsilon})_2 = \frac{1 - \epsilon_{r,2}}{\epsilon_{r,2}} = \frac{1 - 0.5}{0.5} = 1 \quad \text{since } A_{r,1} = A_{r,2}$$

$$A_{r,1}(R_{r,\epsilon})_s = \frac{1 - \epsilon_{r,s}}{\epsilon_{r,s}} = \frac{1 - 0.05}{0.05} = 19 \quad \text{since } A_{r,s} = A_{r,1}$$

$$A_{r,1}(R_{r,F})_s = 1 \quad \text{since } A_{r,s} = A_{r,1}.$$

Then for $q_{r,1\text{-}2}$ we have the following.

(a) Without any shield:

$$q_{r,1\text{-}2} = \frac{\sigma_{\text{SB}}(T_1^4 - T_2^4)}{A_{r,1}[(R_{r,\epsilon})_1 + (R_{r,F})_{1-2} + (R_{r,\epsilon})_2]}$$

$$= \frac{5.670 \times 10^{-8}(\text{W/m}^2\text{-K}^4)[(83.15)^4(\text{K})^4 - (275.15)^4(\text{K})^4]}{1 + 1 + 1}$$

$$q_{r,1\text{-}2} = \frac{-3.222 \times 10^2(\text{W/m}^2)}{3} = -1.074 \times 10^2 \text{ W/m}^2.$$

(a) Physical Model: Evacuated Thermos

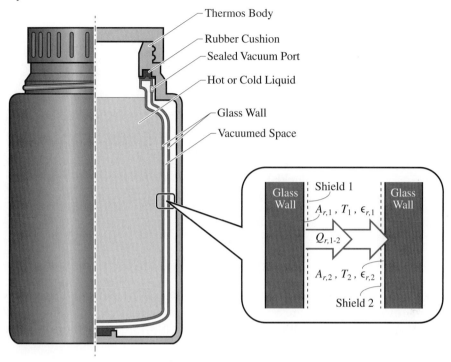

(b) Thermal Circuit Model

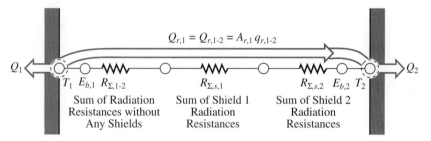

Figure Ex. 4.6. A thermos with two radiation shields. (a) Physical model. (b) Thermal circuit diagram.

(b) One shield:

$$q_{r,1\text{-}2} = \frac{\sigma_{\text{SB}}(T_1^4 - T_2^4)}{A_{r,1}[(R_{r,\epsilon})_1 + (R_{r,F})_{1-s} + (R_{r,\epsilon})_2] + A_{r,1}[2(R_{r,\epsilon})_s + (R_{r,F})_{s-2}]}$$

$$= \frac{-3.222 \times 10^2 (\text{W/m}^2)}{1 + 1 + 1 + (2 \times 19 + 1)}$$

$$q_{r,1\text{-}2} = -7.674 \text{ W/m}^2.$$

(c) Two shields:

$$q_{r,1\text{-}2} = \frac{-3.222 \times 10^2\,(\text{W/m}^2)}{3 + 2 \times 39}$$

$$q_{r,1\text{-}2} = -3.979\ \text{W/m}^2.$$

**COMMENT**

Note that $q_{r,1\text{-}2} < 0$, i.e., heat flows into surface 1. Also note that by using two shields, the amount of heat flow to the inner wall is lowered, compared to that without any shield, by

$$\frac{Q_{r,1\text{-}2}(\text{with 2 shields})}{Q_{r,1\text{-}2}(\text{without any shield})} = \frac{(-3.979)(\text{W/m}^2)A_{r,1}(\text{m}^2)}{(-1.074 \times 10^2)(\text{W/m}^2)A_{r,1}(\text{m}^2)} = 0.0370,$$

which is very substantial reduction.

### 4.4.6 Three-Surface Enclosures

For a three-surface enclosure, with three of the six $Q_i$'s, $\dot{S}_i$'s, and $T_i$'s unknown, a total of six equations are written (two for each surface). Then three of these equations can be considered as being for the radiosities $(q_{r,o})_i$'s. The circuit is shown in Figure 4.14(a).

These six equations are (4.40) and (4.41) written for each surface, i.e.,

$$-Q_1 + \dot{S}_1 = Q_{r,1} = \frac{(q_{r,o})_1 - (q_{r,o})_2}{\dfrac{1}{A_{r,1}F_{1\text{-}2}}} + \frac{(q_{r,o})_1 - (q_{r,o})_3}{\dfrac{1}{A_{r,1}F_{1\text{-}3}}} \tag{4.52}$$

$$-Q_2 + \dot{S}_2 = Q_{r,2} = \frac{(q_{r,o})_2 - (q_{r,o})_1}{\dfrac{1}{A_{r,2}F_{2\text{-}1}}} + \frac{(q_{r,o})_2 - (q_{r,o})_3}{\dfrac{1}{A_{r,2}F_{2\text{-}3}}} \tag{4.53}$$

$$-Q_3 + \dot{S}_3 = Q_{r,3} = \frac{(q_{r,o})_3 - (q_{r,o})_1}{\dfrac{1}{A_{r,3}F_{3\text{-}1}}} + \frac{(q_{r,o})_3 - (q_{r,o})_2}{\dfrac{1}{A_{r,3}F_{3\text{-}2}}} \tag{4.54}$$

$$-Q_1 + \dot{S}_1 = Q_{r,1} = \frac{E_{b,1}(T_1) - (q_{r,o})_1}{\left(\dfrac{1 - \epsilon_r}{A_r \epsilon_r}\right)_1}, \quad E_{b,1}(T_1) = \sigma_{\text{SB}} T_1^4 \tag{4.55}$$

$$-Q_2 + \dot{S}_2 = Q_{r,2} = \frac{E_{b,2}(T_2) - (q_{r,o})_2}{\left(\dfrac{1 - \epsilon_r}{A_r \epsilon_r}\right)_2}, \quad E_{b,2}(T_2) = \sigma_{\text{SB}} T_2^4 \tag{4.56}$$

$$-Q_3 + \dot{S}_3 = Q_{r,3} = \frac{E_{b,3}(T_3) - (q_{r,o})_3}{\left(\dfrac{1 - \epsilon_r}{A_r \epsilon_r}\right)_3}, \quad E_{b,3}(T_3) = \sigma_{\text{SB}} T_3^4. \tag{4.57}$$

In Figure 4.13(b) we have taken $\dot{S}_1 = \dot{S}_2 = \dot{S}_3 = 0$.

(a) Enclosure Radiation: Three-Surface Enclosure, Circuit Diagram

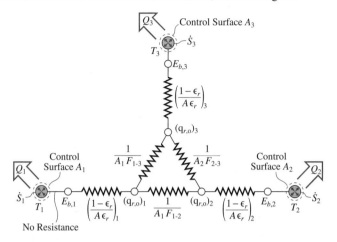

(b) Three-Surface Enclosure with One Surface Ideally Insulated (Reradiating), Circuit Diagram

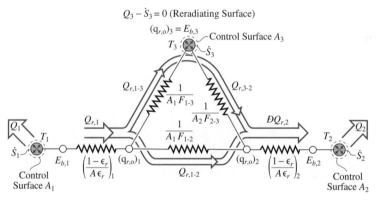

Figure 4.14. Circuit diagram for surface radiation in three-surface enclosure. (a) All three surfaces have nonradiative surface heat flows $Q$. (b) Surface 3 has $\dot{S}_3 - Q_3 = 0$.

Since $E_{b,i} = \sigma_{\mathrm{SB}} T_i^4$, we can consider $E_{b,i}$ as a variable, and therefore, have six linear algebraic equations that are solved for $(q_{r,o})_1$, $(q_{r,o})_2$, $(q_{r,o})_3$, and three of the six $Q_i$'s, $\dot{S}_i$'s, and $E_{b,i}$'s that are unknown.

EXAMPLE 4.7. FAM.S

In an electrically heated, open radiation oven, shown in Figure Ex. 4.7(a), electrical resistance wires are wrapped around a ceramic cylindrical shell. The Joule-heating rate is $\dot{S}_{e,\mathrm{J}}$. The outer surface of the ceramic is ideally insulated, while the inner surface radiates to a cylindrical workpiece (i.e., object to be heated). The ceramic shell and the workpiece have the same length $L$. At a given time, the ceramic is at a temperature $T_2$ and the workpiece is at $T_1$. The surrounding, to which heat is exchanged through the open ends of the oven, is at $T_3$.

(a) Draw the thermal circuit diagram.

(b) Determine the view factor from the relations and compare with the graphical results of Figure 4.11.

(c) Assuming a steady state, determine the fraction of heat transferred to the surroundings.

$\dot{S}_{e,J} = 10^4$ W, $T_1 = 300°$C, $T_3 = 20°$C, $\epsilon_{r,1} = 0.9$, $\epsilon_{r,2} = 0.8$, $D_1 = 3$ cm, $D_2 = 15$ cm, and $l = 60$ cm.

### SOLUTION

(a) This is a three-surface enclosure with surface 3 being the two ends with $\epsilon_{r,3} = 1$ (i.e., an imaginary blackbody surface is used to complete the enclosure). The heat generation $\dot{S}_{e,J}$ leaves surface 2 by radiation, i.e., the outer surface of the ceramic is ideally insulated, $Q_2 = 0$, and there is no energy stored in the ceramic (due to the steady-state assumption). The thermal circuit diagram is shown in Figure Ex. 4.7(b).

(a) Physical Model

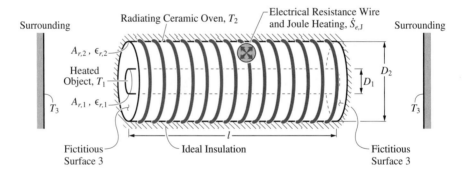

(b) Thermal Circuit Model

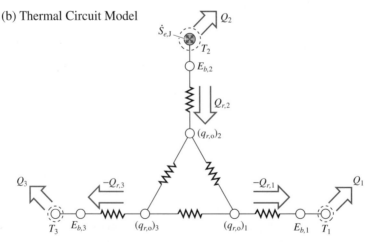

Figure Ex. 4.7. A cylindrical radiation oven heating a cylindrical workpiece placed inside of it. (a) Physical model. (b) Thermal circuit model.

(b) There are seven view factors (because $F_{2\text{-}2} \neq 0$). Using the reciprocity rule (4.34), these are reduced to four. Using the summation rule (4.33), these are yet reduced to two. These two are $F_{2\text{-}2}$ and $F_{2\text{-}1}$. The relations for these are given in Figures 4.11(d) and (e) and are

$$R^* = \frac{D_2}{D_1}, \qquad l^* = \frac{2l}{D_1}$$

$$F_{2\text{-}2} = F_{2\text{-}2}(R^*, l^*) = 1 - \frac{1}{R^*} + \frac{2}{\pi R^*} \tan^{-1} \frac{2(R^{*2} - 1)^{1/2}}{l^*}$$

$$- \frac{l^*}{2\pi R^*} \left\{ \frac{(4R^{*2} + l^{*2})^{1/2}}{l^*} \sin^{-1} \frac{4(R^{*2} - 1) + (l^{*2}/R^{*2})(R^{*2} - 2)}{l^{*2} + 4(R^* - 1)} \right.$$

$$\left. - \sin^{-1} \frac{R^{*2} - 2}{R^{*2}} + \frac{\pi}{2} \left[ \frac{(4R^{*2} + l^{*2})^{1/2}}{l^*} - 1 \right] \right\} \qquad \text{Figure 4.11(d)}$$

$$a = l^{*2} + R^{*2} - 1, \quad b = l^{*2} - R^{*2} + 1$$

$$F_{2\text{-}1} = F_{2\text{-}1}(R^*, l^*) = \frac{1}{R^*} - \frac{1}{\pi R^*} \left\{ \cos^{-1} \frac{b}{a} \right.$$

$$\left. - \frac{1}{2l^*} [(a+2)^2 - (2R^*)^2]^{1/2} \cos^{-1} \frac{b}{aR^*} + b \sin^{-1} \frac{1}{R^*} - \frac{\pi a}{2} \right\} \qquad \text{Figure 4.11(e)}$$

where

$$F_{2\text{-}3} = 1 - F_{2\text{-}2} - F_{2\text{-}1} \qquad F_{1\text{-}3} = 1 - F_{1\text{-}2} \quad \text{summation rule}$$

$$F_{1\text{-}2} = \frac{A_{r,2}}{A_{r,1}} F_{2\text{-}1}, \quad F_{3\text{-}2} = \frac{A_{r,2}}{A_{r,3}} F_{2\text{-}3}, \quad F_{3\text{-}1} = \frac{A_{r,1}}{A_{r,3}} F_{1\text{-}3} \quad \text{reciprocity rule}$$

$$A_{r,2} = \pi D_2 l, \quad A_{r,1} = \pi D_1 l, \quad A_{r,3} = 2 \left[ \frac{\pi}{4} (D_2^2 - D_1^2) \right].$$

Using the numerical values, we have

$$R^* = \frac{D_2}{D_1} = \frac{0.15(\text{m})}{0.03(\text{m})} = 5, \quad l^* = \frac{2l}{D_1} = \frac{2 \times 0.60}{0.03} = 40$$

$$F_{2\text{-}2} = 1 - 0.2 + 0.1273 \times 0.240 - 1.273(1.031 \times 1.180 - 1.16808$$

$$+ 0.04834) = 0.7079$$

$$a = 1,624, \quad b = 1,576$$

$$F_{2\text{-}1} = 0.2 - 0.06366[0.2437 - 0.0125 \times (1.626 \times 10^3 \times 1.376 + 1,576$$

$$\times 0.201 - 2.550 \times 10^3)]$$

$$= 0.2 - 0.06366(0.2437 - 0.0125 \times 4.152) = 0.1867$$

$$F_{2\text{-}3} = 1 - 0.7079 - 0.1878 = 0.1054$$

$$F_{1\text{-}2} = \frac{D_2}{D_1} F_{2\text{-}1} = 0.9337$$

$$F_{1\text{-}3} = 1 - 0.9337 = 0.06630$$

$$F_{3\text{-}1} = \frac{2D_1 l}{D_2^2 - D_1^2} F_{1\text{-}3} = 0.1105$$

$$F_{3\text{-}2} = \frac{2D_2 l}{D_2^2 - D_1^2} F_{2\text{-}3} = 0.8781$$

$$F_{3\text{-}3} = 1 - F_{3\text{-}2} - F_{3\text{-}1} = 0.01140.$$

(c) Now, since $Q_2 = 0$ and $\dot{S}_1 = \dot{S}_3 = 0$, (4.52) through (4.57) become

$$Q_{r,1} = -Q_1 = \frac{(q_{r,o})_1 - (q_{r,o})_2}{\dfrac{1}{\pi D_1 l F_{1\text{-}2}}} + \frac{(q_{r,o})_1 - E_{b,3}(T_3)}{\dfrac{1}{\pi D_1 l F_{1\text{-}3}}}$$

$$Q_{r,2} = \dot{S}_{e,J} = 10^4 (\text{W}) = \frac{(q_{r,o})_2 - (q_{r,o})_1}{\dfrac{1}{\pi D_2 l F_{2\text{-}1}}} + \frac{(q_{r,o})_2 - E_{b,3}(T_3)}{\dfrac{1}{\pi D_2 l F_{2\text{-}3}}}$$

$$-Q_3 = \frac{E_{b,3}(T_3) - (q_{r,o})_1}{\dfrac{1}{\dfrac{\pi}{2}(D_2^2 - D_1^2)F_{3\text{-}1}}} + \frac{E_{b,3}(T_3) - (q_{r,o})_2}{\dfrac{1}{\dfrac{\pi}{2}(D_2^2 - D_1^2)F_{3\text{-}2}}}$$

$$-Q_1 = \frac{E_{b,1} - (q_{r,o})_1}{\dfrac{1 - \epsilon_{r,1}}{\pi D_1 l \epsilon_{r,1}}}$$

$$\dot{S}_{e,J} = \frac{E_{b,2} - (q_{r,o})_2}{\dfrac{1 - \epsilon_{r,2}}{\pi D_2 l \epsilon_{r,2}}}, \qquad E_{b,2} = \sigma_{SB} T_2^4.$$

Note that surface 3 is blackbody and from (4.57) with $(R_{r,\epsilon})_3 = 0$, we have

$$(q_{r,o})_3 = E_{b,3} = \sigma_{SB} T_3^4$$

and have used this above for $(q_{r,o})_3$.

We cannot simply solve for one unknown at a time by successive substitutions. The simultaneous solution requires iterations or use of a solver, for example MATLAB.

We now solve for the six unknowns $(q_{r,o})_1$, $(q_{r,o})_2$, $(q_{r,o})_3$, $T_1$, $Q_2$, and $Q_3$ and the numerical results are

$$(q_{r,o})_1 = 1.789 \times 10^4 \text{ W/m}^2$$

$$(q_{r,o})_2 = 1.327 \times 10^5 \text{ W/m}^2$$

$$(q_{r,o})_3 = 4.179 \times 10^2 \text{ W/m}^2$$

$$T_2 = 983.9 \text{ K}$$

$$Q_1 = 5.994 \times 10^3 \text{ W}$$

$$Q_3 = 4.006 \times 10^3 \text{ W}.$$

**COMMENT**

Note that although $F_{2-2} \neq 0$, it does not appear in (4.53), because the radiosity potential is $(q_{r,o})_2 - (q_{r,o})_2 = 0$.

### 4.4.7 Three-Surface Enclosures with One Surface Having $\dot{S} - Q = 0$

For the special case of one of the surfaces, e.g., surface 3, having $\dot{S}_3 - Q_3 = 0$, i.e., being ideally insulated or adiabatic or having an exact balance between $-Q_3$ and $\dot{S}_3$, this surface completely reradiates the incoming radiation and is called the reradiating surface. Now, consider surface 3 being a reradiating surface. This is shown in Figure 4.14(b), where for surface 3, we have $\dot{S}_3 - Q_3 = 0$. We also have assumed that there are no energy conversions. For this, similar to the two-surface enclosures, we have from Figure 4.14(b)

$$
\begin{aligned}
Q_1 + Q_{r,1} = \dot{S}_1 \quad &\text{or} \quad -Q_1 + \dot{S}_1 = Q_{r,1} \quad &&\text{for node } E_{b,1} \\
Q_2 + Q_{r,2} = \dot{S}_2 \quad & &&\text{for node } E_{b,2} \quad (4.58) \\
Q_3 - Q_{r,1\text{-}3} + Q_{r,3\text{-}2} = \dot{S}_3 \quad &\text{or} \quad Q_{r,1\text{-}3} = Q_{r,3\text{-}2} \quad &&\text{for node } E_{b,3}.
\end{aligned}
$$

From the figure we also have

$$
Q_{r,1} = Q_{r,1\text{-}2} + Q_{r,1\text{-}3} = Q_{r,3\text{-}2} + Q_{r,1\text{-}2} = -Q_{r,2}. \tag{4.59}
$$

Now using $E_{b,1}$ and $E_{b,2}$ as the nodes, we write $Q_{r,1}$ in terms of $E_{b,1}$ and $E_{b,2}$ and the resistances, and from Figure 4.14(b), we have

$$
Q_1 = Q_{r,1} = Q_{r,1\text{-}2} = \frac{E_{b,1}(T_1) - E_{b,2}(T_2)}{\left(\dfrac{1-\epsilon_r}{A_r\epsilon_r}\right)_1 + \dfrac{1}{A_{r,1}F_{1\text{-}2} + \dfrac{1}{\dfrac{1}{A_{r,1}F_{1\text{-}3}} + \dfrac{1}{A_{r,2}F_{2\text{-}3}}}} + \left(\dfrac{1-\epsilon_r}{A_r\epsilon_r}\right)_2}
$$

three surface enclosure with one reradiating surface.

$$
\tag{4.60}
$$

Note that the presence of the reradiating surface reduces the overall view-factor resistance $(R_{r,F})_{1\text{-}2}$. This can be shown by defining an apparent view-factor resistance from (4.60).

$$
[(R_{r,F})_{1\text{-}2}]_{app} \equiv \frac{1}{A_{r,1}F_{1\text{-}2} + \dfrac{1}{\dfrac{1}{A_{r,1}F_{1\text{-}3}} + \dfrac{1}{A_{r,2}F_{2\text{-}3}}}} \qquad \text{apparent view-factor resistance,} \tag{4.61}
$$

as compared to (4.47), where $(R_{r,F})_{1\text{-}2} = 1/A_{r,1}F_{1\text{-}2}$ for the case of no reradiating surface.

Also note that the emissivity of surface 3 does not enter into the relation for $Q_{r,1-2}$. This is because this surface reradiates whatever energy it absorbs. Another observation is that $Q_{r,1-2}$ is not affected by the exact geometry of surface 3, i.e., this surface can be the smallest surface that completes the enclosure or has a much larger surface.

EXAMPLE 4.8. FUN

---

A thermal radiation source is developed by heating one side of a disk of diameter $D_2$ by an acetylene-oxygen torch to a temperature $T_2 = 1,200$ K, as shown in Figure Ex. 4.8. The other side radiates to a smaller, coaxial disk of diameter $D_1$ separated by a distance $l$. In order to direct the greatest amount of heat to the workpiece, an adiabatic (i.e., ideally insulated) conical shell is used.

(a) Draw the thermal circuit diagram.
(b) Determine the rate of radiation heat flowing from surface 2 to 1.
(c) Determine the apparent decrease in the view-factor resistance due to the presence of a reradiating shell.

$$T_1 = 300°C, D_1 = 5 \text{ cm}, \epsilon_{r,1} = 0.8, D_2 = 30 \text{ cm}, \epsilon_{r,1} = 0.9, \text{ and } l = 25 \text{ cm}.$$

**SOLUTION**
(a) The thermal circuit diagram is shown in Figure Ex. 4.8(b). Here $\dot{S}_2 = 0$.
(b) The radiation from surface 1 to 2, in the presence of a reradiating surface 3, is given by (4.60). We need to determine the three view factors. Before any numerical evaluation, we note that

$$F_{1-3} = 1 - F_{1-2} \qquad \text{summation rule}$$

$$F_{2-3} = 1 - F_{2-1} = 1 - \frac{A_{r,1}}{A_{r,2}} F_{1-2} \qquad \text{summation and reciprocity rules.}$$

Then we only need to evaluate $F_{1-2}$. This is done using the relation or the graphical results given in Figure 4.11(a).

$$R_1^* = \frac{D_1}{2l} = \frac{0.05(\text{m})}{2 \times 0.25(\text{m})} = 0.1, \quad R_2^* = \frac{D_2}{2l} = \frac{0.30(\text{m})}{2 \times 0.25(\text{m})} = 0.6$$

$$F_{1-2} = 0.26 \quad \text{for} \quad \frac{1}{R_1^*} = 10 \quad \text{and} \quad R_2^* = 0.6 \quad \text{Figure 4.11(a).}$$

Then

$$F_{1-3} = 1 - 0.26 = 0.74$$

$$F_{2-3} = 1 - \frac{\pi D_1^2/4}{\pi D_2^2/4} F_{1-2} = 1 - \frac{(0.05)^2(\text{m})^2}{(0.30)^2(\text{m})^2} \times 0.26 = 0.993.$$

(a) Physical Model          (b) Thermal Circuit Model

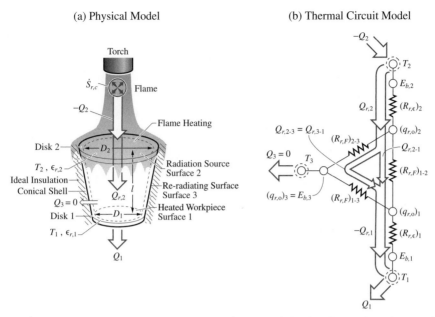

Figure Ex. 4.8. A surface-radiation source (surface 2) for heating surface 1 created by combustion heating and the use of a reradiating surface (surface 3). (a) Physical model. (b) Thermal circuit diagram.

Next, from (4.60) and (4.61), we have

$$Q_{r,2\text{-}1} = \frac{\sigma_{SB}(T_2^4 - T_1^4)}{(R_{r,\epsilon})_1 + [(R_{r,F})_{1\text{-}2}]_{app} + (R_{r,\epsilon})_2}$$

$$(R_{r,\epsilon})_1 = \frac{1 - \epsilon_{r,1}}{\pi R_1^2 \epsilon_{r,1}} = \frac{1 - 0.8}{\pi (0.05)^2 (\text{m})^2 \times 0.8} = 1.273 \text{ m}^2$$

$$(R_{r,\epsilon})_2 = \frac{1 - \epsilon_{r,2}}{\pi R_1^2 \epsilon_{r,2}} = \frac{1 - 0.9}{\pi (0.30)^2 (\text{m})^2 \times 0.9} = 1.572 \text{ m}^{-2}$$

$$[(R_{r,F})_{1\text{-}2}]_{app}$$

$$= \cfrac{1}{\pi R_1^2 F_{1\text{-}2} + \cfrac{1}{\cfrac{1}{\pi R_1^2 F_{1\text{-}3}} + \cfrac{1}{\pi R_2^2 F_{2\text{-}3}}}}$$

$$= \cfrac{1}{\pi (0.025)^2 (\text{m})^2 \times 0.26 + \cfrac{1}{\cfrac{1}{\pi (0.025)^2 (\text{m})^2 \times 0.74} + \cfrac{1}{\pi (0.15)^3 (\text{m})^2 \times 0.993}}}$$

$$= \cfrac{1}{5.060 \times 10^{-4} + \cfrac{1}{\cfrac{1}{1.453 \times 10^{-3}} + \cfrac{1}{7.020 \times 10^{-2}}}}$$

$$= \frac{1}{5.060 \times 10^{-4} + 1.424 \times 10^{-3}} = 5.183 \times 10^2 \text{ m}^{-2}.$$

Now substituting for the surface-grayness and view-factor resistances in the expression for $Q_{r,2\text{-}1}$, we have

$$Q_{r,2\text{-}1} = \frac{5.670 \times 10^{-8} (\text{W/m}^2\text{-K})[(1{,}200)^4(\text{K})^4 - (573.15)^4(\text{K})^4]}{(1.273 \times 10^2 + 1.572 + 5.183 \times 10^2)(\text{m}^{-2})}$$

$$= \frac{1.115 \times 10^5 (\text{W/m}^2)}{6.471 \times 10^2 (1/\text{m}^2)} = 1.723 \times 10^2 \text{ W}.$$

(c) The ratio of $[(R_{r,F})_{1\text{-}2}]_{app}$ to $(R_{r,F})_{1\text{-}2}$ is found noting that

$$(R_{r,F})_{1\text{-}2} = \frac{1}{\pi R_1 F_{1\text{-}2}} = \frac{1}{5.111 \times 10^{-4} (\text{m}^3)} = 1{,}957 \text{ m}^{-2}.$$

Then

$$\frac{[(R_{r,F})_{1\text{-}2}]_{app}}{(R_{r,F})_{1\text{-}2}} = \frac{518.2 (1/\text{m}^2)}{1{,}957 (1/\text{m}^2)} = 0.2648.$$

**COMMENT**

This is a substantial reduction in the view-factor resistance, which is a result of the presence of the reradiating surface. Also, note that emissivity of surface 3, $\epsilon_{r,3}$, does not enter into the radiation heat transfer rate relation. This is because no matter what the extent of its absorption and emission are, this surface emits and reflects whatever it absorbs.

## 4.5 Prescribed Irradiation and Nongray Surfaces*

This section and the sections that follow are on the Web site www.cambridge.org/kaviany. We consider those applications where a spectral or total irradiation is prescribed (as in solar irradiation or irradiation by a laser or another engineered source). In most prescribed irradiations, the dominant wavelength range of irradiation is different than that associated with the surface emission. An example is in solar irradiation on a solar-collector surface where the dominant wavelengths are in the visible range for the irradiation and infrared for the surface emission. Therefore, the total absorptivity and the total emissivity of the surface will not be equal (i.e., the surface behaves as a nongray surface). Here we will consider surface-radiation from a surface without invoking the radiation enclosure. Section 4.5.1 is on Laser Irradiation, 4.5.2 is on Solar Irradiation, 4.5.3 is on Flame Irradiation, and 4.5.4 is on Nongray Surfaces. There are three examples (Examples 4.9 to 4.11).

### 4.5.1 Laser Irradiation $(q_{r,i})_l$*

### 4.5.2 Solar Irradiation $(q_{r,i})_s$*

### 4.5.3 Flame Irradiation $(q_{r,i})_f$*

### 4.5.4 Nongray Surfaces*

## 4.6 Inclusion of Substrate

In our surface-radiation heat transfer so far, we have referred to the substrate only as a possible path for the surface heat flow. A general treatment of surface-radiation with nonuniform (generally perpendicular to the surface) substrate temperature would require inclusion of the conduction heat flux and the integration of the differential-volume energy equation (2.1) over the substrate volume. Here we consider only a few special cases. We begin with a one-dimensional steady-state substrate conduction (i.e., nonuniform substrate temperature). Then we show the conditions under which the substrate temperature may be assumed uniform. Next we consider the substrate sensible heat storage/release (transient) under the assumption of a uniform, time-dependent substrate temperature.

### 4.6.1 Steady-State, Series Conduction-Surface-Radiation, Conduction-Radiation Number $N_r$

The heat supplied/removed from gray, diffuse, and opaque radiating surfaces is conducted through the substrates. Here we address this solid conduction with surface-radiation. We consider steady-state, one-dimensional conduction. As an example, consider the heat supplied from an electrical (i.e., Joule) heater by conduction to a radiating surface, as shown in Figure 4.22(a). In order to inhibit heat flow to the other side of the heater, a conduction thermal insulator is placed on that side. The thermal insulator is not ideal and has a conductivity $k_2$, while the other side has a higher conductivity $k_1$. The insulator has an outside prescribed temperature $T_2$, while at the outside surface of the conductor heat flows out by radiation to another surface that completely encloses surface 1 and is at temperature $T_3$. These are also shown in Figure 4.22(a). The heat generated $\dot{S}_{e,J}$ is mostly removed through the good conductor and this is designated as $Q_{H,1}$, and the small remaining amount $Q_{H,2}$ is removed through nonideal insulator.

The thermal circuit diagram is shown in Figure 4.22(b). Under the steady-state assumption, the energy equation for the heater control volume is the simplified form of (2.9), i.e.,

$$Q \mid_A = \underbrace{\int_A (\mathbf{q} \cdot \mathbf{s}_n) dA}_{\substack{\text{net surface heat flux} \\ \text{heat flux}}} = \underbrace{\dot{S}_{e,J}}_{\substack{\text{electromagnetic} \\ \text{energy conversion}}} \quad \begin{array}{l} \text{integral-volume} \\ \text{energy equation for heater} \end{array} \tag{4.67}$$

or

$$Q_{H,1} + Q_{H,2} = \dot{S}_{e,J}. \tag{4.68}$$

Now we examine the heat flow through the thermal conductor and the thermal insulator.

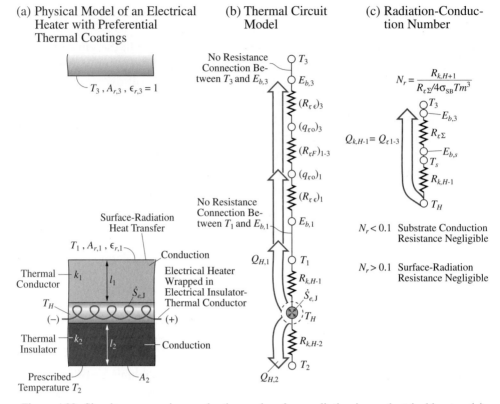

(a) Physical Model of an Electrical Heater with Preferential Thermal Coatings

(b) Thermal Circuit Model

(c) Radiation-Conduction Number

Figure 4.22. Simultaneous series conduction and surface-radiation in an electrical heater. (a) Physical model. (b) Thermal circuit diagram. (c) Radiation-conduction number.

The heat flow through the conductor $Q_{H,1}$ is

$$Q_{H,1} = Q_{k,H\text{-}1} = Q_{r,1\text{-}3} = \frac{T_H - T_1}{R_{k,H\text{-}1}}$$

$$\equiv \frac{E_{b,1} - E_{b,3}}{R_{r,\Sigma}}$$

$$= \frac{E_{b,1} - E_{b,3}}{(R_{r,\epsilon})_1 + (R_{r,F})_{1\text{-}3} + (R_{r,\epsilon})_3} \quad \text{H-1 branch of circuit} \quad (4.69)$$

$$Q_{H,2} = Q_{k,H\text{-}2} = \frac{T_H - T_2}{R_{k,H\text{-}2}} \quad \text{H-2 branch of circuit,} \quad (4.70)$$

where the overall radiation resistance $R_{r,\Sigma}$ is that defined in (4.69).

By choosing a large $R_{k,H\text{-}2}$, the heat loss is reduced.

Now we examine the significance of conduction within the conductor, as compared to the surface-radiation. For this we use (4.69) and for the ratio of the temperature drop through the conductor to that between the radiating surfaces, we then have

$$\frac{T_H - T_1}{T_1 - T_3} = \frac{R_{k,H\text{-}1}}{R_{r,\Sigma}} \frac{E_{b,1} - E_{b,3}}{T_1 - T_3} \quad \begin{array}{l}\text{ratio of temperature drop} \\ \text{for conduction to that for radiation.}\end{array} \quad (4.71)$$

For the right-hand side, we define the radiation mean temperature $T_m(\text{K})$ as

$$\frac{E_{b,1} - E_{b,3}}{T_1 - T_3} = \frac{\sigma_{SB}(T_1^4 - T_3^4)}{T_1 - T_3} = \frac{\sigma_{SB}(T_1^2 + T_3^2)(T_1^2 - T_3^2)}{T_1 - T_3}$$

$$= \sigma_{SB}(T_1^2 + T_3^2)(T_1 + T_3)$$

$$\equiv 4\sigma_{SB}T_m^3, \quad E_{b,1} - E_{b,3} = 4\sigma_{SB}T_m^3(T_1 - T_3) \quad \begin{array}{l}\text{linearized difference} \\ \text{in emission powers,}\end{array} \tag{4.72}$$

where

$$T_m \equiv [\frac{(T_1^2 + T_3^2)(T_1 + T_3)}{4}]^{1/3} \quad \begin{array}{l}\text{radiation mean temperature used} \\ \text{in the coefficient of linearized term.}\end{array} \tag{4.73}$$

We substitute (4.72) into (4.71) and have

$$\frac{T_H - T_1}{T_1 - T_3} = \frac{4\sigma_{SB}T_m^3 R_{k,H\text{-}1}}{R_{r,\Sigma}} = \frac{R_{k,H\text{-}1}}{R_{r,\Sigma}/(4\sigma_{SB}T_m^3)} \equiv N_r \quad \begin{array}{l}\text{conduction-radiation} \\ \text{number } N_r < 0.1; \\ \text{substrate can be} \\ \text{treated as having a} \\ \text{uniform temperature,}\end{array} \tag{4.74}$$

where $N_r$ is the conduction-radiation number.

For small $N_r$, i.e., $N_r < 0.1$, the temperature drop within the solid is insignificant compared to that between the radiating surfaces. This is shown in Figure 4.22(c). Note that the unit for $R_{r,\Sigma}$ is $1/m^2$, while $R_{k,H\text{-}1}$ is in K/W. The unit for $\sigma_{SB}T_m^3$ is $W/m^2$-K.

For the special case of $\epsilon_{r,1} = 1$, $\epsilon_{r,3} = 1$, and $F_{1\text{-}3} = 1$, using the conduction resistance given in Table 3.2, $R_{k,H\text{-}1} = l_1/A_1 k_1$ [Figure 4.22(a)], the conduction-radiation number simplifies to

$$N_r = \frac{R_{k,s}}{R_{r,\Sigma}/4\sigma_{SB}T_m^3} = \frac{4\sigma_{SB}T_m^3 l_1}{k_1} \quad N_r < 0.1 \quad \begin{array}{l}\text{substrate conduction} \\ \text{resistance and temperature} \\ \text{variation are negligible.}\end{array} \tag{4.75}$$

For small $N_r$, i.e., $N_r < 0.1$, this would require a large $k_1$, small $l_1(l)$, or small $T_m$. Then the temperature nonuniformity within the substrate can be neglected.

---

EXAMPLE 4.12. FUN

---

Electrically heated range-top elements glow when they are in high-power settings. The elements have electrical resistance wires that are electrically insulated from the element covers (i.e., casing) by a dielectric filler (e.g., oxide or monoxide ceramic). This ceramic should have a very low electrical conductivity and yet be able to conduct the heat from the heating element to the casing. Oxide ceramics have a low $\sigma_e$ and also low $k$. Nonoxide ceramics (i.e., nitrates such as aluminum nitrite) have low $\sigma_e$, but high $k$. The melting temperature of the resistance wire and the dielectric are also important. The heating element is a coiled wire in a cylindrical form, as shown in Figure Ex. 4.12(a). All the Joule heating leaves through surface 1, i.e., $Q_{k,H\text{-}1} = \dot{S}_{e,J}$.

(a) Draw the thermal circuit diagram.
(b) Determine the temperature $T_H$.
(c) Determine the temperature drop $T_H - T_1$.

$\epsilon_{r,1} = 0.9$, $F_{1-3} = 1$, $T_3 = 20°C$, $D_H = 2$ mm, $D_1 = 10$ mm, $L = 30$ cm, $A_{r,3} \gg A_{r,1}$, $k_1 = 1.5$ W/m-K, and $\dot{S}_{e,J} = 500$ W.

### SOLUTION

(a) The thermal circuit diagram of the heat flow is shown in Figure Ex. 4.12(b). Here we are given the heat flow rate $Q_{H\text{-}1}$. From Figure Ex. 4.12(b), we have

$$Q_{H\text{-}1} = \frac{T_H - T_1}{R_{k,H\text{-}1}} = Q_{r,1\text{-}3} = \frac{E_{b,1} - E_{b,3}}{(R_{r,\epsilon})_1 + (R_{r,F})_{1\text{-}3} + (R_{r,\epsilon})_3} = \dot{S}_{e,J}.$$

(b) Here we first determine $E_{b,1} = \sigma_{SB} T_1^4$, by solving the above for $E_{b,1}$, i.e.,

$$E_{b,1} = \sigma_{SB} T_1^4 = E_{b,3} + \dot{S}_{e,J}[(R_{r,\epsilon})_1 + (R_{r,F})_{1\text{-}3} + (R_{r,\epsilon})_3].$$

The radiation resistances are

$$(R_{r,\epsilon})_1 = \left(\frac{1 - \epsilon_r}{A_r \epsilon_r}\right)_1, \quad A_{r,1} = \pi D_1 L$$

$$(R_{r,F})_{1\text{-}3} = \frac{1}{A_{r,1} F_{1\text{-}3}}, \quad F_{1\text{-}3} = 1$$

$$(R_{r,\epsilon})_3 = \left(\frac{1 - \epsilon_r}{A_r \epsilon_r}\right)_3, \quad A_{r,3} \gg A_{r,1}$$

$$R_{r,\Sigma} = \frac{1}{A_{r,1}}\left(\frac{1 - \epsilon_{r,1}}{\epsilon_{r,1}} + 1 + \frac{A_{r,1}}{A_{r,3}}\frac{1 - \epsilon_{r,3}}{\epsilon_{r,3}}\right) = \frac{1}{A_{r,1}}\left(\frac{1}{\epsilon_{r,1}} - 1 + 1\right) = \frac{1}{A_{r,1}\epsilon_{r,1}}.$$

Then we have

$$\sigma_{SB} T_1^4 = E_{b,3} + \dot{S}_{e,J} R_{r,\Sigma}.$$

Solving for $T_1$, we have

$$T_1 = \left(\frac{E_{b,3} + \dot{S}_{e,J} R_{e,\Sigma}}{\sigma_{SB}}\right)^{1/4}.$$

Next, we use (4.68) and (4.69) again, but this time we solve for $T_H$, i.e.,

$$T_H = T_1 + \dot{S}_{e,J} R_{k,H\text{-}1},$$

where $R_{k,H\text{-}1}$ is for a cylindrical shell and is given in Table 3.2 as

$$R_{k,H\text{-}1} = \frac{\ln(D_1/D_H)}{2\pi L k_1} \quad \text{Table 3.2.}$$

(a) Physical Model of Radiation Cooling of a Wire with Joule Heating    (b) Thermal Circuit Model

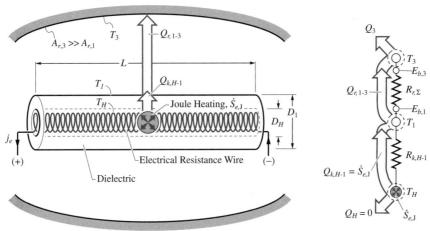

Figure Ex. 4.12. An electrical heating element radiating to its surroundings. (a) Physical model. (b) Thermal circuit diagram ($Q_H = 0$ due to symmetry).

Now substituting for $T_1$ and $R_{k,H\text{-}1}$, in the above expression for $T_H$, we have

$$
T_H = \left[ \frac{E_{b,3} + \dot{S}_{e,J} \dfrac{1}{A_{r,1}\epsilon_{r,1}}}{\sigma_{SB}} \right]^{1/4} + \dot{S}_{e,J} \frac{\ln(D_1/D_H)}{2\pi L k_1}.
$$

(c) The difference in temperature, $T_H - T_1$, is found from the first equation written above, i.e.,

$$
T_H - T_1 = \dot{S}_{e,J} R_{k,H\text{-}1}.
$$

Now using the numerical values, we have

$$
T_H = \left\{ \frac{5.670 \times 10^{-8}(\text{W/m}^2\text{-K}^4) \times (293.15)^4 (\text{K})^4}{5.670 \times 10^{-8}(\text{W/m}^2\text{-K}^4)} \right.
$$

$$
+ \frac{5 \times 10^2(\text{W}) \left[ \dfrac{1}{\pi \times 0.01(\text{m}) \times 0.3(\text{m}) \times 0.9} \right]}{5.670 \times 10^{-8}(\text{W/m}^2\text{-K}^4)} \Bigg\}^{1/4}
$$

$$
+ 5 \times 10^2(\text{W}) \frac{\ln(0.01/0.002)}{2\pi \times 0.3(\text{m}) \times 1.5(\text{W/m-K})}
$$

$$
= \left[ \frac{4.187 \times 10^2 + 5 \times 10^2 \times 117.9(1/\text{m}^2)}{5.670 \times 10^{-8}(\text{W/m}^2\text{-K}^4)} \right]^{1/4} (\text{K})
$$

$$
+ 5 \times 10^2(\text{W}) \times 0.5691(\text{K/W})
$$

or

$$T_H = \left(\frac{4.187 \times 10^2 + 5.895 \times 10^4}{5.670 \times 10^{-8}}\right)^{1/4} (\text{K}) + (5 \times 10^2 \times 0.5691)(\text{K})$$

$$= 1.011 \times 10^3 (\text{K}) + 2.846 \times 10^2 (\text{K}) = 1,296 \text{ K}.$$

Solving for the temperature difference, we have

$$T_H - T_1 = 5 \times 10^2 (\text{W}) \times 0.5691 (\text{K/W}) = 284.6 \text{ K}$$

or

$$T_1 = 1,011 \text{ K}.$$

**COMMENT**
Note that the temperature-drop ratio is given by (4.74) and $N_r < 0.1$ is described by (4.75), i.e.,

$$\frac{T_H - T_1}{T_1 - T_3} = N_r \equiv \frac{R_{k,H\text{-}1}}{R_{r,\Sigma}/\left(4\sigma_{SB}T_m^3\right)} = \frac{284.6}{1,011 - 293.15} = 0.396 > 0.1,$$

i.e., conduction-radiation number $N_r$ is not negligibly small. This supports our treatment, allowing for a nonuniform temperature in the dielectric substrate. Also, note the high heating element temperature $T_H$ and the need for a high melting temperature material.

---

EXAMPLE 4.13. FAM

---

A surface-radiation burner is used to melt solid glass particles (similar to sand particles) for bottle glassmaking. This is shown in Figure Ex. 4.13(a). In order to have a nearly uniform radiation surface temperature $T_1$, the fuel (natural gas) is provided to a combustion chamber in a distributed manner (as compared to using a premixed air-fuel mixture), as shown in the figure. Since the objective is to deliver the heat released from the combustion to the radiating surface, the paths for other heat transfer from the combustion chamber are made substantially more resistive. A large fraction of the heat produced by combustion $\dot{S}_{r,c}$, i.e., $a_1\dot{S}_{r,c}$, flows to the radiating surface (i.e, $a_1$ is close to unity). For a constant combustion gas stream, $Q_{u,1} = Q_{u,2}$. We assume a steady-state heat transfer.

The radiating surface, the sand particles, and the remaining surroundings all have emissivities of unity (blackbody surfaces).

(a) Draw the thermal circuit diagram.
(b) For the conditions given below, determine $T_1$.
(c) Determine $Q_{r,1\text{-}2}$.
(d) Determine $Q_{r,1\text{-}\infty}$.

(a) Physical Model of Surface-Radiation Burner

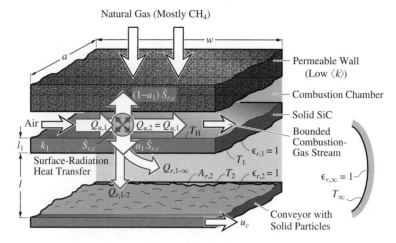

(b) Thermal Circuit Model

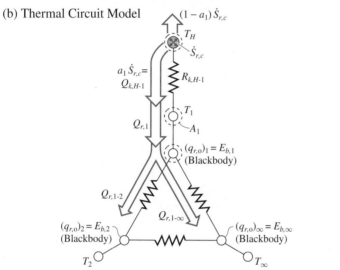

Figure Ex. 4.13. (a) A surface-radiation burner. (b) The thermal circuit model.

(e) Determine $T_H$.

$a_1 \dot{S}_{r,c} = 10$ kW, $w = a = 30$ cm, $l = 5$ cm, $k_1 = 40$ W/m-K, $l_1 = 3$ mm, $T_2 = 500°$C, $T_\infty = 300°$C.

**SOLUTION**

(a) The thermal circuit diagram is shown in Figure Ex. 4.13(b). From this diagram, for the solid ceramic layer (SiC) and for the radiating surface, we have the energy equation

$$Q_{k,\text{H-1}} = \frac{T_H - T_1}{R_{k,\text{H-1}}} = a_1 \dot{S}_{r,c} \quad \text{ceramic-layer conduction resistance}$$

$$Q_{r,1} = Q_{r,1\text{-}2} + Q_{r,1\text{-}\infty} = a_1 \dot{S}_{r,c} \quad \text{radiating surface energy equation.}$$

(b) To determine $T_1$, we need to write $Q_{r,1\text{-}2}$ and $Q_{r,1\text{-}\infty}$ in terms of the temperatures. Because we have all blackbody surfaces, all the surface-grayness resistances vanish. So, all the surface radiosities and emissive powers are equal [this is determined from (4.41)]. Then we use (4.40) for surface 1, and replace the $(q_{r,o})$'s with $E_b$'s. The result is

$$Q_{r,1} = Q_{r,1\text{-}2} + Q_{r,1\text{-}\infty} = \frac{E_{b,1} - E_{b,2}}{\dfrac{1}{A_{r,1}F_{1\text{-}2}}} + \frac{E_{b,1} - E_{b,\infty}}{\dfrac{1}{A_{r,1}F_{1\text{-}\infty}}}.$$

Then the radiating surface energy equation becomes

$$A_{r,1}\sigma_{\text{SB}}[F_{1\text{-}2}(T_1^4 - T_2^4) + F_{1\text{-}\infty}(T_1^4 - T_\infty^4)] = a_1\dot{S}_{r,c}.$$

Solving for $T_1$, we have

$$T_1 = \left[\frac{a_1\dot{S}_{r,c} + A_{r,1}\sigma_{\text{SB}}(F_{1\text{-}2}T_2^4 + F_{1\text{-}\infty}T_\infty^4)}{A_{r,1}\sigma_{\text{SB}}(F_{1\text{-}2} + F_{1\text{-}\infty})}\right]^{1/4}.$$

The view factors $F_{1\text{-}2}$ and $F_{1\text{-}\infty}$ are related through the summation rule (4.37), i.e.,

$$F_{1\text{-}2} + F_{1\text{-}\infty} = 1, \text{ or } F_{1\text{-}\infty} = 1 - F_{1\text{-}2}.$$

The view factor $F_{1\text{-}2}$ is determined from Figure 4.11(b). The input parameters are

$$a^* = \frac{a}{l} \quad \text{and} \quad w^* = \frac{w}{l}.$$

The numerical values are

$$a^* = \frac{30(a)}{5(a)} = 6$$

$$w^* = \frac{30(a)}{5(a)} = 6$$

$$F_{1\text{-}2} \simeq 0.76 \quad \text{Figure 4.11(b)}$$

$$F_{1\text{-}\infty} = 1 - 0.76 = 0.24.$$

Then

$$T_1 = \left\{ \frac{10^4(\text{W}) + 0.3(\text{m}) \times 0.3(\text{m}) \times 5.67 \times 10^{-8}(\text{W/m}^2\text{-K}^4)}{0.3(\text{m}) \times 0.3(\text{m}) \times 5.67 \times 10^{-8}(\text{W/m}^2\text{-K}^4) \times (0.76 + 0.24)} \right.$$

$$\left. \times \frac{[0.76 \times (773.15)^4(\text{K}^4) + 0.24 \times (637.15)^2(\text{K}^4)]}{1} \right\}^{1/4}$$

$$= \left[\frac{10^4 + 5.103 \times 10^{-9}(2.716 \times 10^{11} + 4.928 \times 10^{10})}{5.103 \times 10^{-9}}\right]^{1/4}$$

$$= 1{,}229 \text{ K.}$$

(c) From the energy equation, we have

$$Q_{r,1\text{-}2} = \frac{E_{b,1} - E_{b,2}}{\dfrac{1}{A_{r,1}F_{1\text{-}2}}} = A_{r,1}F_{1\text{-}2}(E_{b,1} - E_{b,2})$$

$$= A_{r,1}F_{1\text{-}2}\sigma_{\text{SB}}(T_1^4 - T_2^4)$$

$$= 0.3(\text{m}) \times 0.3(\text{m}) \times 0.76 \times 5.67 \times 10^{-8}(\text{W/m}^2\text{-K}^4)$$

$$\times \; [(1{,}229)^4(\text{K}^4) - (773.15)^4(\text{K}^4)]$$

$$= 7.459 \times 10^3 \;\; \text{W}.$$

(d) From above

$$Q_{r,1\text{-}\infty} = a_1\dot{S}_{r,c} - Q_{r,1\text{-}2}$$

$$= 10^4(\text{W}) - 7.459 \times 10^3(\text{W}) = 2.541 \; \times 10^3 \text{ W}.$$

(e) From the conduction resistance in the ceramic layer, given above, we have

$$\frac{T_H - T_1}{R_{k,H\text{-}1}} = a_1\dot{S}_{r,c}.$$

From Table 3.2, for the planar slab, we have

$$R_{k,H\text{-}1} = \frac{l_1}{A_k k_1} \quad \text{Table 3.2.}$$

Solving for $T_H$,

$$T_H = T_1 + \frac{(a_1\dot{S}_{r,c})l_1}{A_k k_1},$$

where $A_k = A_{r,1}$. Using the numerical values, we have

$$T_H = 1{,}229\text{ K} + \frac{10^4(\text{W}) \times 3 \times 10^{-3}(\text{m})}{0.3(\text{m}) \times 0.3(\text{m}) \times 40(\text{W/m-K})} = 1{,}237\text{ K}.$$

**COMMENT**

The radiating surface temperature $T_1$ is rather high and as higher radiant power [e.g., $Q_{r,1\text{-}2}/A_{r,1} = 7.459 \times 10^3(\text{W})/0.09(\text{m}^2) = 1.111 \times 10^5$ W/m$^2$] may be demanded, even higher $T_1$ is needed.

The combustion surface temperature $T_H$ is only slightly higher than $T_1$. Then from (4.74 and 4.75), $N_r = 4\sigma_{\text{SB}}T_m^3 l_1/k_1 = (T_H - T_1)/(T_1 - T_2) = 0.01220 < 0.1$. Here we could have justifiably neglected substrate conduction.

Also, 25% of the radiant heat is to the surroundings. This can be used to preheat the sand particles and/or the air-fuel mixture or, if practical, can be reduced by bringing the radiant surface closer to the conveyer.

(a) Thermal Circuit Model

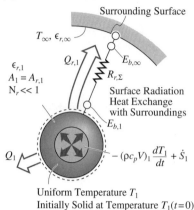

(b) Temperature Variation with
Respect to Time for $\dot{S}_1 - Q_1 = 0$

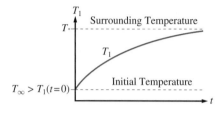

Uniform Temperature $T_1$
Initially Solid at Temperature $T_1(t=0)$

Figure 4.23. Transient heat transfer from a fluid by surface-radiation. A uniform temperature is assumed within the solid. (a) Physical model. (b) Thermal circuit diagram.

### 4.6.2 Transient Heat Transfer: Lumped Capacitance for $N_r < 0.1$

When the conduction-radiation parameter $N_r$ is small, i.e., $N_r < 0.1$, then the temperature variation within the substrate is negligible compared to that occurring outside the solid. Then, in a manner similar to that discussed in Section 3.6, where the temperature nonuniformity within the solid was neglected, we perform a lumped-capacitance (as compared to distributed-capacitance) analysis. For a solid undergoing a temporal change in the temperature and a surface-radiation heat transfer, the integral-volume energy equation (2.9) for constant $\rho$ and $V$ becomes

$$Q|_{A,1} = \int_{A,1} (\mathbf{q}_r \cdot \mathbf{s}_n) dA = -(\rho c_p V)_1 \frac{d}{dt} T_1(t) + \dot{S}_1 \quad \text{for } N_r < 0.1$$

integral-volume energy equation for volume node $T_1$,        (4.76)

where $T_1 = T_1(t)$. Here, for simplicity, we allow only for surface-radiation heat transfer between the solid surface at temperature $T_1$ and its surrounding at temperature $T_\infty$, as shown in Figure 4.23(a). Then for constant $\rho c_p$, the energy equation (4.76) becomes (for no prescribed surface heat transfer, i.e., $Q_1 = 0$)

$$Q\,|_{A,1} = Q_1 + Q_{r,1} = Q_{r,1} = -(\rho c_p V)_1 \frac{dT_1(t)}{dt} + \dot{S}_1, \quad (4.77)$$

where the surface heat transfer by radiation $Q_{r,1}$ is given by (4.74), i.e.,

$$Q\,|_{A,1} = Q_{r,1} = \frac{E_{b,1} - E_{b,\infty}}{(R_{r,\epsilon})_1 + (R_{r,F})_{1-\infty} + (R_{r,\epsilon})_\infty} \equiv \frac{E_{b,1} - E_{b,\infty}}{R_{r,\Sigma}}, \quad (4.78)$$

where $R_{r,\Sigma}$ is the overall resistance (for a serial arrangement of resistances). Note that $R_{r,\Sigma}$ is the sum of the three (two surface-grayness and one view-factor) radiation resistances. The energy equation (4.77) is now written using (4.78) for $Q|_{A,1}$, i.e.,

$$Q_{r,1} = \frac{\sigma_{SB}}{R_{r,\Sigma}}[T_1^4(t) - T_\infty^4] = -(\rho c_p V)_1 \frac{dT_1(t)}{dt} + \dot{S}_1. \tag{4.79}$$

We now consider the case of $\dot{S}_1 = 0$, in order to obtain a closed-form solution. The variables are separated and we have

$$\frac{\sigma_{SB}}{R_{r,\Sigma}(\rho c_p V)_1} dt = \frac{dT_1(t)}{T_\infty^4 - T_1^4(t)}. \tag{4.80}$$

The initial thermal condition for the solid is the prescribed initial temperature $T_1(t = 0)$. Then we have

$$\frac{\sigma_{SB}}{R_{r,\Sigma}(\rho c_p V)_1} \int_0^t dt = \int_{T_1(t=0)}^{T_1} \frac{dT_1(t)}{T_\infty^4 - T_1^4(t)}. \tag{4.81}$$

The result of the integration is

$$\frac{4\sigma_{SB} T_\infty^3}{(\rho c_p V)_1 R_{r,\Sigma}} t \equiv \frac{t}{\tau_1} = \left( \ln \left| \frac{T_\infty + T_1}{T_\infty - T_1} \right| + 2\tan^{-1} \frac{T_1}{T_\infty} \right) \Bigg|_{T_1(t=0)}^{T_1}$$

$$= \left[ \ln \left| \frac{T_\infty + T_1}{T_\infty - T_1} \right| - \ln \left| \frac{T_\infty + T_1(t=0)}{T_\infty - T_1(t=0)} \right| + \right.$$

$$\left. 2\tan^{-1} \frac{T_1}{T_\infty} - 2\tan^{-1} \frac{T_1(t=0)}{T_\infty} \right] \quad \begin{array}{l} \text{for } T_\infty \neq 0, \text{ transient nodal} \\ \text{temperature for } \dot{S}_1 - Q_1 = 0, \\ \tau_1 = \dfrac{(\rho c_p V)_1 R_{r,\Sigma}}{4\sigma_{SB} T_\infty^3}. \end{array}$$

$$\tag{4.82}$$

Figure 4.23(b) shows the anticipated temporal variation of $T_1$, for $T_\infty > T_1(t = 0)$. After a large elapsed time $t$, the solid temperature reaches the surrounding temperature $T_{inf}$.

Note that $(\rho c_p V)_1 R_{r,\Sigma}/4\sigma_{SB} T_\infty^3$ is the time constant $\tau_1(s)$, similar to (3.172). As $(\rho c_p V)_1 R_{r,\Sigma}$ decreases and $T_\infty^3$ increases, $\tau_1$ will decrease. This equation can be used to explicitly solve for the elapsed time $t$ required to reach a desired temperature $T_1(t)$. However, it cannot be explicitly solved for a $T_1(t)$ given an elapsed time $t$ [therefore, it is implicit $T_1(t)$]. If $T_1(t)$ is an unknown, this implicit equation for $T_1(t)$ is solved numerically (i.e., through iteration and/or using solver).

### EXAMPLE 4.14. FAM

In extraterrestrial applications (e.g., in the space station), radiation cooling of liquids is made by forming droplets and injecting the droplets into open space (negligible surface-convection heat transfer), and after a short travel period collecting them. This is shown in Figure Ex. 4.14(a). We use water droplets of diameter $D = 2$ mm, emissivity $\epsilon_{r,1} = 0.8$, initial temperature $T_1(t = 0) = 50°C$

($T_1$ is assumed uniform throughout the droplet, $N_k < 0.1$), and a blackbody surrounding at $T_\infty = 3$ K.

(a) Draw the thermal circuit diagram.
(b) Determine the elapsed time required to cool the droplet to $T_1 = 20°C$.

Use the saturated water thermodynamic properties from Table C.23 and at $T = 313$ K.

**SOLUTION**
(a) The thermal circuit diagram for this problem is given in Figure Ex. 4.14(b).
(b) From (4.82), we have

$$
t = \frac{R_{r,\Sigma}(\rho c_p V)_1}{4\sigma_{SB} T_\infty^3} \left[ \ln \left| \frac{T_\infty + T_1}{T_\infty - T_1} \right| - \ln \left| \frac{T_\infty + T_1(t=0)}{T_\infty - T_1(t=0)} \right| \right.
$$
$$
\left. + 2\tan^{-1} \frac{T_1}{T_\infty} - 2\tan^{-1} \frac{T_1(t=0)}{T_\infty} \right],
$$

(a) Physical Model

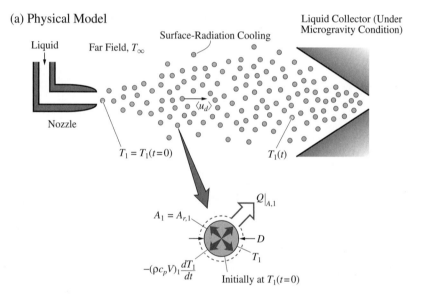

(b) Thermal Circuit Model

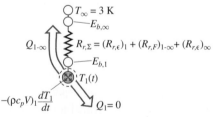

Figure Ex. 4.14. (a) Droplet cooling by surface-radiation in microgravity environment. (b) Thermal circuit model.

where

$$R_{r,\Sigma} = (R_{r,\epsilon})_1 + (R_{r,F})_{1-\infty} + (R_{r,\epsilon})_\infty.$$

The overall radiation resistance is

$$R_{r,\Sigma} = \left(\frac{1-\epsilon_r}{A_r \epsilon_r}\right)_1 + \frac{1}{A_{r,1} F_{1-\infty}} + \left(\frac{1-\epsilon_r}{A_r \epsilon_r}\right)_\infty$$

$$= \left(\frac{1-\epsilon_r}{A_r \epsilon_r}\right)_1 + \frac{1}{A_{r,1}} + 0$$

$$= \frac{1}{A_{r,1}\epsilon_{r,1}},$$

because $\epsilon_{r,\infty} = 1$ and $F_{1-\infty} = 1$ (neglecting the droplet-droplet and droplet-solid surface view factors).

We also note

$$A_{r,1} = 4\pi R^2 = \pi D^2, \quad V_1 = \frac{4}{3}\pi R^3 = \frac{1}{6}\pi D^3.$$

The thermodynamic properties for water at $T_1 = 313$ K are

$$\rho_1 = 994.5 \text{ kg/m}^3, \quad c_{p,1} = 4.178 \times 10^3 \text{ J/kg-K} \quad \text{Table C.23}$$

We now use the numerical values where

$$A_{r,1} = \pi \times (2 \times 10^{-3})^2 (\text{m})^2 = 1.257 \times 10^{-5} \text{ m}^2$$

$$V_1 = \pi \times \frac{(2 \times 10^{-3})^3 (\text{m})^3}{6} = 4.189 \times 10^{-9} \text{ m}^3$$

$$R_{r,\Sigma} = \frac{1}{A_{r,1}\epsilon_{r,1}} = \frac{1}{1.257 \times 10^{-5}(\text{m}^2) \times 0.8} = 9.944 \times 10^4 \text{ 1/m}^2,$$

to determine $t$ as

$$t = \frac{9.944 \times 10^4 (1/\text{m}^2) \times 994.5(\text{kg/m}^3) \times 4.178 \times 10^3 (\text{J/kg-K}) \times 4.189 \times 10^{-9}(\text{m}^3)}{4 \times 5.670 \times 10^{-8}(\text{W/m}^2\text{-K}^4) \times (3)^3(\text{K})^3}$$

$$\times \left[ \ln\left|\frac{3(\text{K}) + 293.15(\text{K})}{3(\text{K}) - 293.15(\text{K})}\right| - \ln\left|\frac{3(\text{K}) + 323.15(\text{K})}{3(\text{K}) - 323.15(\text{K})}\right| \right.$$

$$\left. + 2\tan^{-1}\frac{293.15(\text{K})}{3(\text{K})} - 2\tan^{-1}\frac{323.15(\text{K})}{3(\text{K})} \right]$$

$$= 9.421 \times 10^7 (\text{s}) \times (2.0468052 \times 10^{-2} - 1.8567762$$

$$\times 10^{-2} + 3.1211260 - 3.1230260)$$

$$t = 9.421 \times 10^7 (\text{s}) \times 2.9000000 \times 10^{-7} = 27.31 \text{ s}.$$

**COMMENT**

Since a rather low temperature $T_1$ is desired, this is a long elapsed time. If $T_1 > 20°C$ is chosen, this time will be reduced. Also note that many significant figures must be used in the above arithmetic manipulations, because we are subtracting small and large numbers.

## 4.7 Summary

We have divided the surfaces into gray (radiation properties independent of the wavelength) and nongray (wavelength-dependent properties).

For gray, diffuse, and opaque surfaces making up an enclosure, we have used the concepts of radiosity and irradiation to express the surface-radiation heat transfer rate $Q_r$ and developed the surface-grayness resistance $R_{r,\epsilon}$ and the view-factor resistance $R_{r,F}$. Then depending on the number of surfaces present, various simplifications have been made and are summarized in Chart 4.4. For blackbody surfaces and two-surface enclosures, and also for three-surface enclosures with the third surface having $\dot{S} = Q = 0$, we have expressed the surface-radiation heat transfer $Q_{r,i-j}$ in terms of $E_{b,i} - E_{b,j}$. The overall radiation resistance $R_{r,\Sigma}$ is the sum of $R_{r,\epsilon}$'s and $R_{r,F}$'s. For a larger number of surfaces, we have used the radiosity and the associated equations.

For nongray surfaces and when there is a prescribed irradiation, we have included the surface and volumetric radiation as surface energy conversions.

Finally, the substrate heat transfer and energy storage and conversions have been included. For a small radiation-conduction parameter, we can treat the substrate as having a uniform temperature. Otherwise, we have included the conduction within the substrate.

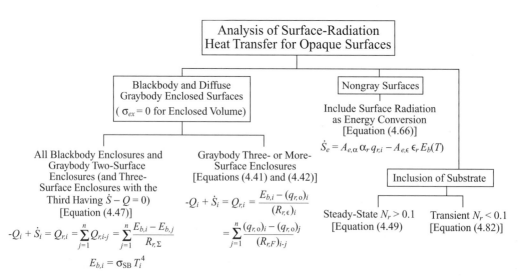

Chart 4.4. Summary of surface-radiation treatments considered.

## 4.8 References*

This section is found on the Web site www.cambridge.org/kaviany.

## 4.9 Problems

### 4.9.1 Volumetric and Surface-Radiation Fundamentals

**PROBLEM 4.1. FUN**

A piece of polished iron, with a surface area $A_r = 1$ m$^2$, is heated to a temperature $T_s = 1,100°$C.

(a) Determine the maximum amount of thermal radiation this surface can emit.
(b) Determine the actual amount of thermal radiation this surface emits. Interpolate the total emissivity from the values listed in Table C.18.
(c) What fraction of the radiation energy emitted is in the visible ($\lambda$ between 0.39 and 0.77 $\mu$m), near infrared ($\lambda$ between 0.77 and 25 $\mu$m), and far infrared ($\lambda$ between 25 and 1,000 $\mu$m) ranges of the electromagnetic spectrum?

**PROBLEM 4.2. FUN**

A blackbody radiation source at $T_i = 500°$C is used to irradiate three different surfaces, namely, (i) aluminum (commercial sheet), (ii) nickel oxide, and (iii) paper. The irradiating surfaces have an area $A_r = 1$ m$^2$ and are assumed gray, diffuse, and opaque. Use the emissivities in Table C.18 [for (ii) extrapolate; for others use the available data].

(a) Sketch the radiation heat transfer arriving and leaving the surface, showing the blackbody emitter and the irradiated surface [see Figure 4.9(b)]. Show heat transfer as irradiation $Q_{r,i}$, absorption $Q_{r,\alpha}$, reflection $Q_{r,\rho}$, emission $Q_{r,\epsilon}$, and radiosity $Q_{r,o}$ (sum of emission and reflection).
(b) If the three surfaces are kept at $T_s = 100°$C, determine the amounts of reflected and absorbed energies.
(c) Determine the rates of energy emitted and the radiosity for each surface.
(d) Determine the net radiation heat transfer rate for each surface. Which surface experiences the highest amount of radiation heating?

**PROBLEM 4.3. FUN**

Three different surfaces are heated to a temperature $T_s = 800°$C. The total radiation heat flux leaving these surfaces (i.e., the radiosity) is measured with a calorimeter and the values $(q_{r,o})_1 = 6,760$ W/m$^2$, $(q_{r,o})_2 = 21,800$ W/m$^2$, and $(q_{r,o})_3 = 48,850$ W/m$^2$ are recorded.

Assume that the reflected radiation is negligible compared to surface emission and that the surfaces are opaque, diffuse, and gray.

(a) Determine the total emissivity for each of these surfaces.
(b) Comment on the importance of considering the surface reflection in the measurement of surface emissivity $\epsilon_r$.

**PROBLEM 4.4. FAM**

In an incandescent lamp, the electrical energy is converted to the Joule heating in the thin-wire filament and this is in turn converted to thermal radiation emission. The filament is at $T = 2{,}900$ K, and behaves as an opaque, diffuse, and gray surface with a total emissivity $\epsilon_r = 0.8$.

Determine the fractions of the total radiant energy and the amount of emitted energy in the (a) visible, (b) near infrared, and (c) remaining ranges of the electromagnetic spectrum.

**PROBLEM 4.5. FUN**

The equation of radiative transfer describes the change in the radiation intensity $I_r$ as it experiences local scattering, absorption, and emission by molecules or larger particles (a medium made of a continuum with dispersed particles). The equation of radiation transfer can be derived from the Boltzmann transport equation, with the inter-photon interactions ignored. In the absence of a significant local emission, and by combining the effects of scattering and absorption into a single volumetric radiation property, i.e., the extinction coefficient $\sigma_{ex}$ (1/m), we can describe the ability of the medium to attenuate radiation transport across it. Under this condition, we can write the equation of radiative transport for a one-dimensional volumetric radiation heat transfer as

$$\frac{dI_\lambda}{dx} = -\sigma_{ex}I_\lambda \quad \text{radiative transport for nonemitting media}$$

or by integrating this over all the wavelength and solid angles (as indicated in the second footnote of Section 4.1.2), we have arrive at the radiation heat flux $q_r$

$$\frac{dq_r}{dx} = -\sigma_{ex}q_r.$$

(a) Integrate this equation of radiative transfer, using $q_r(x = 0) = q_{r,i}(1 - \rho_r)$, where $\rho_r$ is the surface reflectivity, as shown in Figure Pr. 4.5, and show that

$$q_r(x = 0) = q_{r,i}(1 - \rho_r)e^{-\sigma_{ex}x}.$$

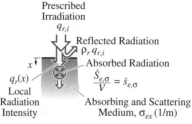

Figure Pr. 4.5. Radiation attenuating (absorbing and scattering) medium with a surface $(x = 0)$ reflection.

(b) Starting from (2.1), and assuming one-dimensional, volumetric radiation heat transfer only, show that

$$q_{r,i}(1 - \rho_r)\sigma_{ex}e^{-\sigma_{ex}x} = -\dot{s}_{e,\sigma} = -\frac{\dot{S}_{e,\sigma}}{V},$$

which is also given by (2.43), when we note that the attenuation of radiation is represented by $\dot{s}_{e,\sigma}$ as a source of energy.

**PROBLEM 4.6. FUN**

When the optical thickness defined by (2.44), $\sigma_{ex}^* = \sigma_{ex}L$, for a heat transfer medium of thickness $L$ and extinction coefficient $\sigma_{ex}$ is larger than 10, the emission and transfer of radiation can be given by the radiation heat flux as (for a one-dimensional heat flow)

$$q_{r,x} = -\frac{16}{3}\frac{\sigma_{SB}T^3}{\sigma_{ex}}\frac{dT}{dx}$$

diffusion approximation for optically thick ($\sigma_{ex}^* > 10$) heat transfer media.

This is called the diffusion approximation.

The equation of radiation (or radiactive) transfer, for an emitting medium with a strong absorption, becomes

$$\frac{dI_{r,b}}{d(x/\cos\theta)} = -\sigma_{ex}I_r + \sigma_{ex}I_{r,b}, \quad \pi I_{r,b} = E_{r,b} = \sigma_{SB}T^4,$$

where $x/\cos\theta$ is the photon path as it travels between surfaces located at $x$ and $x + \Delta x/2$, as shown in Figure Pr. 4.6.

As discussed in Section 4.1.2, the radiation heat flux is found by the integration of $I_r$ over a unit sphere, i.e.,

$$\mathbf{q}_r = \int_0^{2\pi}\int_0^{\pi} \mathbf{s}I_r\cos\theta\sin\theta d\theta d\phi.$$

Note that this integral is over a complete sphere. Also note that $I_{r,b}$ is independent of $\theta$ and $\phi$. For the $x$ direction, using Figure Pr. 4.6, we have $q_{r,x} = \int_0^{2\pi}\int_0^{\pi} I_r\cos\theta\sin\theta d\theta d\phi$.

Using the equation of radiative transfer and the definition of $\mathbf{q}_r$, both given above, derive the above expression for $q_{r,x}$ for the diffusion approximation.

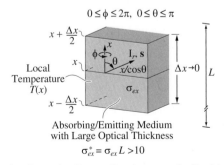

Figure Pr. 4.6. Radiation intensity $I_r$ traveling in an optically thick, emitting medium.

## PROBLEM 4.7. FAM

The human eye is sensitive to the visible range of the photon wavelength and has a threshold for detection of about $E_{b,\lambda} = 0.0936$ W/m$^2$-$\mu$m at wavelength of $\lambda = 0.77$ $\mu$m (largest wavelength in the red band, Figure 4.1). This corresponds to the Draper point in Figure 4.2(a).

(a) The turtle eye is sensitive to the infrared range and if the threshold for detection is the same, i.e., $E_{b,\lambda} = 0.0936$ W/m$^2$-$\mu$m, but at $\lambda = 1.5$ $\mu$m, determine the corresponding temperature at which the turtle can detect blackbody emission.
(b) Using this temperature, at what wavelength would $E_{b,\lambda}$ peak?
(c) Would this turtle eye be able to detect radiation emission from a tank containing liquid water at one atm pressure (Table C.3)?

## PROBLEM 4.8. FUN

Dielectrics, e.g., ceramics such as SiC, have very small spectral extinction index $\kappa_\lambda$ and also small spectral refraction index $n_\lambda$ (optical properties). On the other hand metals have large $\kappa_\lambda$ and $n_\lambda$. Figure Pr. 4.8 shows the interface between two media, 1 and 2, which have different optical properties. The higher is $n_\lambda$, the larger is the reflectivity and the larger is $\kappa_\lambda$, the lower is the transmissivity.

The normal (i.e., $\theta = 0$) emissivity, for the dielectrics and metals, is predicted using these optical properties and a simple relation is given by

$$\epsilon_{r,\lambda} = \frac{4\dfrac{n_{\lambda,2}}{n_{\lambda,1}}}{\left(\dfrac{n_{\lambda,2}}{n_{\lambda,1}} + 1\right)^2 + \kappa_{\lambda,2}^2},$$

where, as shown in Figure 4.6, 1 and 2 (or $i$ and $j$) refer to the two media and here we use air as media 1 (with $n_{\lambda,1} = 1$).

The measured values of $n_\lambda$ and $\kappa_\lambda$ at $\lambda = 5$ $\mu$m are given as

$$\text{SiC:} \quad n_{\lambda,2} = 2.4, \quad \kappa_{\lambda,2} = 0.07, \quad \lambda = 5 \ \mu\text{m}$$

$$\text{Al:} \quad n_{\lambda,2} = 9, \quad \kappa_{\lambda,2} = 65.0, \quad \lambda = 5 \ \mu\text{m}$$

$$\text{air:} \quad n_{\lambda,1} = 1.$$

(a) Determine the spectral hemispherical emissivity $\epsilon_{r,\lambda}$ for SiC and Al in contact with air.
(b) Compare these values with the measured values of total hemispherical emissivity $\epsilon_r$ given in Table C.18.

Air (Optical Properties: $n_{\lambda,1} = 1$)

Medium 1 — Surface Spectral Emissivity, $\epsilon_{r,\lambda}$

Interface

Medium 2

SiC or Al
(Optical Properties: $n_{\lambda,2}, \kappa_{\lambda,2}$)

Figure Pr. 4.8. Interface between two media with different spectral optical properties.

## PROBLEM 4.9. FAM

Some surface materials and coatings are selected for their radiation properties, i.e., their emissivity $\epsilon_r$ and absorptivity $\alpha_r$. Consider the following selections based on surface-radiation properties. All surfaces are at $T_s = 300$ K.

(i) Space suit ($\alpha_r$ for low heat absorption).

(ii) Solar collector surface ($\alpha_r$ for high heat absorption and $\epsilon_r$ for low heat emission).

(iii) Surface of thermos ($\alpha_r$ for low heat absorption).

(a) Choose the materials for the applications (i) to (iii) from Table C.19.

(b) Determine the emissive power for the selected surfaces (i) to (iii).

(c) Determine the surface reflectivity for the selected surfaces (i) to (iii), assuming no transmission (opaque surface).

### 4.9.2 View Factor

## PROBLEM 4.10. FAM

The view factors between two surfaces making up part of an enclosure are given for some geometries in Table 4.2 and Figures 4.11(a) to (e). Determine the view factors ($F_{i-j}$, with $i$ and $j$ specified for each case on top of the figures) for the five surface pairs shown in Figures Pr. 4.10(a) to (e).

(a) Inside Surface of Cylinder (Excluding Top Surface) to Top Surface: $F_{1-2}$

(b) Inside, Side Surfaces of Rectangle to Itself: $F_{1-1}$

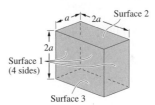

(c) Vertical Side of Right Angle Wedge to Its Horizontal Side: $F_{2-1}$

(d) Surface of Inner Cylinder to Top Opening of Annulus: $F_{1-3}$

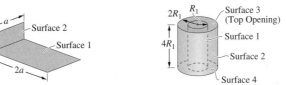

(e) Surface of a Sphere Near a Coaxial Disk to Rest of Its Surroundings: $F_{2-3}$

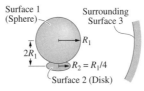

Figure Pr. 4.10. (a) to (e) View factors between surface pairs for five different surface pairs in different geometries.

Note that, for the geometry shown in Figure Pr. 4.10(a), the view factor can be found using only the summation and the reciprocity rules (4.33) and (4.34), and by using simple inspection (i.e., no tables or figures are needed) of the limiting view factor (i.e., surfaces that are completely enclosed by another surface).

### PROBLEM 4.11. FUN

Two planar surfaces having the same area $A = A_1 = A_2$ are to have three different geometries/arrangements, while having nearly the same view factor $F_{1-2}$. These are coaxial circular disks, coaxial square plates, and perpendicular square plates, and are shown in Figure Pr. 4.11.

(a) Determine this nearly equal view factor $F_{1-2}$ (shared among the three geometries).

(b) Under this requirement, are the disks or the plates placed closer together?

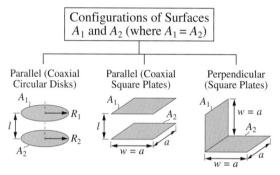

Figure Pr. 4.11. Two planar surfaces having the same area and three different geometries/arrangements.

### 4.9.3 Two-Surface, Opaque, Diffuse Gray Enclosures

### PROBLEM 4.12. FAM

The blackbody surface can be simulated using a large cavity (i.e., an enclosure with a small opening). The internal surfaces of the cavity have a total emissivity $\epsilon_{r,1}$ which is smaller than unity; however, due to the large cavity surface area, compared to its opening (i.e., mouth), the opening appears as a blackbody surface. To show this, consider the cylindrical enclosure shown in Figure Pr. 4.12. The surrounding is assumed to be a blackbody at $T_\infty$.

(a) Equate the net radiation heat transfer $Q_{r,1-\infty}$ from the cavity surface in Figure Pr. 4.12(i) to that in Figure Pr. 4.12(ii). In Figure Pr. 4.12(ii), we use the cavity opening area and an apparent emissivity $\epsilon_{r,1}'$. Then derive an expression for this apparent emissivity.

(b) Show that this apparent emissivity tends to unity, for $L \gg D$.

## (i) Radiation Exchange with Cavity Surface

Surrounding

$\epsilon_{r,\infty} = 1$

$T_\infty$

$A_{r,\infty} \gg A_{r,1'}$
$F_{1'-\infty} = 1$

$Q_{r,1-\infty}$

Cavity

Cavity Mouth $A_{r,1'}$

$D$

$L$

$T_1 \quad \epsilon_{r,1} \quad A_{r,1} = \pi DL + \pi D^2/4$

## (ii) Apparent Radiation Exchange with Cavity Mouth

Surrounding

$\epsilon_{r,\infty} = 1$

$T_\infty$

$A_{r,\infty} \gg A_{r,1'}$
$F_{1'-\infty} = 1$

$Q_{r,1'-\infty} = Q_{r,1-\infty}$

$A_{r,1'} = \pi D^2/4$
$T_{1'} = T_1$

Cavity Mouth

$D$

$\epsilon_{r,1}$

Figure Pr. 4.12. (i) Surface-radiation from a cavity. (ii) The concept of apparent emissivity for the cavity opening.

## PROBLEM 4.13. FAM

Liquid oxygen and hydrogen are used as fuel in space travel. The liquid is stored in a cryogenic tank, which behaves thermally like a thermos. Radiation shields (highly reflecting aluminum or gold foils) are placed over the tank to reduce irradiation to the tank surface (Figure Pr. 4.13). The surface of the tank has a total emissivity of $\epsilon_{r,1} = 0.7$ and the shields have an emissivity of $\epsilon_{r,s} = 0.05$. Consider placing one [Figure Pr. 4.13(i)] and two [Figure Pr. 4.13(ii)] radiation shields on the tank.

### (i) One Shield

### (ii) Two Shields

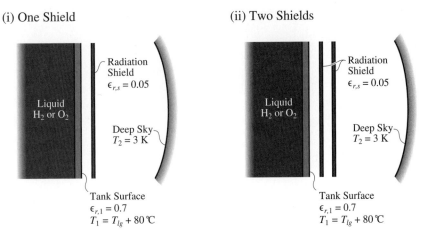

Figure Pr. 4.13. Surface-radiation heat transfer from a cryogenic liquid tank. (i) With one shield. (ii) With two shields.

(a) Draw the thermal circuit diagrams.

(b) Determine the rate of heat flowing out of the tank per unit area for liquid oxygen and liquid hydrogen.

Assume that the surface of the tank is $T_1 = 80°C$ above the saturation temperature of the liquid at one atm pressure, the tank is facing away from the sun, and the deep sky temperature is $T_2 = 3$ K.

## PROBLEM 4.14. DES

A single-junction thermocouple psychrometer is used to measure the relative humidity in air streams flowing through ducts, as shown in Figure Pr. 4.14. In the simplest design, the thermocouple psychrometer consists of a thermocouple bead, which is exposed to the humid air stream, connected simultaneously to a DC power source and a voltmeter. Initially, a voltage is applied to the thermocouple, causing a decrease in the bead temperature, due to the energy conversion from electromagnetic to thermal energy by the Peltier effect. This cooling causes the condensation of the water vapor and the formation of a liquid droplet on the thermocouple bead. When the temperature drops to temperature $T_{t,o}$, the power source is turned off and the voltage generated by the thermocouple is recorded. Since the droplet temperature is lower than the ambient temperature, the droplet receives heat from the ambient by surface convection and surface-radiation. This causes the evaporation of the droplet. The voltage measured between the thermocouple leads is related to the temperature of the thermocouple bead/water droplet. An equilibrium condition is reached when the net heat flow at the droplet surface balances with the energy conversion due to phase change. This equilibrium temperature is called the wet-bulb temperature for the air stream $T_{wb}$.

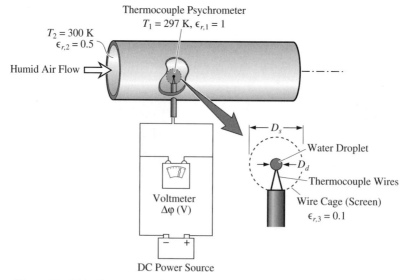

Figure Pr. 4.14. Thermocouple psychrometer with a screen radiation shield.

Figure Pr. 4.14 shows the thermocouple placed in the air stream. The duct diameter is much larger than the thermocouple bead. The water droplet has a diameter $D_d = 0.5$ mm and its surface is assumed to be a blackbody. The tube surface is opaque, diffuse, and gray and has a surface emissivity $\epsilon_{r,2} = 0.5$ and a temperature $T_2 = 300$ K. The evaporation rate of the water is estimated as $\dot{m}_{lg} = 0.00017$ kg/m$^2$-s.

(a) If the droplet temperature is $T_1 = 297$ K, determine the net heat transfer by surface-radiation between the bead and the tube surface and express it as a percentage of the energy conversion due to liquid-vapor phase change.

(b) If the bead is protected by a porous spherical wire cage with diameter $D_s = 3$ mm and the ratio between the open area and total area $a_1 = A_{void}/A_{total} = 0.7$, calculate the reduction in the net heat transfer by surface-radiation $\Delta Q_{r,1}$. The surface of the wires is opaque, diffuse, and gray, and has an emissivity $\epsilon_{r,3} = 0.1$. Using the available results for radiation between two surfaces separated by a screen, and for $A_{r,2} \gg A_{r,3} > A_{r,1}$, the overall radiation resistance is given by

$$(R_{r,\Sigma})_{1\text{-}2} = \frac{1 - \epsilon_{r,1}}{\epsilon_{r,1} A_{r,1}} + \cfrac{1}{a_1 A_{r,1} + \cfrac{1}{\cfrac{1}{A_{r,1}(1 - a_1)} + 2\left(\cfrac{1 - \epsilon_r}{A_r \epsilon_r}\right)_3}}.$$

(c) In order to reduce the amount of heat transfer between the droplet and the tube surface by surface-radiation, should we increase or decrease $a_1 = A_{void}/A_{total}$ and $\epsilon_{r,3}$?

## PROBLEM 4.15. DES
The polymer coating of an electrical wire is cured using infrared irradiation. The wire is drawn through a circular ceramic oven as shown in Figure Pr. 4.15. The polymer coating is thin and the drawing speed $u_w$ is sufficiently fast. Under these conditions, the wire remains at a constant and uniform temperature of $T_1 = 400$ K, while moving through the oven. The diameter of the wire is $d = 5$ mm and its surface is assumed opaque, diffuse, and gray with an emissivity $\epsilon_{r,1} = 0.9$. The oven wall is made of aluminum oxide (Table C.18), has a diameter $D = 20$ cm, and length $L = 1$ m, and its surface temperature is $T_2 = 600$ K. One of the ends of the furnace is closed by a ceramic plate with a surface temperature $T_3 = 600$ K and a surface emissivity

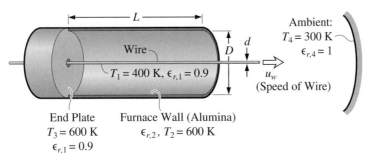

Figure Pr. 4.15. A wire drawn through an oven.

$\epsilon_{r,3} = 0.5$. The other end is open to the ambient, which behaves as a blackbody surface with $T_4 = 300$ K. Ignore the heat transfer by surface convection.

(a) Draw the thermal circuit diagram for the four-surface-radiation enclosure and write all of the relations for determination of the net heat transfer by radiation to the wire surface $Q_{r,1}$.

(b) Assuming that the wire exchanges heat by radiation with the tube furnace surface only (i.e., a two-surface enclosure), calculate the net heat transfer by surface-radiation to the wire surface $Q_{r,1}$.

(c) Explain under what conditions the assumption made on item (b) can be used. Does the net heat transfer by surface-radiation at the wire surface increase or decrease with an increase in the furnace diameter $D$ (all the other conditions remaining the same)? Explain your answer.

## PROBLEM 4.16. FUN

As in the application of radiation shields discussed in Section 4.4.5, there are applications where the surface-radiation through multiple (thin, opaque solid) layers (or solid slabs) is of interest. This is rendered in Figure Pr. 4.16, where these layers are placed between surfaces maintained at $T_h$ and $T_c$. Since for large $N$ (number of layers) the local radiation heat transfer becomes independent of the presence of the far-away layers, we can then use the local (or diffusion) approximation of radiation heat transfer and use the temperature difference (or local temperature gradient) between adjacent layers and write

$$q_{r,x} \equiv -\langle k_r \rangle \frac{dT}{dx} = -\langle k_r \rangle \frac{T_2 - T_1}{l} = -\langle k_r \rangle \frac{T_c - T_h}{L},$$

where $l$ is the spacing between adjacent layers. This radiant conductivity is

$$\langle k_r \rangle = \frac{4\epsilon_e \sigma_{SB} T^3 l}{2 - \epsilon_r}, \quad T = \left[ \frac{(T_1^2 + T_2^2)(T_1 + T_2)}{4} \right]^{1/3}.$$

| (i) Surface-Radiation in Multiple Parallel Layers (Zero Thickness) and Its Representation by Radiant Conductivity $\langle k_r \rangle$ | (ii) Thermal Circuit Model Using Radiant Conductivity $\langle k_r \rangle$ |

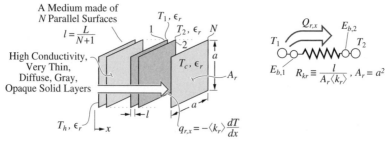

Figure Pr. 4.16. (i) Surface-radiation in a multilayer (each layer opaque) system. (ii) Its thermal circuit representation by radiant conductivity.

(a) Start from (4.47) for radiation between surfaces 1 and 2 and assume that $l \ll a$, such that $F_{1\text{-}2} = 1$. Then show that

$$q_{r,1\text{-}2} = \frac{\epsilon_r \sigma_{SB}(T_1^4 - T_2^4)}{2 - \epsilon_r}.$$

(b) Use that linearization of (4.72) to show that

$$q_{r,1\text{-}2} = \frac{4\epsilon_r \sigma_{SB} T^3(T_1 - T_2)}{2 - \epsilon_r} \quad \text{for } T_1 \to T_2,$$

i.e., for small diminishing difference between $T_1$ and $T_2$.

(c) Then using the definition of $q_{r,x}$ given above, derive the above expression for radiant conductivity $\langle k_r \rangle$.

## PROBLEM 4.17. FAM

A short, one-side closed cylindrical tube is used as a surface-radiation source, as shown in Figure Pr. 4.17. The surface (including the cylindrical tube and the circular closed end) is ideally insulated on its outside surface and is uniformly heated by Joule energy conversion, resulting in a uniform inner surface temperature $T_1 = 800°C$. The heat transfer from the internal surface to the surroundings is by surface-radiation only.

(a) Draw the thermal circuit diagram.
(b) Determine the required Joule heating rate $\dot{S}_{e,J}$.
    $T_\infty = 100°C$, $\epsilon_{r,1} = 0.9$, $D = 15$ cm, $L = 15$ cm.

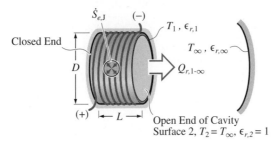

Figure Pr. 4.17. surface-radiation from a one-end closed cavity, to its surroundings.

## PROBLEM 4.18. FAM

A cylindrical piece of wood (length $L$ and diameter $D$) is burning in an oven as shown in Figure Pr. 4.18. The wood can be assumed to be in the central region of the cube furnace (with each oven side having length $a$). The internal oven surface temperature is $T_2$. The burning rate is $\dot{M}_{r,c}$, and the heat of combustion is $\Delta h_{r,c}$. Assume that the only surface heat transfer from the wood is by steady-state radiation.

(a) Draw the thermal circuit diagram.
(b) Determine the wood surface temperature $T_1$.
(c) What would $T_1$ be if $T_2$ were lowered by 80°C?

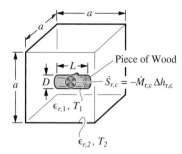

Figure Pr. 4.18. A cylindrical piece of wood burning in an oven.

$T_2 = 80°C$, $\dot{M}_{r,c} = 2.9 \times 10^{-4}$ kg/s, $\Delta h_{r,c} = -1.4 \times 10^7$ J/kg, $\epsilon_{r,1} = 0.9$, $\epsilon_{r,2} = 0.8$, $D = 5$ cm, $L = 35$ cm, $a = 1$ m.

Use geometrical relations (not the tables) to determine the view factors.

## PROBLEM 4.19. FUN

Consider two square (each length $a$) parallel plates at temperatures $T_1$ and $T_2$ and having an equal emissivity $\epsilon_r$. Assume that the distance between them $l$ is much smaller than $a$ ($l \ll a$).

(a) Show that radiative heat flux between surface 1 and 2 is

$$q_{r,1\text{-}2} = \frac{\epsilon_r \sigma_{\text{SB}}(T_1^4 - T_2^4)}{2 - \epsilon_r}.$$

(b) Show that if a radiation shield having the same size and emissivity is placed between them, then

$$q_{r,1\text{-}2} = \frac{\epsilon_r \sigma_{\text{SB}}(T_1^4 - T_2^4)}{2(2 - \epsilon_r)}.$$

## PROBLEM 4.20. FUN

Two very large, parallel plates at maintained temperatures $T_1$ and $T_2$ are exchanging surface-radiation heat. A third large and thin plate is placed in between and parallel to the other plates (Figure Pr. 4.20). This plate has periodic voids (e.g., as in a screen) and the fraction of void area to total surface area is $\epsilon = A_{\text{voids}}/A_{\text{total}}$. The screen is sufficiently thin (or its conductivity is sufficiently large) such that its temperature $T_3$ is uniform across the thickness. All plates have opaque, diffuse, and gray surfaces with the same total emissivity $\epsilon_r$.

(a) Draw the thermal circuit.
(b) Derive the expression for the net heat transfer rate by surface-radiation between surfaces 1 and 2, i.e., $Q_{r,1\text{-}2}$, given by

$$\frac{Q_{r,1\text{-}2}}{A_r} = \frac{E_{b,1} - E_{b,2}}{2\left[\dfrac{1 - \epsilon_r}{\epsilon_r} + \dfrac{1}{2\epsilon + \epsilon_r(1 - \epsilon)}\right]}.$$

(i) Physical Model

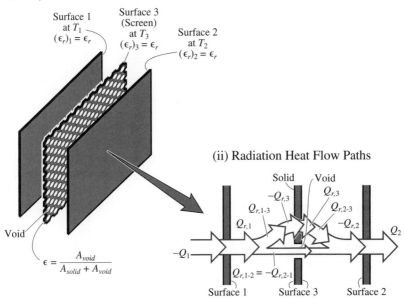

Figure Pr. 4.20. (i) and (ii) Surface-radiation heat transfer between two plates separated by a screen.

Use a three-surface enclosure and allow for heat transfer between surfaces 1 and 2 directly through the screen voids and indirectly through the solid portion of the screen. The screen has radiation exchange on both of its sides, with a zero net heat transfer.

(c) Comment on the limits as $\epsilon \to 0$ and $\epsilon \to 1$.

(d) Would $Q_{r,1\text{-}2}$ increase or decrease with an increase in the emissivity of the screen?

Analyze the above expression for $Q_{r,1\text{-}2}$ in the limits for $\epsilon_r \to 0$ and $\epsilon_r \to 1$.

**PROBLEM 4.21. FUN**

In surface-radiation heat transfer between surfaces 1 and 2, the enclosure geometry dependence of the radiation heat flux $q_{r,2\text{-}1}$ is examined using four different geometries. These are shown in Figures Pr. 4.21(i) through (iv) and are: parallel plates, coaxial cylinders, coaxial spheres, and a disk facing an enclosing hemisphere. The plates are assumed to be placed sufficiently close to each other and the cylinders are assumed to be sufficiently long, such that for all the four enclosure geometries the radiation is only between surfaces 1 and 2 (i.e., two-surface enclosures).

(a) Draw the thermal circuit diagram (one for all geometries).

(b) Determine $q_{r,2\text{-}1} = Q_{r,2\text{-}1}/A_{r,2}$ for the geometries of Figures Pr. 4.21(i)–(iv), for the conditions given below.

$T_1 = 180°C, T_2 = 90°C, \epsilon_{r,1} = \epsilon_{r,2} = 0.8.$

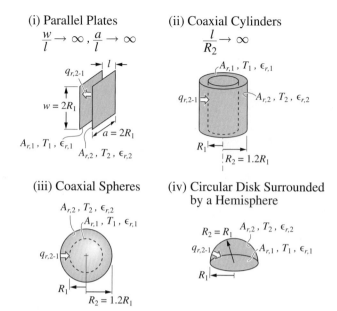

(i) Parallel Plates $\frac{w}{l} \to \infty$, $\frac{a}{l} \to \infty$

$q_{r,2-1}$, $l$, $w = 2R_1$, $a = 2R_1$, $A_{r,1}$, $T_1$, $\epsilon_{r,1}$, $A_{r,2}$, $T_2$, $\epsilon_{r,2}$

(ii) Coaxial Cylinders $\frac{l}{R_2} \to \infty$

$A_{r,1}$, $T_1$, $\epsilon_{r,1}$, $A_{r,2}$, $T_2$, $\epsilon_{r,2}$, $q_{r,2-1}$, $R_1$, $R_2 = 1.2R_1$

(iii) Coaxial Spheres $A_{r,2}$, $T_2$, $\epsilon_{r,2}$, $A_{r,1}$, $T_1$, $\epsilon_{r,1}$, $q_{r,2-1}$, $R_1$, $R_2 = 1.2R_1$

(iv) Circular Disk Surrounded by a Hemisphere $R_2 = R_1$, $A_{r,2}$, $T_2$, $\epsilon_{r,2}$, $q_{r,2-1}$, $A_{r,1}$, $T_1$, $\epsilon_{r,1}$, $R_1$

Figure Pr. 4.21. (i) through (iv) Four enclosure geometries used in determining the dependence of $q_{r,2-1}$ on the enclosure geometry.

### 4.9.4 Three-Surface, Opaque, Diffuse Gray Enclosures

**PROBLEM 4.22. FAM**

A hemispherical Joule heater (surface 1) is used for surface-radiation heating of a circular disk (surface 2). This is shown in Figure Pr. 4.22. In order to make an efficient use of the Joule heating, a hemispherical cap (surface 3) is placed around the heater surface and is ideally insulated.

(a) Draw the thermal circuit diagram.
(b) Determine $Q_{r,1-2}$ for the conditions given below.
(c) Determine $Q_{r,1-2}$ without the reradiator and compare the results with the results in (b).

$\quad R_1 = 5$ cm, $R_2 = 5R_1$, $T_1 = 1,100$ K, $T_2 = 500$ K $\epsilon_{r,1} = \epsilon_{r,2} = 1$.

Assume that $F_{1-2}$ corresponds to that from a sphere to a disk (i.e., assume that the upper hemisphere does not see the disk).

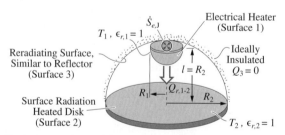

$\dot{S}_{e,J}$, Electrical Heater (Surface 1), $T_1$, $\epsilon_{r,1} = 1$, Reradiating Surface, Similar to Reflector (Surface 3), Ideally Insulated $Q_3 = 0$, Surface Radiation Heated Disk (Surface 2), $l = R_2$, $R_1$, $Q_{r,1-2}$, $R_2$, $T_2$, $\epsilon_{r,2} = 1$

Figure Pr. 4.22. A Joule heater is used for surface-radiation heating of a disk. A reradiating hemisphere is used to improve the heating rate.

## PROBLEM 4.23. FAM

A flat radiation heater is placed along a vertical wall to heat the passing pedestrians who may stop temporarily and face the heater. The heater is shown in Figure Pr. 4.23(i) and is geometrically similar to a full-size mirror. The heater surface is at $T_1 = 600°C$ and the pedestrians have a surface temperature of $T_2 = 5°C$. Assume that the surfaces are opaque, diffuse, and blackbody surfaces (total emissivities are equal to one). Also assume that both the heater and the pedestrian have a rectangular cross section with dimensions $a = 50$ cm and $w = 170$ cm and that the distance between them is $l = 40$ cm, as shown in Figure Pr. 4.23(ii).

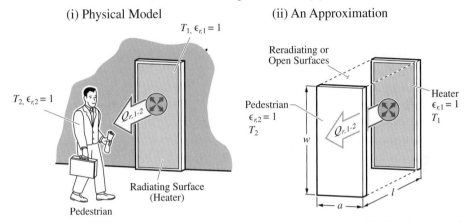

(i) Physical Model      (ii) An Approximation

Figure Pr. 4.23. (i) Physical model of a radiant wall, pedestrian heater. (ii) Idealized model.

(a) Draw the thermal circuit diagram for a three-surface enclosure (including the surroundings as a blackbody surface).
(b) Determine the net radiation heat transfer from the heater to the pedestrian $Q_{r,1\text{-}2}$.
(c) Determine the net radiation heat transfer to the pedestrian, when a reradiating (i.e., ideally insulated) surface is placed around the heater and pedestrian to increase the radiant heat flow $Q_{r,1\text{-}2}$.

## PROBLEM 4.24. FAM

A source for thermal irradiation is found by a Joule heater placed inside a solid cylinder of radius $R_1$ and length $l$. Then a hollow cylinder of radius $R_2$ and length $l$ is placed coaxially around it with this outer cylinder and the top part of the opening ideally insulated. This is shown in Figure Pr. 4.24 with the radiation leaving through the opening at the bottom spacing between the cylinders (surface 2). This results in surface 1 being the high temperature surface with direct and reradiation exchange with surface 2.

(a) Draw the thermal circuit diagram.
(b) Determine the view factors $F_{1'\text{-}2}$, $F_{1\text{-}3}$, and $F_{2\text{-}3}$, using Figures 4.11(d) and (e), and the designations of Figure Pr. 4.24.
(c) Determine the heater surface temperature $T_1$.

   $T_2 = 400$ K, $\epsilon_{r,1} = 0.8$, $\dot{S}_{e,J} = 1,000$ W, $R_1 = 1$ cm, $R_2 = 5$ cm, $l = 10$ cm.

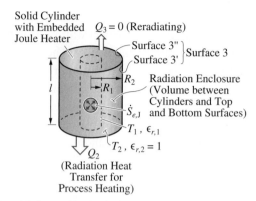

Figure Pr. 4.24. Surface 1 is heated by Joule heating and through direct and reradiation allows for radiation to leave for surface 2.

### PROBLEM 4.25. FUN.S

Consider three opaque, diffuse, and gray surfaces with temperatures $T_1 = 400$ K, $T_2 = 400$ K, and $T_3 = 300$ K, with surface emissivities $\epsilon_{r,1} = 0.2$ and $\epsilon_{r,2} = \epsilon_{r,3} = 0.5$, and areas $A_{r,1} = A_{r,2} = A_{r,3} = 1$ m$^2$.

(a) For (i) surfaces 1 and 2 forming a two-surface enclosure (i.e., $F_{1-2} = 1$), and (ii) surfaces 1, 2, and 3 forming a three-surface enclosure (assume a two-dimensional equilateral triangular enclosure), is there a net radiation heat transfer rate $Q_{r,1-2}$ between surfaces 1 and 2?
(b) If there is a nonzero net heat transfer rate, what is the direction of this heat transfer?
(c) Would this heat transfer rate change if $T_3 = 500$ K?
(d) What is the temperature $T_3$ for which $Q_{r,1-2} = 0$?

### PROBLEM 4.26. FAM

Surface-radiation absorption is used to melt solid silicon oxide powders used for glass making. The heat is provided by combustion occurring over an impermeable surface 1 with dimensions $a = w = 1$ m, as shown in Figure Pr. 4.26. The desired surface temperature $T_1$ is 1,600 K. The silicon oxide powders may be treated as a

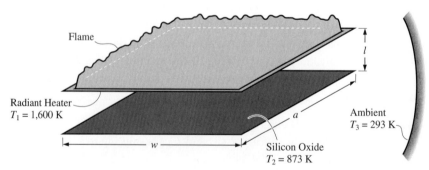

Figure Pr. 4.26. Surface radiant heater heated by a flame over it and forming a three-surface-radiation enclosure.

surface 2, with the same area as the radiant heater, at a distance $l = 0.25$ m away from the heater, and at a temperature $T_2 = 873$ K. The surroundings are at $T_3 = 293$ K. Assume that all surfaces are ideal blackbody surfaces.

(a) Draw the thermal circuit diagram.
(b) Determine the net radiation heat transfer to the silicon oxide surface.

## PROBLEM 4.27. FUN

Surface-radiation emission can be redirected to a receiving surface using reradiating surfaces. Figure Pr. 4.27 renders such a redirection design using a reradiating surface 3. Surface 3 is ideally insulated and is treated as a single surface having a uniform temperature $T_3$.

Surface 1 has a temperature $T_1$ higher than that of surface 2, $T_2$.

(a) Draw the thermal circuit diagram.
(b) Determine $Q_{r,1-2}$ for the given conditions.
(c) Compare this with $Q_{r,1-2}$ without reradiation.
(d) Show the expression for $Q_{r,1-2}$ for the case of $F_{1-2} = 0$ and comment on this expression.

$R_1 = 25$ cm, $R_2 = 25$ cm, $F_{1-2} = 0.1$, $\epsilon_{r,1} = 1.0$, $\epsilon_{r,2} = 1.0$, $T_1 = 900$ K, $T_2 = 400$ K.

Note that since surfaces 1 and 2 are blackbody surfaces, $(q_{r,o})_1 = E_{b,1}$ and $(q_{r,o})_2 = E_{b,2}$.

### Redirection of Radiation
### Using a Reradiating Surface

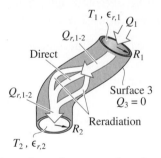

Figure Pr. 4.27. Two blackbody surfaces that are exchanging surface-radiation heat and are completely enclosed by a reradiation surface.

### 4.9.5 Nongray Surfaces and Prescribed Irradiation

## PROBLEM 4.28. FAM

Fire barriers are used to temporarily protect spaces adjacent to fires. Figure Pr. 4.28 shows a suspended fire barrier of thickness $L$ and effective conductivity $\langle k \rangle$ (and $\langle \rho \rangle$ and $\langle c_p \rangle$) subjected to a flame irradiation $(q_{r,i})_f$. The barrier is a flexible, wire-reinforced mat made of a ceramic (high melting temperature, such as $ZrO_2$) fibers.

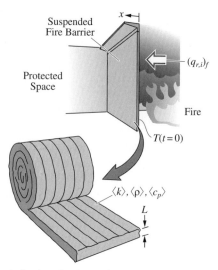

Figure Pr. 4.28. A fire barrier is used to protect a space adjacent to a fire.

The barrier can withstand the high temperatures resulting from the flow of $(q_{r,i})_f$ into the mat until, due to thermal degradation of the fibers and wires, it fails. In some cases the barrier is actively water sprayed to delay this degradation.

The transient conduction through the mat, subject to a constant $(q_{r,i})_f$, can be treated analytically up to the time that thermal penetration distance $\delta_\alpha$ reaches the back of the mat $x = L$. This is done by using the solution given in Table 3.4 for a semi-infinite slab, and by neglecting any surface-radiation emission and any surface convection. Assume that these simplifications are justifiable and the transient temperature $T(x, t)$ can then be obtained, subject to a constant surface flux $q_s = (q_{r,i})_f$ and a uniform initial temperature $T(t = 0)$.

(a) Determine the elapsed time $t$ for the thermal penetration using (3.148).
(b) Determine the surface temperature $T(x = 0, t)$ at this elapsed time.
(c) Using the melting temperature of $ZrO_2$ in Figure 3.8, would the mat disintegrate at this surface?

$L = 3\,\text{cm}$, $\langle k \rangle = 0.2\,\text{W/m-K}$, $\langle \rho \rangle = 600\,\text{kg/m}^3$, $\langle c_p \rangle = 1{,}000\,\text{J/kg-K}$, $(q_{r,i})_f = -10^5$ $\text{W/m}^2$, $T(t = 0) = 40°\text{C}$.

**PROBLEM 4.29. FAM**

Using reflectors (mirrors) to concentrate solar irradiation allows for obtaining very large (concentrated) irradiation flux. Figure Pr. 4.29 shows a parabolic concentrator that results in concentration irradiation flux $(q_{r,i})_c$, which is related to the geometric parameters through the energy equation applied to solar energy, i.e.,

$$(q_{r,i})_s wL = (q_{r,i})_c DL,$$

where $D$ is the diameter, $DL$ is the projected cross-sectional area of the receiving tube, and $wL$ is the projected concentrator cross-sectional area receiving solar irradiation.

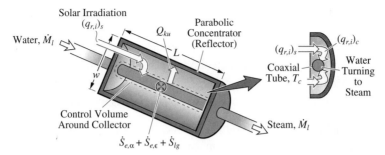

Figure Pr. 4.29. A concentrator-solar collector system used for steam production.

The concentrated irradiation is used to produce steam from saturated (at $T = T_{lg}$) water, where the water mass flow rate is $\dot{M}_l$. The absorptivity of the collector is $\alpha_{r,c}$ and its emissivity $\epsilon_{r,c}$ is lower (nongray surface). In addition to surface emission, the collector loses heat to the ambient through surface convection and is given as a prescribed $Q_{ku}$. Assume that collector surface temperature is $T_c = T_{lg}$.

(a) Draw the thermal circuit diagram.
(b) Determine the stream production rate $\dot{M}_l$.

$(q_{r,i})_s = 200\,\text{W/m}^2, Q_{ku} = 400\,\text{W}, T_c = T_{lg} = 127°\text{C}, \alpha_{r,c} = 0.95, \epsilon_{r,c} = 0.4, D = 5$ cm, $w = 3$ m, $L = 5$ m.

Use Table C.27 for properties of saturated water.

**PROBLEM 4.30. FAM**
Pulsed lasers may be used for the ablation of living-cell membrane in order to introduce competent genes, in gene therapy. This is rendered in Figure Pr. 4.30. The ablation (or scissors) laser beam is focused on the cell membrane using a neodymium yttrium aluminum garnet (Nd:YAG) laser with $\lambda = 532$ nm $= 0.532\ \mu$m, and a focus

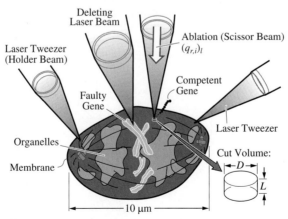

Figure Pr. 4.30. A living cell is ablated at a region on its membrane, for introduction of competent genes.

spot with diameter $D = 500$ $\mu$m. There is a Gaussian distribution of the irradiation across $D$, but here we assume a uniform distribution.

Assume a steady-state heat transfer. Although the intent is to sublimate $\dot{S}_{sg}$ the targeted membrane region for a controlled depth (to limit material removal to the thin, cell membrane), the irradiation energy is also used in some exothermic chemical reaction $\dot{S}_{r,c}$ and in some heat losses presented by $Q$ (this includes surface emission).

(a) Draw the thermal circuit diagram.
(b) Determine the required duration of the laser pulse $\Delta t$, for the following conditions.

$D = 500$ nm, $L = 10$ nm, $\rho = 2 \times 10^3$ kg/m$^3$, $(q_{r,i})_l = 10^{10}$ W/m$^2$, $\Delta h_{sg} = 3 \times 10^6$ J/kg, $\dot{S}_{r,c} = -7 \times 10^{-4}$ W, $Q = 3 \times 10^{-4}$ W, $\alpha_r = 0.9$.

## PROBLEM 4.31. FUN*
This problem is found on the Web site www.cambridge.org/kaviany.

## PROBLEM 4.32. FUN
Semi-transparent, fire-fighting foams (closed cell) have a very low effective conductivity and also absorb radiation. The absorbed heat results in the evaporation of water, which is the main component (97% by weight) of the foam. As long as the foam is present, the temperature of the foam is nearly that of the saturation temperature of water at the gas pressure. A foam covering (i.e., protecting) a substrate while being exposed to a flame of temperature $T_f$ is shown in Figure Pr. 4.32.

The foam density $\langle\rho\rangle$ and its thickness $L$ both decrease as a result of irradiation and evaporation. However, for the sake of simplicity here we assume constant $\langle\rho\rangle$ and $L$. The absorbed irradiation, characterized by the flame irradiation flux $(q_{r,i})_f$ and by the foam extinction coefficient $\sigma_{ex}$, results in the evaporation of foam. The flame is a propane-air flame with a composition given below.

(a) Write the energy equation for the constant-volume foam layer.
(b) Determine the flame irradiation flux $(q_{r,i})_f$ impinging on the foam.
(c) Determine the rate of irradiation absorbed into the foam layer $\dot{S}_{e,\sigma}$. Use (2.43)

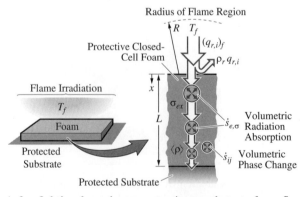

Figure Pr. 4.32. A fire-fighting foam layer protecting a substrate from flame irradiation. A close-up of the closed-cell foam is also shown.

and integrate it over the foam thickness $L$.

(d) Assuming that irradiation heat absorbed results in the foam evaporation, determine the elapsed time for the complete evaporation of the foam, Use (2.25) with $\Delta h_{lg}$ being that of water at $T = 100°C$.

Assume no heat losses.

$\langle \rho \rangle = 30 \text{ kg/m}^3$, $\sigma_{ex} = a_1 \langle \rho \rangle$, $a_1 = 3 \text{ m}^2/\text{kg}$, $L = 10$ cm, $R = 1$ m, $T_f = 1{,}800$ K, $p_{CO_2} = 0.10$ atm, $p_{H_2O} = 0.13$ atm, $\epsilon_s = 10^{-7}$, $\rho_r = 0$.

## PROBLEM 4.33. FAM

Flame radiation from a candle can be sensed by the temperature sensors existing under the thin, skin layer of the human hands. The closer the sensor (or say hand) is, the higher irradiation flux $q_{r,i}$ it senses. This is because the irradiation leaving the approximate flame surface $(q_{r,i})_f A_{r,f}$, $A_{r,f} = 2\pi R_1 L$, is conserved. This is rendered in Figure Pr. 4.33. Then, assuming a spherical radiation envelope $R_2$ (and neglecting the radiation from ends of the cylinder), we have

$$2\pi R_1 L (q_{r,i})_f = 4\pi R_2^2 q_{r,i} \quad \text{for} \quad R_2^2 \gg R_1^2.$$

(a) Determine the flame irradiation flux $(q_{r,i})_f$ at the flame envelope for the heavy-soot condition given below.

(b) Determine the irradiation flux $q_{r,i}$ at $r = R_2$, using the above relation.

$L = 3.5$ cm, $R_1 = 1$ cm, $R_2 = 10$ cm, $T_f = 1{,}100$ K, $p_{CO_2} = 0.15$ atm, $p_{H_2O} = 0.18$ atm, $\epsilon_s = 2 \times 10^{-7}$.

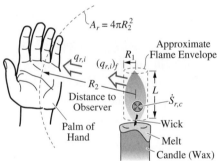

Figure Pr. 4.33. A candle flame and the sensing of its irradiation at a distance $R_2$ from the flame center line.

## PROBLEM 4.34. DES

New coating technologies employ ultraviolet curable coatings and ultraviolet radiation ovens. The coatings contain monomers and oligomers that cross link to form a solid, cured film upon exposure to the ultraviolet radiation. The radiation is produced by a mercury vapor or a gallium UV (ultraviolet) lamp. The intensity of the radiation is selected to suit the type of coating applied, its pigmentation, and its thickness. One advantage of the UV-curable coatings is that a smaller amount of solvent is used and discharged to the atmosphere during curing.

In a wood coating-finishing process, the infrared fraction of the emitted radiation is undesirable. The infrared radiation can heat the wood panels to a threshold temperature where the resins leach into the coating before it cures, thus producing

(i) Oven, lamps, workpiece, and reflectors

(ii) Selective Reflections

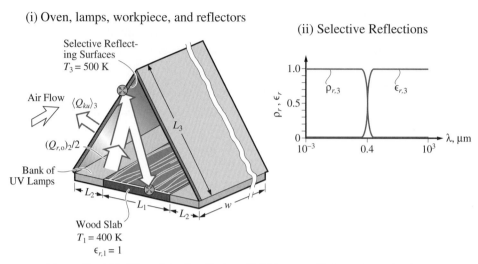

Figure Pr. 4.34.  Ultraviolet irradiation. (i) UV oven. (ii) Selective reflections.

an inferior finish. To prevent this, inclined selective surfaces, which reflect the ultraviolet fraction of the radiation, are used. Figure Pr. 4.34(i) shows a UV oven. A wood panel, with length $L_1 = 80$ cm, and width $w = 1$ m, occupies the central part of the oven and a bank of UV lamps, with length $L_2 = 50$ cm and width $w = 1$ m, are placed on both sides of the workpiece. The top surfaces act as selective reflecting surfaces, i.e., absorb the infrared radiation and reflect the ultraviolet radiation. They are cooled in their back by a low temperature air flow in order to minimize emission of infrared radiation.

The UV lamps emit $(\dot{S}_{e,\epsilon}/A_r)_2 = (q_{r,o})_2 = 7 \times 10^5$ W/m$^2$ which is 95% in the ultraviolet range of the spectrum and 5% in the visible and infrared range of the spectrum. The wood boards have a curing temperature $T_1 = 400$ K and behave as a blackbody surface. The selective surfaces have a temperature $T_3 = 500$ K and the emissivity and reflectivity shown in Figure Pr. 4.34(ii).

(a) Determine the amount of heat transfer by surface convection $Q_{ku}$(W) needed to keep the selective surfaces at $T_3 = 500$ K.
(b) Determine the radiation heat transfer in the ultraviolet range $Q_{r,1}$(UV) and infrared and visible range $Q_{r,1}$(IR + V) reaching the workpiece (surface 1).
(c) Determine the maximum allowed temperature for the selective surfaces $T_{3,max}$, such that the amount of infrared and visible radiation reaching the workpiece is less than 3% of the ultraviolet radiation [i.e., $Q_{r,1}$(IR + V)$/Q_{r,1}$(UV) $< 0.03$].

### 4.9.6  Inclusion of Substrate

The following material can be found on the text Web site www.cambridge.org/kaviany.

**PROBLEM 4.35. DES***

**PROBLEM 4.36. FUN***

**PROBLEM 4.37. FUN\***

**PROBLEM 4.38. DES\***

**PROBLEM 4.39. DES\***

**PROBLEM 4.40. DES.S\***

**PROBLEM 4.41. DES\***

**PROBLEM 4.42. FAM**

A gridded silicon electric heater is used in a microelectromechanical device, as shown in Figure Pr. 4.42. The heater has an electrical resistance $R_e$ and a voltage $\Delta\varphi$ is applied resulting in the Joule heating. For testing purposes, the heater is raised to a steady-state, high temperature (i.e., glowing red). The gridded heater is connected to a substrate through four posts (made of silicon oxide, for low conductivity $k_p$), resulting in conduction heat loss through four support posts. The substrate is at $T_s$ and has an emissivity $\epsilon_{r,s}$. The upper heater surface is exposed to large surface area surroundings at $T_{surr}$. Treat the heater as having a continuous surface (i.e., solid, not gridded), with a uniform temperature $T_1$.

(a) Draw the thermal circuit diagram.
(b) Determine the heater temperature $T_1$.

$T_s = 400°C$, $T_{surr} = 25°C$, $w = 0.5$ mm, $a = 0.01$ mm, $l = 0.01$ mm, $\epsilon_{r,1} = 0.8$, $\epsilon_{r,s} = 1$, $R_e = 1{,}000$ ohm, $\Delta\varphi = 5$ V, $k_p = 2$ W/m-K.

It is not necessary to use tables or figures for the view factors (use $w \gg l$). Use (2.28) for $\dot{S}_{e,J}$.

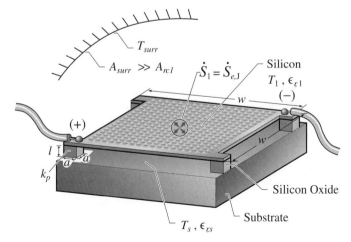

Figure Pr. 4.42. A miniature gridded heater connected to a substrate by four support posts and raised to a glowing temperature.

## PROBLEM 4.43. FAM

A spherical cryogenic (hydrogen) liquid tank has a thin (negligible thickness), double-wall structure with the gap space filled with air, as shown in Figure Pr. 4.43. The air pressure is one atm.

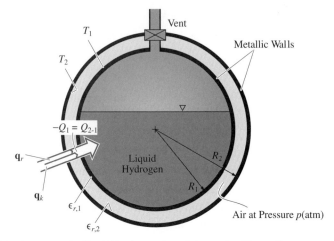

Figure Pr. 4.43. A cryogenic (liquid hydrogen) tank has a double-wall structure with the space between the thin walls filled with atmospheric or subatmospheric pressure air.

(a) Draw the thermal circuit diagram for the heat flow between the outer and inner walls.

(b) Determine the rate of heat transfer to the tank $Q_{2\text{-}1}$.

(c) Would $Q_{2\text{-}1}$ change if the air pressure is reduced to 1/10 atm? How about under ideal vacuum ($p = 0$, $k = 0$)?

$\epsilon_{r,1} = 0.05$, $\epsilon_{r,2} = 0.05$, $R_1 = 1$ m, $R_2 = 1.01$ m, $T_1 = -240°C$, $T_2 = -80°C$.

Use (3.19) for low and moderate gas pressure gases to determine any pressure dependence of the gas conductivity. Use Table C.22 for the atmospheric pressure properties of air and use $T = 150$ K.

## PROBLEM 4.44. FUN

In surface-radiation through multiple, opaque layer systems, such as the one shown in Figure Pr. 4.44, the rate of conduction through the layers can be significant. To include the effect of the layer conductivity, and also the layer spacing indicated by porosity, use the approximation that the local radiation heat transfer is determined by the local temperature gradient, i.e.,

$$q_{r,x} = -\langle k_r \rangle \frac{dT}{dx} = -\langle k_r \rangle \frac{T_2 - T_1}{l_1 + l_2}, \qquad \langle k_r \rangle = \langle k_r \rangle(k_s, \epsilon, \epsilon_r, T_2, l_2),$$

where $\langle k_r \rangle$ is the radiant conductivity.

(a) Using the thermal circuit diagram representing $Q_{r,x}$ in Figure Pr. 4.44(ii), start from (4.47) and use $F_{1'\text{-}2} = 1$ for $l_2 \ll a$, and then use (4.72) to linearize $T_{1'}^4 - T_2^4$ and arrive at

$$q_{r,x} = \frac{4\epsilon_r \sigma_{SB} T^3 (T_1' - T_2)}{2 - \epsilon_r}.$$

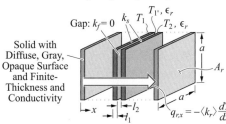
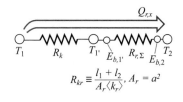

(i) Multiple, Opaque, Finite-
Conductivity Layers (Slabs)

(ii) Thermal Circuit Model
Using Radiant Conductivity $\langle k_r \rangle$

Figure Pr. 4.44. (i) Surface-radiation and solid conduction through a system of parallel slabs.
(ii) Thermal circuit diagram.

(b) Then add $R_k$ and use $T_1 - T_2$ to arrive at the radiant conductivity expression $\langle k_r \rangle$

$$\langle k_r \rangle = \frac{1}{\dfrac{1 - \epsilon}{k_s} + \dfrac{(2 - \epsilon_r)\epsilon}{4\epsilon_r \sigma_{SB} T^3 l_2}}, \quad \epsilon = \frac{l_2}{l_1 + l_2}$$

or

$$\frac{\langle k_r \rangle}{4\sigma_{SB} T^3 l_2} = \frac{1}{\dfrac{4\sigma_{SB} T^3 l_2 (1 - \epsilon)}{k_s} + \dfrac{(2 - \epsilon_r)\epsilon}{\epsilon_r}}$$

or

$$\frac{k_r}{4\sigma_{SB} T^3 l_1} = \frac{\epsilon_r N_r^{-1} (1 - \epsilon)^{-1}}{\epsilon_r + N_r^{-1}(2 - \epsilon_r)}, \quad N_r = \frac{4\sigma_{SB} T^3 l_1}{k_s}.$$

## PROBLEM 4.45. FUN.S

The measured effective thermal conductivity of porous solids, such as that for packed
zirconium oxide fibers given in Figure 3.13(b), does include the radiation contribution. The theoretical prediction can treat the conduction and radiation heat transfer
separately. In a prediction model (derivation for cubic particles is left as an end of the
chapter problem), the effective, combined (total) conductivity for a periodic porous
solid is given by

$$\langle k_{kr} \rangle = \langle k_k \rangle + \langle k_r \rangle$$

$$\langle k_r \rangle = 4D\sigma_{SB} T^3 \frac{\epsilon_r (1 - \epsilon)^{1/3} N_r^{-1}}{\epsilon_r + N_r^{-1}(2 - \epsilon_r)}, \quad N_r = \frac{4\sigma_{SB} T^3 D}{k_s}$$

$$\langle k \rangle = k_f \left( \frac{k_s}{k_f} \right)^{0.280 - 0.757 \log \epsilon + 0.057 \log(k_s / k_f)}$$

Using these, we can compare the predicted and measured results.

Here $\epsilon$ is the porosity, $D$ is the fiber-diameter, $\epsilon_r$ is the fiber emissivity, $k_f$ is the fluid, and $k_s$ is the solid conductivity.

(a) Plot the variation of $\langle k_{kr} \rangle$ with respect to $T$ for $300 \text{ K} \leq T \leq 1{,}000 \text{ K}$, for the zirconium oxide and air system.
(b) Compare the results with the experimental results of Figure 3.13(b).
(c) Is radiation contribution significant in this material?

$D = 10 \ \mu\text{m}$, $\langle \rho \rangle = 1{,}120 \text{ kg/m}^3$.

Use $\langle \rho \rangle = \epsilon \rho_f + (1 - \epsilon)\rho_s = (1 - \epsilon)\rho_s$, for $\rho_f \ll \rho_s$, to determine $\epsilon$. For air use $k_f = 0.0267(\text{W/m-K}) + 5.786 \times 10^{-5}(\text{W/m-K}^4) \times (T - 300)(\text{K})$, and use the only data available for zirconium oxide $k_s$ and $\rho_s$ in Table C.17. For emissivity use $\epsilon_r = 0.9 - 5.714 \times 10^{-4}(T - 300)(\text{K})$ based on Table C.18 for zirconium oxide.

## PROBLEM 4.46. FAM.S
A spherical carbon steel AISI 1010 piece of diameter $D = 1$ cm, initially at $T_1(t = 0) = 1{,}273$ K, is cooled by surface-radiation to a completely enclosing cubic oven made of white refractory brick with each side having a length $L = 10$ cm and a surface temperature $T_2 = 300$ K. Assume that all the surfaces are opaque, diffuse, and gray. For the carbon steel sphere use the higher value for the emissivity of oxidized iron, listed in Table C.18.

(a) Using a software, plot the variation of the piece temperature with respect to time.
(b) Determine the time it takes for the piece to reach $T_1 = 600$ K and compare this result (i.e., the numerical solution) with the one predicted by (4.82) (i.e., the analytic solution).

## PROBLEM 4.47. FUN
An idealized bed of solid particles is shown in Figure Pr. 4.47. The heat is transferred through the bed by surface-radiation and the cubic solid particles have a finite conductivity $k_s$, while conduction through the fluid is neglected ($k_f = 0$). We use the radiant conductivity $\langle k_r \rangle$ defined through

$$q_{r,x} = -\langle k_r \rangle \frac{dT}{dx} = -\langle k_r \rangle \frac{T_2 - T_1}{l_1 + l_2},$$

$$\langle k_r \rangle = \langle k_r \rangle(k_s, \epsilon, \epsilon_r, T, l_2).$$

(a) For the thermal circuit model shown in Figure Pr. 4.44(ii), determine the radiation resistance $R_{r,\Sigma}$ between two adjacent surfaces 1 and 2. Assume that $F_{1'-2} = 1$ and use the linearization given in (4.72).
(b) Add the conduction resistance using series resistances to arrive at

$$\frac{k_r}{4\sigma_{\text{SB}} T^3 l_1} = \frac{\epsilon_r N_r^{-1}(1 - \epsilon)^{1/3}}{\epsilon_r + N_r^{-1}(2 - \epsilon_r)}, \quad N_r = \frac{4\sigma_{\text{SB}} T^3 l_1}{k_s}, \quad \epsilon = 1 - \frac{l_1^3}{(l_1 + l_2)^3}.$$

(i) Surface Radiation and Solid Conduction in a Bed of Cubic Particles

(ii) Thermal Circuit Model Using Radiant Conductivity $\langle k_r \rangle$

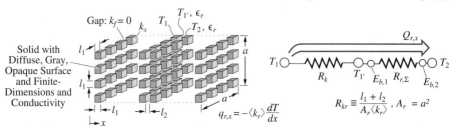

Figure Pr. 4.47. (i) A bed of cubical particles with surface-radiation and solid conduction. (ii) Simplified thermal circuit diagram.

## PROBLEM 4.48. FUN

In the limit of an optically thick medium, $\sigma_{ex}^* = \sigma_{ex}L > 10$, the extinction coefficient $\sigma_{ex}$ and the radiant conductivity are used interchangeably and are related through

$$q_{r,x} \equiv -\langle k_r \rangle \frac{dT}{dx}$$

$$= -\frac{4}{3\sigma_{ex}} \frac{dE_b}{dx} = -\frac{16 \sigma_{SB} T^3}{3} \frac{dT}{\sigma_{ex}} \frac{dT}{dx}.$$

Then

$$\langle k_r \rangle = \frac{16\sigma_{SB} T^3}{3\sigma_{ex}}.$$

For a packed bed of cubical particles of finite conductivity $k_s$, surface emissivity $\epsilon_r$, and linear dimension $l_1$, with interparticle spacing $l_2$, the radiant conductivity can be shown to be approximated by

$$\langle k_r \rangle = \frac{1}{\dfrac{1}{k_s(1 - \epsilon)^{1/3}} + \dfrac{(2 - \epsilon_r)}{4\sigma_{SB}\epsilon_r T^3 l_1}}$$

or

$$\sigma_{ex} = \frac{16\sigma_{SB} T^3}{3k_s(1 - \epsilon)^{1/3}} + \frac{4(2 - \epsilon_r)}{3\epsilon_r l_1 (1 - \epsilon)^{1/3}}.$$

(a) Determine $\sigma_{ex}$ for a bed of alumina cubical particles at $T = 500$ K, with $\epsilon_r = 0.7$ $\epsilon = 0.4$, $k_s = 36$ W/m-K, and (i) $l_2 = 3$ cm, and (ii) $l_2 = 3$ $\mu$m.
(b) Repeat (a) for amorphous silica particles, $\epsilon_r = 0.45$, $k_s = 1.38$ W/m-K, keeping other parameters the same.
(c) Compare these with the results of Figure 2.13 and comment.

**PROBLEM 4.49. FAM**

Spherical, pure, rough-polish aluminum particles of diameter $D_1$ and emissivity $\epsilon_{r,1}$ are heated by surface-radiation while traveling through an alumina ceramic tube kept at a high temperature $T_2$. The tube has an inner diameter $D_2$, a length $l$, and an emissivity $\epsilon_{r,2}$. This is shown in Figure Pr. 4.49. A particle arrives at the entrance to the tube with an initial, uniform temperature $T_1(t=0)$, and exits the tube with a final, uniform temperature $T_1(t=t_f)$. Assume that, throughout the time of travel, the fraction of radiative heat transfer between the particle and the open ends of the tube is negligible (i.e., the view factor $F_{1\text{-ends}}=0$).

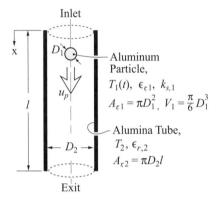

Figure Pr. 4.49. Particles heated in a tube by surface-radiation.

(a) Draw the thermal circuit diagram.
(b) Determine the speed $u_p = l/t_f$ at which the particle must move through the tube in order to exit at the melting temperature $T_1(t=t_f) = T_{sl}$.
(c) Approximate the internal conduction resistance to be $(D_1/2)/(A_{r,1}k_{s,1})$, where $k_{s,1}$ is the thermal conductivity of the solid aluminum, and determine if the assumption of uniform temperature within the particle is valid.

$D_1 = 10\ \mu\text{m}$, $D_2 = 3$ mm, $l = 5$ cm, $T_1(t=0) = 20°\text{C}$, $T_1(t=t_f) = T_{sl}$ (aluminum, Table C.16), $T_2 = 1{,}283$ K. Evaluate the emissivities from Table C.18 and the properties of aluminum at $T = 300$ K (Table C.16).

**PROBLEM 4.50. FUN**

A highly insulated thermos depicted in Figure Pr. 4.50 has five layers of insulation shields on the outside. The wall has two glass layers separated by an evacuated space [Figure Pr. 4.50(i)], or a cork board [Figure Pr. 4.50(ii)].

(a) Draw the thermal circuit diagram.
(b) Determine the heat transfer per unit area associated with each of these designs. Comment on the preference of the cork board or the vacuum. Also, assume that all the surfaces are diffuse and gray. The glass is assumed opaque to radiation.

$T_1 = 90°\text{C}$, $T_3 = 20°\text{C}$, $l = 1$ mm, $L = 7$ mm, $\epsilon_{r,s} = 0.04$, $\epsilon_{r,2} = 0.9$, $\epsilon_{r,3} = 1$.

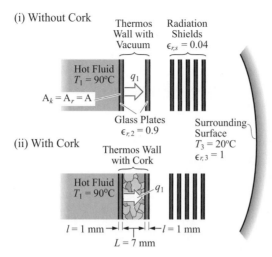

Figure Pr. 4.50. Surface-radiation from a thermos having radiation shields (i) With a cork board wall. (ii) Without a cork board wall.

## PROBLEM 4.51. FAM.S

A person with a surface temperature $T_1 = 31°C$ is standing in a very large room $(A_{r,2} \gg A_{r,1})$ and is losing heat by surface-radiation to the surrounding room surfaces, which are at $T_2 = 20°C$ (Figure Pr. 4.51). Model the person as a cylinder with diameter $D = 0.4$ m and length $L = 1.7$ m placed in the center of the room, as shown in Figure Pr. 4.51. Neglect surface-convection heat transfer and the heat transfer from the ends of the cylinder.

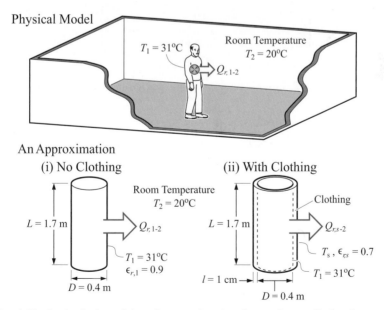

Figure Pr. 4.51. A physical model and approximation for surface-radiation heat exchange between a person and his or her surrounding surfaces, with (i) no clothing and (ii) a layer of clothing.

For a steady-state condition, (a) draw the thermal circuit and (b) determine the rate of heat loss for the case of (i) no clothing covering a body with a surface emissivity $\epsilon_{r,1} = 0.9$, and (ii) for the case of added clothing of thickness $l = 1$ cm with a conductivity $k = 0.1$ W/m-K and a surface emissivity $\epsilon_{r,s} = 0.7$.

Assume that all the surfaces are opaque, diffuse, and gray. Assume negligible contact resistance between the clothing and the body. Comment on the effect of the clothing, for the given temperature difference.

### PROBLEM 4.52. FAM

A thin film is heated with irradiation from a laser source with intensity $q_{r,i} = 10^6$ W/m², as shown in Figure Pr. 4.52. The heat losses from the film are by surface emission and by conduction through the substrate. Assume that the film can be treated as having a uniform temperature $T_1(t)$ and that the conduction resistance through the substrate can be treated as constant.

(a) Draw the thermal circuit diagram.
(b) For an initial temperature $T_1(t = 0) = 20°$C, determine the time needed to raise the temperature of the film $T_1$ to $500°$C.

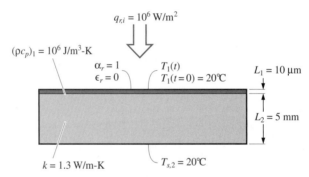

Figure Pr. 4.52. Laser irradiation heating of a thin film on a substrate.

### PROBLEM 4.53. DES

A pipeline carrying cryogenic liquid nitrogen is to be insulated. Two scenarios, shown in Figure Pr. 4.53, are considered. The first one [Figure Pr. 4.53(i)] consists of placing the pipe (tube) concentrically inside a larger diameter casing and filling the space with microspheres insulation material. The microspheres have an effective thermal conductivity of $\langle k \rangle = 0.03$ W/m-K. The tube has an inside diameter $D_1 = 2$ cm and the casing has an outside diameter $D_2 = 10$ cm. Another scenario [Figure Pr. 4.53(ii)] consists of placing a thin polished metal foil between the tube and the casing, thus forming a cylindrical shell with diameter $D_3 = 6$ cm, and then evacuating the spacings. Both the tube and casing have an emissivity $\epsilon_{r,1} = \epsilon_{r,2} = 0.4$ and the thin foil has an emissivity

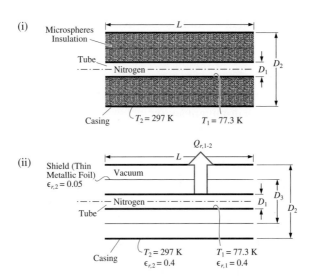

Figure Pr. 4.53. (i) and (ii) Two scenarios for insulation of a cryogenic fluid tube.

$\epsilon_{r,3} = 0.05$. The tube is carrying liquid nitrogen and has a surface temperature $T_1 = 77.3$ K and the casing has a surface temperature $T_2 = 297$ K.

(a) Determine the net heat transfer to liquid nitrogen for the two scenarios using a tube length $L = 1$ m.
(b) How thick should the microsphere insulation be to allow the same heat transfer as that for the evacuated, radiation shield spacing?

## PROBLEM 4.54. FAM

Automatic fire sprinklers, shown in Figure Pr. 4.54, are individually heat activated, and tied into a network of piping filled with pressurized water. When the heat flow from a fire raises the sprinkler temperature to its activation temperature $T_{sl} = 165°F$,

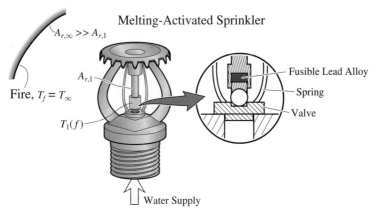

Figure Pr. 4.54. A fire extinguisher actuated by rise in temperature caused by surface-radiation heating.

a lead alloy solder link will melt, and the pre-existing stress in the frame and spring washer will eject the link and retainer from the frame, allowing the water to flow. An AISI 410 stainless steel sprinkler having a mass $M_s = 0.12$ kg and an initial temperature $T_1(t = 0) = 72°$F is used to extinguish a fire having a temperature $T_\infty = 1{,}200°$F and an area $A_{r,\infty}$ much greater than the area of the sprinkler $A_{r,1} = 0.003$ m$^2$. Assume that the dominant source of heat transfer is radiation and that the lumped capacitance analysis is valid.

(a) Draw the thermal circuit diagram.
(b) Determine the elapsed time $t$ needed to raise the sprinkler temperature to the actuation temperature.

## PROBLEM 4.55. FUN

Heat transfer by conduction and surface-radiation in a packed bed of particles with the void space occupied by a gas is approximated using a unit-cell model. Figure Pr. 4.55(i) shows a rendering of the cross section of a packed bed of monosized spherical particles with diameter $D$ and surface emissivity $\epsilon_r$. A two-dimensional, periodic structure with a square unit-cell model is used. The cell has a linear dimension $l$, with the gas and solid phases distributed to allow for an interparticle contact

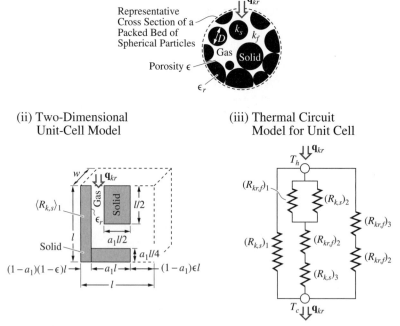

Figure Pr. 4.55. (i) Physical model of a packed bed of spherical particle with the pore space filled with a gas. (ii) A simplified, two-dimensional unit-cell model. (iii) Thermal circuit model for the unit cell.

and also for the presence of the pore space, as shown in Figure Pr. 4.55(ii). The thermal circuit model for this unit cell is shown in Figure Pr. 4.55(iii).

The surface-radiation is approximated by an optically thick medium treatment. This allows for a volumetric presentation of radiation (this is discussed in Section 5.4.6). This uses the concept of radiant (photon) conductivity $\langle k_r \rangle$. One of the models for $\langle k_r \rangle$ (derivation is left as an end of chapter problem) is

$$\langle k_r \rangle = \frac{4\sigma_{\rm SB} T^3 D}{\dfrac{2}{\epsilon_r} - 1} = \frac{4\epsilon_r \sigma_{\rm SB} T^3 D}{2 - \epsilon_r}.$$

Using the geometric parameters shown in Figure Pr. 4.55(ii), show that the total thermal conductivity for the thermal circuit model of Figure Pr. 4.55(iii) is

$$\langle k_{kr} \rangle = \frac{q_{kr}}{(T_h - T_c)l} = (1 - a_1)(1 - \epsilon_r)k_s$$

$$+ \frac{a_1}{\dfrac{1}{(k_f + \langle k_r \rangle) + k_s} + \dfrac{1}{4(k_f + \langle k_r \rangle)} + \dfrac{1}{4k_s}} + (1 - a_1)\epsilon(k_f + \langle k_r \rangle).$$

Here we have combined the surface-radiation with the gas conduction such that in Figure Pr. 4.55(iii), $R_{kr,f}$ uses $k_f + \langle k_r \rangle$ as the conductivity.

### PROBLEM 4.56. FAM.S

The range-top electrical heater has an electrical conductor that carries a current and produces Joule heating $\dot{S}_{e,\rm J}/L$(W/m). This conductor is covered by an electrical insulator. This is shown in Figure Pr. 4.56. An electrical insulator should be a good thermal conductor, in order to a avoid large temperature drop across it. It should also have good wear properties, therefore various ceramics (especially, oxide ceramics) are used. Here we consider alumina ($Al_2O_3$).

Range-Top Electrical Heater

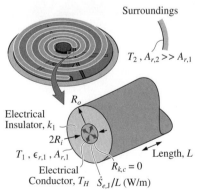

Figure Pr. 4.56. A range-top electrical heater with a cylindrical heating element made of an inner electrical conductor and an outer electrical insulator.

Consider a heater with a circular cross section, as shown in Figure Pr. 4.56. Neglect the conduction contact resistance between the electrical conductor (central cylinder) and the electrical insulator (cylindrical shell). Assume a steady-state surface-radiation heat transfer only (from the heater surface).

(a) Draw the thermal circuit diagram for the heater.
(b) Plot the heater temperature $T_H$ with respect to the outer radius $R_o$, for $2 \le R_o \le 20$ mm, and the conditions given below.
(c) Plot the conduction-radiation number $N_r = R_{k,H-1}/[R_{r,\Sigma}/(4\sigma_{SB}T_m^3)]$, where $T_m$ is defined by (4.73), with respect to $R_o$, for $2 \ll R_o \ll 20$ mm.

$$R_i = 1 \text{ mm}, \dot{S}_{e,J}/L = 5 \times 10^3 \text{ W/m}, T_2 = 30°C, \epsilon_{r,1} = 0.76.$$

## PROBLEM 4.57. FAM

Ice is formed in a water layer as it flows over a cooled surface at temperature $T_c$. Assume that the surface of the water is at the saturation temperature $T_{ls}$ and that the heat transfer across the water layer (thickness $L$) is by steady-state conduction only. The top of the water layer is exposed to the room-temperature surroundings at temperature $T_\infty$, as shown in Figure Pr. 4.57. Assume that water and the surrounding surfaces are opaque, diffuse, and gray (this is a reasonable assumption for water in the near infrared range which is applicable in this problem).

(a) Draw the thermal circuit diagram for the water surface.
(b) Determine the rate of ice formation per unit area $\dot{m}_{ls} = \dot{M}_{ls}/A_l$.

$$T_{ls} = 0°C, T_c = -10°C, T_\infty = 300 \text{ K}, L = 2 \text{ mm}, \epsilon_{r,l} = 1.$$

Assume that the ice is being formed at the top surface of the water layer. Evaluate the water properties at $T = 280$ K (Table C.4 and Table C.13).

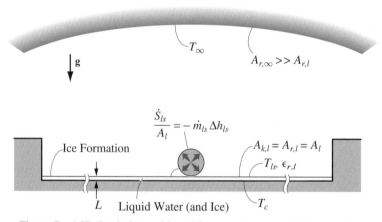

Figure Pr. 4.57. Ice is formed in a thin water layer cooled from below.

### 4.9.7 General*

This section and the Problems that follow are found on the Web site www .cambridge.org/kaviany.

**PROBLEM 4.58. FUN.S***

**PROBLEM 4.59. FUN.S***

**PROBLEM 4.60. FUN.S***

**PROBLEM 4.61. FAM***

# 5

---

# Convection: Unbounded Fluid Streams*

Chapter Five can be found on the text Web site www.cambridge.org/kaviany. In Chapter 5, we begin discussion of convection heat transfer by considering an unbounded fluid stream undergoing a temperature change along the stream flow direction, where this change is caused by an energy conversion. Convection heat transfer is the transfer of heat from one point to another by a net (i.e., macroscopic) fluid motion (i.e., by currents of gas or liquid particles) given by the fluid velocity vector. The fluid motion can be void of random fluctuations, which is called a laminar flow, or it can have these random fluctuations, which is called a turbulent flow. Convection heat transfer examines the magnitude, direction, and spatial and temporal variations of the convection heat flux vector in a fluid stream. In this chapter, we examine convective heat transfer within a fluid stream without directly considering the heat transfer between the fluid and a bounding solid surface (i.e., here we consider only intramedium conduction and convection of unbounded fluid streams). In addition to the occurrence of the intramedium (i.e., bulk) convection, in many practical applications involving cooling and heating of fluids, the review of the intramedium convection allows for an introductory treatment of laminar, steady-state, uniform, one-directional fluid flow without addressing the effect of the fluid viscosity emanating from the bounding surfaces, which would require inclusion of the momentum conservation equation. The interfacial heat transfer between a fluid and a solid surface, i.e., the surface-convection heat flux vector is influenced by the fluid viscosity, since the viscosity results in the nonuniformity of the fluid velocity near the solid surface. Also due to the surface-convection heat transfer, the fluid temperature changes, and this change for semi-bounded fluid stream is confined to a region adjacent to the bounding surface (while the far-field temperature is unchanged); this is treated in Chapter 6. When the fluid stream is bounded, the surface-convection heat transfer alters the fluid temperature across and along the stream, and this is discussed in Chapter 7.

# 6

## Convection: Semi-Bounded Fluid Streams

The surface-convection heat transfer coefficient can be represented in a universal, dimensionless form, similar to the surface friction representation in hydraulics.

– W. Nusselt

A semi-bounded fluid stream in thermal nonequilibrium with its bounding solid surface exchanges heat by surface convection $\mathbf{q}_{ku}$ and this alters its temperature (and $\mathbf{q}_k$ and $\mathbf{q}_u$) only in a region adjacent to the bounding surface (i.e., the far-field temperature is unchanged). In this chapter we consider a steady, semi-bounded fluid stream flowing over a solid surface (e.g., flow over a plate or a sphere) at a far-field velocity $\mathbf{u}_{f,\infty}$ with the solid $T_s$ and the far-field $T_{f,\infty}$ fluid having different temperatures. We initially focus on the fluid and its motion and heat transfer adjacent to the surface (because of this focus, the chapter title, semi-bounded fluid streams, is used. Also as will be shown, there are some similarities with the transient conduction in semi-infinite solid). Later we include the heat transfer with any bounding solid (i.e., the substrate). This is also called external flow heat transfer. In Chapter 7 we will consider steady, bounded fluid streams (e.g., flow in a tube). This is also called internal flow. We are interested in the parameters and conditions influencing the surface-convection heat flux $\mathbf{q}_{ku}$ and its spatial variation. These parameters and conditions include the direction of the fluid flow with respect to the bounding surface (e.g., fluid flow parallel to the surface, perpendicular to the surface, or oblique), flow structure, fluid properties, liquid-gas phase change, and solid-surface conditions. In absence of radiation, within the fluid, the heat flux is given by $\mathbf{q}_k$ and $\mathbf{q}_u$ and the magnitude and direction of both $\mathbf{q}_k$ and $\mathbf{q}_u$ change at and near the surface. In the presence of an impermeable solid bounding surface, due to the fluid viscosity $\mu_f$ resulting in no motion for the fluid stream at the stationary solid surface, we have $\mathbf{u}_f = 0$ at the surface. Then the surface $\mathbf{q}_u = 0$, because $\mathbf{u}_f = 0$. We designate the surface-convection heat flux with $\mathbf{q}_{ku}$ to show that the heat transfer in the fluid at the surface is by conduction, but influenced by fluid motion [i.e., in $-k_f\, \partial T_f/\partial n$ (on surface), $\partial T_f/\partial n$ depends on $\mathbf{u}_{f,\infty}$]. We also define a surface-convection resistance $R_{ku}$, which represents the resistance to this heat conduction through the fluid in

426

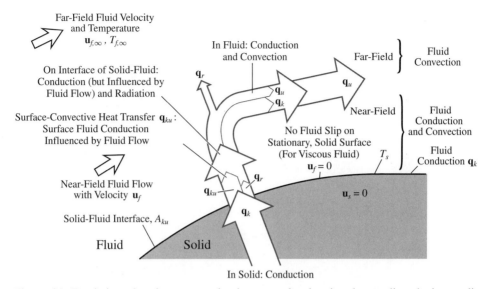

Figure 6.1. Depiction of surface-convection heat transfer showing the no-slip velocity condition on the solid surface. The interfacial heat transfer by conduction on the fluid side depends on the fluid velocity and is designated as $\mathbf{q}_{ku}$, which is called the surface-convection heat flux.

contact with the surface, but includes the effect of the fluid motion (i.e., convection) adjacent to the surface.

## 6.1 Flow and Surface Characteristics

In this chapter we consider a flowing fluid partly bounded by a solid with the condition of thermal nonequilibrium between the far-field fluid and the solid surface (i.e., the far-field fluid and solid are at different temperatures). This would require the inclusion of the fluid momentum (and mass) conservation equation in addition to the energy conservation equation. Here, the semi-bounded fluid is assumed to have a steady and uniform, far-field velocity $u_f = u_{f,\infty}$ and a steady and uniform far-field temperature $T_{f,\infty}$ (for $y \to \infty$, with $y = 0$ at the surface). The fluid has a nonzero viscosity $\mu_f$ (in some special cases, the idealization of zero viscosity is used). For simplicity, stationary solid bodies are assumed, and therefore, due to the fluid no-slip condition (due to the nonzero fluid viscosity), the velocity of the fluid on the solid surface is zero ($\mathbf{u}_f = 0$ on $A_s$). This creates a fluid velocity nonuniformity near the solid surface (a near-field velocity nonuniformity effect). This velocity field is two or three dimensional. Due to the flow instabilities, and if there are surface curvatures, the flow may become time dependent (and then time averaging may be necessary).

The thermal nonequilibrium will result in heat transfer between (intermedium) the fluid and the solid, as well as heat transfer within the fluid and the solid (intramedium), near the solid surface (this is referred to as near-field temperature nonuniformities). These inter- and intramedium heat transfers are depicted in Figure 6.1.

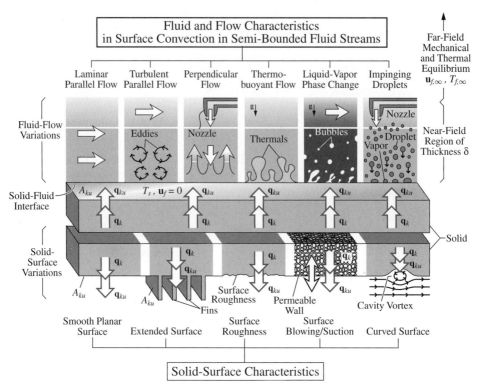

Chart 6.1. Fluid and flow characteristics and solid-surface characteristics influencing $\mathbf{q}_{ku}$.

The near-field temperature and velocity variations in the fluid are similar to the penetration effect in transient conduction discussed in Section 3.5, where a sudden change was introduced in the thermal condition at the bounding surface. Since the fluid is semi-bounded, we cannot simply follow its average temperature or its average convection heat flux as it flows over the solid and exchanges heat. This simply is due to the semi-infinite extent of the fluid, which results in an unchanged, average (over the semi-infinite fluid) fluid temperature. In contrast, for bounded fluids we can readily define an average fluid temperature by integrating the fluid convection heat flux across the known, finite, fluid flow cross section. Therefore, in semi-bounded fluid flows our attention is only on the solid-fluid interfacial (or surface-convection) heat flux $\mathbf{q}_{ku}$.

The fluid and flow, and the solid-surface characteristics influencing $q_{ku}$, are depicted in Chart 6.1. Among the fluid and the flow characteristics are the fluid properties (thermodynamic and transport) and the flow structure. The flow structure includes the direction of the flow with respect to the solid surface. While parallel flows are common, jet-like perpendicular flows are more effective for surface-convection heat transfer. The fluid can also undergo liquid-gas phase change as a result of the surface-convection heat transfer. Other flow structure aspects, such as flow instabilities, turbulence, separation, oscillation, thermobuoyancy, phase-density buoyancy, electromagnetic forces (for charged fluids), and particle suspensions, are also important.

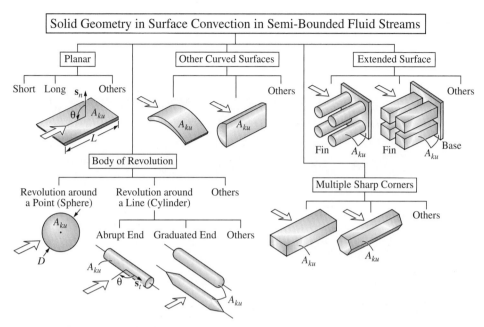

Chart 6.2. A classification of surface geometry for semi-bounded fluids.

The solid-surface characteristics, such as surface roughness, surface extensions (i.e., fins), permeability (allowing for surface blowing or suction), curvature in the surface contour, etc., are also important. The surfaces considered here are all assumed to be smooth. A classification of the surface roughness was given in Section 3.3.5. The heat transfer within the solid (intramedium) may also be important (as in extended surfaces and in solid walls separating two fluids in indirect heat exchange).

We also note that the solid geometry can vary greatly, depending on the application. Chart 6.2 gives a classification of the solid geometry. The solid surface can be planar, or be the surface of a body of revolution, or simply be curved. Again, the solid surface can also have many solid attachments.

The flow structure also varies depending on the surface geometry and the flow inertia. These are shown in Chart 6.3 as various visualized flows. At low velocities, the flow is laminar. After a threshold velocity, a transition from laminar to turbulent flow occurs. Also shown in this chart are thermobuoyancy, phase-density buoyancy, and compressibility effects on the flow structure.

In order to examine the effect of fluid and flow characteristics, we first consider the simple surface geometry of a flat plate. Then we examine other surface geometries. Next we discuss the inclusion of the substrate heat transfer. Finally we discuss surface-convection evaporation cooling, where there is a simultaneous surface transfer of heat and mass. When we neglect the motion of the evaporating liquid, the surface-convection heat and mass become analogous. Chart 6.4 lists aspects of surface-convection heat transfer, with fluid partly bounded by a solid surface, covered in this chapter.

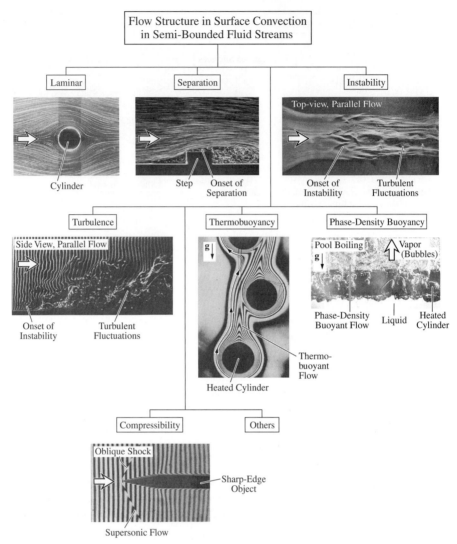

Chart 6.3. A classification of flow structure for semi-bounded fluid.

## 6.2 Semi-Infinite Plate as a Simple Geometry

The semi-infinite plate offers a simple geometry (no surface curvature) and a two dimensionality (as compared to three dimensionality) of the flow. Therefore, it is used to study surface-convection heat transfer and to obtain nearly closed-form solutions for $q_{ku}$ (or $R_{ku}$).

### 6.2.1 Local Surface-Convection Resistance $R_{ku}(°C/W)$

So far we have dealt with the thermal resistance for the intramedium conduction $R_k$, the surface-grayness and the view-factor resistances in enclosure (gray, diffuse, opaque surfaces) surface-radiation exchange $R_{r,\epsilon}$ and $R_{r,F}$, and the axial

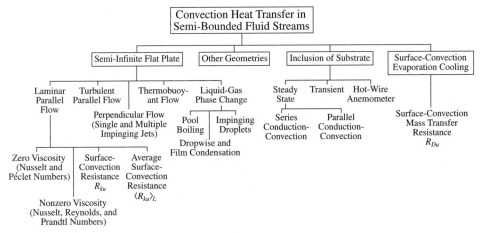

Chart 6.4. Aspects of surface-convection heat transfer, with a semi-bounded fluid stream, considered in Chapter 6.

conduction-convection resistance $R_{k,u}$ in one-dimensional, uniform fluid flow. As will be shown in Section 6.2.3, the heat transfer between a solid surface at a temperature $T_s$ and a semi-bounded flow with a far-field temperature $T_{f,\infty}$ is also represented by a resistance. This is called the surface-convection resistance $R_{ku}$. Note that we have used both subscripts $k$ and $u$ and we have not separated them. The use of $k$ is because we consider stationary, impermeable solid surfaces and use the no-slip condition for the fluid in contact with the solid, then the heat transfer on the surface is by conduction only (assuming for now that radiation is not significant). However, since the rate of this heat transfer depends on the near- and far-field fluid velocity, the subscript $u$ is also used.

We begin by developing some simple relations for $R_{ku}$ for the semi-infinite plate and for simple flow fields (laminar parallel flow, then turbulent parallel fluid flows, then perpendicular, and finally thermobuoyant and phase-density buoyant flows). We will show that $R_{ku}$ varies along the surface. Next we will discuss the average surface-convection resistance $\langle R_{ku}\rangle_L$, where $\langle\,\rangle$ indicates spatial average, which in this case is along the length of the surface $L$. Then we will progressively address more complex flows and geometries. As will be shown, these complexities are most readily dealt with empirically.

### 6.2.2 Viscous versus Thermal Boundary Layer: Prandtl Number Pr

We begin by considering laminar, parallel flow over a semi-infinite plate (this is the simplest semi-bounded fluid flow). To make the analysis yet simpler, we assume that the fluid kinematic viscosity $\nu_f$ is very small compared to its thermal diffusivity $\alpha_f$. In the case of $\nu_f = 0$, the velocity in the near field is also uniform. After developing the needed relation between $q_{ku}$, $A_{ku}R_{ku}$, and $T_s - T_{f,\infty}$, we include the influence of the nonzero viscosity. This will make the velocity field in the near field nonuniform

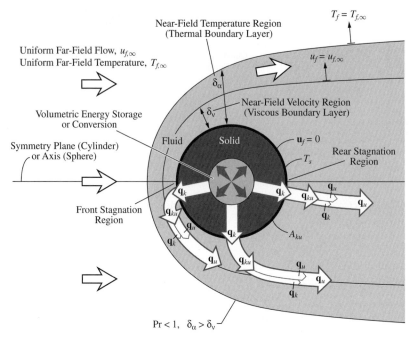

Figure 6.2. Prandtl number Pr represents the ratio of viscous to thermal penetration regions (i.e., ratio of extents of the near-field regions).

and two dimensional. The ratio of $\nu_f$ to $\alpha_f$ is called the Prandtl number Pr ($0 \leq$ Pr $< \infty$)

$$\text{Pr} \equiv \frac{\nu_f}{\alpha_f} \qquad \text{Prandtl number, } 0 \leq \text{Pr} < \infty. \qquad (6.1)$$

As will be shown in Section 6.2.5, Pr compares the extent of the near-field, nonuniform velocity region $\delta_\nu$ to the extent of the near-field, nonuniform temperature region $\delta_\alpha$. In Figure 6.2, $\delta_\nu$ and $\delta_\alpha$ are shown for the case with $\delta_\nu < \delta_\alpha$ and this corresponds to Pr $< 1$. The two near-field regions are also called the viscous $\delta_\nu$ and thermal $\delta_\alpha$ boundary-layer regions or the viscous and thermal penetration regions.

### 6.2.3 Boundary-Layer Flow with Zero Prandtl Number: Axial-Lateral Péclet Number Pe$_L$

We begin by considering the case of zero fluid viscosity. The liquid metals have Pr $\ll 1$ and may be treated as having Pr $\to 0$. Although the case of zero viscosity, $\mu_f = 0$, i.e., Pr $= 0$, which gives a $\delta_\nu = 0$, does not allow for a general treatment, it is instructive due to its simplicity. The simplicity is in not requiring the inclusion of fluid

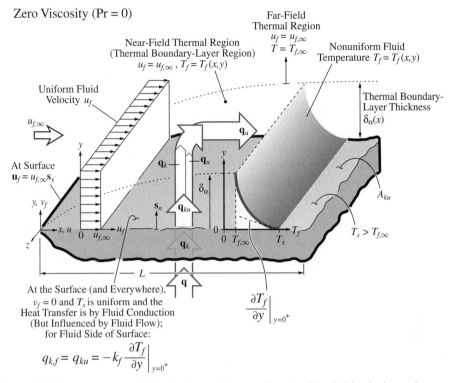

Figure 6.3. Flow over a semi-infinite plate with zero viscosity. The fluid velocity and temperature distributions are also shown.

momentum (and mass) conservation equation. In Section 6.2.4, we will consider the case of nonzero viscosity and include the fluid momentum conservation equation.

Here, we consider laminar (i.e., no turbulent fluctuations), parallel, uniform (near- and far-field) flow $u_{f,\infty}$, over a semi-infinite, impermeable plate. The plate is at a uniform surface temperature at $T_s$, and the fluid far-field temperature $T_{f,\infty}$. This is rendered in Figure 6.3.

In order to determine the relation between $q_{ku}$ and $T_s - T_{f,\infty}$, i.e., to determine $R_{ku}$, we begin with the differential-volume ($\Delta V \to 0$) energy equation (2.1), i.e.,

$$\nabla \cdot \mathbf{q} = \frac{\displaystyle\int_{\Delta A}(\mathbf{q}\cdot\mathbf{s}_u)dA}{\Delta V} = -\frac{\partial}{\partial t}(\rho c_p)_f T_f + \dot{s} \quad \text{fluid energy equation.} \quad (6.2)$$

We assume that $u_{f,\infty}$, $T_{f,\infty}$, and $T_s$ are constant, and therefore, we have a steady state, and (6.2) becomes

$$\nabla \cdot \mathbf{q} = \dot{s}. \quad (6.3)$$

Next, we assume that there is no energy conversion and (6.3) becomes

$$\nabla \cdot \mathbf{q} = 0 \quad \text{(divergence of } \mathbf{q} \text{ is zero) fluid energy equation.} \quad (6.4)$$

This is the case where the direction and contribution of the different mechanisms of heat transfer change within the fluid thermal boundary layer. We assume that radiation heat transfer is negligible and write (6.4) as

$$\nabla \cdot (\mathbf{q}_k + \mathbf{q}_u) = \nabla \cdot [-k_f \nabla T_f + (\rho c_p)_f \mathbf{u}_f T_f] = 0. \tag{6.5}$$

For the uniform, far-field and near-field fluid flow, we have $\mathbf{u}_f = u_{f,\infty} \mathbf{s}_x$. We also have a two-dimensional near-field temperature distribution $T_f = T_f(x, y)$. Then assuming constant $k_f$ and $\rho_f$ ($c_p$ is already assumed constant), (6.5) becomes

$$\frac{\partial}{\partial x} q_{k,x} \quad + \quad \frac{\partial}{\partial y} q_{k,y} \quad + \quad \frac{\partial}{\partial x} q_{u,x} \quad =$$

$$-k_f \frac{\partial^2 T_f}{\partial x^2} \quad - \quad k_f \frac{\partial^2 T_f}{\partial y^2} \quad + \quad (\rho c_p)_f u_{f,\infty} \frac{\partial T_f}{\partial x} = 0. \tag{6.6}$$

| spatial rate of change of $\mathbf{q}_k$ in $x$ direction | spatial rate of change of $\mathbf{q}_k$ in $y$ direction | spatial rate of change of $\mathbf{q}_u$ in $x$ direction |

### (A) Boundary-Layer Approximation

Next we make the assumption that the thermal penetration (or boundary-layer) thickness $\delta_\alpha$, shown in Figure 6.3, is much smaller than $L$, where $L$ is the distance along the $x$ direction over which we are interested in the surface convection $q_{ku}$. This condition $\delta_\alpha/L < 1$ occurs when the fluid moves fast enough to confine the effect of surface-convection heat transfer to a small region near the surface. At $x = 0$ (called the leading edge), $\delta_\alpha = 0$. Then we write this as

$$\frac{\delta_\alpha}{L} < 1 \quad \text{lateral penetration is smaller than axial distance,} \tag{6.7}$$

for $x > 0$ (also called downstream of the leading edge). Since $T_f$ varies between $T_s$ and $T_{f,\infty}$, both in the $x$ and in the $y$ directions, we can use (6.7) to write the following relations for $k_f \partial^2 T_f/\partial x^2$ and $k_f \partial^2 T_f/\partial y^2$, both appearing in the energy equation (6.6). The result is

$$k_f \frac{\partial^2 T_f}{\partial x^2} \simeq k_f \frac{T_s - T_{f,\infty}}{L^2}, \quad k_f \frac{\partial^2 T_f}{\partial y^2} \simeq k_f \frac{T_s - T_{f,\infty}}{\delta_\alpha^2} \quad \begin{array}{l} \text{approximation of} \\ \text{second derivatives.} \end{array} \tag{6.8}$$

By comparing these and using (6.7), we have

$$k_f \frac{\partial^2 T_f}{\partial x^2} \ll k_f \frac{\partial^2 T_f}{\partial y^2} \quad \begin{array}{l} \text{spatial rate of change of axial} \\ \text{conduction is negligible compared} \\ \text{to lateral conduction.} \end{array} \tag{6.9}$$

This is called the boundary-layer approximation.

Using (6.9), we simplify (6.6) to

$$-k_f \frac{\partial^2 T_f}{\partial y^2} \quad + \quad (\rho c_p)_f u_{f,\infty} \frac{\partial T_f}{\partial x} = \frac{\partial}{\partial y} \mathbf{s}_y \cdot \mathbf{q}_k + \frac{\partial}{\partial x} \mathbf{s}_x \cdot \mathbf{q}_u = 0 \quad \begin{array}{l} \text{boundary-layer} \\ \text{energy equation} \\ \text{for Pr = 0.} \end{array}$$

| spatial rate of change of $\mathbf{q}_k$ in $y$ direction | spatial rate of change of $\mathbf{q}_u$ in $x$ direction |

$$\tag{6.10}$$

(a) Zero Viscosity, Parallel Flow: No Lateral Convection

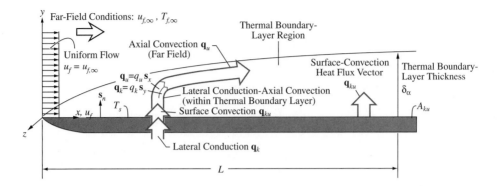

(b) Thermal Boundary-Layer Thickness $\delta_\alpha$

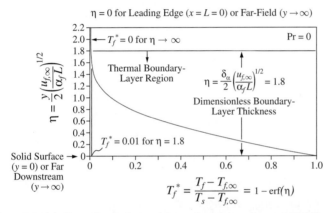

Figure 6.4. (a) Growth of thermal boundary layer for parallel flow. For the case of zero viscosity no lateral fluid motion (i.e., no lateral convection) is present. (b) The dimensionless temperature distribution with the spatial variables $x$ and $y$ combined into the similarity variable $\eta$. The thermal boundary-layer thickness is also shown.

The vectorial form of the energy equation is reintroduced to show that the dominant conduction is lateral (perpendicular to the surface), while the only (because of the one-dimensional fluid flow) convection is axial (parallel to the surface). This is also rendered in Figure 6.4(a), where the lack of lateral convection (due to zero viscosity and thus the uniformity of flow) should be noted.

We note that the axial temperature variation in the convection terms allows for affecting the temperature via the upstream changes, but the changes occurring downstream cannot affect the upstream. This is a characteristic of the so-called initial-value problem. Therefore, this boundary-layer treatment is such that any $x$ location downstream of a location $L$ does not influence the heat transfer at $L$.

Now using the following scales, we make $T_f$, $x$, and $y$ dimensionless (an asterisk indicates a dimensionless quantity), i.e.,

$$T_f^* = \frac{T_f - T_{f,\infty}}{T_s - T_{f,\infty}}, \qquad x^* = \frac{x}{L}, \qquad y^* = \frac{y}{L} \qquad \text{dimensionless variables.} \qquad (6.11)$$

Then (6.10) becomes

$$-\frac{\partial^2 T_f^*}{\partial y^{*2}} \quad + \quad \mathrm{Pe}_L \frac{\partial T_f^*}{\partial x^*} \quad = 0 \qquad \begin{array}{l} \text{dimensionless} \\ \text{energy equation,} \end{array}$$

$$\underbrace{\phantom{-\frac{\partial^2 T_f^*}{\partial y^{*2}}}}_{\substack{\text{lateral rate of change} \\ \text{of conduction}}} \qquad \underbrace{\phantom{\mathrm{Pe}_L \frac{\partial T_f^*}{\partial x^*}}}_{\substack{\text{axial rate of change} \\ \text{of convection}}} \tag{6.12}$$

where, similar to Section 5.2, the Péclet number $\mathrm{Pe}_L$ is

$$\mathrm{Pe}_L = \frac{u_{f,\infty} L}{\alpha_f}, \qquad \alpha_f = \left(\frac{k}{\rho c_p}\right)_f \qquad \text{axial-lateral Péclet number.} \tag{6.13}$$

Note that, as compared to Chapter 5, here $\mathrm{Pe}_L$ is the ratio of the axial convection to lateral conduction. The heat flows from the surface by conduction (but influenced by convection), and for the case of an isothermal surface this heat flow is perpendicular to the surface. As we move away from the surface, the heat flux vector changes direction and once we are at the edge of the boundary layer, only convection heat transfer occurs (and is parallel to the surface).

It can be shown that for (6.7) to hold, i.e., for the fluid to move fast enough to confine the temperature nonuniformity to a small region adjacent to the surface, we need to have a large $\mathrm{Pe}_L$ (say larger than 100).

Noting the order of the derivatives, to integrate (6.12), we need one thermal boundary condition for the $x$ direction (leading-edge condition), and two thermal boundary conditions for the $y$ direction (surface and far-field conditions). The bounding-surface and far-field thermal conditions are (2.63) and (2.65), along with the mechanical conditions (uniform velocity along $x$); these are

$$v_f(x, y \to \infty) = 0 \quad \text{or} \quad v_f^*(x^*, y^*) = 0$$
$$u_f(x, y) = u_{f,\infty} \quad \text{or} \quad u_f^*(x^*, y^* \to \infty) = \frac{u_f}{u_{f,\infty}} = 1$$

$$\begin{aligned} T_f(x = 0, y) &= T_{f,\infty} & \text{or} & \quad T_f^*(x^* = 0, y^*) = 0 \\ T_f(x, y = 0) &= T_s & \text{or} & \quad T_f^*(x^*, y^* = 0) = 1 \\ T_f(x, y \to \infty) &= T_{f,\infty} & \text{or} & \quad T_f^*(x^*, y^* \to \infty) = 0 \end{aligned}$$

$$\text{thermal and mechanical conditions, at various locations.} \tag{6.14}$$

Note that we have stated that the far-field condition occurs far from the surface. This is because we have not yet determined $\delta_\alpha$. The solution shown below will determine $\delta_\alpha$, once we choose a practical marking between the near- and the far-field regions. This is similar to the transient conduction treatment in Section 3.5.

### (B) Fluid Temperature Distribution

Note that both $x$ and $y$ have the same range, $0 \le x < \infty$ and $0 \le y < \infty$. This and the properties of the differential equation (6.12), which is based on the boundary-layer approximations, will allow us to define a similarity variable $\eta$ as

$$\eta = \frac{y}{2}\left(\frac{u_{f,\infty}}{\alpha_f L}\right)^{1/2} \qquad \text{dimensionless similarity variable,} \tag{6.15}$$

where $L$ is a location along the $x$ axis. As in Section 3.5.1 using (3.136), the similarity variable allows us to combine the two independent variables, and then the partial differential equation reduces to an ordinary differential equation. Here using (6.15), (6.12) becomes (after several steps)

$$\frac{d^2 T_f^*}{d\eta^2} + 2\eta \frac{dT_f^*}{d\eta} = 0 \quad \text{dimensionless energy equation.} \tag{6.16}$$

The thermal boundary conditions given in (6.14) are then reduced from three to two, i.e.,

$$T_f^*(\eta = 0) = 1, \qquad T_f^*(\eta \to \infty) = 0. \tag{6.17}$$

The solution to (6.16), after applying the conditions (6.17), is

$$T_f^* = \frac{T_f - T_{f,\infty}}{T_s - T_{f,\infty}} = 1 - \text{erf}(\eta) \quad \begin{array}{l}\text{solution for dimensionless} \\ \text{fluid temperature distribution,}\end{array} \tag{6.18}$$

where $\text{erf}(\eta)$ is the Gaussian error function.

The variation of $T^*$ with respect to $\eta$ is shown in Figure 6.4(b). Note that (6.12) is similar to the transient conduction equation (3.140). The roles of $x$ here and $t$ there are similar. (Also the range for $x$ and $t$ are the same, i.e., $0 \le x < \infty$, $0 \le t < \infty$.) Therefore, the solution found for $T^*$ for this conduction-convection problem, (6.18), is similar to that for conduction-storage problem, i.e., (3.140). Again, the error function is defined as

$$\text{erf}(\eta) = \frac{2}{\pi^{1/2}} \int_0^\eta e^{-z^2} dz \quad \text{error function.} \tag{6.19}$$

In Section 3.5, the function $\text{erf}(\eta)$ was plotted in Figure 3.33(a) and listed in Table 3.5, where it was shown that $\text{erf}(0) = 0$ and $\text{erf}(\infty) = 1$.

We choose $\delta_\alpha$ as the location where $T_f^* = 0.01$ (instead of $T_f^* = 0$, which is for $y \to \infty$, since $T_f^* = 0$ is reached asymptotically). This is shown in Figure 6.4(b). From the variation of $T^*$ with respect to $\eta$ given by (6.18), and the magnitude of $\text{erf}(\eta)$ in Table 3.5, we find that

$$T_f^* = 0.01 \quad \text{for} \quad \eta = 1.8 \quad \begin{array}{l}\eta = 1.8 \text{ marks separation between near and} \\ \text{far fields (criterion for marking the edge} \\ \text{of the boundary layer).}\end{array} \tag{6.20}$$

Now we determine $\delta_\alpha$ by using the definition of $\eta$ in (6.15) and the value $\eta = 1.8$ (as the marking between the near- and far-field regions). From (6.20), for any location $x = L$, we have

$$\eta(\delta_\alpha) = \frac{\delta_\alpha^*}{2} \text{Pe}_L^{1/2} \equiv 1.8 \quad \text{or} \quad \delta_\alpha \equiv 3.6 \left(\frac{\alpha_f L}{u_{f,\infty}}\right)^{1/2} \quad \begin{array}{l}\text{thickness of thermal} \\ \text{boundary layer for} \\ \text{laminar flow with Pr} = 0.\end{array} \tag{6.21}$$

This shows that $\delta_\alpha$ increases (grows) as $L^{1/2}$, increases as $\alpha_f$ increases, and decreases as $u_{f,\infty}$ increases.

The temperature distribution given by (6.18) and the uniform velocity distribution are shown in Figure 6.3. The thermal boundary-layer thickness $\delta_\alpha = \delta_\alpha(x \text{ or } L)$ and its growth along the $x$ axis (or as $L$ becomes larger) are also shown. Within the boundary-layer region $0 \le y \le \delta_\alpha$, the temperature changes monotonically (i.e., with no change in the sign of the slope). For the condition rendered in Figure 6.3, i.e., $T_{f,\infty} < T_s$, this corresponds to a drop in temperature as $y$ approaches $\delta_\alpha$.

### (C) Surface-Convection Heat Transfer

We now evaluate the surface heat flux $q_{ku}$. Since we have assumed a uniform surface temperature $T_s$, the heat flux vector $\mathbf{q}_{ku}$ is perpendicular to the surface. Due to the assumed zero viscosity, the fluid slips over the surface and is parallel to the surface. We use the bounding-surface energy equation (2.63), and neglect the radiation on the fluid side. Note that $\mathbf{u}_f \cdot \mathbf{s}_n = 0$ at $y = 0$, i.e., no fluid flows perpendicular to surface. Then, referring to Figure 6.3, the surface-convection heat flow rate $Q_{k,f}(y = 0) = Q_{ku} = A_{ku}q_{ku}$ (which is the same as $A_{ku}q_f$) at the surface is found from (2.63) as

$$-Q_{k,s} + Q_{k,f} = A_{ku}k_s\frac{\partial T_s}{\partial y}\Big|_{y=0^-} -A_{ku}k_f\frac{\partial T_f}{\partial y}\Big|_{y=0^+} \equiv A_{ku}k_s\frac{\partial T_s}{\partial y}\Big|_{y=0^-} +A_{ku}\,q_{ku} = 0$$

bounding-surface energy equation                                    (6.22)

or

$$q_{k,f}(y = 0) \equiv \mathbf{q}_{ku} \cdot \mathbf{s}_n = q_{ku} = -k_f\frac{\partial T_f}{\partial y}\Big|_{y=0^+} \quad \begin{array}{l}\text{definition of surface-}\\\text{convection heat flux,}\end{array} \quad (6.23)$$

where we note that the surface-convection heat flux $q_{k,f}(y = 0) \equiv q_{ku}$ is by conduction through the fluid and at the surface where $\mathbf{u}_f = 0$, but it is influenced by the fluid motion through $\partial T_f/\partial y/\big|_{y=0^+}$. The surface-convection heat flux vector is normal to the surface. Note that here, $v_f = 0$. Since we have assumed negligible surface radiation, in (6.22) $q_{ku}$ is balanced by the solid-side conduction. By differentiating (6.18) and evaluating the derivative at $y = 0$, (6.22) becomes

$$q_{ku} = \frac{k_f}{L}\frac{\mathrm{Pe}_L^{1/2}}{\pi^{1/2}}(T_s - T_{f,\infty}), \quad \frac{\partial T_f}{\partial y}\Big|_{y=0^+} = -\frac{\mathrm{Pe}_L^{1/2}}{\pi}\frac{T_s - T_{f,\infty}}{L} \quad \begin{array}{l}\text{relation for surface-}\\\text{convection heat flux.}\\\text{for } \mu_f = 0.\end{array}$$

(6.24)

Therefore, $\partial T_f/\partial y\big|_{y=0^+}$ is related to $u_{f,\infty}$ through $\mathrm{Pe}_L$.

Now by examining (6.24), similar to (3.50), we define the local surface-convection thermal resistance through

$$R_{ku} \equiv \frac{T_s - T_{f,\infty}}{Q_{ku}}, \quad Q_{ku} = A_{ku}q_{ku} \quad \begin{array}{l}\text{definition of local,}\\\text{surface-convection resistance.}\end{array} \quad (6.25)$$

Table 6.1. *Average surface-conduction resistance for semi-bounded fluid streams*

| Heat flow rate | Potential | Resistance |
|---|---|---|
| $\langle Q_{ku} \rangle_L = A_{ku} \langle q_{ku} \rangle_L$ $= (T_s - T_{f,\infty})/\langle R_{ku} \rangle_L$, W $\langle Q_{ku} \rangle_D = A_{ku} \langle q_{ku} \rangle_D$ $= (T_s - T_{f,\infty})/\langle R_{ku} \rangle_D$, W | $T_s - T_{f,\infty}$, °C | $\langle R_{ku} \rangle_L = L/A_{ku} \langle Nu \rangle_L k_f$, °C/W, semi-infinite plate $\langle R_{ku} \rangle_D = D/A_{ku} \langle Nu \rangle_D k_f$, finite objects |

This is referred to as the Newton law of cooling.

Then using (6.24), the local surface-convection resistance $R_{ku}$ becomes

$$A_{ku} R_{ku} = \frac{L}{k_f} \frac{\pi^{1/2}}{\text{Pe}_L^{1/2}} = \frac{L^{1/2}}{k_f^{1/2}} \frac{\pi^{1/2}}{(\rho c_p)_f^{1/2} u_{f,\infty}^{1/2}} = \frac{\pi^{1/2} L^{1/2}}{(\rho c_p k)_f^{1/2} u_{f,\infty}^{1/2}} \quad \begin{array}{l} \text{local, surface-convection} \\ \text{resistance for laminar} \\ \text{flow with Pr} = 0. \end{array}$$

$$(6.26)$$

Equation (6.26) shows that $R_{ku}$ increases as $L$ increases. This is because as heat is transferred to the boundary-layer region and the thermal boundary-layer thickness $\delta_\alpha$ grows, the heat has to be transferred laterally (by conduction) through a larger distance. The resistance decreases as the product $(k_f \rho c_p)_f u_{f,\infty}$ increases. Note that, as in (3.144), the product $(\rho c_p k)^{1/2}$ is the thermal effusivity.

Similar to (3.144), we can combine (6.24) and (6.25) and write for $q_{ku}$

$$q_{ku} = \frac{T_s - T_{f,\infty}}{A_{ku} R_{ku}} = \frac{k_f}{L} \frac{\text{Pe}_L^{1/2}}{\pi^{1/2}} (T_s - T_{f,\infty}) = \frac{(\rho c_p k)_f^{1/2} u_{f,\infty}^{1/2}}{\pi^{1/2} L^{1/2}} (T_s - T_{f,\infty}). \quad (6.27)$$

Note that similar to the transient, distributed heat conduction through a semi-infinite medium discussed in Section 3.5.1, the heat transfer rate is proportional to the fluid effusivity, $(\rho c_p k)_f^{1/2}$, i.e., (3.145) and (6.27) are similar. As expected, (6.27) shows that $R_{ku}$ is not a function of $T_s - T_{f,\infty}$, and therefore, it can be used similar to the conduction resistance. Table 6.1 shows the resistance analogy of the surface-convection heat flow. In Table 6.1, the resistance is given in terms of the average, dimensionless conductance, which is described below.

### 6.2.4 Dimensionless Surface-Convection Conductance: Nusselt Number $Nu_L \equiv N_{ku,f}$, and Heat Transfer Coefficient $h$

It is customary to normalize $A_{ku} R_{ku}$[°C/(W/m²)], i.e., making it dimensionless, by scaling it with respect to the conduction resistance. Here we use the same area for conduction (i.e., $A_k = A_{ku}$) and use the axial conduction resistance as $A_k R_{k,f} = L/k_f$ [°C/(W/m²)]. In general, $A_k R_k$ is larger than $A_{ku} R_{ku}$. These two resistances are shown in Figure 6.5. The ratio of $A_k R_{k,f} = L/k_f$ to $A_{ku} R_{ku}$ is called the Nusselt number $Nu_L$ (subscript $L$ indicates the length scale used in the normalization), and is given as

Dimensionless Local Surface-Convection Conductance (Nusselt Number $\mathrm{Nu}_L$):

$$\mathrm{Nu}_L \equiv \frac{1}{R_{ku}{}^*} \equiv \frac{R_k}{R_{ku}}$$

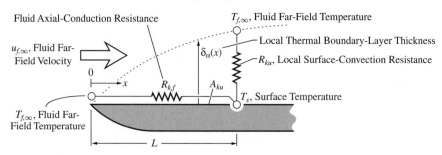

Figure 6.5. A description of the dimensionless surface-convection conductance, the Nusselt number.

$$\mathrm{Nu}_L \equiv N_{ku,f} = \frac{1}{R_{ku}^*} = \frac{R_{k,f}}{R_{ku}} = \frac{A_k R_{k,f}}{A_{ku} R_{ku}} = \frac{L}{A_{ku} R_{ku} k_f} = \frac{q_{ku} L}{(T_s - T_{f,\infty}) k_f} \qquad (6.28)$$

$$\text{definition of local Nusselt number } \mathrm{Nu}_L, \, 0 \le \mathrm{Nu}_L < \infty.$$

or

$$A_{ku} R_{ku} = \frac{L}{\mathrm{Nu}_L k_f} \qquad \text{relation between surface-convection resistance} \qquad (6.29)$$
$$\text{and Nusselt number.}$$

The Nusselt number $\mathrm{Nu}_L$ is the dimensionless surface-convection conductance (i.e, $N_{ku,f}$, the ratio of surface-convection conductance to conduction conductance). Also we note that $0 \le \mathrm{Nu}_L < \infty$.

Note that because the axial location $L$ is used in $R_k$, this can be called the axial-lateral Nusselt number, indicating that $R_k$ is the axial conduction resistance and $R_{ku}$ is lateral surface-convection resistance.

Using (6.27) and (6.28), we have

$$\mathrm{Nu}_L = \frac{\mathrm{Pe}_L^{1/2}}{\pi^{1/2}} \qquad \text{local Nusselt number for laminar, parallel} \qquad (6.30)$$
$$\text{flow over semi-infinite plate for } \mathrm{Pr} = 0.$$

The inverse of $A_{ku} R_{ku}$ is called the heat transfer coefficient $h(\mathrm{W/m^2\text{-}{}^\circ C})$, i.e.,

$$h \equiv \frac{1}{A_{ku} R_{ku}} \quad \text{or} \quad A_{ku} h \equiv \frac{1}{R_{ku}} \qquad \text{definition of heat transfer coefficient } h. \qquad (6.31)$$

Therefore, $h$ is the surface-convection conductance per unit area. The convenience in the use of the heat transfer coefficient is evident when, using (6.25), we express $q_{ku}$ in terms of $h$ from (6.31), i.e.,

$$Q_{ku} = A_{ku} q_{ku} = A_{ku} h (T_s - T_{f,\infty}) \qquad \text{surface convection in terms of } h. \qquad (6.32)$$

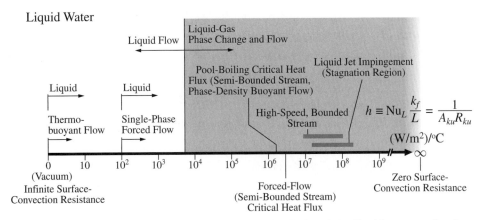

Figure 6.6. Typical values of the local heat transfer coefficient $h$ for liquid water under thermobuoyant and single- and two-phase (phase change) forced flow conditions.

As an alternative to (6.28), the Nusselt number can be written in terms of $h$ as

$$\mathrm{Nu}_L = \frac{q_{k,u}L}{(T_s - T_{f,\infty})k_f} \equiv \frac{hL}{k_f} \quad \text{or} \quad h \equiv \mathrm{Nu}_L\frac{k_f}{L} \qquad \begin{array}{l}\text{relation between}\\ \text{Nusselt number and}\\ \text{heat transfer coefficient.}\end{array} \qquad (6.33)$$

Finally, we can also write (6.28) as

$$q_{ku} = \frac{\mathrm{Nu}_L k_f}{L}(T_s - T_{f,\infty}) \qquad \begin{array}{l}\text{local surface-convection heat}\\ \text{transfer in terms of local } \mathrm{Nu}_L.\end{array} \qquad (6.34)$$

Note that in (6.30), for $\mathrm{Pe}_L = 0$ (i.e., $u_{f,\infty} = 0$), $\mathrm{Nu}_L$ (and $h$) are also zero, i.e., no surface-convection heat transfer occurs from the surface when the far-field velocity is zero. This is because, under steady state, no conduction heat transfer is possible for the semi-infinite fluid medium bounded by a planar surface. This was discussed in Section 3.3.1(B) and in Section 3.3.8. In practice and under gravity, buoyancy caused by the temperature difference in the fluid can cause a thermobuoyant motion and then surface-convection heat transfer occurs. We will discuss this in Section 6.5.

The heat transfer coefficient $h$ is zero for a vacuum. Since the thermobuoyant motions generally result in smaller fluid velocities, compared to forced flows, the thermobuoyant-flow heat transfer coefficient is generally smaller than that for forced flows. The range of $h$ is shown in Figure 6.6, for liquid water. As phase change occurs (evaporation and condensation) $h$ becomes yet larger (Section 6.6). The liquid-gas, two-phase flow can be caused by phase-density difference buoyancy (e.g., pool boiling) or may be forced (but under normal gravity, the phase-density buoyancy remains significant even when the flow is also forced).

To summarize, we write $\mathrm{Nu}_L$ in terms of $\mathrm{Pe}_L$, or other dimensionless fluid and surface parameters (such as those discussed next for the case $\mathrm{Pr} \neq 0$), e.g., (6.30). These $\mathrm{Nu}_L$ relations are determined by analysis or by experiment (i.e., empirically). Then the surface-convection resistance $R_{ku}$ is found from its relation to $\mathrm{Nu}_L$ as given by (6.29). Finally $Q_{ku}$ is determined using (6.25). We now proceed to include the influence of the fluid viscosity $\mu_f$ on $Nu_L$.

---

### EXAMPLE 6.1. FAM

---

The surface of a solid nuclear reactor is cooled by forced, parallel flow of mercury, as shown in Figure Ex. 6.1(a). Assume that the flow can be idealized as a zero-viscosity ($\mu_f = 0$) flow depicted in Figure 6.3.

(a) Draw the thermal circuit diagram.
(b) Determine the boundary-layer thickness at the tail end of the reactor surface.
(c) Determine the surface-convection heat flow at locations (i) $x = L/4$, (ii) $L/2$, and (iii) $L$.
(d) Determine the Nusselt number based on $L$.
(e) Determine the heat transfer coefficient $h$.
(f) Determine the surface-convection resistance per unit area $A_{ku}R_{ku}$ at location $x = L$.
(g) As we will show, for laminar flow with $Pr > 0.6$, the local Nusselt number is given by $Nu_L = 0.332 Re_L^{1/2} Pr^{1/3}$, where $Re_L = Pe_L Pr^{-1}$ and $Re_L$ is the Reynolds number based on the $L$. Determine $Nu_L$ from this expression and compare it with that for zero viscosity.

$L = 6$ cm, $w = 2$ cm, $T_s = 314°C$, $T_{f,\infty} = 20°C$, and $u_{f,\infty} = 20$ cm/s. The mercury properties can be evaluated at 420 K from Table C. 24.

Surface-Convection Cooling of a Planar Reactor Surface

(a) Physical Model                    (b) Thermal Circuit Model

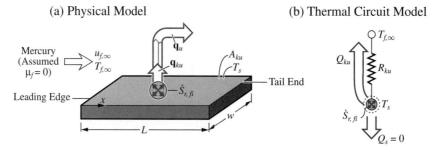

Figure Ex. 6.1. The surface of a solid nuclear reactor cooled by flowing mercury. (a) Physical model. (b) Thermal circuit model.

### SOLUTION

(a) The thermal circuit diagram is shown in Figure Ex. 6.1(b).
(b) The thermal boundary-layer thickness (the only boundary layer, since $\mu_f = 0$) is given by (6.21) as

$$\delta_\alpha = 3.6 \left( \frac{\alpha_f L}{u_{f,\infty}} \right)^{1/2}.$$

Here, we have from Table C.24 for Hg at $T = 420$ K

$$\alpha_f = \left(\frac{k}{\rho c_p}\right)_f = \frac{11.4(\text{W/m-K})}{13,270(\text{kg/m}^3) \times 130(\text{J/kg-K})} = 6.608 \times 10^{-6} \text{ m}^2/\text{s} \quad \text{Table C.24,}$$

and $L = 6$ cm $= 0.06$ m, $u_{f,\infty} = 20$ cm/s $= 0.2$ m/s.

Using these for the evaluation of $\delta_\alpha$, we have

$$\delta_\alpha = 3.6 \times \left[\frac{6.607 \times 10^{-6}(\text{m}^2/\text{s}) \times 0.06(\text{m})}{0.20(\text{m})}\right]^{1/2} = 5.069 \times 10^{-3} \text{ m} = 5.069 \text{ mm}.$$

(c) The surface-convection heat flux is given by (6.27) as

$$q_{ku} = \frac{k_f}{L} \frac{\text{Pe}_L^{1/2}}{\pi^{1/2}}(T_s - T_{f,\infty})$$

$$\text{Pe}_L = \frac{u_{f,\infty}L}{\alpha_f}.$$

Here we have

$$T_s - T_{f,\infty} = 314(°\text{C}) - 20(°\text{C}) = 294°\text{C}$$

$$\text{Pe}_L = \frac{u_{f,\infty}}{\alpha_f}L = \frac{0.20(\text{m/s})}{6.609 \times 10^{-6}(\text{m}^2/\text{s})}L = (3.027 \times 10^4)L.$$

Then

$$q_{ku} = \frac{11.4(\text{W/m-K})}{L(\text{m})} \times \left[\frac{0.20(\text{m/s})}{6.607 \times 10^{-6}(\text{m}^2/\text{s})}L(\text{m})\right]^{1/2} \frac{1}{\pi^{1/2}} \times 294(°\text{C}).$$

    (i) For    $L = \dfrac{0.06(\text{m})}{4} = 0.015$ m,    $q_{ku} = 2.687 \times 10^6$ W/m$^2$

    (ii) for    $L = \dfrac{0.06(\text{m})}{2} = 0.03$ m,    $q_{ku} = 1.900 \times 10^6$ W/m$^2$

    (iii) for    $L = 0.06$ m,            $q_{ku} = 1.344 \times 10^6$ W/m$^2$.

(d) The Nusselt number based on $L$ is defined by relation (6.28) and for this flow is given by (6.30), i.e.,

$$\text{Nu}_L = \frac{L}{A_{ku}R_{ku}k_f} = \frac{\text{Pe}_L^{1/2}}{\pi^{1/2}}.$$

Here, for $L = 0.06$ m we have

$$\text{Nu}_L = \left[\frac{0.20(\text{m/s}) \times 0.06(\text{m})}{6.609 \times 10^{-6}(\text{m}^2/\text{s})}\right]^{1/2} \frac{1}{\pi^{1/2}} = (1.816 \times 10^3)^{1/2}\frac{1}{\pi^{1/2}} = 23.83.$$

(e) The heat transfer coefficient is given by (6.33) as

$$h = \text{Nu}_L \frac{k_f}{L}.$$

Here we have for $L = 0.06$ m

$$h = 23.83 \times \frac{11.4(\text{W/m-}^\circ\text{C})}{0.06(\text{m})} = 4.529 \times 10^3 \text{ W/m}^2\text{-}^\circ\text{C}.$$

(f) The surface-convection resistance can be determined from (6.26), or since $h$ is already calculated, we can use (6.31), i.e.,

$$A_{ku}R_{ku} = \frac{1}{h} = 2.208 \times 10^{-4} {}^\circ\text{C}/(\text{W/m}^2).$$

(g) From Table C.24, we have Pr = 0.0136. Then

$$\text{Re}_L = \text{Pe}_L \text{Pr}^{-1} = \frac{0.20(\text{m/s}) \times 0.06(\text{m})}{6.609 \times 10^{-6}(\text{m}^2/\text{s})} \times (0.0136)^{-1} = 1.335 \times 10^5$$

$$\text{Nu}_L = 0.332 \text{Re}_L^{1/2} \text{Pr}^{1/3} = 0.332(1.335 \times 10^5)^{1/2}(0.0136)^{1/3} = 28.96.$$

Using the zero viscosity results given above, we had $\text{Nu}_L = 24.42$. This results in a difference of 19.42%.

**COMMENT**

Note that $\delta_\alpha$ is very small, i.e., $\delta_\alpha/L = 0.0434$. The mercury flown to cool the surface, is affected only in a small region (less than 3 mm thick) adjacent to the chip surface.

Also note that $q_{ku}$ decreases by a factor of 2 between locations $L/4$ and $L$ (as the boundary-layer thickness increases).

Laminar, parallel (as compared to perpendicular) flows are not the most effective surface-convection heat transfer flow arrangements. This will be explored further as we consider turbulent parallel flows and perpendicular flows.

### 6.2.5 Nonzero Prandtl Number: Reynolds Number $\text{Re}_L$

When both thermal and viscous boundary layers are significant, i.e., $\text{Pr} \neq 0$, the near-field velocity nonuniformity influences the surface-convection resistance $R_{ku}$. Figure 6.7(a) renders the presence of both viscous and thermal boundary layers, for $\text{Pr} > 1$, i.e., $\delta_v > \delta_\alpha$.

Here we will first consider forced, laminar parallel flow. This flow occurs when the flow inertial force, compared to the viscous force, is below a threshold value. This condition will be discussed in Section 6.3.

The two-dimensional, steady-state solution to the velocity field $\mathbf{u}_f = (u_f, v_f)$ is determined from the momentum and mass conservation equations given in Appendix B. The $x$- and $y$-direction velocity components $u_f$ and $v_f$ are determined subject to the surface and far-field mechanical (as compared to thermal) conditions. The boundary-layer approximations, similar to those discussed in the last section, are made.

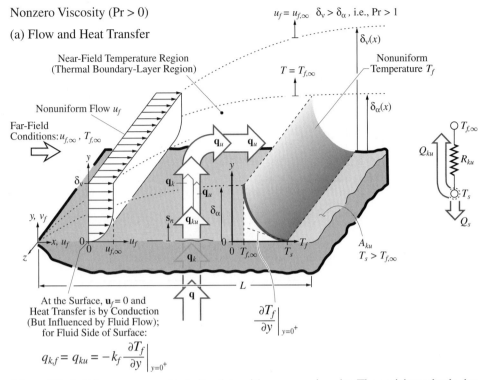

Nonzero Viscosity (Pr > 0)

(a) Flow and Heat Transfer

$u_f = u_{f,\infty}$  $\delta_v > \delta_\alpha$, i.e., Pr > 1

Figure 6.7. (a) Flow over a semi-infinite plate with nonzero viscosity. The anticipated velocity and temperature distributions are also shown.

### (A) Boundary-Layer Approximation

As with (6.10), the vectorial form of the energy equation is written below to show that the dominant conduction heat flow is lateral (perpendicular to the surface), while the convection heat flow is both lateral and axial (perpendicular and parallel to the surface). This is due to the two-dimensionality of the velocity field (as compared to one-dimensional uniform flow considered in Section 6.2.3, i.e., where we used $v_f = 0$). After the bounding-layer approximations [similar to that given by (6.9)], the energy equation (6.5) becomes

$$\frac{\partial}{\partial y}\mathbf{s}_y \cdot \mathbf{q}_k \;+\; \frac{\partial}{\partial x}\mathbf{s}_x \cdot \mathbf{q}_u \;+\; \frac{\partial}{\partial y}\mathbf{s}_y \cdot \mathbf{q}_u \;=$$

$$-k_f\frac{\partial^2 T_f}{\partial y^2} \;+\; (\rho c_p)_f u_f \frac{\partial T_f}{\partial x} \;+\; (\rho c_p)_f v_f \frac{\partial T_f}{\partial y} = 0 \qquad \text{energy equation.} \qquad (6.35)$$

| spatial rate of change of $\mathbf{q}_k$ in $y$ direction | spatial rate of change of $\mathbf{q}_u$ in $x$ direction | spatial rate of change of $\mathbf{q}_u$ in $y$ direction |

Note that since $(\rho c_p)_f$ is assumed constant, the mass conservation equation (given later) is used to simplify the convection terms. For the two-dimensional, laminar, steady-state flow considered here, the thermal boundary conditions remain those given by (6.14).

(b) Velocity Distribution for Parallel Component

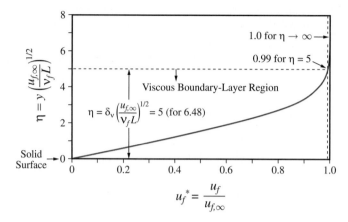

(c) Velocity Distribution for Perpendicular Component

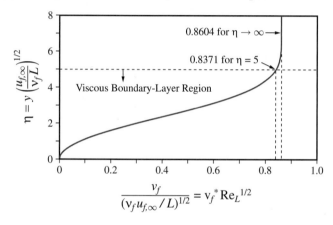

(d) Temperature Distribution

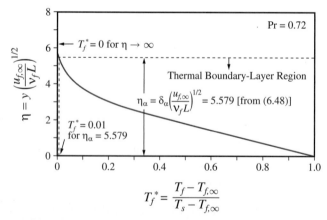

Figure 6.7. (b) and (c) Predicted dimensionless velocity distributions. (d) Predicted dimensionless, temperature distribution for Pr = 0.72.

The corresponding $x$-direction momentum conservation equation is given by (B.51). We consider the case of steady, two-dimensional, incompressible flow with no external-pressure gradient, no buoyancy force, no electromagnetic force, and a Newtonian fluid behavior. Using the appropriate boundary-layer approximations, the $x$-direction momentum conservation equation (B.52) becomes (with subscript $f$ added)

$$\underbrace{\rho_f \left( u_f \frac{\partial u_f}{\partial x} + v_f \frac{\partial u_f}{\partial y} \right)}_{\substack{\text{spatial rate of change} \\ \text{of inertial force} \\ \text{in } x \text{ direction}}} = \underbrace{\mu_f \frac{\partial^2 u_f}{\partial y^2}}_{\substack{\text{spatial rate of change} \\ \text{of viscous stress} \\ \text{in } x \text{ direction}}} \qquad \begin{array}{l} x\text{-direction} \\ \text{momentum equation.} \end{array} \qquad (6.36)$$

The associated mass conservation (also called the continuity) equation (B.49) becomes (again for incompressible fluid flow)

$$\underbrace{\frac{\partial u_f}{\partial x} + \frac{\partial v_f}{\partial y}}_{\substack{\text{spatial rate of} \\ \text{change of mass flux}}} = 0 \quad \text{continuity equation.} \qquad (6.37)$$

These two conservation equations, (6.36) and (6.37), are solved for $u_f$ and $v_f$, subject to the mechanical boundary conditions

$$u_f(x = 0, y) = u_{f,\infty}$$

$$u_f(x, y = 0) = 0$$

$$u_f(x, y \to \infty) = u_{f,\infty}$$

$$v_f(x = 0, y) = v_f(x, y = 0) = 0. \qquad (6.38)$$

### (B) Fluid Temperature Distribution

We now normalize the energy (6.35), momentum (6.36), and continuity (6.37) equations. We use the length scale $L$ and the velocity scale $u_{f,\infty}$. Again the temperature is scaled using $T_s - T_{f,\infty}$. Then (6.35) to (6.37) become

$$-\frac{\partial^2 T_f^*}{\partial y^{*2}} + \mathrm{PrRe}_L \left( u_f^* \frac{\partial T_f^*}{\partial x^*} + v_f^* \frac{\partial T_f^*}{\partial y^*} \right) = 0 \quad \text{dimensionless energy equation} \qquad (6.39)$$

$$-\frac{\partial^2 u_f^*}{\partial y^{*2}} + \mathrm{Re}_L \left( u_f^* \frac{\partial u_f^*}{\partial x^*} + v_f^* \frac{\partial u_f^*}{\partial y^*} \right) = 0 \quad \begin{array}{l} \text{dimensionless } x\text{-direction} \\ \text{momentum equation} \end{array} \qquad (6.40)$$

$$\frac{\partial u_f^*}{\partial x^*} + \frac{\partial v_f^*}{\partial y^*} = 0 \quad \text{dimensionless continuity equation,} \qquad (6.41)$$

where

$$T_f^* = \frac{T_f - T_{f,\infty}}{T_s - T_{f,\infty}}, \quad x^* = \frac{x}{L}, \quad y^* = \frac{y}{L}, \quad u_f^* = \frac{u_f}{u_{f,\infty}}, \quad v_f^* = \frac{v_f}{u_{f,\infty}}$$

$$\mathrm{Pr} = \frac{\nu_f}{\alpha_f}, \quad \mathrm{Re}_L = \frac{u_{f,\infty}L}{\nu_f} \qquad \text{dimensionless variables and parameters.} \qquad (6.42)$$

The dimensionless mechanical boundary conditions (6.38) and thermal conditions (6.14) become

$$u_f^*(x^*, y^* = 0) = v_f^*(x^*, y^* = 0) = 0, \quad T_f^*(x^*, y^* = 0) = 1$$

$$u_f^*(x^*, y^* \to \infty) = 1, \quad T_f^*(x^*, y^* \to \infty) = 0 \quad \begin{array}{l}\text{dimensionless thermal and}\\ \text{mechanical conditions}\\ \text{at various locations.}\end{array}$$

$$u_f^*(x^* = 0, y^*) = 1, \quad v_f^*(x^* = 0, y^*) = T_f^*(x^* = 0, y^*) = 0$$

$$(6.43)$$

First $u_f^*$ and $v_f^*$ are determined by solving (6.40) and (6.41). This is done by defining a stream function that satisfies (6.41) and a similarity variable that reduces (6.40) to an ordinary differential equation[†] (left as an end of chapter problem). Then using these, the solution for $T^* = T^*(x, y)$ is found from (6.39). Finally, using these, $q_{k,f}(y = 0) \equiv q_{ku}$ is determined from (6.23). Note that $\mathbf{u}_f = 0$ on the impermeable solid surface with a viscous fluid flow [the impermeable solid makes $v_f^*(x, y = 0)$, while the viscosity makes $u_f^*(x, y = 0) = 0$]. Figures 6.7(b) and (c) show the predicted velocity distributions. The results are presented as $u_f/u_{f,\infty}$ and $v_f/(\nu_f u_{f,\infty}/L)^{1/2}$ as a function of the similarity variable $\eta = y(u_{f,\infty}/\nu_f x)^{1/2}$. Note that for $\eta \to \infty$ (i.e., $y \to \infty$), for the streamwise velocity $u_f^*$ the expected uniform far-field velocity is obtained. The lateral velocity $v_f^*$ initially increases with $y$ (or $\eta$), and then reaches an asymptotic limit. This lateral flow is needed due to the streamwise velocity near the surface. Figure 6.7(d) shows the dimensionless temperature distribution in the thermal boundary-layer region, for $\mathrm{Pr} = 0.72$.

---

[†] The stream function is related to the components of the two-dimensional velocity field through

$$u = \frac{\partial \psi}{\partial y} \quad \text{and} \quad v = -\frac{\partial \psi}{\partial x}.$$

The dimensionless stream function is defined as

$$\psi^* = \frac{\psi}{(\nu_f u_{f,\infty} L)^{1/2}}$$

and the dimensionless similarity variable is defined as

$$\eta = y \left( \frac{u_{f,\infty}}{\nu_f L} \right)^{1/2},$$

which is similar to (6.15), but uses the kinematic viscosity instead of thermal diffusivity. Using $\psi^*$ eliminates the need to include (6.41), and use of $\eta$ leads to an ordinary differential equation with $\psi^*$ as the dependent and $\eta$ as the independent variable. The steps are left as an end of chapter problem.

### (C) Surface-Convection Heat Transfer

The solution to the fluid temperature distribution is left as an end of chapter problem. The result for $q_{ku}$ and $\mathrm{Nu}_L$ are

$$
\begin{aligned}
\mathrm{Nu}_L &= \frac{q_{ku}L}{(T_s - T_{f,\infty})k_f} \equiv -k_f \left.\frac{\partial T}{\partial y}\right|_{y=0^+} \frac{L}{(T_s - T_{f,\infty})k_f} = \frac{L}{A_{ku}R_{ku}k_f} \\
&= 0.332\mathrm{Re}_L^{1/2}\mathrm{Pr}^{1/3} \quad \begin{array}{l}\text{laminar, parallel flow over} \\ \text{semi-infinite plate for } \mathrm{Pr} > 0.6.\end{array}
\end{aligned} \tag{6.44}
$$

The Reynolds number $\mathrm{Re}_L = F_u/F_\mu$ is defined in (6.42) as

$$
\mathrm{Re}_L \equiv \frac{F_u}{F_\mu} \equiv \frac{u_{f,\infty}L}{\nu_f}, \qquad \nu_f = (\mu/\rho)_f \quad \text{definition of Reynolds number.} \tag{6.45}
$$

The Reynolds number is the ratio of the inertial force to viscous forces. As will be discussed, for $\mathrm{Re}_L$ below a threshold value of $\mathrm{Re}_{L,t} = 5 \times 10^5$ (called the transition Reynolds number, discussed in Section 6.3.4), this parallel flow is laminar. This threshold value is designated as $\mathrm{Re}_{L,t}$, where subscript $t$ stands for transition. Note that the three dimensionless numbers $\mathrm{Pe}_L$, $\mathrm{Re}_L$, and $\mathrm{Pr}$ are related as $\mathrm{Pe}_L = \mathrm{Re}_L\mathrm{Pr}$. Using (6.44), we solve for $A_{ku}R_{ku}$ [similar to (6.26)], i.e.,

$$
\begin{aligned}
A_{ku}R_{ku} &= \frac{L}{k_f\mathrm{Nu}_L} = \frac{L}{k_f}\frac{1}{0.332\mathrm{Re}_L^{1/2}\mathrm{Pr}^{1/3}} \\
&= \frac{1}{0.332}\frac{\nu_f^{1/6}L^{1/2}}{k_f^{2/3}(\rho c_p)^{1/3}u_{f,\infty}^{1/2}} \quad \begin{array}{l}\text{local, surface-convection resistance} \\ \text{for laminar flow with } \mathrm{Pr} > 0.6.\end{array}
\end{aligned} \tag{6.46}
$$

Note that (6.46) is similar to (6.26) in the $L$ and $u_{f,\infty}$ dependence. Next we use (6.44) again and we write for $q_{ku}$

$$
\begin{aligned}
q_{ku} &= \frac{(T_s - T_{f,\infty})}{A_{ku}R_{ku}} = \frac{k_f}{L}0.332\mathrm{Re}_L^{1/2}\mathrm{Pr}^{1/3}(T_s - T_{f,\infty}) \\
&= 0.332\frac{\rho_f^{1/2}k_f u_{f,\infty}^{1/2}}{\mu_f^{1/2}L^{1/2}}\mathrm{Pr}^{1/3}(T_s - T_{f,\infty}).
\end{aligned} \tag{6.47}
$$

Compared to (6.24), here $\partial T_f/\partial y\,|_{y=0^+} = -0.332\mathrm{Re}_L^{1/2}\mathrm{Pr}^{1/3}(T_s - T_{f,\infty})/L$ and $\mathrm{Re}_L$ contains $u_{f,\infty}$. Note that as $L \to 0$, $q_{ku} \to \infty$, i.e., the local resistance diminishes as the edge of the boundary layer is approached. Also note that comparing to (6.27), the role of fluid heat capacity is included in $\mathrm{Pr}$, but otherwise a similar dependence on $u_{f,\infty}$ and $L$ exists.

The $\mathrm{Pr}^{1/3}$ dependence does not directly appear in the solution (as compared to the $\mathrm{Re}_L^{1/2}$ relation), but is the result of a curve fit to the numerical results

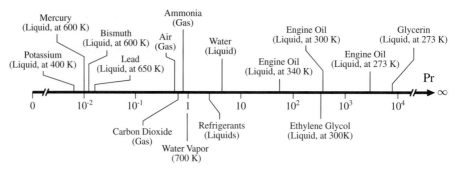

At 300 K and 1 atmosphere pressure, unless stated otherwise.

Figure 6.8. Magnitude of Prandtl number for some gaseous and liquid phases of some substances.

for $0.6 \leq \mathrm{Pr} \leq 60$. The two boundary-layer thicknesses, $\delta_v$ and $\delta_\alpha$, are related through

$$\frac{\delta_v}{\delta_\alpha} = \mathrm{Pr}^{1/3}, \quad \delta_v = 5.0 \left( \frac{\nu_f L}{u_{f,\infty}} \right)^{1/2} = 5.0 L \mathrm{Re}_L^{-1/2}$$

$$\delta_\alpha = 5.0 \left( \frac{\nu_f L}{u_{f,\infty}} \right)^{1/2} \left( \frac{\alpha_f}{\nu_f} \right)^{1/3} \qquad \begin{array}{l} \text{viscous and thermal, laminar} \\ \text{boundary-layer thicknesses,} \qquad (6.48) \\ \mathrm{Re}_L < \mathrm{Re}_{L,t} = 5 \times 10^5, \end{array}$$

i.e., as $\mathrm{Pr} \to 0$, $\delta_v \ll \delta_\alpha$ and as $\mathrm{Pr} \to \infty$ $\delta_v \gg \delta_\alpha$. The criterion for the viscous boundary-layer thickness $\delta_v$ is $u_f(\delta_v) = 0.99 u_{f,\infty}$. This is shown in Figure 6.7(b).

Figure 6.8 gives the range of the Prandtl numbers for the gaseous and liquid phases of some substances. Liquid metals have the smallest Pr (i.e., $\nu_f \ll \alpha_f$) and oils and other large-chain organic liquids have very large Pr (i.e., $\nu_f \gg \alpha_f$). Air and most other gases have a Pr of near unity. Water has a Pr of near ten. The Prandtl number of various fluids are listed in Tables C.22 to C.24 and C.26 in Appendix C, along with their other thermodynamic and transport properties.

Since the lateral component of the velocity $v_f$ is positive (flow moves away from the surface as the viscous boundary layer grows), this assists (i.e., enhances) the lateral heat transfer.

Since no allowance is made for the variation of the fluid properties ($\rho_f$, $c_{p,f}$, $k_f$, and $\mu_f$) with temperature, as a good approximation, the properties are evaluated at $(T_s + T_{f,\infty})/2$. This is called the film temperature.

### 6.2.6 Average Nusselt Number $\langle \mathrm{Nu} \rangle_L$ and Average Surface-Convection Resistance $\langle R_{ku} \rangle_L$

In general, we are interested in the integrated surface-convection heat transfer $\langle Q_{ku} \rangle_L$, integrated over a given surface area $A_{ku}$. For the semi-infinite plate considered, and for a length $L$ along the flow direction, this area has a length $L$ along the plate and has a width $w$. Then we define the integrated heat transfer rate $\langle Q_{ku} \rangle_L$, average resistance $\langle R_{ku} \rangle_L$, and Nusselt number $\langle \mathrm{Nu} \rangle_L$ as

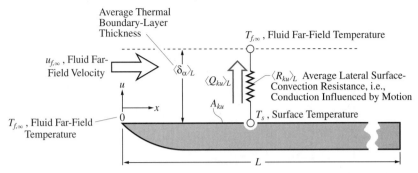

Figure 6.9. The average surface-convection and the average boundary-layer thickness for surface-convection heat transfer from a semi-infinite plate.

$$A_{ku} = Lw, \quad \langle Q_{ku} \rangle_L = \frac{T_s - T_{f,\infty}}{\langle R_{ku} \rangle_L} = A_{ku} \langle \mathrm{Nu} \rangle_L \frac{k_f}{L}(T_s - T_{f,\infty}) \quad \text{integrated surface-convection heat transfer rate,} \quad (6.49)$$

where the surface area here is $A_{ku} = Lw$. The definition of spatial average is given by (2.68). The average of $\mathrm{Nu}_x$ (replacing subscript $L$ with subscript $x$ in the integrand) taken over the distance $0 \le x \le L$, i.e., $\langle \mathrm{Nu} \rangle_L$, is defined, starting from (6.28), as the surface-average Nusselt number

$$\langle \mathrm{Nu} \rangle_L \equiv \frac{A_{ku} \langle R_{k,f} \rangle}{A_{ku} \langle R_{ku} \rangle} = \left( \frac{L}{k_f} \right) \frac{1}{L} \int_0^L \frac{k_f \mathrm{Nu}_x}{x} dx = \int_0^L \frac{\mathrm{Nu}_x}{x} dx \quad \text{definition of average Nusselt number.} \quad (6.50)$$

For laminar, parallel flow with $\mathrm{Nu}_L$ given by (6.44), we have

$$\langle \mathrm{Nu} \rangle_L = 0.332 \mathrm{Pr}^{1/3} \left( \frac{u_{f,\infty}}{\nu_f} \right)^{1/2} \int_0^L x^{-1/2} dx = 2\,\mathrm{Nu}_L$$

$$= 0.664 \mathrm{Re}_L^{1/2} \mathrm{Pr}^{1/3} = \frac{L}{A_{ku} \langle R_{ku} \rangle_L k_f}$$

$$\mathrm{Re}_L \le \mathrm{Re}_{L,t} = 5 \times 10^5, \text{ average Nusselt number for laminar flow, } \mathrm{Pr} > 0.6. \quad (6.51)$$

The average surface-convection resistance $\langle R_{ku} \rangle_L$ is rendered in Figure 6.9.

Using this, we write the surface-averaged heat transfer rate $\langle Q_{ku} \rangle_L$ in (6.49) as

$$\langle Q_{ku} \rangle_L = A_{ku} \frac{k_f}{L}[0.664 \mathrm{Re}_L^{1/2} \mathrm{Pr}^{1/3}](T_s - T_{f,\infty})$$

$$= Lw \frac{k_f}{L}\left[ 0.664 \left( \frac{u_{f,\infty}L}{\nu_f} \right)^{1/2} \mathrm{Pr}^{1/3} \right](T_s - T_{f,\infty}). \quad (6.52)$$

Variation of Surface-Convection Resistance $R_{ku}(x)$
[and Dimensionless Conductance $Nu_L(x)$] Along the Surface

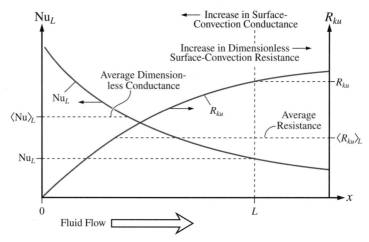

Figure 6.10. Variation of the local $Nu_L(x)$ and $R_{ku}(x)$ and their average values $\langle Nu \rangle_L$ and $\langle R_{ku} \rangle_L$, with respect to location $L$.

Note that the average heat transfer coefficient $\langle h \rangle_L$ is defined similar to (6.31) as

$$\langle h \rangle_L \equiv \langle Nu \rangle_L \frac{k_f}{L} \qquad \text{definition of average heat transfer coefficient.} \qquad (6.53)$$

From (6.51), we note that since the local Nusselt number increases with $L$ in $L^{1/2}$ proportionality, then the average $\langle Nu_L \rangle_L$ is twice the local $Nu_L$ (because of the $Re_L^{1/2}$ relation). Figure 6.10 demonstrates this relationship between $\langle Nu_L \rangle_L$ (or $\langle h \rangle_L$) and $Nu_L$ (or $h$), and $\langle R_{ku} \rangle_L$ and $R_{ku}$.

Note that $R_{ku}$ is zero at the leading edge ($x = 0$). However, the boundary-layer approximation discussed in the last section is not valid near and at $x = 0$. In practice, a finite resistance exists near the leading edge (heat is axially and laterally conducted through the fluid). As $L$ increases, $\langle R_{ku} \rangle_L$ also increases; then as $L$ increases there is more resistance to heat transfer (compared to that near the leading edge).

EXAMPLE 6.2. FAM

As part of a paper fabrication process, after sheet forming of the pulp, the sheet, while extended and rolling between an upper and a lower roller, is heated and partially dried by surface convection (and in general also by surface radiation, but neglected here) using the hot gaseous products of a combustion. This is shown in Figure Ex. 6.2(a). This heat transfer to the pulp results in the pulp drying. Consider a pulp surface temperature $T_s$, and treat the combustion products as air at temperature $T_{f,\infty}$ flowing parallel to the surface at a far-field velocity $u_{f,\infty}$. Assume that the pulp sheet is stationary relative to the flowing gas and that a flat, semi-infinite approximation is valid for the pulp sheet, with the leading edge, $x = 0$, marked as in Figure Ex. 6.2.

(a) Draw the thermal circuit diagram.
(b) Determine the rate of surface-convection heat transfer $\langle Q_{ku} \rangle_L$.
(c) Determine the boundary-layer thickness at $x = L$, $\delta_\alpha(L)$.

   $T_s = 35°C$, $T_{f,\infty} = 900°C$, $u_{f,\infty} = 0.5$ m/s, $L = 2$ m, $w = 1$ m.

   Determine the thermophysical properties of air at $T = 700$ K from Table C.22.

Surface-Convection Heating (and Drying) of a Pulp Sheet

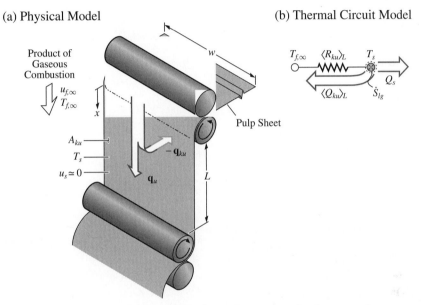

Figure Ex. 6.2. Pulp sheet drying by surface-convection heat transfer.

**SOLUTION**
(a) The thermal circuit diagram is shown in Figure Ex. 6.2(b).
(b) The surface-convection heat transfer rate, from the hot gas to the pulp $\langle Q_{ku} \rangle_L$, is given by (6.49), i.e.,

$$\langle Q_{ku} \rangle_L = A_{ku} \langle Nu \rangle_L \frac{k_f}{L}(T_s - T_{f,\infty}).$$

The Nusselt number depends on the Reynolds number $\mathrm{Re}_L$, defined by (6.45) as

$$\mathrm{Re}_L = \frac{u_{f,\infty} L}{\nu_f}.$$

From Table C.22, and at $T = 700$ K, we have

$$\nu_f = 6.515 \times 10^{-5} \text{ m}^2/\text{s} \quad \text{Table C.22,}$$

where we note that (Pa-s) = (N-s/m$^2$) = (kg-m-s/m$^2$-s) = (kg/m-s). Then

$$\mathrm{Re}_L = \frac{0.5(\text{m/s}) \times 2(\text{m})}{6.515 \times 10^{-5}(\text{m}^2/\text{s})} = 1.535 \times 10^4 < \mathrm{Re}_{L,t} = 5 \times 10^5 \quad \text{laminar flow.}$$

Then we can use (6.51) for $\langle \mathrm{Nu} \rangle_L$, i.e.,

$$\langle \mathrm{Nu} \rangle_L = 0.664 \mathrm{Re}_L^{1/2} \mathrm{Pr}^{1/3}.$$

From Table C.22 and at $T = 700$ K, we have

$$\mathrm{Pr} = 0.70 \qquad\qquad\qquad \text{Table C.22}$$

$$k_f = 0.0513 \text{ W/m-K} \qquad\qquad \text{Table C.22.}$$

Then

$$\langle \mathrm{Nu} \rangle_L = 0.664 \times (1.535 \times 10^4)^{1/2} \times (0.7)^{1/3} = 73.05.$$

With $A_{ku} = Lw$, the heat transfer rate $\langle Q_{ku} \rangle_L$ becomes

$$\langle Q_{ku} \rangle_L = 2(\text{m}) \times 1(\text{m}) \times 73.05 \times \frac{0.0513(\text{W/m-K})}{2(\text{m})} \times (900 - 35)(\text{K})$$

$$= 3.242 \times 10^3 \text{ W.}$$

(c) For laminar flow $\delta_\alpha(L)$ is given by (6.48) as

$$\delta_\alpha = 5 \left( \frac{\nu_f L}{u_{f,\infty}} \right)^{1/2} \mathrm{Pr}^{-1/3} \quad \text{thermal boundary-layer thickness}$$

$$\delta_\alpha = 5 \left[ \frac{6.515 \times 10^{-5}(\text{m}^2/\text{s}) \times 2(\text{m})}{0.5(\text{m/s})} \right]^{1/2} (0.7)^{-1/3} = 0.0909 \text{ m} = 9.09 \text{ cm.}$$

**COMMENT**

In practice, higher heating rates are needed. Also, the portion of the gas stream affected by this heat transfer is confined to a small region adjacent to the pulp sheet. An alternative, effective, and efficient method is the surface-radiation heat transfer. There, the combustion occurs over a radiating porous surface that faces the pulp sheet.

### 6.3 Parallel Turbulent Flow: Transition Reynolds Number $\mathrm{Re}_{L,t}$

Turbulent flow is marked by random fluctuations in the fluid velocity $\mathbf{u}_f$. For a boundary-layer flow, with the Cartesian coordinate system and for $u_f$ being the velocity in the $x$ direction, Figure 6.11(a) renders this turbulent velocity function. Figures 6.11(c) and (d) show the structures of the viscous and thermal boundary layers for laminar and turbulent flows. The distributions of the velocity and temperature are shown in terms of the dimensionless velocity $u_f^*$ and temperature $T_f^*$. As with the laminar flow, the near-field nonuniformities are confined to regions adjacent

(a) Fluid Particle Turbulent Velocity Fluctuations in the $x$ direction, $u_f'$

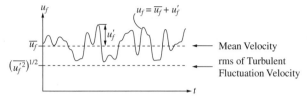

(b) Turbulent Mixing Length, $\lambda_t$

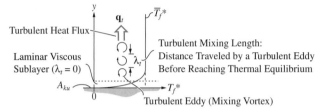

(c) Viscous Laminar and Turbulent Boundary Layers, $\delta_v$ and $\bar{\delta}_v$

   (i) Laminar          (ii) Turbulent

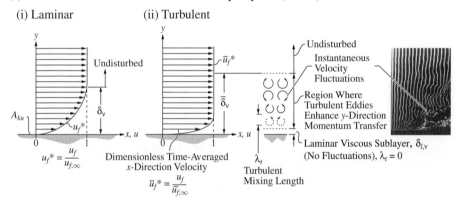

(d) Thermal Laminar and Turbulent Boundary Layers, $\delta_\alpha$ and $\bar{\delta}_\alpha$

   (i) Laminar          (ii) Turbulent

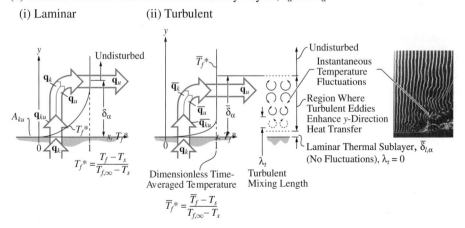

Figure 6.11. (a) Turbulent velocity fluctuations showing the mean and fluctuating components of the $x$-direction velocity $u_f'$. (b) A rendering of turbulent mixing length. (c) Structure of the viscous laminar and turbulent boundary layers. (d) Structure of the thermal laminar and turbulent boundary layers. (c) and (d) include renderings of the turbulent eddies and the turbulent mixing lengths that contribute to the lateral ($y$ direction) heat transfer.

to the surface and are marked by the viscous and thermal boundary-layer thicknesses $\bar{\delta}_v$ and $\bar{\delta}_\alpha$. For turbulent boundary layers, the fluctuations vanish far away and very close to the surface. The region adjacent to the surface is called the laminar sublayer. The laminar viscous sublayer has a thickness $\bar{\delta}_{l,v}$ and the laminar thermal sublayer has a thickness $\bar{\delta}_{l,\alpha}$. These are marked in Figures 6.11(c) and (d). In turbulent boundary layers, although the disturbed region is larger than its laminar counterpart, most of the changes in the velocity and the temperature occur very close to the surface. As such, adjacent to the surface the gradients of the velocity and the temperature are much larger than those in the laminar boundary layers.

As we discussed in Section 5.4.7, it is customary to decompose the fluid velocity into a time-averaged (or mean) $\bar{\mathbf{u}}_f$ and a fluctuating component $\mathbf{u}'_f$, i.e.,

$$\mathbf{u}_f = \bar{\mathbf{u}}_f + \mathbf{u}'_f, \quad \bar{\mathbf{u}}_f \equiv \frac{1}{\tau} \int_0^\tau \mathbf{u}_f \, dt \quad \text{mean and fluctuating components of } \mathbf{u}_f, \quad (6.54)$$

where $\tau(s)$ is the time period for averaging and is taken to be long enough such that $\bar{\mathbf{u}}_f$ no longer changes by any further increase in $\tau$.

The temperature is also decomposed as

$$T_f = \overline{T}_f + T'_f, \quad \overline{T}_f \equiv \frac{1}{\tau} \int_0^\tau T_f \, dt \quad \text{mean and fluctuating components of } T_f. \quad (6.55)$$

Turbulence can be produced within the boundary layer when the flow becomes unstable. This is called the boundary-layer turbulence. The boundary-layer turbulence, in an otherwise parallel laminar flow (far-field), occurs at a transition Reynolds number $\mathrm{Re}_{L,t}$. This boundary-layer transition Reynolds number is $\mathrm{Re}_{L,t} = 5 \times 10^5$.

Turbulence can also be present in the far-field flow and this is called the free-stream turbulence. This can be caused by propellers, the presence of grid nets, other interactions with solid surfaces, or by instabilities in the far-field flow. This was discussed in Section 5.5.6. Our discussion here is limited to boundary-layer turbulence.

Although on the solid surface $\mathbf{u}_f = 0$, the surface-convection heat transfer is influenced by the velocity and temperature fluctuations near to the surface, similar to the laminar-flow case. The turbulent velocity fluctuations are three dimensional and the component perpendicular to the surface makes the largest contribution to the surface-convection heat transfer.

### 6.3.1 Turbulent Convection Heat Flux*

This section and the sections that follow are found on the Web site www .Cambridge.org/kaviany. In this section we discuss the turbulent fluctuations contribution to the heat transfer in a way similar to the molecular fluctuations. Here, the combined effects of the turbulent velocity fluctuation, the ability of the fluid to store/release heat, and the ability of the fluctuations to carry the fluid content a short distance before reaching thermal equilibrium also result in heat transfer in the presence of a temperature nonuniformity.

### 6.3.2 Microscale Turbulent Convection: Turbulent Conductivity $k_t$ and Turbulent Mixing Length $\lambda_t$*

### 6.3.3 Variation of Turbulent Mixing Length Near Solid Surface*

### 6.3.4 Averaged, Laminar-Turbulent Nusselt Number

For the fluid flow that begins as a laminar flow at $x = 0$, the presence of the solid surface causes a flow instability at a location $L_i$. Shortly after this distance the flow becomes turbulent at a location $L_t > L_i$. This location is marked with the criterion

$$\text{Re}_{L,t} = \frac{u_{f,\infty} L_t}{\nu_f} \simeq 5 \times 10^5 \quad \text{transition to boundary-layer turbulence,} \qquad (6.64)$$

where $\text{Re}_{L,t}$ is called the transition (from laminar flow to unstable, transitional flow) Reynolds number.

For a turbulent flow over a semi-infinite plate, the empirical results for the local, turbulent Nusselt number have been correlated as

$$\text{Nu}_L \equiv \overline{\text{Nu}}_L = 0.0296 \text{Re}_L^{4/5} \text{Pr}^{1/3} \text{ for } \quad 5 \times 10^5 < \text{Re}_L < 10^8 \text{ and } 0.6 < \text{Pr} < 60$$

$$\text{local, turbulent parallel flow over semi-infinite plate,} \qquad (6.65)$$

where for simplicity we have dropped the overbar indicating time averaging. A similar form of $\text{Nu}_L$ can be derived using the mixing length model. This is left as an end of chapter problem.

The empirically obtained turbulent viscous and thermal boundary-layer thicknesses $\overline{\delta}_\nu$ and $\overline{\delta}_\alpha$, similar to (6.48), are

$$\overline{\delta}_\nu \equiv \delta_\nu = 0.37 L \text{Re}_L^{-1/5} = 0.37 \left( \frac{\nu_f}{u_{f,\infty}} \right)^{1/5} L^{4/5}, \quad \frac{\overline{\delta}_\nu}{\overline{\delta}_\alpha} \simeq 1 \qquad (6.66)$$

viscous and thermal turbulent boundary-layer thicknesses.

While for laminar flow the relative thicknesses of the viscous and thermal boundary layers are related to the Prandtl number, in the turbulent boundary layer the transport of momentum and thermal energy are dominated by the turbulent fluctuations, which results in the two boundary layers having approximately the same overall thickness.

Now, to find the average Nusselt number for $L > L_t$, we include the initial laminar portion of the flow and use the average $\langle \text{Nu} \rangle_L$ as defined by (6.50). The result is

$$\langle\overline{Nu}\rangle_L \equiv \langle Nu\rangle_L = (0.037\mathrm{Re}_L^{4/5} - 871)\mathrm{Pr}^{1/3} \quad \text{for } 5 \times 10^5 < \mathrm{Re}_L < 10^8,$$

and $0.6 < \mathrm{Pr} < 60$

average laminar-turbulent, parallel flow over semi-infinite plate. (6.67)

The derivation of (6.67) is left as an end of chapter problem. As before, the temperature used to determine the fluid thermodynamic and transport properties (i.e., $\rho, c_p, \mu$, and $k$) in these Nusselt number relations should be the arithmetic average of the surface and the far-field temperatures, i.e., $(T_s + T_{f,\infty})/2$ (the film temperature).

Detailed discussion of turbulent flow heat transfer is deferred to advanced studies [7, 9].

### EXAMPLE 6.3. FAM

In rapid cooling of plastic sheets formed by molding, water is used as the coolant. Consider a thin sheet with a square cross section $L \times L$, with water flowing over one side of it with a far-field velocity $u_{f,\infty}$ parallel to the sheet. The surface temperature is $T_s$ and the far-field temperature is $T_{f,\infty}$. This is depicted in Figure Ex. 6.3(a).

(a) Draw the thermal circuit diagram.
(b) Determine the rate of surface-convection cooling.

(a) Semi-Infinite Plate with a Uniform Far-Field Temperature and Velocity

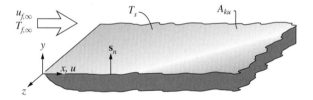

(b) Surface-Convection Heat Transfer: Forced Parallel Flow

(i) Physical Model  (ii) Thermal Circuit Model

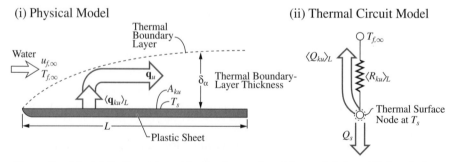

Figure Ex. 6.3. (a) Surface-convection heat transfer with parallel flow. (b) Rendering of surface-convection heat transfer with parallel flow.

(c) Determine the thermal boundary-layer thickness at the tail edge.

(d) Determine the average surface-convection resistance $\langle R_{ku}\rangle_L$.

$L = 20$ cm, $T_s = 95°$C, $T_{f,\infty} = 20°$C, and (i) $u_{f,\infty} = 0.5$ m/s, (ii) $u_{f,\infty} = 5$ m/s. Determine the thermophysical properties from Table C.23 and at $T = 330$ K.

We will continue to use this example, but with different fluid flow arrangements and phase change, in the following sections to allow for a comparative examination of the surface-convection heat transfer methods.

**SOLUTION**

(a) The thermal circuit diagram is shown in Figure Ex. 6.3(b). The energy equation is the surface energy equation (2.66), which reduces to

$$Q\mid_A = Q_s + \langle Q_{ku}\rangle_L = 0.$$

(b) The surface-convection heat transfer rate is given by (6.49), i.e.,

$$\langle Q_{ku}\rangle_L = A_{ku}\langle Nu\rangle_L \frac{k_f}{L}(T_s - T_{f,\infty}),$$

where $A_{ku} = L^2$.

The Nusselt number is related to the Reynolds number $Re_L$. The Reynolds number is given by (6.45) as

$$Re_L = \frac{u_{f,\infty}L}{\nu_f}.$$

For water we use Table C.23, and find that for $T = 330$ K

$$\nu_f = 505 \times 10^{-9} \text{m}^2\text{/s} \quad \text{Table C.23}$$
$$k_f = 0.648 \times \text{W/m-K} \quad \text{Table C.23}$$
$$Pr = 3.22 \quad \text{Table C.23}.$$

Then

(i) $Re_L = \dfrac{0.5(\text{m/s}) \times 0.2(\text{m})}{505 \times 10^{-9}(\text{m}^2\text{/s})} = 1.980 \times 10^5 < Re_{L,t} = 5 \times 10^5$ laminar-flow regime

(ii) $Re_L = \dfrac{5(\text{m/s}) \times 0.2(\text{m})}{505 \times 10^{-9}(\text{m}^2\text{/s})} = 1.980 \times 10^6 > Re_{L,t} = 5 \times 10^5$ turbulent-flow regime.

Based on this magnitude for $Re_L$, we use (6.51) and (6.67) for $\langle Nu\rangle_L$, i.e.,

(i) $\langle Nu\rangle_L = 0.664 Re_L^{1/2} Pr^{1/3} = 0.664(1.980 \times 10^5)^{1/2}(3.22)^{1/3}$

$= 436.1$

(ii) $\langle Nu\rangle_L = (0.037 Re_L^{4/5} - 871)Pr^{1/3}$

$= [(0.037(1.980 \times 10^6)^{4/5} - 871)](3.22)^{1/3}$

$= 4.666 \times 10^3$ average laminar-turbulent Nusselt number.

Now returning to $\langle Q_{ku}\rangle_L$, we have

(i) $\langle Q_{ku}\rangle_L = (0.2)^2(\text{m}^2) \times 436.1 \times 10^3 \times \dfrac{0.648(\text{W/m-K})}{0.2(\text{m})} \times (95 - 20)(\text{K})$

$= 4.239 \times 10^3 \text{ W}$

(ii) $\langle Q_{ku}\rangle_L = (0.2)^2(\text{m}^2) \times 4.666 \times 10^3 \times \dfrac{0.648(\text{W/m-K})}{0.2(\text{m})} \times (95 - 20)(\text{K})$

$= 4.535 \times 10^4 \text{ W}.$

(c) The thermal boundary-layer thickness is given by (6.48) and (6.66) as

(i) $\delta_\alpha = 5\left(\dfrac{\nu_f L}{u_{f,\infty}}\right)^{1/2} Pr^{-1/3}$

$= 5\left[\dfrac{505 \times 10^{-9}(\text{m/s}) \times 0.2(\text{m})}{0.5(\text{m/s})}\right]^{1/2} (3.22)^{-1/3}$

$= 1.522 \times 10^{-3} \text{ m} = 1.522 \text{ mm}$

(ii) $\delta_\alpha = \delta_\nu = 0.37\left(\dfrac{\nu_f}{u_{f,\infty}}\right)^{1/5} L^{4/5}$

$= 0.37\left[\dfrac{505 \times 10^{-9}(\text{m}^2/\text{s})}{5(\text{m/s})}\right]^{1/5} (0.2)^{4/5}(\text{m})^{4/5} = 4.073 \times 10^{-3} \text{ m}$

$= 2.758 \text{ mm}.$

(d) The average surface-convection resistance $\langle R_{ku}\rangle_L$ is found from (6.49), i.e.,

(i) $\langle R_{ku}\rangle_L = \dfrac{T_s - T_{f,\infty}}{\langle Q_{ku}\rangle_L} = \dfrac{(95 - 20)(^\circ\text{C})}{4.239 \times 10^3(\text{W})} = 0.01769^\circ\text{C/W}$

$A_{ku}\langle R_{ku}\rangle_L = (0.2)^2(\text{m}^2) \times 0.01769(^\circ\text{C/W}) = 7.077 \times 10^{-4\,\circ}\text{C/(W/m}^2)$

(ii) $\langle R_{ku}\rangle_L = \dfrac{T_s - T_{f,\infty}}{\langle Q_{ku}\rangle_L} = \dfrac{(95 - 20)(^\circ\text{C})}{4.535 \times 10^4(\text{W})} = 0.001654^\circ\text{C/W}$

$A_{ku}\langle R_{ku}\rangle_L = (0.2)^2(\text{m}^2) \times 0.001654(^\circ\text{C/W}) = 6.615 \times 10^{-5\,\circ}\text{C/(W/m}^2).$

Figure Ex. 6.3(b) shows the average heat flux vector $\langle \mathbf{q}_{ku}\rangle_L$, which is perpendicular to the surface (between $T_s$ is uniform over the surface). The thermal circuit diagram is also shown.

**COMMENT**

Note that for (ii), turbulent-flow regime, the average surface-convection heat flux is rather substantial, i.e.,

$$\langle q_{ku}\rangle_L = \dfrac{\langle Q_{ku}\rangle_L}{A_{ku}} = \dfrac{4.535 \times 10^4(\text{W})}{0.2^2(\text{m}^2)} = 1.134 \times 10^5 \text{ W/m}^2.$$

However, this requires a high velocity. Also note that $\delta_\alpha$ is rather small, and for the water outside this small boundary layer the temperature does not change.

## 6.4 Perpendicular Flows: Impinging Jets

In flows parallel to an isothermal surface, the heat flux **q** begins flowing perpendicular to the surface at the surface ($y = 0$) and then turns parallel to the surface, as the edge of the boundary layer is approached. This keeps the convected heat within the boundary layer and flowing parallel to the surface. These result in the growth of the thermal boundary layer and the deterioration of the heat transfer, as the distance from the leading edge increases. One method of avoiding this deterioration is to flow the fluid perpendicular to the surface, as in the multiple impinging jets. This results in a rather small surface-convection resistance in the areas of impingement and the fluid affected by the heat transfer leaves the heated (or cooled) surface region by flowing out through the spacing in between the nozzles. This is shown in Figures 6.12(a) and (b).

To show the surface convection in the region of jet impingement, we first examine a single circular jet. Then we address the surface-convection heat transfer using multiple impinging jets. Figure 6.12(a) shows a schematic of a circular jet impinging on a flat plate. Here rendered is the case of a jet cooling a surface (heat flow from the solid surface), but with assumed negligible thermobuoyancy, the results apply equally to jets heating a surface. We assume that the jet fluid is the same as the ambient fluid (in general different fluids and different phases can be used). The jet can be at the temperature of the fluid surrounding the jet (ambient fluid) or can have a different temperature. Here we consider the same temperature for the exiting jet fluid and the ambient fluid. In addition, we assume that the ambient fluid is otherwise quiescent (no motion).

The jet flow is generally turbulent, even within the nozzle (i.e., before leaving the nozzle). The nozzle diameter is $D$, the distance to the surface is $L_n$, the time- and area-averaged nozzle velocity is $\langle \overline{u}_f \rangle_A = \langle u_f \rangle_A = \langle u_f \rangle$, and the nozzle exit temperature is $T_{f,\infty}$. The solid planar surface is assumed to be isothermal and at temperature $T_s$. For a perpendicular (as compared to the tilted) round jet, there is a symmetry around the nozzle axis. The nozzle flow area is divided into three regions. The free-jet region, where the flow changes from an inside tube-like velocity profile (with zero velocity at the tube surface) to a free-surface jet flow (where there is no solid surface and at some radial distance the fluid velocity gradually becomes that of the ambient fluid, which is assumed to be stagnant here). The flow impinges on the surface and then turns flowing parallel to the surface. This is called the stagnation-flow region. The parallel flow portion is called the wall-jet region.

In general, the surface-convection resistance is the smallest (largest Nusselt number) in the stagnation-flow region and increases in the wall-jet region. The wall-jet region has some similarity to the parallel flow discussed before, except that the far-field velocity is zero for the single impinging jet.

The local surface-convection resistance $R_{ku}(r = L)$ depends on the radial location $r = L$ measured from the intersection of the nozzle axis with the solid surface. This local resistance depends on the nozzle Reynolds number $\text{Re}_D$

$$\text{Re}_D = \frac{\langle u_f \rangle D}{\nu_f} \quad \text{nozzle Reynolds number,} \quad (6.68)$$

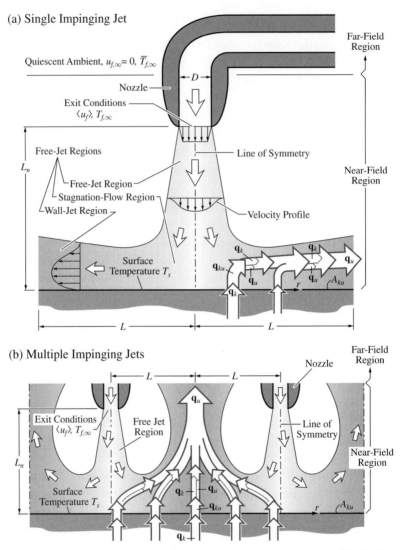

Figure 6.12. Fluid flow and heat transfer. (a) Single, round impinging jet. (b) Multiple round impinging jets.

and on Pr, $L_n/D$, and $L/D$. The surface-convection heat transfer is represented in terms of the local dimensionless surface-convection conductance, i.e., the local Nusselt number $Nu_L$. Typical variation of $Nu_L$ with respect to $L$ is shown in Figure 6.13 for air. Note the general trend of decay in $Nu_L$ with the distance $L$, except when $L_n/D$ is relatively small (i.e., $L_n/D \leq 5$). The local averaged Nusselt number is defined as

$$Nu_L = \frac{A_{ku}R_k}{A_{ku}R_{ku}} = \frac{L}{A_{ku}R_{ku}k_f} = \frac{q_{ku}L}{(T_s - T_{f,\infty})k_f} \quad \text{or} \quad A_{ku}R_{ku} = \frac{L}{Nu_L k_f}. \quad (6.69)$$

Taking the average of $Nu_L$ over a circle of diameter $2L$ on the surface, we have the average Nusselt number as

$$\langle Nu \rangle_L = \frac{A_{ku}R_k}{A_{ku}\langle R_{ku}\rangle_L} = \frac{2}{L^2}\int_0^L Nu_x x dx \quad \begin{array}{l}\text{average Nusselt number} \\ \text{over an area of radius } L.\end{array} \quad (6.70)$$

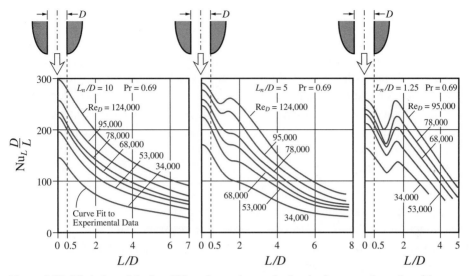

Figure 6.13. Variation of the local Nusselt number under impinging, round nozzle, with respect to the normalized radial distance from the nozzle axis. The results are for air with three different normalized nozzle distances and for several nozzle Reynolds numbers. (From Martin, H., reproduced by permission © 1977, Academic.)

One of the existing correlations for $\langle \text{Nu} \rangle_L$, for a single round jet, is [15]

$$\langle \text{Nu} \rangle_L = 2\text{Re}_D^{1/2}\text{Pr}^{0.42}(1 + 0.005\text{Re}_D^{0.55})^{1/2}\frac{1 - 1.1D/L}{1 + 0.1(L_n/D - 6)D/L} \quad (6.71)$$

$$\text{for } 7.5 > \frac{L}{D} > 2.5, \quad \text{correlation for } \langle \text{Nu} \rangle_L \text{ for a single, round jet.}$$

For a square array of round jets with a distance $2L$ between the center of the adjacent nozzles [as shown in Figure 6.12(b)], the correlation for $\langle \text{Nu} \rangle_L$ is

$$\langle \text{Nu} \rangle_L = \text{Re}_D^{2/3}\text{Pr}^{0.42}\frac{2L}{D}\left\{1 + \left[\frac{L_n/D}{0.6/(1 - \epsilon)^{1/2}}\right]^6\right\}^{-0.05}$$

$$\times (1 - \epsilon)^{1/2}\frac{1 - 2.2(1 - \epsilon)^{1/2}}{1 + 0.2(L_n/D - 6)(1 - \epsilon)^{1/2}}$$

$$\text{for } \frac{2L}{D} > 2.5 \quad \text{correlation for } \langle \text{Nu} \rangle_L \text{ for a square array of round jets,} \quad (6.72)$$

where

$$\epsilon = 1 - \frac{\pi D^2}{16L^2}. \quad (6.73)$$

Note that $\epsilon$ is the void-area fraction in the plane of the nozzle axis. The above correlations are best suited for $2 \leq L_n/D \leq 12$, and $2{,}000 \leq \text{Re}_D < 100{,}000$.

EXAMPLE 6.4. FAM

Consider the rapid cooling of the plastic sheet discussed in Example 6.3, again using water, but here with a perpendicular high-speed, jet flow. Use a single, impinging jet with a jet nozzle diameter $D$ and average nozzle flow velocity $\langle u_f \rangle$, with nozzle exit temperature $T_{f,\infty}$ and surface at $T_s$. This is depicted in Figure Ex. 6.4(b). In the calculation of $\langle Nu \rangle_L$, use a surface with a square area $2L \times 2L$. As before, $2L = 20$ cm, $T_s = 95°C$, and $T_{f,\infty} = 20°C$. The nozzle diameter is $D = 2.5$ cm and it is at $L_n = 10$ cm from the surface. The average nozzle flow velocity $\langle u_f \rangle = 25$ m/s.

(a) Draw the thermal circuit diagram.
(b) Determine the rate of surface-convection cooling.
(c) Determine the average surface-convection resistance $\langle R_{ku} \rangle_L$.
(d) Compare the results with that of the parallel flow cooling of Example 6.3 and comment on the difference.

   Evaluate properties at $T = 330$ K.

Surface-Convection Heat Transfer:<br/>Comparison of Forced Parallel and Perpendicular Flows

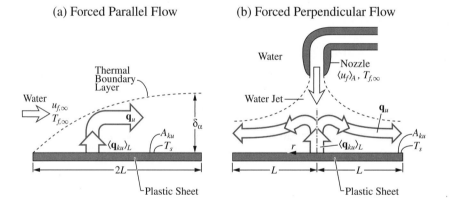

(c) Thermal Circuit Model

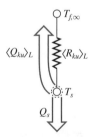

Figure Ex. 6.4. Surface-convection heat transfer by a single, round nozzle and comparison with parallel flow. (a) Parallel flow. (b) Perpendicular flow. (c) Thermal circuit model.

**SOLUTION**

(a) The thermal circuit diagram is shown in Figure Ex. 6.4(c).

(b) The rate of surface-convection heat transfer is that given by (6.49), i.e.,

$$\langle Q_{ku}\rangle_L = A_{ku}\langle \mathrm{Nu}\rangle_L \frac{k_f}{L}(T_s - T_{f,\infty}),$$

where

$$A_{ku} = 4L^2.$$

For a single, normal jet, the Nusselt number is found from (6.71), i.e.,

$$\langle \mathrm{Nu}\rangle_L = 2\mathrm{Re}_D^{1/2}\mathrm{Pr}^{0.42}(1 + 0.005\mathrm{Re}_D^{0.55})^{1/2}\frac{1 - 1.1D/L}{1 + 0.1(L_n/D - 6)(D/L)},$$

where from (6.68)

$$\mathrm{Re}_D = \frac{\langle u_f\rangle D}{\nu_f}.$$

Note that here $L$ is the radius of the circle on the surface. As in Example 6.3, the thermophysical properties are obtained from Table C.23, and we have

$$\nu_f = 505 \times 10^{-9} \text{ m}^2/\text{s} \quad \text{Table C.23}$$
$$k_f = 0.648 \text{ W/m-K} \quad \text{Table C.23}$$
$$\mathrm{Pr} = 3.22 \quad \text{Table C.23.}$$

Then

$$\mathrm{Re}_D = \frac{25(\text{m/s}) \times 0.025(\text{m})}{505 \times 10^{-9}(\text{m/s}^2)} = 1.238 \times 10^6.$$

Note that this $\mathrm{Re}_D$ is outside the range of the data used in the correlation (6.71). Due to lack of an alternative correlation, we will use (6.71). Now we evaluate $\langle \mathrm{Nu}\rangle_L$ and we note that $2L = 20$ cm.

$$\langle \mathrm{Nu}\rangle_L = 2(1.238 \times 10^6)^{1/2}(3.22)^{0.42}[1 + 0.005(1.238 \times 10^6)^{0.55}]^{1/2} \times$$
$$\frac{1 - 1.1[0.025(\text{m})/0.10(\text{m})]}{1 + 0.1[0.10(\text{m})/0.025(\text{m}) - 6][0.025(\text{m})/0.10(\text{m})]}$$
$$= 3.637 \times 10^9 \times 3.496 \times 0.7632$$
$$= 9.702 \times 10^4.$$

The surface-convection heat transfer rate is

$$\langle Q_{ku}\rangle_L = (0.2)^2(\text{m}^2) \times 9.702 \times 10^4 \times \frac{0.648(\text{W/m-K})}{0.1(\text{m})}(95 - 20)(\text{K})$$
$$= 1.886 \times 10^6 \text{ W.}$$

(c) The average surface-convection resistance $\langle R_{ku} \rangle_L$ is given by (6.49), i.e.,

$$\langle R_{ku} \rangle_L = \frac{T_s - T_{f,\infty}}{\langle Q_{ku} \rangle_L} = \frac{(95 - 20)(^\circ C)}{1.886 \times 10^6 (W)} = 3.976 \times 10^{-5} {}^\circ C/W$$

$$A_{ku} \langle R_{ku} \rangle_L = (0.2)^2 (m^2) \times 3.976 \times 10^5 (^\circ C/W) = 1.590 \times 10^{-6} {}^\circ C/(W/m^2).$$

(d) Although the amount of water used in the impinging-jet cooling is much less than that for parallel flow, for the conditions used, the heat transfer rate is larger for the single jet, i.e.,

$$\langle Q_{ku} \rangle_L = 4.535 \times 10^4 \text{ W} \quad \text{forced, turbulent parallel flow}$$

$$\langle Q_{ku} \rangle_L = 1.886 \times 10^6 \text{ W} \quad \text{forced, turbulent single jet.}$$

**COMMENT**

Note that we used (6.71) outside the range of $Re_D$ used in the construction of the correlation. Further improvement in jet cooling can be made by using multiple nozzles.

## 6.5 Thermobuoyant Flows

Consider that a fluid under the gravitational influence is designated by the gravity vector **g** with the gravitational acceleration constant $g$. The fluid is quiescent, i.e., $\mathbf{u}_{f,\infty} = 0$, in the far-field region and has a temperature $T_{f,\infty}$. When this fluid is partly bounded by a solid surface at a temperature $T_s$, different than $T_{f,\infty}$, the fluid may undergo a thermobuoyant motion. This is due to the interaction of the gravity vector $g$ and the local change in the fluid density $\Delta\rho_f$ caused by the temperature nonuniformity. This motion, which is confined to the region affected by heat transfer (i.e., confined to the nonuniform temperature region), is called the heat transfer-controlled buoyant motion or thermobuoyant motion (or natural convection or free convection). This motion, and its effect on the surface-convection heat transfer, are significant under a significant gravity (i.e., at or near the surfaces of massive objects such as the earth). Under microgravity (for example, in the space-shuttle orbital motion around the earth, where a near balance exists between the gravitational force and the centrifugal force, such that the net acceleration is of the order of $10^{-3}g_o$, where $g_o = 9.807$ m/s$^2$), the thermobuoyant motion becomes less significant. Table C.7 lists the magnitude of $g$ as a function of the distance from the earth surface.

The fluid volumetric thermal expansion coefficient $\beta_f (1/K)$, discussed in Appendix A, is a measure of the relative change in the fluid specific volume $v_f$ (or density $\rho_f$) with respect to a change in the fluid temperature $T_f$. This volumetric thermal expansion coefficient is defined by (A.10) as

$$\beta_f \equiv \frac{1}{v_{f,o}} \frac{\partial v_f}{\partial T_f} \Big|_{p_o} = -\frac{1}{\rho_{f,o}} \frac{\partial \rho_f}{\partial T_f} \Big|_{p_o} \qquad \text{volumetric expansion coefficient,} \qquad (6.74)$$

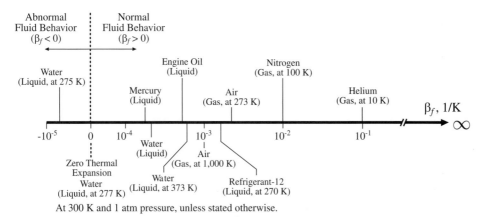

Figure 6.14. Typical values of the volumetric thermal expansion coefficient $\beta_f$ for various gases and liquids.

where $\rho_{f,o} = 1/v_{f,o}$, $T_o$, and $p_o$ are the reference fluid density, temperature, and pressure. Typical values for $\beta_f$ for various gases and liquids are shown in Figure 6.14. A normal fluid is the fluid that has a positive volumetric thermal expansion coefficient, $\beta_f > 0$, i.e., its specific volume increases with temperature. Water around the temperature of $4°C$ has a negative $\beta_f$ and is referred to as an abnormal fluid in this temperature range.

For a fluid with $\beta_f > 0$, Figures 6.15(a) to (c) show the thermobuoyant flow around cooled or heated flat plates. In these figures, grouping is made by cooling $(T_s < T_{f,\infty})$ versus heating $(T_s > T_{f,\infty})$, by vertical versus horizontal orientation of the plate, and by the extent of the plate (finite, semi-infinite, or infinite). There is also a distinction made for a plume-type motion, where a warm (rising plume) or cold (falling plume) fluid flows in a given direction until it comes into thermal equilibrium with the ambient. There is also a cellular motion in which the fluid circulates. Other directions, i.e., the oblique (or tilted) arrangement, and geometries, are also possible, but not shown in Figures 6.15(a) to (c).

With $\beta_f > 0$, when the temperature in a fluid varies in the horizontal direction (i.e., perpendicular to the gravity vector $\mathbf{g}$), the lighter region of the fluid rises and the heavier region of the fluid descends. When a temperature variation exists in the vertical direction (along or opposite to $\mathbf{g}$), and if the heavier fluid is at the bottom, there will not be a thermobuoyant flow and heat transfer within the fluid is by conduction only. If the lighter fluid is at the bottom and the fluid is semi-bounded, there will be a thermobuoyant motion. For fluids bounded at a top and bottom by infinite horizontal plates, the presence of the heavier fluid on the top does not guarantee a motion, unless a threshold of this unstable fluid buoyancy is exceeded. Then it is said that there is an onset of the thermobuoyant motion.

The volumetric buoyancy force in the differential-volume momentum conservation equation, discussed in Appendix B, is given in (B.32) as $\rho\mathbf{g}$. For simple, compressible fluids, the density $\rho_f$ depends on the pressure $p$ and temperature $T_f$,

(a) Isothermal, Horizontal Finite Plate

(i) Heated Plate, $T_s > T_{f,\infty}$              (ii) Cooled Plate, $T_s < T_{f,\infty}$

Figure 6.15. (a) Thermobuoyant flow over upper surface of an isothermal, horizontal finite plate, heated or cooled.

i.e., $\rho_f = \rho_f(p, T_f)$. Now we expand (Taylor series expansion) $\rho_f$ around $\rho_{f,o}$ (a reference density at $T_o$ and $p_o$) at constant $p$ and take only the first term with the derivative and write

$$\rho_f = \rho_{f,o} + \frac{\partial \rho_f}{\partial T_f}\Big|_{p_o} (T_f - T_o) + 0, \qquad T_f \to T_o \quad \begin{array}{l}\text{Taylor series expansion of}\\ \rho_f(p_o, T_f) \text{ around } \rho_{f,o}.\end{array}$$
$$(6.75)$$

Now we use the definition of $\beta_f$ in (6.74) and replace $\partial \rho_f / \partial T_f|_{p_o}$ with $-\rho_{f,o}\beta_f$ that becomes

$$\rho_f = \rho_{f,o} - \rho_{f,o}\beta_f(T_f - T_o) \quad \text{linear } \rho_f\text{-}T_f \text{ relation.} \qquad (6.76)$$

In addition to the typical values of $\beta_f$ shown in Figure 6.14, in Appendix C, Table C.23 lists $\beta_f$ for some liquids along with other thermophysical properties. For gases, we note that for an ideal gas $\partial \rho_f / \partial T_f|_p$ is found by the appropriate differentiation and we have for $\beta_f$

$$\beta_f = \frac{1}{T_f}, \quad p = \rho_f \frac{R_g}{M} T_f \quad \text{ideal gas.} \qquad (6.77)$$

The temperature $T_f$ used is in absolute scale, i.e., K, and as $T_f \to 0$, the thermal expansion coefficient increases substantially, tending to infinity as $T_f \to 0$ K is approached. But no gas is expected to exist as $T_f \to 0$.

(b) Isothermal, Horizontal Infinite Flat Plate

(i) Heated Plate, $T_s > T_{f,\infty}$ : Turbulent, Cellular Thermobuoyant Flow on Top of Plate

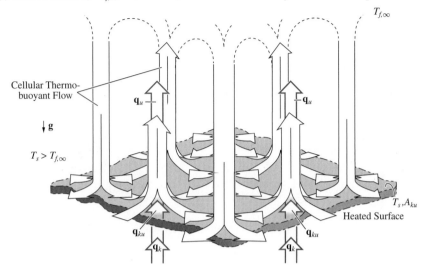

(ii) Cooled Plate, $T_s < T_{f,\infty}$ : No Thermobuoyant Motion on Top of Plate

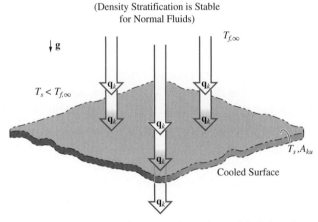

Figure 6.15. (b) Thermobuoyant flow over isothermal, horizontal infinite plate, heated or cooled.

### 6.5.1 Laminar Flow Adjacent to Vertical Plates: Grashof and Rayleigh Numbers $Gr_L$, $Ra_L$

Consider an isothermal, semi-infinite vertical surface at temperature $T_s$ and a fluid far-field temperature $T_{f,\infty}$, with $L$ being the length along the vertical surface ($L$ being less than $L_i$, where a boundary-layer instability occurs in the thermobuoyant flow). Figure 6.16 shows a schematic of the problem considered for $T_s > T_{f,\infty}$, i.e., heated plate. Here $\mathbf{q}$ is conducted through the substrate and then removed from the surface by surface convection $\mathbf{q}_{ku}$ and then conducted and convected $\mathbf{q}_k + \mathbf{q}_u$ inside the boundary layer and finally convected $\mathbf{q}_u$ away from the surface. As with

(c) Isothermal, Vertical Semi-Infinite Plate: Thermobuoyant Flow
   (Boundary-Layer Flow and Heat Transfer)

(i) Heated Plate, $T_s > T_{f,\infty}$          (ii) Cooled Plate, $T_s < T_{f,\infty}$

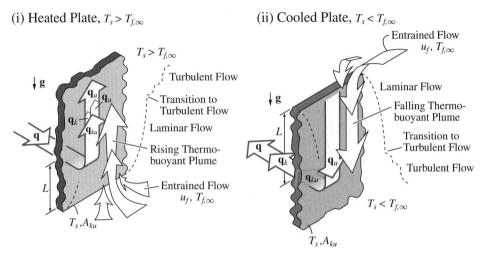

Figure 6.15. (c) Thermobuoyant flow adjacent to isothermal, vertical semi-infinite plate, heated or cooled.

the forced, parallel flows (Section 6.2.5), here this parallel flow changes the direction of **q** from perpendicular to the surface at $y = 0$ to parallel to surface near the edge of the boundary layer (the edge is marked by $y = \delta_\alpha$).

    The fluid flow (momentum equation), i.e., $\mathbf{u}_f(\mathbf{x}, t)$, is influenced by the temperature distribution in the fluid $T_f(\mathbf{x}, t)$, which in turn depends on the heat transfer (energy equation) within the fluid as again influenced by $\mathbf{u}_f$. Therefore, the momentum and energy equations are coupled. This requires the simultaneous solution of both the energy and momentum (and mass) conservation equations in order to find a simultaneous solution for $T_f(\mathbf{x}, t)$ and $\mathbf{u}_f(\mathbf{x}, t)$. The case of steady-state heat transfer and fluid flow, with two-dimensional $T_f = T_f(x, y)$ and $\mathbf{u}_f = (u_f, v_f)$ fields, i.e., semi-infinite plates, has been studied with great detail.

    For zero external pressure, and subject to the boundary-layer and other approximations (some related to the variations of the properties $\rho_f$, $c_{p,f}$, $\mu_f$, and $k_f$ with respect to temperature), the steady-state, two-dimensional energy (2.1), momentum (B.51), and mass (B.49) conservation equations become

$$\frac{\partial}{\partial y}\mathbf{s}_y \cdot \mathbf{q}_k \quad + \quad \frac{\partial}{\partial x}\mathbf{s}_x \cdot \mathbf{q}_u \quad + \quad \frac{\partial}{\partial y}\mathbf{s}_x \cdot \mathbf{q}_u \quad = 0 \quad \text{energy equation}$$

spatial rate of    spatial rate of    spatial rate of
change in $\mathbf{q}_k$      change in $\mathbf{q}_u$      change in $\mathbf{q}_u$                 (6.78)
in $y$ direction     in $x$ direction     in $y$ direction

or

$$-k_f\frac{\partial^2 T_f}{\partial y^2} + (\rho c_p)_f\left(u_f\frac{\partial T_f}{\partial x} + v_f\frac{\partial T_f}{\partial y}\right) = 0 \quad \text{energy equation} \qquad (6.79)$$

(a) Thermobuoyant Flow and Heat Transfer

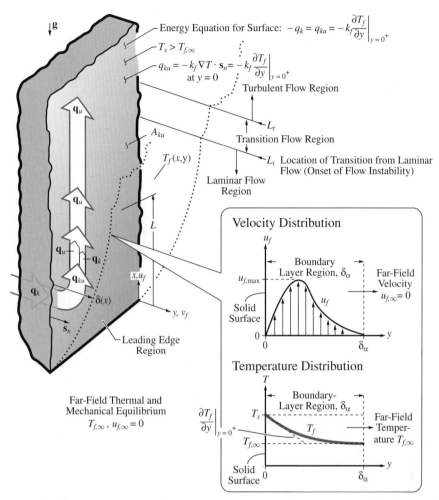

Figure 6.16. (a) Thermobuoyant flow and heat transfer adjacent to a semi-infinite, vertical plate. The anticipated distributions of the velocity $u_f$ and temperature $T_f$ are also shown.

$$-\mu_f \frac{\partial^2 u_f}{\partial y^2} + \rho_f \left( u_f \frac{\partial u_f}{\partial x} + v_f \frac{\partial u_f}{\partial y} \right) = rho_f g \beta_f (T_f - T_{f,\infty}) \quad \begin{array}{l} x\text{-direction momentum} \\ \text{equation} \end{array}$$

$$\begin{array}{ccc} \text{spatial rate of} & \text{spatial rate of} & \text{volumetric} \\ \text{change of} & \text{change of} & \text{gravitational} \\ \text{viscous stress} & \text{inertial force} & \text{force} \end{array} \qquad (6.80)$$

$$\frac{\partial u_f}{\partial x} + \frac{\partial v_f}{\partial y} = 0 \quad \text{continuity equation.}$$

$$\text{spatial rate of change of mass flux} \qquad (6.81)$$

The expected velocity and temperature distributions $u_f = u_f(x, y)$, $T_f = T_f(x, y)$, at a location $x$, are also shown in Figure 6.16. The other component of velocity $v_f = v_f(x, y)$ is also important, but is not shown in the figure.

**(b) Velocity Distribution for Parallel Component**

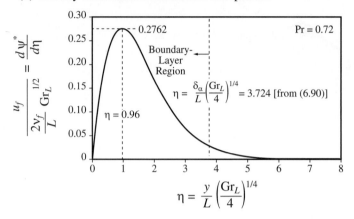

**(c) Velocity Distribution for Perpendicular Component**

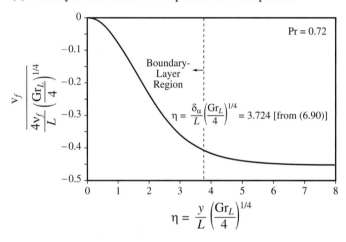

**(d) Temperature Distribution**

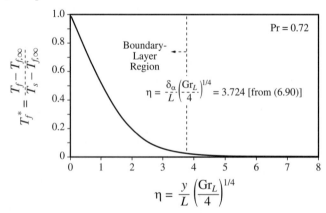

Figure 6.16. (b) Predicted dimensionless streamwise fluid velocity. (c) Predicted dimensionless lateral fluid velocity. (d) Predicted temperature. All are shown as functions of the similarity variable, for $Pr = 0.72$.

We now scale (i.e., normalize) the energy, momentum, and continuity equations. We use the length scale $L$ and, since the maximum velocity $u_{f,max}$ is not known a priori, we choose $v_f/L$ as the velocity scale. The choice of $[g\beta_f(T_s - T_{f,\infty})L/(1 + \mathrm{Pr})]^{1/2}$ as the velocity scale is more meaningful [13], but for the sake of simplicity, we will use $v_f/L$. For the temperature, we use $T_s - T_{f,\infty}$. Then (6.79) to (6.81) become

$$-\frac{\partial^2 T_f^*}{\partial y^{*2}} + \mathrm{Pr}\left(u_f^* \frac{\partial T_f^*}{\partial x^*} + v_f^* \frac{\partial T_f^*}{\partial y^*}\right) = 0 \qquad \text{dimensionless energy equation} \quad (6.82)$$

$$-\frac{\partial^2 u_f^*}{\partial y^{*2}} + \left(u_f^* \frac{\partial u_f^*}{\partial x^*} + v_f^* \frac{\partial v_f^*}{\partial y^*}\right) = \mathrm{Gr}_L T^* \qquad \begin{array}{l}\text{dimensionless } x\text{-direction}\\ \text{momentum equation}\end{array} \quad (6.83)$$

$$\frac{\partial u_f^*}{\partial x^*} + \frac{\partial v_f^*}{\partial y^*} = 0 \qquad \text{dimensionless continuity equation,}$$

$$(6.84)$$

where

$$T_f^* = \frac{T_f - T_{f,\infty}}{T_s - T_{f,\infty}}, \quad x^* = \frac{x}{L}, \quad y^* = \frac{y}{L}, \quad u_f^* = \frac{u_f}{v_f/L}, \quad v_f^* = \frac{v_f}{v_f/L}, \quad \mathrm{Pr} = \frac{v_f}{\alpha_f},$$

$$\mathrm{Gr}_L = \frac{F_{\Delta T}}{F_\mu} = \frac{g\beta_f(T_s - T_{f,\infty})L^3}{v_f^2} \qquad \text{dimensionless variables and parameters.}$$

$$(6.85)$$

The Grashof number $\mathrm{Gr}_L = F_{\Delta T}/F_\mu$ is the ratio of buoyant force to viscous force. Since we assume that $\mathrm{Gr}_L \geq 0$, whenever $T_{f,\infty} < T_s$, we rewrite the expression for $\mathrm{Gr}_L$ as $T_{f,\infty} - T_s$.

Note that these conservation equations are similar to (6.43) used for forced flow.

The mechanical and thermal bounding-surface and far-field conditions are also similar to (6.14) and (6.38), except here the far-field velocity $u_{f,\infty}$ is zero.

$$\begin{array}{ll} u_f^*(x^*, y^* = 0) = v_f^*(x^*, y^* = 0) = 0, & T_f^*(x^*, y^* = 0) = 1 \\ u_f^*(x^*, y^* \to \infty) = 0, \quad T_f^*(x^*, y^* \to \infty) = 0 \\ u_f^*(x^* = 0, y^*) = v_f^*(x^* = 0, y^*), & T_f^*(x^* = 0, y^*) = 0, \end{array} \qquad \begin{array}{l}\text{dimensionless thermal}\\ \text{and mechanical conditions}\\ \text{at various locations.}\end{array}$$

$$(6.86)$$

Comparing the two-dimensional, boundary-layer conservation equations (6.82) to (6.84) with those for the forced flow, i.e., (6.39) to (6.41), we note that here $\mathrm{Pr}$ and $\mathrm{Gr}_L$ play roles similar to $\mathrm{Pr}$ and $\mathrm{Re}_L$ in the forced flow.

A similarity solution exists for this thermobuoyant flow. This combines $y$ and $x$ into one variable $\eta$ and reduces (6.82) to (6.84) to ordinary, differential equations.[†]

---

[†] The dimensionless stream function is defined as

$$\psi^* = \frac{\psi}{4v_f(\mathrm{Gr}_L/4)^{1/4}}$$

and the dimensionless similarity variable is defined as

$$\eta = \frac{y}{x}(\mathrm{Gr}_L/4)^{1/4}.$$

The details of the solution are not given here, but are left as an end of chapter problem. Figures 6.16(b) and (c) show the predicted velocity distributions for Pr = 0.72 (e.g., air near room temperature). The results are presented as $u_f/(2v_f\,\mathrm{Gr}_L^{1/2}/L)$, and $v_f/[4v_f/L(\mathrm{Gr}_L/4)^{1/4}]$ as a function of the similarity variable $\eta = y(\mathrm{Gr}_L/4)^{1/4}/L$. Note that for $\eta \to \infty$ (i.e., far from the surface), the streamwise velocity $u_f$ vanishes (as expected for an otherwise quiescent ambient). However, the lateral velocity $v_f$ (which is pointing toward the surface, and therefore is negative) reaches an asymptotic value far from the boundary layer. This lateral flow is needed to provide the mass flow rate added to the boundary layer as it grows along the wall. Figure 6.16(d) shows the dimensionless temperature distribution $T_f^*$ within the boundary layer.

The solution for the local dimensionless surface-convection conductance $\mathrm{Nu}_L$ is [6]

$$\mathrm{Nu}_L = \frac{1}{R_{ku}^*} = \frac{L}{A_{ku}R_{ku}k_f} = \frac{q_{ku}L}{(T_s - T_{f,\infty})k_f} = \frac{-k_f\dfrac{\partial T_f}{\partial y}\big|_{y=0^+}\,L}{(T_s - T_{f,\infty})k_f}$$

$$= \frac{0.503}{[1 + (0.492/\mathrm{Pr})^{9/16}]^{4/9}}\mathrm{Ra}_L^{1/4}, \qquad \mathrm{Ra}_L < 10^9 \quad\begin{array}{l}\text{laminar, thermobuoyant}\\ \text{flow over vertical,}\\ \text{semi-infinite plate.}\end{array}$$

$$(6.87)$$

Compared to (6.47), $k_f\partial T_f/\partial y\,|_{y=0^+} = -0.503\mathrm{Ra}_L^{1/4}(T_s - T_{f,\infty})/L[1 + (0.492/\mathrm{Pr})^{9/16}]^{4/9}$.

The product $\mathrm{Gr}_L\mathrm{Pr}$ is named the Rayleigh number $\mathrm{Ra}_L$, i.e.,

$$\boxed{\mathrm{Ra}_L \equiv \mathrm{Gr}_L\mathrm{Pr} = \frac{g\beta_f(T_s - T_{f,\infty})L^3}{v_f\alpha_f} \qquad \text{Rayleigh number.} \qquad (6.88)}$$

Note that here the resistance $R_{ku}$ depends on the temperature difference $T_s - T_{f,\infty}$. This is referred to as the nonlinearity of the relation between $q_{ku}$ and $T_s - T_{f,\infty}$. This nonlinearity is characteristic of thermobuoyant flows.

The power of 1/4 on the Rayleigh number is a characteristic of thermobuoyant laminar flow and heat transfer. The average Nusselt number $\langle\mathrm{Nu}\rangle_L$ is defined in (6.50) and here using (6.87) we have

$$\langle\mathrm{Nu}\rangle_L = \frac{\langle q_{ku}\rangle_L L}{(T_s - T_{f,\infty})k_f} = \int_0^L \frac{\mathrm{Nu}_x}{x}dx$$

$$= \frac{4}{3}\mathrm{Nu}_L$$

$$= \frac{4}{3}\frac{0.503}{[1 + (0.492/\mathrm{Pr})^{9/16}]^{4/9}}\mathrm{Ra}_L^{1/4} \quad\begin{array}{l}\text{average Nusselt number for}\\ \text{laminar, thermobuoyant flow}\\ \text{adjacent to vertical, semi-}\\ \text{infinite plate, } \mathrm{Ra}_L < 10^9.\end{array} \qquad (6.89)$$

$$\equiv a_1\mathrm{Ra}_L^{1/4}$$

The boundary-layer approximation that allows neglecting the axial (along $x$ direction) conduction, is valid when $\text{Ra}_L$ is large (i. e. $\text{Ra}_L > 10^4$). Then as $\text{Ra}_L \to 0$, the axial conduction tends to increase the value of $\text{Nu}_L$.

The magnitude of $g$ as a function of the distance from the sea level $r(\text{m})$ is given in Table C.7. For $r = 0$, we have $g = 9.807 \text{ m/s}^2$ (generally designated as the standard gravitational constant $g_o$).

The laminar, boundary-layer thickness $\delta_\alpha = \delta_\nu$ is approximated as

$$\delta_\alpha = \delta_\nu = 3.93L \left( \text{Pr} + \frac{20}{21} \right)^{1/4} (\text{Gr}_L^{1/2} \text{Pr})^{-1/2} \quad \begin{array}{l} \text{laminar thermal and viscous} \\ \text{boundary-layer thicknesses,} \\ \text{Ra}_L < 10^9. \end{array}$$

$$(6.90)$$

The boundary-layer thickness $\delta_\alpha$ grows as $L^{1/4}$ and the surface-convection heat flux $q_{ku}$ decreases with this growth (i.e., $R_{ku}$ increases as $L^{1/4}$).

In practice, surface-averaged Nusselt number relations usable over a large of $\text{Ra}_L$ are used. Such correlations are listed in Table 6.4 and will be discussed with the summary of correlations in Section 6.7.

### 6.5.2 Turbulent Thermobuoyant Flow over Vertical Plate

The transition to turbulent thermobuoyant flow is represented empirically by

$$\text{Ra}_{L,t} = \frac{g\beta(T_s - T_{f,\infty})L_t^3}{\nu_f \alpha_f} = 10^9 \quad \text{transition to turbulent, thermobuoyant flow.}$$

$$(6.91)$$

A characteristic of turbulent thermobuoyant flow and heat transfer is the lack of dependence of $q_{ku}$ on $L$. This would require that the Nusselt number $\text{Nu}_L$ be proportional to $\text{Ra}_L^{1/3}$. One of the correlations for the semi-infinite, vertical plate that is applicable across the laminar- and turbulent-flows regimes is [16]

$$\langle \text{Nu} \rangle_L = ((\langle \text{Nu}_{L,l} \rangle)^6 + \langle \text{Nu}_{L,t} \rangle^6)^{1/6}$$

$$\langle \text{Nu}_{L,l} \rangle = \frac{2.8}{\ln[1 + 2.8/(a_1 \text{Ra}_L^{1/4})]} \quad \text{laminar-flow contribution}$$

$$\langle \text{Nu}_{L,t} \rangle = \frac{0.13 \text{Pr}^{0.22}}{(1 + 0.61 \text{Pr}^{0.81})^{0.42}} \text{Ra}_L^{1/3} \quad \text{turbulent-flow contribution}$$

$$a_1 = \frac{4}{3} \frac{0.503}{[1 + (0.492/\text{Pr})^{9/16}]^{4/9}}$$

$$\begin{array}{l} \text{correlation for laminar and turbulent, thermobuoyant} \\ \text{flow over vertical, semi-infinite plate, } 0 \leq \text{Ra}_L \leq \infty, \end{array}$$

$$(6.92)$$

where $Ra_L$ is defined in (6.88). This laminar flow contribution allows for the effect of the axial conduction, which causes higher Nusselt number at low values of $Ra_L$. This is done using natural logarithmic function that tends towards (6.90), when $Ra_L$ is larger than $10^8$ (this is discussed in an end of chapter problem).

Other correlations for thermobuoyant-flow Nusselt number, for different surface geometries, will be given in Tables 6.3 to 6.6 and will be discussed in Section 6.7.2.

### 6.5.3 Combined Thermobuoyant-Forced Parallel Flows

In general, the presence of a fluid density nonuniformity resulting from a fluid temperature nonuniformity can assist (i.e., same direction) or oppose (i.e., opposite direction) a forced flow. A criterion for assessing the significance of the thermobuoyant flow in a parallel forced flow is the ratio $Gr_L/Re_L^2$. For $Gr_L/Re_L^2 \ll 1$ the thermobuoyant flow is negligible, and for $Gr_L/Re_L^2 \gg 1$ the thermobuoyant flow is dominant, compared to forced parallel flow.

### EXAMPLE 6.5. FAM

Again, consider the cooling of the square plastic sheet of Examples 6.3 and 6.4. Here we place the hot plastic sheet vertically, and on one side of it we have a body of water that is otherwise quiescent, i.e., zero far-field velocity $u_{f,\infty} = 0$, and with a far-field temperature $T_{f,\infty}$. This is depicted in Figure Ex. 6.5(a). The plastic surface is at $T_s$. The sheet dimension is again $L \times L$. Assume that the sheet can be treated as a vertical, semi-infinite plate (i.e., neglecting the end effect due to finite length) and treat the thermobuoyant flow and heat transfer accordingly.

(a) Draw the thermal circuit diagram.
(b) Determine the rate of surface-convection cooling.
(c) Determine the average surface-convection resistance $\langle R_{ku} \rangle_L$.
(d) Comment on the comparison between this method of cooling versus parallel and perpendicular forced flows.

$L = 20$ cm, $T_s = 95°C$, $T_{f,\infty} = 20°C$.
Evaluate the water properties at $T = 330$ K.

**SOLUTION**
(a) The thermal circuit diagram is shown in Figure Ex. 6.5(b).
(b) The rate of surface-convection heat transfer is again given by (6.49) as

$$\langle Q_{ku} \rangle_L = A_{ku} \langle q_{ku} \rangle_L = A_{ku} \langle Nu \rangle_L \frac{k_f}{L} (T_s - T_{f,\infty}),$$

where

$$A_{ku} = L^2.$$

Surface-Convection Heat Transfer: Thermobuoyant Parallel Flow

(a) Physical Model  (b) Thermal Circuit Model

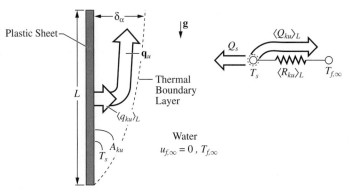

Figure Ex. 6.5. Surface-convection heat transfer by thermobuoyant flow. (a) Physical model. (b) Thermal circuit model.

For an isothermal, semi-infinite plate, we have the relation for $\langle \text{Nu} \rangle_L$ given by (6.92) and the Rayleigh number $\text{Ra}_L$, from (6.88) is

$$\text{Ra}_L = \frac{g\beta_f(T_s - T_{f,\infty})L^3}{\nu_f \alpha_f}.$$

Using Table C.23 for water (at $T = 330$ K), and $g = 9.807$ m/s$^2$, we have

$\beta_f = 0.000273$ 1/K   Table C.23 (at $T = 310$ K, the closest data available)
$k_f = 0.648$ W/m-K   Table C.23
$\alpha_f = 1.54 \times 10^{-7}$ m$^2$/s Table C.23
$\nu_f = 5.05 \times 10^{-7}$ m$^2$/s Table C.23
$\text{Pr} = 3.22$   Table C.23.

Then

$$\text{Ra}_L = \frac{9.807(\text{m/s}^2) \times 0.000273(1/\text{K}) \times (95-20)(\text{K}) \times (0.2)^3(\text{m}^3)}{5.05 \times 10^{-7}(\text{m}^2/\text{s}) \times 1.54 \times 10^{-7}(\text{m}^2/\text{s})}$$

$$= 2.066 \times 10^{10} \quad \text{turbulent, thermobuoyant flow.}$$

The $\langle \text{Nu} \rangle_L$ relation (6.92) is

$$\langle \text{Nu} \rangle_L = [(\text{Nu}_{L,l})^6 + (\text{Nu}_{L,t})^6]^{1/6}$$

$$\langle \text{Nu}_{L,l} \rangle = \frac{2.8}{\ln(1 + 2.8/a_1 \text{Ra}^{1/4})}$$

$$\langle \text{Nu}_{L,t} \rangle = \frac{0.13\text{Pr}^{0.22}}{(1 + 0.61\text{Pr}^{0.81})^{0.42}}\text{Ra}_L^{1/3}$$

$$a_1 = \frac{4}{3}\frac{0.503}{[1 + (0.492/\text{Pr})^{9/16}]^{4/9}}.$$

Then

$$a_1 = 0.5874$$

$$\langle \text{Nu}_{L,l} \rangle = \frac{2.8}{\ln\left[1 + 2.8/0.5874(2.066 \times 10^{10})^{1/4}\right]}$$

$$= 224.1$$

$$\langle \text{Nu}_{L,t} \rangle = \frac{0.13 \times (3.22)^{0.22}}{[1 + 0.61(3.22)^{0.81}]^{0.42}}(2.066 \times 10^{10})^{1/3}$$

$$= \frac{0.1681}{1.487} \times 2.744 \times 10^3 = 310.0$$

$$\langle \text{Nu} \rangle_L = [(224.1)^6 + (310.0)^6]^{1/6}$$

$$= 317.0.$$

The surface-convection heat transfer rate is

$$\langle Q_{ku} \rangle_L = (0.2)^2 (\text{m}^2) \times 317.0 \times \frac{0.648 (\text{W/m-K})}{0.2(\text{m})} \times (95 - 20)(\text{K})$$

$$= 3,081 \text{ W}.$$

(c) The average surface-convection resistance $\langle R_{ku} \rangle_L$ is given by (6.49), i.e.,

$$\langle R_{ku} \rangle_L = \frac{T_s - T_{f,\infty}}{\langle Q_{ku} \rangle_L} = \frac{(95 - 20)(^\circ\text{C})}{3.081 \times 10^3 (\text{W})} = 0.02434^\circ\text{C/W}$$

$$A_{ku}\langle R_{ku} \rangle_L = (0.2)^2 (\text{m}^2) \times 0.02434(^\circ\text{C/W}) = 9.737 \times 10^{-4} \,^\circ\text{C/(W/m}^2).$$

(d) Comparing thermobuoyant, forced-parallel, and forced-perpendicular flows, the thermobuoyant flow is the least effective ($\langle R_{ku} \rangle_L = 0.02434^\circ\text{C/W}$), followed by laminar and turbulent forced parallel ($\langle R_{ku} \rangle_L = 0.001769^\circ\text{C/W}$ and $\langle R_{ku} \rangle_L = 0.001654^\circ\text{C/W}$), with the single-jet perpendicular flow being the most effective ($\langle R_{ku} \rangle_L = 0.00003976^\circ\text{C/W}$) among them.

**COMMENT**
Parallel, thermobuoyant flows assist in passive cooling and heating heat transfer from surfaces. But for moderate temperature differences $|T_s - T_{f,\infty}|$, it is not very effective.

## 6.6 Liquid-Gas Phase Change

Here we consider pure fluid substances such that the saturation temperature $T_{lg}$ and the saturation pressure $p_g$ are uniquely related and given by the Clausius-Clapeyron relation, $T_{lg} = T_{lg}(p_g)$, which is given in Appendix A, i.e., (A.14).

When the heat transfer liquid in contact with a solid is heated with the solid surface temperature $T_s$ exceeding the saturation temperature $T_{lg} = T_{lg}(p_g)$, then surface-bubble nucleation, i.e., boiling can occur. Depending on the surface condition and the solid-fluid pairs, a threshold value of $T_s - T_{lg}$, which is called the surface superheat, must be exceeded for the surface bubble nucleation to occur. This may be a few °C for rough surfaces or a few tens of °C for smooth surfaces.

A similar phenomenon occurs when the heat transfer gas (i.e. vapor) is cooled on a solid surface with a $T_s$ below $T_{lg}(p_g)$ and then condensation can occur. This is after a threshold $T_{lg} - T_s$, which is called the surface subcooling, is exceeded, and then surface-droplet nucleation occurs.

The surface-convection heat flux $q_{ku}$, with liquid-gas phase change, in addition to the surface superheat/subcooling depends on a combination of the fluid (liquid and gas) velocity, fluid thermophysical properties, and surface characteristics (orientation with respect to gravity, roughness, coatings, extended surface, etc.). As before these are all represented through $R_{ku}$.

When the fluid (gas or liquid) is in the saturated state far away from the heat transfer surface, then the addition/removal of heat results in a phase change, i.e., for saturated fluids we have $q_{ku} = \dot{S}_{ij}/A_{ku} = -\dot{m}_{ij}\Delta h_{ij}$, where $\dot{m}_{ij}$ is the rate of phase change per unit area of the heated/cooled surface. When the fluid is not saturated (i.e., subcooled liquid or superheated vapor), then we assume that the surface-convection heat transfer is first used as the sensible heat to bring the fluid to the saturated state and then it is used for phase change.

Among the fluid properties that influence the surface-convection resistance $R_{ku}$ during boiling or condensation are the heat of evaporation $\Delta h_{lg}$(J/kg), the surface-tension coefficient (or surface tension) $\sigma$(N/m), the contact angle $\theta_c$, and the phase-density difference $\Delta\rho_{lg}$(kg/m$^3$) $= \rho_l - \rho_g$, in addition to those already discussed for the single-phase surface convection. The surface tension, heat of evaporation, and phase-density difference all vanish as the critical state $(T_c, p_c, \rho_c)$ is reached. For example, for water the surface tension $\sigma$ is correlated as a function of $T$(K) as

$$\sigma = \sigma_o \left(1 - \frac{T - 273.16}{T_c - \Delta T_m - 273.16}\right)^{1.2} \quad \begin{array}{l}\text{variation of surface tension}\\ \text{of water with temperature,}\end{array} \quad (6.93)$$

where $\sigma_o = 0.0755$ N/m (surface tension at $T = 273.16$ K, which is the triple point of water), $T_c = 647.13$ K ($p_c = 2.206 \times 10^7$ Pa), and $\Delta T_m = 6.14$ K. Note that using this correlation $\sigma$ vanishes at $T_c - \Delta T_m$; this is because a simple correlation such as this cannot accurately predict $\sigma = \sigma(T)$, especially near $T_c$. The surface tensions of some organic liquid-vapor interfaces (liquids with their own vapors) are listed in Table C.25, as a function of temperature. For comparison, water is also included. Table C.26 is for some other liquids (including liquid mercury) with their vapors but for one atm pressure only. Tables C.28 and C.29 list liquid-vapor saturated thermophysical properties, including $\sigma$, for R-134 and sodium.

When a gas and liquid, separated by their common interface, come in contact with a solid, the contact line between the three phases is called the common line. The contact angle $\theta_c$ is the angle between the liquid-solid and liquid-gas interface, measured within the liquid. The contact angle is zero for a perfectly wetting liquid and equal to 180° for a perfectly nonwetting liquid. Most liquids are in between these limits. A bubble or droplet nucleating (i.e., forming) on a solid surface (i.e., a third

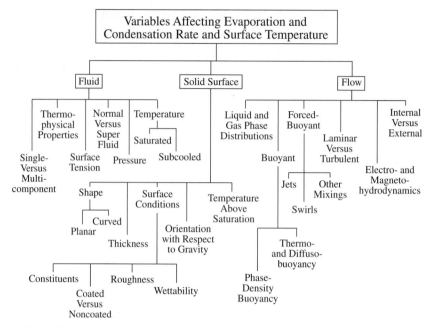

Chart 6.5. A list of fluid, solid surface, and flow variables affecting the evaporation and condensation rate and the surface temperature in surface convection with liquid-vapor phase change.

phase or a dissimilar, immiscible substance) is referred to as heterogeneous nucleation. This is in contrast to a bubble or droplet forming without a solid surface, which is called a homogeneous nucleation (e.g., droplets occurring when the vapor pressure is increased at constant pressure or bubbles forming when the liquid pressure is decreased at constant pressure). Here in our discussion of the surface-convection heat transfer with phase change, we consider only heterogeneous nucleations.

In the following, we will discuss the mechanisms of heat transfer in surface convection with liquid-gas phase change. Chart 6.5 gives a classification of the surface evaporation and condensation. The variables are related to the fluid, the solid surface, and the flow. Surface roughness and orientation, as well as forced versus buoyant flows, influence the rate of surface-convection heat transfer. The liquid subcooling increases $q_{ku}$ by allowing for the collapse of bubbles a short distance from where they are formed. When the phases are arranged in layers, the phenomenon is less complicated (as in film boiling and film condensation). When surface bubble or droplet nucleation occurs, the phenomenon is more complex. We will begin by examining the maximum possible heat flux in evaporation and condensation and show the resistances that prevent reaching this theoretical limit.

### 6.6.1 Theoretical Maximum Heat Flux for Evaporation and Condensation $q_{max}$ (Kinetic Limit)*

This section is found on the Web site www.cambridge.org/kaviany. In this section we discuss how the continuous vapor (or condensate) production rate is controlled by the rate of heat delivered (or removed) to the liquid-vapor interface and by the rate of vapor (or condensate) removal. We discuss and give results for this theoretical kinetic limit for evaporation/condensation rate. This theoretical maximum has not yet been observed in any experiment due to limits on liquid/vapor supply/removal and heat delivery. There is also an example (Example 6.6).

### 6.6.2 Bubble and Film Evaporation: Pool Boiling Regimes

As the solid surface temperature $T_s$ is raised above the saturation temperature $T_{lg}$ and beyond a threshold superheat $T_s - T_{lg}$, the surface vapor-bubble nucleation occurs. The bubble nucleation occurs at discrete sites that are randomly distributed over conventionally prepared (including polished) surfaces. This incipient solid surface superheat can be very small for rough surfaces and wetting liquids with a slow heating rate of the surface. However, for smooth surfaces and perfectly wetting liquids, and when the surface is rapidly heated (such as in the inkjet printer discussed in Example 8.2), this superheat can be large. The magnitude of this incipient superheat $(T_s - T_{lg})_{inc}$ depends on the surface conditions, such as the surface roughness (including the statistics of size, geometry, angle of orientation with respect to the normal to the surface, etc.), the surface cavities, the surface coating, and noncondensible gas trapped in the cavities. It also depends on the fluid and the hydrodynamics conditions. The hydrodynamics and the subcooling (the far-field liquid being at a temperature lower than saturation) influence the temperature distribution on the surface as well as in the liquid adjacent to the surface.

### (A) Pool-Boiling Regimes

Pool boiling is the evaporation of a liquid bounded by a heated solid surface with the vapor and liquid motions caused primarily by the density difference between the vapor and the liquid phase. The thermobuoyant flow within the vapor is generally negligible and that within the liquid is only secondary. Here, normal gravity (at the earth surface) or a significant net acceleration is assumed to be present. Figure 6.20(a) shows the various regimes occurring in pool boiling. As the surface superheat $T_s - T_{lg}$ is increased, evaporation does not occur until a threshold superheat for the surface-bubble nucleation inception $(T_s - T_{lg})_{inc}$ is reached. This single-phase (liquid) heat transfer regime is also shown in the figure.

As the surface superheat $T_s - T_{lg}$ is further increased, the surface-bubble nucleation regime is encountered, corresponding to a sharp increase in the surface heat flux $q_{ku}$ (for a given rise in the surface temperature). This is called the surface-nucleation (or nucleate-boiling) regime and has several subregimes. As shown in Figure 6.20, at low surface superheats, isolated bubbles are formed at active nucleation sites and this is called the isolated-bubbles regime. At higher $q_{ku}$ (or $T_s - T_{lg}$), the frequency of the bubble departure increases and the number of active sites also increases. This is referred to as the combined isolated-bubbles and jets regime. Next, these neighboring jets coalesce and the vapor-mushroom regime is found. As $T_s - T_{lg}$ (or $q_{ku}$) is further increased, the dry-patches regime appears. At a surface superheat

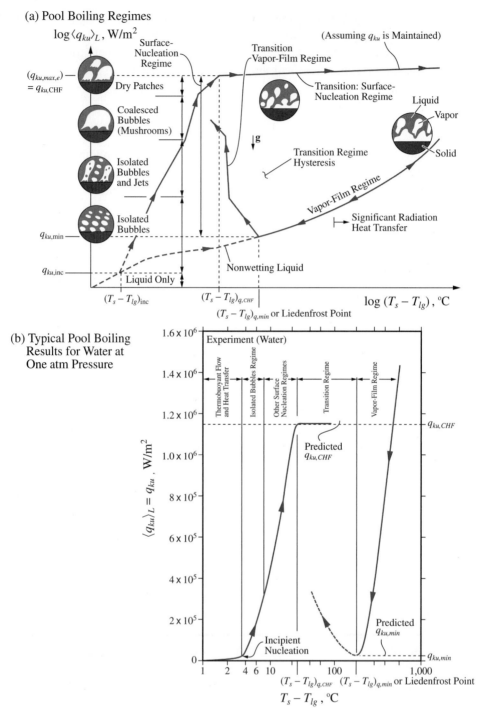

Figure 6.20. (a) Pool boiling, i.e., heat transfer and evaporation caused by raising the bounding, solid surface temperature to a temperature $T_s$ above the liquid saturation temperature $T_{lg}$. The various regimes and subregimes are shown along with the incipient nucleation and transition regime hysteresis. The variation of the surface heat flux is shown versus the surface superheat. (b) A typical pool boiling result for water at one atm pressure. (From Lienhard, J.H., reproduced by permission. © 1987, Prentice-Hall.)

$(T_s - T_{lg})_{q,CHF}$, a maximum in $q_{ku}$ is found. This $q_{ku}$ is called the critical heat flux, is designated by $q_{ku,CHF}$, and occurs at $(T_s - T_{lg})_{q,CHF}$. Any further increase in $q_{ku}$ corresponds to a large increase in the surface temperature or $T_s - T_{lg}$.

During quenching (i.e., placing high surface temperature solids in liquids), at sufficiently larger surface superheat $T_s - T_{lg}$, a vapor film (vapor-film regime in Figure 6.22) is formed at the surface. In steady state, this vapor film is maintained by a steady vaporization and the liquid is prevented from wetting the solid surface (intermittent surface wetting has been observed in thin vapor films). Now, starting from the vapor-film regime and decreasing the superheat, it is possible to reduce the superheat to a minimum value $(T_s - T_{lg})_{q,min}$, while maintaining a vapor film. This is called the Liedenfrost point. There is a corresponding surface heat flux at $(T_s - T_{lg})_{q,min}$ that is called the minimum heat flux $q_{ku,min}$. In the transition regime, bubble nucleation occurs, but the vapor removal mechanism is hindered. In the vapor-film regime, liquid is mostly not in contact with the bounding surface and heat is transferred by conduction and radiation across the vapor film to the vapor-liquid interface where the vapor production occurs.

Figure 6.20(b) shows typical pool boiling results for water at one atm pressure [12]. Note that $q_{ku,CHF}$ is about $1.2 \times 10^6$ W/m$^2$ or 1.2 MW/m$^2$. For this surface, the incipient nucleation superheat $(T_s - T_{lg})_{inc}$ is about 4°C and $q_{ku,CHF}$ occurs at $(T_s - T_{lg})_{q,CHF}$ of about 30°C. This low incipient nucleation superheat is characteristic of nonpolished surfaces. Also, note that the Liedenfrost superheat is about $(T_s - T_{lg})_{q,min} = 200$°C.

From Figure 6.19(a), and also in Example 6.6, we find that for water at one atm pressure, the theoretical maximum heat flux, $q_{max} = 2.195 \times 10^8$ W/m$^2$. Comparing this with pool boiling in Figure 6.20(b), we note that only less than 1% of this theoretical limit is achieved. A rendering of the surface-bubble nucleation, growth, and bubble departure on a heated solid surface is given in Figure 6.21(a). The surface-bubble nucleation number per unit area $N_b/A$ is influenced by the surface, fluid, and hydrodynamic parameters.

The growth of the bubbles on the surface is influenced by the liquid motion around the bubble as well as the heat transfer through the liquid and the solid surface. The bubble departure is caused by a combination of the phase-density buoyancy force, the hydrostatic pressure gradient, the interfacial drag force, the liquid flow inertia, and the surface forces.

A rendering of the formation and departure of bubble mushrooms on a heated surface is given in Figure 6.21(b). As the surface heat flux $q_{ku}$ (or surface superheat $T_s - T_{lg}$) increases, the bubbles coalesce and large mushroom-shape bubbles appear on the surface and periodically depart. Observations have shown that the liquid reaches the solid surface inside these vapor mushrooms and evaporates, causing their growth. When the phase-density buoyant and interfacial viscous drag forces can overcome the surface tension forces, the bubble departs, as depicted in Figure 6.21(b). Many models have been proposed for the evaporation within the vapor mushrooms. Most of them assume a discontinuous liquid layer, called the macrolayer, that surrounds the vapor columns or chimneys, which are envisioned as being on the nucleation sites.

(a) Isolated-Bubble Regime in Pool Boiling (with Liquid Microlayer)

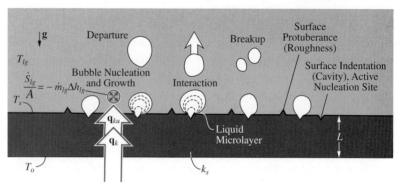

(b) Coalesced-Bubble Regime in Pool Boiling (with Liquid Macrolayer)

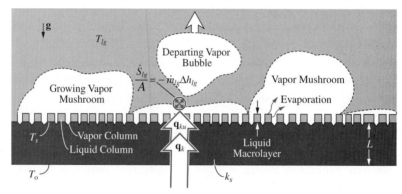

Figure 6.21. Surface-bubble nucleation regime in pool boiling. (a) With isolated bubbles. (b) With coalesced bubbles. The heat flow through the bounding surface is also shown.

### (B) Correlations for Pool Boiling

As expected, boiling and condensation correlations include many variables. Here we list a few correlations, namely fully-developed nucleate boiling, critical heat flux, and minimum heat flux, for pool boiling from an infinitely large horizontal surface facing up.

The fully-developed nucleate boiling is correlated and given in [17] as

$$
A_{ku}\langle R_{ku}\rangle_L = \frac{T_s - T_{lg}}{\langle q_{ku}\rangle_L}
$$

$$
= a_s^3 \frac{\Delta h_{lg}^2}{\mu_l c_{p,l}^3 (T_s - T_{lg})^2}\left(\frac{\sigma}{g\Delta\rho_{lg}}\right)^{1/2}\mathrm{Pr}_l^n \quad
\begin{array}{l}\text{fully-developed nucleate}\\ \text{boiling, on horizontal}\\ \text{surface facing up}\end{array}
$$

$$
n = 3 \text{ (water)}, \quad n = 5.1 \text{ (all other fluids)}, \tag{6.98}
$$

Film Boiling Adjacent to a Vertical Plate

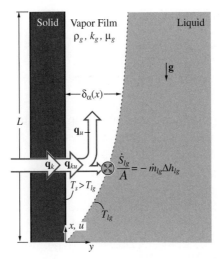

Figure 6.22. Film evaporation adjacent to a vertical surface.

where the Nusselt number becomes

$$\langle \mathrm{Nu} \rangle_L = \frac{L}{A_{ku}\langle R_{ku} \rangle_L k_l} = \frac{\mathrm{Ja}_l^2 \mathrm{Bo}_L^{1/2}}{a_s^3 \mathrm{Pr}_l^n}, \quad \mathrm{Ja}_l = \frac{c_{p,l}(T_s - T_{lg})}{\Delta h_{lg}}, \quad \mathrm{Bo}_L = \frac{g \Delta \rho_{lg} L^2}{\sigma}, \quad (6.99)$$

where $\mathrm{Ja}_l$ is the Jakob number (also used in Section 5.1), and $\mathrm{Bo}_L = F_{\Delta\rho}/F_\sigma$ is the Bond number based on $L$. The Jakob number is the ratio of the sensible heat of the liquid to the heat of evaporation. The Bond number is the ratio of the phase-density buoyancy force to the surface tension. In (6.98) $n = 3$ for water and $n = 5.1$ for other fluids. The constant $a_s$ depends on the surface condition and the fluid-solid pair. Table 6.2 gives a listing for several pairs.

Table 6.2. *Empirical constant $a_s$ in the pool nucleate boiling correlation*

| Fluid-Solid Pair | $a_s$ |
|---|---|
| water-nickel | 0.006 |
| water-platinum | 0.013 |
| water-copper | 0.013 |
| water-brass | 0.006 |
| carbon tetrachloride-copper | 0.013 |
| benzene-chromium | 0.010 |
| *n*-pentane-chromium | 0.015 |
| ethanol-chromium | 0.0027 |
| isopropyl alcohol-copper | 0.0025 |
| 35% potassium carbonate-copper | 0.0054 |
| 50% potassium carbonate-copper | 0.0027 |
| *n*-butyl alcohol-copper | 0.0030 |

The critical heat flux $q_{ku,CHF}$ is correlated and as a dimensionless quantity (designated with an asterisk) is given by

$$q^*_{ku,CHF} = \langle q^*_{ku,CHF} \rangle_L = \frac{q_{ku,CHF}}{\rho_g \Delta h_{lg}} \left( \frac{\rho_g^2}{\sigma g \Delta \rho_{lg}} \right)^{1/4} = \text{Ku} = \frac{\pi}{24} = 0.13 \qquad (6.100)$$

critical heat flux in pool boiling on horizontal surface facing up.

This dimensionless critical heat flux is also called the Kutateladze number Ku or boiling number. During quenching of solid surfaces ($T_s - T_{lg}$ very large) in an otherwise stagnant saturated liquid, the minimum surface-convection heat flux $q_{ku,min}$ occurs, in film boiling just before the transition boiling begins. The dimensionless $q_{ku,min}$ is correlated with the fluid properties as

$$q^*_{ku,min} = \langle q^*_{ku,min} \rangle_L = \frac{q_{ku,min}}{\rho_g \Delta h_{lg}} \left[ \frac{(\rho_l + \rho_g)^2}{\sigma g \Delta \rho_{lg}} \right]^{1/4} = 0.09 \qquad \begin{array}{l} \text{minimum heat flux in film,} \\ \text{pool boiling on horizontal} \\ \text{surface facing up.} \end{array}$$
$$(6.101)$$

The above relations are for pool boiling from plain, smooth horizontal surfaces. By forming random and ordered surface roughness and using various surface attachments, it is possible to decrease the surface superheat $(T_s - T_{lg})_{q,CHF}$ and increase the critical heat flux $q_{ku,CHF}$. These improve the number of nucleation sites and/or increase the surface area (extended surface), and/or assist in liquid flow toward the surface by capillarity [13]. Significant reduction in $(T_s - T_{lg})_{q,CHF}$ and more than three times improvement in $q_{ku,CHF}$ are achievable.

### (C) Correlation for Film Boiling on Vertical Plate

As another phase-density buoyant flow, consider the film boiling (i.e, evaporation) adjacent to heated vertical plates. The surface temperature $T_s$ is much larger than $T_{lg}$. This is shown in Figure 6.22. The vertical plate is assumed semi-infinite. At a location $L$ along the plate, the vapor film has a thickness $\delta_g(L)$.

By first neglecting the effect of radiation, the vapor-film thickness $\delta_g(L)$, the Nusselt number $\text{Nu}_L$, and averaged Nusselt number $\langle \text{Nu} \rangle_L$, for film boiling, are predicted as [18]

$$\delta_g(L) = \left[ \frac{4 k_g \nu_g (T_s - T_{lg}) L}{g \Delta \rho_{lg} \Delta h_{lg}} \right]^{1/4} \qquad (6.102)$$

$$\text{Nu}_L = \frac{L}{\delta_g} = \left[ \frac{g \Delta \rho_{lg} \Delta h_{lg} L^3}{4 k_g \nu_g (T_s - T_{lg})} \right]^{1/4}, \qquad (6.103)$$

and using (6.50), we have

$$\langle Nu \rangle_L = \frac{\langle q_{ku} \rangle_L L}{(T_s - T_{lg})k_g} = 0.9428 \left[ \frac{g \Delta \rho_{lg} \Delta h_{lg} L^3}{k_g v_g (T_s - T_{lg})} \right]^{1/4} \quad \begin{array}{l} \text{laminar, nonwavy film} \\ \text{boiling on vertical,} \\ \text{semi-infinite plate.} \end{array}$$
$$(6.104)$$

Then

$$\langle q_{ku} \rangle_L = \langle Nu \rangle_L \frac{k_g}{L} (T_s - T_{lg}). \tag{6.105}$$

The radiation contribution is the special case of two-surface enclosure and (4.49) applying with $F_{s-lg} = 1$, i.e.,

$$\langle q_r \rangle_L = \frac{\sigma_{SB}(T_s^4 - T_{lg}^4)}{\left( \dfrac{1 - \epsilon_r}{\epsilon_r} \right)_s + 1 + \left( \dfrac{1 - \epsilon_r}{\epsilon_r} \right)_{lg}} = \frac{\sigma_{SB}(T_s^4 - T_{lg}^4)}{\dfrac{1}{\epsilon_{r,s}} + \dfrac{1}{\epsilon_{r,lg}} - 1} \quad \begin{array}{l} \text{radiation heat flux} \\ \text{for film boiling,} \end{array} \tag{6.106}$$

where in general $\epsilon_{r,lg} \simeq 1$.

Then by adding the surface convection and radiation heat transfer we have

$$\langle Q \rangle_L = A_{ku}(\langle q_{ku} \rangle_L + \langle q_r \rangle_L) \quad \text{surface convection and radiation.} \tag{6.107}$$

### EXAMPLE 6.7. FAM

Again, consider the plastic sheet of Example 6.5, and here allow for the surface-convection cooling by evaporation by exposing one side of the plastic (placed horizontally) sheet to a saturated water pool, as shown in Figure Ex. 6.7(a). At one atmosphere pressure, the boiling temperature of the water is found from Table C.26 and is $T_{f,\infty} = T_{lg} = 373.15$ K. The plastic sheet is at $T_s = 115°C = 388.15$ K. Assume that the surface superheat $T_s - T_{lg} = 15°C$ is sufficient to cause nucleate boiling of the plastic surface (which is a reasonable assumption).

(a) Draw the thermal circuit diagram.
(b) Use (6.98) to determine the rate of surface-convection cooling.
(c) Again using (6.98), determine the average surface-convection resistance $\langle R_{ku} \rangle_L$.

$L = 0.2$ m, $p_g = 1.013 \times 10^5$ Pa.

For lack of specific data, take $a_s$ from Table 6.2, that corresponding to the water-copper pair.

### SOLUTION
(a) The thermal circuit diagram is shown in Figure 6.7(b).
(b) The rate of surface-convection heat transfer is given by (6.98) as

$$\langle Q_{ku} \rangle_L = \frac{T_s - T_{lg}}{\langle R_{ku} \rangle_L} = (T_s - T_{lg})\langle Nu \rangle_L \frac{k_l}{L} A_{ku} = \frac{\mu_l c_{p,l}^3 (T_s - T_{lg})^3}{a_s^3 \Delta h_{lg}^2 [\sigma/(g \Delta \rho_{lg})]^{1/2} Pr_l^3} A_{ku},$$

## Surface-Convection Heat Transfer: Pool Boiling

(a) Physical Model  (b) Thermal Circuit Model

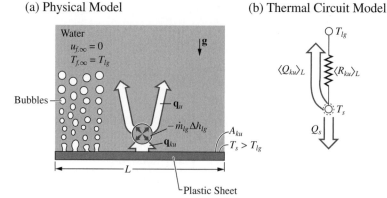

Figure Ex. 6.7. Surface-convection heat transfer by pool boiling. (a) Physical model. (b) Thermal circuit model.

where

$$A_{ku} = L^2.$$

From Table C.26, we have

| | | |
|---|---|---|
| $\mu_l$ | $= 277.5 \times 10^{-6}$ Pa-s | Table C.26 |
| $c_{p,l}$ | $= 4{,}220$ J/kg-K | Table C.26 |
| $\Delta h_{lg}$ | $= 2{,}257 \times 10^3$ J/kg | Table C.26 |
| $\sigma$ | $= 0.05891$ N/m | Table C.26 |
| $\Delta\rho_{lg}$ | $= \rho_l - \rho_g = (958.3 - 0.596)(\text{kg/m}^3) = 957.7$ kg/m$^3$ | Table C.26 |
| $\text{Pr}_l$ | $= 1.72$ | Table C.26. |

We take the standard gravitational constant, $g = g_o = 9.807$ m/s$^2$. Also, from Table 6.2, we have $a_s = 0.013$. Then

$$\langle Q_{ku}\rangle_L = \frac{277.53 \times 10^{-6}(\text{kg/m-s}) \times (4{,}220)^3 (\text{J/kg-K})^3 (388.15 - 373.15)^3 (\text{K})^3}{(0.013)^3 (2{,}256.7 \times 10^3)^2 (\text{J/kg})^2 \left[\dfrac{(0.05891)(\text{N/m})}{(9.807)(\text{m/s}^2) \times (957.7)(\text{kg/m}^3)}\right]^{1/2} (1.72)^3}$$

$$= \frac{7.039 \times 10^{11}}{1.119 \times 10^7 \times (6.272 \times 10^{-6})^{1/2}(1.72)^3} \times (0.2)^2$$

$$= 1.971 \times 10^4 \text{ W}.$$

Note that the unit of $(\sigma/g\Delta\rho_{lg})^{1/2}$ is m.

(c) The average surface-convection resistance is given by (6.98), i.e.,

$$\langle R_{ku}\rangle_L = \frac{T_s - T_{lg}}{\langle Q_{ku}\rangle_L} = \frac{(388.15 - 373.15)(^\circ\text{C})}{1.971 \times 10^4 (\text{W})} = 7.611 \times 10^{-4} \,^\circ\text{C/W}$$

$$A_{ku}\langle R_{ku}\rangle_L = (0.2)^2 (\text{m}^2) \times 7.611 \times 10^{-4} (^\circ\text{C/W}) = 3.044 \times 10^{-5} \,^\circ\text{C/(W/m}^2).$$

**COMMENT**

Assuming that our fully-developed nucleate boiling condition is justifiable, the surface-convection resistance with phase change is smaller than that for forced, parallel, and thermobuoyant flows. We can compare this with the kinetic-theory limit of $q_{ku}$. From (6.96), we have

$$q_{max} = \left(\frac{R_g}{2\pi M}\right)^{1/2} \Delta h_{lg}(\rho_g T_{lg}^{1/2} - \rho_{g,\infty} T_{lg,\infty}^{1/2}).$$

We will use $T_{lg}$ as the surface temperature and $T_{lg,\infty}$ as the far-field temperature. Then from Table C.27, we have

$$q_{max} = \left[\frac{8,314.5(\text{J/kmole-K})}{2\pi \times 18.015(\text{kg/kmole})}\right]^{1/2} \times 2.216 \times 10^6(\text{J/kg}) \times [1.028(\text{kg/m}^3)$$
$$\times (388.15)^{1/2}(\text{K}^{1/2}) - 0.596(\text{kg/m}^3) \times (373.15)^{1/2}(\text{K}^{1/2})$$
$$= 1.708 \times 10^8 \text{ W/m}^2.$$

$$\frac{\langle q_{ku}\rangle_L}{q_{max}} = \frac{1.971 \times 10^5(\text{W})/(0.2)^2(\text{m})^2}{1.708 \times 10^8(\text{W/m}^2)} = 0.002885,$$

i.e., 1% of the limit is achieved. As was mentioned, active liquid supply and vapor removal (as compared to the passive, buoyant supply and removal) assist in increasing $\langle q_{ku}\rangle_L$. As shown in Figure 6.19(b), higher values of $\langle q_{ku}\rangle_L$ have been achieved (by forcing the liquid flow).

### 6.6.3 Dropwise and Film Condensation*

This section is found on the Web site www.cambridge.org/kaviany. In this section, we examine the surface-droplet nucleation, growth, and departure on a vertical, cooled surface bounding a saturated vapor. Most existing analyses of droplet formation and dynamics, on cooled surfaces bounding a vapor, are for an otherwise quiescent vapor. We consider both filmwise and dropwise condensation.

### 6.6.4 Impinging-Droplets Surface Convection with Evaporation*

This section is found on the Web site www.cambridge.org/kaviany. In this section we discuss how in order to continuously remove large heat fluxes from a surface while using a small amount of liquid, forced droplet spray cooling is used. During droplet contact with the surface, contact conduction occurs and this results in evaporation and vapor convection. Droplets also entrain (cause to move) the local gas (e.g., air and vapor) and thus flowing gas exchanges heat with the surface and this is designated as bulk convection. Radiation heat transfer between the droplets and the surface can also become significant. We give a correlation for surface-convection resistance for impinging, evaporating droplets and give an example (Example 6.8).

## 6.7  Summary of Nusselt Number Correlations

In Chart 6.2, a classification of the solid geometries was given. Among objects used with semi-bounded fluid flow are a single cylinder, a collection of cylinders (in cross flow), and a single sphere. The single-phase flow and heat transfer for these objects have been well studied. The fluid flow over objects is influenced by the relative strength of the inertial, viscous, and gravity forces, and the imposed pressure gradient. We do not intend a complete catalogue of the Nusselt numbers, but here we give some typical examples. We begin by considering a single sphere. Next we will discuss other geometries and then list some of the available Nusselt number correlations in tables.

### 6.7.1  Sphere: Forced-Flow Regimes and Correlations

We examine the surface-convection heat transfer from a single sphere to demonstrate the effect of various flow regimes (these regimes are described below) occurring as the fluid far-field velocity increases, on the surface-convection resistance.

The local surface-convection resistance varies over the surface of the sphere. This variation is influenced by the structure of the fluid flow around it. These variations are divided into regimes, marked by the magnitude of the Reynolds number based on the sphere diameter $D$, $Re_D$. Because of the unsteadiness, i.e., vortex shedding and turbulence, the complete prediction of the interfacial heat transfer for various regimes has not yet been achieved. Therefore, the high $Re_D$ regimes are only treated experimentally.

### *(A) Conduction Regime,* $Re_D = 0$

In this regime, there is no motion, i.e., $u_{f,\infty} = 0$. The result is the conduction results presented in Section 3.3.1. The local Nusselt number (which is independent of $\theta$ for this case) is found from (3.66) as

$$\langle Nu \rangle_D = Nu_D = \frac{q_k D}{(T_s - T_{f,\infty})k_f} = \frac{q_{ku} D}{(T_s - T_{f,\infty})k_f} = 2, \quad \text{for}$$

$$Re_D = \frac{u_{f,\infty} D}{\nu_f} = 0 \quad \text{conduction regime}, \tag{6.121}$$

i.e., there is a steady-state conduction heat transfer from a sphere to its surrounding given by (6.120).

### *(B) Attached-Wake Regime,* $Re_D < 400$

Numerical solutions have been obtained using the continuity, momentum, and energy equations for Reynolds numbers less than those corresponding to the onset of vortex shedding, i.e., $Re_D < 400$. Figure 6.27(a) shows those results for $Pr = 0.71$ and several values of $Re_D$. The variation of the local Nusselt number with respect to the polar angle ($\theta = 0$ is at the front stagnation point) is shown. The $Re_D \to 0$ asymptote in which $Nu_D$ does not change with $\theta$ is also shown. For $Re_D \geq 20$, where the flow separation occurs, the separation angle $\theta_s$ corresponding to the $Re_D$ is also

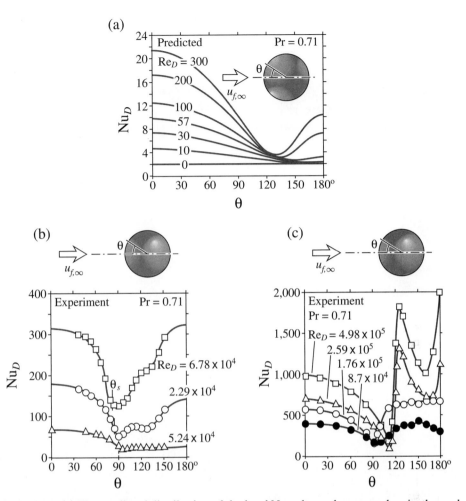

Figure 6.27. (a) The predicted distribution of the local Nusselt number around an isothermal sphere for various (but low) Reynolds numbers and $Pr = 0.71$. (b) The measured distribution of the local Nusselt number for some subcritical Reynolds numbers. The predictions are based on the boundary-layer approximations. The separation angles are also shown. (c) The measured local Nusselt number distribution for sub- and supercritical Reynolds numbers.

indicated. For $Re_D < 20$, the local Nusselt number decreases monotonically with increase in $\theta$. For $30 < Re_D < 57$, a minimum in $Nu_D$ appears and moves upstream as $Re_D$ increases. The minimum is behind the separation point, and the recirculation causes an increase in the local $Nu_D$ in the wake region. The heat transfer from the recirculating region to the free stream is by conduction. The front (or upstream) stagnation region has the largest local $Nu_D$.

### (C) Subcritical Regime, $Re_D < 10^5$

For $Re_D > 400$, vortices are shed and the flow behind the sphere oscillates. For $Re_D > 3,000$ the flow in the front stagnation region approaches a boundary-layer behavior. This boundary-layer flow and heat transfer have been analyzed using the laminar boundary-layer analysis of the variable far-field velocity.

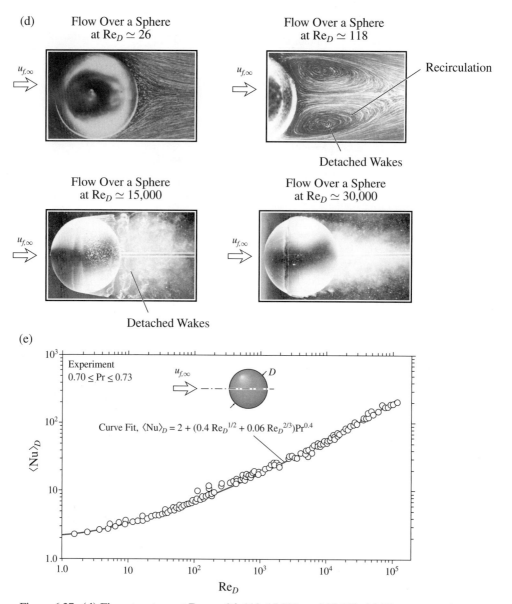

Figure 6.27. (d) Flow structure at $Re_D = 26$, 118, 15,000, and 30,000. (e) The measured area-averaged Nusselt number as a function of the Reynolds number and for $0.70 < Pr < 0.73$.

The front stagnation boundary-layer analysis predicts a $Re_D^{1/2}$ dependence for $Nu_D$ and an angular variation that is in good agreement with the experimental results for the stagnation region. Figure 6.27(b) shows the experimental results for $Pr = 0.71$, and $Re_D = 5.24 \times 10^3$, $2.29 \times 10^4$, and $6.78 \times 10^4$. The predictions are also shown along with the separation angle $\theta_s$ corresponding to each $Re_D$. Note that the boundary-layer theory predicts $Nu_D = 0$ at $\theta_s$, because of the neglected angular conduction. The local $Nu_D$ begins to increase rapidly with $Re_D$ for $5 \times 10^3 < Re_D$ and for $Re_D = 6.78 \times 10^4$ the front and near maxima are nearly equal.

### (D) Critical Regime, $10^5 < \text{Re}_D$

For yet larger $\text{Re}_D$, the behavior in the near region begins to change even further. Figure 6.27(c) shows the experimental results for $\text{Pr} = 0.71$ and $\text{Re}_D = 8.7 \times 10^4$, $1.76 \times 10^5$, $2.59 \times 10^5$, and $4.89 \times 10^5$. At $\text{Re}_D = 2.59 \times 10^5$ transition has occurred and a second minimum appears. The turbulent flow in the rear follows a $\text{Re}_D^{4/5}$ relation, while in the front stagnation region it follows a $\text{Re}_D^{1/2}$ relation, for $\text{Nu}_D$. Therefore, at the rear $\text{Nu}_D$ becomes larger than the front, as the $\text{Re}_D$ increases beyond the transition wave. Figure 6.27(d) shows the structure of flow (visualized) for four different $\text{Re}_D$.

The experimental results for the polar-angle averaged Nusselt number, defined as

$$\langle \text{Nu} \rangle_D \equiv \frac{1}{180°} \int_0^{180°} \text{Nu}_D(\theta) \, d\theta \quad \text{surface-averaged Nusselt number,} \quad (6.122)$$

is shown in Figure 6.27(e) for $1 \leq \text{Re}_D \leq 10^5$, and for $0.70 \leq \text{Pr} \leq 0.73$. The results are for several experiments.

The effect of the Prandtl number is included in the correlation [22]

$$\langle \text{Nu} \rangle_D = 2 + (0.4\text{Re}_D^{1/2} + 0.06\text{Re}_D^{2/3})\text{Pr}^{0.4} \quad \text{for single sphere, all regimes.} \quad (6.123)$$

For $2 \times 10^3 < \text{Re}_D$, the correlation can be interpreted as combining the front-stagnation region and the wake-turbulent region contributions. From Figure 6.27(e), note that the average surface-convection conductance is rather insensitive to the hydrodynamic details described above.

### 6.7.2 Tabular Presentation of Correlations for Some Flows and Geometries

Tables 6.3 to 6.6 summarize the correlations for the all range Nusselt number for single-phase forced flow, thermobuoyant flow, and phase change, for semi-bounded fluid flowing over objects with different geometries. The various dimensionless quantities are listed in Table C.1(c).

Most of the correlations are curve fits to the experimental results (similar to Figure 6.27) and are accurate to generally within $\pm 20\%$. Some of the forced-flow correlations for Nusselt number for semi-bounded fluid streams over flat plates are given in Table 6.3. These are valid within a particular range of the Reynolds number. The correlations give $\langle \text{Nu} \rangle_L$ or $\langle \text{Nu} \rangle_D$ as a function of $\text{Re}_L$ or $\text{Re}_D$ and Pr. For the case of perpendicular flows, other geometrical parameters are also used.

The correlations for cross flow over cylinders of various cross-sectional geometries, and for spheres, are given in Table 6.4. A general set of correlations for the average, thermobuoyant-flow Nusselt numbers is suggested in Table 6.4 [16]. These use a natural logarithmic relation to correct for the thin boundary-layer approximations and a laminar-turbulent correlation based on a power-law relation. Therefore the resulting correlations are valid over both laminar- and turbulent-flow regimes.

Table 6.3. *Correlations for Nusselt number for semi-bounded fluid streams, for single-phase forced flows over flat plates*

$$\langle Q_{ku}\rangle_{L(\,or\,D)} = A_{ku}(T_s - T_{f,\infty})\frac{\langle Nu\rangle_{L(\,or\,D)}k_f}{L(\,or\,D)}, \quad \langle R_{ku}\rangle_{L(\,or\,D)} = \frac{L(\,or\,D)}{A_{ku}\langle Nu\rangle_{L(\,or\,D)}k_f}.$$

All properties evaluated at average fluid temperature of $(T_s + T_{f,\infty})/2$.

---

## Semi-Bounded Fluid Streams: Flows Parallel and Perpendicular to Semi-Infinite Plates

Parallel Laminar and Turbulent Flows

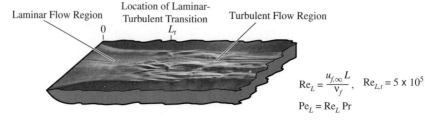

$$\mathrm{Re}_L = \frac{u_{f,\infty}L}{\nu_f}, \quad \mathrm{Re}_{L,t} = 5 \times 10^5$$

$$\mathrm{Pe}_L = \mathrm{Re}_L\,\mathrm{Pr}$$

(a) Laminar Flow

(b) Combined, Laminar-Turbulent Flow

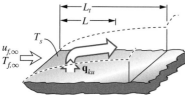

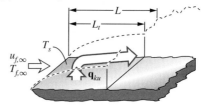

$\mathrm{Re}_L \leq 5 \times 10^5,\ 0.6 < \mathrm{Pr}$

$\langle Nu\rangle_L = 0.664\,\mathrm{Re}_L^{1/2}\,\mathrm{Pr}^{1/3}$

$\langle Nu\rangle_L = 2\dfrac{\mathrm{Pe}_L^{1/2}}{\pi^{1/2}}\ ,\ \mathrm{Pr} = 0$

$\mathrm{Re}_L > 5 \times 10^5,\ 0.6 < \mathrm{Pr} < 60$

$\langle Nu\rangle_L = (0.037\,\mathrm{Re}_L^{4/5} - 871)\,\mathrm{Pr}^{1/3}$

---

Perpendicular Turbulent Jet Flow

(a) Single Nozzle

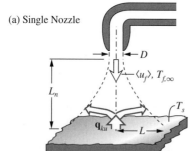

Single Impinging Jet

$$\langle Nu\rangle_L = 2\mathrm{Re}_D^{1/2}\,\mathrm{Pr}^{0.42}\,(1 + 0.005\,\mathrm{Re}_D^{0.55})^{1/2}\frac{1 - 1.1\,D/L}{1 + 0.1(L_n/D - 6)D/L}$$

$$\frac{L}{D} > 2.5, \quad \mathrm{Re}_D = \frac{\langle u_f\rangle D}{\nu_f}$$

(b) Arrays of Nozzles

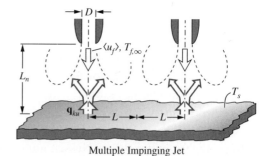

Multiple Impinging Jet

$$\frac{2L}{D} > 2.5, \quad \mathrm{Re}_D = \frac{\langle u_f\rangle D}{\nu_f}, \quad \varepsilon = 1 - \frac{\pi D^2}{16L^2}$$

$$\langle Nu\rangle_L = \mathrm{Re}_D^{2/3}\,\mathrm{Pr}^{0.42}\frac{2L}{D}\left\{1 + \left[\frac{L_n/D}{0.6/(1-\varepsilon)^{1/2}}\right]^6\right\}^{-0.05}(1-\varepsilon)^{1/2}\frac{1 - 2.2(1-\varepsilon)^{1/2}}{1 + 0.2(L_n/D - 6)(1-\varepsilon)^{1/2}}$$

Table 6.4. *Correlations for Nusselt number for semi-bounded fluid streams, for cross flow over cylinders (various cross-section geometry) and spheres. Also listed is the rotating disc*

## Semi-Bounded Fluid Streams: Forced Flows Across Various Objects

### Cylinders

$$\langle Nu \rangle_D = a_1\, Re_D^{a_2}\, Pr^{1/3}, \quad Re_D = \frac{u_{f,\infty} D}{\nu_f}$$

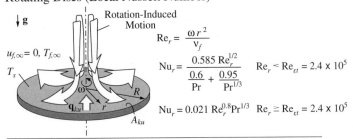

| | $a_1$ | $a_2$ | $Re_D$ |
|---|---|---|---|
| | 0.683 | 0.466 | $0 - 4 \times 10^3$ |
| | 0.193 | 0.618 | $4 \times 10^3 - 4 \times 10^4$ |
| | 0.027 | 0.805 | $4 \times 10^4 - 4 \times 10^5$ |
| | 0.246 | 0.588 | $5 \times 10^3 - 10^5$ |
| | 0.102 | 0.675 | $5 \times 10^3 - 10^5$ |
| | 0.160 | 0.638 | $5 \times 10^3 - 2 \times 10^4$ |
| | 0.0385 | 0.782 | $2 \times 10^3 - 10^5$ |
| | 0.153 | 0.638 | $5 \times 10^3 - 10^5$ |
| | 0.228 | 0.731 | $4 \times 10^3 - 2 \times 10^4$ |

### Spheres

$$\langle Nu \rangle_D = 2 + (0.4\, Re_D^{1/2} + 0.06\, Re_D^{2/3})\, Pr^{0.4}, \quad Re_D = \frac{u_{f,\infty} D}{\nu_f}$$

### Rotating Discs (Local Nusselt Number)

$$Re_r = \frac{\omega r^2}{\nu_f}$$

$$Nu_r = \frac{0.585\, Re_r^{1/2}}{\dfrac{0.6}{Pr} + \dfrac{0.95}{Pr^{1/3}}} \qquad Re_r < Re_{r,t} = 2.4 \times 10^5$$

$$Nu_r = 0.021\, Re_r^{0.8} Pr^{1/3} \qquad Re_r \geq Re_{r,t} = 2.4 \times 10^5$$

Table 6.5. *Correlations for Nusselt number for semi-bounded fluid streams, for single-phase, thermobuoyant flows. Note that when $T_s < T_{f,\infty}$, $Ra_L$ should be written in terms of $T_{f,\infty} - T_s$, i.e., $Ra_L \geq 0$. The thermal expansion for gases is idealized with $\beta_f = 1/T_f$, and for some liquids $\beta_f$ is listed in Table C.23, where the average fluid temperature of $(T_s + T_{f,\infty})/2$ is used for this and other properties. The correlations are for both laminar and turbulent flows*

---

### Semi-Bounded Fluid Streams: Thermobuoyant Flows

**Vertical Plates**

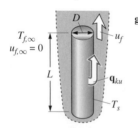

Laminar    Turbulent

$$\langle Nu \rangle_L = [\langle Nu_{L,l} \rangle^6 + \langle Nu_{L,t} \rangle^6]^{1/6}$$

$$\langle Nu_{L,l} \rangle = \frac{2.8}{\ln[1 + 2.8/(a_1 Ra_L^{1/4})]}$$

1/4 Dependency for Laminar Flow

$$\langle Nu_{L,t} \rangle = \frac{0.13 Pr^{0.22}}{(1 + 0.61 Pr^{0.81})^{0.42}} Ra_L^{1/3}$$

1/3 Dependency for Turbulent Flow

Prandtl Number Dependency for $Nu_{L,l}$

$$a_1 = \frac{4}{3} \frac{0.503}{[1 + (0.492/Pr)^{9/16}]^{4/9}}$$

$$Ra_L = \frac{g\beta_f (T_s - T_{f,\infty}) L^3}{\nu_f \alpha_f}$$

$$\langle Nu \rangle_L = \frac{\langle q_{ku} \rangle L}{(T_s - T_{f,\infty}) k_f}$$

---

**Horizontal Plates (Rectangular or Circular Plates and Rings)**

$A_{ku}$, $P_{ku}$
($P_{ku}$ is Perimeter)
$L = \dfrac{A_{ku}}{P_{ku}}$

$$\langle Nu \rangle_L = [\langle Nu_{L,l} \rangle^{10} + \langle Nu_{L,t} \rangle^{10}]^{1/10}$$

$$\langle Nu_{L,l} \rangle = \frac{1.4}{\ln[1 + 1.4/(0.835 a_1 Ra_L^{1/4})]}$$

$$\langle Nu_{L,t} \rangle = 0.14 Ra_L^{1/3}$$

$$a_1 = \frac{4}{3} \frac{0.503}{[1 + (0.492/Pr)^{9/16}]^{4/9}}$$

$$Ra_L = \frac{g\beta_f (T_s - T_{f,\infty}) L^3}{\nu_f \alpha_f}$$

$$\langle Nu \rangle_L = \frac{\langle q_{ku} \rangle L}{(T_s - T_{f,\infty}) k_f}$$

---

**Horizontal Circular Cylinders**

$L >> D$

$$\langle Nu \rangle_D = [\langle Nu_{D,l} \rangle^{3.3} + \langle Nu_{D,t} \rangle^{3.3}]^{1/3.3}$$

$$\langle Nu_{D,l} \rangle = \frac{1.6}{\ln[1 + 1.6/(0.772 a_1 Ra_D^{1/4})]}$$

$$\langle Nu_{D,t} \rangle = \frac{0.13 Pr^{0.22}}{(1 + 0.61 Pr^{0.81})^{0.42}} Ra_D^{1/3}$$

$$a_1 = \frac{4}{3} \frac{0.503}{[1 + (0.492/Pr)^{9/16}]^{4/9}}$$

$$Ra_D = \frac{g\beta_f (T_s - T_{f,\infty}) D^3}{\nu_f \alpha_f}$$

$$\langle Nu \rangle_D = \frac{\langle q_{ku} \rangle D}{(T_s - T_{f,\infty}) k_f}$$

---

**Vertical Circular Cylinders**

Same as vertical plate provided $\delta_\alpha << D$ can be justified;

this is when $\dfrac{D}{L} \geq \dfrac{35}{Gr_L^{1/4}}$

---

**Spheres**

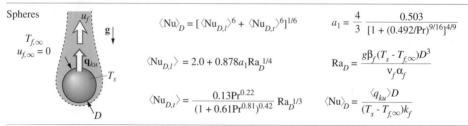

$$\langle Nu \rangle_D = [\langle Nu_{D,l} \rangle^6 + \langle Nu_{D,t} \rangle^6]^{1/6}$$

$$\langle Nu_{D,l} \rangle = 2.0 + 0.878 a_1 Ra_D^{1/4}$$

$$\langle Nu_{D,t} \rangle = \frac{0.13 Pr^{0.22}}{(1 + 0.61 Pr^{0.81})^{0.42}} Ra_D^{1/3}$$

$$a_1 = \frac{4}{3} \frac{0.503}{[1 + (0.492/Pr)^{9/16}]^{4/9}}$$

$$Ra_D = \frac{g\beta_f (T_s - T_{f,\infty}) D^3}{\nu_f \alpha_f}$$

$$\langle Nu \rangle_D = \frac{\langle q_{ku} \rangle D}{(T_s - T_{f,\infty}) k_f}$$

Table 6.6. *Correlations for Nusselt number for semi-bounded fluid streams, for phase-density buoyant flows: pool boiling, film evaporation, and film condensation, and for impinging droplets*

## Semi-Bounded Fluid Streams: Phase Change

Pool Boiling (Horizontal Plates)

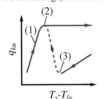

(1) Fully-Developed Nucleate Boiling for Saturated Liquid

$$\langle Nu \rangle_L = \frac{L}{k_l} \frac{1}{A_{ku} \langle R_{ku} \rangle_L} = \frac{L}{a_s^3 k_l} \frac{\mu_l c_{p,l}^3 (T_s - T_{lg})^2}{\Delta h_{lg}^2} \left( \frac{g \Delta \rho_{lg}}{\sigma} \right)^{1/2} Pr_l^{-n}$$

$a_s$ is listed in connection to Table 6.2
$n$ is 3 for water and 5.1 for all other fluids

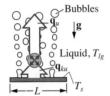

(2) Critical Heat Flux

$$q_{ku,CHF}^* = \frac{q_{ku,CHF}}{\rho_g \Delta h_{lg}} \left( \frac{\rho_g^2}{\sigma g \Delta \rho_{lg}} \right)^{1/4} = 0.13$$

(3) Minimum Heat Flux

$$q_{ku,min}^* = \frac{q_{ku,min}}{\rho_g \Delta h_{lg}} \left[ \frac{(\rho_l + \rho_g)^2}{\sigma g \Delta \rho_{lg}} \right]^{1/4} = 0.09$$

Vapor-Film Regime Boiling (Vertical Plates)

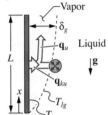

Laminar, Nonwavy Vapor Film

$$\langle Nu \rangle_L = \frac{L}{k_g} \frac{1}{(A_{ku} \langle R_{ku} \rangle_L)} = 0.943 \left[ \frac{g \Delta \rho_{lg} \Delta h_{lg} L^3}{k_g \nu_g (T_l - T_{lg})} \right]^{1/4}$$

Filmwise Condensation (Vertical Plates)

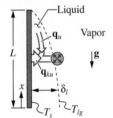

Laminar, Nonwavy Liquid Film

$$\langle Nu \rangle_L = \frac{L}{k_l} \frac{1}{(A_{ku} \langle R_{ku} \rangle_L)} = 0.943 \left[ \frac{g \Delta \rho_{lg} \Delta h_{lg} L^3}{k_l \nu_l (T_{lg} - T_s)} \right]^{1/4}$$

Forced Impinging-Droplets Vapor-Film Regime (Horizontal Plate)

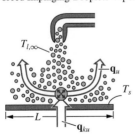

$$\langle Nu \rangle_L \frac{k_l}{L} = (A_{ku} \langle R_{ku} \rangle_L)^{-1} = \frac{\langle q_{ku} \rangle_L}{T_s - T_{l,\infty}}$$

$$= \rho_{l,\infty} \Delta h_{lg,\infty} \frac{\langle \dot{m}_d \rangle}{\rho_{l,\infty}} \eta_d \left[ 1 - \frac{\langle \dot{m}_d \rangle / \rho_{l,\infty}}{(\langle \dot{m}_d \rangle / \rho_l)_o} \right] \frac{1}{T_s - T_{l,\infty}}$$

$$+ 1,720 \, (T_s - T_{l,\infty})^{-0.088} \langle D \rangle^{-1.004} \langle u_d \rangle^{0.764} \frac{(\langle \dot{m}_d \rangle / \rho_{l,\infty})^2}{(\langle \dot{m}_d \rangle / \rho_l)_o}$$

$$\left( \frac{\langle \dot{m}_d \rangle}{\rho_l} \right)_o \equiv 5 \times 10^{-3} \text{ m/s, for } 180^\circ C < T_s - T_{l,\infty} < 380^\circ C$$

$$\eta_d \equiv \frac{3.68 \times 10^4}{\rho_{l,\infty} \Delta h_{lg,\infty}} (T_s - T_{l,\infty})^{1.691} \langle D \rangle^{-0.062}$$

$$\Delta h_{lg,\infty} \equiv (c_{p,l})_\infty (T_{lg} - T_{l,\infty}) + \Delta h_{lg}$$

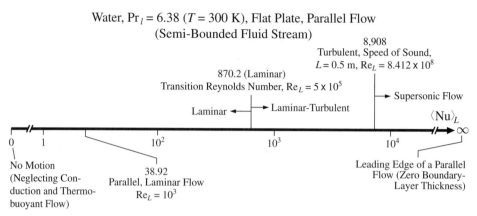

Figure 6.28. Range of Nusselt number $\langle Nu \rangle_L$ for parallel, semi-bounded water stream over a semi-infinite plate (with no phase change).

The laminar thermobuoyant flow is characterized by the 1/4 power appearing with $Ra_L$, and the turbulent flow is signified by $Ra_L$ to the power 1/3 power relation. The results are for flat plates (vertical and horizontal), horizontal cylinders, and spheres. The definition of $Ra_L$ and $\langle Nu \rangle_L$ (or $Ra_D$ and $\langle Nu \rangle_D$, etc.), are also listed in the table. The vertical cylinder is treated as a vertical plate, when the condition given in Table 6.4 is satisfied. The local Nusselt number correlation for rotating disks (in a stagnant ambient) is also listed for both laminar and turbulent flow regimes. Table 6.6 lists some of the correlations for the surface-convection conductance with phase change. The list is only suggestive and more comprehensive listings can be found in [17, 18].

Depending on the planar or radial surface geometry, $\langle Nu \rangle_L$ or $\langle Nu \rangle_D$ is used, and the surface-convection heat transfer rate is given by (6.49), i.e.,

$$\langle Q_{ku} \rangle_{L(\text{or } D)} = \frac{T_s - T_{f,\infty}}{\langle R_{ku} \rangle_{L(\text{or } D)}} = A_{ku} \frac{k_f}{L(\text{or } D)} \langle Nu \rangle_{L(\text{or } D)}(T_s - T_{f,\infty})$$

$$\text{Re}_{L(\text{or } D)} = \frac{u_{f,\infty} L(\text{or } D)}{\nu_f}. \tag{6.124}$$

As an example, Figure 6.28 shows the range of $\langle Nu \rangle_L$ for parallel flow of water over flat plates. The results for a plate with $L = 0.5$ m and a velocity equal to the speed of sound are also given. The average Nusselt number is 870.2 where the transition to turbulent flow ($\text{Re}_L = 5 \times 10^5$) occurs.

### EXAMPLE 6.9. FAM.S

As discussed, phase change offers the smallest resistance to surface convection, followed by forced and thermobuoyant flows. Consider a cylinder cooled by water. Determine the product $A_{ku}\langle R_{ku} \rangle_D = (T_s - T_{f,\infty})/q_{ku}$ for a long cylinder with diameter $D = 0.02$ m and $D = 0.08$ m, cooled in liquid water. Use (a) forced cross flow, (b) thermobuoyant flow, and (c) phase-density buoyant (with nucleate pool boiling) flow.

Use the correlations for $\langle \text{Nu} \rangle_D$ listed in Tables 6.4, 6.5, and 6.6. For the nucleate boiling use the results for a flat plate facing up (this is a good approximation when the diameter is large).

Determine the thermophysical properties for water at one atm pressure and at $T = 330$ K for thermobuoyant and forced flows, and at $T = 373$ K for pool boiling.

**SOLUTION**

(a) For the forced cross flow, we use the correlation in Table 6.4, i.e.,

$$\langle \text{Nu} \rangle_D = a_1 \text{Re}^{a_2} \text{Pr}^{1/3}, \quad \text{Re}_D = \frac{u_{f,\infty} D}{\nu_f}.$$

The constants $a_1$ and $a_2$ depend on $\text{Re}_D$ as shown in Table 6.4.

From (6.124), we have for $A_{ku}\langle R_{ku} \rangle_D$

$$A_{ku}\langle R_{ku} \rangle_D = \frac{T_s - T_{f,\infty}}{q_{ku}} = \frac{D}{\langle \text{Nu} \rangle_D k_f}.$$

For water, at $T = 330$ K, from Table C.23 we have

$$\nu_f = 5.05 \times 10^{-7} \text{ m}^2/\text{s}$$

$$k_f = 0.648 \text{ W/m-K}$$

$$\text{Pr} = 3.22.$$

(b) For the thermobuoyant flow, we use the correlation in Table 6.5, i.e.,

$$\langle \text{Nu} \rangle_D = (\langle \text{Nu}_{D,l} \rangle^{3.3} + \langle \text{Nu}_{D,t} \rangle^{3.3})^{1/3.3}$$

$$\langle \text{Nu}_{D,l} \rangle = \frac{1.6}{\ln[1 + 1.6(0.772 a_1 \text{Ra}_D^{1/4})]}$$

$$\langle \text{Nu}_{D,t} \rangle = \frac{0.13 \text{Pr}^{0.22}}{(1 + 0.61 \text{Pr}^{0.81})^{0.42}} \text{Ra}_D^{1/3}$$

$$a_1 = \frac{4}{3} \frac{0.503}{[1 + (0.492/\text{Pr})^{9/16}]^{4/9}}$$

$$\text{Ra}_D = \frac{g \beta_f (T_s - T_{f,\infty}) D^3}{\nu_f \alpha_f}.$$

Again

$$A_{ku}\langle R_{ku} \rangle_D = \frac{D}{\langle \text{Nu} \rangle_D k_f}.$$

For water, at $T = 325$ K, from Table C.23, we have

$$\beta_f = 0.000203 \text{ 1/K} \quad \text{available only at } T = 290 \text{ K}$$

$$\alpha_f = 1.57 \times 10^{-7} \text{ m}^2/\text{s}.$$

Comparison Among Surface-Convection Resistances for
Horizontal Cylinders Cooled by Water

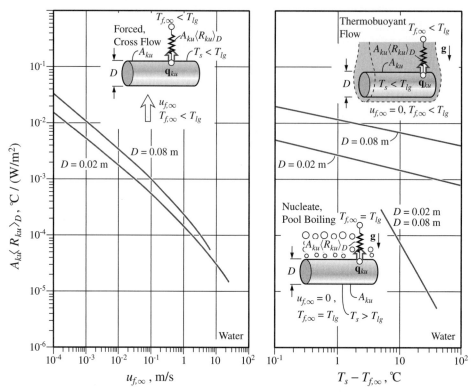

Figure Ex. 6.9. Surface-convection resistance for a horizontal cylinder cooled by a water
stream. Forced cross flow, thermobuoyant flow, and nucleate pool boiling are used.

(c) For the pool nucleate boiling, we use the correlation in Table 6.5. Here we
directly write for $A_{ku}\langle R_{ku}\rangle_D$ as

$$\frac{1}{A_{ku}\langle R_{ku}\rangle_D} = \frac{1}{a_s^3}\frac{\mu_l c_{p,l}^3 (T_s - T_{lg})^2}{\Delta h_{lg}^2}\left(\frac{g\Delta\rho_{lg}}{\sigma}\right)^{1/2}\mathrm{Pr}_l^{-n}.$$

From Table 6.5, for water we have $n = 3$, and we choose a water-copper pair
that gives $a_s = 0.013$ from Table 6.2.

For water at $T = 373$ K, from Table C.26 we have

$$\mu_l = 2.775 \times 10^{-4}\ \text{Pa-s}$$

$$c_{p,l} = 4{,}220\ \text{J/kg-K}$$

$$\Delta h_{lg} = 2.567 \times 10^6\ \text{J/kg}$$

$$\Delta\rho_{lg} = (958.3 - 0.597)(\text{kg/m}^3) = 957.7\ \text{kg/m}^3$$

$$\sigma = 0.05891\ \text{N/m}$$

$$\mathrm{Pr}_l = 1.73$$

$$g = 9.807\ \text{m/s}^2.$$

Figure Ex. 6.9 shows the variation of $A_{ku}\langle R_{ku}\rangle_D$ for the three flows. The forced cross flow is shown to the left with $u_{f,\infty}$ as the variable. The thermobuoyant and pool nucleate boiling results are shown to the right, with $T_s - T_{f,\infty}$ as the variable. The results for both diameters are shown.

Note that forced flow and nucleate boiling both result in a $A_{ku}\langle R_{ku}\rangle_D$ that is small compared to the thermobuoyant flow. For forced flow, a yet lower resistance can be obtained by further increasing the water velocity (the sonic velocity in water is about 1,500 m/s). The highest velocity used is based on the range of validity of the correlation.

**COMMENT**

The maximum surface superheat, $T_s - T_{lg}$ used is that corresponding to the critical heat flux for water. Note that the cylinder with the larger diameter has a larger product $A_{ku}\langle R_{ku}\rangle_D$. The pool nucleate boiling results are independent of $D$, because the correlation for flat plate was used.

---

EXAMPLE 6.10. FAM

---

The wind-chill index is the rate of surface-convection (and surface-radiation) heat transfer per unit area $\langle q_{ku}\rangle_D$ from a person with a surface temperature $T_s$ in a crossing wind [1]. This is depicted in Figure Ex. 6.10(a). The heat loss is characterized by the ambient air velocity $u_{f,\infty}$ (m/s) and temperature $T_{f,\infty}$ (°C). The body is approximated as a long cylinder. The average Nusselt number $\langle Nu\rangle_D$ is determined empirically as a function of the wind velocity. The correlation used is slightly different than that listed in Table 6.2.

The equivalent wind-chill temperature (also called the wind-chill factor) is the apparent ambient temperature $(T_{f,\infty})_{app}$ that results in the same heat loss

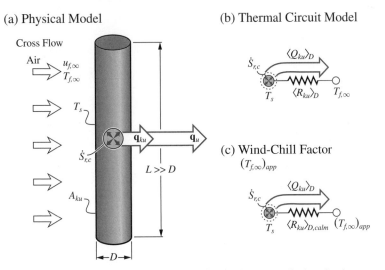

Surface-Convection Heat Transfer in Cross Flow over a Long Cylinder

Figure Ex. 6.10. Physical modeling of a human body as a cylinder. Surface-convection heat transfer from a cylinder in cross flow. (a) Physical model. (b) Thermal circuit model. (c) Wind-chill factor.

but under a calm wind condition (the calm wind velocity is considered to be 6.4 km/hr), $\langle Nu \rangle_{D,calm}$.

(a) Draw the thermal circuit diagram.
(b) Determine the wind-chill index $\langle q_{ku} \rangle_D$.
(c) Determine the wind-chill factor $(T_{f,\infty})_{app}$.
    $D = 35$ cm, $u_{f,\infty} = 10$ km/hr, $T_s = 33°$C, and $T_{f,\infty} = 10°$C.
    Evaluate the air properties from Table C.22 at $T = 300$ K.

**SOLUTION**
(a) The thermal circuit diagram is shown in Figure Ex. 6.10 (b). The energy equation is the surface energy equation (2.7), which reduces to $Q|_A = \langle Q_{ku} \rangle_D = \dot{S}_r$.
(b) The surface-convection heat flux is given by (6.124) as

$$\langle q_{ku} \rangle_D = \frac{\langle Q_{ku} \rangle_D}{A_{ku}} = \langle Nu \rangle_D \frac{k_f}{D}(T_s - T_{f,\infty}).$$

In order to evaluate $\langle Nu \rangle_D$ from Table 6.3, we need to first determine the Reynolds number given by (6.124), $Re_D = u_{f,\infty}D/v_f$. From Table C.22, we have for $T = 300$ K

$$v_f = 15.66 \times 10^{-6} \text{ m}^2/\text{s} \quad \text{Table C.22}$$
$$k_f = 0.0267 \text{ W/m-K} \quad \text{Table C.22}$$
$$Pr = 0.69 \quad \text{Table C.22.}$$

Then the Reynolds number is

$$Re_D = \frac{u_{f,\infty}D}{v_f} = \frac{\dfrac{10 \times 10^3}{3,600}(\text{m/s}) \times 0.35(\text{m})}{15.66 \times 10^{-6}(\text{m}^2/\text{s})} = 6.208 \times 10^4.$$

From Table 6.4, we have

$$\langle Nu \rangle_D = a_1 Re_D^{a_2} Pr^{1/3}, \quad a_1 = 0.027, \quad a_2 = 0.805$$

$$\langle Nu \rangle_D = 0.027(6.208 \times 10^4)^{0.805}(0.69)^{1/3} = 172.2.$$

For $\langle q_{ku} \rangle_D$, we have

$$\langle q_{ku} \rangle_D = 172.2 \times \frac{0.06267(\text{W/m-K})}{0.35(\text{m})}(33 - 10)(°\text{C})$$

$$= 302.1 \text{ W/m}^2 \quad \text{wind-chill index.}$$

(c) The wind-chill factor is defined as

$$\langle q_{ku} \rangle_D = \frac{k_f}{D}\langle Nu \rangle_{D,calm}[T_s - (T_{f,\infty})_{app}],$$

or

$$(T_{f,\infty})_{app} = T_s - \frac{\langle q_{ku} \rangle_D D}{k_f \langle Nu \rangle_{D,calm}},$$

where $\langle Nu \rangle_{D,calm}$ is for $u_{f,\infty} = 6.4$ km/hr. We need to determine $\langle Nu \rangle_{D,calm}$ and first we calculate $Re_D$, i.e.,

$$Re_D = \frac{\dfrac{6.4 \times 10^3}{3,600}(m/s) \times 0.35(m)}{15.66 \times 10^{-6}(m^2/s)} = 3.974 \times 10^4 \simeq 4 \times 10^4.$$

From Table 6.3, this is in the same range of $Re_D$ as above, i.e.,

$$\langle Nu \rangle_D = 0.027(3.947 \times 10^4)^{0.805}(0.69)^{1/3} = 120.2.$$

Then

$$(T_{f,\infty})_{app} = 33(°C) - \frac{302.1(W/m^2) \times 0.35(m)}{0.0267(W/m\text{-}K) \times 120.2} = 33(°C) - 32.94(°C)$$

$$= 0.06183°C \quad \text{wind-chill factor (apparent ambient temperature).}$$

**COMMENT**

The actual wind-chill index and factor use radiation and an empirical relation that when combined lowers the effect of $\langle Nu \rangle_D$. Therefore, both the wind-chill index and the apparent ambient temperature are higher than the values obtained above [1].

---

### EXAMPLE 6.11. FAM.S

---

In the incandescent lamp, the electrical energy converted into the Joule heating in the thin-wire filament is again converted into thermal radiation by surface emission. This is shown in Figure Ex. 6.11(a). The filament is at $T_2 = 2,900$ K and has

(a) Physical Model

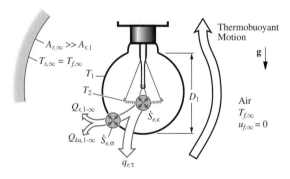

(b) Thermal Circuit Model

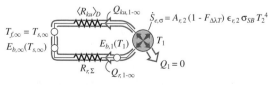

Figure Ex. 6.11. Surface-convection and surface-radiation heat transfer for a light bulb. (a) Physical model. (b) Thermal circuit model.

a total emissivity $\epsilon_{r,2} = 0.8$. This thermal radiation emission is mostly absorbed in the glass bulb (approximated as a sphere of diameter $D_1 = 6$ cm) and the remainder is transmitted through the bulb. In problem 4.13 at $T_2 = 2,900$ K, the calculated fraction of the radiation in the visible range is $F_{\Delta\lambda T}(\text{visible}) = 0.1012$. Assume that the remainder of the emission is absorbed in the glass bulb, i.e., $\dot{S}_{e,\sigma} = A_{r,2}(1 - F_{\Delta\lambda T})\epsilon_{r,2}\sigma_{\rm SB}T_2^4$. Here the filament is a wire with $D_2 = 0.2$ mm, $L_2 = 1$ m. This heat absorbed by the glass bulb is removed by a thermobuoyant-flow surface convection and by surface radiation. The ambient and the surrounding surfaces are at $T_{f,\infty} = T_{s,\infty} = 300$ K. The bulb emissivity is that of smooth glass listed in Table C.18.

(a) Draw the thermal circuit diagram for the glass bulb heat transfer.
(b) Determine the temperature of the glass bulb $T_1$.
  Determine the properties of air at $T = 300$ K.

**SOLUTION**
(a) Figure 6.11(b) shows the thermal circuit diagram.
(b) Applying the integral-volume energy equation (2.72), and assuming a steady-state heat transfer, we have

$$Q_1 + Q_{r,1\text{-}\infty} + Q_{ku,1\text{-}\infty} = \dot{S}_{e,\sigma}$$
$$= A_{r,2}(1 - F_{\Delta\lambda T})\epsilon_{r,2}\sigma_{\rm SB}T_2^4.$$

Here $Q_1 = 0$, and for the surface convection we have

$$Q_{ku,1\text{-}\infty} = \frac{T_1 - T_{f,\infty}}{\langle R_{ku}\rangle_D},$$

$$\langle R_{ku}\rangle_D = \frac{D}{A_{ku}\langle \mathrm{Nu}\rangle_D k_f}, \quad A_{ku} = \pi D^2$$

From Table 6.5, for the sphere we have

$$\langle \mathrm{Nu}\rangle_D = (\langle \mathrm{Nu}_{D,l}\rangle^6 + \langle \mathrm{Nu}_{D,t}\rangle^6)^{1/6}$$

$$\langle \mathrm{Nu}_{D,l}\rangle = 2.0 + 0.878 a_1 \mathrm{Ra}_D^{1/4}$$

$$\langle \mathrm{Nu}_{D,t}\rangle = \frac{0.13\mathrm{Pr}^{0.22}}{(1 + 0.61\mathrm{Pr}^{0.81})^{0.42}\mathrm{Ra}_D^{1/2}}$$

$$a_1 = \frac{4}{3}\frac{0.503}{[1 + (0.492/\mathrm{Pr})^{9/16}]^{4/19}}$$

$$\mathrm{Ra}_D = \frac{g\beta_f(T_1 - T_{f,\infty})D^3}{\nu_f\alpha_f}.$$

For radiation, we have a two-surface enclosure with both surfaces being gray, diffuse opaque surfaces with $A_{r,\infty} \gg A_{r,2}$. Then we use (4.49), because we also have $F_{1-\infty} = 1$, i.e.,

$$Q_{r,1} = Q_{r,1-\infty} = A_{r,1}\epsilon_{r,1}[E_{b,1}(T_1) - E_{b,\infty}(T_{s,\infty})]$$

$$A_{r,1} = A_{ku}.$$

We now substitute for $Q_{r,1-\infty}$ and $Q_{ku,1-\infty}$ in the energy equation, i.e.,

$$A_{ku}\langle Nu\rangle_D \frac{k_f}{D}(T_1 - T_{f,\infty}) + A_{r,1}\epsilon_{r,1}\sigma_{SB}(T_1^4 - T_{s,\infty}^4) = A_{r,2}(1 - F_{\Delta\lambda T})\epsilon_{r,2}\sigma_{SB}T_2^4.$$

This equation is solved for $T_1$ and is nonlinear in $T_1$. A software is used for the solution. The numerical values for the parameters are

$\epsilon_{r,1} = 0.94$  Table C.18

$\epsilon_{r,2} = 0.8$, $T_2 = 2{,}900$ K, $\sigma_{SB} = 5.67 \times 10^{-6}$ W/m²-K⁴  Table C.18

$D_1 = 0.06$ m, $T_{f,\infty} = T_{s,\infty} = 300$ K

$k_f = 0.0267$ W/m-K  Table C.22

$Pr = 0.69$  Table C.22

$\beta_f = 1/T_{f,\infty} = 1/300$ 1/K  ideal gas (6.77)

$\nu_f = 1.566 \times 10^{-5}$ m²/s  Table C.22

$\alpha_f = 2.257 \times 10^{-5}$ m²/s  Table C.22

$g = 9.807$ m/s²

$A_{r,2} = \pi D_2 L_2 = \pi \times 2 \times 10^{-4}(\text{m}) \times 10^{-2}(\text{m}) = 6.280 \times 10^{-6}$ m²

$F_{\Delta\lambda T} = 0.1012$.

From the solver, we have the results

$$T_1 = 410.2 \text{ K}$$

$$= 137.1°\text{C}.$$

The intermediate quantities computed are

$$\dot{S}_{e,\sigma} = 21.28 \text{ W}, \quad Ra_D = 2.202 \times 10^6, \quad \langle Nu\rangle_D = 19.68,$$

$$Q_{r,1-\infty} = 10.37 \text{ W}, \quad Q_{ku,1-2} = 10.91 \text{ W}.$$

**COMMENT**

A fraction of the near infrared radiation is also absorbed in the glass bulb. This results in a higher light bulb temperature $T_2$.

## 6.8 Inclusion of Substrate

So far, we have assumed that the thermal condition at the solid-fluid interface is known a priori (and considered mostly as a specified surface temperature). The heat transfer at the fluid-solid interface is a result of (or results in) conduction heat transfer in the solid. Here we address this solid heat conduction occurring simultaneously with the surface-convection (this is called the conjugate conduction-surface convection heat transfer). Here we consider layered fluid-solid composites. First we consider layered composite solid-fluid media with steady-state heat flowing perpendicular to the layers. Next we examine heat flow parallel to the layer. Next we consider transient substrate heat transfer. The general case of distributed, transient conduction with surface convection is not discussed for space economy. The lumped-capacitance approximation (i.e., uniform substrate temperature) treatment is given. Finally we consider the hot-wire anemometer (i.e., velocity) sensor as an application of coupling between the surface convection and substrate energy conversion and heat transfer.

### 6.8.1 Steady-State, Series Conduction-Surface Convection: Conduction-Convection or Biot Number $N_{ku,s} \equiv \mathrm{Bi}_L$

The perpendicular flow of heat through the solid-fluid composites is a series arrangement of conduction and surface-convection resistances. We demonstrate this series arrangement through an example.

Consider the layered structure in an electrical heater shown in Figure 6.29. Assuming a steady-state, one-dimensional heat flow, the heat generated by the Joule heating $\dot{S}_e(\mathrm{W})$ is removed from the top by passing through a thermal conductor and then through the interface of this conductor and a liquid undergoing pool boiling (surface convection with phase change). This branch of heat flow is designated as $Q_{H-1}$. The remainder of the heat generated is removed as a heat loss $Q_{H-2}$ through the lower path where a thermal insulator is placed to reduce this heat and then this heat loss flows from the interface of the insulator to the ambient air through a thermobuoyant surface convection. This is similar to the process considered in Section 4.6.1, except there we dealt with surface radiation.

We write the integral-volume energy equation (2.9) for the heater, for the control volume shown in the Figure 6.29, and $Q_H = 0$, as

$$Q \mid_A = \underset{\substack{\text{net surface} \\ \text{heat flow}}}{\int_A (\mathbf{q} \cdot \mathbf{s}_n) dA} = Q_{H-1} + Q_{H-2} = \underset{\substack{\text{electromagnetic} \\ \text{energy conversion,}}}{\dot{S}_{e,\mathrm{J}},} \tag{6.125}$$

where

$$Q_{H-1} = \frac{T_H - T_{f,\infty,1}}{R_{k,H-1} + R_{ku,H-1}} = \frac{T_H - T_{s,1}}{R_{k,H-1}} = \frac{T_{s,1} - T_{f,\infty,1}}{R_{ku,H-1}} \tag{6.126}$$

$$Q_{H-2} = \frac{T_H - T_{f,\infty,2}}{R_{k,H-2} + R_{ku,H-2}} = \frac{T_H - T_{s,2}}{R_{k,H-2}} = \frac{T_{s,2} - T_{f,\infty,2}}{R_{ku,H-2}}. \tag{6.127}$$

Note that for the $H$-1 branch of the heat flow, the ratio of the temperature drop in the thermal conductor $T_H - T_{s,1}$ to that in the fluid boundary layer $T_{s,1} - T_{f,\infty,1}$ is

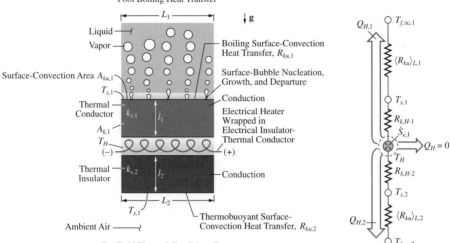

(a) Physical Model
(b) Thermal Circuit Model

(c) Biot Number

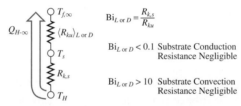

Figure 6.29. Simultaneous conduction and surface convection in an electrical heater. (a) Physical model. (b) Thermal circuit model. (c) Definition of Biot number.

$$\mathrm{Bi}_{L,1} \equiv N_{ku,s} \equiv \frac{R_{k,H-1}}{R_{ku,H-1}} = \frac{R_{k,H-1}}{\langle R_{ku}\rangle_{L-1}} = \frac{T_H - T_{s,1}}{T_{s,1} - T_{f,\infty,1}}$$

$$= \langle \mathrm{Nu}\rangle_{L,1} \frac{l_1 k_{f,1}}{L_1 k_{s,1}} \quad \begin{array}{l}\text{conduction-surface convection} \\ \text{(or Biot) number,}\end{array} \qquad (6.128)$$

where the inverse of the surface-convection conductance, $\langle R_{ku,1}\rangle_{L-1} = A_{ku,1}\langle \mathrm{Nu}\rangle_{L,1}k_{f,1}/L_1$, is the surface-convection conductance for surface 1 (note that as before, we have used the length along the surface, i.e., $L_1$, for the dimensionless surface-convection conductance) and $A_{k,1}k_{s,1}/l_1$ is the conduction conductance for layer 1 (where as before we have used $l_1$ as the length).

For the $H$-2 branch of the heat flow, with $A_{ku,1} = A_{k,1}$ we have

$$\frac{T_H - T_{s,2}}{T_{s,2} - T_{f,\infty,2}} = \frac{R_{k,H-2}}{R_{ku,H-2}} = \frac{R_{k,H-2}}{\langle R_{ku}\rangle_{L-2}} = \langle \mathrm{Nu}\rangle_{L,2}\frac{l_2 k_{f,2}}{L_2 k_{s,2}} \equiv \mathrm{Bi}_{L,2} \quad \begin{array}{l}\text{Biot number} \\ \text{for } H\text{-2 branch.}\end{array} \qquad (6.129)$$

In practice, $\text{Bi}_L < 0.1$ is considered low and $\text{Bi}_L > 10$ is considered high.

For an optimum heat flow, i.e., minimum heat loss (or $Q_{H-2} \to 0$), we need to choose $k_{s,1} \to \infty$ and $k_{s,2} \to 0$. Then $\text{Bi}_{L,1} \to 0$ and $\text{Bi}_{L,2} \to \infty$. This would correspond to negligible temperature drop due to conduction in the $H$-1 branch and a dominant temperature drop due to conduction in the $H$-2 branch.

In general, the Biot number $\text{Bi}_{L \text{ or } D} \equiv R_{k,s}/\langle R_{ku}\rangle_{L \text{ or } D}$, i.e.,

$$\text{Bi}_{L \text{ or } D} = N_{ku,s} = \frac{R_{k,s}}{\langle R_{ku}\rangle_{L \text{ or } D}} \quad \begin{matrix} \text{Bi}_{L \text{ or } D} < 0.1, \text{ negligible substrate} \\ \text{temperature nonuniformity, } \text{Bi}_{L \text{ or } D} > 10, \\ \text{negligible fluid temperature nonuniformity,} \end{matrix} \quad (6.130)$$

where $s$ stands for solid or substrate. The Biot number $\text{Bi}_{L \text{ or } D}$ is a measure of the significance of the temperature variation within the solid, as compared to that in the fluid boundary layer. This is shown in Figure 6.29(c). Note that $\text{Bi}_L$, for surface convection, is similar to $N_r$ for surface radiation (Section 4.6.1). As $\text{Bi}_L \to 0$, the solid temperature drop becomes insignificant and as $\text{Bi}_L \to \infty$, the fluid temperature drop becomes insignificant.

Also, note that for $\text{Bi}_L > 10$, the surface-convection resistance is negligible and then the surface temperature becomes nearly that of the fluid far field, i.e., $T_s \simeq T_{f,\infty}$.

When the substrate heat transfer is not in one dominant direction, then a characteristic conduction path length is used.

### EXAMPLE 6.12. FAM.S

A high speed liquid flowing inside a tube results in maintaining the internal surface of the tube at temperature $T_1$. The tube is made of (i) copper or (ii) PVC (polyvinylchloride). These are examples of metallic and polymeric solids and

(a) Physical Model                    (b) Thermal Circuit Model

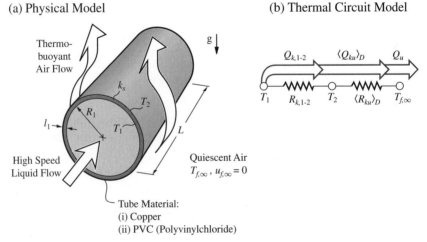

Figure Ex. 6.12. (a) Heat transfer from a fluid moving in a tube, to the ambient air. (b) Thermal circuit model of the problem.

represent high and low thermal conductivities. From Table C.14, at $T = 350$ K, we find $k_s$(copper) $= 397$ W/m-K and $k_s$(PVC) $= 0.35$ W/m-K.

The surface of the tube is warmer than the ambient, quiescent air, and therefore, a thermobuoyant flow occurs, resulting in the surface-convection heat transfer. This is shown in Figure Ex. 6.12(a).

(a) Draw the thermal circuit diagram.
(b) Write the expression for the heat flowing through this composite system.
(c) Determine $\langle R_{ku} \rangle_D$ for the two tube materials. Note that $\langle Nu \rangle_D$ depends on the tube outside temperature $T_2$, which is not known a priori.
(d) Determine the $Bi_D$ for the two tube materials and comment on the significance or negligibility of the temperature drop across the tube wall.

$R_1 = 15$ cm, $l_1 = 1$ cm, $L = 50$ cm, $T_{f,\infty} = 20°$C, and $T_1 = 80°$C.
Evaluate air properties at $T = 350$ K.

**SOLUTION**
(a) The thermal circuit diagram is shown in Figure Ex. 6.12(b).
(b) The heat flowing through the composite layers is

$$Q_{k,1\text{-}2} = \frac{T_1 - T_2}{R_{k,1\text{-}2}}$$

$$= \langle Q_{ku} \rangle_D = \frac{T_2 - T_{f,\infty}}{\langle R_{ku} \rangle_D}$$

$$= \frac{T_1 - T_{f,\infty}}{R_{k,1\text{-}2} + \langle R_{ku} \rangle_D} \equiv \frac{T_1 - T_{f,\infty}}{R_\Sigma}, \quad D = 2(R_1 + l_1).$$

(c) The conduction resistance is found from Table 3.2, i.e.,

$$R_{k,1\text{-}2} = \frac{\ln R_2/R_1}{2\pi L k_s} = \frac{\ln(R_1 + R_2)/R_1}{2\pi L k_s}.$$

The resistance $\langle R_{ku} \rangle_D$ is related to $\langle Nu \rangle_D$ through (6.124), i.e.,

$$\langle R_{ku} \rangle_D = \frac{D}{A_{ku} \langle Nu \rangle_D k_f}, \quad A_{ku} = \pi D L.$$

Now $\langle Nu \rangle_D$ is given by the correlation in Table 6.5 as

$$\langle Nu \rangle_D = (\langle Nu_{D,l} \rangle^{3.3} + \langle Nu_{D,t} \rangle^{3.3})^{1/3.3},$$

where $Nu_{D,l}$ and $Nu_{D,t}$ are listed in Table 6.5 and are related to the Rayleigh number

$$Ra_D = \frac{g\beta_f(T_2 - T_{f,\infty})D^3}{\nu_f \alpha_f}.$$

Here we note that $\langle R_{ku} \rangle_D$, through $\langle Nu \rangle_D$, depends on $T_2$, which is not known. We need to determine $T_2$ and $\langle R_{ku} \rangle_D$ simultaneously. For this we write

$\langle R_{ku}\rangle_D = \langle R_{ku}\rangle_D(T_2)$ and from the above heat flow rate equation, we rewrite

$$\frac{T_1 - T_2}{R_{k,1\text{-}2}} = \frac{T_1 - T_{f,\infty}}{R_{k,1\text{-}2} + \langle R_{ku}\rangle_D(T_2)}$$

and then solve this for $T_2$. This is a nonlinear algebraic equation and is solvable by hand (through iteration) or by using a software.

We now write the expressions for $\langle \mathrm{Nu}_{D,l}\rangle$ and $\langle \mathrm{Nu}_{D,t}\rangle$ from Table 6.5, i.e.,

$$\langle \mathrm{Nu}_{D,l}\rangle = \frac{1.6}{\ln[1 + 1.6/(0.772a_1\,\mathrm{Ra}_D^{1/4})]}$$

$$\langle \mathrm{Nu}_{D,t}\rangle = \frac{0.13\mathrm{Pr}^{0.22}}{(1 + 0.61\mathrm{Pr}^{0.81})^{0.42}}\,\mathrm{Ra}_D^{1/3}$$

$$a_1 = \frac{4}{3}\frac{0.503}{[1 + (0.492/\mathrm{Pr})^{9/16}]^{4/9}}.$$

For the properties for air at $T = 350$ K, from Table C.22, we have

$$\mathrm{Pr} = 0.69$$

$$k_f = 0.0300 \text{ W/m-K}$$

$$\nu_f = 2.030 \times 10^{-5} \text{ m}^2/\text{s}$$

$$\alpha_f = 2.944 \times 10^{-5} \text{ m}^2/\text{s}.$$

Also

$$\beta = \frac{1}{T} = \frac{1}{350 \text{ (K)}} \quad \text{ideal gas}$$

$$g = 9.807 \text{ m/s}^2$$

$$A_{ku} = \pi(R_1 + l_1)L = \pi(0.15 + 0.001)(\text{m}) \times 0.5(\text{m}) = 0.2512 \text{ m}^2$$

$$D = 2(R_1 + l_1) = 2(0.15 + 0.01)(\text{m}) = 0.32 \text{ m}.$$

Then

$$\mathrm{Ra}_D = \frac{9.8(\text{m/s}^2) \times 2.857 \times 10^{-3}(1/\text{K}) \times (T_2 - 293.16)(\text{K}) \times (0.32)^3(\text{m}^3)}{2.030 \times 10^{-5}(\text{m}^2/\text{s}) \times 2.944 \times 10^{-5}(\text{m}^2/\text{s})}$$

$$a_1 = 0.5132.$$

Now solving for $T_2$, we find

$$\text{(i) copper}: \ T_2 = 353.14 \text{ K} = 79.99°\text{C}$$

$$\text{(ii) PVC}: \ T_2 = 345.72 \text{ K} = 72.57°\text{C}.$$

The two surface-convection resistances are

$$(i) \text{ copper}: \text{ Ra}_D = 9.216 \times 10^7, \quad \langle\text{Nu}\rangle_D = 53.12$$

$$\langle R_{ku}\rangle_D(T_2 = 353.14 \text{ K}) = 0.3996°\text{C/W}$$

$$(ii) \text{ PVC}: \text{ Ra}_D = 7.667 \times 10^7, \quad \langle\text{Nu}\rangle_D = 51.07$$

$$\langle R_{ku}\rangle_D(T_2 = 345.72 \text{ K}) = 0.4157°\text{C/W},$$

and the two conduction resistances are

$$(i) \text{ copper}: R_{k,1\text{-}2} = 5.177 \times 10^{-5}°\text{C/W}$$

$$(ii) \text{ PVC}: R_{k,1\text{-}2} = 0.05872°\text{C/W}.$$

(d) From (6.130), the two Biot numbers are

$$\text{Bi}_D(\text{copper}) = \frac{R_{k,1\text{-}2}}{\langle R_{ku}\rangle_D} = \frac{5.177 \times 10^{-5}°\text{C/W}}{0.3996°\text{C/W}} = 1.296 \times 10^{-4} < 0.1 \qquad \begin{matrix}\text{substrate}\\\text{temperature}\\\text{drop is}\\\text{negligible}\end{matrix}$$

$$\text{Bi}_D(\text{PVC}) = \frac{0.05872}{0.4157} = 0.1413 > 0.1 \qquad \begin{matrix}\text{substrate temperature drop}\\\text{is significant.}\end{matrix}$$

**COMMENT**

The outside surface temperature $T_2$ depends on the heat flow rate through the tube wall, which in turn depends on $T_2$ through the thermobuoyant flow. Therefore, a simultaneous solution for $Q_{1\text{-}\infty}$ and $T_s$ is sought. Here for copper we have a temperature drop $T_1 - T_2 = 0.01°\text{C}$ and for PVC we have $T_1 - T_2 = 7.430°\text{C}$, with $T_1 - T_{f,\infty} = 60°\text{C}$. From the above results we have

$$\text{copper}: Q_{1\text{-}\infty} = \frac{T_1 - T_2}{R_{k,1\text{-}2}} = \frac{(353.15 - 353.14)(\text{K})}{5.17 \times 10^{-5}(\text{K/W})} = 150.1 \text{ W}$$

$$\text{PVC}: Q_{1\text{-}\infty} = \frac{(353.15 - 345.72)(\text{K})}{0.05872(\text{K/W})} = 126.5 \text{ W},$$

i.e., lower $Q_{1\text{-}\infty}$ for tube with PVC material.

### EXAMPLE 6.13. FUN

Electrical conducting wires carrying current are electrically insulated by a dielectric (i.e., electrical insulation) jacket (i.e., shell). Polymers (e.g., polyvinylchloride, called PVC) are dielectric and also have a relatively low conductivity.

Due to the electrical resistance of the wire $R_e(\text{ohm})$, the current flow $J_e(\text{A})$ results in the Joule heating $\dot{S}_{e,J}$ and this heat generation has to be removed. Figure Ex. 6.13(a) depicts the energy conversion and heat transfer. The heat flows by conduction through the wire (which, because of the high thermal conductivity, is assumed to have a uniform temperature). Then by conduction it flows through the dielectric material and by surface convection to the ambient fluid (here, air) and is finally convected away. The surface-convection surface area can be optimized for the maximum heat dissipation, i.e., for minimum wire temperature.

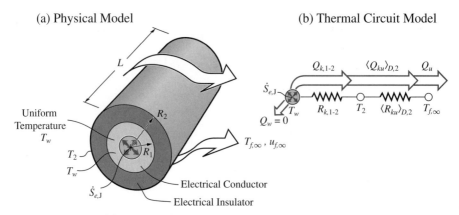

(a) Physical Model             (b) Thermal Circuit Model

Figure Ex. 6.13. (a) Joule heating and surface-convection heat transfer from a wire coated with an electrical insulation, i.e., a dielectric shell. (b) Thermal circuit model for heat flow.

(a) Draw the thermal diagram for the heat flow.

(b) Write the energy equation for the control surface enclosing the wire. Write the expression for the surface heat flow rate $Q|_A$.

(c) Minimize $R_\Sigma$ (the overall resistance) with respect to $R_2$, the other radius of the dielectric shield, and find the general relation for $R_2$, corresponding to the minimum $R_\Sigma$.

(d) Assume that $k_f \langle Nu \rangle_{D,2}/D_2 = h$ is constant, where $D_2 = 2R_2$, i.e., $h$ is independent of $D_2$. Then derive an explicit relation for this optimum $R_2$ (this is called the critical radius).

(e) Use Table 6.3 to determine the optimum $R_2$.

**SOLUTION**

(a) The thermal circuit diagram is shown in Figure Ex. 6.13(b).

(b) The integral-volume energy equation for the wire is (6.125), i.e.,

$$Q\,|_{A,w} = \dot{S}_{e,J} \quad \text{energy equation for wire,}$$

where, using $Q_w = 0$, we have

$$Q\,|_{A,w} = \frac{T_w - T_{f,\infty}}{R_{k,1-2} + \langle R_{ku}\rangle_{D,2}} = \frac{T_w - T_{f,\infty}}{R_\Sigma} \quad \text{surface heat transfer rate,}$$

where from Table 3.1

$$R_{k,1-2} = \frac{\ln(R_2/R_1)}{2\pi L k_s} \quad \text{Table 3.1}$$

and from (6.124) and noting that $A_{ku} = \pi D_2 L = 2\pi R_2 L$, we have

$$
\begin{aligned}
\langle R_{ku} \rangle_{D,2} &= \frac{D_2}{A_{ku} \langle Nu \rangle_{D,2} k} \\
&= \frac{2R_2}{2\pi R_2 L \langle Nu \rangle_{D,2} k_f} = \frac{1}{\pi L \langle Nu \rangle_{D,2} k_f},
\end{aligned}
$$

with $D_2 = 2R_2$.

(c) To minimize $R_\Sigma$, we differentiate $R_\Sigma$ and set the resultant equal to zero, i.e.,

$$
\frac{d}{dR_2} R_\Sigma = \frac{d}{dR_2} \left[ \frac{\ln(R_2/R_1)}{2\pi L k_s} + \frac{1}{\pi L k_f \langle Nu \rangle_{D,2}} \right] = 0.
$$

We then substitute the $h$, which is independent of $R_2$, and write (6.33) as the relation between $\langle Nu \rangle_{D,2}$ and $D_2$, and finally perform differentiation.

(d) If we assume that

$$
k_f \langle Nu \rangle_{D,2} = h D_2 \quad \text{assuming that } h \text{ is independent of } D_2,
$$

where $h$ is the heat transfer coefficient, substituting for $\langle Nu \rangle_{D,2}$ and differentiating we have,

$$
\frac{d}{dR_2} \left[ \frac{\ln(R_2/R_1)}{2\pi L k_s} + \frac{1}{2\pi L h R_2} \right] = 0.
$$

The result is

$$
\frac{1}{2\pi L k_s R_2} - \frac{1}{2\pi L h R_2^2} = 0.
$$

Solving for $R_2$, we have

$$
R_2 = \frac{k_s}{h} = R_c \quad \text{for } h \text{ independent of } D_2,
$$

where $R_c$ is called the critical radius (i.e., corresponding to minimum $R_\Sigma$).

(e) We note that, in general, $h$ depends on $D_2$. For example, in forced flow, from Table 6.4, we have for cylinders

$$
\frac{h D_2}{k_f} = \langle Nu \rangle_{D,2} = a_1 Re_{D,2}^{a_2} Pr^{1/3}, \quad a_2 < 1.
$$

Note that from Table 6.4, for example for $Re_{D,2} < 4 \times 10^3$, $a_2$ is 0.466. Then $h$ is proportional to $D_2^{a_2-1}$.

Now if we use the relation

$$
h = \frac{k_f}{D_2} a_1 Re_{D,2}^{a_2} Pr^{1/3} \equiv a_R D_2^{a_2-1}, \quad a_R = k_f a_1 \frac{u_{f,\infty}^{a_2}}{\nu_f^{a_2}} Pr^{1/3},
$$

where $a_R$ contains all the other constants (i.e., $a_1$, $a_2$, Pr, etc.), then we can write

$$k_f \langle \mathrm{Nu} \rangle_{D,2} = a_R D_2^{1+(a_2-1)}$$

$$= a_R D_2^{a_2}.$$

By substituting for $k_f \langle \mathrm{Nu} \rangle_{D,2}$, we have

$$\frac{d}{dR_2} \left[ \frac{\ln R_2/R_1}{2\pi L k_s} + \frac{1}{2^{a_2}\pi L a_R R_2^{a_2}} \right].$$

After differentiation, we have

$$\frac{1}{2\pi L k_s R_2} + \frac{-a_2}{2^{a_2}\pi L a_R R_2^{a_2+1}} = 0$$

or

$$\frac{1}{2k_s} - \frac{a_2}{2^{a_2} a_R R_2^{a_2-1}} = 0.$$

Solving for $R_2$, we have

$$R_2 = R_c = \left( \frac{a_2 k_s}{2^{a_2-1} a_R} \right)^{1/a_2} = \left( \frac{a_2 k_s v_f^{a_2}}{2^{a_2-1} k_f a_1 u_{f,\infty}^{a_2} \mathrm{Pr}^{1/3}} \right)^{1/a_2}.$$

For the case of $a_2 = 1$, which gives $a_R = h$, we obtain $R_2 = R_c = k_s/h$, as before. For $0 < a_2 < 1$, a different critical radius is found. Note that the unit of $a_R$ depends on $a_2$.

### COMMENT

The surface-convection resistance decreases as $R_2$ increases, while the conduction resistance increases with $R_2$. The minimum in $R_\Sigma$ occurs at the critical radius $R_c$. This $R_2$ depends on the Reynolds number $\mathrm{Re}_{d,2}$. This is demonstrated in an end of chapter problem, where it is shown that the region near this minimum is rather flat, i.e., the minimum is not very pronounced.

### 6.8.2 Steady-State Simultaneous Conduction-Surface Convection: Extended Surface and Fin Efficiency $\eta_f$

Extended surfaces are layered solid surface extensions that provide for large surface area for surface-convection heat transfer. The heat flowing to or from a solid surface flows in parallel through areas with solid-extension attachments (called the fin area), and also through areas with no such attachments (called the bare area). Figure 6.30 shows two extended surfaces in the form of rectangular cross-section fins and circular cross-section fins (also called pin fins). In practice, many such fins are placed on the heat transfer surface with an optimum spacing. The distinction between the surface convection from surfaces of various geometries and that from the extended surfaces is that for the extended surfaces the surface temperature changes along the surface,

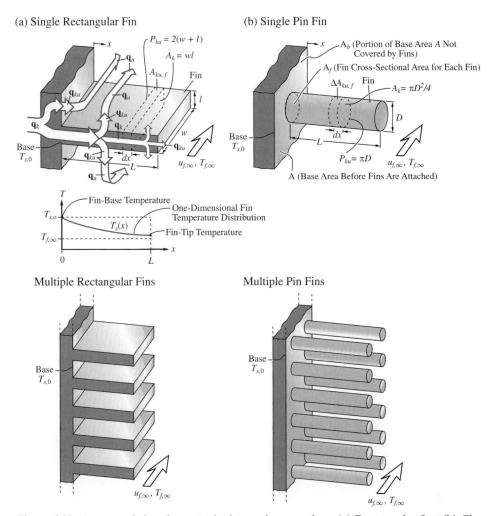

Figure 6.30. An extended surface attached to a planar surface. (a) Rectangular fins. (b) Circular (pin) fins.

as a result of the surface-convection heat transfer. Therefore, only the temperature at the base of the extended surfaces is known a priori.

In order to determine the heat transfer from these extended surfaces, their surface temperature $T_s = T_s(x)$ needs to be determined. The heat flows from the base of the fin through the fin by conduction and simultaneously leaves the surface by surface convection and radiation. Here we assume that the radiation heat transfer is negligible. It can be shown that effective fins have a length $L$ to thickness ($l$ or $D$) ratio that is much larger than unity; therefore, the temperature variation within the fin is approximated (and justified) as one dimensional and along the fin, i.e., $T_s = T_s(x)$.

The control volume (and the associated energy equation) is selected as a combined integral- and differential-length, as shown in Figure 6.30. The energy equation is similar to (2.11). The geometrical variables and the energy equation are

$$\Delta V = A_k \Delta x = wl \Delta x, \ A_{ku,f} = P_{ku,f} \Delta x = 2(w+l)\Delta x, \ A_k = wl \quad \begin{array}{l}\text{for rectangular}\\ \text{cross-section fin}\end{array}$$

$$\Delta V = \frac{A_k \Delta x \pi D^2}{4}\Delta x, \ A_{ku,f} = P_{ku,f} \Delta x = \pi D \Delta x, \ A_k = \frac{\pi D^2}{4} \quad \begin{array}{l}\text{for circular}\\ \text{cross-section fin}\end{array}$$

$$(6.131)$$

$$\underbrace{\frac{\int_{\Delta A}(\mathbf{q}\cdot\mathbf{s}_n)dA}{\lim\limits_{\Delta V\to 0}\Delta V}}_{\substack{\text{integral}\\ \text{surface-}\\ \text{convection}\\ \text{heat transfer}}} = \underbrace{\frac{P_{ku,f}}{A_k}q_{ku}}_{} + \underbrace{\frac{d}{dx}q_k}_{\substack{\text{spatial rate}\\ \text{of change of}\\ \mathbf{q}_k \text{ in}\\ x \text{ direction}}} = \underbrace{\frac{2(w+l)}{wl}q_{ku} + \frac{d}{dx}q_k = 0}_{\text{for rectangular fins}}$$

$$(6.132)$$

or

$$P_{ku,f}q_{ku} + A_k\frac{d}{dx}q_k = 0 \tag{6.133}$$

or

$$P_{ku,f}q_{ku} + A_k\frac{d}{dx}\left(-k_s\frac{dT_s}{dx}\right) = 0 \quad \begin{array}{l}\text{combined integral- and}\\ \text{differential-length energy equation.}\end{array} \tag{6.134}$$

Note that $P_{ku}$ is the surface-convection perimeter (perpendicular to the $x$-axis) for each fin.

Now we represent the surface-convection heat flux $q_{ku}$ using the dimensionless conductance and we use a conductance over the fin surface, i.e., $\langle \text{Nu}\rangle_w$ or $\langle \text{Nu}\rangle_D$. For the rectangular fin, from (6.124) we have

$$\langle \text{Nu}\rangle_w = \langle \text{Nu}\rangle_w(\text{Re}_w, \text{Pr}, \text{geometry of fin}), \tag{6.135}$$

$$q_{ku,f} = \langle \text{Nu}\rangle_w\frac{k_f}{w}[T_s(x) - T_{f,\infty}]. \tag{6.136}$$

Here we state that $T_s = T_s(x)$ and for simplicity use only $T_s$. Using (6.136) in (6.134) and assuming that $k_s$ is constant, we have

$$P_{ku,f}\langle \text{Nu}\rangle_w\frac{k_f}{w}(T_s - T_{f,\infty}) - A_k k_s\frac{d^2 T_s}{dx^2} = 0 \tag{6.137}$$

or

$$\frac{d^2 T_s}{dx^2} - \frac{P_{ku,f}\langle \text{Nu}\rangle_w\dfrac{k_f}{w}}{A_k k_s}(T_s - T_{f,\infty}) = 0. \tag{6.138}$$

One boundary condition for this second-order differential equation is that $T_s$ is a prescribed temperature at the base, i.e.,

$$T_s = T_{s,0} \quad \text{at } x = 0. \tag{6.139}$$

The second boundary condition is the surface energy equation (2.67) applied at $x = L$. In obtaining the solution, it is much easier (and without much loss of accuracy) to account for the surface convection from the tip area (i.e., $x = L$) by adding a correction length to the fin length and then assuming that the tip is insulated (no heat flows out of the tip). This correction leads to the corrected length $L_c$, defined as

$$L_c = L + \frac{l}{2} \quad \text{corrected length for rectangular fin} \tag{6.140}$$

$$L_c = L + \frac{D}{4} \quad \text{corrected length for circular fin.} \tag{6.141}$$

Then the bounding-surface energy equation (2.67), for the case of no energy conversion, no surface radiation, no surface convection, and with conduction only in the solid side, becomes

$$k_s \frac{dT_s}{dx} = 0 \quad \text{at} \quad x = L_c \quad \text{surface energy equation at the tip surface.} \tag{6.142}$$

The temperature distribution $T_s = T_s(x)$ is found by twice integrating (6.138) and determining the constants. The solution is

$$\frac{T_s(x) - T_{f,\infty}}{T_{s,0} - T_{f,\infty}} = \frac{T_s(x) - T_{f,\infty}}{\Delta T_{max}} = \frac{\cosh[m(L_c - x)]}{\cosh(mL_c)} \quad \begin{array}{l} \text{temperature} \\ \text{distribution} \\ \text{along fin,} \end{array} \tag{6.143}$$

where

$$
\begin{aligned}
m &= \left( \frac{P_{ku,f} \langle Nu \rangle_w k_f}{A_k k_s w} \right)^{1/2} \\
&= \left( \frac{P_{ku,f} \langle Nu \rangle_w \dfrac{k_f}{w} \dfrac{L_c^2}{L_c^2}}{A_k k_s} \right)^{1/2} = \left( P_{ku,f} L_c \frac{\langle Nu \rangle_w \dfrac{k_f}{w}}{A_k} \frac{1}{k_s / L_c} \right)^{1/2} \frac{1}{L_c} = \left( \frac{R_{k,s}}{\langle R_{ku} \rangle_w} \right)^{1/2} \frac{1}{L_c} \\
&\equiv Bi_w^{1/2} \frac{1}{L_c} \qquad \text{ratio of solid-conduction to surface-convection resistance.}
\end{aligned}
\tag{6.144}
$$

Here $m(1/m)$ is a measure of extinction of the fin temperature. This decay or drop in fin temperature is related to the square root of the ratio of axial, solid conduction resistance $R_{k,s} = L_c / A_k k_s$ to the surface-convection resistance $\langle R_{ku} \rangle_w = P_{ku,f} L_c \langle Nu \rangle_w k_f / w$, i.e., $Bi_w^{1/2} = (R_{k,s} / \langle R_{ku} \rangle_w)^{1/2}$. When $m$ is large, the conduction resistance is much larger than the surface-convective resistance and the temperature drop along the fin becomes noticeable over only a short distance from the base. Since the surface-convection heat transfer rate depends on the difference between the fin and the far-field fluid temperatures, then the local surface-convection heat transfer rate decreases as the fin temperature drops. Therefore, it is desirable to have a relatively low value for $m$.

The magnitude of the hyperbolic cosine of argument $mL_c$, $\cosh(mL_c)$, is listed for discrete values of $mL_c$ in Table 6.7.

Table 6.7. *Hyperbolic cosine and tangent with argument* $mL_c = \mathrm{Bi}_{w\ or\ D}^{1/2}$, *as a function of* $mL_c$

| $mL_c$ | $\cosh(mL_c)$ | $\tanh(mL_c)$ | $mL_c$ | $\cosh(mL_c)$ | $\tanh(mL_c)$ |
|------|------|------|------|------|------|
| 0.00 | 1.0000 | 0.00000 | 1.20 | 1.8107 | 0.83365 |
| 0.10 | 1.0050 | 0.09967 | 1.40 | 2.1509 | 0.88535 |
| 0.20 | 1.0201 | 0.19738 | 1.60 | 2.5775 | 0.92167 |
| 0.30 | 1.0453 | 0.29131 | 1.80 | 3.1075 | 0.94681 |
| 0.40 | 1.0811 | 0.37995 | 2.00 | 3.7622 | 0.96403 |
| 0.50 | 1.1276 | 0.46212 | 3.00 | 10.068 | 0.99505 |
| 0.60 | 1.1855 | 0.53705 | 4.00 | 27.308 | 0.99933 |
| 0.70 | 1.2552 | 0.60437 | 5.00 | 74.210 | 0.99991 |
| 0.80 | 1.3374 | 0.66404 | 7.00 | 548.32 | 1.00000 |
| 0.90 | 1.4331 | 0.71630 | 9.00 | 4051.5 | 1.00000 |
| 1.00 | 1.5431 | 0.76159 | 10.00 | 11013 | 1.00000 |

The heat flowing into the fin at $x = 0$ by conduction is equal to that which leaves from the fin surface through surface convection. This heat flowing into the fin is found by differentiating $T_s = T_s(x)$, from (6.143), and evaluating it at $x = 0$, and then multiplying the resultant by $-A_k k_s$. This gives

$$A_k q_k(x = 0) = -A_k k_s \frac{dT_s}{dx}|_0 = \left( \langle \mathrm{Nu} \rangle_w \frac{k_f}{w} P_{ku,f} A_k k_s \right)^{1/2} (T_{s,0} - T_{f,\infty}) \tanh(mL_c)$$

heat flow into a fin. (6.145)

It is customary to rewrite (6.145) in terms of the fin efficiency $\eta_f$, that is the measure of the deviation of the fin temperature $T_s = T_s(x)$ from the base temperature $T_{s,0}$. This gives

$$A_k q_k(x = 0) = A_{ku,f} \langle \mathrm{Nu} \rangle_w \frac{k_f}{w} \eta_f (T_{s,0} - T_{f,\infty}), \quad A_{ku,f} = P_{ku,f} L_c, \quad (6.146)$$

where

$$\eta_f = \frac{\tanh(mL_c)}{mL_c}, \quad 0 \le \eta_f \le 1 \quad \text{fin efficiency}$$

$$= \frac{\tanh[(R_{k,s}/\langle R_{ku}\rangle_w)^{1/2}]}{(R_{k,s}/\langle R_{ku}\rangle_w)^{1/2}} = \frac{\tanh(\mathrm{Bi}_w^{1/2})}{\mathrm{Bi}_w^{1/2}}. \quad (6.147)$$

When $R_{k,s} \ll \langle R_{ku}\rangle_w$, then $\eta_f = 1$, because the fin is at a uniform temperature $T_{s,0}$. This can be shown by expanding the $\tanh(z)$ in terms of $z$ and then taking only the leading term (left as an end of chapter problem).

The variation of $\eta_f$ with respect to $mL_c = \mathrm{Bi}_w^{1/2}$ is shown in Figure 6.31(a). The high and low $\mathrm{Bi}_w$ limits are also shown. Note that from (6.146), the product of $A_{ku,f}$ and $\eta_f$ should be kept large for an effective fin usage. While small $\mathrm{Bi}_w$ makes $\eta_f$ tend toward unity, it also results in smaller $A_{ku,f}$. So a trade-off is made between $A_{ku,f}$ and $q_f$.

(a) Variation of Fin Efficiency $\eta_f$ with Respect to $mL_c = \mathrm{Bi}_w^{1/2}$

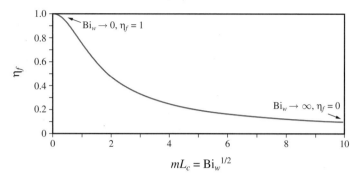

(b) Thermal Circuit Model

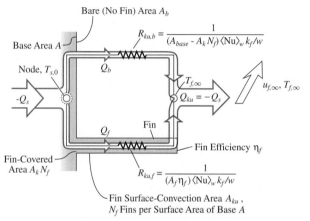

Figure 6.31. (a) Variation of fin efficiency $\eta_f$ with respect to $mL_c = \mathrm{Bi}_w^{1/2}$. (b) Thermal circuit model for heat transfer for a base surface through its base and fined surfaces.

The magnitude of $\tanh(mL_c)$ is also listed in Table 6.7 for discrete values of $mL_c$. For $mL_c \to 0$, the fin will be at a uniform temperature $T_s$. For $mL_c > 3$, the numerator in (6.147) does not increase noticeably compared to the denominator. Then $\eta_f$ begins to decrease substantially as $mL_c$ increases beyond 3.

Now for a base surface of area $A_{base}$ (before any fins are attached) that has $N_f$ fins attached to it, with each fin having a cross-sectional area $A_k$, we have $A_b = A_{base} - N_f A_k$ as the total bare surface area $A_b$. This bare area also undergoes surface convection. We also have $A_{ku,f} = P_{ku,f} L_c$ for each fin and $A_f = N_f P_{ku,f} L_c = N_f A_{ku,f}$ as the total fin surface area $A_f$ for surface convection.

Under steady-state condition, with no surface energy conversion and with the surface heat transfer given by $-Q_s$ and $Q_{ku}$, as shown in Figure 6.31(b), surface-integral energy equation (2.71) is applicable. Then for the total heat transfer from the base $Q_{ku}$ (sum of heat flow for bare surface $Q_b$ and heat flow for finned surface $Q_f$), we have

$$- Q_s = Q_{ku} = Q_b + Q_f = \frac{T_{s,0} - T_{f,\infty}}{R_\Sigma} \qquad \begin{array}{l}\text{total surface-convection} \\ \text{heat transfer.}\end{array} \qquad (6.148)$$

Here we have used two resistances in parallel. We can also write this as

$$Q_{ku} = \langle Q_{ku}\rangle_{s-\infty} = (A_b + A_f\eta_f)\langle Nu\rangle_w \frac{k_f}{w}(T_{s,0} - T_{f,\infty}), \qquad (6.149)$$

where

$$A_b = A_{base} - N_f A_k, \quad A_f = N_f A_{ku,f} = N_f P_{ku,f} L_c \quad \text{base and single fin surface areas}$$

$$(6.150)$$

$$\frac{1}{R_\Sigma} = \sum_{i=1}^{2} \frac{1}{R_{ku,i}} = \frac{1}{R_{ku,b}} + \frac{1}{R_{ku,f}} \quad \text{overall resistance} \qquad (6.151)$$

$$R_{ku,b}^{-1} = A_b \langle Nu\rangle_w \frac{k_f}{w} \quad \text{base-surface resistance} \qquad (6.152)$$

$$R_{ku,f}^{-1} = A_f \eta_f \langle Nu\rangle_w \frac{k_f}{w} \quad \text{fin-surface resistance.} \qquad (6.153)$$

Note that we have assumed and used the same $\langle Nu\rangle_w$ for the bare surface and the fin surface. Note that $Q_s$ can in turn be given in terms of other resistances (e.g., substrate conduction). $\langle Nu\rangle_w$ is based on $w$.

This heat transfer is viewed as taking parallel paths through the base surface and the fin surface, as depicted in Figure 6.31. Other fin geometries and the optimization of fins are discussed in [10,18].

The fin effectiveness $\Gamma_f$ is defined as the ratio of $Q_{ku}$ with fins to $Q_{ku}$ without fins. Then assuming the same $\langle Nu\rangle_w$ for all surfaces, $\Gamma_f \equiv Q_{ku}(\text{with fins})/Q_{ku}(\text{without fins}) = (A_b + \eta_f A_f)/A_{base}$, note again that $A_{base}$ is the base area (before fin placement), and $\Gamma_f > 10$ is considered a substantial surface convection enhancement by addition of the fin. As fins are added to the surface, to maintain a high $\langle Nu\rangle_w$, the fluid stream velocity must increase. For thermobuoyant flows, fins are spaced as to not significantly reduce this limited motion.

### EXAMPLE 6.14. FAM

A square base at $T_{s,0} = 80°C$ [rendered in Figure Ex. 6.14(a)], with each side having a dimension $a = 10$ cm, has pure aluminum pin fins of diameter $D = 7$ mm and length $L = 70$ mm, attached to it. There are a total of $N_f = 10 \times 10 = 100$ fins. The air flows across the fins at a velocity of $u_{f,\infty} = 1$ m/s, and temperature $T_{f,\infty} = 20°C$. Assume that the isolated cylinder $\langle Nu\rangle_D$ relation in Table 6.4 is applicable. (We can make this assumption since the fluid flow through the fins can be treated as bounded fluid. A more accurate correlation will be given in Section 7.4.4, where we will discuss interacting discontinuous solid surfaces.)

Surface-Convection with Extended Surface

(a) Physical Model                    (b) Thermal Circuit Model

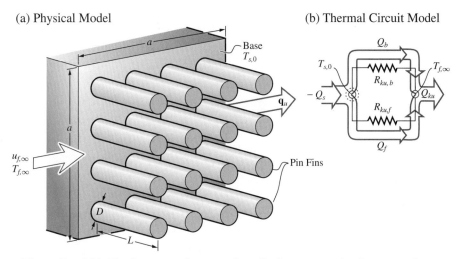

Figure Ex. 6.14. Pin fins on a planar surface. Surface-convection heat transfer occurs from the bare base surface and the fin surface. (a) Physical model. (b) Thermal circuit model.

(a) Draw the thermal circuit diagram.
(b) Determine the Nusselt number $\langle Nu \rangle_D$.
(c) Determine the fin efficiency.
(d) Determine the total heat flow rate from the base $-Q_s = Q_{ku}(W)$.
   Determine the thermophysical properties at $T = 300$ K.

**SOLUTION**
(a) The thermal circuit diagram is given in Figure Ex. 6.14(b). The rate of surface-convection heat transfer from the bare and finned areas of the base is given by (6.149). Here we use the circular fins and replace $\langle Nu \rangle_w k_f / w$ with $\langle Nu \rangle_D k_f / D$, i.e.,

$$-Q_s = Q_{ku} = [(A - N_f A_k) + N_f A_{ku,f} \eta_f] \langle Nu \rangle_D \frac{k_f}{D} (T_{s,0} - T_{f,\infty}).$$

(b) To determine the Nusselt number $\langle Nu \rangle_D$, the correlation we use (an approximation) is from Table 6.4, i.e.,

$$\langle Nu \rangle_D = a_1 Re_D^{a_2} Pr^{1/3} \quad \text{Table 6.4,}$$

where, as defined in (6.124),

$$Re_D = \frac{u_{f,\infty} D}{\nu_f}.$$

The thermophysical properties for air are obtained from Table C.22 at $T = 300$ K, i.e.,

$$v_f = 15.66 \times 10^{-6} \ \text{m}^2/\text{s} \qquad \text{Table C.22}$$

$$k_f = 0.0267 \ \text{W/m-K} \qquad \text{Table C.22}$$

$$\text{Pr} = 0.69 \qquad \text{Table C.22.}$$

Also from Table C.14, we have for aluminum (pure) at $T = 300$ K

$$k_s = 237 \ \text{W/m-K} \qquad \text{Table C.14.}$$

Then

$$\text{Re}_D = \frac{1(\text{m/s}) \times 7 \times 10^{-3}(\text{m})}{15.66 \times 10^{-6}(\text{m}^2/\text{s})} = 447.0.$$

From Table 6.4

$$a_1 = 0.683, \quad a_2 = 0.466 \quad \text{Table 6.4}$$

$$\langle \text{Nu} \rangle_D = 0.683(447.0)^{0.466}(0.69)^{1/3} = 10.37.$$

(c) The fin efficiency $\eta_f$ is given by (6.147) as

$$\eta_f = \frac{\tanh(mL_c)}{mL_c},$$

where $L_c$ and $m$ are given by (6.141) and (6.144), i.e.,

$$L_c = L + \frac{D}{4}$$

$$m = \left( \frac{P_{ku,f} \langle \text{Nu} \rangle_D k_f}{A_k k_s D} \right)^{1/2},$$

where $P_{ku,f} = \pi D$, and $A_k = \pi D^2/4$. Then

$$m = \left[ \frac{\pi D \langle \text{Nu} \rangle_D k_f}{(\pi D^2/4) k_s D} \right]^{1/2}$$

or

$$m = \left( \frac{4 \langle \text{Nu} \rangle_D k_f}{D^2 k_s} \right)^{1/2} = \left( \frac{4 \langle \text{Nu} \rangle_D}{D^2} \frac{k_f}{k_s} \right)^{1/2}$$

$$= \left[ \frac{4 \times 10.37}{(7 \times 10^{-3})^2 (\text{m}^2)} \frac{0.0267(\text{W/m-K})}{237(\text{W/m-K})} \right]^{1/2} = 9.766 \ 1/\text{m}.$$

Then

$$mL_c = 9.766(1/\text{m}) \times [7 \times 10^{-2}(\text{m}) + (7 \times 10^{-3}/4)(\text{m})] = 0.7006.$$

From Table 6.7, tanh(0.7006) = 0.6044, and for $\eta_f$ we have

$$\eta_f = \frac{\tanh(mL_c)}{mL_c} = \frac{0.6044}{0.7006} = 0.8627 \quad \text{fin efficiency.}$$

(d) The relation for $Q_{ku}$ was listed earlier, and we have

$$A_{ku,f} = P_{ku,f}L_c = \pi D(L + D/4)$$

$$Q_{ku} = [(a \times a - N_f \pi D^2/4) + N_f \pi D(L + D/4) \times \eta_f]\langle Nu \rangle_D \frac{k_f}{D}(T_{s,0} - T_{f,\infty})$$

$$Q_{ku} = [(0.1)^2(\mathrm{m})^2 - 10^2 \times \pi \times (7 \times 10^{-3})^2(\mathrm{m})^2/4 + 10^2 \times \pi \times 7 \times 10^{-3}(\mathrm{m})$$

$$\times (7 \times 10^{-2} + 7 \times 10^{-3}/4)(\mathrm{m}) \times 0.8627] \times 10.37$$

$$\times \frac{0.0267(\mathrm{W/m\text{-}K})}{7 \times 10^{-3}(\mathrm{m})} \times (80 - 20)(^\circ\mathrm{C})$$

$$= [0.01(\mathrm{m})^2 - 3.847 \times 10^{-3}(\mathrm{m})^2 + 0.1361(\mathrm{m})^2] \times 39.55(\mathrm{W/m\text{-}K}) \times 60(^\circ\mathrm{C})$$

$$= 345.2 \text{ W.}$$

**COMMENT**

Note that the fins have added an effective surface area of

$$N_f A_{ku,f} \eta_f - N_f A_k = 0.1361(\mathrm{m})^2 - 3.847 \times 10^{-3}(\mathrm{m})^2 = 0.1323 \text{ m}^2,$$

for surface-convection heat transfer. This should be compared to the base area $A_{base} = a^2 = 0.01 \text{ m}^2$. Also note that the fin effectiveness is

$$\Gamma_f = \frac{A_b + A_f \eta_f}{A_{base}}$$

$$= \frac{a^2 - N_f \pi \dfrac{D^2}{4} + \eta_f N_f \pi D \left(L + \dfrac{D}{4}\right)}{a^2}$$

$$= \frac{0.01 - 0.003847 + 0.1323}{0.01} = 14.23.$$

This is an increase of 14.23 times in the surface area for surface convection. The efficiency is relatively large, indicating that the entire fin is nearly at the base temperature. The high concentration of fins allows for a large surface area; this high concentration may reduce the fluid velocity and $\langle Nu \rangle_D$ would decrease due to this resistance to the fluid flow. Also, $\langle Nu \rangle_D$ may not be the same for the base surface and for the fins.

### 6.8.3 Transient: Lumped Capacitance for $Bi_L < 0.1$

The general case of distributed, transient substrate conduction with surface convection, including the penetration front, is not treated here for space economy. The results can be found in [19] and the graphical presentation is referred to as the Heisler graphs. For the case of large conduction-convection number Bi, i.e., $Bi_{L \text{ or } D} > 10$, the temperature variation within the substrate is dominant over that in the fluid. Then the surface of the substrate will be nearly at the far-field fluid temperature $T_{f,\infty}$ and we can use the results of Section 3.5 for the nonuniform transient substrate temperature. When the Fourier number $Fo_{L \text{ or } R}$ is large enough to allow for the penetration of the surface temperature change to travel through the substrate and cause a near uniform temperature (Chart 3.7), then we assume uniform substrate temperature. For example, for spheres this would require that $Fo_{L \text{ or } R} > 0.5$.

When the conduction-convection parameter $Bi_{L \text{ or } D}$ is small, then the temperature variation within the substrate is negligible compared to that occurring outside the solid. As with the transient surface-radiation heat transfer, discussed in Section 4.6.2, this is referred to as the lumped-capacitance analysis. Our treatment here is similar to that in Section 3.5.2, except we use a surface-convection resistance $\langle R_{ku} \rangle_{L \text{ or } D}$. For a solid undergoing temporal change in its temperature and a surface-convection heat transfer as shown in Figure 6.32(a), the integral-volume energy equation (2.9) becomes

$$Q|_{A,1} = \int_{A_1} (\mathbf{q}_{ku} \cdot \mathbf{s}_n) dA = -(\rho c_p V)_1 \frac{dT_1}{dt} + \dot{S}_1 \quad \text{for } Bi_L < 0.1$$

integral-volume energy equation for volume node $T_1$, (6.154)

where $T_1 = T_1(t)$, i.e., $T$ is assumed uniform within the solid. Here we allow for surface convection as shown in Figure 6.32(a). We use a surface-average resistance $\langle R_{ku} \rangle_{L \text{ or } D}$, such that $Q_{ku,1-\infty} = [T_1(t) - T_{f,\infty}]/\langle R_{ku} \rangle_{L \text{ or } D}$. We will allow for other surface heat flow $Q_1$ in addition to the surface convection. By assuming a constant $\rho c_p$, (6.154) becomes

$$Q|_{A,1} = Q_1 + Q_{ku,1-\infty} = Q_1 + \frac{T_1(t) - T_{f,\infty}}{\langle R_{ku} \rangle_{L \text{ or } D}} = -(\rho c_p V)_1 \frac{dT_1(t)}{dt} + \dot{S}_1. \quad (6.155)$$

Here we assume that $Q_1$ is constant to allow us to analytically integrate (6.155). The initial thermal condition is a prescribed temperature $T_1(t = 0)$. The results presented here are similar to those given in Section 3.6.2. The solution to (6.155), using this initial condition, is

$$T_1(t) = T_{f,\infty} + [T_1(t = 0) - T_{f,\infty}]e^{-t/\tau_1} + a_1\tau_1(1 - e^{-t/\tau_1})$$

transient solid temperature for time-dependent surface heat transfer rate,

(6.156)

where

$$\tau_1 = (\rho c_p V)_1 \langle R_{ku} \rangle_{L \text{ or } D}, \quad a_1 = \frac{\dot{S}_1 - Q_1}{(\rho c_p V)_1}. \quad (6.157)$$

(a) Thermal Circuit Model

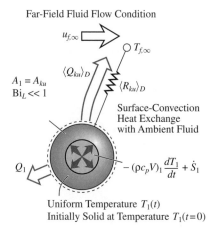

Far-Field Fluid Flow Condition

$u_{f,\infty}$

$T_{f,\infty}$

$\langle Q_{ku}\rangle_D$

$A_1 = A_{ku}$
$Bi_L \ll 1$

$\langle R_{ku}\rangle_D$

Surface-Convection
Heat Exchange
with Ambient Fluid

$Q_1$

$-(\rho c_p V)_1 \dfrac{dT_1}{dt} + \dot{S}_1$

Uniform Temperature $T_1(t)$
Initially Solid at Temperature $T_1(t=0)$

(b) Temperature Variation with Respect to Elapsed Time

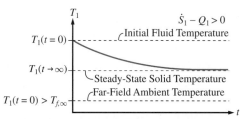

$T_1$

$\dot{S}_1 - Q_1 > 0$

Initial Fluid Temperature

$T_1(t=0)$

$T_1(t \to \infty)$

Steady-State Solid Temperature

Far-Field Ambient Temperature

$T_1(t=0) > T_{f,\infty}$

$t$

(c) Dimensionless Temperature Versus Dimensionless Time, for $\dot{S}_1 - Q_1 = 0$

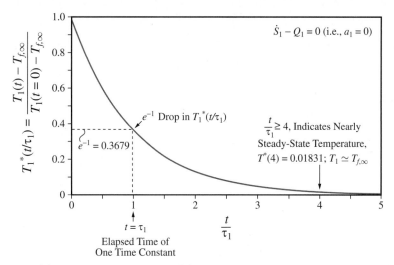

$T_1^*(t/\tau_1) = \dfrac{T_1(t) - T_{f,\infty}}{T_1(t=0) - T_{f,\infty}}$

$\dot{S}_1 - Q_1 = 0$ (i.e., $a_1 = 0$)

$e^{-1}$ Drop in $T_1^*(t/\tau_1)$

$e^{-1} = 0.3679$

$\dfrac{t}{\tau_1} \geq 4$, Indicates Nearly
Steady-State Temperature,
$T^*(4) = 0.01831; T_1 \simeq T_{f,\infty}$

$t = \tau_1$

$\dfrac{t}{\tau_1}$

Elapsed Time of
One Time Constant

Figure 6.32. (a) Thermal circuit model. (b) Transient, lumped temperature variation of an object subject to surface-convection heat transfer. (c) Variation of dimensionless temperature with respect to dimensionless time, for $\dot{S}_1 - Q_1 = 0$.

Here $\tau_1$(s) is the time constant or the relaxation time for the time response of the solid to the sudden surface-convection heat exchange.[†]

For $\dot{S}_1 - Q_1 = 0$, the dimensionless temperature $T^*(t/\tau_1)$ is plotted in Figure 6.32(c) versus dimensionless time $t/\tau_1$. Note that for $t/\tau = 1$, $T^*(t/\tau)$ drops to $e^{-1}$ or $T^*(1) = 0.3679$. For $t/\tau_1 > 4$, $T_1 \simeq T_{f,\infty}$, i.e, $T_1^*(4) = 0.01839 \simeq 0$.

Figure 6.32(b), similar to Figure 3.38, also shows the anticipated temporal variation of temperature, $T_1 = T_1(t)$, for the case of $\dot{S}_1 - Q_1 \neq 0$ and $T_{f,\infty} < T_1(t = 0)$. When $\dot{S}_1 - Q_1 \neq 0$, the steady-state solid temperature $T_1(t \to \infty)$ will be different than $T_{f,\infty}$ and is found by setting $t \to \infty$. Then (6.155) becomes

$$T_1(t \to \infty) = T_{f,\infty} + a_1\tau_1 = T_{f,\infty} + (\dot{S}_1 - Q_1)\langle R_{ku}\rangle_{L \text{ or } D} \quad \begin{array}{l}\text{steady-state}\\\text{temperature.}\end{array} \quad (6.158)$$

The surface-convection resistance is given in terms of the Nusselt number (6.28), using the flow characteristic length $L$ or $D$, i.e., $\langle R_{ku}\rangle_{L \text{ or } D}^{-1} = k_f A_{ku}\langle Nu\rangle_{L \text{ or } D}/L$ (or $D$).

---

### EXAMPLE 6.15. FAM

---

In a powder-based surface-coating process, aluminum particles are heated in a very hot stream of argon gas, $T_{f,\infty} = 1,500$ K, before deposition on the substrate. The spherical powder particles of average diameter $D = 100$ $\mu$m are injected into the stream at a temperature $T_1(t = 0) = 20°$C. The particles will reach a temperature $T_1 = T_{sl} - 30°$C, where $T_{sl}$ is the melting temperature of aluminum, before deposition. The relative velocity of the gas and particle (which is used in the evaluation of $\langle Nu\rangle_D$) is $u_{f,\infty} - u_1 = 5$ m/s. These are depicted in Figure Ex. 6.15(a).

(a) Draw the thermal circuit model.
(b) Determine the Nusselt number $\langle Nu\rangle_D$.
(c) Determine the time constant $\tau_1$.
(d) Determine the time of flight required for the particles to reach $T_{sl} - 30°$C.
(e) Examine the Biot number $Bi_D$.

Estimate the internal conduction resistance as $D/4A_{ku}k_s$ for validation of the lumped capacitance treatment. Evaluate the argon thermophysical properties at $T_{f,\infty} = 1,500$ K from Table C.22.

---

[†] Note that when using (6.157), (6.155) becomes

$$\frac{dT_1}{dt} + \frac{T_1 - T_{f,\infty}}{\tau_1} = a_1.$$

With $a_1 = 0$, this shows how $T_1$ returns to its thermal equilibrium value $T_{f,\infty}$ during the relaxation period $\tau_1$. Under steady state, $dT_1/dt = 0$ and we have $T_1(t \to \infty) = T_{f,\infty} + a_1\tau_1$.

(a) Physical Model                                          (b) Thermal Circuit Model

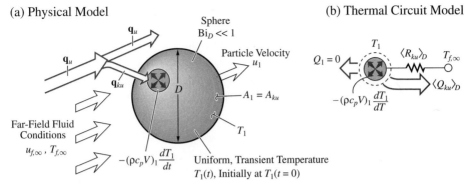

Figure Ex. 6.15. (a) An aluminum particle is heated by surface convection and its temperature rises to near the melting temperature. (b) Thermal circuit model of the problem.

**SOLUTION**

(a) The thermal circuit model is shown in Figure Ex. 6.15(b).

(b) The Nusselt number for a sphere in forced flow is given in Table 6.4, i.e.,

$$\langle Nu \rangle_D = 2 + (0.4 Re_D^{1/2} + 0.06 Re_D^{2/3}) Pr^{0.4} \qquad \text{Table 6.4,}$$

where, as defined in (6.124),

$$Re_D = \frac{u_{f,\infty} D}{\nu_f}.$$

For argon at $T = 1{,}500$ K, we have from Table C.22

$$k_f = 0.0551 \text{ W/m-K} \quad \text{Table C.22}$$

$$\nu_f = 2.18 \times 10^{-4} \text{ m}^2/\text{s} \quad \text{Table C.22}$$

$$Pr = 0.67 \quad \text{Table C.22.}$$

Then

$$Re_D = \frac{5(\text{m/s}) \times 1 \times 10^{-4}(\text{m})}{2.18 \times 10^{-4}(\text{m}^2/\text{s})} = 2.294$$

$$\langle Nu \rangle_D = 2 + [0.4 \times (2.294)^{1/2} + 0.06 \times (2.294)^{2/3}] \times (0.67)^{0.4}$$

$$= 2 + 1.807 = 3.807.$$

(c) The time constant $\tau_1$ is given by (6.157) as

$$\tau_1 = (\rho c_p V)_1 \langle R_{ku} \rangle_D$$

$$\langle R_{ku} \rangle_D = \frac{D}{A_{ku} \langle Nu \rangle_D k_f}, \qquad A_{ku} = 4\pi R^2 = \pi D^2.$$

From Table C.16, we have for aluminum (for $T = 300$ K)

$$T_{sl} = T_s = 933 \text{ K} \quad \text{Table C.16}$$
$$\rho = 2{,}702 \text{ kg/m}^3 \quad \text{Table C.16}$$
$$c_p = 903 \text{ J/kg-K} \quad \text{Table C.16}$$
$$k_s = 237 \text{ W/m-K} \quad \text{Table C.16.}$$

Also

$$V = \frac{4}{3}\pi R^3 = \frac{1}{6}\pi D^3.$$

Then

$$\tau_1 = \frac{(\rho c_p)_1 \frac{1}{6}\pi D^3 D}{\pi D^2 \langle \text{Nu} \rangle_D k_f} = \frac{1}{6}\rho c_p \frac{D^2}{\langle \text{Nu} \rangle_D k_f}$$

$$= \frac{1}{6}(2{,}702)(\text{kg/m}^3) \times 903(\text{J/kg-K}) \times \frac{(10^{-4})^2 (\text{m})^2}{3.807 \times 0.0551(\text{W/m-K})}$$

$$= 0.01939 \text{ s} = 19.39 \text{ ms.}$$

(d) The transient temperature is given by (6.156), i.e.,

$$T_1(t) = T_{f,\infty} + [T_1(t=0) - T_{f,\infty}]e^{-t/\tau_1} + a_1\tau_1(1 - e^{-t/\tau_1}),$$

where, since $Q_1 = \dot{S}_1 = 0$, we have

$$a_1 = \frac{\dot{S}_1 - Q_1}{(\rho c_p V)_1} = 0.$$

Then we write the temperature distribution as

$$\frac{T_1(t) - T_{f,\infty}}{T_1(t=0) - T_{f,\infty}} = e^{-t/\tau_1}.$$

Since we know all the temperatures on the left-hand side, we take the natural logarithm of this expression and solve for $t$. The result is

$$-\frac{t}{\tau_1} = \ln \frac{T_1(t) - T_{f,\infty}}{T_1(t=0) - T_{f,\infty}}$$

$$t = -\tau \ln \frac{T_1(t) - T_{f,\infty}}{T_1(t=0) - T_{f,\infty}}$$

$$= -(0.01939)(\text{s}) \ln \frac{(933 - 30)(\text{K}) - 1{,}500(\text{K})}{(20 + 273.15)(\text{K}) - 1{,}500(\text{K})}$$

$$= -(0.01939)(\text{s}) \ln \frac{-597(\text{K})}{-1{,}206.85(\text{K})} = 0.01365 \text{ s} = 13.65 \text{ ms.}$$

(e) The Biot number $Bi_D$ is defined by (6.128) as

$$Bi_D = \frac{R_k}{\langle R_{ku} \rangle_D} = \frac{A_{ku} \langle Nu \rangle_D k_f / D}{4 A_{ku} k_s / D} = \langle Nu \rangle_D \frac{k_f}{4 k_s}$$

$$= 3.807 \times \frac{0.0551 (\text{W/m-K})}{4 \times 237 (\text{W/m-K})} = 2.128 \times 10^{-4}$$

$$< 0.1 \quad \text{solid temperature variation is negligible.}$$

Here we have used the same areas for conduction and surface convection.

**COMMENT**

Note that $\langle Nu \rangle_D$ is very small (nearly that of conduction only, which is 2) resulting in $Bi_D \ll 1$. Therefore, we can justifiably neglect the internal resistance, i.e., the temperature variation within the solid, and use the lumped-capacitance analysis.

The time constant is proportional to $D^2$. Here $t$ for reaching $T_1 = T_{sl} - 30°C$ is nearly one time constant.

EXAMPLE 6.16. FAM.S

During the period when hot water is not flowing through the faucet, it cools in the pipes by surface convection and surface-radiation heat transfer from the pipe surface. This is shown in Figure Ex. 6.16(a). Assume that the vertical portion of the pipe shown in the figure and its water content have a uniform temperature

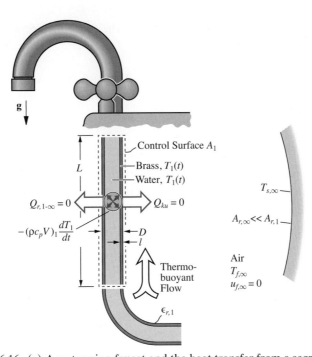

Figure Ex. 6.16. (a) A water pipe-faucet and the heat transfer from a segment of pipe.

(b) Thermal Circuit Model

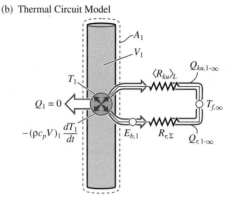

(c) Decay in Pipe-Water Temperature

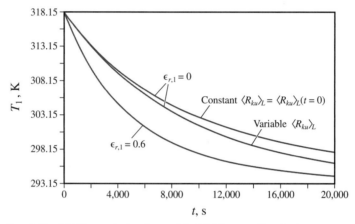

Figure Ex. 6.16. (b) Thermal circuit model. (c) The predicted pipe-water temperature history, for the pipe-water system.

$T_1(t)$. Initially, we have $T_1(t = 0) = 45°C$. The surface-convection heat transfer is due to a thermobuoyant motion. Use the correlation for vertical flat plate for the $\langle Nu \rangle_L$ for pipe surface.

(a) Draw the thermal circuit diagram.

(b) Plot the temperature history $T_1(t)$ for the set of parameters given below. Since the energy equation is an ordinary differential equation, and since for combined surface-convection and radiation heat transfer it cannot be analytically integrated, use the software.

(c) For the case of no radiation, compare the results for $T(t)$ to that predicted by (6.156).

$T_{f,\infty} = 20°C$, $T_{s,\infty} = 20°C$, $D = 8$ cm, $l = 3$ mm, and $L = 30$ cm

The pipe material is oxidized brass (Tables C.16 and C.18). Determine the air and water properties at $T = 300$ K.

**SOLUTION**

(a) The thermal circuit diagram is shown in Figure Ex. 6.16(b).

(b) The energy equation for the uniform temperature pipe-water segment is (6.154), with $Q_1 = \dot{S}_1 = 0$. Then we have

$$Q|_{A,1} = Q_{ku,1\text{-}\infty} + Q_{r,1\text{-}\infty} = -(\rho c_p V)_1 \frac{dT_1}{dt}.$$

The volume $V$ is occupied by water $(w)$ and brass $(b)$, i.e.,

$$(\rho c_p V)_1 = (\rho c_p)_w \pi \frac{(D - 2l)^2}{4} L + (\rho c_p)_b \pi \frac{[D^2 - (D - 2l)^2]}{4} L.$$

For surface radiation, for a two-surface enclosure, with $A_{r,\infty} \ll A_{r,1}$ and $F_{1\text{-}\infty} = 1$, we have (4.49), i.e.,

$$Q_{r,1\text{-}\infty} = A_{r,1} \epsilon_{r,1} \sigma_{\mathrm{SB}} (T_1^4 - T_{s,\infty}^4), \quad A_{r,1} = \pi DL = A_{ku}.$$

The surface convection is given by (6.124), i.e.,

$$Q_{ku,1\text{-}\infty} = A_{ku} \langle \mathrm{Nu} \rangle_L \frac{k_f}{L} (T_1 - T_{f,\infty}).$$

The Nusselt number is found from Table 6.5, where it is suggested that we use the correlation for the vertical plates, i.e.,

$$\langle \mathrm{Nu} \rangle_L = [(\langle \mathrm{Nu}_{L,l} \rangle)^6 + (\langle \mathrm{Nu}_{L,t} \rangle)^6]^{1/6}$$

$$\langle \mathrm{Nu}_{L,l} \rangle = \frac{2.8}{\ln[(1 + 2.8/(a_1 \mathrm{Ra}^{1/4})]}$$

$$\langle \mathrm{Nu}_{L,t} \rangle = \frac{0.13 \mathrm{Pr}^{0.22}}{(1 + 0.61 \mathrm{Pr}^{0.81})^{0.42}} \mathrm{Ra}_L^{1/3}$$

$$a_1 = \frac{4}{3} \frac{0.503}{[1 + (0.492/\mathrm{Pr})^{9/16}]^{4/9}}$$

$$\mathrm{Ra}_L = \frac{g \beta_f (T_f - T_{f,\infty}) L^3}{\nu_f \alpha_f}.$$

The thermophysical properties are, at $T = 300$ K,

|  |  |  |
|---|---|---|
| water : | $\rho_w = 997$ kg/m$^3$ | Table C.27 |
|  | $c_{p,w} = 4{,}179$ J/kg-K | Table C.27 |
| brass : | $\rho_b = 8{,}933$ kg/m$^3$ | Table C.16 |
|  | $c_{p,b} = 355$ J/kg-K | Table C.16 |
|  | $\epsilon_{r,1} = 0.60$ | Table C.18 |
| air : | $k_f = 0.0267$ W/m-K | Table C.22 |
|  | $\beta_f = \dfrac{1}{300}$ 1/K | ideal gas (6.77) |
|  | $\nu_f = 1.566 \times 10^{-5}$ m$^2$/s | Table C.22 |
|  | $\alpha_f = 2.257 \times 10^{-5}$ m$^2$/s | Table C.22 |
|  | Pr $= 0.69$ | Table C.22 |
| other parameters : | $T_{f,\infty} = T_{s,\infty} = 293.15$ K |  |
|  | $D = 0.06$ m, $l = 10^{-3}$ m, $L = 0.3$ m. |  |

The results for $T_1(t)$ are plotted in Figure Ex. 6.16(c). The results with and without the surface radiation are shown. The surface radiation accelerates the cooldown. Note that within about one hour, the water temperature drops to near 30°C. This is the average surface temperature for the human skin. Any further drop in the temperature results in a cold sensation when this water flows over the human hands.

(c) From the case of no surface radiation, $Q_{r,1\text{-}\infty} = 0$, and (6.156), for $\dot{S}_1 - Q_1 = 0$ ($a_1 = 0$), we have

$$T_1(t) = T_{f,\infty} + [T_1(t = 0) - T_{f,\infty}]e^{-t/\tau_1}$$

$$\tau_1 = (\rho c_p V)_1 \langle R_{ku} \rangle_L.$$

The solution is valid for a constant $\langle R_{ku} \rangle_L$. Here we use $\langle R_{ku} \rangle_L(t = 0) = 4.353°$C/W.

$$\tau_1 = 2307.2 \times 3.997 = 9,222 \text{ s}.$$

Now, after $t = 2$ hr $= 7{,}200$ s, we have

$$T_1(t = 2 \text{ hr}) = 293.15(\text{K}) + (318.15 - 293.15)(\text{K}) \, e^{-[7,200 \text{ (s)}]/[9,222 \text{ (s)}]}$$

$$= 304.6 \text{ K} = 31.45°\text{C}.$$

This is what is found from the use of a solver (e.g., MATLAB) by setting $\epsilon_{r,1} = 0$, and using a constant $\langle R_{ku} \rangle_L(t = 0)$, as shown in Figure Ex. 6.15(c). The effect of variable $\langle R_{ku} \rangle_L$ is less profound compared to that of surface emissivity.

**COMMENT**

The Rayleigh number at $t = 0$ is

$$\text{Ra}_L = \frac{9.807(\text{m/s}^2) \times \dfrac{1}{300}(1/\text{K}) \times 25(\text{K}) \times (0.3)^3(\text{m}^3)}{1{,}566 \times 10^{-5}(\text{m/s}^2) \times 2.257 \times 10^{-5}(\text{m/s}^2)}$$

$$= 6.243 \times 10^7.$$

Then

$$\text{Gr}_L = \frac{\text{Ra}_L}{\text{Pr}} = \frac{6.243 \times 10^7}{0.69} = 9.048 \times 10^7.$$

Then, we have

$$\frac{D}{L} = \frac{0.06}{0.30} = \frac{1}{5} = 0.2, \quad \text{which is smaller than} \quad \frac{35}{\text{Gr}_L^{1/4}} = \frac{35}{97.53} = 0.3589.$$

Table 6.4 requires that $D/L$ be larger than 0.3589. This shows that the condition for using the vertical plate correlation is not satisfied and a Nusselt number relation for the vertical cylinders should be used [16]. However, a large error is not expected.

Surface-radiation heat transfer results in more heat loss and a shorter elapsed time for reaching near equilibrium.

### 6.8.4 Hot-Wire Anemometry*

This section is found on the Web site www.cambridge.org/kaviany. An anemometer is an instrument used for the measurement of the velocity of moving fluids. In this section we introduce the hot-wire anemometer which uses the surface-convection heat transfer from a heated wire (and its relationship to the far-field fluid velocity) to indirectly measure the far-field velocity. The hot-wire anemometer probe is made of a probe body, which has two supporting needles, a wire sensor at the one end, and electrodes at the other end. We give the relevant relation and an example (Example 6.17).

## 6.9 Surface-Convection Evaporation Cooling*

This section is found on the Web site www.cambridge.org/kaviany. The resistance to vapor flow in a gas mixture is not zero and the vapor pressure is not uniform, because the gas contains other species (called inert or noncondensable, because of their higher boiling temperatures). A common example is water vapor in air (mostly nitrogen and oxygen). Then this mass diffusion limit has to be at rest. In this section we consider the combined heat and mass transfer limits to surface-convection evaporation. We give the relevant relations and give an example (Example 6.18).

## 6.10 Summary

We have examined flow and heat transfer of a semi-bounded fluid flowing over solid objects of various geometries. We have developed the concept of local surface-convection resistance $R_{ku}$ and showed how the fluid flow arrangement (with respect to solid surface) and flow structure influence $R_{ku}$. We also showed that in practice the surface-average resistance $\langle R_{ku}\rangle_L$ or $\langle R_{ku}\rangle_D$ is used. Then in Tables 6.3 to 6.6 we listed some of the available correlations for $\langle R_{ku}\rangle_L$ and $\langle R_{ku}\rangle_D$ (or the dimensionless conductance $\langle Nu\rangle_L$ or $\langle Nu\rangle_D$). In these correlations, we found dimensionless parameters such as $Re_L$ and $Pr$ in forced flows, $Ra_L$ and $Pr$ in thermobuoyant flows, and $Bo_L$ and $Ja_l$ in phase-density buoyant flows with phase change. By comparing $\langle R_{ku}\rangle_L$ (or $\langle R_{ku}\rangle_D$) for various flows with and without phase change, the desired flow and phase change strategy can be found.

We also examined various surface geometries, those for which the surface temperature $T_s$ is uniform, and those with a nonuniform surface temperature. For extended surfaces we used the concept of fin efficiency. We also considered transient heat transfer for objects with surface-convection heat transfer.

Finally, we addressed the simultaneous heat and mass transfer in surface-convection evaporation cooling and introduced the average surface-convection mass transfer resistance $\langle R_{Du} \rangle_L$. This resistance is related to its heat transfer counterpart $\langle R_{ku} \rangle_L$ through (6.180).

## 6.11 References*

This section is found on the Web site www.cambridge.org/kaviany.

## 6.12 Problems

### 6.12.1 Laminar Parallel-Flow Surface Convection

**PROBLEM 6.1. FUN**

Surface-convection heat transfer refers to heat transfer across the boundary separating a fluid stream and a condensed-phase (generally solid) volume, as rendered in Figure Pr. 6.1 for a stationary solid. Assume a uniform solid surface temperature $T_s$ and a far-field temperature $T_{f,\infty} \neq T_s$.

(a) Draw the heat flux vector tracking starting from $\mathbf{q}_{k,s}$ and ending with $\mathbf{q}_u$ away from the surface. At the surface use $\mathbf{q}_{k,f} = \mathbf{q}_{ku}$ and in the thermal boundary layer show both conduction and convection. Neglect radiation heat transfer.
(b) Qualitatively draw the fluid temperature distribution $T_f(y)$ at the location shown in Figure Pr. 6.1.
(c) Show the thermal boundary-layer thickness $\delta_\alpha$ on the same graph. Show the viscous boundary-layer thickness for $\mathrm{Pr} > 1$ and $\mathrm{Pr} < 1$.
(d) Draw the thermal circuit diagram for the solid surface and write the expression for the average heat transfer rate $\langle Q_{ku} \rangle_L$.
(e) What should the average surface-convection resistance $\langle R_{ku} \rangle_L$ be for $T_s$ to be the same as $T_{f,\infty}$, i.e., for the solid surface temperature to be made equal to the fluid stream far-field temperature?
(f) If $T_{f,\infty} \neq T_s$ what should $\langle R_{ku} \rangle_L$ be so there is no surface-convection heat transfer (ideal insulation)? vspace*-6pt

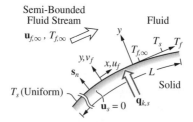

Figure Pr. 6.1. A rendering of a semi-bounded fluid stream passing over a solid surface with $T_{f,\infty} \neq T_s$.

**PROBLEM 6.2. FUN**

A surface, treated as a semi-infinite plate and shown in Figure Pr. 6.2, is to be heated with a forced, parallel flow. The fluids of choice are (i) mercury, (ii) ethylene glycol (antifreeze), and (iii) air.

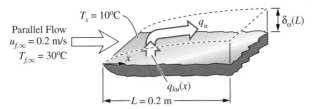

Figure Pr. 6.2. A semi-bounded fluid stream exchanging heat with its semi-infinite plate bounding surface.

For the tailing edge of the plate $x = L$, do the following.

(a) Determine the local rate of heat transfer per unit area $q_{ku}(W/m^2)$.
(b) Determine the thermal boundary-layer thickness $\delta_\alpha(mm)$. Use the Nusselt number relation for $Pr \neq 0$.
(c) For mercury, also use the relation for Nusselt number for a zero viscosity (i.e., $Pr = 0$) and compare the results with that obtained from the nonzero viscosity relations.

$T_s = 10°C$, $T_{f,\infty} = 30°C$, $u_{f,\infty} = 0.2$ m/s, $L = 0.2$ m.
Evaluate the properties at $T = 300$ K, from Tables C.22 and C.23.

## PROBLEM 6.3. FAM

The top surface of a microprocessor chip, which is modeled as a semi-infinite plate, is to be cooled by forced, parallel flow of (i) air, or (ii) liquid Refrigerant-12. The idealized surface is shown in Figure Pr. 6.3. The effect of the surfaces present upstream of the chip can be neglected.

(a) Determine the surface-convection heat transfer rate $\langle Q_{ku} \rangle_L(W)$.
(b) Determine the thermal boundary-layer thickness at the tail edge of the chip $\delta_\alpha(mm)$.

Evaluate the properties at $T = 300$ K.

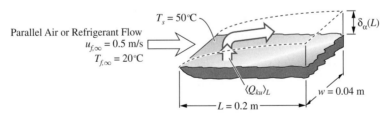

Figure Pr. 6.3. Surface of a microprocessor is cooled by a semi-bounded fluid stream.

## PROBLEM 6.4. FUN

As discussed in Section 6.2.5, the stream function $\psi$, for a two-dimensional, laminar fluid flow $(u_f, v_f)$, expressed in the Cartesian coordinate $(x, y)$, is defined through

$$u_f \equiv \frac{\partial \psi}{\partial y}, \quad v_f \equiv -\frac{\partial \psi}{\partial x}.$$

Show that this stream function satisfies the continuity equation (6.37).

**PROBLEM 6.5. FUN**

As discussed in Section 6.2.5, the two-dimensional, $(x, y)$, $(u_f, v_f)$, laminar steady viscous, boundary-layer momentum equation (6.36) can be reduced to an ordinary differential equation using a dimensionless similarity variable

$$\eta \equiv y \left( \frac{u_{f,\infty}}{v_f x} \right)^{1/2}$$

and a dimensionless stream function

$$\psi^* \equiv \frac{\psi}{(v_f u_{f,\infty} x)^{1/2}}, \quad u_f \equiv \frac{\partial \psi}{\partial y}, \quad v_f \equiv -\frac{\partial \psi}{\partial x}.$$

(a) Show that the momentum equation (6.36) reduces to

$$2 \frac{d^3 \psi^*}{d\eta^3} + \psi^* \frac{d^2 \psi^*}{d\eta^2} = 0.$$

This is called the Blasius equation.

(b) Show that energy equation (6.35) reduces to

$$\frac{d^2 T_f^*}{d\eta^2} + \frac{1}{2} \mathrm{Pr} \psi^* \frac{d T_f^*}{d\eta} = 0, \quad T_f^* = \frac{T_f - T_{f,\infty}}{T_s - T_{f,\infty}}.$$

**PROBLEM 6.6. FUN**

Use a solver to integrate the dimensionless transformed boundary-layer momentum equation, i.e., the third-order, ordinary Blasius differential equation

$$2 \frac{d^3 \psi^*}{d\eta^3} + \psi^* \frac{d^2 \psi^*}{d\eta^2} = 0,$$

subject to surface and far-field mechanical conditions

$$\text{at } \eta = 0: \quad \frac{d\psi^*}{d\eta} = \psi^* = 0$$

$$\text{for } \eta \rightarrow \infty: \quad \frac{d\psi^*}{d\eta} = 1.$$

Plot $\psi^*$, $d\psi^*/d\eta = u_f/u_{f,\infty}$, and $d^2\psi^*/d\eta^2$, with respect to $\eta$.

Note that with an initial-value problem solver, such as MATLAB, the second derivative of $\psi^*$ at $\eta = 0$ must be guessed. This guess is adjusted till $d\psi^*/d\eta$ becomes unity for large $\eta$.

Hint: $d^2\psi^*/d\eta^2(\eta = 0)$ is between 0.3 to 0.4.

## 6.12.2 Turbulent Parallel-Flow Surface Convection

**PROBLEM 6.7. FAM**

During part of the year, the automobile windshield window is kept at a temperature significantly different than that of the ambient air. Assuming that the flow and heat transfer over the windshield can be approximated as those for parallel flow over a semi-infinite, flat plate, examine the role of the automobile speed on the surface-convection heat transfer from the window. These are shown in Figure Pr. 6.7.

(a) To that end, determine the average Nusselt number $\langle Nu \rangle_L$.
(b) Determine the average surface-convection thermal resistance $A_{ku} \langle R_{ku} \rangle_L [°C/(W\text{-}m^2)]$.
(c) Determine the surface-averaged rate of surface-convection heat transfer $\langle Q_{ku} \rangle_L (W)$.

Consider automobile speeds of 2, 20, and 80 km/hr. Comment on the effects of the flow-regime transition and speed on the surface-convection heat transfer.

The ambient air is at $-10°C$ and the window surface is at $10°C$. The window is 1 m long along the flow direction and is 2.5 m wide.

Use the average temperature between the air and the window surface to evaluate the thermophysical properties of the air.

(i) Physical Model

(ii) An Approximation of Heat Transfer from Windshield

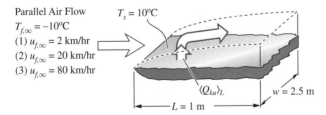

Figure Pr. 6.7. (i) Fluid flow and heat transfer over an automobile windshield window. (ii) Its idealization as parallel flow over a semi-infinite plate.

**PROBLEM 6.8. FAM.S**

A square flat surface with side dimension $L = 40$ cm is at $T_s = 120°C$. It is cooled by a parallel air flow with far-field velocity $u_{f,\infty}$ and far-field temperature $T_{f,\infty} = 20°C$.

(a) Use a solver (such as MATLAB) to plot the variation of the averaged surface-convection heat transfer rate $\langle Q_{ku} \rangle_L$(W) with respect to $u_{f,\infty}$(m/s) from zero up to the sonic velocity. Use (3.20) to find the sonic velocity.

(b) Determine the air velocity needed to obtain $\langle q_{ku} \rangle_L = 1{,}200$ W/m$^2$.

## PROBLEM 6.9. FAM

On a clear night, a water layer formed on a paved road can freeze due to radiation heat losses to the sky. The water and the pavement are at the freezing temperature $T_1 = 0°$C. The water surface behaves as a blackbody and radiates to the sky at an apparent temperature of $T_{sky} = 250$ K. The ambient air flows parallel and over the water layer at a speed $u_{f,\infty} = 9$ m/s and temperature $T_{f,\infty}$, which is greater than the water temperature. Assume that the surface convection is modeled using a surface that has a length $L = 2$ m along the flow and a width $w = 1$ m (not shown in the figure) perpendicular to the flow. These are shown in Figure Pr. 6.9.

(a) Draw the thermal circuit diagram.

(b) Determine the maximum ambient temperature below which freezing of the water layer occurs.

(c) When a given amount of salt is added to the water or ice, the freezing temperature drops by 10°C. If the water surface is now at $-10°$C, will freezing occur? Use the property values found for (b), the given $u_{f,\infty}$ and $T_{sky}$, and the freestream air temperature found in (b).

Neglect the heat transfer to the pavement and evaluate the air properties at $T = 273.15$ K (Table C.22).

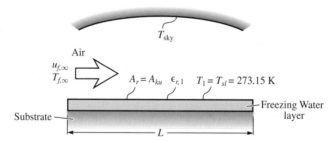

Figure Pr. 6.9. Radiation cooling of a thin water film and its surface-convection heating.

### 6.12.3 Impinging-Jet Surface Convection

## PROBLEM 6.10. DES

A square flat plate, with dimensions $a \times a$, is being heated by a thermal plasma (for a coating process) on one of its sides. To prevent meltdown and assist in the coating process, the other side is cooled by impinging air jets. This is shown in Figure Pr. 6.10. In the design of the jet cooling, a single, large-diameter nozzle [Figure Pr. 6.10(i)], or nine smaller diameter nozzles [Figure Pr. 6.10(ii)] are to be used.

For each design, do the following.

(a) Determine the average Nusselt number $\langle Nu \rangle_L$.

(b) Determine the average surface-convection thermal resistance $A_{ku}\langle R_{ku}\rangle_L$ [°C/(W/m²)].

(c) Determine the rate of surface-convection heat transfer $\langle Q_{ku}\rangle_L(W)$.

$\quad a = 30$ cm, $T_s = 400$°C, $T_{f,\infty} = 20$°C.

$\quad$ Single nozzle: $D = 3$ cm, , $L_n = 6$ cm, $L = 15$ cm, $\langle u_f \rangle_A = 1$ m/s.

$\quad$ Multiple nozzles: $D = 1$ cm, $L_n = 2$ cm, $L = 5$ cm, $\langle u_f \rangle_A = 1$ m/s.

Use the average temperature between the air and the surface to evaluate the thermophysical properties of the air.

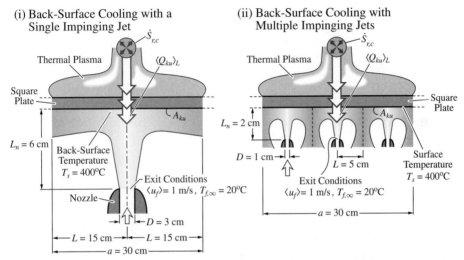

**(i) Back-Surface Cooling with a Single Impinging Jet**

**(ii) Back-Surface Cooling with Multiple Impinging Jets**

Figure Pr. 6.10. (i) A single impinging jet used for surface cooling. (ii) Multiple impinging jets.

## PROBLEM 6.11. FUN

Permanent damage occurs to the pulp of a tooth initially at $T(t = 0) = 37$°C, when it reaches a temperature $T_p = 41$°C. Therefore, to prevent nerve damage, a water coolant must be constantly applied during many standard tooth drilling operations. In one such operation, a drill, having a frequency $f = 150$ Hz, a burr diameter $D_b = 1.2$ mm, a tooth contact area $A_c = 1.5 \times 10^{-7}$ m², and a coefficient of friction between the drill burr and the tooth $\mu_F = 0.4$, is used to remove an unwanted part of the tooth. The contact force between the drill burr and the tooth is $F = 0.05$ N. During the contact time, heat is generated by surface friction heating. In order to keep the nerves below their threshold temperature, the tooth surface must be maintained at $T_s = 45$°C by an impinging jet that removes 80% of the generated heat. The distance between the jet and the surface $L_n$ is adjustable. Use the dimensions shown in Figure Pr. 6.11.

(a) Draw the thermal circuit diagram.

(b) Write the surface energy equation for the tooth surface.

(c) Determine the location $L_n$ of the jet that must be used in order to properly cool the tooth. $T_{f,\infty} = 20°C$, $\langle u_f \rangle = 0.02$ m/s, $D = 1.5$ mm, $L = 4$ mm, $\mu_F = 0.4$, $f = 150$ 1/s, $\Delta u_i = 2\pi f R_b$, $p_c = F_c/A_c$, $F_c = 0.05$ N, $A_c = 1.5 \times 10^{-7}$ m$^2$, $D_b = 2R_b = 1.2$ mm.

Use the same surface area for heat generation and for surface convection (so surface area $A_{ku}$ will not appear in the final expression used to determine $L_n$). Determine the water properties at $T = 293$ K.

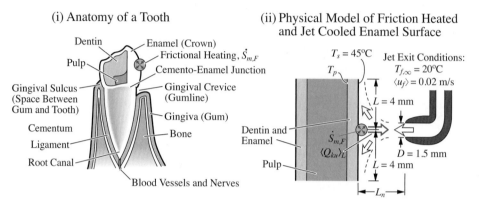

(i) Anatomy of a Tooth                  (ii) Physical Model of Friction Heated and Jet Cooled Enamel Surface

Figure Pr. 6.11. (i) Anatomy of a tooth. (ii) Physical model of cooling of a tooth during drilling.

## PROBLEM 6.12. FAM

Heat-activated, dry thermoplastic adhesive films are used for joining surfaces. The adhesive film can be heated by rollers, hot air, radio-frequency and microwaves, or ultrasonics. Consider a flat fabric substrate to be coated with a polyester adhesive film with the film, heated by a hot air jet, as shown in Figure Pr. 6.12. The film is initially at $T_1(t = 0)$. The thermal set temperature is $T_{sl}$. Assume that the surface-convection heat transfer results in the rise in the film temperature with no other heat transfer.

(a) Draw the thermal circuit diagram.
(b) Determine the elapsed time needed to reach $T_{sl}$, for the conditions given below.

$L_n = 4$ cm, $L = 10$ cm, $D = 2$ cm, $\langle u_f \rangle = 1$ m/s, $l = 0.2$ mm, $T_{f,\infty} = 200°C$, $T_{sl} = 120°C$, $T_1(t = 0) = 20°C$.

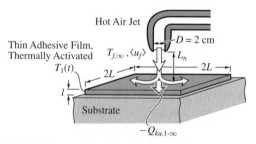

Figure Pr. 6.12. A heat-activated adhesive film heated by a hot air jet.

Determine the air properties at $T = 350$ K. For polyester, use Table C.17 and the properties of polystyrene.

## PROBLEM 6.13. DES

A pure aluminum plate is to be rapidly cooled from $T_1(t = 0) = 40°C$ to $T_1(t) = 20°C$. The plate has a half-length $L = 12$ cm and thickness $w = 0.2$ cm. The plate is to be cooled using water by placing it at a distance $L_n = 10$ cm from a faucet with diameter $D = 2$ cm. The water leaves the faucet at a temperature $T_{f,\infty} = 5°C$ and velocity $\langle u \rangle_A = 1.1$ m/s. There are two options for the placement of the plate with respect to the water flow. The plate can be placed vertically, so the water flows parallel and on both sides of the plate. Then the water layer will have a thickness $l = 2$ mm on each side of the plate [shown in Figure Pr. 6.13(i)]. Alternately, it can be placed horizontally with the water flowing as a jet impingement [shown in Figure Pr. 6.13(ii)].

(a) Draw the thermal circuit diagram.
(b) Determine which of the orientations gives the shorter cooling time.

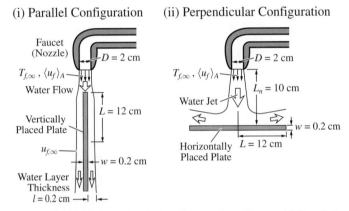

Figure Pr. 6.13. Cooling of an aluminum plate under a faucet. (i) Parallel configuration. (ii) Perpendicular configuration.

Assume that the results for impinging jets can be used here, even though the jet fluid (water) is not the same as the ambient (air) fluid. Use the water as the only fluid present. Assume a uniform plate temperature. Estimate the parallel, far-field velocity $u_{f,\infty}$ using the mass flow rate out of the faucet. This approximate flux is assumed uniform over the rectangular flow cross section ($l \times 2L$) and is assumed to be flowing over a square surface ($2L \times 2L$).

### 6.12.4 Thermobuoyant-Flow Surface Convection

## PROBLEM 6.14. FAM

A bottle containing a cold beverage is awaiting consumption. During this period, the bottle can be placed vertically or horizontally, as shown in Figure Pr. 6.14. Assume that the bottle can be treated as a cylinder of diameter $D$ and length $L$.

We wish to compare the surface-convection heat transfer to the bottle when it is (i) standing vertically, or (ii) placed horizontally.

(a) Determine the average Nusselt numbers $\langle Nu \rangle_L$ and $\langle Nu \rangle_D$.

(b) Determine the average surface-convection thermal resistances $A_{ku}\langle R_{ku}\rangle_L$ $[°C/(W/m^2)]$ and $A_{ku}\langle R_{ku}\rangle_D[°C/(W/m^2)]$.

(c) Determine the rates of surface-convection heat transfer $\langle Q_{ku}\rangle_L(W)$ and $\langle Q_{ku}\rangle_D(W)$.

For the vertical position, the surface-convection heat transfer is approximated using the results of the vertical plate, provided that the boundary-layer thickness $\delta_\alpha$ is much less than the bottle diameter $D$.

$D = 10$ cm, $L = 25$ cm, $T_s = 4°C$, $T_{f,\infty} = 25°C$.

Neglect the end areas. Use the average temperature between the air and the surface to evaluate the thermophysical properties of the air.

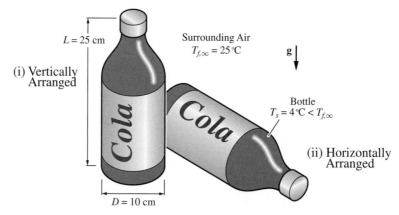

Figure Pr. 6.14. Thermobuoyant flow and heat transfer from beverage bottles. (i) Standing vertically. (ii) Placed horizontally.

## PROBLEM 6.15. FAM

The fireplace can provide heat to the room through surface convection and surface radiation from that portion of the fireplace wall heated by the combustion products exiting through a chimney behind the wall. This heated area is marked on the fireplace wall in Figure Pr. 6.15. Assume this portion of the wall (including the fireplace) is maintained at a steady, uniform temperature $T_s$. The surface convection is by a thermobuoyant flow that can be modeled as the flow adjacent to a heated vertical plate with length $L$. The surface-radiation exchange is between this heated portion of the wall and the remaining surfaces in the room. Assume that all the remaining wall surfaces are at a steady uniform temperature $T_w$.

(a) Draw the thermal circuit diagram.

(b) Determine the surface-convection heat transfer rate.

(c) Determine the surface-radiation heat transfer rate.

(d) Assume the fire provides an energy conversion rate due to combustion of $\dot{S}_{r,c} = 19,500$ W. (This would correspond to a large 5 kg log of wood burning at a constant rate to total consumption in one hour.) Determine the efficiency of the fireplace as a room-heating system through the heated wall only (efficiency is defined as the ratio of the total surface heat transfer rate to the rate of energy conversion $\dot{S}_{r,c}$).

$T_s = 32°C$, $T_{f,\infty} = 20°C$, $T_w = 20°C$, $\epsilon_{r,s} = 0.8$, $\epsilon_{r,w} = 0.8$, $w = 3$ m, $L = 4$ m, $a = 6$ m.

Determine the air properties at 300 K (Table C.22).

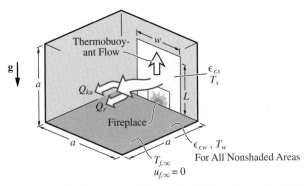

Figure Pr. 6.15. Surface convection and radiation from a heated wall.

## PROBLEM 6.16. FUN

As discussed in Section 6.5.1, the two-dimensional $(x, y)$, $(u_f, v_f)$, laminar viscous, thermobuoyant boundary-layer (for vertical, uniform surface temperature plate) momentum equation (6.80) can be reduced to an ordinary differential equation using a dimensionless similarity variable

$$\eta = \frac{y}{x} \left( \frac{\text{Gr}_x}{4} \right)^{1/4},$$

and a dimensionless stream function

$$\psi^* = \frac{\psi}{4v_f \left( \dfrac{\text{Gr}_x}{4} \right)^{1/4}}, \quad u_f = \frac{\partial \psi}{\partial y}, \quad v_f = -\frac{\partial \psi}{\partial x}, \quad \text{Gr}_x = \frac{g\beta_f (T_s - T_{f,\infty})x^3}{v_f^2}.$$

(a) Show that the momentum equation (6.80) reduces to

$$\frac{d^3\psi^*}{d\eta^3} + 3\psi^* \frac{d^2\psi^*}{d\eta^2} - 2\left( \frac{d\psi^*}{d\eta} \right)^2 + T_f^* = 0 \quad T_f^* = \frac{T_f - T_\infty}{T_s - T_{f,\infty}}.$$

(b) Show that the energy equation (6.79) reduces to

$$\frac{d^2 T_f^*}{d\eta^2} + 3\text{Pr}\psi^* \frac{dT_f^*}{d\eta} = 0 \quad \text{Pr} = \frac{v_f}{\alpha_f}.$$

## PROBLEM 6.17. FUN

Use a solver to integrate the dimensionless, transformed coupled boundary-layer momentum and energy equations for thermobuoyant flow, i.e.,

$$\frac{d^3 \psi^*}{d\eta^3} + 3\psi^* \frac{d^2 \psi^*}{d\eta^2} - 2 \left( \frac{d\psi^*}{d\eta} \right)^2 + T_f^* = 0$$

$$\frac{d^2 T_f^*}{d\eta^2} + 3\Pr\psi^* \frac{dT_f^*}{d\eta} = 0,$$

subject to the surface and far-field thermal and mechanical conditions

$$\text{at} \quad \eta = 0 : \frac{d\psi^*}{d\eta} = \psi^* = 0, \quad T_f^* = 1$$

$$\text{for} \quad \eta \to \infty : \frac{d\psi^*}{d\eta} = 0, \quad T_f^* = 0.$$

Use $\Pr = 0.72$ and plot $\psi^*$, $d\psi^*/d\eta$, $d^2\psi^*/d\eta^2$, $T_f^*$, and $dT_f^*/d\eta$, with respect to $\eta$.

Note that with an initial-value problem solver such as MATLAB, the second derivative of $\psi^*$ and first derivative of $T_f^*$ at $\eta = 0$ must be guessed. The guesses are adjusted till $d^2\psi^*/d\eta^2$ becomes zero for large $\eta$.

Use $d^2\psi^*/d\eta^2(\eta = 0) = 0.6760$ and $dT_f^*/d\eta^*(\eta = 0) = -0.5064$.

## PROBLEM 6.18. FAM

An aluminum flat sheet, released from hot pressing, is to be cooled by surface convection in an otherwise quiescent air, as shown in Figure Pr. 6.18. The sheet can be placed vertically [Figure Pr. 6.18(i)] or horizontally [Figure Pr. 6.18(ii)]. Both sides of the sheet undergo heat transfer. For the horizontal arrangement, treat the lower surface the same as the top, using the Nusselt number relations listed in Table 6.5 for the top surface. This is just a rough approximation.

(a) Draw the thermal circuit diagram.
(b) Determine the surface-convection heat transfer rate $\langle Q_{ku} \rangle_L$ for the two arrangements and for the conditions given below.

$w = L = 0.4$ m, $T_{f,\infty} = 25°$C, $T_s = 430°$C.
Use air properties at $\langle T_f \rangle_\delta = (T_s + T_{f,\infty})/2$.

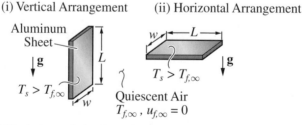

Figure Pr. 6.18. A sheet of aluminum is cooled in an otherwise quiescent air. (i) Vertical placement. (ii) Horizontal placement.

### 6.12.5  Boiling and Condensation Surface Convection

**PROBLEM 6.19. FAM**

Water, initially at $T = 12°C$, is boiled in a portable heater at one atm pressure, i.e., it has its temperature raised from $12°C$ to $100°C$. The heater has a circular, nickel surface with $D = 5$ cm and is placed at the bottom of the water, as shown in Figure Pr. 6.19. The amount of water is 2 kg (which is equivalent to 8 cups) and the water is to be boiled in 6 min.

(a) Determine the time-averaged (constant with time) electrical power needed $\dot{S}_{e,J}(W)$ assuming no heat losses.

(b) Determine the critical heat flux $q_{ku,CHF}(W/m^2)$ for this fluid and then comment on whether the required electrical power per unit area is greater or smaller than this critical heat flux. Note that the surface-convection heat transfer rate (or the electrical power) per unit area should be less than the critical heat flux; otherwise, the heater will burn out.

(c) Determine the required surface temperature $T_s$, assuming nucleate boiling. Here, assume that the effect of the liquid subcooling on the surface-convection heat transfer rate is negligible. When the subcooling is not negligible (i.e., the water is at a much lower temperature than the saturation temperature $T_{lg}$), the larger temperature gradient between the surface and the liquid and the collapse of the bubbles away from the surface, will increase the rate of heat transfer.

(d) Determine the average surface-convection thermal resistance $A_{ku}\langle R_{ku}\rangle_L$ $[°C/(W/m^2)]$ and the average Nusselt number $\langle Nu\rangle_L$.

The properties for water are given in Table C.23.

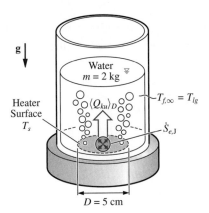

Figure Pr. 6.19. An electric water heater using boiling surface-convection heat transfer.

**PROBLEM 6.20. FAM**

Steam is produced by using the flue gas from a burner to heat a pool of water, as shown in Figure Pr. 6.20. The water and the flue gas are separated by a plate. On the flue-gas side (modeled as air), the measurements show that the flue-gas, far-field temperature is $T_{f,\infty} = 977°C$ and flows parallel to the surface at $u_{f,\infty} = 2$ m/s, while the flue-gas side surface of the plate is at $T_{s,2} = 110°C$. The heat flows through the plate (having a length $L = 0.5$ m and a width $w$) into the water (water is at the saturated temperature $T_{lg} = 100°C$ and is undergoing nucleate boiling).

(a) Draw the thermal circuit diagram.
(b) Determine the surface temperature of the plate on the water side $T_{s,1}$. For the nucleate boiling Nusselt number correlation, use $a_s = 0.013$.

Evaluate the flue-gas properties as those of air at the flue-gas film temperature (i.e., at the average temperature between the flue-gas side surface temperature of the plate and the flue-gas, far-field temperature). For water, use the saturation liquid-vapor properties given in Table C.26.

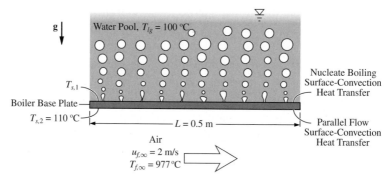

Figure Pr. 6.20. A solid surface separating flue gas and water with a large difference between the far-field temperatures causing the water to boil and produce steam.

**PROBLEM 6.21. FAM**
A glass sheet is vertically suspended above a pan of boiling water and the water condensing over the sheet and raises its temperature. This is shown in Figure Pr. 6.21.

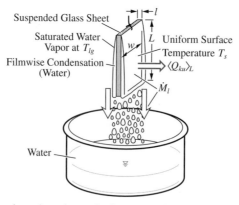

Figure Pr. 6.21. A glass sheet is vertically suspended in a water-vapor ambient and the heat released by condensation raises the sheet temperature.

Filmwise condensation and uniform sheet temperature $T_s$ are assumed. Note that the condensate is formed on both sides of the sheet. Also assume a steady-state heat transfer.

(a) Draw the thermal circuit diagram.

(b) Determine the heat transfer rate $\langle Q_{ku}\rangle_L$, for the conditions given below, for each side.

(c) Determine the condensate flow rate $\dot{M}_l = -\dot{M}_{lg}$, for each side.

(d) Is this a laminar film condensate flow?

$l = 1$ mm, $L = 15$ cm, $w = 15$ cm, $T_{lg} = 100°$C, $T_s = 40°$C.

Use the saturated water properties at $T_{lg}$.

## PROBLEM 6.22. FAM

To boil water by electrical resistance heating would require a large electrical power per unit area of the heater surface. For a heater having a surface area for surface convection $A_{ku}$, this power from (2.28) is

$$\dot{S}_{e,J} = \frac{\Delta\varphi^2}{R_e}, \qquad A_{ku} = \pi DL,$$

where $\Delta\varphi$ is the applied voltage, $R_e$ is the electrical resistance, and $D$ and $L$ are the diameter and length of the heater. Consider the water-boiler shown in Figure Pr. 6.22. Using Figure 6.20(b), assume a surface superheat $T_s - T_{lg} = 10°$C is needed for a significant nucleate boiling. Then use the nucleate boiling correlation of Table 6.6.

(a) Draw the thermal circuit diagram for the heater.

(b) Determine the surface-convection heat transfer rate $\langle Q_{ku}\rangle_L$, for the conditions given below.

(c) For an electrical resistance of $R_e = 20$ ohm, what should be the applied voltage $\Delta\varphi$, and the electrical current $J_e$ ?

$a_s = 0.01$, $D = 0.5$ cm, $L = 12$ cm, $T_{lg} = 100°$C.

Use saturated water properties at $T = T_{lg}$.

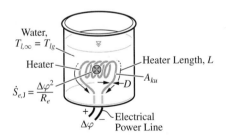

Figure Pr. 6.22. A Joule heater is used to boil water.

### 6.12.6 Impinging-Droplets Surface Convection

## PROBLEM 6.23. FAM

To reduce the air conditioning load, the roof of a commercial building is cooled by a water spray. The roof is divided into segments with each having a dedicated sprinkler, as shown in Figure Pr. 6.23. Assume that the impinging-droplet film evaporation relation of Table 6.6 can be used here.

(a) Draw the thermal circuit diagram for the panel surface.
(b) Using the conditions given below, determine the rate of surface-convection heat transfer $\langle Q_{ku} \rangle_L$ from the roof panel.

$L = 4$ m, $T_{f,\infty} = 30°$C, $T_s = 210°$C, $\langle D \rangle = 100$ $\mu$m, $\langle u_d \rangle = 2.5$ m/s, $\langle \dot{m}_d \rangle / \rho_{l,\infty} = 10^{-3}$ m/s.

Evaluate the water properties at $T = 373$ K.

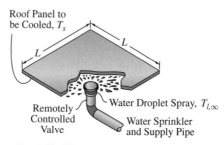

Figure Pr. 6.23. Water-spray cooling of a roof panel.

## PROBLEM 6.24. FAM

In using water evaporation in surface-convection heat transfer, compare pool boiling by saturated water ($T_{l,\infty} = T_{lg}$), as shown in Figure 6.20(b), and droplet impingement by subcooled water droplets ($T_{l,\infty} < T_{lg}$) as shown in Figure 6.26.

Select between water pool boiling [Figure Pr. 6.24(i)] and droplet impingement [Figure Pr. 6.24(ii)], by comparing peak, and minimum, surface-convection heat flux.

(i) Pool Boiling: Vapor-Film Regime          (ii) Impinging Droplets: Vapor-Film Regime

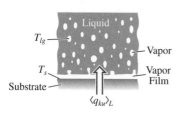   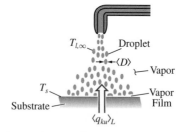

Figure Pr. 6.24. Selection of a water evaporation surface cooling method, between (i)pool boiling and (ii)impinging droplets.

For pool boiling, use the peak in $\langle Q_{ku} \rangle_L$ as given by the critical heat flux, i.e., (6.100), and for the minimum use (6.101). For impinging droplets, use the peak shown in Figure 6.26, which is nearly independent of the droplet mass flux, in the high mass flux regime, and for the minimum use the approximate correlation (6.116). The correlations are also listed in Table 6.6.

Pool boiling: $T_{lg} = 100°$C, $T_s = 300°$C.

Impinging droplets: $\langle \dot{m}_d \rangle = 1.43 \; kg/m^2\text{-s}$, $\langle u_d \rangle = 3.21 \; m/s$, $\langle D \rangle = 480 \; \mu m$, $T_{l,\infty} = 20°C$, $T_{lg} = 100°C$, $T_s = 300°C$, evaluate properties at 310 K.

Note that not all these conditions are used in every case considered.

### 6.12.7 Surface Convection with Other Flows and Geometries

**PROBLEM 6.25. FAM**

A person caught in a cold cross wind chooses to curl up (crouching as compared to standing up) to reduce the surface-convection heat transfer from his clothed body. Figure Pr. 6.25 shows two idealized geometries for the person while crouching [Figure Pr. 6.25(i)] and while standing up [Figure Pr. 6.25(ii)].

(a) Draw the thermal circuit diagram and determine the heat transfer rate for the idealized spherical geometry.

(b) Draw the thermal circuit diagram and determine the heat transfer rate for the idealized cylindrical geometry. Neglect the heat transfer from the ends of the cylinder.

(c) Additional insulation (with thermal conductivity $k_i$) is to be worn by the standing position to reduce the surface-convection heat transfer to that equal to the crouching position. Draw the thermal circuit diagram and determine the necessary insulation thickness $L$ to make the two surface convection heat transfer rates equal. Assume that $T_1$ and the surface-convection resistance for the cylinder will remain the same as in part (b).

(d) What is the outside-surface temperature $T_2$ of the added insulation?

$D_s = 50 \; cm$, $D_c = 35 \; cm$, $L_c = 170 \; cm$, $T_1 = 12°C$, $T_{f,\infty} = -4°C$, $u_{f,\infty} = 5 \; m/s$, $k_i = 0.1 \; W/m\text{-K}$.

Use air properties (Table C.22) at $T = 300$ K.

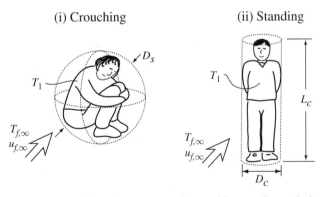

Figure Pr. 6.25. Two positions by a person in a cold cross flow of air. (i) Crouching position. (ii) Standing position.

**PROBLEM 6.26. FUN**

A thermocouple is placed in an air stream to measure the stream temperature, as shown in Figure Pr. 6.26. The steady-state temperature of the thermocouple bead of diameter $D$ is determined through its surface-convection (as a sphere in a semi-bounded fluid stream) and surface-radiation heat transfer rates.

(a) Draw the thermal circuit diagram.
(b) Determine the thermocouple bead temperature $T_1$.
(c) Comment on the difference between $T_1$ and $T_{f,\infty}$. How can the difference (measurement error) be reduced?

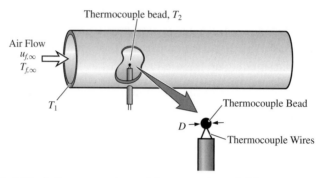

Figure Pr. 6.26. A thermocouple used for measuring a fluid stream temperature.

$u_{f,\infty} = 2$ m/s, $T_{f,\infty} = 600°$C, $T_1 = 400°$C, $\epsilon_{r,w} = 0.9$, $\epsilon_{r,s} = 0.8$, $D = 1$ mm.

Neglect the heat transfer to and from the wires and treat the surface-convection heat transfer to the thermocouple bead as a semi-bounded fluid flow over a sphere.

Assume the tube length $L$ is large (i.e, $L \rightarrow \infty$). Evaluate the fluid properties at $T = 350$ K (Table C.22).

Numerical hint: The thermocouple bead temperature should be much closer to the air stream temperature than to the tube surface temperature. For iterations, start with a a guess of $T = 820$ K.

## PROBLEM 6.27. FAM

To prevent the flame from blow-off by a cross wind, a lighter is desired with a flame anchor (i.e., flame holder) in the form of a winding wire placed in the air-fuel stream undergoing combustion. This is shown in Figure Pr. 6.27(i). The wire retains (through its heat storage) a high temperature and will maintain the flame around it, despite a large, intermittent cross flow. Assume that the combustion of the $n$-butane in air is complete before the gas stream at temperature $T_{f,\infty}$ and velocity $u_{f,\infty}$ reach the flame holder. Treat the flame holder as a long cylinder with steady-state, surface-convection heating and surface-radiation cooling. The simplified heat transfer model is also shown, in Figure Pr. 6.27(ii).

(a) Draw the thermal circuit for the flame holder.
(b) Determine the flame holder temperature $T_s$ for the following conditions.

$T_{f,\infty} = 1,300°$C, $D = 0.3$ mm, $T_{surr} = 30°$C, $\epsilon_{r,s} = 0.8$.

Use the adiabatic, laminar flame speed, for the stoichiometric $n$-butane in air, from Table C.21(a), for the far-field fluid velocity $u_{f,\infty}$.

You do not need to use tables or graphs for the view factors. Assume the properties of the combustion products are those of air at $T = 900$ K.

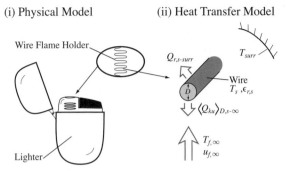

(i) Physical Model (ii) Heat Transfer Model

Figure Pr. 6.27. (i) A winding-wire flame holder used in a lighter. (ii) A simplified heat transfer model for the wire.

## PROBLEM 6.28. FUN

Consider measuring the temperature $T_{f,\infty}$ of an air stream using a thermocouple. A thermocouple is a junction made of two dissimilar materials (generally metals, as discussed in Section 2.3.2), as shown in Figure Pr. 6.28. The wires are electrically (and if needed, thermally) insulated. The wire may not be in thermal equilibrium with the stream. This can be due to the nonuniformity of temperature along the wire. Then the temperature of the thermocouple bead (i.e., its tip) $T_{s,L}$ may not be close enough to $T_{f,\infty}$, for the required accuracy. Consider an air stream with a far-field temperature $T_{f,\infty} = 27°C$ and a far-field velocity $u_{f,\infty} = 5$ m/s. Assume that the bare (not insulated) end of the wire is at temperature $T_{s,0} = 15°C$. Consider one of the thermocouple wires made of copper, having a diameter $D = 0.2$ mm and a bare-wire length $L = 5$ mm.

Using the extended surface analysis, determine the expected uncertainty $T_{f,\infty} - T_{s,L}$. Evaluate the properties of air at 300 K.

(i) Idealized Thermocouple Wire Model    (ii) A Thermocouple for Measurement of $T_{f,\infty}$

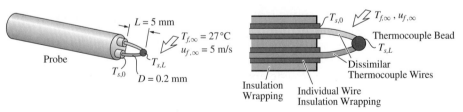

Figure Pr. 6.28. A thermocouple junction used for temperature measurement in an air stream. (i) Idealized thermocouple wire model. (ii) Thermocouple for measurement of far-field fluid temperature $T_{f,\infty}$.

## PROBLEM 6.29. FUN

Shape-memory actuation devices are capable of recovering a particular shape upon heating above their transformation temperature. These alloys can be made of nickel and titanium and display two types of material properties. When at a temperature below their transformation temperature $T_t$, they display the properties of martensite and, when above this temperature they display the properties of austenite. The NiTi alloy shown in Figure Pr. 6.29 is shaped as a spring and deforms from a compressed state [Figure Pr. 6.29(i)] to an extended state [Figure Pr. 6.29(ii)] when heated above its transformation temperature.

(i) Martensite $T_1 < T_t$

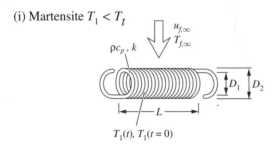

(ii) Austenite $T_1 \geq T_t$

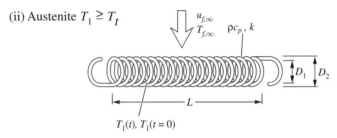

Figure Pr. 6.29. Springs made of shape-memory alloy and used for thermal actuation.

This spring is being tested for its suitability for use in the closing of heating ducts within a desired elapsed time. In order to close the duct, the spring must extend a lightweight-beam induced closing mechanism within 20 s. The air flow within the heating duct has a velocity $u_{f,\infty}$ and temperature $T_{f,\infty}$, near the spring. The spring has a length $L$, an outer diameter $D_2$, an inner diameter $D_1$, and an initial temperature $T_1(t = 0)$.

Assume that the entire spring is at a uniform temperature $T_1(t)$ and the dominant surface heat transfer is by surface convection.

(a) Draw the thermal circuit diagram.
(b) Use the properties of the low-temperature form of NiTi listed below and determine if the spring will activate during the time allowed.

Martensite: $\rho = 6{,}450$ kg/m$^3$, $k = 8.6$ W/m-°C, $c_p = 837.36$ J/kg-°C, $T_1(t = 0) = 21$°C, $T_t = 50$°C, $L = 4$ cm, $D_2 = 0.5$ cm, $D_1 = 0.4$ cm, $u_{f,\infty} = 5$ m/s, $T_{f,\infty} = 77$°C.
Evaluate the properties of air at $T = 350$ K.

**PROBLEM 6.30. FAM**
A steel cylindrical rod is to be cooled by surface convection using an air stream, as shown in Figure Pr. 6.30. The rod can be placed perpendicular [Figure Pr. 6.30(i)] or

parallel [Figure Pr. 6.30(ii)] to the stream. The Nusselt number for the parallel flow can be determined by assuming a flat surface. This is justifiable when the viscous boundary-layer thickness $\delta_v$ is smaller than $D$.

(a) Draw the thermal circuit diagram.
(b) Determine the heat transfer rates $\langle Q_{ku} \rangle_D$ and $\langle Q_{ku} \rangle_L$, for the conditions given below (i.e., $\delta_\alpha < D$).
(c) Is neglecting the surface curvature and using a flat surface, for the parallel flow, justifiable?

   $D = 1.5$ cm, $L = 40$ cm, $u_{f,\infty} = 4$ m/s, $T_{f,\infty} = 25°$C, $T_s = 430°$C.
   Determine the air properties at $\langle T_f \rangle_\delta = (T_s + T_{f,\infty})/2$.
   Neglect the heat transfer from the end surfaces.

(i) Cross (Perpendicular) Flow        (ii) Parallel Flow

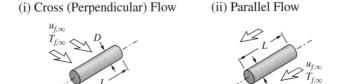

Figure Pr. 6.30. A steel rod is cooled in an air stream with the choice of placing it perpendicular or parallel to the stream. (i) Cross (perpendicular) flow. (ii) Parallel flow.

## PROBLEM 6.31. FUN
An automobile brake rotor is idealized as a solid disc, as shown in Figure Pr. 6.31. In a laboratory test the rotor is friction heated at a rate $\dot{S}_{m,F}$, under steady-state heat transfer and its assumed uniform temperature becomes $T_s$. Assume that the heat transfer is by surface convection only and that the fluid (air) motion is only due to the rotation (rotation-induced motion) (Table 6.4).

Rotating Disc

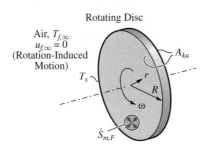

Figure Pr. 6.31. A rotating disc is heated by friction and cooled by surface convection.

(a) Draw the thermal circuit diagram.
(b) Determine the radial location $r_{tr}$ that the flow regime changes from laminar to turbulent ($Re_{r,tr} = 2.4 \times 10^5$).
(c) Integrate the local surface convection over the entire surface area (two sides, neglect the edge) to find $\langle R_{ku} \rangle_L^{-1}$.
(d) Determine the rotor temperature $T_s$.

$\omega = 130$ rad/s, $R = 30$ cm, $T_{f,\infty} = 20°C$, $\dot{S}_{m,F} = 2 \times 10^4$ W.
Determine the air properties at $T = 400$ K.

## PROBLEM 6.32. FAM

A person remaining in a very cold ambient will eventually experience a drop in body temperature (i.e., experience hypothermia). This occurs when the body no longer converts sufficient chemical-bond energy to thermal energy to balance the heat losses. Consider an initial uniform temperature of $T_1(t = 0) = 31°C$ and a constant energy conversion of $\dot{S}_{r,c} = 400$ W. The body may be treated as a cylinder made of water with a diameter of 40 cm and a length of 1.7 m, as shown in Figure Pr. 6.32(ii). Assuming that the lumped-capacitance analysis is applicable, determine the elapsed time for a drop in the body temperature of $\Delta T_1 = 10°C$.

(a) Consider the ambient to be air with a temperature of $T_{f,\infty} = -10°C$, blowing at $u_{f,\infty} = 30$ km/hr across the body (i.e., in cross flow).
(b) Consider the ambient to be water with a temperature of $T_{f,\infty} = 0°C$ with a thermobuoyant motion $(u_{f,\infty} = 0$ km/hr$)$ in the water along the length of the cylinder. For the thermobuoyant motion, use the results for a vertical plate and assume that the body temperature is the time-averaged body temperature (i.e., an average between the initial and the final temperature). This results in a constant surface-convection resistance.

Evaluate the properties at the average temperature between the initial temperature and the far-field fluid temperature.

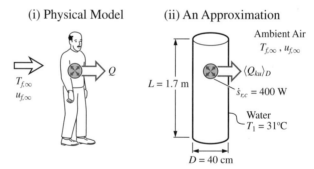

Figure Pr. 6.32. (i) Surface-convection heat transfer from a person. (ii) Its geometric presentation.

## PROBLEM 6.33. FAM

A methane-air mixture flows inside a tube where it is completely reacted generating a heating rate of $\dot{S}_{r,c} = 10^4$ W. This heat is removed from the tube by a cross flow of air, as shown in Figure Pr. 6.33.

(a) Draw the thermal circuit diagram and determine the tube surface temperature with no surface radiation.
(b) Repeat part (a) with surface radiation included.
   Evaluate the properties of air at $T = 300$ K.

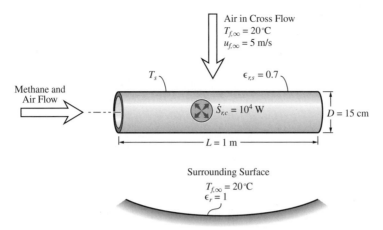

Figure Pr. 6.33. Heat removal from a combustion tube.

### 6.12.8 Inclusion of Substrate

**PROBLEM 6.34. FUN**

Consider surface-convection heat transfer from a sphere of radius $R$ and initial temperature $T_s(t = 0)$ as rendered in Figure Pr. 6.34. The time dependence of the uniform sphere temperature, with surface convection, is given by (6.156) and is valid for $\mathrm{Bi}_R < 0.1$. Also for $\mathrm{Bi}_R > 10$ the surface-convection resistance becomes negligible and then the constant surface temperature, distributed transient temperature given in Figure 3.33(b)(ii) becomes valid. In the Biot number regime $10 < \mathrm{Bi}_R < 0.1$, numerical or series, closed-form solutions are used. In an existing series solution, when the elapsed time is sufficiently large (i.e., $\mathrm{Fo}_R > 0.2$) such that the penetration distance has reached and passed the center of the spheres, a single term from this series solution may be used to obtain $T_s = T_s(r, t)$. This solution for the center of the sphere, i.e., $r = 0$, is

$$T_s^*(r = 0, t) = \frac{T_s(r = 0, t) - T_{f,\infty}}{T_s(t = 0) - T_{f,\infty}} = a_1 e^{-a_2^2 \mathrm{Fo}_R}, \quad \mathrm{Fo}_R = \frac{t\alpha_s}{R^2} > 0.2,$$

where the constants $a_1$ and $a_2$ depend on $\mathrm{Bi}_R$ and are listed for some values of $\mathrm{Bi}_R$ in Table Pr. 6.34. From (6.128), we have

$$\mathrm{Bi}_R = \frac{R_{k,s}}{\langle R_{ku}\rangle_D} = \frac{\langle \mathrm{Nu}\rangle_D k_f / D}{k_s / R}.$$

(a) Show that (6.156) can be written as

$$T_s^*(t) = \frac{T_s(t) - T_{f,\infty}}{T_s(t = 0) - T_{f,\infty}} = e^{-3\mathrm{Fo}_R \mathrm{Bi}_R}, \quad \mathrm{Bi}_R < 0.1.$$

(b) Plot $T_s^*(t)$ with respect to $\mathrm{Fo}_R$, for $0.01 \leq \mathrm{Fo}_R \leq 1$, and for $\mathrm{Bi}_R = 0.01, 0.1, 1, 10$, and 100.

(c) On the above graph, mark the center temperature $T_s(r = 0, t)$ for $\mathrm{Fo}_R = 0.2$ and 1.0, and for the Biot numbers listed in part (b).

Table Pr. 6.34. *The constants in the one-term solution*

|  | $Bi_R$ | $a_1$ | $a_2$ | $(3Bi_R)^{1/2}$ |
|---|---|---|---|---|
|  | 0 | 1.000 | 0 | 0 |
|  | 0.01 | 1.003 | 0.1730 | 0.1732 |
| Lumped | 0.10 | 1.030 | 0.5423 | 0.5477 |
|  | 1.0 | 1.273 | 1.571 | 1.414 |
|  | 10 | 1.943 | 2.836 | 4.472 |
| Constant Surface | 100 | 1.999 | 3.110 | 14.14 |
| Temperature | $\infty$ | 2.000 | $3.142 = \pi$ | $\infty$ |

(d) For $Fo_R = 0.2$ and 1.0, also mark the results found from Figure 3.33(b)(ii), noting that this corresponds to $Bi_R \rightarrow \infty$.
(e) Comment on the regime of a significant difference among the results of the lumped-capacitance treatment, the distributed-capacitance treatment with $Bi_R \rightarrow \infty$, and the single-term solution for distributed capacitance with finite $Bi_R$.

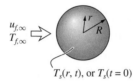

$T_s(r, t)$, or $T_s(t = 0)$

Figure Pr. 6.34. A sphere of initial temperature $T_s(t = 0)$ is placed in a fluid stream with far-field conditions $T_{f,\infty}$, and $u_{f,\infty}$.

## PROBLEM 6.35. FAM
In a rapid solidification-coating process, a liquid metal is atomized and sprayed onto a substrate. The atomization is by gas injection into a spray nozzle containing the liquid-metal stream. The injected gas is small compared to the gas (assume air) entrained by the droplet spray stream. This entrained gas quickly cools the droplets such that at the time of impingement on the substrate the droplets contain a threshold amount of liquid that allows for them to adhere to each other and to the substrate surface. This is shown in Figure Pr. 6.35, where a plastic balloon is coated with a tin layer and since the droplets are significantly cooled by surface convection, the balloon is unharmed. Assume that each droplet is independently exposed to a semi-bounded air stream.

(a) Draw the thermal circuit diagram for a tin droplet cooled from the initial temperature $T_1(t = 0)$ to its solidification temperature $T_{sl}$. Assume a uniform temperature $T_1(t)$.
(b) By neglecting any motion within the droplet, determine if a uniform droplet temperature can be assumed; use $R_{k,s} = D/4A_{ku}k_s$.
(c) Determine the time of flight $t$, for the following conditions.
$$T_1(t = 0) = 330°C, \; T_{f,\infty} = 40°C, \; u_{f,\infty}(\text{relative velocity}) = 5 \text{ m/s}, \; D = 50 \; \mu m.$$

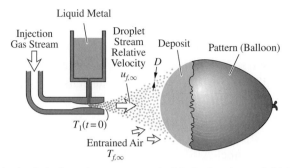

Figure Pr. 6.35. A plastic balloon is spray coated with tin droplets solidifying on its surface.

The Nusselt number can be determined (Table 6.4) using the relative velocity and the properties of tin are given in Tables C.5 and C.16. Determine the air properties at $T = 400$ K.

## PROBLEM 6.36. FAM
In order to enhance the surface-convection heat transfer rate, fins (i.e., extended surfaces) are added to a planar surface. This is shown in Figure Pr. 6.36. The surface has a square geometry with dimensions $a = 30$ cm and $w = 30$ cm and is at $T_{s,o} = 80°C$. The ambient is air with a far-field velocity of $u_{f,\infty} = 1.5$ m/s and a temperature of $T_{f,\infty} = 20°C$ flowing parallel to the surface. There are $N = 20$ rectangular fins made of pure aluminum and each is $l = 2$ mm thick and $L = 50$ mm long.

(a) Determine the rate of heat transfer for the plate without the fins.
(b) Determine the rate of heat transfer for the plate with the fins. Treat the flow over the fins as parallel along the width $w$, thus neglecting the effect of the base and the neighboring fins on the flow and heat transfer [i.e., use the same Nusselt number for parts (a) and (b)].

Assume that the Nusselt number is constant and evaluate the properties at the average temperature between the plate temperature and the far-field fluid temperature.

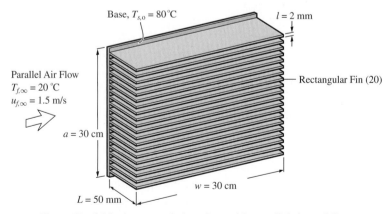

Figure Pr. 6.36. An extended surface with parallel, forced flow.

## PROBLEM 6.37. FAM

An automobile disc-brake converts mechanical energy (kinetic energy) to thermal energy. This thermal energy is stored in the disc and is transferred to the ambient by surface convection and surface radiation and is also transferred to other mechanical components by conduction (e.g., the wheel, axle, suspension, etc). The rate of energy conversion decreases with time due to the decrease in the automobile speed. Here, assume that it is constant and is $(\dot{S}_{m,F})_o = 6 \times 10^4$ W. Assume also that the heat loss occurs primarily by surface-convection heat transfer from the disc surface. The disc is made of carbon steel AISI 4130 (Table C.16) and its initial temperature is $T_1(t = 0) = T_{f,\infty} = 27°C$. The disc surface-convection heat transfer is from the two sides of disc of diameter $D = 35$ cm, as shown in Figure Pr. 6.37, and the disc thickness is $l = 1.5$ cm. The Nusselt number is approximated as that for parallel flow over a plate of length $D$ and determined at the initial velocity. The average automobile velocity is $u_{f,\infty} = 40$ km/hr and the ambient air is at $T_{f,\infty}$.

(a) Assuming that the lumped-capacitance analysis is applicable, determine the temperature of the disc after 4 s $[T_1(t = 4 \text{ s})]$.
(b) Using this temperature [i.e., $T_1(t = 4 \text{ s})$] as the initial temperature and setting the heat generation term equal to zero (i.e., the brake is released), determine the time it takes for the disc temperature to drop to $t_1 = 320$ K.
(c) Evaluate the Biot number and comment on the validity of the lumped-capacitance assumption. For the Biot number, the conduction resistance is based on the disc thickness $l$, while the surface convection resistance is based on the disc diameter $D$.
    Evaluate the air properties at $T_{f,\infty}$.

(i) Physical Model          (ii) An Approximation

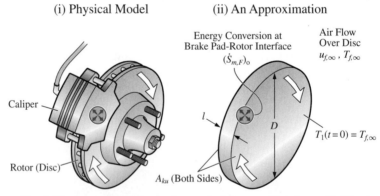

Figure Pr. 6.37. An automobile brake cooled by parallel flow. (i) Physical model. (ii) Approximate model.

## PROBLEM 6.38. FAM

In a portable, phase-change hand warmer, titanium bromide (TiBr$_4$, Table C.5) liquid is contained in a plastic cover (i.e., encapsulated) and upon solidification at the freezing temperature $T_{sl}$ (Table C.5) releases heat. A capsule, which has a thin, rectangular shape and has a cross-sectional area $A_k = 0.04$ m$^2$, as shown in

Figure Pr. 6.38, is placed inside the pocket of a spectator watching an outdoor sport. The capsule has a planar surface area of $A_k = 0.04 \text{ m}^2$.

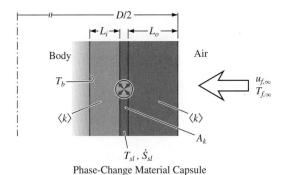

Figure Pr. 6.38. Simplified, physical model for heat transfer from a hand warmer.

The pocket has a thick insulation layer on the outside of thickness $L_o = 2$ cm (toward the ambient air) and a thinner insulation layer on the inside of thickness $L_i = 0.4$ cm (toward the body). The effective conductivity for both layers is $\langle k \rangle = 0.08$ W/m-K. The body temperature is $T_b = 32°$C. The outside layer is exposed to surface convection with a wind blowing as a cross flow over the body (diameter $D$) at a speed $u_{f,\infty} = 2$ m/s and temperature $T_{f,\infty} = 2°$C. For the surface-convection heat transfer, use cross flow over a cylinder of diameter $D = 0.4$ m. Assume that the heat flow is steady and that the temperatures are constant. Treat the conduction heat transfer as planar and through a cross-sectional surface area $A_k$.

(a) Draw the thermal circuit diagram.
(b) Determine all the thermal resistances that the heat flow from the capsule encounters. Use planar geometry for the conduction resistances.
(c) Determine the heat rate toward the body and toward the ambient air.
(d) Determine the total energy conversion rate $\dot{S}_{sl}(\text{W})$.
    Determine the air properties at $T = 300$ K from Table C.22.

**PROBLEM 6.39. FAM.S**
An automobile disc-brake is shown in Figure Pr. 6.39. Nearly all the kinetic energy of the automobile is converted to friction heating during a complete stop. The friction energy conversion is the product of the friction force and the velocity of the disc. The friction force is nearly equal to the deceleration force of the automobile.

Then, using the automobile mass $M_a$ and velocity $u_a$, we can express the energy conversion as

$$\dot{S}_{m,\text{F}} = \mathbf{F}_\text{F} \cdot \mathbf{u}_a = \left( -M_a \frac{du_a}{dt} \right) u_a,$$

where we have used (1.27) and a balance between inertial and friction forces and then replaced the friction force with the inertial force (i.e., acceleration).

Automobile Disc-Brake: Physical Model

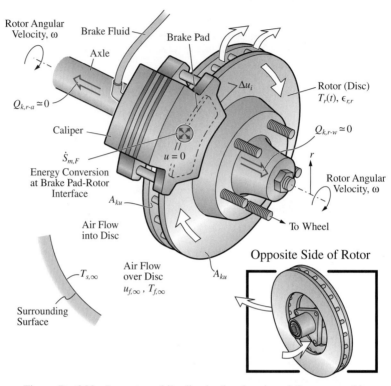

Figure Pr. 6.39. An automobile disc-brake showing airflow around it.

During the stop that takes $\tau(s)$, the velocity is approximated as

$$u_a = u_{a,o} \left(1 - t/\tau\right).$$

Then,

$$\dot{S}_{m,\mathrm{F}} = M_a \frac{u_{a,o}^2}{\tau} \left(1 - t/\tau\right), \quad t \leq \tau.$$

Generally, the front wheels absorb 65% of the total energy. Then

$$\dot{S}_{m,\mathrm{F}} = \frac{0.65}{2} M_a \frac{u_{a,o}^2}{\tau} \left(1 - t/\tau\right) \quad \text{for each front wheel for } t \leq \tau.$$

(a) Draw the thermal circuit diagram for the disc-brake assuming lumped capacitance. Assume that the conduction losses to the pad, axle, and wheel are negligible.
(b) Plot the disc temperature with the disc initially at 20°C, for elapsed times up to 100 s.
(c) Comment on the time variation of the disc temperature.

Assume that the disc can be modeled as a solid disc made of carbon steel AISI 1010 with diameter $D = 35$ cm and thickness $l = 1.5$ cm. The surface emissivity is $\epsilon_{r,r} = 0.4$ and the surroundings are assumed to have blackbody surfaces at

$T_{s,\infty} = 20°C$. Consider heat transfer from the outside surface of the rotor and evaluate the Nusselt number using parallel flow with $L = D$ and at the initial velocity of $u_{f,\infty} = 80$ km/hr. The ambient air is also at $T_{f,\infty} = 20°C$. The automobile stop occurs over $\tau = 4$ s, and an average midsize automobile has a mass of $M_a = 1,500$ kg.

Evaluate properties at $T = 300$ K.

## PROBLEM 6.40. FAM

A microprocessor chip generates Joule heating and needs to be cooled below a damage threshold temperature $T_{p,max} = 90°C$. The heat transfer is by surface convection from its top surface and by conduction through the printed-circuit-board substrate from its bottom surface. The surface convection from the top surface is due to air flow from a fan that provides a parallel flow with a velocity of $u_{f,\infty}$. The conduction from the bottom surface is due to a temperature drop across the substrate of $T_p - T_s$. The substrate is fabricated from a phenolic composite and has a thermal conductivity of $k_s$.

(a) Pentium Pro Microprocessor

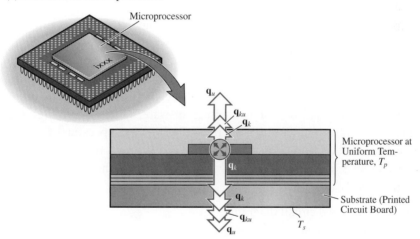

(b) Two Different Surface-Convection Designs

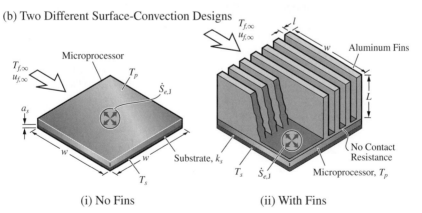

(i) No Fins                    (ii) With Fins

Figure Pr. 6.40. Surface-convection cooling of a microprocessor. (a) Physical Model. (b) Two different surface-convection designs (i) without, and (ii) with back fins.

An idealized schematic of the system is shown in Figure Pr. 6.40.

(a) Draw the thermal circuit diagram for the steady-state heat transfer from the processor.
(b) Determine $T_p$ for the case with no fins and with $T_p - T_s = 10°\,C$.
(c) Determine $T_p$ for the case with aluminum fins and with $T_p - T_s = 1°\,C$.
(d) Comment on the difference between the two cases with respect to a damage threshold temperature.

$\dot{S}_{e,J} = 35$ W, $T_{f,\infty} = 25°$ C, $u_{f,\infty} = 0.5$ m/s, $w = 7$ cm, $a_s = 1.5$ mm, $k_s = 0.3$ W/m-K.

Fin data: $L = 3.5$ cm, $l = 1$ mm, $N_f = 16$; evaluate the properties of aluminum at $T = 300$ K.

Evaluate the properties of air at $T = 300$ K. Neglect the contact resistance between the processor and the substrate. Neglect the edge heat losses. Assume the processor is at a uniform temperature $T_p$. Assume that the energy conversion occurs uniformly within the microprocessor chip.

## PROBLEM 6.41. FAM
To analyze the heat transfer aspects of the automobile rear-window defroster, the window and the very thin resistive heating wires can be divided into identical segments. Each segment consists of an individual wire and an $a \times L \times l$ volume of glass affected by this individual wire/heater. Each segment has a uniform, transient temperature $T_1(t)$. This is shown in Figure Pr. 6.41. In the absence of any surface phase change (such as ice or snow melting, or water mist evaporating), the Joule heating results in a temperature rise from the initial temperature $T_1(t = 0)$, and in a surface heat loss to the surroundings. The surface-convection heat loss to the surroundings is represented by a resistance $\langle R_{ku} \rangle_a$. The surrounding far-field temperature is $T_{f,\infty}$.

(a) Draw the thermal circuit diagram.
(b) Determine the surface-convection resistance.
(c) Show that the lumped capacitance approximation is valid using $l$ for the conduction resistance.

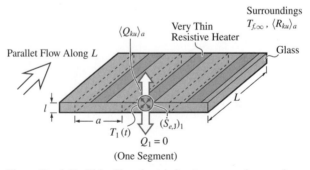

Figure Pr. 6.41. Thin-film electric heaters on a glass surface.

(d) Assuming no surface phase change occurs, determine the steady-state temperature of the glass.

(e) Still assuming no surface phase change occurs, determine the glass temperature after an elapsed time $t = 5$ min.

$T_1(t = 0) = -15°C$, $T_{f,\infty} = -15°C$, $u_{f,\infty} = 6$ m/s, $l = 3$ mm, $a = 2$ cm, $L = 1.5$ m, $(\dot{S}_{e,J})_1 = 15$ W.

Determine the glass plate properties from Table C.17, and properties of air at 270 K.

## PROBLEM 6.42. FAM

In particle spray surface coating using impinging-melting particles, prior to impingement the particles are mixed with a high temperature gas as they flow through a nozzle. The time of flight $t$ (or similarly the nozzle-to-surface distance) is chosen such that upon arrival at the surface the particles are heated (i.e., their temperature is raised) close to their melting temperature. This is shown in Figure Pr. 6.42. The relative velocity of the particle-gas, which is used in the determination of the Nusselt number, is $\Delta u_p$. Consider lead particles of diameter $D$ flown in an air stream of $T_{f,\infty}$. Assume that the particles are heated from the initial temperature of $T_1(t = 0)$ to the melting temperature $T_{sl}$ with surface-convection heat transfer only (neglect radiation heat transfer).

(a) Draw the thermal circuit diagram.

(b) Determine the Biot number $Bi_D$, based on the particle diameter $D$. Can the particles be treated as lumped capacitance?

(c) Determine the time of flight $t$ needed to reach the melting temperature $T_{sl}$.

$T_1(t = 0) = 20°C$, $T_{f,\infty} = 1{,}500$ K, $D = 200$ $\mu$m, $\Delta u_p = 50$ m/s.

Determine the air properties at $T = 1{,}500$ K (Table C.22), and the lead properties at $T = 300$ K (Table C.16).

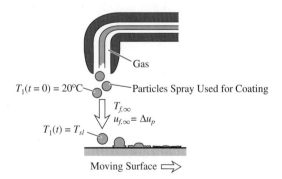

Figure Pr. 6.42. A particle spray surface-coating process using impinging-melting particles.

## PROBLEM 6.43. FAM

A rectangular (square cross section) metal workpiece undergoing grinding, shown in Figure Pr. 6.43, heats up and it is determined that a surface-convection cooling is needed. The fraction of the energy converted by friction heating $\dot{S}_{m,F}$, that results in this heating of the workpiece, is $a_1$. This energy is then removed from the top of the workpiece by surface convection. A single, round impinging air jet is used. Assume steady-state heat transfer and a uniform workpiece temperature $T_s$.

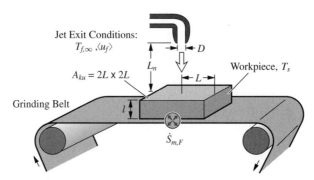

Figure Pr. 6.43. Grinding of a metal workpiece.

(a) Draw the thermal circuit diagram.
(b) Determine the workpiece temperature $T_s$.
(c) What should the ratio of the workpiece thickness $l$ to its conductivity $k_s$ be for the uniform temperature assumption to be valid?

$\dot{S}_{m,F} = 3,000$ W, $a_1 = 0.7$, $T_{f,\infty} = 35°C$, $\langle u_f \rangle = 30$ m/s, $D = 1.5$ cm, $L = 15$ cm, $L_n = 5$ cm.

Evaluate properties of air at $T = 300$ K.

## PROBLEM 6.44. FAM

A microprocessor with the Joule heating $\dot{S}_{e,J}$ is cooled by surface convection for one of its surfaces. An off-the-shelf surface attachment is added to this surface and has a total of $N_f$ square-cross-sectional aluminum pin fins attached to it, as shown in Figure Pr. 6.44. Air is blown over the fins and we assume that the Nusselt number can be approximated using the far-field air velocity $u_{f,\infty}$ and a cross flow over each

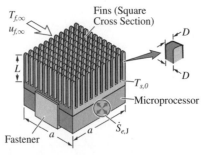

Figure Pr. 6.44. A microprocessor with the Joule heating and a surface-convection cooling. There is an attached extended surface for reduction of the microprocessor temperature.

square-cross-sectional cylinder fin (i.e., the Nusselt number is not affected by the presence of the neighboring fins). This is only a rough approximation.

(a) Draw the thermal circuit diagram.
(b) Determine the fin efficiency.
(c) Determine the steady-state surface temperature $T_{s,0}$.
(d) Determine the fin effectiveness.

$T_{f,\infty} = 35°C$, $u_{f,\infty} = 2$ m/s, $\dot{S}_{e,J} = 50$ W, $D = 2$ mm, $a = 5$ cm, $N_f = 121$, $L = 2$ cm.

Evaluate the air and aluminum properties at $T = 300$ K. Assume that the $\langle Nu \rangle_D$ correlation of Table 6.4 is valid.

### PROBLEM 6.45. FUN

Desiccants (such as silica gel) are porous solids that adsorb moisture (water vapor) on their large interstitial surface areas. The adsorption of vapor on the surface results in formation of an adsorbed water layer. This is similar to condensation and results in liberation of energy. The heat of adsorption, similar to the heat of condensation, is negative and is substantial. Therefore, during adsorption the desiccant heats up. The heat of adsorption for some porous solids is given in Table C.5(b). Consider a desiccant in the form of pellets and as an idealization consider a spherical pellet of diameter $D$ in a mist-air stream with far-field conditions $T_{f,\infty}$ and $u_{f,\infty}$. Assume that the released energy is constant.

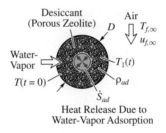

Figure Pr. 6.45. A desiccant pellet in a cross, moist air flow.

(a) Draw the thermal circuit diagram for the pellet.
(b) Determine the pellet temperature after an elapsed time $t_o$.

$\dot{S}_1 = \dot{S}_{ad} = \Delta h_{ad} \rho_{ad} V / t_o$, $D = 5$ mm, $\rho_{ad} = 200$ kg/m$^3$, $T_1(t = 0) = 10°C$, $T_{f,\infty} = 10°C$, $(\rho c_p)_1 = 10^6$ J/m$^3$-K, $u_{f,\infty} = 3$ cm/s, $\Delta h_{ad} = 3.2 \times 10^6$ J/kg, $t_o = 1$ hr.

Evaluate properties of air at $T = 300$ K.

### PROBLEM 6.46. FUN.S

Consider the concept of the critical radius discussed in Example 6.13. An electrical-current conducting wire is electrically insulated using a Teflon layer wrapping, as shown in Figure Pr. 6.46. Air flows over the wire insulation and removes the Joule heating. The thermal circuit diagram is also shown.

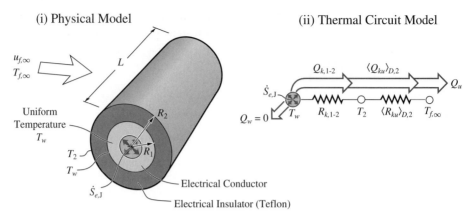

Figure Pr. 6.46. (i) An electrical-current carrying wire is electrically insulated with a Teflon layer wrapping. (ii) Thermal circuit diagram.

(a) Plot the variation of $R_\Sigma = R_{k,1\text{-}2} + \langle R_{ku}\rangle_{D,2}$ with respect to $R_2$, for $R_1 \le R_2 \le 3R_1$.
(b) Determine $R_2 = R_c$ (where $R_\Sigma$ is minimum).
(c) Show the contributions due to $R_{k,1\text{-}2}$ and $\langle R_{ku}\rangle_{D,2}$ at $R_2 = R_c$.
 (d) Also determine $R_c$ from the expression given in Example 6.13, i.e.,

$$R_c = \left(\frac{a_2 k_s}{2^{a_2-1} a_R}\right)^{1/a_2}.$$

$L = 1$ m, $R_1 = 3$ mm, $u_{f,\infty} = 0.5$ m/s. Evaluate the air properties at $T = 300$ K.

Thermal conductivity of Teflon is given in Table C.17.

## PROBLEM 6.47. FUN
In designing fins, from (6.149) we note that a combination of high fin surface area $A_f$ and high fin efficiency $\eta_f$ are desirable. Therefore, while high $\eta_f (\eta_f \to 1)$ is desirable, $\eta_f$ decreases as $A_f$ increases. From (6.149) the case of $\eta_f \to 1$ corresponds to $\mathrm{Bi}_w \to 0$.

Show that in the limit of $mL_c \to 0$, the fin efficiency tends to unity. Note that

$$\tanh(z) = \frac{\sinh(z)}{\cosh(z)}, \quad \sinh(z) = \frac{e^z - e^{-z}}{2},$$

$$\cosh(z) = \frac{e^z + e^{-z}}{2}, \quad e^z = 1 + z + \frac{z^2}{2} + \cdots .$$

## PROBLEM 6.48. FUN
The body of the desert tortoise (like those of other cold-blooded animals) tends to have the same temperature as its ambient air. During daily variations of the ambient temperature, this body temperature also varies, but due to the sensible heat storage, thermal equilibrium (i.e., the condition of being at the same temperature) does not exist at all times. Consider the approximate model temperature variation given in Figure Pr. 6.48(i), which is based on the ambient temperature measured in

early August, 1992, near Las Vegas, Nevada. Assume that a desert tortoise with a uniform temperature is initially at $T_1(t = 0) = 55°C$. It is suddenly exposed to an ambient temperature $T_{f,\infty} = 35°C$ for 6 hours, after which the ambient temperature suddenly changes to $T_{f,\infty} = 55°C$ for another 12 hours before suddenly dropping back to the initial temperature. The heat transfer is by surface convection only (for accurate analysis, surface radiation, including solar radiation, should be included). The geometric model is given in Figure Pr. 6.48(ii), with surface convection through the upper (hemisphere) surface and no heat transfer from the bottom surface. For the Nusselt number, use that for forced flow over a sphere.

(a) Draw the thermal circuit diagram.
(b) Determine the body temperature after an elapsed time of 12 hours, i.e., $T_1(t = 12 \text{ hr})$ for $R_1 = 20$ cm.
(c) Repeat for $R_1 = 80$ cm.

$\rho_1 = 1,000 \text{ kg/m}^3$, $c_{p,1} = 900 \text{ J/kg-K}$, $u_{f,\infty} = 2 \text{ m/s}$.
Evaluate the air properties at $T = 320$ K.

(i) Ambient Air Temperature

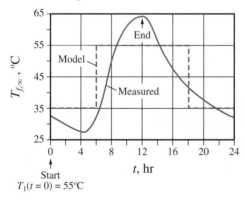

(ii) Geometric Model

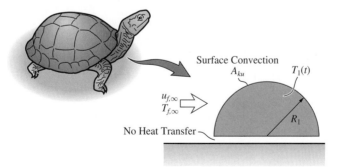

Figure Pr. 6.48. (i) A measured daily ambient air temperature variation over 24 hr and an approximation (model) to the temperature variation. (ii) A geometric model for a tortoise in forced cross flow.

**PROBLEM 6.49. FUN**

When humans experience hypothermia (exposure to extreme low temperatures resulting in lower temperature over part or the entire body), the body provides intensified metabolic reactions to supply more heat. Glucose ($C_6H_{12}O_6$) is the primary body fuel and the magnitude of its heat of oxidation $\Delta h_{r,c}$ is rather large; therefore, the body prepares a less energetic fuel called ATP (a combination of adenine, ribose, and three phosphate radicals, see Problem 2.41). Here for simplicity we assume that glucose oxidation results in thermal energy release as given by

$$C_6H_{12}O_6 + 6O_2 \rightarrow 6CO_2 + 6H_2O.$$

During hypothermia, more energy conversion requires a larger oxygen consumption rate (in chemical-bond energy conversion $\dot{S}_{r,c}$).

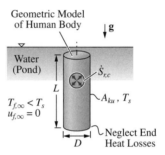

Figure Pr. 6.49. Hypothermia resulting from excessive heat loss, as experienced during submergence in very cold water.

Consider a human fallen into a cold water pond as shown in Figure Pr. 6.49. Assume steady-state heat transfer with surface convection due to the thermobuoyant motion (neglect end heat losses).

(a) Draw the thermal circuit diagram.
(b) Determine $\dot{S}_{r,c}$.
(c) Determine the oxygen consumption rate $\dot{M}_{O_2}$.
(d) Compare the rate calculated in part (c) with that associated with the rest condition $\dot{S}_{r,c} = 45$ W [Figure Ex. 1.3(d)].

$D = 0.45$ m, $L = 1.70$ m, $T_s = 15°C$, $T_{f,\infty} = 4°C$, $\Delta h_{r,c} = -1.6 \times 10^7$ J/kg. Evaluate the water properties at $T = 290$ K.

6.12.9 Surface-Convection Evaporation Cooling*

This section and the Problems that follow are found on the Web site www .cambridge.org/kaviany.

**PROBLEM 6.50. FUN.S\***

**PROBLEM 6.51. FUN.S\***

**PROBLEM 6.52. FAM.S\***

**PROBLEM 6.53. FAM\***

6.12.10 General\*

**PROBLEM 6.54. FUN\***

**PROBLEM 6.55. FUN\***

**PROBLEM 6.56. FUN\***

**PROBLEM 6.57. FUN\***

**PROBLEM 6.58. FAM\***

**PROBLEM 6.59. FAM\***

**PROBLEM 6.60. FAM\***

# Convection: Bounded Fluid Streams

As the fluid passes along the wall in turbulent motion, fluid particles coalesce into lumps, which move bodily and cling together retaining their
momentum for a given transverse length (a mixture or mixing length) and
this is analogous to the mean-free path in the kinetic theory of gases.

– L. Prandtl

A bounded fluid stream in thermal nonequilibrium with its bounding solid surface exchanges heat, and this alters its temperature across and along the stream. That this change occurs throughout the stream is what distinguishes it from the semi-bounded streams where the far-field temperature is assumed unchanged (i.e., the change was confined to the thermal boundary layer). We now consider the magnitude, direction, and spatial and temporal variations of $\mathbf{q}_u$ in a bounded fluid stream (also called an internal flow) that is undergoing surface-convection heat transfer $q_{ku}$. As in Chapter 6, initially our focus is on the fluid and its motion and heat transfer (and this is the reason for the chapter title, bounded fluid streams); later we include the heat transfer in the bounding solid. We examine $\mathbf{q}_u$ within the fluid and adjacent to the solid surface. As in Chapter 6, at the solid surface, where $\mathbf{u}_f = 0$, we use $\mathbf{q}_{ku}$ to refer to the surface-convection heat flux vector. We will develop the average convection resistance $\langle R_u \rangle_L$, which represents the combined effect of surface-convection resistance and the change in the average fluid temperature as it exchanges heat with the bounding surface. This change in the average fluid temperature was not included when we considered semi-bounded fluids (there, no change in the average temperature occurs, due to the semi-infinite extent of the fluid). In addition to single, bounded fluid streams, we consider heat transfer between two bounded, fluid streams separated by a bounding surface. Chart 7.1 renders these bounded, single- and two-stream heat transfers.

## 7.1 Flow and Surface Characteristics

In this chapter, we consider a flowing fluid bounded by a solid surface. In this definition, we include flow through solid surfaces that are continuous (e.g., flow

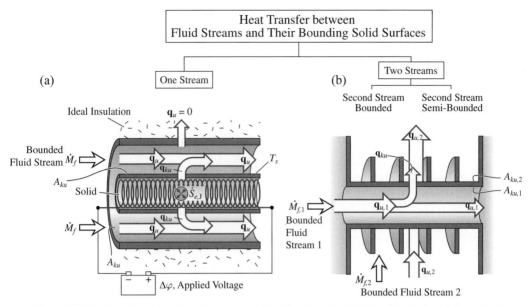

Chart 7.1. A classification of surface convection for bounded fluids. (a) A single, bounded fluid stream exchanging heat with its bounding surface, and (b) heat exchange between two, bounded fluid streams, separated by a bounding surface.

through a tube) and solid surfaces that are discontinuous (e.g., flow across or along a bank of tubes). Thermal nonequilibrium exists between the fluid, which has a local temperature $T_f$, and the bounding solid, which has a surface local temperature $T_s$. Since the fluid is bounded and flows at a finite mass flow rate $\dot{M}_f$ (kg/s), or at a mass flux $\dot{m}_f$ (kg/m$^2$-s), its temperature changes as it exchanges heat with the bounding solid. We will determine this change in the fluid temperature. Because of the no-slip condition at the solid surface, the fluid velocity is zero and varies across the flow cross section $A_u$. For a one-dimensional flow, a flow cross-section averaged velocity $\langle u_f \rangle$ is used. Similarly, because the fluid temperature changes across the flow cross section, a velocity and flow cross-section averaged fluid temperature $\langle T_f \rangle$ is also used.

The local rate of heat transfer between the solid and fluid depends on the local surface-convection resistance $R_{ku}$ and the difference in the local temperatures, i.e., $T_s - \langle T_f \rangle$. This resistance is influenced by solid-surface and flow characteristics.

Among the solid-surface characteristics is the surface roughness (on the side where the fluid flows; note that in some cases, e.g., heat exchangers, two different fluids flow inside and outside of a tube). Extended surfaces (fins) are also used with gases or whenever $R_{ku}$ for a plain surface is large. Chart 7.2 shows the solid geometry variations. The cross-sectional geometry and the geometry of the solid (continuous solid or discontinuous solid) are also among solid geometry variations. In order to obtain a large surface-convection area $A_{ku}$, discontinuous solid surfaces are used. For the two-stream heat exchange, a bounding surface separates these streams. The flow and solid geometry varies among the different two-stream heat exchangers. These

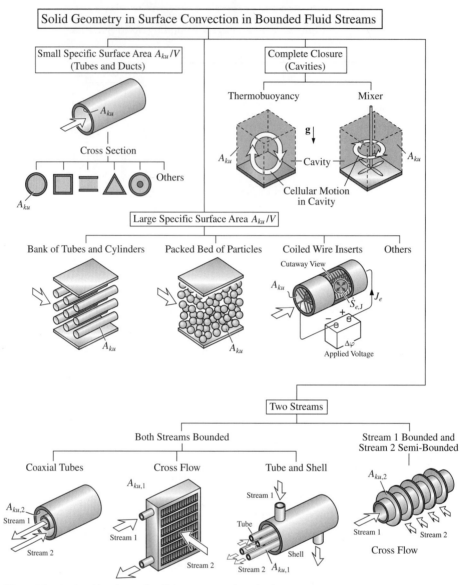

Chart 7.2. A classification of solid geometry for single- and two-stream heat exchange of bounded flow streams.

include cross-flow, tube and shell, and coaxial tube-type heat exchangers. These are shown in Chart 7.2.

Among the fluid characteristics are the fluid properties (thermodynamic and transport) and the flow structure. Flow structure includes fluid dynamics aspects such as instabilities (laminar to turbulent) and thermobuoyancy. The fluid can also undergo liquid-gas phase change as a result of heat transfer; therefore, the phase-density buoyancy becomes very significant (and the flow orientation with respect to the gravity vector becomes a significant parameter). These are shown in Chart 7.3.

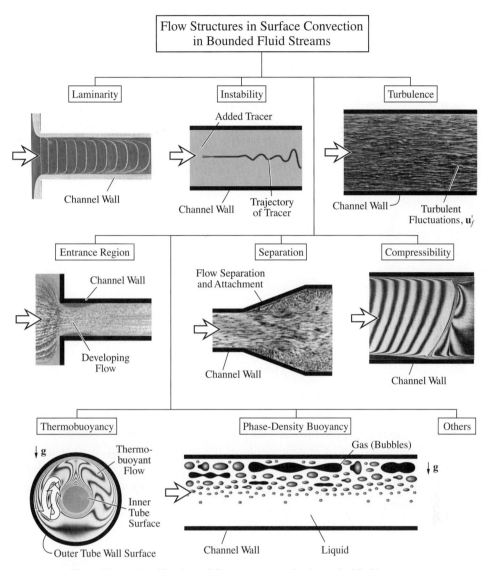

Chart 7.3. A classification of flow structures for bounded fluid streams.

Chart 7.4 gives a summary of the various aspects of surface-convection heat transfer of bounded fluid streams considered in this chapter. First, we consider flow and heat transfer in circular tubes and define the bounded fluid stream heat transfer characteristics, such as average fluid velocity, average local fluid temperature, and local and average surface-convection resistance and Nusselt number. Then we introduce the effectiveness and number of transfer units used in the evaluation of the fluid exit temperature and the heat transfer rate. Next we examine the factors influencing the Nusselt number, such as laminar versus turbulent flow and the distance from the tube entrance. The correlations for various solid geometries are reviewed next. The two-stream heat exchange ends the topics in this chapter.

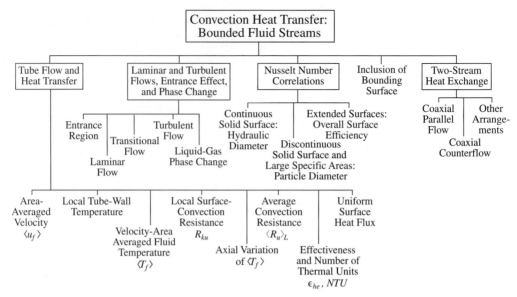

Chart 7.4. Various aspects of surface-convection heat transfer, for bounded fluids, considered in Chapter 7.

## 7.2 Tube Flow and Heat Transfer

Consider steady-state fluid flow into a tube of circular cross section of diameter $D$, with the incoming fluid temperature $T_f$ different than the tube surface temperature $T_s$. Figure 7.1 shows a typical tube flow and heat transfer. We consider a unidirectional flow, i.e., having only an axial component of the velocity $\mathbf{u}_f = (u_f, 0, 0)$. This axial component is designated by $u_f$ and is given as $u_f = u_f(r)$, where $r$ is the radial location. Initially we assume a laminar flow, which has a smooth velocity distribution. This is depicted in Figure 7.1.

The cross-sectional averaged fluid velocity $\langle u_f \rangle$ is obtained by averaging the fluid velocity over the tube flow cross-sectional area $A_u$, i.e.,

$$\langle u_f \rangle_A \equiv \langle u_f \rangle \equiv \frac{1}{A_u} \int_{A_u} u_f \, dA = \frac{8}{D^2} \int_0^{D/2} u_f(x, r) r \, dr \quad \begin{array}{l} \text{area-averaged} \\ \text{fluid velocity,} \end{array} \quad (7.1)$$

where $\langle \ \rangle_A$ indicates an area-averaged quantity and $D$ is the tube diameter. Here for brevity, we have used $\langle u_f \rangle \equiv \langle u_f \rangle_A$.

From the equation for conservation of mass (1.26), for steady-state flow we have (for the control surface $A$ applied over any segment of the tube containing the fluid)

$$\int_A (\dot{\mathbf{m}}_f \cdot \mathbf{s}_n) dA = \int_{A_{ku}+A_u} (\dot{\mathbf{m}}_f \cdot \mathbf{s}_n) dA = \int_{A_{ku}} (\dot{\mathbf{m}}_f \cdot \mathbf{s}_n) dA + \int_{A_u} (\dot{\mathbf{m}}_f \cdot \mathbf{s}_n) dA = 0,$$

$$\dot{m}_f = \rho_f \langle u_f \rangle = \text{constant} \qquad \text{continuity equation,} \qquad (7.2)$$

Radial Velocity and Temperature Distributions and Average
Velocity and Temperature in an impermeable wall Tube

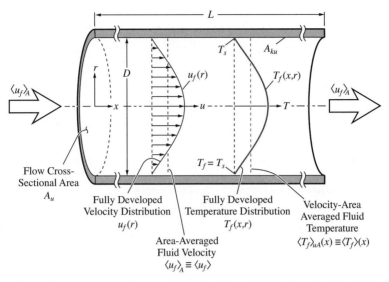

Figure 7.1. Flow and heat transfer in an impermeable wall circular tube. Typical radial fluid velocity and temperature distributions are shown. The local and average fluid velocity and temperature are also shown.

where on $A_{ku}$, we have $\dot{\mathbf{m}}_f = \rho_f \mathbf{u}_f = 0$. On $A_u$, the net mass flow is zero, but the mass flux $\dot{m}_f$ (kg/m²-s) is constant. This indicates that flow entering the control surface $A_u$ is equal to that leaving it because the solid surface area $A_{ku}$ is impermeable.

Upon integration over the flow cross section $A_u$, we have the mass flow rate $\dot{M}_f$ (kg/s)

$$\dot{M}_f = \int_{A_u} \rho_f u_f \, dA = \rho_f \langle u_f \rangle A_u = A_u \dot{m}_f = \quad \begin{array}{l} \text{constant bounded stream} \\ \text{mass flow rate.} \end{array} \quad (7.3)$$

### 7.2.1 Velocity-Area Averaged Fluid Temperature

The axial (i.e., along the $x$ direction) convection heat flux $q_u = (\rho c_p)_f u_f T_f$ varies across the cross section of the tube, because $u_f$ and $T_f$ both vary. Figure 7.1 shows typical variations of $u_f$ and $T_f$ across the tube cross section. To obtain the cross-section averaged heat flow rate $Q_u$, through the tube cross-sectional area $A_u$, we integrate $(\rho c_p)_f u_f T_f$ across $A_u$ to have

$$Q_u = \int_{A_u} q_u \, dA = \int_{A_u} (\rho c_p)_f u_f T_f \, dA \quad \text{convection heat flow rate.} \quad (7.4)$$

We have been assuming a constant $c_{p,f}$ and here we also assume that $\rho_f$ is constant. We now define the velocity-area averaged fluid temperature $\langle T_f \rangle$ and average convection heat flux $\langle q_u \rangle$ through

$$Q_u = \langle q_u \rangle A_u \equiv (\rho c_p)_f \langle u_f \rangle \langle T_f \rangle A_u = \int_{A_u} (\rho c_p)_f u_f T_f \, dA \quad \begin{array}{l} \text{velocity-area} \\ \text{averaged fluid,} \end{array} \quad (7.5)$$

or define $\langle T_f \rangle (x)$ at location $x$ along the flow as

$$\langle T_f \rangle_{uA}(x) \equiv \langle T_f \rangle (x) \equiv \frac{\displaystyle \int_{A_u} u_f T_f \, dA}{\langle u_f \rangle A_u} \qquad \text{local velocity-area (or mixed)} \atop \text{averaged fluid temperature.} \qquad (7.6)$$

The velocity-area averaged fluid temperature $\langle T_f \rangle (x)$ is also called the local bulk, or mixed fluid temperature. Here again, for brevity we have used $\langle T_f \rangle \equiv \langle T_f \rangle_{uA}$. This velocity-area averaged temperature varies along the tube axis, i.e., $\langle T_f \rangle = \langle T_f \rangle (x)$.

### 7.2.2 Tube-Surface Temperature $T_s$: Uniform or Varying

The solid temperature varies across the wall thickness to allow for heat transfer across the tube wall. Here we have designated the inner-surface tube temperature as $T_s$ and, depending on the direction of the heat flow, the outer-surface tube temperature is higher or lower than $T_s$.

In general, $T_s$ varies along the tube, i.e., $T_s = T_s(x)$. This variation is influenced by the fluid temperature, which also varies along the tube axis (because of heat transfer with the tube surface). For example, if there is a second fluid flowing outside and along the tube with a temperature different than the fluid inside the tube, then the temperature of the two fluids and the wall would vary along the tube axis.

The variation of $T_s$ along the tube does influence the local surface-convection resistance $A_{ku} R_{ku}$ (between the tube surface and the fluid flowing inside the tube). However, some general and relatively accurate results are obtained for $R_{ku}$, when it is assumed that either $T_s$ is uniform along the tube or $q_s$ is uniform along the tube. We note that these uniform $T_s$ and $q_s$ need not be prescribed and can be determined from the analysis of heat flow through a multitude of heat transfer media. This is done by considering the complete path of the heat flow starting from the fluid flowing through the tube and then following the heat flow through the tube wall and then from the outside of the tube to a second fluid flowing over the tube, etc.

We first discuss the case of a uniform $T_s$, which is the condition most used in the determination of $A_{ku} R_{ku}$. We discuss the case of a uniform $q_s$ in Section 7.2.7.

### 7.2.3 Local and Average Surface-Convection Resistance: Nusselt Number $\mathrm{Nu}_D$, $\langle \mathrm{Nu} \rangle_D$

The local surface-convection resistance to the heat transfer between the fluid and the tube surface is defined through

$$q_{k,f}(r = R) = -k_f \frac{\partial T}{\partial r}\Big|_{r=R} \equiv \mathbf{q}_{ku} \cdot \mathbf{s}_n \equiv q_{ku}$$

$$\equiv \frac{T_s - \langle T_f \rangle}{A_{ku} R_{ku}} \qquad \text{local surface-convection resistance,} \qquad (7.7)$$

where $q_{ku}$ is taken to be positive (i.e., directed along $\mathbf{s}_n$) when the local surface temperature $T_s$ is higher than the local mixed fluid temperature. Here $A_{ku}$ is the surface-convection area for the fluid-surface heat transfer (inner surface area of the tube).

The local surface-convection resistance $R_{ku}$ depends on the fluid properties, fluid velocity $u_f = u_f(r)$, and the surface thermal condition (uniform $T_s$ or $q_s$).

For laminar flow with uniform $T_s$ (or $q_s$), $R_{ku}$ can be predicted from the energy conservation equation, in a manner similar to that done in Section 6.2.5 for the semi-bounded fluid. Similar to the discussion in Section 6.2.5, the heat transfer between the surface and the fluid is by conduction on both sides of the surface ($\mathbf{u}_f = 0$, due to solid impermeability and fluid viscosity), because $u_f = 0$ on the surface. Now referring to Figure 7.1, with $R = D/2$, we have

$$q_{k,f}(r = R, x) \equiv q_{ku}(x) = (\mathbf{q} \cdot \mathbf{s}_n)_{r=R,x} = -k_f \frac{\partial T_f}{\partial r}\Big|_{r=R,x} = \frac{T_s - \langle T_f\rangle(x)}{A_{ku}R_{ku}(x)}, \quad (7.8)$$

$$u_f = 0 \text{ at } r = R \quad \begin{array}{l}\text{relation between derivative of fluid temperature}\\\text{and the local surface-convection heat transfer rate.}\end{array}$$

Here we have shown that $R_{ku}$ can vary along the flow direction $x$.

For some special cases, e.g., laminar flow in a tube under the conditions of fully-developed temperature and velocity fields and a constant surface heat flux (these will be discussed in Sections 7.3 and 7.4) closed-form solutions can be found for $\langle Nu\rangle_D$. This is done starting with the differential-volume energy equation, similar to Sections 6.2.3 and 6.2.5. An example is given as an end of chapter problem.

The normalized or dimensionless local surface-convection conductance $Nu_D$ is used here, similar to that in Section 6.2.4. The length scale conventionally used is the diameter $D$. Then we have the local Nusselt number $Nu_D$ as

$$Nu_D \equiv \frac{1}{R_{ku}^*} \equiv \frac{R_{k,f}}{R_{ku}(x)} = \frac{A_{ku}R_{k,f}}{A_{ku}R_{ku}(x)} = \frac{D}{A_{ku}R_{ku}(x)k_f} = \frac{q_{ku}D}{[T_s - \langle T_f\rangle(x)]k_f} \quad (7.9)$$

local Nusselt number $Nu_D$.

Note that because the lateral length $D$ is used in $R_{k,f}$, this can be called the lateral-lateral Nusselt number. The definition of $Nu_D$ is depicted in Figure 7.2.

In general, $Nu_D$ is not very large, as compared to $Nu_L$ for semi-bounded fluids. This is because here the length scale $D$ is used in $R_{k,f}$, and $D$ correctly signifies the lateral (or radial) conduction (as compared to $L$ used for the semi-bounded fluid).

Since $Nu_D$ can vary along the surface (as the velocity and temperature fields vary), i.e., $Nu_D = Nu_D(x)$, the average Nusselt number along the tube length $L$ is defined through the average resistance $\langle R_{ku}\rangle_D$ as,

$$\langle R_{ku}\rangle_D \equiv \frac{1}{L}\int_0^L R_{ku}(x)dx, \quad \langle Nu\rangle_D = \frac{k_f}{\langle R_{ku}\rangle_D D}\int_0^L Nu_D(x)dx. \quad (7.10)$$

Note that, in contrast to (6.50), $Nu_D$ is scaled here with the length scale $D$, which does not change along $x$. This variation of $Nu_D$ along the tube, $Nu_D = Nu_D(x)$, is discussed in Sections 7.3.2 and 7.3.3.

### Local Nusselt Number $\mathrm{Nu}_D$ of a Tube

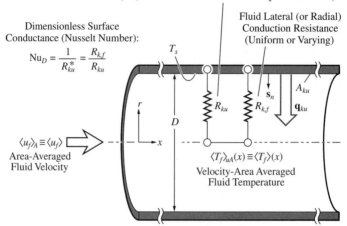

Figure 7.2. Definition of Nusselt number and designation of local conduction and surface-convection resistances.

When there is no change in the stream temperature, i.e., $\langle T_f \rangle$ is constant (such as during liquid/vapor phase with no significant pressure drop), then (7.9) is used with $\langle \mathrm{Nu} \rangle_D$ and the treatment will be similar to that for the semi-bounded fluid streams. When $\langle T_f \rangle = \langle T_f \rangle(x)$, we need to determine the temperature variation along the flow and this is now discussed.

### 7.2.4 Negligible Axial Conduction: Large Péclet Number

Heat is convected by the fluid flowing along the tube axis, while the heat exchange is made with the tube surface. Then, convection heat flux $q_u$ varies along the tube axis, due to this heat exchange. In most cases, the fluid conduction along the tube axis (i.e., axial conduction) is assumed negligible. This is justifiable for relatively large velocities, as discussed in Section 5.2 (i.e., for high Péclet number, $\mathrm{Pe}_L = \langle u_f \rangle L/\alpha_f \gg 1$, where the length here is the tube length). Therefore, we use the axial fluid convection $q_u$, along with surface-convection $q_{ku}$, in describing the bounded fluid heat transfer. The surface convection includes the effect of lateral conduction through the use of $R_{ku}$, i.e., the lateral resistance to heat flow between the surface and the flowing fluid.

### 7.2.5 Axial Variation of Fluid Temperature for Uniform $T_s$: Effectiveness $\epsilon_{he}$ and Number of Transfer Units $NTU$

Now consider a control volume $\Delta V = \pi R^2 \Delta x$ taken along the tube, which has differential length in the $x$ direction only and contains the fluid at location $x$. The fluid has a velocity-area averaged fluid temperature $\langle T_f \rangle = \langle T_f \rangle(x)$. This is depicted in Figure 7.3. The tube inner surface temperature is $T_s$.

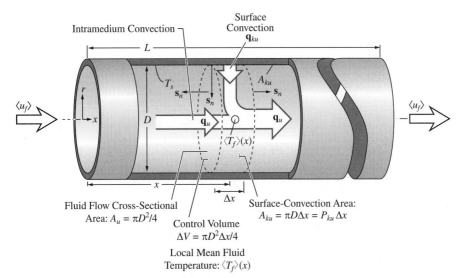

Figure 7.3. Combined differential-integral control volume for energy equation for flow in a circular tube. The surface temperature $T_s$ is uniform.

For the fluid control volume shown in Figure 7.3, the energy equation in the absence of any volumetric terms is that discussed in Section 2.2.2 for a control volume that combines the integral and differential lengths. We use (2.11), when $q_{ku}$ is given by (7.9). As shown in Figure 7.3, we have $A_u = \pi D^2/4$ and $\Delta A_{ku} = \pi D \Delta x$. Then for $\Delta V = \pi D^2 \Delta x/4 \to 0$, as $\Delta x \to 0$, we have

$$\lim_{\Delta x \to 0} \frac{\int_{\Delta A} (\mathbf{q} \cdot \mathbf{s}_n) dA}{\Delta V} = -\frac{4}{D} q_{ku} + \frac{d}{dx} \langle q_u \rangle = 0 \qquad (7.11)$$

or

$$-P_{ku} q_{ku} + A_u \frac{d}{dx} \langle q_u \rangle = 0 \qquad \text{combined integral-differential} \atop \text{length energy equation.} \qquad (7.12)$$

Now using $\langle q_u \rangle = (\rho c_p)_f \langle u_f \rangle \langle T_f \rangle$ from (7.5), (7.12) becomes

$$P_{ku} \frac{\langle T_f \rangle - T_s}{A_{ku} R_{ku}} + A_u \frac{d}{dx} (\rho c_p)_f \langle u_f \rangle \langle T_f \rangle = 0, \qquad (7.13)$$

where $P_{ku}$ is the perimeter for the surface-convection heat transfer surface, which is the inner perimeter of the tube, i.e., $P_{ku} = \pi D$.

Since we have already assumed that $\rho_f$ and $c_{p,f}$ are constant, then from (7.2), $\langle u_f \rangle$ is also constant. Then (7.13) becomes

$$-\frac{P_{ku}}{A_u (\rho c_p)_f \langle u_f \rangle A_{ku} R_{ku}} (T_s - \langle T_f \rangle) + \frac{d}{dx} \langle T_f \rangle = 0. \qquad (7.14)$$

We now use the constant surface temperature $T_s$ to rewrite this as

$$\frac{d}{dx} (T_s - \langle T_f \rangle) + \frac{P_{ku} \mathrm{Nu}_D (k_f/D)}{A_u (\rho c_p)_f \langle u_f \rangle} (T_s - \langle T_f \rangle) = 0 \qquad \text{energy equation.} \qquad (7.15)$$

This ordinary differential equation is integrated over $0 \le x \le L$, where $T_s - \langle T_f \rangle$ at $x = 0$ is prescribed as the entrance condition

$$(T_s - \langle T_f \rangle)_{x=0} = T_s - \langle T_f \rangle_0, \quad \text{at } x = 0, \tag{7.16}$$

or

$$\langle T_f \rangle(x = 0) \equiv \langle T_f \rangle_0 \quad \text{entrance thermal condition.} \tag{7.17}$$

The integration of (7.15) with use of the entrance condition (7.16) gives the temperature at any location $L$ along $x$ as

$$\frac{\langle T_f \rangle_L - T_s}{\langle T_f \rangle_0 - T_s} = \exp\left[-\frac{P_{ku}(k_f/D)\int_0^L \mathrm{Nu}_D dx}{A_u(\rho c_p)_f \langle u_f \rangle}\right]$$

$$= \exp\left[-\frac{P_{ku}(k_f/D)L\langle \mathrm{Nu}\rangle_D}{A_u(\rho c_p)_f \langle u_f \rangle}\right]$$

$$= \exp\left[-\frac{A_{ku}\langle \mathrm{Nu}\rangle_D(k_f/D)}{A_u(\rho c_p)_f \langle u_f \rangle}\right]$$

$$\equiv \exp\left[-\frac{1}{\langle R_{ku}\rangle_D(\dot{M}c_p)_f}\right] \quad \begin{array}{l}\text{fluid temperature}\\ \text{distribution along} \\ \text{tube axis,}\end{array} \tag{7.18}$$

where $\langle T_f \rangle_L$ is referred to as the fluid exit temperature. Here we have used the average surface-convection resistance as

$$\langle R_{ku}\rangle_D = \frac{D}{A_{ku}\langle \mathrm{Nu}\rangle_D k_f}, \tag{7.19}$$

where $A_{ku} = P_{ku}L$, and $\dot{M}_f = \dot{m}_f A_u$ as defined by (7.2) and (7.3).

In (7.18) we have used the average dimensionless surface conductance $\langle \mathrm{Nu}\rangle_D$ defined by (7.10).

In correlations to be listed in Section 7.4, this average conductance is given as a function of the flow and surface variables.

The number of transfer units $NTU$ is defined as

$$NTU \equiv \frac{R_{u,f}}{\langle R_{ku}\rangle_D} = \frac{1}{\langle R_{ku}\rangle_D(\dot{M}c_p)_f} = \frac{A_{ku}\langle \mathrm{Nu}\rangle_D k_f/D}{(\dot{M}c_p)_f} \quad \begin{array}{l}\text{number of transfer}\\ \text{units } 0 \le NTU < \infty.\end{array} \tag{7.20}$$

Note that $NTU$ is the ratio of the surface-convection conductance $1/\langle R_{ku}\rangle_D$, to the ability of the fluid stream to convect heat, i.e., the fluid stream convection conductance $(\dot{M}c_p)_f$. Now we rewrite (7.18), using $NTU$, as

$$\frac{T_s - \langle T_f \rangle_L}{T_s - \langle T_f \rangle_0} = \frac{\langle T_f \rangle_L - T_s}{\langle T_f \rangle_0 - T_s} = e^{-NTU} \quad \text{fluid temperature at location } L, \langle T_f \rangle_L. \tag{7.21}$$

The variation of $\langle T_f \rangle_L$ with respect to $NTU$ (or $L$) is shown in Figure 7.4(a).

(a) Variation of Average Fluid Temperature with Respect to *NTU*

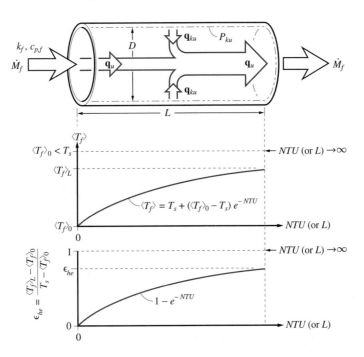

(b) $\epsilon_{he} - NTU$ Relation for a Single Stream

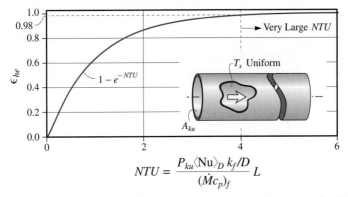

$$NTU = \frac{P_{ku}\langle Nu\rangle_D \, k_f/D}{(\dot{M}c_p)_f} L$$

Figure 7.4. (a) The variation of average fluid temperature (and effectiveness) with respect to *NTU*. (b) Variation of $\epsilon_{he}$ with respect to *NTU*, for a single stream.

A heat exchange effectiveness $\epsilon_{he}$ is defined as

$$\epsilon_{he} \equiv \frac{\langle T_f\rangle_L - \langle T_f\rangle_0}{T_s - \langle T_f\rangle_0} = \frac{\langle T_f\rangle_0 - \langle T_f\rangle_L}{\langle T_f\rangle_0 - T_s} = 1 - e^{-NTU} \qquad \begin{array}{l}\text{heat exchange} \\ \text{effectiveness} \\ 0 \le \epsilon_{he} \le 1.\end{array} \qquad (7.22)$$

This is called the $\epsilon_{he}$-*NTU* relation. Then for a very large *NTU* (i.e., $NTU > 4$), the effectiveness $\epsilon_{he}$ is unity. As *NTU* decreases, $\epsilon_{he}$ approaches zero. Figure 7.4(a)

### (c) Average Convection Resistance $\langle R_u \rangle_L$

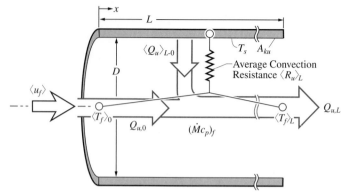

Figure 7.4. (c) Average convection resistance $\langle R_u \rangle_L$ for bounded flow streams. The thermal node presentation uses one surface node and two fluid nodes, but $\langle R_u \rangle_L$ is based on $T_s - \langle T_f \rangle_0$.

shows the variation of $\epsilon_{he}$ with respect to $NTU$. Note that $NTU$ can be made large by increasing $P_{ku}\langle Nu \rangle_D k_f / D$, or by increasing $L$, or by decreasing $(\dot{M}c_p)_f$. Note the exponential change in $\langle T_f \rangle_L$ with an asymptotic approach to $T_s$ for large $NTU$.

As marked in Figure 7.4(b), for the exit fluid temperature $\langle T_f \rangle_L$ to approach $T_s$, i.e., $\epsilon_{he} \to 0.98$, $NTU$ should be 4 or larger. However, in practice a smaller value is used, resulting in a $\langle T_f \rangle_L$ different than $T_s$. It may also be desirable for $\langle T_f \rangle_L$ to be substantially different than $T_s$; then a much smaller $NTU$ is used.

As we will discuss in Section 7.4, (7.22) describes the temperature $\langle T_f \rangle_L$ at a location $L$ for a bounded fluid exchanging heat with any constant surface temperature $T_s$ (continuous or discontinuous solid surface). In this equation, the areas $A_{ku}$ and $A_u$, and the average dimensionless surface-convection conductance $\langle R_{ku} \rangle_L$, vary with the surface geometry and are listed in Section 7.4.

The integrated heat transfer rate $\langle Q_{ku} \rangle_L$ or $\langle Q_u \rangle_{L-0}$ is found by writing the integral-volume energy equation (2.9) over the integral volume $V = A_u L$, i.e.,

$$\int_A (\mathbf{q} \cdot \mathbf{s}_n) dA = \underbrace{\int_{A_{ku}} q_{ku} dA}_{\substack{\text{surface} \\ \text{convection}}} + \underbrace{\int_{A_u} q_u dA}_{\substack{\text{convection at} \\ x = L \text{ and } x = 0}} = 0 \qquad \substack{\text{integral-volume} \\ \text{energy equation}}$$

$$= -\int_0^L P_{ku} q_{ku} dx + A_u(\langle q_u \rangle_L - \langle q_u \rangle_0) = 0$$

$$\equiv -\langle Q_{ku} \rangle_L + A_u(\dot{m}c_p)_f(\langle T_f \rangle_L - \langle T_f \rangle_0) = 0$$

$$\equiv -\langle Q_{ku} \rangle_L + \langle Q_u \rangle_{L-0} = 0. \qquad (7.23)$$

Then the convection and the surface-convection heat transfer are

$$\langle Q_u \rangle_{L-0} = A_u(\dot{m}c_p)_f(\langle T_f \rangle_L - \langle T_f \rangle_0)$$

$$= (\dot{M}c_p)_f(\langle T_f \rangle_L - \langle T_f \rangle_0) = \langle Q_{ku} \rangle_L$$

$$= (\dot{M}c_p)_f(T_s - \langle T_f \rangle_0)\epsilon_{he} \qquad \substack{\text{change in fluid convection} \\ \text{heat flow rate,}} \qquad (7.24)$$

Table 7.1. *Average convection resistance for bounded fluid streams*

| Heat flow rate | Potential | Resistance |
|---|---|---|
| $\langle Q_u \rangle_{L-0} = \langle Q_{ku} \rangle_L =$ | $T_s - \langle T_f \rangle_0, °C$ | $\langle R_u \rangle_L = 1/(\dot{M}c_p)_f \epsilon_{he}, °C/W$ |
| $(\dot{M}c_p)_f(\langle T_f \rangle_L - \langle T_f \rangle_0) =$ | | $\epsilon_{he} = 1 - e^{-NTU}$ |
| $(T_s - \langle T_f \rangle_0)/\langle R_u \rangle_L, W$ | | $NTU \equiv \dfrac{1}{\langle R_{ku} \rangle_D (\dot{M}c_p)_f} = \dfrac{A_{ku}\langle Nu \rangle_D k_f/D}{(\dot{M}c_p)_f}$ |

where we have used (7.22) and $\langle Q_u \rangle_{L-0}$ is positive when heat flows from the tube to the fluid.[†]

### 7.2.6 Average Convection Resistance $\langle R_u \rangle_L (°C/W)$

In thermal circuit analysis, it is more convenient to give $\langle Q_u \rangle_{L-0}$ in terms of $T_s - \langle T_f \rangle_0$. This eliminates the direct determination of $\langle T_f \rangle_L$ from (7.22). Here we define an average convection resistance $\langle R_u \rangle_L$ as

$$\langle Q_u \rangle_{L-0} \equiv \frac{T_s - \langle T_f \rangle_0}{\langle R_u \rangle_L} \quad \text{surface-averaged heat transfer rate} \quad (7.25)$$

or

$$\langle R_u \rangle_L \equiv \frac{T_s - \langle T_f \rangle_0}{\langle Q_u \rangle_{L-0}} \quad \text{average convection resistance in terms of } T_s - \langle T_f \rangle_0. \quad (7.26)$$

Now setting (7.24) equal to (7.25) and substituting for $\langle T_f \rangle_L$ from (7.22) gives

$$\langle R_u \rangle_L = \frac{T_s - \langle T_f \rangle_0}{(\dot{M}c_p)_f(\langle T_f \rangle_L - \langle T_f \rangle_0)} = \frac{1}{(\dot{M}c_p)_f \epsilon_{he}} = \frac{1}{(\dot{M}c_p)_f(1 - e^{-NTU})} \quad \begin{array}{l}\text{average}\\\text{convection}\\\text{resistance.}\end{array} \quad (7.27)$$

Again the proportion of $\langle Q_u \rangle_{L-0}$ to the temperature difference $T_s - \langle T_f \rangle_0$ suggests the use of an electrical circuit analogy. Table 7.1 shows the use of this analogy.

Figure 7.4(c) shows the physical significance of $\langle R_u \rangle_L$. Note that for $A_{ku} = P_{ku}L \to 0$, this resistance becomes equal to $\langle R_{ku} \rangle_L$ or $D/(A_{ku}\langle Nu \rangle_D k_f)$. This can

---

[†] Note that by taking the natural log of both sides of (7.18) we have

$$\ln\left(\frac{T_s - \langle T_f \rangle_L}{T_s - \langle T_f \rangle_0}\right) = -\frac{A_{ku}}{A_u}\frac{1}{(\dot{m}c_p)_f}\frac{\langle Nu \rangle_D k_f}{D} = -\frac{A_{ku}\langle Nu \rangle_D k_f/D}{\langle Q_u \rangle_{L-0}/(\langle T_f \rangle_L - \langle T_f \rangle_0)},$$

where we have used $\langle Q_u \rangle_{L-0}$ from (7.24).

From this, it is customary to solve for $\langle Q_u \rangle_{L-0}$ and use

$$\langle Q_u \rangle_{L-0} \equiv -\frac{1}{\langle R_{ku} \rangle_D}\Delta T_{lm} = -\frac{A_{ku}\langle Nu \rangle_D k_f}{D}\Delta T_{lm},$$

where the log-mean-temperature difference $\Delta T_{lm}$ is defined as

$$\Delta T_{lm} \equiv \frac{\langle T_f \rangle_L - \langle T_f \rangle_0}{\ln[(T_s - \langle T_f \rangle_L)/(T_s - \langle T_f \rangle_0)]}.$$

Comparison Between Surface-Convection Heat Transfer
to Semi-Bounded and Bounded Fluids

(a) Semi-Bounded Fluid

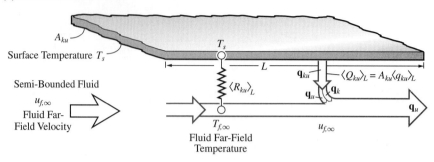

(b) Bounded Fluid

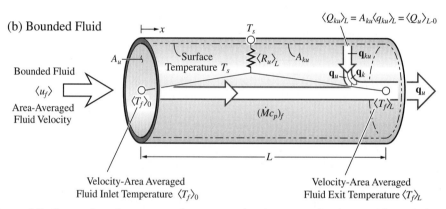

Figure 7.5. Comparison between surface-convection heat transfer for semi-bounded and bounded fluids. (a) Semi-bounded fluid, where $T_{f,\infty}$ does not change along $L$. (b) Bounded fluid, where $\langle T_f \rangle$ changes along $L$.

be shown by expanding the exponent and taking the leading term as $L \to 0$ (this is left as an end of chapter problem).

Figure 7.5 compares the length-averaged, surface-convection resistance, for the semi-bounded fluid $\langle R_{ku} \rangle_L$ and $\langle R_u \rangle_L$ for the bounded fluid. In the semi-bounded fluid flow, the far-field fluid temperature $T_{f,\infty}$ does not change, while for the bounded fluid temperature $\langle T_f \rangle$ changes with $L$. Note that because of this, in rendering $\langle R_u \rangle_L$, the bounded fluid connects to two fluid thermal nodes $\langle T_f \rangle_0$ and $\langle T_f \rangle_L$. However, the average convection resistance $\langle R_u \rangle_L$ is defined based on $\langle T_f \rangle_0$ only. This allows for direct determination of $\langle Q_u \rangle_{L-0}$.

### 7.2.7 Prescribed Uniform Surface Heat Flux $q_s$

When a uniform $q_s$ is prescribed, then the energy equation is written similar to (7.23). For $V = \pi D^2 L/4$, with $A_{ku} = \pi D L$ and $A_u = \pi D^2/4$, we have

$$\int (\mathbf{q} \cdot \mathbf{s}_n) dA = -A_{ku} q_s + A_u (\langle q_u \rangle_L - \langle q_u \rangle_0) = 0 \qquad \begin{array}{l} \text{integral-volume} \\ \text{energy equation.} \end{array} \qquad (7.28)$$

Note that $L$ is any location along the tube axis and when it is the end of the tube, it signifies the exit condition. The above can be written as

$$A_{ku}q_s = Q_s = \langle Q \rangle_{L-0} = A_u(\dot{m}c_p)_f(\langle T_f \rangle_L - \langle T_f \rangle_0), \tag{7.29}$$

or solving for $\langle T_f \rangle_L$, we have

$$\langle T_f \rangle_L = \langle T_f \rangle_0 + \frac{A_{ku}q_s}{(\dot{m}c_p)_f A_u} = \langle T_f \rangle_0 + \frac{A_{ku}q_s}{(\dot{M}c_p)_f}. \tag{7.30}$$

Note that $q_s$ is positive when it is flowing to the fluid.

The surface temperature $T_s$ also varies along the tube, and $T_s(L)$ is influenced by the surface-convection resistance $\langle R_{ku} \rangle_D$. The surface temperature is found from the definition of surface-convection resistance given by (7.9), i.e.,

$$q_s = \frac{T_s(L) - \langle T_f \rangle_L}{A_{ku}R_{ku}}$$

$$= \mathrm{Nu}_D \frac{k_f}{D}[T_s(L) - \langle T_f \rangle_L] \tag{7.31}$$

or

$$T_s(L) = \langle T_f \rangle_L + \frac{q_s}{\langle \mathrm{Nu} \rangle_D k_f / D} \quad \begin{array}{l} \text{surface temperature} \\ \text{at location } L. \end{array} \tag{7.32}$$

The surface temperature at the inlet $T_s(x = 0)$ is found in a similar manner and is

$$q_s = A_{ku}\mathrm{Nu}_D \frac{k_f}{D}[T_s(0) - \langle T_f \rangle_0] \tag{7.33}$$

or

$$T_s(0) = \langle T_f \rangle_0 + \frac{q_s}{\langle \mathrm{Nu} \rangle_D k_f / D} \quad \begin{array}{l} \text{surface temperature} \\ \text{at the entrance.} \end{array} \tag{7.34}$$

The correlations for $\langle \mathrm{Nu} \rangle_D$ under the condition of uniform $q_s$ will be listed in Section 7.4.5.

When the bounding-surface thermal condition is neither uniform temperature $T_s$ nor uniform heat flux $q_s$, generally division of the bounding surface into small segments is used. This will be discussed in Section 7.5.

### EXAMPLE 7.1. FAM

Engine oil is cooled by flowing through a tube at a rate of $\dot{M}_f = 0.05$ kg/s with an inlet temperature $\langle T_f \rangle_0 = 80°$C. The surface of the tube is cooled and maintained at a constant temperature of $T_s = 30°$C (by, e.g., removing the heat from this surface to another fluid). This is rendered in Figure Ex. 7.1(a).

The tube has a diameter $D = 2$ cm, and a length $L = 20$ m (this can be a coiled tube instead of a straight tube). The Nusselt number is $\langle \mathrm{Nu} \rangle_D = 3.66$.

(a) Draw the thermal circuit diagram and show the expected fluid temperature distribution.

(b) Determine the number of transfer units $NTU$.

(c) Determine the effectiveness $\epsilon_{he}$.

(d) Determine the fluid exit temperature $\langle T_f \rangle_L$.

(e) Determine the average convection resistance $\langle R_u \rangle_L$.

(f) Determine the average convection heat transfer rate $\langle Q_u \rangle_{L\text{-}0}$.

Evaluate the oil thermophysical properties from Table C.23 and at $T = 330$ K.

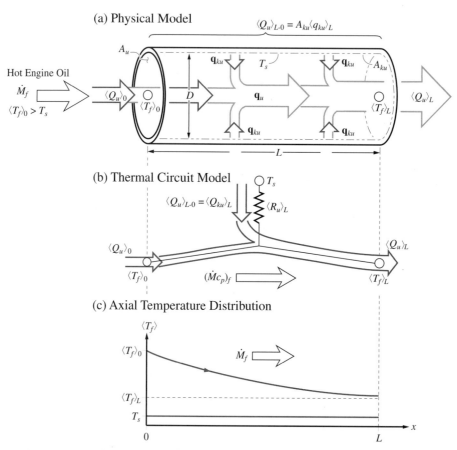

Figure Ex. 7.1. (a) Flow and heat transfer aspects of hot oil flowing through a tube with the tube surface cooled and maintained at $T_s$. (b) Thermal circuit model. (c) Axial distribution of fluid temperature.

**SOLUTION**

(a) The thermal circuit diagram is shown in Figure Ex. 7.1(b). The anticipated decay of the fluid temperature is shown in Figure Ex. 7.1(c).

(b) The definition of the number of thermal units $NTU$ is given by (7.20), i.e.,

$$NTU = \frac{A_{ku} \langle \mathrm{Nu} \rangle_D k_f / D}{(\dot{M} c_p)_f},$$

where

$$A_{ku} = \pi DL.$$

Then

$$NTU = \frac{\pi L \langle \mathrm{Nu} \rangle_D k_f}{\dot{M}_f c_{p,f}}.$$

From Table C.23 and at $T = 330$ K, we have

$$c_{p,f} = 2{,}040 \text{ J/kg-K} \quad \text{Table C.23}$$

$$k_f = 0.14 \text{ W/m-K} \quad \text{Table C.23.}$$

Using these, we have

$$NTU = \frac{\pi \times 20(\mathrm{m}) \times 3.66 \times 0.14(\mathrm{W/m\text{-}K})}{0.05(\mathrm{kg/s}) \times 2{,}040(\mathrm{J/kg\text{-}K})} = 0.3156.$$

(c) The heat exchanger effectiveness $\epsilon_{he}$ is defined by (7.22), i.e.,

$$\epsilon_{he} = 1 - e^{-NTU} = 1 - e^{-0.3156} = 0.2707.$$

(d) The fluid exit temperature $\langle T_f \rangle_L$ is found from the expression for $\epsilon_{he}$, i.e., (7.22),

$$\epsilon_{he} = \frac{\langle T_f \rangle_0 - \langle T_f \rangle_L}{\langle T_f \rangle_0 - T_s} = \frac{\langle T_f \rangle_L - \langle T_f \rangle_0}{T_s - \langle T_f \rangle_0}.$$

Solving for $\langle T_f \rangle_L$, we have

$$\langle T_f \rangle_L = \langle T_f \rangle_0 - \epsilon_{he}(\langle T_f \rangle_0 - T_s)$$

$$= 80(^\circ\mathrm{C}) - 0.2707(80 - 30)(^\circ\mathrm{C}) = 66.47^\circ\mathrm{C}.$$

(e) The average convection resistance $\langle R_u \rangle_L$ is given by (7.27), i.e.,

$$\langle R_u \rangle_L = \frac{1}{(\dot{M}c_p)_f (1 - e^{-NTU})}$$

$$= \frac{1}{(0.05)(\mathrm{kg/s}) \times 2{,}040(\mathrm{J/kg\text{-}{}^\circ C})(1 - e^{-0.3156})} = 0.03622^\circ\mathrm{C/W}.$$

(f) The surface-averaged heat transfer rate $\langle Q_u \rangle_{L\text{-}0} = \langle Q_{ku} \rangle_L$ is given by (7.25), i.e.,

$$\langle Q_u \rangle_{L\text{-}0} = \frac{T_s - \langle T_f \rangle_0}{\langle R_u \rangle_L}$$

$$= \frac{(30 - 80)(^\circ\mathrm{C})}{0.03622(^\circ\mathrm{C/W})} = -1.380 \times 10^3 \text{ W}.$$

**COMMENT**

Note that the magnitude of $NTU$ used here is not large. This is due to the small value used for $\langle \mathrm{Nu} \rangle_D$. The value $\langle \mathrm{Nu} \rangle_D = 3.66$ is for laminar flows. Laminar flows occur at small fluid velocities, or for high fluid viscosities, or small tube diameters. High molecular weight liquids, such as oils, have a relatively high viscosity. We will discuss the correlation for $\langle \text{Nu} \rangle_D$ in Section 7.4.

The thermophysical properties of the fluid change along the tube as its temperature changes. A more accurate calculation would take this into account by using the predicted fluid exit temperature to correct for the fluid properties [by using $(\langle T_f \rangle_0 + \langle T_f \rangle_L)/2$ for fluid properties]. This would require an iteration.

### 7.3 Laminar and Turbulent Flows, Entrance Effect, Thermobuoyant Flows, and Phase Change

The local dimensionless, surface-convection conductance $\text{Nu}_D$ is defined by (7.9). In determining the heat transfer over a finite tube length $L$, we use the tube-length average Nusselt number $\langle \text{Nu} \rangle_D$ as defined by (7.10). Here we discuss some factors influencing $\text{Nu}_D$ and $\langle \text{Nu} \rangle_D$. In general $\text{Nu}_D$ depends on the tube Reynolds number $\text{Re}_D$, Prandtl number Pr, flow regime (laminar, transitional, or fully-turbulent), distance from tube entrance $L$, surface roughness, cooled versus heated fluid (i.e., $T_s < \langle T_f \rangle$ versus $T_s > \langle T_f \rangle$), etc. Liquid-gas phase change also influences $\text{Nu}_D$. These can be summarized as

$$\text{Nu}_D = \text{Nu}_D(\text{Re}_D, \text{Pr, flow regime, distance from entrance } L, \text{ surface} \atop \text{roughness, cooled versus heated fluid, phase change, etc.}), \quad (7.35)$$

where

$$\text{Re}_D = \frac{\langle u_f \rangle D}{\nu_f}, \quad \text{Pr} = \frac{\nu_f}{\alpha_f}. \quad (7.36)$$

We now discuss some of these factors.

### 7.3.1 Laminar versus Turbulent Flow: Transition Reynolds Number $\text{Re}_{D,t}$

For tubes, the laminar, transitional, or fully-turbulent flow regimes are determined based on the tube Reynolds number $\text{Re}_D$, i.e.,

$$\text{Re}_{D,t} \equiv \begin{cases} \text{Re}_D < 2{,}300 & \text{laminar flow regime} \\ 2{,}300 < \text{Re}_D < 10{,}000 & \text{transitional flow regime} \\ \text{Re}_D > 10{,}000 & \text{fully-turbulent flow regime.} \end{cases} \quad (7.37)$$

Here, $\text{Re}_{D,t} = 2{,}300$ marks the transition from laminar flow to an unstable, transitional (between laminar and turbulent) flow. When there is a lack of heat transfer correlations for the transitional flow regime, the fully-turbulent flow correlations may be used.

Thermal and Viscous Entrance Lengths

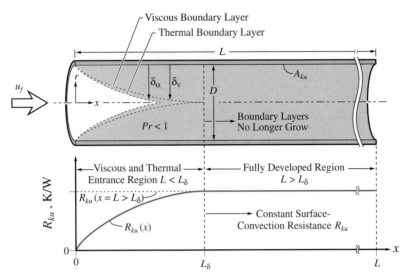

Figure 7.6. Developing (i.e., growing) boundary layers (viscous and thermal) along axial $x$-direction in the entrance region of a tube. After a length $L$, the surface-convection resistance no longer varies.

The presence of surface roughness reduces the surface-convection resistance $R_{ku}$. The cooled versus heated fluid condition influences the fluid viscosity, which in turn influences the hydrodynamics (i.e., velocity distribution). Also, the fluid properties are taken into account in separate and various forms, e.g., including the evaluation of properties at the average (between the inlet and the outlet) fluid temperature, $(\langle T_f \rangle_0 + \langle T_f \rangle_L)/2$.

### 7.3.2 Developing versus Fully Developed Region, Entrance Length $L_\delta$: Laminar Flow

For flow entering a tube with a uniform temperature (across the flow cross section), the local surface-convection resistance $A_{ku}R_{ku}$ is zero at the entrance to the tube (the boundary-layer thickness is zero). The vanishing of the local surface-convection resistance at the entrance (which corresponds to an infinitely large local $Nu_D$ and $q_{ku}$) is similar to that which we found for parallel flows over a semi-infinite plate (Sections 6.2.3 and 6.2.5). Along the tube, the surface-convection resistance increases and reaches an asymptotic value when the boundary layers emanating from the tube internal surface reach and meet at the tube centerline. This is shown in Figure 7.6. The location along the tube at which the boundary layers meet (i.e., where the fully-developed region is reached) is designated as the entrance length $x = L_\delta$(m).

The fully developed region is also marked by a fluid temperature distribution $T_f(r, x)$, which is normalized as $T_f^*(r) = (T_s - T_f)/(T_s - \langle T_f \rangle)$, and this does not change along the tube. This allows for $T_s$, $T_f$, and $\langle T_f \rangle$ to vary along $x$, while $T_f^*$ only changes radially.

The magnitude of $\mathrm{Nu}_D$ for laminar flow in a circular tube can be obtained without many simplifying assumptions and the results are available for the entrance region and for the asymptotically reached, fully-developed region. These results will be given in Section 7.4. The entrance length $L_\delta$ for laminar flow is given by

$$\frac{L_\delta}{D} = 0.06\mathrm{Re}_D\mathrm{Pr} = 0.06\mathrm{Pe}_D \quad \begin{array}{l} \text{entrance length for laminar,} \\ \text{circular-tube flow, } \mathrm{Re}_D < \mathrm{Re}_{D,t} = 2,300. \end{array} \quad (7.38)$$

Here $\mathrm{Pe}_D$ is the axial-lateral Péclet number. In Section 5.2, the physical significance of Pe was discussed. This relation shows that as $\mathrm{Re}_D$ increases from 0 to 2,300, $L_\delta/D$ also increases. For oils (Pr $\gg$ 1) $L_\delta/D$ is much larger than that for mercury (Pr is about 0.003 at room temperature at $p = 1$ atm), air (Pr = 0.7 at atmospheric STP), and water (Pr of about 7 at room temperature and $p = 1$ atm). The quantity $(L/D\mathrm{Pe}_D)$ is called the Graetz number.

When the developing region is more than 1/10 of the tube length, we include the entrance effect, i.e., when $L_\delta \geq 0.1L$, where $L$ is the tube length. Then we use a length-averaged Nusselt number. When $L_\delta \leq 0.1L$, i.e., $L/D \geq 0.6\mathrm{Pe}_D$, then the entrance effect can be neglected.

### 7.3.3 Entrance Length $L_\delta$: Turbulent Flow

Because $\mathrm{Nu}_D$ is small for laminar flows, turbulent flow is preferred for the surface-convection heat transfer purposes, unless the fluid viscosity is very large (such as oils and other high viscosity organic liquids), which then requires an excessive pressure drop to be maintained along the tube. So, it is advantageous to have a turbulent flow ($\mathrm{Re}_D > 2,300$, and preferably $\mathrm{Re}_D > 10,000$). The small-scale thermal mixing that occurs in the turbulent-flow structure enhances the lateral (e.g., radial) heat transfer. This mixing is by the turbulent eddies discussed in Section 6.3.

The entrance length $L_\delta$ for fully-turbulent flow is a weaker function of $\mathrm{Re}_D$, as compared to the laminar flow, as is independent of Pr. One of the relations is [6]

$$\frac{L_\delta}{D} \simeq 4.4\mathrm{Re}_D^{1/6} \quad \begin{array}{l} \text{entrance length for fully-turbulent,} \\ \text{circular-tube flow, } \mathrm{Re}_D \geq 10^4. \end{array} \quad (7.39)$$

Again, for a given tube length $L$, the entrance effect is neglected when $L_\delta \leq 0.1L$. Because of the generally shorter entrance lengths (compared to laminar flows), the effect of entrance length on $\langle \mathrm{Nu} \rangle_D$ is generally neglected for turbulent flows.

### 7.3.4 Thermobuoyant Flows

As in Section 6.5, under gravitational acceleration and a significant spatial variation of the fluid temperature, a thermobuoyant motion occurs. Among the common geometries for bounded fluids with potentially significant thermobuoyant flows are the through flows between vertical plates, cellular motion in enclosures, and cellular motion in horizontal fluid layers. The fluid motion in the first geometry is monotonic in direction, while the motion in the last two are recirculating (i.e., cellular).

Section 7.4.5 contains a list of correlations for thermobuoyant-flow Nusselt numbers for the above three geometries.

One of the applications in which the thermobuoyant motion becomes important is in fluid layers introduced for insulating purposes. If the thermobuoyant flow is not significantly reduced, the convection across the layer can become undesirably significant.

In some enclosed-fluid, surface-convection cooling applications, a large thermobuoyant motion is desirable for heat removal without forced flow.

### 7.3.5 Liquid-Gas Phase Change

In Chapter 6, we considered evaporation and condensation adjacent to a solid surface submerged in a semi-bounded liquid or vapor. We used far-field, single-phase liquid or vapor conditions. In internal (i.e., bounded) flows, these far-field conditions become the entrance conditions. As an example, we consider boiling of a subcooled liquid, $\langle T_f \rangle_0 < T_{lg,0}(p_g)$, in a tube. The entrance vapor quality is zero, i.e., $x = 0$. As an initially subcooled, single-phase liquid stream with velocity $\langle u_f \rangle = \langle u_l \rangle$ enters, it is heated eventually and undergoes phase change (evaporation); the generated vapor (if not collapsed in a subcooled liquid) flows downstream. Then the liquid-vapor interface will not be continuous and simple.

Figure 7.7 depicts an internal, upward flow of an initially (entrance condition) subcooled liquid through a tube heated with a constant heat flux $q_s = q_{ku}$. Right at the entrance to the tube, the solid surface temperature $T_s$ is nearly the inlet subcooled liquid temperature $\langle T_f \rangle_0$, because no thermal boundary layer is yet developed. With the increase in the distance from the entrance $x$, the surface temperature increases. When the temperature at the surface reaches the bubble-nucleation inception temperature, bubble nucleation begins on the surface. These bubbles collapse in the subcooled liquid and heat the liquid until a distance downstream where the liquid subcooling diminishes and the bubbles no longer collapse and begin to coalesce. As the amount of vapor in the cross section of the tube increases with the distance $z$, i.e., the void (or vapor) volume fraction increases, the liquid will flow as a liquid film and as dispersed droplets. Finally, the liquid film disappears and only dispersed droplets flow, which then completely evaporate. After that, the flow consists of superheated vapor.

Many regimes of two-phase flow and heat transfer can be identified in this upward liquid flow and evaporation process. A simple classification identifying seven regimes is shown in Figure 7.7. Starting from the entrance to the tube, after the regime of subcooled-liquid flow with solid-surface bubble nucleation, the regime of bubbly flow and solid-surface bubble nucleation is encountered. As the void (or vapor) volume fraction increases, the plug- and churn-flow regimes, with the continued

Boiling and Vapor Superheating in Bounded, Inlet-Subcooled Liquid Flow

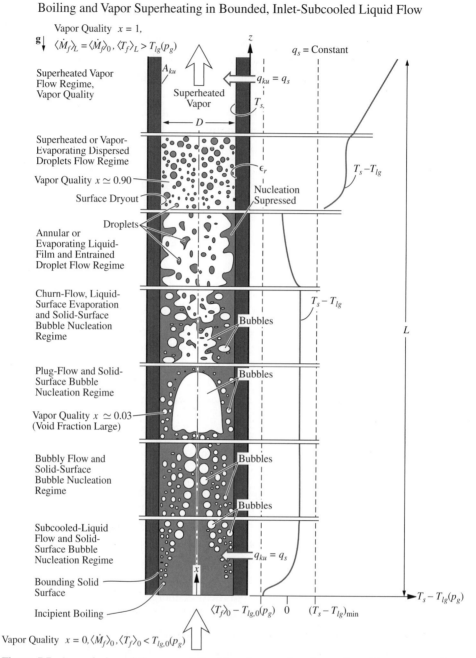

Figure 7.7. A rendering of the upward liquid and vapor flow in a tube with a constant heat flux. The inlet liquid is subcooled $\langle T_f \rangle_0 < T_{lg,0}(p_g)$. The tube surface superheat distribution $T_s - T_{lg}$, as well as the various regimes encountered along the tube axis, are also shown. The vapor quality $x$ is also shown.

solid-surface bubble nucleation, are observed. Further downstream, with the liquid fraction decreasing, the thin liquid film (in the annular-flow regime) evaporation occurs and the solid-surface bubble nucleation is suppressed. As a result of this change in the mechanism of vapor formation, the surface temperature decreases. The surface temperature increases once the film totally evaporates. In the dispersed-droplet superheated-vapor flow regime, the evaporation of the droplets prevents large surface temperatures. Large solid-surface temperatures occur in the superheated-vapor flow regime, where $x = 1$. The anticipated surface temperature distribution along the tube is also shown in Figure 7.7. Due to the decrease in the local pressure along the tube, the saturation temperature also decreases. The beginning of the dispersed-droplet regime is generally marked by a minimum in the surface superheat $(T_s - T_{lg})_{min}$.

Note that because of the substantial difference between the vapor quality $x$ and void fraction (the void fraction being much larger, because the vapor occupies a larger volume), a vapor quality $x$ of only a few percent can result in void fractions as large as 50% (depending on how far the state is from the critical state).

Correlations for boiling in horizontal and vertical tubes will be discussed in Section 7.4.5.

In contrast to evaporation, heat removal from a bounded vapor stream results in condensation. Condensation in turbulent vapor flow through a tube is fairly well correlated using an annular liquid-gas flow regime model. This correlation will be discussed in Section 7.4.5.

## 7.4 Summary of Nusselt Number Correlations

In addition to circular tubes, tubes with other cross-sectional geometries (e.g., square, annulus) are used. The surfaces can also have added surface extensions (fins). In order to further increase the surface area, discontinuous surfaces with large surface areas per unit volume (called the specific surface area $A_{ku}/V$) are used. Here we first discuss the continuous solid surfaces. Then we summarize the correlations for $\langle Nu \rangle_D$ given in the tables. Finally, we will discuss discontinuous solid surfaces and the extended surfaces.

### 7.4.1 Laminar Flow: Tube Cross-Sectional Geometry Dependence

The results for fully-developed, laminar $\langle Nu \rangle_D$ for various tube cross-sectional geometries are listed in Table 7.2 (the tables of correlations are grouped together and presented in Section 7.4.5). These include circular, square, and triangular cylinders, parallel plates, and annuli. The results for $T_s$ or $q_s$ specified at the surface are given. Note that $\langle Nu \rangle_D$ depends on the particular cross-sectional geometry (this is in contrast to the turbulent flow discussed below). The results for developing, laminar flow in circular tubes are also given in Table 7.2, for the uniform $T_s$ condition. Since the penetrative lateral conduction is significant in the entrance region, the Péclet

Table 7.2. *Correlations for Nusselt number for bounded fluid streams, laminar and forced flows*

---

### Bounded Fluid Streams: Laminar Flows

---

Laminar, Fully-Developed Fields, Forced Flows, $\mathrm{Re}_{D,h} \leq 2{,}300$,

$$\frac{L}{D_h} \geq 0.6\,\mathrm{Re}_{D,h}\,\mathrm{Pr}, \quad D_h = \frac{4A_u}{P_{ku}}$$

Circular Cylinders

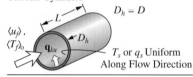

$D_h = D$

$T_s$ or $q_s$ Uniform Along Flow Direction

$T_s$ Uniform: $\langle \mathrm{Nu} \rangle_{D,h} = 3.66$

$q_s$ Uniform: $\langle \mathrm{Nu} \rangle_{D,h} = 4.36$

---

Rectangular Cylinders

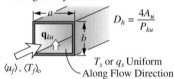

$D_h = \dfrac{4A_u}{P_{ku}}$

$T_s$ or $q_s$ Uniform Along Flow Direction

| $a/b$ | $\langle \mathrm{Nu} \rangle_{D,h}$ | |
|---|---|---|
| | $T_s$ Uniform | $q_s$ Uniform |
| 1 | 2.98 | 3.61 |
| 2 | 3.39 | 4.12 |
| 3 | 4.79 | 3.96 |
| 8 | 5.60 | 6.49 |
| $\infty$ | 7.54 | 8.23 |

---

Triangular Cylinders

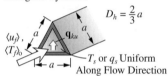

$D_h = \dfrac{2}{3}a$

$T_s$ or $q_s$ Uniform Along Flow Direction

$T_s$ Uniform: $\langle \mathrm{Nu} \rangle_{D,h} = 2.47$

$q_s$ Uniform: $\langle \mathrm{Nu} \rangle_{D,h} = 3.11$

---

Parallel Plates

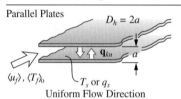

$D_h = 2a$

$T_s$ or $q_s$ Uniform Flow Direction

$T_s$ Uniform: $\langle \mathrm{Nu} \rangle_{D,h} = 7.54$

$q_s$ Uniform: $\langle \mathrm{Nu} \rangle_{D,h} = 8.23$

---

Circular Annuli: Inner-Surface Convection

$D_h = D_2 - D_1$

Outer Surface: Ideal Insulation $q = 0$

$T_s, \langle \mathrm{Nu} \rangle_{D,h}$

| $D_1/D_2$ | $\langle \mathrm{Nu} \rangle_{D,h}$ |
|---|---|
| 0.05 | 17.81 |
| 0.10 | 11.91 |
| 0.20 | 8.499 |
| 0.40 | 6.583 |
| 0.60 | 5.912 |
| 0.80 | 5.580 |
| Parallel Plates → 1.00 | 5.385 |

---

Laminar, Developing Fields (Entrance Effect), Forced Flows

$$\frac{L}{D_h} \leq 0.6\,\mathrm{Re}_{D,h}\,\mathrm{Pr}, \quad \mathrm{Pe}_{D,h} = \mathrm{Re}_{D,h}\,\mathrm{Pr}$$

For Circular Cross Sections: Hydraulic Diameter $D_h = D$, $T_s$ Uniform

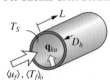

$$\langle \mathrm{Nu} \rangle_{D,h} = 2.409 \left( \frac{L/D_h}{\mathrm{Pe}_{D,h}} \right)^{-1/3} - 0.7 \qquad 0 < \frac{L/D_h}{\mathrm{Pe}_{D,h}} \leq 0.03$$

$$\langle \mathrm{Nu} \rangle_{D,h} = 3.66 + \frac{0.0499}{(L/D_h)/\mathrm{Pe}_{D,h}} \qquad 0.03 \leq \frac{L/D_h}{\mathrm{Pe}_{D,h}}$$

Table 7.3. *Correlations for Nusselt number for bounded fluid streams, for transitional-flow and turbulent-flow regime forced flows*

---

**Bounded Fluid Streams: Turbulent Flows**

---

**Turbulent, Fully Developed Fields, Forced Flow, $Re_{D,h} \gtrsim 10^4$**

| For All Cross-Sectional Geometries (Use Hydraulic Diameter) | $\dfrac{L}{D_h} > 44\,Re_{D,h}^{1/6},\ \ D_h = 4A_u/P_{ku}$ |
|---|---|

---

| $0.7 < Pr < 160$ | Uniform $T_s$ Along Flow Direction |
|---|---|
| $\langle Nu \rangle_{D,h} = 0.023\,Re_{D,h}^{4/5}\,Pr^n$ | $n = 0.3$ for $T_s < \langle T_f \rangle_0$, $\quad n = 0.4$ for $T_s > \langle T_f \rangle_0$ |

---

**$Pr < 0.1$ (Liquid Metals)**

| $\langle Nu \rangle_{D,h} = 4.8 + 0.0156\,Re_{D,h}^{0.85}\,Pr^{0.93}$ | Uniform $T_s$ |
|---|---|
| $\langle Nu \rangle_{D,h} = 6.3 + 0.0167\,Re_{D,h}^{0.85}\,Pr^{0.93}$ | Uniform $q_s$ |

---

Rough Surface, Circular Tube, Using the Surface Friction Coefficient $c_f$

$$\langle Nu \rangle_{D,h} = \frac{(c_f/2)\,(Re_{D,h}-1{,}000)\,Pr}{1 + 12.7(c_f/2)^{1/2}\,(Pr^{2/3}-1)}, \qquad \frac{1}{c_f^{1/2}} = -2.0\log\left(\frac{\langle \delta^2 \rangle^{1/2}}{3.7\,D} + \frac{2.51}{Re_{D,h}\,c_f^{1/2}}\right)$$

$\langle \delta^2 \rangle^{1/2}$ is the Root-Mean Square of Surface Roughness

---

**Transitional Flow $2{,}300 < Re_{D,h} < 10^4$**

$$\langle Nu \rangle_{D,h}^{10} = (\langle Nu \rangle_{D,h})_l^{10} + \left(\frac{e^{(2{,}200-Re_{D,h})/365}}{(\langle Nu \rangle_{D,h})_l^2} + \frac{1}{(\langle Nu \rangle_{D,h})_t^2}\right)^{-5}$$

$(\langle Nu \rangle_{D,h})_l$ and $(\langle Nu \rangle_{D,h})_t$ are the laminar and turbulent flow-regime Nusselt number correlations given in Table 7.2 and Table 7.3 (above).

---

(axial-lateral) number $Pe_D$, discussed in Section 6.2, appears in the relation. Note that for $L/D \to \infty$, the fully developed result of $\langle Nu \rangle_D = 3.66$ is recovered.

### 7.4.2 Turbulent Flow: Geometric Universality of Using Hydraulic Diameter $D_h$

In turbulent flows, the presence of small-scale eddies causes thermal mixing, and therefore, at a short distance from the surface the geometry of the surface plays a less significant role. The concept of hydraulic diameter $D_h$, defined as

$$D_h = \frac{4 \times \text{flow cross-sectional area}}{\text{perimeter for surface convection}} = \frac{4A_u}{P_{ku}} \quad \text{hydraulic diameter,} \quad (7.40)$$

allows for the unification of $\langle Nu \rangle_D$ correlations, for fully-developed turbulent flow, and for circular and noncircular tubes and annuli. This is unlike the results for fully-developed laminar flow $\langle Nu \rangle_D$ listed in Table 7.2. Table 7.3 lists the $\langle Nu \rangle_{D,h}$ correlations for fully-developed turbulent flows. Note that for liquid metal, a separate correlation is recommended. This table also includes correlations for bounded, thermobuoyant flows. Further correlations are found in [6]. Note that from (7.40), for circular cross sections we have $D_h = D$, when we use $P_{ku} = \pi D$ and $A_u = \pi D^2/4$.

Packed Beds of Monosize Spherical Particles

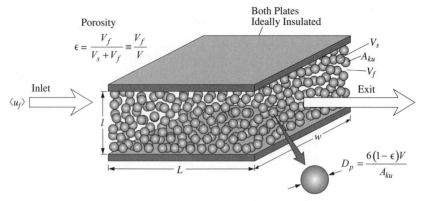

Figure 7.8. A packed bed of monosize spherical particles.

Subject to some assumptions and simplifications, a semi-empirical correlation for $\langle Nu \rangle_{D,h}$ is derived.[†]

### 7.4.3  Discontinuous Solid Surfaces and Large Specific Surface Areas: Geometric Universality of Particle Diameter $D_p$

#### (A) Large Specific Surface Area

In order to increase the surface area for surface-convection heat transfer $A_{ku}$, or in particular the surface area per unit volume $A_{ku}/V\,(\text{m}^2/\text{m}^3$ or $1/\text{m})$, also called the specific surface area, banks of tubes, or packed beds of spherical particles (and other particle geometries) are used. For example, consider the packed bed of spherical particles, shown in Figure 7.8. The solid particles have an average diameter $D_p$ and the bed has a porosity $\epsilon$, defined by (3.27) as the ratio of the fluid volume $V_f$ to the

---

[†] The surface-convection heat transfer to a fluid flowing turbulently in a smooth pipe, with $T_s$ uniform, can be predicted and expressed by the dimensionless surface-convection conductance, i.e., the Nusselt number $Nu_D$, (7.9). This is done by using measured velocity distributions (in order to define the radial distribution of the turbulent thermal diffusivity), in an analysis proposed by Seban and Shimazaki ("Heat Transfer to a Fluid Flowing Turbulently in a Smooth Pipe with Walls at Constant Temperature," 1951, *ASME Transactions*, pp. 803–809).

   The analysis is based on an iterative method, where a radial distribution of fluid mean temperature, as $(T_s - T_f)/[T_s - T_f(r = 0)] = (1 - r/R)^{1/7}$, found from the constant wall heat flux $(\overline{q}_s)$ case is used as the first approximation for the determination of the fluid temperature distribution. The temperature distribution obtained from the first iteration serves as the next approximation in the iterative procedure, until the temperature distribution converges.

   The analysis is divided into following steps. In step one, an expression is found for the radial distribution of fluid temperature $\overline{T}_f - \overline{T}_f(r = 0)$, as a function of the dimensionless temperature distribution $T^* = (T_s - T_f)/[T_s - T_f(r = 0)]$, and the turbulent thermal diffusivity distribution $\alpha_t(r)$. In the second step, the radial dependence of the turbulent thermal diffusivity is determined, and then, it is used along with the expression for the radial temperature distribution for constant heat flux. Then a new temperature distribution is found. When convergence of the temperature profile is obtained, the Nusselt number is determined. This Nusselt number is compared with the Dittus and Boelter correlation (given in Table 7.3). These steps are divided into three end of chapter problems.

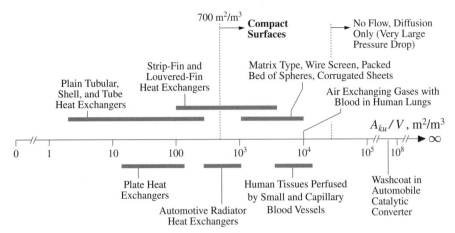

Figure 7.9. Examples of large specific surface area $A_{ku}/V$.

total (fluid plus solid) volume $V$. We can give $A_{ku}/V$ in terms of $\epsilon$ and $D_p$ as

$$\epsilon \equiv \frac{V_f}{V} \equiv \frac{V_f}{V_s + V_f}, \text{ or } \frac{V_s}{V_s + V_f} = 1 - \epsilon \qquad \begin{array}{l} \text{definition of porosity,} \\ \text{and solid fraction} \end{array} \qquad (7.41)$$

$$\frac{A_{ku}}{V_s} = \frac{\pi D_p^2}{\pi D_p^3/6} = \frac{6}{D_p}, \text{ or } \frac{A_{ku}}{V} = \frac{6(1-\epsilon)}{D_p} \qquad \begin{array}{l} \text{specific surface area for} \\ \text{a packed bed of spheres.} \end{array} \qquad (7.42)$$

Then, as $D_p$ decreases, $A_{ku}/V$ becomes very large. For example, the random arrangement of monosized (i.e., same size) spherical particles gives a porosity of near 0.40 (independent of $D_p$). Using (7.42) we have

$$\frac{A_{ku}}{V} = \frac{3.6}{D_p} \qquad \text{specific surface area for randomly packed bed of spheres.} \qquad (7.43)$$

For $D_p = 10^{-6}, 10^{-4}$, and $10^{-3}$ m, we have $A_{ku}/V_s$ equal to $3.6 \times 10^6, 3.6 \times 10^4$, and $3.6 \times 10^3$ m$^2$/m$^3$. While flow through beds of particles with $D = 10^{-6}$ m (one $\mu$m) requires a large pressure drop, beds of $D = 10^{-3}$ m (one mm) particles give a surface area of 2,400 m$^2$/m$^3$ and offer a significant surface area for surface-convection heat transfer with a reasonable pressure drop (especially for gases).

Figure 7.9 gives a classification of surface-convection specific surface area into the compact and noncompact surfaces. Typical heat exchanger applications are also shown. Note the large specific surface area of the human lungs.

### (B) Correlations for $\langle Nu \rangle_{D,p}$

Table 7.5 lists $\langle Nu \rangle_{D,p}$ for a bank of tubes or a packed bed of spheres (also applies to nonspherical particles with an equivalent particle diameter $D_p$), or any porous medium [12]. A general equivalent particle diameter $D_p$ is defined through (7.42) as

$$D_p \equiv \frac{6V_s}{A_{ku}} = \frac{6(1-\epsilon)V}{A_{ku}} \qquad \text{particle diameter based on volume and surface area.} \qquad (7.44)$$

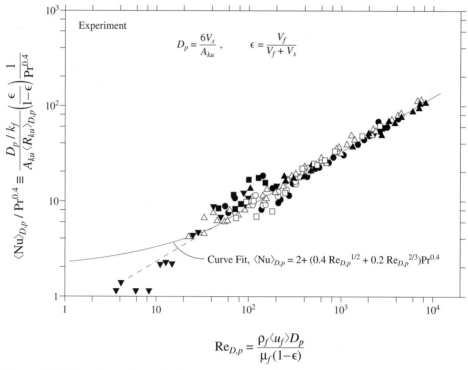

Figure 7.10. Experimental results (for different fluids) and curve fit for surface-convection heat transfer for packed bed of particles and bank of tubes. (From Whitaker, S., reproduced by permission ©1973, Krieger.)

Then the particle Reynolds number $\mathrm{Re}_{D,p}$ and the particle Nusselt number $\langle \mathrm{Nu} \rangle_{D,p}$, are defined as

$$\mathrm{Re}_{D,p} \equiv \frac{\rho_f \langle u_f \rangle D_p}{\mu_f (1-\epsilon)} \quad \text{Reynolds number for discontinuous solids} \tag{7.45}$$

$$\langle \mathrm{Nu} \rangle_{D,p} \equiv \frac{D_p}{A_{ku} \langle R_{ku} \rangle_{D,p} k_f} \left( \frac{\epsilon}{1-\epsilon} \right) \quad \begin{array}{l}\text{Nusselt number for} \\ \text{discontinuous solids,}\end{array} \tag{7.46}$$

where $L$ is the bed linear dimension, and $\langle u_f \rangle$ is the average velocity, which is related to the fluid mass flow rate through

$$\dot{M}_f = A_u \dot{m}_f = A_u \rho_f \langle u_f \rangle. \tag{7.47}$$

The flow area $A_u$ is taken to be the cross-sectional area (including solid and void) through which the apparent stream flows. Therefore, $\langle u_f \rangle$ is the area-averaged velocity, averaged over the combined solid and fluid cross-sectional areas. This is done for the sake of convenience.

One of the correlations for $\langle \mathrm{Nu} \rangle_{D,p}$, as a curve fit to the experimental results, is given in Figure 7.10. This correlation allows for a laminar, stagnation-flow regime signified by $\mathrm{Re}_{D,p}^{1/2}$ and an inertial-flow regime signified by $\mathrm{Re}_{D,p}^{2/3}$. It also includes a

Prandtl number dependence. Note that for $\mathrm{Re}_{D,p} = 0$, we have $\langle \mathrm{Nu} \rangle_{D,p} = 2$ which is the conduction limit discussed in Section 6.7.1(A).

The analyses of Sections 7.2.5 and 7.2.6 also apply here, and $A_{ku}$ will be the total surface area available for surface-convection heat transfer. Again, we assume that $A_{ku}$ is maintained at a uniform temperature $T_s$. This can be done in the cases of a bank of tubes, for example, by flowing a fluid through them. For the case of spherical particles, heat generation or large heat storage can maintain $T_s$ uniform.

The area for the fluid flow $A_u$ is defined as the cross-sectional area through which the fluid enters and exits (this includes both solid and fluid). In practice, the mass flow rate $\dot{M}_f$ flowing into the packed bed or the bank of tubes is used and includes the area $A_u$.

### 7.4.4 Extended Surfaces: Overall Surface Efficiency

Similar to extended surfaces used with the external flows (semi-bounded fluid), as discussed in Section 6.8.2, when the internal flow (bounded fluid) surface-convection resistance is large, extended surfaces are used. The extended surfaces offer large $A_{ku}$, but due to conduction resistance, the surface temperature over the entire extended surface is not $T_s$. We will allow for the nonuniformity of the surface temperature of extended surfaces using the base temperature as $T_s$ with the extended surfaces having a temperature nonuniformity represented by the fin efficiency $\eta_f$. Then using the bare surface area $A_b$, the total fin area $A_f$, and the fin efficiency $\eta_f$, the total surface-convection resistance [similar to (6.151)] is

$$\frac{1}{\langle R_{ku} \rangle_D} = \sum \frac{1}{R_{ku,i}} = \frac{1}{R_{ku,b}} + \frac{1}{R_{ku,f}} \quad \begin{array}{l} \text{combined bare-} \\ \text{and fin-surface resistances} \end{array} \qquad (7.48)$$

$$R_{ku,b}^{-1} = A_b \langle \mathrm{Nu} \rangle_D \frac{k_f}{D}, \quad R_{ku,f}^{-1} = A_f \eta_f \langle \mathrm{Nu} \rangle_D \frac{k_f}{D}. \qquad (7.49)$$

Then in the expression for $NTU$, (7.20), we simply use the modified surface area

$$
\begin{aligned}
A_{ku} &= A_b + A_f \eta_f \\
&\equiv (A_b + A_f) \eta_{f,o}
\end{aligned}
\quad \begin{array}{l} \text{modified surface-convection area} \\ \text{and overall fin efficiency } \eta_{f,o}, \end{array} \qquad (7.50)
$$

where in (7.49) we have assumed and used the same $\langle \mathrm{Nu} \rangle_D$ for the bare and fin surfaces and also have defined an overall fin efficiency $\eta_{f,o}$ for the surface-convection surface area.

### 7.4.5 Tabular Presentation of Correlation for Some Geometries

Tables 7.2 to 7.6 list the correlation for $\langle \mathrm{Nu} \rangle_D$ for bounded fluids. The correlations for the bounded fluid streams in the laminar-flow regime are given in Table 7.2. These include the various tube cross-sectional geometries and the entrance effect. Note that the entrance effect is negligible for $L/(D_h \mathrm{Re}_{D,h} \mathrm{Pr}) > 0.6$.

Table 7.4. *Correlations for Nusselt number for bounded fluid streams, for thermobuoyant flows. The thermal expansion for gases is idealized with $\beta_f = 1/T_f$, and for some liquids $\beta_f$ is listed in Table C.23, where the average fluid temperature of $(T_s + T_{f,\infty})/2$ or $(T_{s,1} + T_{s,2})/2$ is used for this and other properties. The correlations are for both laminar and turbulent flows*

**Bounded Fluid Streams: Thermobuoyant Flows**

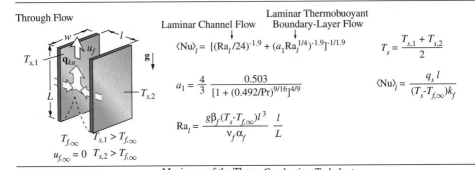

Through Flow

Laminar Channel Flow → Laminar Thermobuoyant Boundary-Layer Flow →

$$\langle Nu \rangle_l = [(Ra_l/24)^{-1.9} + (a_1 Ra_l^{1/4})^{-1.9}]^{-1/1.9}$$

$$T_s = \frac{T_{s,1} + T_{s,2}}{2}$$

$$a_1 = \frac{4}{3} \frac{0.503}{[1 + (0.492/Pr)^{9/16}]^{4/9}}$$

$$\langle Nu \rangle_l = \frac{q_s \, l}{(T_s - T_{f,\infty})k_f}$$

$$Ra_l = \frac{g\beta_f(T_s - T_{f,\infty})l^3}{\nu_f \alpha_f} \frac{l}{L}$$

$T_{f,\infty} \quad T_{s,1} > T_{f,\infty}$
$u_{f,\infty} = 0 \quad T_{s,2} > T_{f,\infty}$

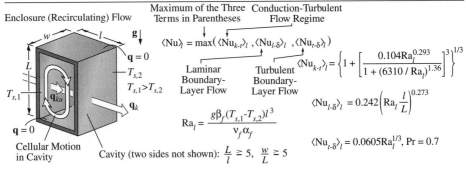

Enclosure (Recirculating) Flow

Maximum of the Three Terms in Parentheses → Conduction-Turbulent Flow Regime →

$$\langle Nu \rangle = max(\langle Nu_{k\text{-}t} \rangle_l, \langle Nu_{l\text{-}\delta} \rangle_l, \langle Nu_{l\text{-}\delta} \rangle_l)$$

Laminar Boundary-Layer Flow → Turbulent Boundary-Layer Flow →

$$\langle Nu_{k\text{-}t} \rangle_l = \left\{ 1 + \left[ \frac{0.104 Ra_l^{0.293}}{1 + (6310/Ra_l)^{1.36}} \right]^3 \right\}^{1/3}$$

$$Ra_l = \frac{g\beta_f(T_{s,1} - T_{s,2})l^3}{\nu_f \alpha_f}$$

$$\langle Nu_{l\text{-}\delta} \rangle_l = 0.242 \left( Ra_l \frac{l}{L} \right)^{0.273}$$

$$\langle Nu_{l\text{-}\delta} \rangle_l = 0.0605 Ra_l^{1/3}, \quad Pr = 0.7$$

Cellular Motion in Cavity

Cavity (two sides not shown): $\frac{L}{l} \geq 5, \quad \frac{w}{L} \geq 5$

Walls 1 and 2 have $q \neq 0$; all other walls have $q = 0$.

Cellular Horizontal Fluid Layer (Heated From Below) Flow

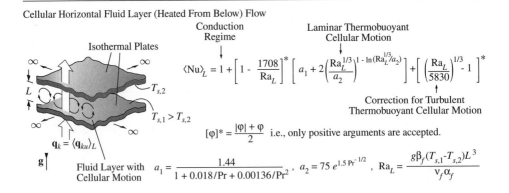

Conduction Regime → Laminar Thermobuoyant Cellular Motion →

$$\langle Nu \rangle_L = 1 + \left[ 1 - \frac{1708}{Ra_L} \right]^* \left[ a_1 + 2\left( \frac{Ra_L^{1/3}}{a_2} \right)^{1 - \ln(Ra_L^{1/3}/a_2)} + \left[ \left( \frac{Ra_L}{5830} \right)^{1/3} - 1 \right] \right]^*$$

Correction for Turbulent Thermobuoyant Cellular Motion →

$$[\varphi]^* = \frac{|\varphi| + \varphi}{2} \text{ i.e., only positive arguments are accepted.}$$

Isothermal Plates

$T_{s,2}$
$T_{s,1} > T_{s,2}$

$q_k = \langle q_{ku} \rangle_L$

Fluid Layer with Cellular Motion

$$a_1 = \frac{1.44}{1 + 0.018/Pr + 0.00136/Pr^2}, \quad a_2 = 75 \, e^{1.5 \, Pr^{-1/2}}, \quad Ra_L = \frac{g\beta_f(T_{s,1} - T_{s,2})L^3}{\nu_f \alpha_f}$$

Most of the results are curve fit to experimental results (such as Figure 7.10). The expected accuracy is generally ±20.

Table 7.3 is for transitional and turbulent forced flows. For turbulent flows, Table 7.3 also uses the concept of hydraulic diameter. The transitional-flow regime (2,300 < $Re_{D,h} < 10^4$) correlation, uses the laminar- and turbulent-flow regime correlations

Table 7.5. *Correlations for Nusselt number for bounded fluid streams, for discontinuous solid surfaces and large specific surface areas, including packed bed of particles*

---

Bounded Fluid Streams: Packed Beds of Particles, Porous Solids, and Collection of Cylinders

Iso-diametral Spherical Particles

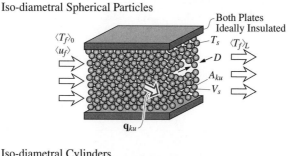

Iso-diametral Cylinders

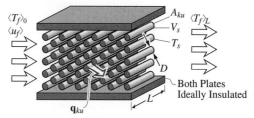

---

Other Particle Geometries

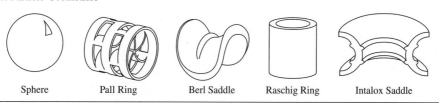

| Sphere | Pall Ring | Berl Saddle | Raschig Ring | Intalox Saddle |

---

listed in Tables 7.2 and 7.3. A correlation is given for the liquid metals ($Pr \ll 1$) for the turbulent-flow regime, in Table 7.3.

Table 7.4 lists the Nusselt number correlations for three thermobuoyant flows. These are thermobuoyant flow between two isothermal vertical plates (the plate temperatures are different than the far-field fluid temperature), thermobuoyant flow in an enclosed fluid with two of its vertical walls maintained at uniform, but different, temperatures (all other surfaces ideally insulated), and horizontal fluid layers heated from below. Further correlations are listed in [9].

Table 7.5 is for discontinuous solid surfaces, surfaces with large specific surface areas, including packed beds of particles, and other porous media, and uses the concept of equivalent particle diameter.

A correlation for condensation of turbulent vapor flow in tubes is given in Table 7.6. Examples of boiling correlations for horizontal and vertical tubes are also listed [11]. These correlations use the vapor quality $x$ and either the Reynolds or the Froude number. The boiling correlations also use the Kutateladze number.

The lists are not extensive and further relations are given in [5, 6, 10, 11].

As an example of the range of $\langle Nu \rangle_D$, water flow in circular tubes is considered in Figure 7.11. At the speed of sound in water ($a_s = 1,497$ m/s at STP), and for

Table 7.6. *Correlations for Nusselt number for bounded fluid streams, for forced flow with phase change*

---

### Bounded Fluid Streams: Phase Change

---

Condensation: Annular Liquid-Film Regime

$$\text{Re}_{l,eq} = \frac{\dot{m}_{eq} D_h}{\mu_l}$$

$$\langle \text{Nu} \rangle_{D,h} = 5.03 \, \text{Re}_{l,eq}^{1/3} \, \text{Pr}_l^{1/3} \quad , \quad \text{Re}_{l,eq} < 50{,}000$$

$$\langle \text{Nu} \rangle_{D,h} = 0.0265 \, \text{Re}_{l,eq}^{0.8} \, \text{Pr}_l^{1/3} \quad \text{Re}_{l,eq} > 50{,}000$$

$$\dot{m}_{eq} = \dot{m}_f \left[ (1-x) + x \left( \frac{\rho_l}{\rho_g} \right)^{1/2} \right] \qquad x \text{ is the vapor quality, } x = \frac{\dot{m}_g}{\dot{m}_f}.$$

---

Boiling

(i) Vertical Tubes

$$\frac{\langle \text{Nu} \rangle_{D,h}}{(\langle \text{Nu} \rangle_{D,h})_l} = 1 + 3{,}000 \, \text{Ku}^{0.86} + 1.12 \left( \frac{x}{1-x} \right)^{0.75} \left( \frac{\rho_l}{\rho_g} \right)^{0.41}$$

(ii) Horizontal Tubes

$$\frac{\langle \text{Nu} \rangle_{D,h}}{(\langle \text{Nu} \rangle_{D,h})_l} = \left[ \frac{\langle \text{Nu} \rangle_{D,h}}{(\langle \text{Nu} \rangle_{D,h})_l} \right]_{\text{Vertical Tube}} \times \text{Fr}_{D,h}^{(0.1 - 2 \text{Fr}_{D,h})}$$

$$\text{Fr}_{D,h} = \frac{\dot{m}_f^2}{\rho_l^2 g D_h} \quad \text{is the Froude number.}$$

$(\langle \text{Nu} \rangle_{D,h})_l$ is the Nusselt number using the total mass flow rate, but assuming liquid flow only and is defined using Tables 7.2 and 7.3.

$\text{Ku} = \frac{\langle q_{ku} \rangle_{D,h}}{\dot{m}_f \Delta h_{lg}}$ is the Kutateladze or boiling number. $x$ is the vapor quality.

---

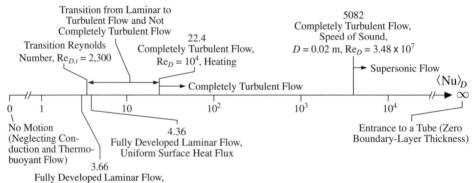

Figure 7.11. Range of $\langle \text{Nu} \rangle_D$ for bounded water stream in circular tubes (with no phase change).

$D = 0.02$ m, we have $\langle \mathrm{Nu} \rangle_D = 5{,}082$. For laminar flow, we have $\langle \mathrm{Nu} \rangle_D$ as low as 3.66. The correlations in Tables 7.3 to 7.5 apply under listed constraints, such as a range of $\mathrm{Re}_D$.

### EXAMPLE 7.2. FAM

Water is heated by flowing through a tube (constant surface temperature $T_s$) of diameter $D = 1$ cm. It can flow using a small power pump with a velocity of $\langle u_f \rangle = 0.08$ m/s or, using a larger power pump, with a velocity of $\langle u_f \rangle = 0.5$ m/s.

(a) Determine $\dot{M}_f$ for both velocities.
(b) Determine $\langle \mathrm{Nu} \rangle_D$ for both velocities.
(c) Determine the tube length required if the desired heat exchanger effectiveness $\epsilon_{he}$ is 0.5, for both velocities.

Evaluate the water properties at $T = 330$ K, and neglect the entrance effect.

### SOLUTION

(a) The mass flow rate, for each velocity, is given by (7.3) as

$$\dot{M}_f = A_u \rho_f \langle u_f \rangle,$$

where for a circular tube

$$A_u = \pi \frac{D^2}{4}.$$

From Table C.23, at $T = 330$ K, we have

| | |
|---|---|
| $\rho_f = 986.8$ kg/m$^3$ | Table C.23 |
| $c_{p,f} = 4{,}183$ J/kg-K | Table C.23 |
| $v_f = 505 \times 10^{-9}$ m$^2$/s | Table C.23 |
| $k_f = 0.648$ W/m-K | Table C.23 |
| $\mathrm{Pr} = 3.22$ | Table C.23. |

For $\langle u_f \rangle = 0.08$ m/s

$$\dot{M}_f = \frac{\pi D^2}{4} \rho_f \langle u_f \rangle$$

$$= \frac{\pi \times (10^{-2})^2 (\mathrm{m})^2}{4} \times 986.8 (\mathrm{kg/m}^3) \times 0.08 (\mathrm{m/s}) = 6.201 \times 10^{-3} \ \mathrm{kg/s}.$$

For $\langle u_f \rangle = 0.5$ m/s

$$\dot{M}_f = \frac{\pi \times (10^{-2})^2}{4} \times 986.4 \times 0.5 = 3.875 \times 10^{-2} \ \mathrm{kg/s}.$$

(b) The Nusselt number $\langle Nu \rangle_D$ depends on the Reynolds number $Re_D$. For circular tubes, the correlations are listed in Table 7.2 and Table 7.3. The Reynolds number is given by (7.36), i.e.,

$$Re_D = \frac{\langle u_f \rangle D}{\nu_f}.$$

For $\langle u_f \rangle = 0.8$ m/s, we have

$$Re_D = \frac{0.08(\text{m/s}) \times 10^{-2}(\text{m})}{5.05 \times 10^{-7}(\text{m}^2/\text{s})}$$

$$= 1{,}584 < Re_{D,t} = 2{,}300 \quad \text{laminar flow regime.}$$

For $\langle u_f \rangle = 0.5$ m/s,

$$Re_D = \frac{0.5(\text{m/s}) \times 10^{-2}(\text{m})}{5.05 \times 10^{-7}(\text{m}^2/\text{s})}$$

$$= 9{,}901 > Re_{D,t} = 2{,}300 \quad \text{turbulent flow.}$$

For $\langle u_f \rangle = 0.08$ m/s, and under uniform temperature $T_s$ and fully developed conditions, we have from Table 7.2

$$Re_D = 1{,}584 < Re_{D,t} = 2{,}300, \quad \langle Nu \rangle_D = 3.66 \quad \text{Table 7.2.}$$

For $\langle u_f \rangle = 0.5$ m/s, and under fully developed conditions, we have from Table 7.3

$$Re_D = 9{,}901 > Re_{D,t} = 2{,}300, \quad \langle Nu \rangle_D = 0.023 Re_D^{4/5} Pr^n \quad \text{Table 7.3.}$$

Here we have $T_s > \langle T_f \rangle$ and from Table 7.3, $n = 0.4$. Then

$$\langle Nu \rangle_D = 0.023 \times (9{,}901)^{4/5}(3.22)^{0.4} = 57.73.$$

Note the larger Nusselt number associated with turbulent flows.

(c) The effectiveness for the heat exchange between a bounded fluid and its constant-temperature bounding surface is given by (7.22) as

$$\epsilon_{he} = \frac{\langle T_f \rangle_0 - \langle T_f \rangle_L}{\langle T_f \rangle_0 - T_s} = 1 - e^{-NTU}.$$

Here we are given $\epsilon_{he} = 0.5$, or

$$0.5 = 1 - e^{-NTU}$$

$$e^{-NTU} = 0.5$$

$$NTU = -\ln 0.5 = 0.6931.$$

From (7.20), the definition of $NTU$ is

$$NTU = \frac{1}{(\dot{M}c_p)_f \langle R_{ku} \rangle_D} = \frac{A_{ku} \langle Nu \rangle_D k_f}{D(\dot{M}c_p)_f},$$

where

$$A_{ku} = \pi DL.$$

Then

$$NTU = \frac{\pi L k_f \langle Nu \rangle_D}{\dot{M}_f c_{p,f}}.$$

Solving for $L$, we have

$$L = \frac{NTU \dot{M}_f c_{p,f}}{\pi k_f \langle Nu \rangle_D} \qquad \text{required length.}$$

For $\langle u_f \rangle = 0.08$ m/s

$$L = \frac{0.6931 \times 4{,}183\text{(J/kg-K)} \times 6.201 \times 10^{-3}\text{(kg/s)}}{\pi \times 0.648\text{(W/m-K)} \times 3.66} = 2.413 \text{ m.}$$

For $\langle u_f \rangle = 0.5$ m/s

$$L = \frac{0.6931 \times 4{,}183 \times 3.875 \times 10^{-2}}{\pi \times 0.648 \times 57.73} = 0.9560 \text{ m.}$$

Note the shorter length required for the turbulent flow.

**COMMENT**

The required length is directly proportional to $\dot{M}_f$ and inversely proportional to $\langle Nu \rangle_D$. Here the increase in $\langle Nu \rangle_D$ more than compensates for the increase in $\dot{M}_f$, and therefore, a shorter length is needed for the higher mass flow rate. This increase in $\langle Nu \rangle_D$ is due to the change in the flow regime (change from laminar to turbulent regime).

EXAMPLE 7.3. FUN

Ethylene glycol (antifreeze) liquid is cooled by passing it through a bank of tubes, as shown in Figure Ex. 7.3(a). The tubes are maintained at a surface temperature $T_s = 45°C$ by an internal flow of a phase changing fluid. The tubes are arranged within a cubic space ($L \times l \times w$) with $L = l = w = 15$ cm. The tube diameter is $D = 1$ cm and there are 11 tubes along $L$ and 11 tubes along $l$. The ethylene glycol mass flow rate is $\dot{M}_f = 5$ kg/s and its inlet temperature is $\langle T_f \rangle_0 = 90°C$.

(a) Draw the thermal circuit diagram.
(b) Determine the particle Reynolds number, $Re_{D,p}$.
(c) Determine the particle Nusselt number $\langle Nu \rangle_{D,p}$.
(d) Determine $NTU$.

(e) Determine the heat transfer rate $\langle Q_u \rangle_{L-0}$.

(f) Determine the exit temperature of the ethylene glycol stream $\langle T_f \rangle_L$.

Evaluate the properties of the ethylene glycol at $T = 350$ K, using Table C.23.

**SOLUTION**

(a) The thermal circuit diagram is shown in Figure Ex. 7.3(b).

(b) We will use the correlation in Table 7.5 for discontinuous solid surfaces. First we need to evaluate the particle diameter from (7.44), i.e.,

$$D_p = \frac{6V_s}{A_{ku}}.$$

Here the tubes should be treated as solid (because the surface convection under study occurs on the outside surface), and we have for each tube

$$V_s = \pi D^2 w / 4$$

$$A_{ku} = \pi D w.$$

Then

$$D_p = \frac{6\pi D^2 w}{4\pi D w} = \frac{6}{4} D = \frac{3}{2} D = 1.5 \times (0.01)(\text{m})$$

$$= 1.5 \times 10^{-2} \text{ m}.$$

The porosity $\epsilon$ is defined by (7.41) as

$$\epsilon = \frac{V_f}{V_s + V_f} = \frac{Llw - N_t \pi D^2 w / 4}{Llw}$$

$$= 1 - \frac{N_t \pi D^2}{4Ll},$$

where $N_t$ is the number of tubes, which here is $N_t = 11 \times 11 = 121$. Then

$$\epsilon = 1 - \frac{121 \times \pi \times (10^{-2})^2 (\text{m})^2}{4 \times (0.015)^2 (\text{m})^2} = 0.5776.$$

The particle Reynolds number is defined by (7.45) as

$$\text{Re}_{D,p} = \frac{\rho_f \langle u_f \rangle D_p}{\mu_f (1 - \epsilon)},$$

where $\langle u_f \rangle$ is found from $\dot{M}_f$ in (7.47), i.e.,

$$\dot{M}_f = A_u \rho_f \langle u_f \rangle$$

or

$$\langle u_f \rangle = \frac{\dot{M}_f}{A_u \rho_f}, \quad A_u = l \times w.$$

Surface-Convection Heat Transfer:
Bounded Flow over a Bank of Tubes

(a) Physical Model

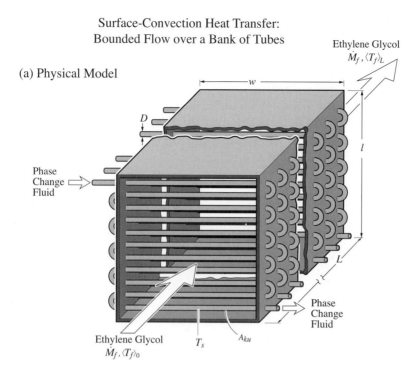

(b) Thermal Circuit Model

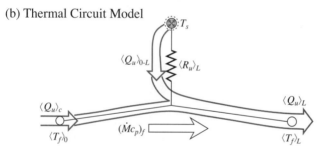

Figure Ex. 7.3. Flow and surface-convection heat transfer in a bank of tubes. (a) Physical model. (b) Thermal circuit model.

Then, using this in the expression for $\text{Re}_{D,p}$, we have

$$\text{Re}_{D,p} = \frac{\dot{M}_f D_p}{\mu_f (1 - \epsilon) l w}.$$

From Table C.23 and at $T = 350$ K, for ethylene glycol we have

$$
\begin{aligned}
\rho_f &= 1{,}079 \text{ kg/m}^3 && \text{Table C.23} \\
c_{p,f} &= 2{,}640 \text{ J/kg-K} && \text{Table C.23} \\
\nu_f &= 3.25 \times 10^{-6} \text{ m}^2/\text{s} && \text{Table C.23} \\
k_f &= 0.261 \text{ W/m-K} && \text{Table C.23} \\
\text{Pr} &= 35.2 && \text{Table C.23.}
\end{aligned}
$$

Then

$$\text{Re}_{D,p} = \frac{5(\text{kg/s}) \times 1.5 \times 10^{-2}(\text{m})}{3.25 \times 10^{-6}(\text{m}^2/\text{s}) \times 1{,}079(\text{kg/m}^3)(1 - 0.5776)(0.15)^2(\text{m})^2}$$

$$= 2.251 \times 10^3.$$

(c) The correlation for $\langle \text{Nu} \rangle_{D,p}$ is given in Table 7.5 as

$$\langle \text{Nu} \rangle_{D,p} = 2 + (0.4\text{Re}_{D,p}^{1/2} + 0.2\text{Re}_{D,p}^{2/3})\text{Pr}^{0.4}.$$

Then

$$\langle \text{Nu} \rangle_{D,p} = 2 + [0.4 \times (2.251 \times 10^3)^{1/2} + 0.2 \times (2.251 \times 10^3)^{2/3}](35.2)^{0.4}$$

$$= 223.6.$$

Note the rather large $\langle \text{Nu} \rangle_{D,p}$.

(d) The *NTU* is defined by (7.20) and, using the definition of $\langle \text{Nu} \rangle_{D,p}$ (i.e., the relation between $\langle \text{Nu} \rangle_{D,p}$ and $\langle R_{ku} \rangle_L$), as given by (7.46), we have

$$NTU = \frac{1}{(\dot{M}c_p)_f \langle R_{ku} \rangle_L} = \frac{A_{ku}}{(\dot{M}c_p)_f} \frac{\langle \text{Nu} \rangle_{D,p} k_f}{D_p} \frac{1 - \epsilon}{\epsilon},$$

where

$$A_{ku} = N_t \pi D w.$$

Then

$$NTU = \frac{11 \times 11 \times \pi \times 0.01(\text{m}) \times 0.15(\text{m}) \times 223.6 \times 0.261(\text{W/m-K}) \times (1 - 0.5776)}{5(\text{kg/s}) \times 2{,}640(\text{J/kg-°C}) \times 0.015(\text{m}) \times 0.5776}$$

$$= 0.1231.$$

(e) The heat transferred rate is given by (7.25) and (7.27), i.e.,

$$\langle Q_u \rangle_{L\text{-}0} = \frac{T_s - \langle T_f \rangle_0}{\langle R_u \rangle_L}$$

$$= -(\langle T_f \rangle_0 - T_s)\dot{M}c_p \epsilon_{he}$$

$$= -(\langle T_f \rangle_0 - T_s)\dot{M}c_p (1 - e^{-NTU})$$

$$= -[90(°\text{C}) - 45(°\text{C})] \times 5(\text{kg/s}) \times 2{,}640(\text{J/kg-°C}) \times (1 - e^{-0.1231})$$

$$= -6.878 \times 10^4 \text{ W}$$

(f) The exit temperature is determined from (7.22), i.e.,

$$\frac{\langle T_f \rangle_0 - \langle T_f \rangle_L}{\langle T_f \rangle_0 - T_s} = 1 - e^{-NTU}$$

or

$$\langle T_f \rangle_L = \langle T_f \rangle_0 - (1 - e^{-NTU})(\langle T_f \rangle_0 - T_s)$$

$$= 90(°C) - (1 - e^{-0.1231})(90 - 45)(°C)$$

$$= 90(°C) - 5.212(°C) = 84.79°C.$$

**COMMENT**

Note that over a short length of $L = 0.15$ m, a $6.342°$C temperature drop occurred for the fluid. This noticeable change over a rather short distance is due to the large surface area and the relatively large Nusselt number. To reduce $\langle T_f \rangle_L$ further, a larger $A_{ku}$ is needed and this is done by using more tubes $N_t$ or by using a larger length $L$.

## 7.5 Inclusion of Bounding Solid

So far, the bounding-surface thermal boundary condition we have mostly been concerned with has been a prescribed inside surface temperature $T_s$, which has been assumed to be uniform (except for Section 7.2.7) and constant (i.e., steady-state heat transfer). A general treatment of a bounding solid can be made by considering various thermal interactions. Figure 7.12 shows these interactions.

Figure 7.12 renders (i) surface-convection heat transfer from the outside surface of the solid (at temperature $T_{s,0}$), the heat transfer by (ii) conduction across a bounding surface (lateral solid conduction), (iii) prescribed heat transfer rate $Q_s$, (iv) sensible heat storage/release within this bounding surface, and (v) conduction along this bounding surface (axial solid conduction). These can be significant to the problem considered and should be included in the analysis. Also shown are

Inclusion of Bounding Surface: Bounded Fluid

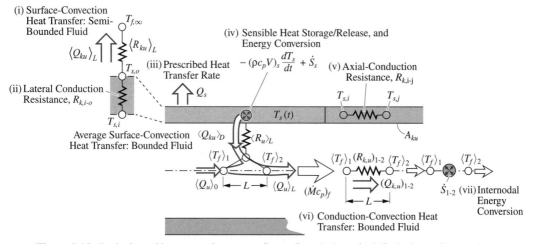

Figure 7.12. Inclusion of heat transfer across (lateral) and along (axial) the bounding surface.

(vi) conduction-convection heat transfer, and (vii) internodal energy conversion within the bounded fluid stream. While the simultaneous inclusion of all these requires a numerical solution, inclusion of the lateral conduction or the lumped-capacitance treatment of heat storage/release can be readily done, leading to closed-form solutions. For example, for bounding-surface energy storage/release with uniform, time-dependent solid temperature $T_s(t)$, the solution given by (6.156) is applicable.

---

### EXAMPLE 7.4. FUN

---

There are many surface-convection heat transfer problems in which the surface-convection surface area per unit volume $A_{ku}/V$ is very large (e.g., Figure 7.8). Then due to this large specific surface area (or large $NTU$), a fluid having a temperature $\langle T_f \rangle_0$, and flowing into this porous solid will come to thermal equilibrium with the solid over a short distance. Show that by using the fluid flow per unit solid volume $\dot{n}_f$, the energy equation for a porous solid of volume $V_s$, with a uniform temperature $T_s$, is given by

$$(\dot{n}c_p)_f V_s (T_s - \langle T_f \rangle_0) = -(\rho c_p V)_s \frac{dT_s}{dt} + \dot{S}_s$$

integral-volume energy equation for porous solid with very large $A_{ku}/V_s$.

Start from (2.72) and allow for the surface-convection heat transfer mechanism only.

### SOLUTION
Starting from (2.72) and allowing for surface-convection heat transfer mechanism only, we have

$$\langle Q_u \rangle_{L\text{-}0} = -(\rho c_p V)_s \frac{dT_s}{dt} + \dot{S}_s,$$

where from (7.25), (7.27), and (7.20), we have

$$\langle Q_u \rangle_{L\text{-}0} = \frac{T_s - \langle T_f \rangle_0}{\langle R_u \rangle_L} = (\dot{M}c_p)_f (1 - e^{-NTU})(T_s - \langle T_f \rangle_0)$$

$$NTU = \frac{A_{ku} \langle \mathrm{Nu} \rangle_D k_f / D}{(\dot{M}c_p)_f}$$

$$= \frac{A_{ku}}{V_s} \frac{\langle \mathrm{Nu} \rangle_D k_f / D}{(\dot{n}c_p)_f}.$$

Now for $A_{ku}/V_s \gg 1 (1/\mathrm{m})$, we have

$$NTU \to \infty.$$

This limit of very large number of transfer units results in $\langle T_f \rangle_L = T_s$, where $T_s$ is the uniform porous solid temperature. Then

$$\langle Q_u \rangle_{L\text{-}0} = (\dot{M}c_p)_f(1 - e^{-\infty})(T_s - \langle T_f \rangle_0)$$
$$= (\dot{M}c_p)_f(T_s - \langle T_f \rangle_0).$$

Now using $(\dot{M}c_p)_f = (\dot{n}c_p)_f V_s$, we have

$$\langle Q_u \rangle_{L\text{-}0} = (\dot{n}c_p)_f V_s(T_s - \langle T_f \rangle_0).$$

Using this in the above energy equation, we have

$$(\dot{n}c_p)_f V_s(T_s - \langle T_f \rangle_0) = -(\rho c_p V)_s \frac{dT_s}{dt} + \dot{s}_s V_s$$

integral-volume energy equation for porous solid with very large $A_{ku}/V_s$

or

$$(\dot{n}c_p)_f(T_s - \langle T_f \rangle_0) = -(\rho c_p)_s \frac{dT_s}{dt} + \dot{s}_s.$$

**COMMENT**

The heat transfer in blood flow through the arterioles (the very small arteries supplying blood to tissues) is described by the above equation (and the conduction heat transfer is added) and this is used to predict the local tissue temperature. This is called the bioheat energy equation model. Typical values for $\dot{n}_f \,(\text{kg/m}^3\text{-s})$ are about 6 kg/m$^3$-s. In the kidney $\dot{n}_f$ is as large as 10 kg/m$^3$-s and in the prostate $\dot{n}_f$ is as low as 1 kg/m$^3$-s.

Conduction heat transfer can be added to the above energy equation. In Example 3.23, the simplified form of the bioheat energy equation was used with ultrasonic energy conversion.

EXAMPLE 7.5. FUN

A flame arrester is a heat storage, high melting temperature, porous solid (also called filter) placed upstream of a gaseous premixed fuel-oxidant stream to cool the flame (i.e., to quench or extinguish it) and prevent it from reaching the fuel source during the flashback. A stainless steel 316 flame arrester element is intended for use with methane as the fuel and is shown in Figure Ex. 7.5(a), with the flashback flame passing through it. The arrester casing is not shown. During the travel through the arrester, the flame is extinguished. Treat the flow of hot stream as steady and assume a uniform solid temperature.

(a) Draw the thermal circuit diagram.
(b) Write the energy equation for the solid and by inspection show that (6.156) applies to this problem.
(c) Determine the Nusselt number.
(d) Determine the $NTU$ and $\langle R_u \rangle_L$.

(a) Physical Model

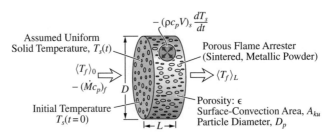

Figure Ex. 7.5. (a) A flame arrester element (placed inside a casing) used for cooling (and quenching) flashback flame in premixed fuel-oxidant stream.

(e) Using (6.156), determine the rise in the temperature of the flame arrester after an elapsed time $t_o$. Estimate the outlet temperature of the stream after an elapsed time $t_o$.

$T_s(t = 0) = 40°C$, $\langle T_f \rangle_0 = 1,600°C$, $\dot{M}_f = 1.0$ g/s, $\epsilon = 0.25$, $D = 3$ cm, $L = 2$ cm, $A_{ku} = 0.46$ m$^2$, $D_p = 0.2$ mm, $t_o = 5$ s.

Use constant air properties at $T = 1,500$ K.

**SOLUTION**

(a) Figure Ex. 7.5(b) shows the thermal circuit diagram.

(b) The energy equation for the solid, from Figure 7.5(b), is

$$\langle Q_u \rangle_{L\text{-}0} = -(\rho c_p V)_s \frac{dT_s(t)}{dt}$$

$$\frac{T_s - \langle T_f \rangle_0}{\langle R_u \rangle_L} = -(\rho c_p V)_s \frac{dT_s(t)}{dt}.$$

The average convection resistance $\langle R_u \rangle_L$ is defined in Table 7.1, i.e,

$$\langle R_u \rangle_L = \frac{1}{(\dot{M} c_p)_f (1 - e^{NTU})}$$

$$NTU = \frac{1}{\langle R_{ku} \rangle_D (\dot{M} c_p)_f}.$$

(b) Thermal Circuit Diagram

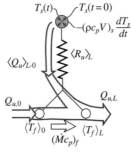

Figure Ex. 7.5. (b) Thermal circuit diagram.

The energy equation is similar to (6.155), i.e., with $Q_s$ and $\dot{S}_s$ both equal to zero, then (6.156) is the solution,

$$T_s(t) = \langle T_f \rangle_0 + (T_s(t = 0) - \langle T_f \rangle_0)e^{-t/\tau_s}$$

$$\tau_s = (\rho c_p V)_s \langle R_u \rangle_L.$$

(c) The Nusselt number for porous media is given in Table 7.5 as

$$\langle \mathrm{Nu} \rangle_{D,p} = 2 + (0.4 \mathrm{Re}_{D,p}^{1/2} + 0.2 \mathrm{Re}_{D,p}^{2/3}) \mathrm{Pr}^{0.4}$$

$$\mathrm{Re}_{D,p} = \frac{\rho_f \langle u_f \rangle D_p}{\mu_f(1-\epsilon)} = \frac{\dot{M}_f D_p}{A_u \mu_f(1-\epsilon)}.$$

From Table C.22, at $T = 1{,}500$ K, we have

$\mu_f = \rho_f \nu_f = 0.235(\mathrm{kg/m^3}) \times 2.29 \times 10^{-4}(\mathrm{m^2/s}) = 5.382 \times 10^{-5}$ Pa-s    Table C.22

$k_f = 0.0870$ W/m-K    Table C.22

$c_p = 1{,}202$ J/kg-K    Table C.22

$\mathrm{Pr} = 0.70$    Table C.22.

Then

$$\mathrm{Re}_{D,p} = \frac{10^{-3}(\mathrm{kg/s}) \times 2 \times 10^{-4}(\mathrm{m})}{\dfrac{\pi \times (3 \times 10^{-2})^2(\mathrm{m^2})}{4} \times 5.382 \times 10^{-5}(\mathrm{Pa\text{-}s})(1 - 0.25)}$$

$$= 7.014$$

$$\langle \mathrm{Nu} \rangle_{D,p} = 2 + [0.4 \times (7.014)^{1/2} + 0.2 \times (7.014)^{2/3}] \times (0.7)^{0.4}$$

$$= 3.554.$$

(d) From Table 7.4, or from (7.46), we have

$$\langle R_{ku} \rangle_D = \frac{D_p}{A_{ku} \langle \mathrm{Nu} \rangle_{D,p} k_f} \left( \frac{\epsilon}{1-\epsilon} \right).$$

Then

$$NTU = \frac{A_{ku} \langle \mathrm{Nu} \rangle_{D,p} k_f (1 - \epsilon)}{D_p \epsilon (\dot{M} c_p)_f}$$

$$= \frac{0.46(\mathrm{m^2}) \times 3.554 \times 0.0870(\mathrm{W/m\text{-}K}) \times (1 - 0.25)}{2 \times 10^{-4}(\mathrm{m}) \times 0.25 \times 10^{-3}(\mathrm{kg/s}) \times 1{,}202(\mathrm{J/kg\text{-}K})} = 1{,}775.$$

(e) The average convection resistance $\langle R_u \rangle_L$ is

$$\langle R_u \rangle_L = \frac{1}{(\dot{M} c_p)_f (1 - e^{-NTU})}$$

$$= \frac{1}{10^{-3}(\mathrm{kg/s}) \times 1{,}202(\mathrm{J/kg\text{-}K})(1 - e^{-1.775})} = 0.8319^\circ \mathrm{C/W}.$$

The time constant $\tau_s$ is

$$\tau_s = (\rho c_p V)_s \langle R_u \rangle_L = (\rho c_p)_s \left( \frac{\pi D^2}{4} L \right) (1 - \epsilon) \langle R_u \rangle_L,$$

where we have used the volume of the solid. From Table C.16, for stainless steel 316, we have

$$\text{stainless steel 316:} \quad \rho_s = 8{,}238 \text{ kg/s} \quad \text{Table C.16}$$

$$c_{p,s} = 486 \text{ J/kg-K} \quad \text{Table C.16.}$$

Then

$$\tau_s = 8{,}238 (\text{kg/m}^3) \times 486 (\text{J/kg-K}) \times \frac{\pi \times (3 \times 10^{-2})^2 (\text{m}^2)}{4} \times 2 \times 10^{-2} (\text{m})$$

$$\times (1 - 0.25) \times 0.8319 (^\circ \text{C/W}) = 35.30 \text{ s.}$$

Then

$$T_s(t = t_o = 5 \text{ s}) = 1{,}600^\circ \text{C} + [40^\circ \text{C} - 1{,}600^\circ \text{C}] e^{-5/35.50} = 246.0^\circ \text{C}$$

The instantaneous fluid stream outlet temperature is determined from (7.21), i.e.,

$$\langle T_f \rangle_L (t = t_o = 5 \text{ s}) = T_s(t) + [\langle T_f \rangle_0 - T_s(t_o)] e^{-NTU}$$

$$= 246.0 (^\circ \text{C}) + [1{,}600 (^\circ \text{C}) - 246.0 (^\circ \text{C})] e^{-1.775} = 246.0^\circ \text{C.}$$

**COMMENT**

The large $NTU$ ensures that the hot stream is cooled to the solid temperature. The large time constant ensures that the arrester will not warm up quickly. Both $NTU$ and $\tau_s$ are reduced by reducing the arrester volume.

## 7.6 Heat Exchange between Two Bounded Streams

In many applications, the heat content of one fluid stream is exchanged with that of another fluid stream, without allowing the two streams to mix or have any direct contact (this is called an indirect heat exchange). The two streams designated as hot and cold streams are separated by an impermeable surface, and the equipment containing the two streams and this impermeable surface is called a heat exchanger. A practical heat exchanger is designed to allow for the least overall thermal resistance $R_\Sigma$ to the heat flow between the two streams. The parasitic heat exchange with the ambient of the heat exchanger is also minimized and in the analysis it is assumed the heat exchange is only between the two streams (i.e., no external heat exchange). Heat exchangers were briefly shown in Chart 7.2 and are discussed in detail in [3,12]. The common heat exchanger types shown in the chart are coaxial (or parallel axes), cross flow, and tube and shell. Both streams can be bounded or one of them may be

## Coaxial Parallel-Flow Heat Exchanger

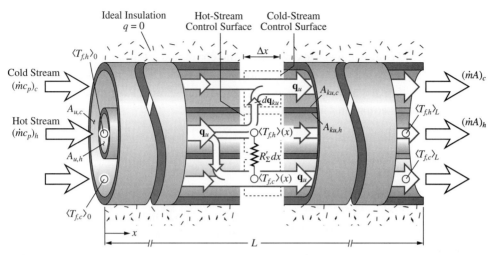

Figure 7.13. Coaxial, parallel-flow heat exchanger showing the overall thermal resistance between the two fluid streams over a length $\Delta x$. The inlet and outlet temperatures for the hot and cold streams are also shown.

semi-bounded. Here we examine the coaxial heat exchanger and briefly refer to the others.

### 7.6.1 Exit Temperatures in Coaxial Heat Exchangers

The simplest two-stream heat exchanger is that shown in Figure 7.13. It is called the coaxial heat exchanger, where the two streams can flow parallel or counter (i.e., opposite) to each other. In Figure 7.13, a parallel-flow coaxial circular tube heat exchanger is shown.

For an incremental distance $dx$ along the heat exchanger, the heat flows from the hot stream to the cold stream, overcoming the resistances due to the surface convection (on the hot- and cold-stream sides), and the wall conduction. These resistances are encountered in series and their sum is called the overall resistance $R_\Sigma$. The resistance for a unit length is designated as $R'_\Sigma$ ($^\circ$C/W-m). Then the overall resistance over a small length $\Delta x$ is

$$R'_\Sigma dx = R'_{ku,c}\Delta x + R'_{k,c-h}\Delta x + R'_{ku,h}\Delta x + \cdots , \qquad (7.51)$$

where $R'_{ku,c}$ is the surface-convection resistance for the cold stream for a unit length, $R'_{k,c-h}$ is the wall-conduction resistance for a unit length, etc.

Then the overall resistance over the length $L$, covering from the entrance $x = 0$ to the exit $x = L$ of the heat exchanger, is found by integration of (7.51), i.e.,

$$R_\Sigma = \int_0^L R'_\Sigma dx = R_{ku,c} + R_{k,c-h} + R_{ku,h} + \cdots \quad \begin{array}{l}\text{overall heat}\\ \text{exchanger resistance.}\end{array} \qquad (7.52)$$

Referring to Figure 7.13, we begin the analysis of the heat exchangers by identifying the differential volumes and surfaces, i.e.,

$$\Delta V = A_u \Delta x, \qquad A_{ku} = P_{ku} \Delta x, \tag{7.53}$$

for the hot and cold streams. Next we write the combined integral- and differential-length energy equation (7.11) for each stream. We note that $d\langle q_{u,c}\rangle/dx$ is positive, while $d\langle q_{u,h}\rangle/dx$ is negative. We use also $q_{ku}$ as positive when it leaves the hot stream. Then the energy equation for the cold stream is

$$\lim_{\Delta x \to 0} \frac{\displaystyle\int_{\Delta A_c} (\mathbf{q}\cdot\mathbf{s}_n)dA}{\Delta V_c} = -\frac{P_{ku,c}}{A_{u,c}}q_{ku,c} + \frac{d}{dx}\langle q_{u,c}\rangle = 0 \tag{7.54}$$

or

$$-P_{ku,c}q_{ku,c} + A_{u,c}\frac{d}{dx}\langle q_{u,c}\rangle = 0 \qquad \begin{array}{l}\text{combined integral- and differential-length,}\\ \text{energy equation for cold stream.}\end{array} \tag{7.55}$$

Similarly, for the hot stream we have

$$\lim_{\Delta x \to 0} \frac{\displaystyle\int_{\Delta A_h} (\mathbf{q}\cdot\mathbf{s}_n)dA}{\Delta V_h} = \frac{P_{ku,h}}{A_{u,h}}q_{\langle ku,h\rangle} + \frac{d}{dx}q_{u,h} = 0 \tag{7.56}$$

or

$$P_{ku,h}q_{ku,h} + A_{u,h}\frac{d}{dx}\langle q_{u,h}\rangle = 0 \qquad \begin{array}{l}\text{combined integral- and differential-length,}\\ \text{energy equation for hot stream.}\end{array} \tag{7.57}$$

Now we use $q_u = (\dot{m}c_p)_f\langle T_f\rangle$ from (7.5), and multiply (7.55) and (7.57) by the differential volume $A_{u,h}dx$ and $A_{u,c}dx$, where we now use $dx = \Delta x$. The results are

$$-P_{ku,c}q_{ku,c}dx + A_{u,c}(\dot{m}c_p)_c d\langle T_{f,c}\rangle = 0 \quad \text{energy equation for cold stream} \tag{7.58}$$

$$P_{ku,h}q_{ku,h}dx + A_{u,h}(\dot{m}c_p)_h d\langle T_{f,h}\rangle = 0 \quad \text{energy equation for hot stream.} \tag{7.59}$$

The surface-convection heat transfer across the incremental length $dx$, $dQ_{ku}$, is given in terms of the overall resistance, i.e., local resistance $R'_\Sigma dx$ and the local temperature difference between the two streams $\langle T_{f,h}\rangle - \langle T_{f,c}\rangle$, i.e.,

$$dQ_{ku} = P_{ku,h}q_{ku,h}dx = -P_{ku,c}q_{ku,c}dx \equiv \frac{\langle T_{f,h}\rangle - \langle T_{f,c}\rangle}{R'_\Sigma dx} \quad \begin{array}{l}\text{heat exchanged over}\\ \text{incremental length } dx.\end{array} \tag{7.60}$$

This heat transfer across the wall separating the two streams is rendered in Figure 7.13.

From (7.58) and (7.59), we note that $dQ_{ku}$ is also equal to

$$dQ_{ku} = dQ_u = -(A_u\dot{m}c_p)_h d\langle T_{f,h}\rangle = (A_u\dot{m}c_p)_c d\langle T_{f,c}\rangle. \tag{7.61}$$

We now rearrange this as

$$d(\langle T_{f,h}\rangle - \langle T_{f,c}\rangle) = -dQ_{ku}\left[\frac{1}{(A_u\dot{m}c_p)_h} + \frac{1}{(A_u\dot{m}c_p)_c}\right]. \tag{7.62}$$

Next substituting in this for $dQ_{ku}$ from (7.60), and using $A_u(\dot{m}c_p)_f \equiv \dot{M}c_p$, we have

$$d(\langle T_{f,h}\rangle - \langle T_{f,c}\rangle) = -\frac{\langle T_{f,h}\rangle - \langle T_{f,c}\rangle}{R'_\Sigma dx}\left[\frac{1}{(\dot{M}c_p)_h} + \frac{1}{(\dot{M}c_p)_c}\right] \qquad (7.63)$$

or

$$\frac{d(\langle T_{f,h}\rangle - \langle T_{f,c}\rangle)}{\langle T_{f,h}\rangle - \langle T_{f,c}\rangle} = \frac{1}{R'_\Sigma dx}\left[\frac{1}{(\dot{M}c_p)_h} + \frac{1}{(\dot{M}c_p)_c}\right]. \qquad (7.64)$$

We now integrate this along $x$, from $x = 0$ to $x = L$, with the temperatures designated as

$$\langle T_{f,h}\rangle = \langle T_{f,h}\rangle_0, \quad \langle T_{f,c}\rangle = \langle T_{f,c}\rangle_0 \qquad \text{at } x = 0 \qquad (7.65)$$

$$\langle T_{f,h}\rangle = \langle T_{f,h}\rangle_L, \quad \langle T_{f,c}\rangle = \langle T_{f,c}\rangle_L \qquad \text{at } x = L. \qquad (7.66)$$

The integration of (7.64), subject to these end conditions, gives

$$\int_{\langle T_{f,h}\rangle_0 - \langle T_{f,c}\rangle_0}^{\langle T_{f,h}\rangle_L - \langle T_{f,c}\rangle_L} \frac{d(\langle T_{f,h}\rangle - \langle T_{f,c}\rangle)}{\langle T_{f,h}\rangle - \langle T_{f,c}\rangle} = -\int_0^L \left[\frac{1}{(\dot{M}c_p)_h} + \frac{1}{(\dot{M}c_p)_c}\right]\frac{1}{R'_\Sigma dx} \qquad (7.67)$$

$$\ln\frac{(\langle T_{f,h}\rangle_L - \langle T_{f,c}\rangle_L)}{\langle T_{f,h}\rangle_0 - \langle T_{f,c}\rangle_0} = -\frac{1}{R_\Sigma}\left[\frac{1}{(A_u\dot{m}c_p)_h} + \frac{1}{(A_u\dot{m}c_p)_c}\right]. \qquad (7.68)$$

We now rearrange this, similar to (7.21), as

$$\boxed{\frac{\langle T_{f,h}\rangle_L - \langle T_{f,c}\rangle_L}{\langle T_{f,h}\rangle_0 - \langle T_{f,c}\rangle_0} = \exp\left\{-\frac{1}{R_\Sigma}\left[\frac{1}{(\dot{M}c_p)_h} + \frac{1}{(\dot{M}c_p)_c}\right]\right\} \qquad \begin{array}{l}\text{relation between}\\\text{inlet and exit}\\\text{temperatures.}\end{array}} \qquad (7.69)$$

This relationship is valid for both parallel-flow and counterflow arrangements of the two streams. For the parallel-flow heat exchange, $\langle T_{f,h}\rangle_0$ and $\langle T_{f,c}\rangle_0$ are specified as the inlet conditions, i.e., at $x = 0$. For the counterflow heat exchange, the inlet conditions are split between the two ends, i.e., at $x = 0$ and $x = L$. The temperature distribution for both arrangements are shown in Figure 7.14.

We now integrate (7.61) between the inlet and the exit. For parallel flow, we have the rate of heat exchange between the two streams $\langle Q_u\rangle_{L\text{-}0}$ as

$$\langle Q_u\rangle_{L\text{-}0} = (\dot{M}c_p)_c(\langle T_{f,c}\rangle_L - \langle T_{f,c}\rangle_0) \qquad (7.70)$$

$$\langle Q_u\rangle_{L\text{-}0} = (\dot{M}c_p)_h(\langle T_{f,h}\rangle_0 - \langle T_{f,h}\rangle_L) \qquad \text{rate of heat exchange for parallel flow.} \qquad (7.71)$$

For the counterflow, depending on which stream arrives at $x = 0$ and which at $x = L$, the subscripts of the temperatures, in the above equations, are altered accordingly.

Coaxial Heat Exchangers

(a) Parallel-Flow Heat Exchanger

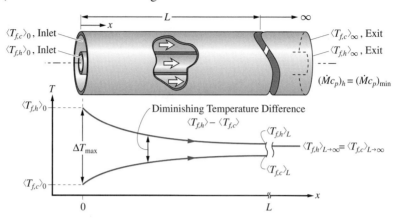

(b) Counterflow Heat Exchanger

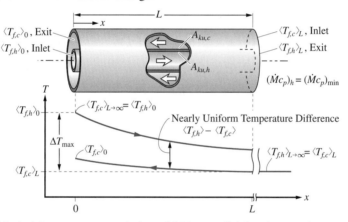

Figure 7.14.  Axial temperature variations. (a) For parallel-flow heat exchanger. (b) For counterflow heat exchanger. For both, $(\dot{M}c_p)_h < (\dot{M}c_p)_{min}$.

### 7.6.2  Heat Exchanger Effectiveness $\epsilon_{he}$ and Number of Transfer Units $NTU$

For parallel flow, the maximum temperature difference in the system is the difference between the inlet of the hot stream $\langle T_{f,h}\rangle_0$ and the inlet of the cold stream $\langle T_{f,c}\rangle_0$. Similar to Section 7.2.5, the effectiveness of the heat exchanger $\epsilon_{he}$ is defined by noting that in (7.70) and (7.71), the fluid with the smaller $(\dot{M}c_p)_f$, designated with $(\dot{M}c_p)_{min}$, will undergo the larger temperature change. Also, the largest temperature difference available is $\Delta T_{max}$, and for a parallel-flow arrangement we have $\Delta T_{max} = \langle T_{f,h}\rangle_0 - \langle T_{f,c}\rangle_0$. Then for the parallel flow, we have

$$\epsilon_{he} \equiv \frac{\Delta\langle T_f\rangle\,|_{(\dot{M}c_p)_{min}}}{\Delta T_{max}} = \frac{\Delta\langle T_f\rangle\,|_{(\dot{M}c_p)_{min}}}{\langle T_{f,h}\rangle_0 - \langle T_{f,c}\rangle_0} \quad \begin{array}{l}\text{parallel-flow heat}\\ \text{exchanger effectiveness.}\end{array} \qquad (7.72)$$

Using this, (7.69) can be written as the $\epsilon_{he}$-$NTU$ relation

$$\epsilon_{he} = \frac{1 - e^{-NTU(1+C_r)}}{1 + C_r} \qquad \epsilon_{he}\text{-}NTU \text{ relation for the parallel-flow heat exchanger,} \quad (7.73)$$

where, as in Section 7.2.5 the number of transfer units $NTU$ is defined similar to (7.20), and here the ratio of thermal capacitances $C_r$ is also defined, i.e.,

$$NTU \equiv \frac{1}{R_\Sigma (\dot{M}c_p)_{min}}, \qquad 0 \le NTU < \infty \qquad \begin{array}{l} NTU \text{ for two-stream} \\ \text{heat exchange,} \end{array} \quad (7.74)$$

and

$$C_r \equiv \frac{(\dot{M}c_p)_{min}}{(\dot{M}c_p)_{max}}, \qquad 0 \le C_r \le 1 \quad \text{ratio of thermal capacitances.} \quad (7.75)$$

For counterflow with the cold stream entering at $x = 0$ and the hot stream entering at $x = L$, we have $\Delta T_{max} = \langle T_{f,h}\rangle_0$, and for the effectiveness, we have

$$\epsilon_{he} \equiv \frac{\Delta\langle T_f\rangle \mid_{(\dot{M}c_p)_{min}}}{\Delta T_{max}} = \frac{\Delta\langle T_f\rangle \mid_{(\dot{M}c_p)_{min}}}{\langle T_{f,h}\rangle_L - \langle T_{f,c}\rangle_0} \qquad \begin{array}{l} \text{counterflow heat exchanger} \\ \text{effectiveness.} \end{array} \quad (7.76)$$

Using this, (7.69) can be written as

$$\epsilon_{he} = \frac{1 - e^{-NTU(1-C_r)}}{1 - C_r e^{-NTU(1-C_r)}}, \qquad C_r < 1 \qquad \begin{array}{l} \epsilon\text{-}NTU \text{ relation for the} \\ \text{counterflow heat exchanger,} \end{array} \quad (7.77)$$

with the case of $C_r = 1$ given by

$$\epsilon_{he} = \frac{NTU}{1 + NTU}, \qquad C_r = 1 \quad \text{counterflow heat exchange.} \quad (7.78)$$

In (7.73) and (7.77), note that when one of the streams is semi-bounded (so its far-field temperature does not change) or when one of the streams undergoes a liquid-gas phase change (such that its temperature remains at the saturation temperature), then $C_r = 0$. This is because $(\dot{M}c_p)_{max} \to \infty$, due to apparent heat capacity tending to infinity $c_p = c_{p,app} \to \infty$ during phase change. The apparent heat capacity was defined by (3.8) in Section 3.1.4. Then (7.73) and (7.77) become

$$\epsilon_{he} = 1 - e^{-NTU}, \qquad C_r = 0 \qquad \begin{array}{l} \text{one stream undergoing phase} \\ \text{change for all heat exchangers.} \end{array} \quad (7.79)$$

This is the same as (7.22) for the constant surface temperature fluid-solid surface heat exchange. Similar to (7.22), (7.69) shows that as $NTU \to \infty$, $\epsilon_{he} = 1$.

Also note that for the parallel flow, the effectiveness given by (7.73) does not have an upper limit $\epsilon_{he} = 1$, when $C_r > 0$. This is the case even when $NTU \to \infty$, i.e., $R_\Sigma(\dot{M}c_p)_{min} \to 0$. This is shown in Figure 7.14, where the temperature distributions along the heat exchanger, for both arrangements, are shown. For the parallel flow, as the hot stream loses heat and the cold stream absorbs heat, they both approach their exit temperatures, which in the limit of $NTU \to \infty$, from (7.73), is

$$\epsilon_{he} = \frac{\Delta\langle T_f \rangle \mid_{(\dot{M}c_p)_{min}}}{\langle T_{f,h}\rangle_0 - \langle T_{f,c}\rangle_0} = \frac{1}{1 + C_r} \quad \text{parallel flow with } NTU \to \infty. \tag{7.80}$$

However for the counterflow, from (7.77), we have the limit for $NTU \to \infty$ as

$$\epsilon_{he} = \frac{\Delta\langle T_f \rangle \mid_{(\dot{M}c_p)_{min}}}{\langle T_{f,h}\rangle_L - \langle T_{f,c}\rangle_0} = 1 \quad \text{counterflow with } NTU \to \infty. \tag{7.81}$$

The variation of $\epsilon_{he}$ with respect to $NTU$ is plotted in Figures 7.15(a) and (b) for both stream arrangements and for various values of $C_r$. Note that the counterflow arrangement allows for achieving $\epsilon_{he} = 1$ for any $C_r$ and for $NTU \to \infty$, as given by (7.81). The maximum $\epsilon_{he}$ for the parallel-flow arrangement is limited by $C_r$, for $NTU \to \infty$, as given by (7.80). When it is desired for both streams to achieve nearly the same $\Delta T$'s, then a $C_r$ of near unity is used. When it is desired that one of the streams achieves a $\Delta T$ larger than the other, then a smaller $C_r$ is used.

### 7.6.3 $\epsilon_{he}$-$NTU$ Relation for Other Heat Exchangers

When additional surface area is needed, other heat exchangers are used. Among these are the cross-flow and the tube and shell heat exchangers (shown in Chart 7.2). The cross-flow heat exchanger is used in many gas-liquid heat exchangers, where the gas-side surface generally has extended surfaces. The tube and shell heat exchanger is used in many process heat exchangers and allows for over-atmospheric and below-atmospheric operating pressures, for both streams. These are discussed in [3, 8, 11].

Table 7.7 lists the $\epsilon_{he}$-$NTU$ relations for some heat exchangers. The effectiveness is the same used in (7.72) and (7.76), i.e.,

$$\epsilon_{he} \equiv \frac{\Delta\langle T_f \rangle \mid_{(\dot{M}c_p)_{min}}}{\Delta T_{max}} \quad \text{general definition of effectiveness,} \tag{7.82}$$

where $\Delta T_{max}$ is the maximum available temperature difference between the two streams.

The rate of heat exchange is [similar to (7.70) and (7.71)]

$$\langle Q_u\rangle_{L-0} = (\dot{M}c_p)_c(\langle T_{f,c}\rangle_L - \langle T_{f,c}\rangle_0) = (\dot{M}c_p)_c \Delta T_c \tag{7.83}$$

$$\langle Q_u\rangle_{L-0} = (\dot{M}c_p)_h(\langle T_{f,h}\rangle_0 - \langle T_{f,h}\rangle_L) = -(\dot{M}c_p)_h \Delta T_h, \tag{7.84}$$

where $\Delta T_c$ and $\Delta T_h$ are the temperature changes of the two streams.

(a) $\epsilon_{he}$ - *NTU* for Parallel-Flow Heat Exchanger

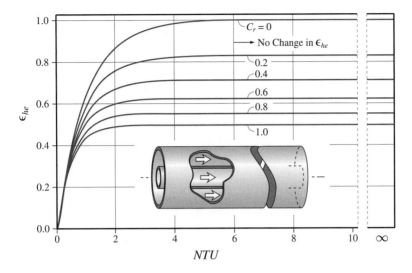

(b) $\epsilon_{he}$ - *NTU* for Counterflow Heat Exchanger

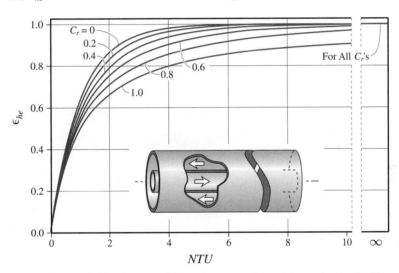

Figure 7.15. Variation of effectiveness with respect to number of thermal units. (a) For parallel-flow coaxial tube flow. (b) For counterflow coaxial tube flow.

Table 7.7 lists the $\epsilon_{he}$-*NTU* relations for the three types of heat exchangers discussed [8]. For the parallel axes (e.g., coaxial) heat exchanger, the two arrangements are parallel-flow and counterflow. For the cross-flow heat exchanger, distinction is made for the hot and cold streams to mix as they exchange (this tends to destroy any nonuniformity of temperature along the stream path). There are four possible cases for such mixing and they are all listed in Table 7.7. For the tube and shell heat exchanger, the number of tube passes and the number of shells are included in the relations.

Table 7.7. $\epsilon_{he}$-*NTU relation for some exchanger types. Some common heat exchanger types are shown in Chart 7.2*

| Arrangement | $\epsilon_{he}$-NTU Relation |
|---|---|
| **Single streams:** (and all exchangers when $C_r = 0$) | $\epsilon_{he} = 1 - \exp(-NTU)$ |
| **Two streams, with parallel axes:** | |
| parallel-flow | $\epsilon_{he} = \dfrac{1 - \exp[-NTU(1 + C_r)]}{1 + C_r}$ |
| counterflow | $\epsilon_{he} = \dfrac{1 - \exp[-NTU(1 - C_r)]}{1 - C_r \exp[-NTU(1 - C_r)]}, \quad C_r \neq 1$ |
| | $\epsilon_{he} = \dfrac{NTU}{1 + NTU}, \quad C_r = 1$ |
| **Two streams, in cross flow:** | |
| both fluids unmixed | $\epsilon_{he} = 1 - \exp\left\{\dfrac{NTU^{0.22}}{C_r}[\exp(-C_r NTU^{0.78}) - 1]\right\}$ |
| both fluids mixed | $\epsilon_{he} = \left[\dfrac{1}{1 - \exp(-NTU)} + \dfrac{C_r}{1 - \exp(-C_r NTU)} - \dfrac{1}{NTU}\right]^{-1}$ |
| $(\dot{M}c_p)_{max}$ mixed, $(\dot{M}c_p)_{min}$ unmixed | $\epsilon_{he} = \dfrac{1}{C_r}\{1 - \exp[C_r(e^{-NTU} - 1)]\}$ |
| $(\dot{M}c_p)_{max}$ unmixed, $(\dot{M}c_p)_{min}$ mixed | $\epsilon_{he} = 1 - \exp\left\{-\dfrac{1}{C_r}[1 - e^{-C_r NTU}]\right\}$ |
| **Two streams, one in shell and other in tubes:** | |
| one shell pass; 2,4,6,... tube passes[1] | $\epsilon_{he} = \epsilon_{he,1} = 2\left\{1 + C_r + (1 + C_r^2)^{1/2}\right.$ $\left. \times \dfrac{1 + \exp[-NTU(1 + C_r^2)^{1/2}]}{1 - \exp[-NTU(1 + C_r^2)^{1/2}]}\right\}^{-1}$ |
| $n$ shell passes; $2n$, $4n$, ... tube passes[†] | $\epsilon_{he} = \dfrac{\left[\left(\dfrac{1 - \epsilon_{he,1} C_r}{1 - \epsilon_{he,1}}\right)^n - 1\right]}{\left[\left(\dfrac{1 - \epsilon_{he,1} C_r}{1 - \epsilon_{he,1}}\right)^n - C_r\right]}$ |

[†] In calculating $\epsilon_{he,1}$, the $NTU$ per shell pass is used (i.e., $NTU/n$).

### 7.6.4 Heat Exchanger Analysis

#### (A) Average Heat Exchanger Convection Resistance $\langle R_u \rangle_L$

We can write (7.83) and (7.84) in terms of the $(\dot{M}c_p)_{min}$ and $\epsilon_{he}$ using (7.82) for use in the thermal circuit models, similar to (7.25). Then we define an average convection resistance $\langle R_u \rangle_L$ using the two stream inlet temperatures as

$$\langle Q_u \rangle_{L\text{-}0} \equiv \frac{\Delta T_{max}}{\langle R_u \rangle_L} = \Delta T_{max}(\dot{M}c_p)_{min}\epsilon_{he} \quad \begin{array}{l}\text{heat transfer rate and} \\ \text{average convection resistance} \\ \text{for heat exchangers,}\end{array} \quad (7.85)$$

where $\Delta T_{max}$ is the maximum temperature between the hot and cold streams. This is the inlet temperature of the hot stream minus the inlet temperature of the cold stream. The average convection resistance for two-stream heat exchange $\langle R_u \rangle_L$ is shown in Figures 7.16(a) and (b).

From above, we note that

$$\langle R_u \rangle_L \equiv \frac{1}{(\dot{M}c_p)_{min}\epsilon_{he}}, \quad \epsilon_{he} = \epsilon_{he}(NTU, C_r, \text{heat exchanger type}) \qquad (7.86)$$

$\langle R_u \rangle_L$ average convection resistance for heat exchangers.

Note that (7.85) and (7.86) are similar to (7.25) and (7.27), except here the maximum available temperature is used in place of $\langle T_{f,h} \rangle_0 - \langle T_{f,c} \rangle_0$ (for parallel flow). The $\epsilon_{he}$-$NTU$ relations, used in determining $\langle R_u \rangle_L$, are given in Table 7.7.

The average convection resistance for two-stream heat exchangers is similar to that for the constant wall temperature heat exchanger, and for the case of $C_r \to 0$ (i.e., one of the streams being semi-bounded or undergoing a phase change, thus having a constant second fluid temperature), (7.86) is identical to (7.27).

The thermal circuit model for a two-stream heat exchanger is shown in Figure 7.16(b). There, following the convention we used for the constant surface temperature heat exchange, we have connected the average convection resistance $\langle R_u \rangle_L$ between the inlet and exit temperatures of each stream. However, we use the difference between the two inlet temperatures (i.e., the maximum temperature difference) in the expression for $\langle Q_u \rangle_{L-0}$, i.e., (7.85).

Finally, note that $NTU$ contains the overall resistance $R_\Sigma$, as given by (7.74).

### (B) Summary of Relations

In summary, there are four fundamental relations used in the heat exchanger analysis.

(i), (ii) The first two are the relations for $\langle Q_u \rangle_{L-0}$ in terms of the temperatures and the mass flow rates of the two streams, i.e., (7.83) and (7.84).
(iii) The third relation is the definition of the effectiveness $\epsilon_{he}$ given by (7.72) or (7.76).
(iv) The last is the $\epsilon_{he}$-$NTU$ relation as listed in Table 7.7.

From these four relations, we generally need to determine $\langle Q_u \rangle_{L-0}$ plus three other unknowns. One of the unknowns can be $\epsilon_{he}$. Since at least one temperature is known for each stream, the two unknowns can be the other two temperatures, or one unknown can be the $NTU$ and the other can be one of the temperatures, etc.

Alternatively, we can use (7.85), (7.86), (7.82), and Table 7.7. These would readily give $\langle Q_u \rangle_{L-0}$ and $\langle R_u \rangle_L$, while the other four equations and relations give $\langle Q_u \rangle_{L-0}$ and the exit temperatures.

(a) Average Convection Resistance for Two-Stream Heat Exchanger

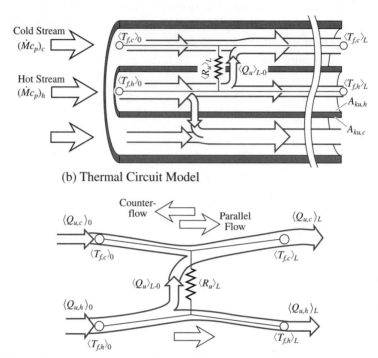

(b) Thermal Circuit Model

Figure 7.16. (a) Average convection resistance $\langle R_u \rangle_L$ for a coaxial, two-stream heat exchanger. (b) Average heat exchanger convection resistance $\langle R_u \rangle_L$ thermal circuit model. Note that subscript 0 indicates the port at $x = 0$ and this can be the inlet or the exit port, depending on the stream direction.

EXAMPLE 7.6. FAM

In a coaxial counterflow heat exchanger [such as the one shown in Figure 7.15(b)], liquid water is the cold stream and flows in the inner cylinder with a mass flow rate $\dot{M}_c = 0.5$ kg/s and an inlet temperature $\langle T_{f,c} \rangle_0 = 20°$C. Liquid Dowtherm J is the hot stream and flows in the annulus with a mass flow rate $\dot{M}_h = 0.25$ kg/s, and an inlet temperature $\langle T_{f,h} \rangle_L = 120°$C. The overall resistance is $R_\Sigma = 0.001°$C/W.

(a) Show the expected temperature distribution for the hot and cold streams and draw the thermal circuit diagram.
(b) Determine the number of transfer units $NTU$ [using $(\dot{M}c_p)_{min}$].
(c) Determine the ratio of the thermal capacitances $C_r$.
(d) Determine the effectiveness $\epsilon_{he}$.
(e) Determine the exit temperature of the hot stream $\langle T_{f,h} \rangle_0$.
(f) Determine the exit temperature of the cold stream $\langle T_{f,c} \rangle_L$.
(g) Determine the average heat exchanger convection resistance $\langle R_u \rangle_L$.

Evaluate the thermophysical properties for the saturated states at $p = 1.013 \times 10^5$ Pa, as given in Table C.26.

**SOLUTION**

(a) The temperature distributions along the heat exchanger, for the hot and cold streams, are shown in Figure Ex. 7.6(a). The results are for the quantitative analysis performed below. The thermal circuit diagram is shown in Figure Ex. 7.6(b).

(a) Temperature Distribution along a Counterflow Heat Exchanger

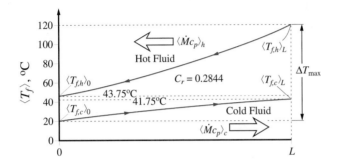

(b) Thermal Circuit Model

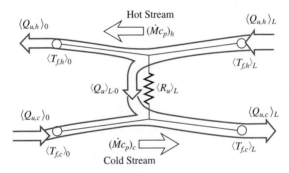

Figure Ex. 7.6. (a) Temperature distribution along the counterflow heat exchanger. (b) Thermal circuit diagram for the heat exchanger.

(b) The definition of $NTU$ for the two-stream heat exchange is given by (7.74), i.e.,

$$NTU \equiv \frac{1}{R_\Sigma (\dot{M} c_p)_{min}}.$$

To determine $(\dot{M} c_p)_{min}$, we need to compare the two streams, i.e.,

$$(\dot{M} c_p)_c = \dot{M}_c c_{p,c} \text{ liquid water}$$
$$(\dot{M} c_p)_h = \dot{M}_c c_{p,h} \text{ liquid Dowtherm J.}$$

From Table C.26, we have

liquid water $\qquad c_{p,c} = 4,220$ J/kg-°C Table C.26
liquid Dowtherm J $\quad c_{p,h} = 2,400$ J/kg-°C Table C.26.

Then

$$(\dot{M}c_p)_c = 0.5(\text{kg/s}) \times 4{,}220(\text{J/kg-}°\text{C}) = 2{,}110 \text{ W/}°\text{C}$$

$$(\dot{M}c_p)_h = 0.25(\text{kg/s}) \times 2{,}400(\text{J/kg-}°\text{C}) = 600 \text{ W/}°\text{C}.$$

Then

$$(\dot{M}c_p)_{min} = (\dot{M}c_p)_h = 600 \text{ W/}°\text{C}.$$

Next

$$NTU = \frac{1}{0.001(°\text{C/W}) \times 600(\text{W/}°\text{C})} = 1.667.$$

(c) The ratio of the heat capacitances $C_r$ is defined by (7.75), i.e.,

$$C_r \equiv \frac{(\dot{M}c_p)_{min}}{(\dot{M}c_p)_{max}} = \frac{(\dot{M}c_p)_h}{(\dot{M}c_p)_c} = \frac{600(\text{W/}°\text{C})}{2{,}110(\text{W/}°\text{C})} = 0.2844.$$

(d) The $\epsilon_{he}$-$NTU$ relation for the counterflow arrangement is given in Table 7.7, i.e.,

$$\epsilon_{he} = \frac{1 - e^{-NTU(1-C_r)}}{1 - C_r e^{-NTU(1-C_r)}} = \frac{1 - e^{-1.667}(1 - 0.2844)}{1 - 0.2844 e^{-1.667(1-0.2844)}}$$

$$= \frac{1 - e^{-1.193}}{1 - 0.2844 e^{-1.193}} = 0.7625.$$

(e) The hot-stream exit temperature $\langle T_{f,h}\rangle_0$ is determined from the definition of $\epsilon_{he}$, for the counterflow arrangement, given by (7.82)

$$\epsilon_{he} = \frac{\langle T_{f,h}\rangle_L - \langle T_{f,h}\rangle_0}{\langle T_{f,h}\rangle_L - \langle T_{f,c}\rangle_0}.$$

Solving this for $\langle T_{f,h}\rangle_0$, we have

$$\langle T_{f,h}\rangle_0 = \langle T_{f,h}\rangle_L - \epsilon_{he}(\langle T_{f,h}\rangle_L - \langle T_{f,c}\rangle_0)$$

$$= 120(°\text{C}) - 0.7625(120 - 20)(°\text{C}) = 43.75°\text{C}.$$

(f) The heat transfer rate $\langle Q_u\rangle_{L\text{-}0}$ is determined from (7.84), i.e.,

$$\langle Q_u\rangle_{L\text{-}0} = -(\dot{M}c_p)_h(\langle T_{f,h}\rangle_L - \langle T_{f,h}\rangle_0) = (\dot{M}c_p)_c(\langle T_{f,c}\rangle_L - \langle T_{f,c}\rangle_0).$$

Solving for $\langle T_{f,c}\rangle_L$, we have

$$\langle T_{f,c}\rangle_L = \langle T_{f,c}\rangle_0 - \frac{(\dot{M}c_p)_h(\langle T_{f,h}\rangle_L - \langle T_{f,h}\rangle_0)}{(\dot{M}c_p)_c} = \langle T_{f,c}\rangle_0 - \frac{\langle Q_u\rangle_{L\text{-}0}}{(\dot{M}c_p)_c}$$

$$= 20(°\text{C}) - \frac{600(\text{W/}°\text{C}) \times 76.25(°\text{C})}{2{,}110(\text{W/}°\text{C})} = 20(°\text{C}) - \frac{4.575 \times 10^4(\text{W})}{2{,}110(\text{W/}°\text{C})}$$

$$= 41.79°\text{C}.$$

The inlet and exit temperatures are shown in Figure Ex. 7.6(a).

(g) The average heat exchanger convection resistance $\langle R_u \rangle_L$ is given by (7.86), i.e.,

$$\langle R_u \rangle_L = \frac{1}{(\dot{M} c_p)_{min} \epsilon_{he}}$$

$$= \frac{1}{600 (\text{W/}^\circ\text{C}) \times 0.7625} = 0.002186 \,^\circ\text{C/W}.$$

**COMMENT**

Note that from division of (7.70) by (7.71) we have

$$\frac{\langle T_{f,c} \rangle_L - \langle T_{f,c} \rangle_0}{\langle T_{f,h} \rangle_L - \langle T_{f,h} \rangle_0} = \frac{(\dot{M} c_p)_h}{(\dot{M} c_p)_c} = C_r,$$

i.e., $\Delta \langle T_{f,c} \rangle = C_r \Delta \langle T_{f,h} \rangle = 0.2844 \Delta \langle T_{f,h} \rangle$. Also note that $\langle R_u \rangle_L > R_\Sigma$. This is because $\langle R_u \rangle_L$ is based on $\Delta T_m$. The actual temperature difference is smaller and changes as the two streams exchange heat.

### 7.6.5 Overall Thermal Resistance in Heat Exchangers $R_\Sigma$

As was given by (7.52), for the case of thermal resistances due to the surface convection and wall conduction, we have the overall thermal resistance

$$R_\Sigma = R_{ku,c} + R_{k,c-h} + R_{ku,h} + \cdots \quad \text{overall resistance.} \qquad (7.87)$$

In turn the individual surface-convection resistances are given in terms of the $\langle \text{Nu} \rangle_D$ correlations of Section 7.4, and the conduction resistance is given by the relations of Section 3.3.1. Other resistances due to contact (Section 3.3.5) or surface deposits (called scaling or fouling) are also added when they are significant [1, 3, 11].

It is desirable that the overall resistance $R_\Sigma$ be minimized and that the resistance contributions from the cold-stream surface convection, the wall conduction, and the hot-stream surfac convection, be nearly equal. In practice, by choosing metallic walls, the wall conduction resistance is reduced. For the surface-convection resistances, the resistance is lowest for flow with phase change (liquid-gas), followed by single-phase liquid phase flow and then by single-phase gas phase flow. Therefore, when one of the fluids is a gas, the surface-convection resistance for this stream dominates, unless extended surfaces are used on the gas-side surface.

The overall resistance $R_\Sigma$ given by (7.87) can be written in terms of the $\langle \text{Nu} \rangle_D$ as

$$R_\Sigma = R_{ku,c} + R_k + R_{ku,h} + \cdots$$

$$= \left[ A_{ku,c} (\eta_{f,o})_c \langle \text{Nu} \rangle_{D,c} \frac{k_{f,c}}{D_c} \right]^{-1} + R_{k,c-h} + \left[ A_{ku,h} (\eta_{f,o})_h \langle \text{Nu} \rangle_{D,h} \frac{k_{f,h}}{D_h} \right]^{-1} + \cdots,$$
$$(7.88)$$

where we have used an overall efficiency $\eta_{f,o}$ for the cold and hot stream surface-convection surface areas. The overall efficiency was discussed in Section 7.4.3 and defined by (7.50). The wall resistance includes any added contact resistance due to attachment of extended surfaces (if the fins are not fabricated as an integrated part of the wall) and any resistance due to corrosion and formation of deposits.

As can be recognized from Figures 7.14(a) and (b), neither a uniform surface temperature nor a uniform surface heat flux condition exists in the two-stream heat exchanger. The local surface temperature takes a value between the local fluid temperatures $\langle T_{f,h} \rangle$ and $\langle T_{f,c} \rangle$. The local surface heat flux is proportional to a local temperature difference $\langle T_{f,h} \rangle - \langle T_{f,c} \rangle$. In practice, the more readily available correlations for $\langle \mathrm{Nu} \rangle_D$, which are for a constant surface temperature, are used (Tables 7.2 to 7.6). When the constant heat flux results are available, they should be used.

## 7.6.6 Dielectric and Inert Heat-Transfer Fluids*

This section is found on the Web site www.cambridge.org/kaviany. When a heat transfer fluid circulates in a heat exchanger (i.e., undergoes heating and cooling) in a closed cycle, then this fluid can be selected based on its melting and freezing temperatures, its chemical inertness (i.e., reactivity with the heat exchanger surfaces), as well as its thermodynamic and transport properties. In this section, we introduce the heat-transfer fluids as synthetic organic fluids especially designed for their noncorrosiveness when used in metallic heat exchangers. We also give an example (Example 7.7).

## 7.7 Summary

In this chapter, we have examined the heat exchange between a fluid stream and its bounding surface, and between two fluid streams separated by a solid membrane. We have defined the average fluid velocity $\langle u_f \rangle$ and the average fluid temperature $\langle T_f \rangle$. We defined the average convection resistance $\langle R_u \rangle_L$ given by (7.27) and based on the entrance fluid temperature. This resistance includes the change in the fluid temperature as it exchanges heat with the surface. Therefore, this average convection resistance includes both the surface-convection resistance $\langle R_{ku} \rangle_D$ and the change in the fluid temperature. This is done through the number of transfer units $NTU$, given by (7.20), which includes the surface-convection resistance $\langle R_{ku} \rangle_D$ and the fluid convection conductance $(\dot{M}c_p)_f$. The effectiveness $\epsilon_{he}$ defined by (7.22) indicates how close the fluid temperature has become at the exit, to the surface temperature $T_s$. Various correlations for $\langle \mathrm{Nu} \rangle_D = D/A_{ku} \langle R_{ku} \rangle_D k_f$ are listed in Tables 7.2 to 7.6. These include continuous and discontinuous solid surfaces. The latter allows for very large specific surface area $A_{ku}/V$. When $\langle \mathrm{Nu} \rangle_D$ for both constant temperature and constant heat flux are available, the constant heat flux results may be more appropriate.

For a two-stream heat exchange, we have extended the definition of $NTU$ to (7.74), since the fluid with the smaller $(\dot{M}c_p)_f$, i.e., $(\dot{M}c_p)_{min}$, offers the most convection resistance (or undergoes the most temperature change). We have also extended the definition of $\epsilon_{he}$ to (7.82), which includes the temperature of both streams. The average convection resistance $\langle R_u \rangle_L$ is given by (7.86). The overall resistance $R_\Sigma$ includes the surface-convection resistances $R_{ku,c}$ and $R_{ku,h}$, and the wall-conduction resistance $R_{k,c-h}$, as given in (7.88). The $\epsilon_{he}$-$NTU$ relation depends on the heat exchanger type and arrangement. Some of these relations are listed in Table 7.7.

## 7.8 References*

This section is found on the Web site www.cambridge.org/kaviany.

## 7.9 Problems

### 7.9.1 Average Convection Resistance and $\epsilon_{he}$-$NTU$

**PROBLEM 7.1. FAM**
Air is heated while flowing in a tube. The tube has a diameter $D = 10$ cm and a length $L = 4$ m. The inlet air temperature is $\langle T_f \rangle_0 = 20°C$ and the tube surface is at $T_s = 130°C$. The cross-sectional averaged air velocity is $\langle u_f \rangle = 2$ m/s. These are shown in Figure Pr. 7.1.

(a) Draw the thermal circuit diagram.
(b) Determine the Nusselt number $\langle Nu \rangle_D$.
(c) Determine the number of transfer units $NTU$.
(d) Determine the heat transfer effectiveness $\epsilon_{he}$.
(e) Determine the average convection resistance $\langle R_u \rangle_L (°C/W)$.
(f) Determine the convection heat transfer rate $\langle Q_u \rangle_{L-0}(W)$.
(g) Determine the air exit temperature $\langle T_f \rangle_L (°C)$.

Evaluate the properties of air at $T = 300$ K.

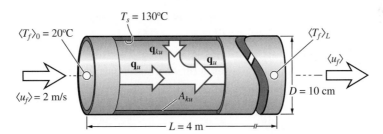

Figure Pr. 7.1. Constant surface temperature tube heating bounded air stream.

**PROBLEM 7.2. FAM.S**
A thermoelectric cooler maintains the surface temperature of a metallic block at $T_s = 2°C$. The block is internally carved to form a connected, circular channel with diameter $D = 0.8$ cm and length $L = 40$ cm. This is shown in Figure Pr. 7.2. Through this channel, a water stream flows and is cooled. The inlet temperature for the water is $\langle T_f \rangle_0 = 37°C$.

(a) Determine and plot (i) the number of transfer units $NTU$, (ii) the thermal effectiveness $\epsilon_{he}$, (iii) the water exit temperature $\langle T_f \rangle_L (°C)$, and (iv) the convection

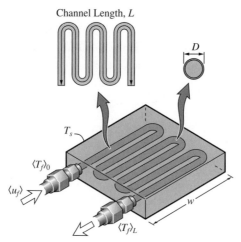

Figure Pr. 7.2. A water-stream cooling block with internal channels.

heat transfer rate $\langle Q_u \rangle_{L-0}$ (W), as a function of the water velocity for $0 < \langle u_f \rangle < 2$ m/s.
(b) At what water velocity is the exit temperature $\langle T_f \rangle_L$, 28°C above $T_s$?

Evaluate the properties at the inlet temperature.

### PROBLEM 7.3. FUN
A fluid enters a tube of uniform surface temperature $T_s$ with temperature $\langle T_f \rangle_0$ and
exits at $\langle T_f \rangle_L$. The tube has a length $L$, a cross-sectional area $A_u$, a surface-convection
area $A_{ku}$, a surface temperature $T_s$, and a mass flow rate $\dot{M}_f$.

(a) Show that for $NTU \to 0$, $\langle T_f \rangle_L$ becomes a linear function of $NTU$.
(b) For a tube with a circular cross section, obtain an expression for $NTU$ as a
function of the tube diameter $D$ and length $L$.
(c) How can the length $L$ and the diameter $D$ be changed such that $NTU \to 0$?

### PROBLEM 7.4. FUN
The blood flow through human tissues is by very small arteries called the arterioles,
which have diameters in the range of 5 to 50 $\mu$m and a length of a few centimeters.
These are fed by small arteries, which in turn are fed by the aorta. Each arteriole
empties into 10 to 100 capillaries, which have porous walls and are the sites of the
exchange between the blood and interstitial tissue fluid. These are shown in Figure
Pr. 7.4. There are about $10^{10}$ capillaries in peripheral tissue. This cascading of blood
vessels results in a large increase in the total flow cross section $A_u$, as listed in
Table Pr. 7.4 along with the cross-sectional area and the time-area averaged blood
velocity $\langle \overline{u_f} \rangle$.

As the total flow cross-sectional area $A_u$ increases, $\langle \overline{u_f} \rangle$ decreases (because the
mass flow rate is conserved).

In Example 7.4, we showed that for a very large specific surface area, i.e.,
$A_{ku}/V \to \infty$, we have $NTU \to \infty$ and that any fluid entering a porous solid with an
inlet temperature $\langle T_f \rangle_0$ will leave with its exit temperature reaching the local solid
temperature, i.e., $\langle T_f \rangle_L = T_s$.

Table Pr. 7.4. *Cross-sectional area (total) and time-area averaged blood velocity through various segments of blood pathways*

|                | $A_u$, cm$^2$ | $\langle \overline{u_f} \rangle$, cm/s |
|----------------|-----------|--------------------------|
| aorta          | 2.5       | 33                       |
| small arteries | 20        | 4.1                      |
| arterioles     | 40        | 2.1                      |
| capillaries    | 2,500     | 0.033                    |
| venules        | 250       | 0.33                     |
| small veins    | 80        | 1.0                      |
| venue cavao    | 8         | 10                       |

Blood Arterioles and Capillaries

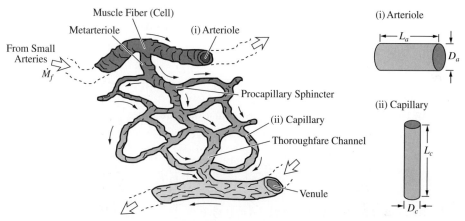

Figure Pr. 7.4. Blood supply to tissue by arterioles feeding the capillaries.

(a) For the conditions given below and in Table Pr. 7.4, determine $NTU$ for (i) an arteriole, and (ii) a capillary. Address the entrance effect [note that from Table 7.2, for $(L/D)/\mathrm{Pe}_D > 0.6$, the entrance effect is negligible].

(b) Using (7.21), (7.22), and (7.24), comment on the ability of these blood streams to control the local tissue temperature $T_s$.

$L_a = 2$ mm, $D_a = 50$ $\mu$m, $L_c = 30$ $\mu$m, $D_c = 3$ $\mu$m, $\rho_f = 1,000$ kg/m$^3$, $c_{p,f} = 3,000$ J/kg-K, $\mu_f = 10^{-3}$ Pa-s, $k_f = 0.6$ W/m-K.

**PROBLEM 7.5. FUN**

The thermally fully developed regime is defined in terms of dimensionless temperature $T^*$ as $\partial T^*/\partial x \equiv \partial[(T_s - T_f)/(T_s - \langle T_f \rangle)]/\partial x = 0$, where $x$ is along the tube and $T_f = T_f(r, x)$, $\langle T_f \rangle = \langle T_f \rangle(x)$, and $T_s(x)$ all change with $x$. For laminar, fully-developed temperature and velocity fields in a tube flow, the differential energy equation for the cylindrical coordinate system is the simplified form of (B.62). Due to the assumed angular symmetry, the $\phi$ dependence is omitted and because of the fully developed fields, the axial conduction and radial convection are omitted. Then for a steady-state heat transfer, we have (using the coordinates of Figure 7.1) from (B.62)

$$-k_f \frac{1}{r} \frac{\partial}{\partial r}\left(r \frac{\partial T_f}{\partial r}\right) + \frac{\partial}{\partial x}(\rho c_p)_f u_f(r) T_f = 0,$$

where

$$u_f(r) = 2\langle u_f \rangle \left[1 - \left(\frac{2r}{D}\right)^2\right].$$

These are used to determine the fluid temperature $T_f(r, x)$, along with the condition of uniform heat flux $q_s$ on the tube wall, which results in the combined integral-differential length energy equation (7.12), i.e.,

$$-P_{ku}q_s + A_u \frac{d}{dx}(\rho c_p)_f \langle u_f \rangle \langle T_f \rangle = 0.$$

Here $q_s$ is taken to be positive when it flows into the fluid.

Then the Nusselt number is given by (7.19), i.e.,

$$\mathrm{Nu}_D = \langle \mathrm{Nu} \rangle_D = \frac{q_s D}{[T_s(x) - \langle T_f \rangle(x)]k_f}.$$

(a) Show that $\partial T_f/\partial x = d\langle T_f \rangle/dx$ is uniform along the tube.
(b) Derive the expression for $T_f = T_f(r, x)$, i.e.,

$$T_f(r, x) = T_s(x) - \frac{2q_s D}{k_f}\left[\frac{3}{16} + \frac{1}{16}\left(\frac{2r}{D}\right)^4 - \frac{1}{4}\left(\frac{2r}{D}\right)^2\right].$$

(c) Derive the expression for $T_s(x) - \langle T_f \rangle(x)$.
(d) Using (7.9) show that $\langle \mathrm{Nu} \rangle_D = 4.364$, for uniform $q_s$.

### 7.9.2 Tubes and Ducts: Hydraulic Diameter

### PROBLEM 7.6. FAM

A rectangular channel used for heating a nitrogen stream, as shown in Figure Pr. 7.6, is internally finned to decrease the average convection resistance $\langle R_u \rangle_L$. The channel wall is at temperature $T_s$ and nitrogen gas enters at a velocity $\langle u_f \rangle$ and temperature $\langle T_f \rangle_0$. The flow is turbulent (so the general hydraulic-diameter based Nusselt number of Table 7.3 can be used). The channel wall and the six fins are made of pure aluminum. Assume the same $\langle \mathrm{Nu} \rangle_{D,h}$ for channel wall and fin surfaces and assume a fin efficiency $\eta_f = 1$.

(a) Draw the thermal circuit diagram.
(b) For the conditions given below, determine the exit temperature $\langle T_f \rangle_L$.
(c) Determine the heat flow rate $\langle Q_u \rangle_{L\text{-}0}$ for the same conditions.

$\langle u_f \rangle = 25$ m/s, $\langle T_f \rangle_0 = -90°C$, $T_s = 4°C$, $a = 20$ mm, $w = 8$ mm, $L = 20$ cm, $L_f = 4$ mm, $l = 1$ mm.

Determine the nitrogen properties at $T = 250$ K.

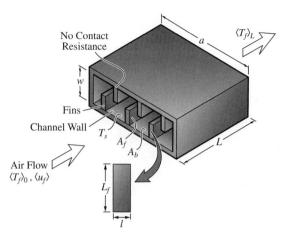

Figure Pr. 7.6. Heat transfer between a bounded fluid stream and channel wall with extended surface.

## PROBLEM 7.7. FUN

Many small devices, such as computer components, can malfunction when exposed to high temperatures. Fans and other such devices are used to cool these components constantly. To aid in the cooling process, extremely small heat exchangers are integrated with the parts being cooled. These heat exchangers are microfabricated through a co-extrusion process such that complex cross-sectional geometries can be formed using densely packed arrays. An example is shown in Figure Pr. 7.7. A small compressor is used to force air at temperature $\langle T_f \rangle_0 = 20°C$ and velocity $\langle u_f \rangle = 5$ m/s through the copper channel having a surface temperature $T_s = 90°C$, and the dimensions are as shown in Figures Pr. 7.7(i) and (ii).

(i) Physical Model for Microchannel      (ii) Idealized Extended Surface Geometry
    with Large Internal Extended Surface        (One of Four Rows)

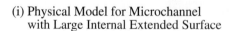

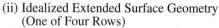

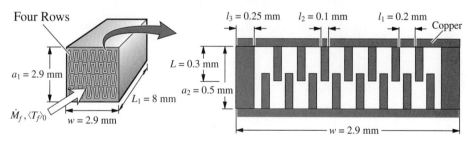

Figure Pr. 7.7. (i) A small heat exchanger. (ii) Cross section of each of the four rows.

(a) Draw the thermal circuit diagram.

(b) Assuming the flow can be approximated as flow through parallel plates, in which $D_h = 2l_1$ and $l_1 = 0.2$ mm, determine the $NTU$ and the heat transfer $\langle Q_u \rangle_{L\text{-}0}$ from the copper heat exchanger to the passing fluid.

(c) Now treating the flow as flow through a packed bed of particles, determine the $NTU$ and the heat transfer $\langle Q_u \rangle_{L\text{-}0}$ from the copper surface to the passing fluid.

Comment on how the predicted Nusselt numbers obtained from the two treatments differ.

Evaluate the air properties at $T = 300$ K.

## PROBLEM 7.8. FAM

In a research nuclear (fission) reactor, a 17 channel element core, with each element being a rectangular cylinder of cross-sectional area $a \times w$ and length $L$, is used. This is shown in Figure Pr. 7.8.

The nuclear fission energy conversion rate $\dot{S}_{r,fi}$ occurs in the channel walls. The coolant flow rate $\dot{M}_f$ per channel is designed for a desired channel wall temperature $T_s$. When for some reason, this flow rate is reduced, $T_s$ can raise to hazardous levels.

(a) Draw the thermal circuit diagram.
(b) Determine the fluid exit temperature $\langle T_f \rangle_L$ and the channel surface temperature $T_s$ for the conditions given below.
(c) If the mass flow rate is reduced by one half, what will $\langle T_f \rangle_L$ and $T_s$ be? Since $\langle T_f \rangle_0$ is below $T_{lg}$, for $T_s > T_{lg}$ any bubble formed will collapse as it departs from the surface (this is called subcooled boiling, Section 7.3.5). Comment on how bubble nucleation affects $T_s$.

Figure Pr. 7.8. A nuclear (fission) reactor cooled by a bounded water stream.

$L = 100$ cm, $a = 2.921$ cm, $w = 7.5$ cm, $\dot{S}_{r,fi} = 4.5$ kW, $\dot{M}_f = 0.15$ kg/s, $\langle T_f \rangle_0 = 45°$C. Determine the water properties at $T = 310$ K.

## PROBLEM 7.9. FAM

The bounded fluid (air) stream surface-convection heat transfer from an automobile brake is by the flow induced by the rotor rotation. There is surface-convection heat transfer from the outside surfaces of the rotor and also from the vane between the two rotor surfaces (called the ventilation area or vent), as shown in Figure Pr. 7.9.

The air flow rate through the vent is given by an empirical relation for the average velocity $\langle u_f \rangle$ as

$$\langle u_f \rangle (\text{m/s}) = rpm \times 0.0316(R_2^2 - R_1^2)^{1/2},$$

where $R_2(\text{m})$ and $R_1(\text{m})$ are outer and inner radii shown in Figure Pr. 7.9. The air enters the vent at temperature $\langle T_f \rangle_0$ and the rotor is at a uniform temperature $T_s$. Assume a uniform flow cross-sectional area (although in practice it is tapered) and a total flow area $A_u = N_c a w$ and a total vent surface-convection area $A_{ku}$. Allow for the entrance effect using circular cross-section Nusselt number. Here $a$ and $w$ are for the rectangular cross section of each channel in the $N_c$ vents.

(a) Draw the thermal circuit diagram.
(b) Determine the vent surface heat transfer rate $\langle Q_u \rangle_{L\text{-}0}$.
(c) Determine the air exit temperature $\langle T_f \rangle_L$.

(i) Automobile Disc-Brake: Physical Model

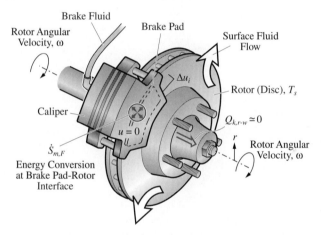

(ii) Rotation-Induced Flow Through Vanes (Ventilated Disc)

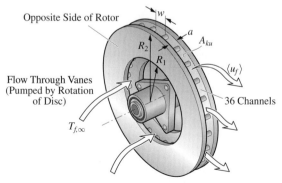

Figure Pr. 7.9. A ventilated automobile brake with rotation induced vent air flow. (i) Physical model. (ii) Rotation-induced flow.

$a = 0.5$ cm, $w = 1.5$ cm, $A_{ku} = 700$ cm$^2$, $rpm = 750$, $N_c = 36$, $R_1 = 10$ cm, $R_2 = 15$ cm, $\langle T_f \rangle_0 = 20°$C, $T_s = 400°$C.

Evaluate air properties at $T = 300$ K. For simplicity, consider only the bounded air stream heat transfer, i.e., do not include the semi-bounded air stream heat transfer from the rotor surface.

## PROBLEM 7.10. FAM

A refrigerant R-134a liquid-vapor stream is condensed, while passing thorough a compact condenser tube, as shown in Figure Pr. 7.10. The stream enters at a mass flow rate $\dot{M}_f$, thermodynamic quality $x_0$, and a temperature $\langle T_f \rangle_0(p_{g,0})$ and here for simplicity assume that the exit conditions are the same as the inlet conditions. The condenser wall is at temperature $T_s$. Assume that the liquid (condensate) flow regime is annular and use the applicable correlation given in Table 7.6.

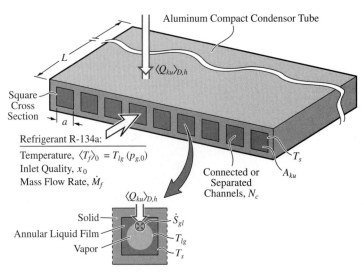

Figure Pr. 7.10. A compact condenser tube is used to condense a refrigerant stream. The tube is a multichannel, extruded aluminum tube. There are $N_c$ square cross-sectional channels in the tube.

(a) Draw the thermal circuit diagram.
(b) Based on the constant $x$ and $\langle T_f \rangle$ assumptions, determine the heat transfer rate, $\langle Q_{ku} \rangle_{D,L}$, for the conditions given below.
(c) From the results of (b), determine the condensation rate $\dot{M}_{gl}$ and estimate the exit quality $x_L$.
(d) Comment on how an iteration may be used to improve on the accuracy of these predictions.

$N_c = 10$, $a = 1.5$ mm, $L = 15$ cm, $p_{g,0} = 1.681$ MPa, $x_0 = 0.5$, $T_s = 58°$C, $\dot{M}_f = 10^{-3}$ kg/s.

Use the saturation properties of Table C.28 at $p_{g,0}$.

**PROBLEM 7.11. FAM**

A compact evaporation tube, made of $N_c$ circular channels and placed vertically, is used for the evaporation of a stream mixture of liquid and vapor of refrigerant R-134a. This is shown in Figure Pr. 7.11, where the inlet quality is $x_0$ and the inlet temperature $\langle T_f \rangle_0$ is determined from the inlet pressure $p_{g,0}$. The liquid-vapor mass flow rate is $\dot{M}_f$. For simplicity assume that along the evaporator $\langle T_f \rangle$ and $x$ remain constant and use the correlation given in Table 7.6.

(a) Draw the thermal circuit diagram.
(b) Based on contact $x$ and $\langle T_f \rangle$ assumptions, determine the heat transfer rate, $\langle Q_{ku} \rangle_D$.
(c) From the results of (b), determine the evaporation rate $\dot{M}_{lg}$ and the exit quality $x_L$.
(d) Comment on how an iteration may be used to improve on the accuracy of predictions.

$\qquad N_c = 8$, $D = 2$ mm, $L = 15$ cm, $p_{g,0} = 0.4144$ MPa, $x_0 = 0.4$, $T_s = 12°C$, $\dot{M}_f = 10^{-3}$ kg/s.

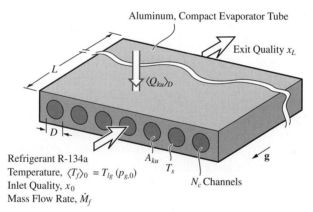

Figure Pr. 7.11. A compact evaporation tube, having $N_c$ circular channel, and placed vertically, is used to evaporate a refrigerant R-134a stream.

### 7.9.3 High Specific Surface Area: Particle Diameter

**PROBLEM 7.12. FAM**

A convection air heater is designed using forced flow through a square channel that has electrically heated wires running across it, as shown in Figure Pr. 7.12.

(a) Draw the thermal circuit diagram.
(b) Determine the average wire temperature for a heating rate (i.e., Joule energy conversion) $\dot{S}_{e,J} = 1,500$ W. Model the wires as a bank of tubes.

Evaluate the properties of air at 300 K.

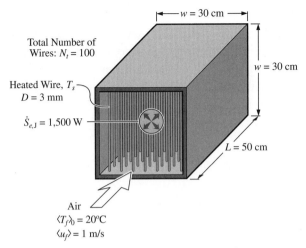

Figure Pr. 7.12.  An electric, air-stream heater.

### PROBLEM 7.13. FAM

The monolith automobile catalytic converter is designated by the number of channels (square geometry) per square inch. Current designs are between 400 and 600 channels per square inch (62 and 93 channels per square centimeter). Each channel has dimensions $a \times a$ and the channel wall thickness is $l$. Assume that the converter has a square cross section of $w \times w$ and a length $L$. This is shown in Figure Pr. 7.13.

The exhaust gas mass flow rate $\dot{M}_f$ (kg/s) is related to the rpm by

$$\dot{M}_f = \frac{1}{2}\frac{rpm}{60}\rho_{f,o}V_d\eta_V,$$

where the density $\rho_{f,o}$ is that of air at the intake condition, $V_d$ is the total displacement volume, and $\eta_V$ is the volumetric efficiency.

For simplicity, during the start-up, assume a uniform channel wall temperature $T_s$.

(a) Draw the thermal circuit diagram.
(b) Determine the exit temperature of the exhaust gas for both the 400 and 600 channels per square inch designs.

$\langle T_f \rangle_0 = 500°C$, $T_s = 30°C$, $w = 10$ cm, $L = 25$ cm, $l = 0.25$ mm, $V_d = 2.2 \times 10^{-3}$ m$^3$, $rpm = 2,500$, $\eta_V = 0.9$, $\rho_{f,o} = 1$ kg/m$^3$.

Evaluate the exhaust gas properties using air at $T = 700$ K.

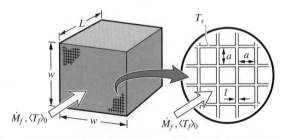

Figure Pr. 7.13.  A monolith automobile catalytic converter.

## PROBLEM 7.14. DES

In many cutting operations it is imperative that the cutting tool be maintained at an operating temperature $T_s$, well below the tool melting point. The cutting tool is shown in Figure Pr. 7.14(a). Two designs for the cooling of the cutting tool are considered. The coolant is liquid nitrogen and the objective is to maintain the tool temperature at $T_s = 500°C$. The nitrogen stream is a saturated liquid at one atm pressure and flows with a mass flow rate of $\dot{M}_{N_2} = 1.6 \times 10^{-3}$ kg/s. These designs are also shown in Figure 7.14(a). The first design uses direct liquid nitrogen droplet impingement. The liquid nitrogen is at temperature $T_{lg,\infty} = 77.3$ K. The average droplet diameter is $\langle D \rangle = 200 \times 10^{-6}$ m, and the average droplet velocity is $\langle u_d \rangle = 3$ m/s. The second design uses the mixing of liquid nitrogen and air, where, after mixing, the nitrogen becomes superheated. Then the mixture is flown internally through the cutting tool. The air enters the mixer at a temperature of $T_a = 20°C$ and flows with a mass flow rate of $\dot{M}_a = 3.2 \times 10^{-3}$ kg/s. The mixture enters the permeable tool with a temperature $\langle T_f \rangle_0$ and flows through the cutting tool where a sintered-particle region forms a permeable-heat transfer medium. The average particle diameter is $D_p = 1$ mm and the porosity is $\epsilon = 0.35$, as depicted in Figure Pr. 7.14(b).

(a) Draw the thermal circuit diagram for the two designs.
(b) Assuming that both designs have the same liquid nitrogen mass flow rate and that (6.116) is valid, determine the amount of surface-convection cooling for surface droplet impingement cooling.
(c) Making the same assumptions, determine the internal transpiration cooling.

Take the mixture conductivity to be $k_f = 0.023$ W/m-K, and evaluate the mixture specific heat using the average of the air specific heat at $T = 300$ K and the

(a) Cutting-Tool Energy Conversion $\dot{S}_{m,F}$ and Two Cooling Methods

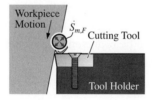

(i) Impinging Droplet Cooling

(ii) Single-Phase Bounded-Fluid Stream Cooling

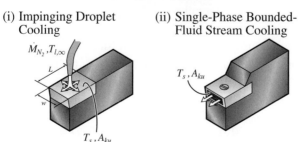

Figure Pr. 7.14. (a) Cooling of a cutting tool by (i) a semi-bounded stream, and (ii) a bounded coolant stream.

(b) Physical Model of Cooling in Second Methods

(i) Mixer

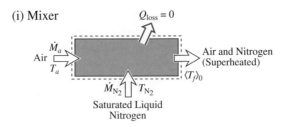

(ii) Bounded-Fluid Stream Through
     Porous Cutting Tool

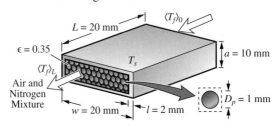

Figure Pr. 7.14. (b) Physical model of cooling of a permeable cutting tool by a bounded coolant stream.

superheated nitrogen specific heat from Table C.26. Also, use the properties of air at $T = 300$ K for part (b). For part (c) use $\langle T_f \rangle_0$, determined from the energy equation for the adiabatic mixture, to determine the properties of air for the mixture.

### PROBLEM 7.15. FAM

An electrical (Joule) heater is used for heating a stream of air $\dot{M}_f$ from temperature $\langle T_f \rangle_0$ to $\langle T_f \rangle_L$. The heater is made of a coiled resistance wire placed inside a tube of diameter $D$. The heater temperature is assumed uniform and at $T_s$. This is shown in Figure Pr. 7.15. The coiled wires can be represented as a porous medium of porosity $\epsilon$ with an equivalent particle diameter $D_p$.

(a) Draw the thermal circuit diagram.
(b) Determine the particle Nusselt number $\langle Nu \rangle_{D,p}$.
(c) Determine the number of transfer units, $NTU$.
(d) Determine the wire surface temperature $T_s$, as a function of $\dot{S}_{e,J}$ (W).
(e) Determine the air exit temperature as a function of $\dot{S}_{e,J}$ (W).
(f) Comment on the safety features that must be included to avoid heater meltdown (melting temperature $T_{sl}$).

$\dot{M}_f = 10^{-3}$ kg/s, $\langle T_f \rangle_0 = 20°$C, $D_p = 1$ mm, $\epsilon = 0.95$, $D = 1.9$ cm, $L = 30$ cm, $T_{sl} = 1,200$ K.

Evaluate air properties at $T = 500$ K.

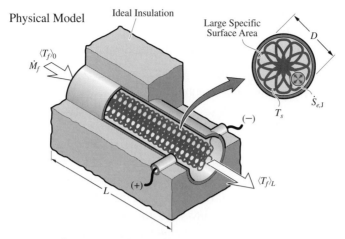

Figure Pr. 7.15. An electric air-stream heater.

## PROBLEM 7.16. FUN.S

Sensible heat storage in packed beds is made by flow of a hot fluid stream through the bed and heat transfer by surface convection. A heat-storage bed of carbon-steel AISI 1010 spherical particles is shown in Figure Pr. 7.16. Hot air is flowing through the bed with an inlet temperature $\langle T_f \rangle_0$ and a mass flow rate $\dot{M}_f$. The initial bed temperature is $T_s(t = 0)$. When the assumption of $\mathrm{Bi}_L < 0.1$ is valid, then (6.156) can be used to predict the uniform, time-dependent bed temperature.

(a) Draw the thermal circuit diagram.
(b) Determine the bed effective thermal conductivity $\langle k \rangle$.
(c) Determine $\langle \mathrm{Nu} \rangle_{D,p}$, $\langle R_{ku} \rangle_{D,p}$, and $NTU$.
(d) Show that the Biot number $\mathrm{Bi}_L = R_{k,s}/\langle R_{ku} \rangle_{D,p}$ is not less than 0.1 (where $R_{k,s} = L/A_k k = 1/L\langle k \rangle$), and therefore, that assuming a uniform bed temperature is not justifiable (although that assumption makes the analysis here much simpler).

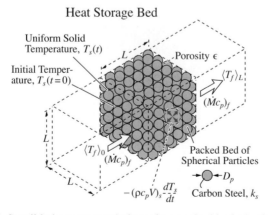

Figure Pr. 7.16. Sensible heat storage/release in a packed bed of spherical particles.

(e) Assume a uniform bed temperature and use (6.156) to plot $T_s(t)$ and $\langle T_f \rangle_L(t)$ for up to four time constants $\tau_s$.

(f) Determine the amount of heat stored during this period.

$T_s(t = 0) = 30°C$, $\langle T_f \rangle_0 = 190°C$, $\dot{M}_f = 4$ kg/s, $L = 2$ m, $D_p = 8$ cm, $\epsilon = 0.40$. Evaluate air properties at $T = 400$ K.

### PROBLEM 7.17. FUN

To improve the surface-radiation heat transfer from a fireplace, metallic chains are suspended above the flame, as shown in Figure Pr. 7.17. The hot, thermobuoyant flue gas flows through and around the chains. The mass flow rate through the chains can be estimated from the fluid flow friction and the available thermobuoyant force and is assumed known here. In steady-state, this surface-convection heat transfer is balanced by the surface-radiation exchange with the surroundings, which is at $T_\infty$. The radiating surface can be modeled as the front surface area of a solid rectangle of dimension $w \times w \times L$ (as shown in Figure Pr. 7.17) scaled by the solid fraction term $(1 - \epsilon)$, where $\epsilon$ is the porosity. Only the surface radiation from the surface facing the surrounding is considered. The entire chain is assumed to have a uniform temperature $T_s$.

(a) Draw the thermal circuit diagram.

(b) Determine the chain temperature $T_s$.

(c) Determine the surface-radiation heat transfer rate $Q_{r,s-\infty}$.

$\dot{M}_f = 0.007$ kg/s, $\langle T_f \rangle_0 = 600°C$, $\epsilon = 0.7$, $A_{ku} = 3$ m$^3$, $D_p = 4$ mm, $\epsilon_{r,s} = 1$, $w = 30$ cm, $L = 50$ cm, $T_\infty = 20°C$.

Treat the flue gas as air and evaluate the properties at $T = 500$ K.

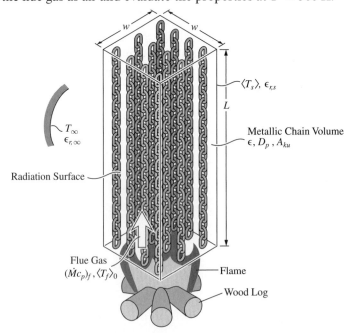

Figure Pr. 7.17. Chains suspended above flame for enhanced surface radiation to the surroundings.

### 7.9.4 Inclusion of Bounding Solid and Other Flows

**PROBLEM 7.18. FAM**

Two printed circuit boards containing Joule heating components are placed vertically and adjacent to each other for a surface-convection cooling by the thermobuoyant motion of an otherwise quiescent air. This is shown in Figure Pr. 7.18. Consider surface-convection from both sides of each board.

(a) Draw the thermal circuit diagram.
(b) Determine the Joule heating rate $\dot{S}_{e,J}$, per board, for the following conditions.

$w = 10$ cm, $L = 15$ cm, $l = 4$ cm, $T_s = 65°C$, $T_{f,\infty} = 30°C$.
Determine the air properties at $\langle T_f \rangle_\delta = (T_s + T_{f,\infty})/2$.

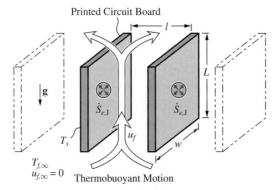

Printed Circuit Board

Figure Pr. 7.18. Two printed circuit boards are placed vertically and adjacent to each other for a surface-convection cooling by the thermobuoyant air motion.

**PROBLEM 7.19. FAM**

A solar collector is placed horizontally, and in order to reduce the heat losses to ambient air, a glass sheet is placed on top of it with the gap occupied by air. This is shown in Figure Pr. 7.19. The energy absorbed by the collector surface $\dot{S}_{e,\alpha}$ is used to heat a water stream flowing underneath it, $Q_c$, or lost to air above it through $\langle Q_{ku} \rangle_L$. Assume a unit surface area of $A_c = 1$ m$^2$ and otherwise treat the collector surface as being infinity large in both directions. The heat transfer between the collector surface and the glass surface is by cellular thermobuoyant motion of the enclosed air. The Nusselt number correlation for this motion is given in Table 7.4 with the gap distance designated as $L$.

(a) Draw the thermal circuit diagram for the collector.
(b) Determine the rate of heat transfer per unit collector surface to the coolant (water) stream $Q_c/A_c$.

$\dot{S}_{e,\alpha}/A_c = \alpha_{r,c}(q_{r,i})_s = 400$ W/m$^2$, $L = 2$ cm, $T_{s,1} = 60°C$, $T_{s,2} = 35°C$.
Determine the air properties at $\langle T_f \rangle = (T_{s,1} + T_{s,2})/2$.

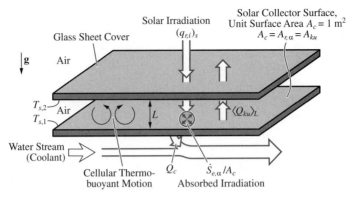

Figure Pr. 7.19. A glass cover is used to reduce heat loss from a solar collector surface to ambient air. There is a cellular motion in the air gap between the two surfaces.

### PROBLEM 7.20. FUN

In order to melt the winter ice formed on a concrete surface, hot-water carrying tubes are embedded in the concrete, near its surface. This is shown in Figure Pr. 7.20. Assume a steady-state heat transfer. The heat flows from the hot-water stream, through the tube wall and through the concrete to the surface (at temperature $T_c$). There it melts the ice with the phase change energy conversion rate designated with $\dot{S}_{sl}$.

(a) Draw the thermal circuit diagram for the heat transfer between the hot-water stream and the concrete surface.
(b) Determine the Nusselt number $\langle Nu \rangle_D$.
(c) Determine the surface-convection resistance $\langle R_{ku} \rangle_D$.
(d) Determine the concrete conduction resistance between the tube surface and concrete surface $R_{k,s\text{-}1}$, using Table 3.3(a). Divide the per-tube resistance by the number of tubes $N_t = a/w$ to obtain the total resistance.
(e) Determine the rate of heat transfer $\langle Q_u \rangle_{L\text{-}0}$ and the rate of ice melting $\dot{M}_{sl}$.

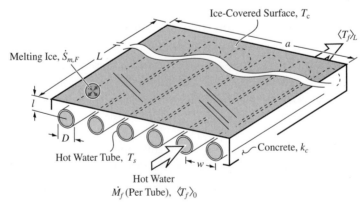

Figure Pr. 7.20. Hot-water carrying tubes are placed near a concrete surface to melt the winter ice formed on the surface.

$D = 3$ cm, $l = 5$ cm, $w = 10$ cm, $\langle T_f \rangle_0 = 25°C$, $T_c = 0°C$, $k_c$(concrete) $= 1.0$ W/m-K, $\langle u_f \rangle = 0.5$ m/s, $a = L = 5$ m.

Assume that the tubes have a negligible thickness. Determine the water properties at $T = 290$ K. The heat of melting for water is given in Table C.4. Note that for a uniform melting, a small $NTU$ is used.

## PROBLEM 7.21. FUN

In an internal combustion engine, for the analysis of the surface-convection heat transfer on the inner surface of the cylinder, $\langle Nu \rangle_D$ must be determined for the cylinder conditions. Figure Pr. 7.21 gives a rendering of the problem considered. The Woschni [15] correlation for this Nusselt number uses the averaged cylinder velocity and cylinder pressure and is

$$\langle Nu \rangle_D = 0.035 Re_D^m, \quad Re_D = \frac{\rho_f D \langle u_f \rangle}{\mu_f},$$

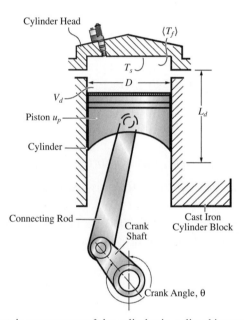

Figure Pr. 7.21. Geometric parameters of the cylinder in a diesel internal combustion engine.

where the averaged fluid velocity in the cylinder $\langle u_f \rangle$ is given by

$$\langle u_f \rangle = a_1 (\overline{u_p^2})^{1/2} + a_2 \frac{V_f (p_f - p_o)}{M_f (R_g / M)},$$

$$(\overline{u_p^2})^{1/2} = 2 L_d (rpm) \frac{2\pi}{60(s/min)}, \quad p_o = \frac{M_f}{V_f} \frac{R_g}{M} T_o.$$

Here $p_o$ is motored pressure used as a reference pressure, which is determined with the intake manifold air temperature $T_o$. In the second term in the averaged velocity $\langle u_f \rangle$ expression, the pressure rise $p_f - p_o$ is caused by combustion (and is used only during the combustion period).

(a) Determine $\langle Nu_D \rangle$ and $\langle q_{ku} \rangle_D$, for (i) the combustion period and (ii) the intake period.

(b) Comment on the magnitude of $\langle Nu \rangle_D$ (i) during the combustion period ($a_1 = 2.28$, $a_2 = 3.24 \times 10^{-3}$ m/s-K, $M_f = 6.494 \times 10^{-3}$ kg, $\langle T_f \rangle = 1{,}700$ K, $p_f = 23$ MPa, $V_f = 1.3966 \times 10^{-4}$ m$^3$) and (ii) during the intake period ($a_2 = 0$, $p_f = 0.2$ MPa, $\langle T_f \rangle = 400$ K).

$m = 0.8$, $D = 0.125$ m, $L_d = 0.14$ m, $rpm = 1{,}600$ rpm, $T_o = 329$ K, $T_s = 800$ K, $R_g/M = 290.7$ J/kg-K.

Note that the pressure and temperature given above are selected for a diesel engine. Evaluate the properties of air for (i) at $T = \langle T_f \rangle = 1{,}700$ K, and for (ii) at $T = \langle T_f \rangle = 400$ K, using Table C.22. Use the ideal gas law and given pressure and temperature to evaluate the density of the gas (air). Use dynamic viscosity (not kinematic viscosity).

### PROBLEM 7.22. FAM.S

Placing an air gap between brick walls can reduce the heat transfer across the composite, when the thermobuoyant motion in the air gap is not significant. Consider the one-dimensional heat flow through a composite of two brick walls and an air gap between them, as shown in Figure Pr. 7.22.

(a) Draw the thermal circuit diagram.
(b) Determine the steady-state heat flow rate $Q_{1-2}$ for the case of $l_a$ equal to (i) 1 cm, (ii) 2 cm, and (iii) 4 cm.
(c) Comment on the minimum $Q_{1-2}$ and its corresponding air gap size $l_a$.

$T_1 = 40°C$, $T_2 = 15°C$, $l_1 = l_2 = 10$ cm, $w = 6$ m, $L = 3$ m, $\langle k_1 \rangle = \langle k_2 \rangle = 0.70$ W/m-K.

Evaluate the air properties at $T = 300$ K and use $\beta_f = 2/[(T_1 + T_2)(K)]$. Since $T_{s,1}$ and $T_{s,2}$ depend on $Q_{1-2}$, and in turn the overall resistance $R_{\Sigma,1-2}$ depends on $T_{s,1}$ and $T_{s,2}$, a solver should be used.

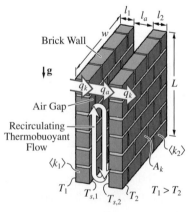

Figure Pr. 7.22. An air gap placed between two brick walls to reduce heat transfer. The thermobuoyant flow in the air gap enclosure is also shown.

## PROBLEM 7.23. FUN

In arriving at the axial temperature distribution of a bounded fluid stream (entering at a temperature of $\langle T_f \rangle_0$ and at a surface temperature of $T_s$), i.e., (7.22), the axial fluid conduction was neglected. This is valid for high Péclet number ($\mathrm{Pe}_L = \langle u_f \rangle L / \alpha_f$) streams where $L$ is the length along the flow. For low $\mathrm{Pe}_L$, i.e., for low velocities or high $\alpha_f$, this axial conduction may become significant.

The axial conduction can be added to the energy equation (7.11) for flow through a tube of diameter $D$, and this gives

$$- P_{ku} q_{ku} + A_{ku} \frac{d}{dx} q_u + A_{ku} \frac{d}{dx} q_k = 0$$

or

$$- P_{ku} q_{ku} + A_{ku} (\rho c_p)_f \langle u_f \rangle \frac{d \langle T_f \rangle}{dx} - A_{ku} k_f' \frac{d^2 \langle T_f \rangle}{dx} = 0,$$

where $k_f'$ is the sum of fluid conduction and a contribution due to averaging of the nonuniform fluid temperature and velocity over tube cross-sectional area. This contribution is called the dispersion (or Taylor dispersion).

Here constant thermophysical properties are assumed. This axial conduction-convection and lateral surface-convection heat transfer in a bounded fluid stream is shown in Figure Pr. 7.23.

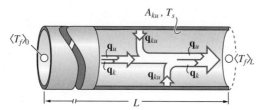

Figure Pr. 7.23. Axial conduction-convection and lateral surface-convection heat transfer in a bounded fluid stream.

Using the surface-convection resistance $R_{ku}$ and the Nusselt number, we can write this energy equation as

$$- \frac{d^2 \langle T_f \rangle}{dx^2} + \frac{\langle u_f \rangle}{\alpha_f} \frac{d \langle T_f \rangle}{dx} + \frac{P_{ku} \langle \mathrm{Nu} \rangle_D k_f}{A_{ku} D k_f} (\langle T_f \rangle - T_s) = 0 \qquad \alpha_f = \frac{k_f'}{(\rho c_p)_f},$$

where $T_s$ is assumed constant.

The fluid thermal conditions at $x = 0$ and $x = L$ are

$$\langle T_f \rangle (x = 0) = \langle T_f \rangle_0$$

$$\langle T_f \rangle (x = L) = \langle T_f \rangle_L.$$

(a) Using the length $L$ and the temperature difference $T_s - \langle T_f \rangle_0$, show that the energy equation becomes

$$\frac{d^2 \langle T_f^* \rangle}{dx^{*2}} - \mathrm{Pe}_L \frac{d \langle T_f^* \rangle}{dx^*} - \mathrm{Pe}_L NTU \frac{D}{4L} \langle T_f^* \rangle = 0,$$

$$x^* = \frac{x}{L}, \quad \langle T_f^* \rangle = \frac{\langle T_f \rangle - T_s}{\langle T_f \rangle_0 - T_s}, \quad \mathrm{Pe}_L = \frac{\langle u_f \rangle L}{\alpha_f},$$

$$NTU = \frac{A_{ku} \langle \mathrm{Nu} \rangle_D k_f}{(\dot{M} c_p)_f D}, \quad (\dot{M} c_p)_f = A_u \rho_f \langle u_f \rangle c_{p,f},$$

$$A_{ku} = P_{ku} L, \quad \langle T_f^* \rangle (x^* = 0) = 1, \quad \langle T_f^* \rangle (x^* = 1) = \langle T_f^* \rangle_L.$$

(b) Show that the solution to this ordinary second-order differential equation gives the fluid axial temperature distribution as

$$\langle T_f^* \rangle (x^*) = \frac{\langle T_f \rangle - T_s}{\langle T_f \rangle_0 - T_s} = \frac{\langle T_f^* \rangle_L (e^{m_2 x^*} - e^{m_1 x^*}) + e^{m_2} e^{m_1 x^*} - e^{m_1} e^{m_2 x^*}}{e^{m_2} - e^{m_1}}.$$

Note that the solution to

$$\frac{d^2 \langle T_f^* \rangle}{dx^{*2}} - b \frac{d \langle T_f^* \rangle}{dx^*} - c \langle T_f^* \rangle = 0$$

is

$$\langle T_f^* \rangle (x^*) = a_1 e^{m_1 x^*} + a_2 e^{m_2 x^*}, \quad m_{1,2} = \frac{b \pm (b^2 + 4c)^{1/2}}{2}.$$

(c) What would $\langle T_f^* \rangle (x^*)$ be for the case of adiabatic end conduction, i.e., $d \langle T_f \rangle / dx |_{x=L} = 0$?

### 7.9.5 Heat Exchangers

**PROBLEM 7.24. FAM**
Plate-type heat exchangers are corrugated thin metallic plates held together in a frame, as shown in Figure Pr. 7.24. Gasket or welding is used for sealing. The flow arrangement is counterflow with the periodic alternation shown in the figure.

Consider an assembly of $N$ plates, each having a surface area $w \times L$ (making a total surface area $NwL$). The heat exchanger transfers heat between a combustion flue-gas stream with an inlet temperature $\langle T_{f,h} \rangle_0$ and a room temperature air stream with $\langle T_{f,c} \rangle_L$ and a mass flow rate $\dot{M}_{f,c} = \dot{M}_{f,h}$. The rate of heat transfer is $\langle Q_u \rangle_{L\text{-}0}$ and the overall resistance is $R_\Sigma$.

(a) Draw the thermal circuit diagram.
(b) Determine the cold fluid exit temperature.
(c) Determine the number of plates $N$ required.

(i) Plate-Type Heat Exchanger

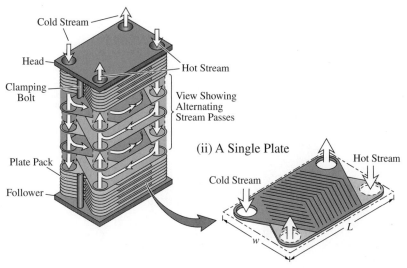

Figure Pr. 7.24. (i) A plate-type heat exchanger. (ii) A single plate in a plate-type heat transfer.

$\langle T_{f,h}\rangle_0 = 900°C$, $\langle T_{f,c}\rangle_L = 20°C$, $\dot{M}_{f,c} = \dot{M}_{f,h} = 70$ g/s, $\langle Q_u\rangle_{L\text{-}0} = 45$ kW, $R_\Sigma = (10^{-3}/NwL)°C/W$, $w = 0.08$ m, $L = 0.35$ m.

Evaluate the thermophysical properties using air at $T = 700$ K.

## PROBLEM 7.25. FAM

An auxiliary, fuel-fired automobile heater uses Diesel fuel and heats the water circulating through the radiator or the water circulating through the heater core (i.e., the heat exchanger for heating the air flowing through the passenger compartment). This heat exchanger is shown in Figure Pr. 7.25 along with the heat flux vector track showing the direction of heat transfer.

The flue gas (products of the combustion of the diesel-air mixture) flows through a counterflow heat exchanger with the flue gas flowing through the inner annulus and the water flowing through the outer annulus. The overall thermal resistance is $R_\Sigma = 0.5°C/W$ and the heat exchanger is $L = 25$ cm long. The mass flow rates of the fuel, air, and water are $\dot{M}_F = 0.03$ g/s, $\dot{M}_a = 0.75$ g/s, and $\dot{M}_w = \dot{M}_c = 5$ g/s. The water inlet temperature is $\langle T_{f,c}\rangle_0 = 15°C$ and the fuel and air combustion chamber inlet temperature is $T_{f,\infty} = 20°C$. Assume complete combustion. The heat of combustion is found in Table 5.2.

(a) Draw the thermal circuit diagram.
(b) Determine the flue gas temperature leaving the combustion chamber $\langle T_{f,h}\rangle_L$.
(c) Determine the exit temperatures for the flue gas $\langle T_{f,h}\rangle_0$ and water $\langle T_{f,c}\rangle_L$.
(d) Determine the amount of heat exchanged between the two streams and the heater efficiency, defined as $\langle Q_u\rangle_{L\text{-}0}/\dot{S}_{r,c}$.

Fuel-Fired Auxiliary Automobile Heater

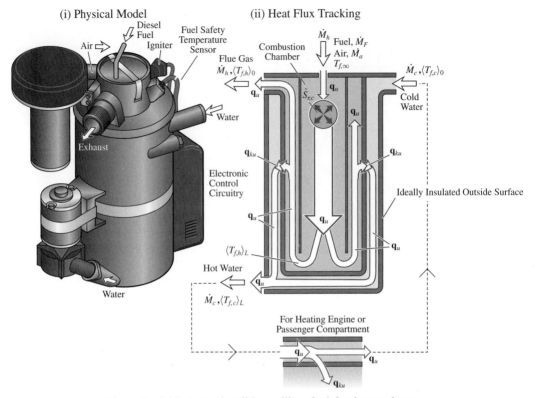

Figure Pr. 7.25. A combustible auxiliary fuel-fired water heater.

This is a counterflow heat exchanger. Use (5.34) to determine $\langle T_{f,h}\rangle_L$. Use $\langle T_{f,c}\rangle_0$, and $\langle T_{f,h}\rangle_L$ as the inlet temperatures. For the gas, use the properties of air at $T = 1{,}000$ K and for the water, evaluate the properties at $T = 310$ K.

### PROBLEM 7.26. FAM

During cardiopulmonary bypass, in open-heart surgery, the blood is cooled by an external heat exchanger to lower the body temperature. This lowering of the body temperature (called whole-body hypothermia) reduces metabolic demand and protects the vital organs during the operation. The heat exchanger is part of the extra-corporeal circulation circuit shown in Figure Pr. 7.26(i). Special pumps (e.g., roller pumps, which use compression of elastic tubes to move the liquid) are used to protect the blood cells from mechanical damage. A shell and tube heat exchanger (with one shell pass) is used to cool (and later heat) the bloodstream, using a water stream, as shown in Figure Pr. 7.26(ii), to the hypothermic temperature $\langle T_{f,h}\rangle_L$.

(a) Draw the thermal circuit diagram.
(b) Determine the bloodstream exit temperature $\langle T_{f,h}\rangle_L$.
(c) Determine $\langle Q_u\rangle_{L\text{-}0}$.

(i) Extracorporeal Circulation Circuit Used in Open-Heart Surgery

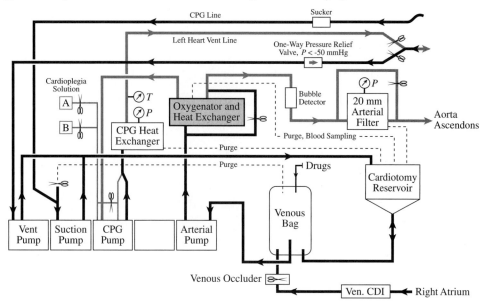

(ii) Oxygenator and Blood Heat Exchanger

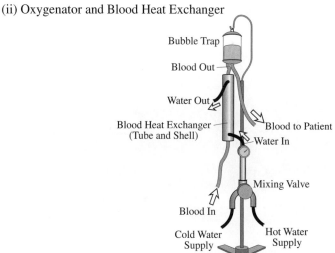

Figure Pr. 7.26. (i) An extracorporeal circulation circuit used in open-heart surgery. (ii) A tube and shell heat exchanger used for cooling (and heating) the blood stream.

$\langle T_{f,h}\rangle_0 = 37°C$, $\langle T_{f,c}\rangle_0 = 15°C$, $(\dot{M}_f/\rho_f)_h = 250$ ml/min, $(\dot{M}_f/\rho_f)_c = 200$ ml/min, $R_\Sigma = 5 \times 10^{-2}$ °C/W.

Use properties of water at $T = 300$ K, for both blood and water.

**PROBLEM 7.27. FAM**

Gray whales have counterflow heat exchange in their tongues to preserve heat. The tip of the tongue is cooled with the cold sea water. The heat exchange is between the incoming warm bloodstream (entering with the deep-body temperature) flowing through the arteries and the outgoing cold bloodstream (leaving the tongue surface region) flowing through the veins. This is shown in Figure Pr. 7.27. In each heat exchanger unit, nine veins of diameter $D_c$ completely encircle (no heat loss to the surroundings) the central artery of diameter $D_h$. The length of the heat exchange region is $L$.

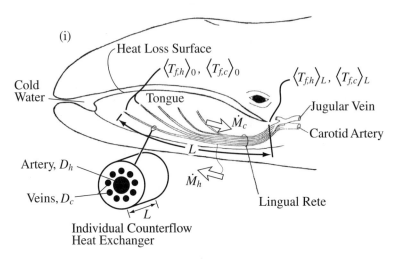

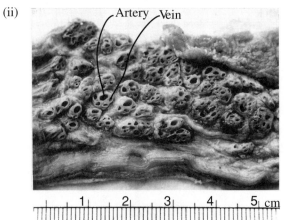

Figure Pr. 7.27. (i) A schematic of the vascular heat exchanger in the gray whale tongue. (ii) The cross section of the lingual rete showing several vascular heat exchangers.

(a) Draw the thermal circuit diagram.
(b) Determine $R_{ku,c}$ and $R_{ku,h}$.
(c) Determine the exit temperature of the cold bloodstream $\langle T_{f,c} \rangle_L$.

$\langle T_{f,c} \rangle_0 = 2°C$, $\langle T_{f,h} \rangle_L = 36°C$, $L = 55$ cm, $D_h = 3$ mm, $D_c = 1$ mm, $R_{k,h\text{-}c} = 5°C/W$, $\langle u_{f,h} \rangle = 1$ mm/s, $\langle u_{f,c} \rangle = 1$ mm/s.

Use water properties at $T = 290$ K for blood. Note that for $C_r = 1$, (7.78) should be used for the counterflow heat exchanger. Also use uniform $q_s$ results for the Nusselt numbers.

## PROBLEM 7.28. FAM

The fan-coil furnace used in domestic air heaters uses a cross-flow heat exchange (both fluids unmixed) between a stream of products of combustion and a stream of air, as shown in Figure Pr. 7.28. The thermal resistance on the air side is $R_{ku,c}$, that on the combustion products side is $R_{ku,h}$, and the conduction resistance of the separating wall is negligible.

(a) Draw the thermal circuit diagram.
(b) Determine the rate of heat exchange between the two streams.
(c) Determine the efficiency of this air heater (defined as the ratio of the rate of heat exchange through the heat exchanger $\langle Q_u \rangle_{L-0}$ to the rate of energy conversion by combustion in the hot fluid stream preheater $\dot{S}_{r,c}$).

$\langle T_{f,c} \rangle_0 = 15°C$, $\langle T_{f,h} \rangle_0 = 800°C$, $\dot{M}_c = 0.1$ kg/s, $\dot{M}_h = 0.01$ kg/s, $R_{ku,c} = 2 \times 10^{-2}°C/W$, $R_{ku,h} = 3 \times 10^{-2}°C/W$, $\dot{S}_{r,c} = 7{,}300$ W.

Treat the combustion products as air and determine the air properties at $T = 300$ K (Table C.22). The $\epsilon_{he}$-$NTU$ relation for this heat exchanger is given in Table 7.7.

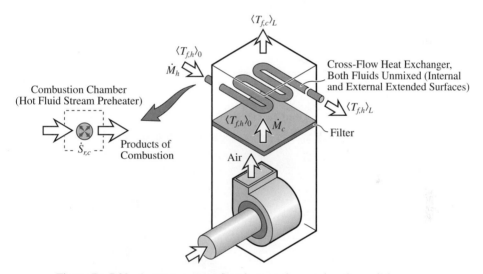

Figure Pr. 7.28. A gas-gas, cross-flow heat exchanger in a fan-coil furnace.

## PROBLEM 7.29. FAM

The automobile passenger compartment heater uses a cross-flow heat exchanger (called the heater core), as shown in Figure Pr. 7.29. Due to the presence of fins, the air (cold stream) is unmixed as it flows through the heater exchanger. The water (hot stream) flowing through flat tubes is also unmixed.

(a) Draw the thermal circuit diagram.
(b) Determine the number of thermal units $NTU$.
(c) Determine the ratio of thermal capacitance $C_r$.
(d) Determine the effectiveness $\epsilon_{he}$.
(e) Determine the exit temperature of the cold stream (air) $\langle T_{f,c} \rangle_L$.

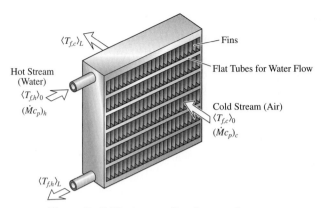

Figure Pr. 7.29.  A cross-flow heat exchanger.

(f) Determine the exit temperature of the hot stream (water) $\langle T_{f,h} \rangle_L$.

(g) Determine the amount of heat exchanged $\langle Q_u \rangle_{L\text{-}0}$.

$\dot{M}_c = 0.03$ kg/s, $\dot{M}_h = 0.10$ kg/s, $\langle T_{f,c} \rangle_0 = 4°C$, $\langle T_{f,h} \rangle_0 = 50°C$, $R_\Sigma = 3 \times 10^{-2}\,°C/W$.

Evaluate the properties at $T = 300$ K.

### 7.9.6  Overall Thermal Resistance in Heat Exchangers $R_\Sigma$

**PROBLEM 7.30. FAM**

In a shell and tube heat exchanger, two fluid streams exchange heat. This is shown in Figure Pr. 7.30. The hot stream is a saturated steam at $\langle T_{f,h} \rangle_0 = 400$ K

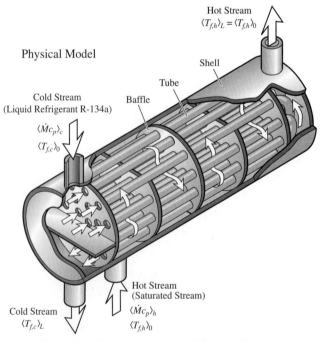

Figure Pr. 7.30.  A tube and shell heat exchanger.

($p_g = 0.2455$ MPa, Table C.27) and it loses heat and condenses. The cold stream is the subcooled liquid Refrigerant R-134a.

(a) Draw the thermal circuit diagram.
(b) Determine the number of thermal units $NTU$.
(c) Determine the effectiveness $\epsilon_{he}$.
(d) Determine the exit temperature of the cold stream $\langle T_{f,c}\rangle_L$.
(e) Determine the heat exchange rate $\langle Q_u\rangle_{L-0}$.

$\dot{M}_c = 3$ kg/s, $\langle T_{f,h}\rangle_0 = \langle T_{f,h}\rangle_L = 400$ K, $\langle T_{f,c}\rangle_0 = 300$ K, $R_\Sigma = 3 \times 10^{-3}\,^\circ$C/W.

Evaluate the refrigerant R-134a saturation properties at $T = 303.2$ K (Table C.28).

## PROBLEM 7.31. FAM

Many hot-water heaters consist of a large reservoir used to store the hot water. This may have a batch-type processing that results in a low efficiency, because the hot water must be constantly heated to make up for heat losses. Alternatively,

(i) Hot Water Heater

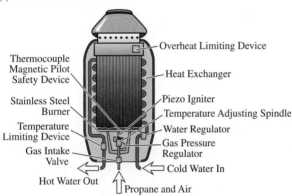

(ii) Adiabatic Flame Temperature

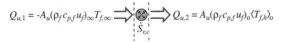

$$Q_{u,1} = -A_u(\rho_f c_{p,f} u_f)_\infty T_{f,\infty} \Rightarrow \otimes \Rightarrow Q_{u,2} = A_u(\rho_f c_{p,f} u_f)_o \langle T_{f,h}\rangle_o$$

(iii) Heat Exchanger Dimensions

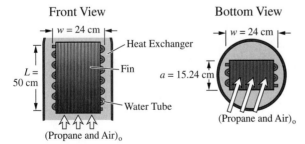

(iv) Fin Dimensions

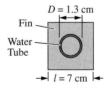

Linear Fin Density = 8 fins/cm
Fin Thickness $l_f = 0.5$ mm

Figure Pr. 7.31. (i) A wall-mounted, on-demand, hot-water heater. (ii) Adiabatic flame temperature. (iii) Heat exchanger dimensions. (iv) Fin dimensions.

no-storage, on-demand, high-efficiency crossflow heat exchangers can provide the hot water needed. One such design, along with its dimensions, is shown in Figure Pr. 7.31. In this design a mixture of air and propane, initially at a temperature $T_{f,\infty} = 25°C$ and with a fuel mass fraction of $(\rho_F/\rho_f)_1 = 0.015$, undergoes combustion with no heat loss (i.e., $Q_{loss} = 0$) with a generation of $\dot{S}_{r,c} = 12{,}900$ W. The flue gas then flows over a tube of diameter $D = 1.3$ cm that is curved as shown in Figure Pr. 7.31, such that the total length is $5w$. The tube contains water with an inlet temperature of $\langle T_{f,c}\rangle_0 = 20°C$ and a volumetric flow rate of $1(\text{gal/min}) = 6.3 \times 10^{-5}(\text{m}^3/\text{s})$. In order to increase the heat transfer, the tube is surrounded by fins with a density of 8 fins per cm and a fin efficiency $\eta_f = 1$.

(a) Draw the thermal circuit diagram.
(b) Determine the overall efficiency of the heat exchanger, defined as $\eta = \langle Q_u\rangle_{L\text{-}0}/\dot{S}_{r,c}$.

    Use the properties of water at $T = 310$ K and treat the combustion products as air with the properties evaluated at $T = 300$ K.

### PROBLEM 7.32. FAM

Water is heated in a heat exchanger where the hot stream is a pressurized fluid undergoing phase change (condensing). The pressure of the hot stream is regulated such that $\Delta T_{max} = \langle T_{f,h}\rangle_0 - \langle T_{f,c}\rangle_0$ remains constant. The parallel-flow heat exchanger is shown in Figure Pr. 7.32. The heat exchanger is used for the two cases where the average cold stream temperature (i) $\langle T_{f,c}\rangle = 290$ K, and (ii) $\langle T_{f,c}\rangle = 350$ K. These influence the thermophysical properties.

(a) Draw the thermal circuit diagram.
(b) Determine the $NTU$ for cases (i) and (ii) and briefly comment on the effect of cold stream average temperature on the capability of the heat exchanger.
(c) Determine the heat transfer rates (W) for cases (i) and (ii).
(d) Determine the rates of condensation [i.e., mass flow rates of condensed fluid in the hot stream (kg/s)] for cases (i) and (ii) assuming $\Delta h_{lg}$ remains constant.

    $\Delta T_{max} = 20°C$, $\dot{M}_{f,c} = 0.5$ kg/s, $L = 5$ m, $D_i = 2$ cm, $\Delta h_{lg} = 2.2 \times 10^6$ J/kg.

    Evaluate the water properties (Table C.23), for the cold stream, at the cold stream average temperature for each case. Neglect the effect of coiling (bending) of the tube on the Nusselt number. Neglect the wall and hot stream thermal resistances.

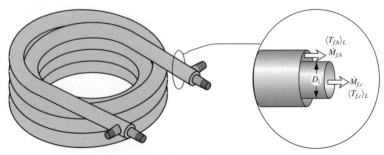

Figure Pr. 7.32. A coaxial heat exchanger.

## PROBLEM 7.33. FUN

A water boiler using natural gas combustion is shown in Figure Pr. 7.33. Consider the evaporator section, where the water temperature is assumed constant and at $T_{lg}(p_g)$. The conduction through the walls separating the combustion flue gas and the water stream is assumed negligible. Also, assume that water-side surface convection resistance $R_{ku,c}$ is negligibly small.

(a) Draw the thermal circuit diagram.
(b) Determine the length $L$ needed to transfer 70% of $\dot{S}_{r,c}$ to the water.

$\dot{M}_{f,h} = 0.02$ kg/s, $\dot{S}_{r,c} = 30$ kW, $D = 15$ cm, $\langle T_{f,h}\rangle_0 = 1{,}200°$C, $\langle T_{f,c}\rangle_L = \langle T_{f,c}\rangle_0 = T_{lg} = 100°$C.

Evaluate flue gas properties using air at $T = 1{,}000$ K.

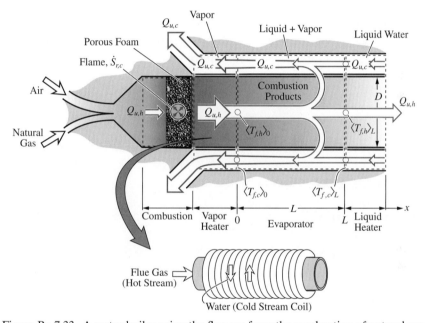

Figure Pr. 7.33. A water boiler using the flue gas from the combustion of natural gas.

## PROBLEM 7.34. FAM.S

Air and nitrogen (gas) stream exchange heat in a parallel-flow, coaxial tube heat exchanger. The inner diameter is $D_1 = 3$ cm and the outer diameter is $D_2 = 5$ cm. This is shown in Figure Pr. 7.34. Nitrogen flows through the inside tube at an average velocity of $\langle u_f\rangle_c = 1$ m/s and enters at $\langle T_{f,c}\rangle_0 = 4°$C. Air flows in the outside tube at an average velocity of $\langle u_f\rangle_h = 2$ m/s and enters at $\langle T_{f,h}\rangle_0 = 95°$C. Neglect the tube wall thickness and assume heat exchange only between these streams.

(a) For a nitrogen exit temperature of $40°$C, determine the amount of heat transferred $\langle Q_u\rangle_{L-0}$.
(b) Also, determine the length of heat exchanger $L$ needed.

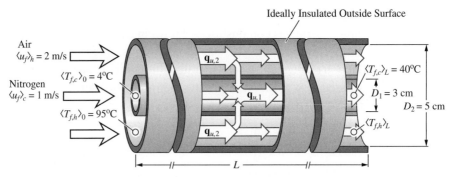

Figure Pr. 7.34. A parallel-flow heat exchanger.

(c) Assuming that the average velocities and inlet temperatures remain the same, what would be the maximum increase in the nitrogen stream temperature $\Delta\langle T_{f,c}\rangle_{max}$ that could be achieved?

For simplicity, evaluate the air and nitrogen properties at their inlet temperatures. Use the constant surface temperature condition to determine the Nusselt numbers.

### 7.9.7 General*

The following Problems are found on the Web site www.cambridge.org/kaviany.

**PROBLEM 7.35. FUN***

**PROBLEM 7.36. FUN***

**PROBLEM 7.37. FUN***

**PROBLEM 7.38. FUN.S***

**PROBLEM 7.39***

**PROBLEM 7.40. FAM***

**PROBLEM 7.41. FUN.S***

**PROBLEM 7.42. FUN***

**PROBLEM 7.43. FAM***

# 8

---

# Heat Transfer in Thermal Systems*

Chapter 8 is found on the Web site www.cambridge.org/kaviany. Chapter 8 is about heat transfer analysis and addresses control of heat transfer, in conjunction with energy storage and conversion, for innovative applications and optimized performance in thermal systems. This heat transfer can occur at various, cascading length scales within the system. The analysis is done by modeling the transport, storage, and conversion of thermal energy as thermal circuits. The elements of these circuits, i.e., mechanisms and models of resistances, storage, and energy conversion, have been discussed in Chapters 2 to 7. In this chapter we consider their combined usage in some innovative, significant, and practical thermal systems. We begin by addressing the primary thermal functions of heat transfer media and bounding surfaces. Then we summarize the elements of thermal engineering analysis. These include the need for assumptions and approximations for reduction of the physical models and conceptual thermal processes to thermal circuit models. Next we give five examples (Examples 8.1 to 8.5), in detail, to demonstrate the application of the fundamentals and relations developed in the text, along with the use of software. The examples are selected for their innovative potentials. There are also five end-of-chapter problems (Problems 8.1 to 8.5).

## 8.1  Primary Thermal Functions*

### 8.1.1  Primary Functions of Heat Transfer Media*

### 8.1.2  Primary Thermal Functions of Bounding Surfaces*

### 8.1.3  Heat Transfer Material*

## 8.2 Thermal Engineering Analysis*

### 8.2.1 Simplifications, Approximations, and Assumptions: Modeling*

### 8.2.2 Nodal Energy Equation*

### 8.2.3 Simultaneous Inclusion of Various Heat Transfer Mechanisms*

### 8.2.4 Optimization*

### 8.3 Examples*

### 8.4 Summary*

### 8.5 References*

### 8.6 Problems*

# Nomenclature*

All the symbols, superscripts and subscripts used in this textbook are found on the Web site www.cambridge.org/kaviany.

# Glossary*

The definitions for the main terms used in this textbook are found on the Web site www.cambridge.org/kaviany.

# Answers to Problems

Life's battles don't always go
To the stronger or faster man;
But soon or late the man who wins
Is the man who thinks he can.
$-$ W. Wintle

**1.13**   (b)   $t = 247$ s

1.14   (b)   $dT/dt = 1.889 \times 10^{-3}\,°\text{C/s}$

1.15   (b)   $T_g(t_f) = 5{,}200°\text{C}$, (c) $\eta = 2.5\%$

1.16       $Q_{ku} + Q_r = 12$ kW

1.17   (c)   $u_\delta = 0.5904$ mm/hr

1.18   (b)   $\langle T \rangle_A = 13.55°\text{C}$

1.19       $\tau = 744$ s $= 0.207$ hr

1.20   (b)   $dT_r/dt = 8.696°\text{C/s}$

**2.1**   (a)   $\lim_{\Delta V \to 0} \int_A \mathbf{q} \cdot \mathbf{s}_n dA / \Delta V \equiv \nabla \cdot \mathbf{q} = 8a$

2.3   (a)   $\dfrac{2}{R_o^2 - R_i^2}(q_{ku,o}R_o + q_{ku,i}R_i) + \dfrac{dq_{k,z}}{dz} = 0$

2.4   (b)   $L = 7.362$ m

2.5       brake pad region: $T(t = 4\,\text{s}) = 93.50°\text{C}$; entire rotor: $T(t = 4\,\text{s}) = 66.94°\text{C}$

2.6   (a)   $\dot{m}_{ph} = 5.353 \times 10^{35}$ photon/m$^2$-s, (b) $\int_{-\infty}^{\infty} \dot{S}_{e,\sigma} dt = 0.1296$ J
      (c)   $\dot{S}_{e,\sigma}/V = 2.972 \times 10^{12}$ J/m$^3$

2.7   (a)   $\dot{s}_{e,J} = 2.576 \times 10^8$ W/m$^3$, (b) $\rho_e(T) = 1.610 \times 10^{-5}$ ohm-m

2.8   (a)   $\dfrac{\partial T}{\partial t} = 0.5580°\text{C/s}$, (b) $\dfrac{\partial T}{\partial t} = 0.7259°\text{C/s}$

2.10   (a)   $\nabla \cdot \mathbf{q} = 10^3$ W/m$^3$

2.11   (a)   $\mathbf{q}_k = -[1.883 \times 10^3\,(°\text{C/m}) \times k(\text{W/m-}°\text{C})]\mathbf{s}_x$

      (c)   $q_{ku}\dfrac{2(L_y + L_z)}{L_y L_z} - k\dfrac{d^2 T}{dx^2} = 0$

2.12      $\dot{s}_{m,\mu} = 1.66 \times 10^{13}$ W/m$^3$

2.13  (a)  $\rho_{CH_4} = 8.427 \times 10^{-5}$ g/cm$^3$, $\rho_{O_2} = 3.362 \times 10^{-4}$ g/cm$^3$

     (b)  $\dot{m}_{r,CH_4}$(without Pt) $= 1.598 \times 10^{-7}$ g/cm$^2$-s,

          $\dot{m}_{r,CH_4}$(with Pt) $= 1.754 \times 10^{-3}$ g/cm$^2$-s

2.14      $T_s = 1{,}454$ K

2.15  (a)  $\alpha_e$(Ta) $= 1.035 \times 10^{-3}$ 1/K, $\alpha_e$(W) $= 1.349 \times 10^{-3}$ 1/K

     (b)  $\rho_e$(Ta) $= 1.178 \times 10^{-6}$ ohm-m, $\rho_e$(W) $= 7.582 \times 10^{-7}$ ohm-m

     (c)  $R_e$(Ta) $= 7.500$ ohm, $R_e$(W) $= 4.827$ ohm

     (d)  $J_e$(Ta) $= 3.651$ A, $J_e$(W) $= 4.551$ A

     (e)  $\Delta\varphi$(Ta) $= 27.38$ V, $\Delta\varphi$(W) $= 21.97$ V

2.16  (a)  $J_e = 3.520$ A, (b) $P_e = 29.74$ W

2.17  (c)  $T_2 = 2{,}195$ K

2.18  (a)  $\Delta t_1 = 0.4577$ hr, (b) $\Delta t_2 = 9.928$ hr

2.19  (a)  $\dot{s}_{m,p} = -2 \times 10^8$ W/m$^3$, (c) $T_o - T_i = -42.27°$C

2.20  (a)  $\langle\dot{s}_{m,\mu}\rangle_A = 3.660 \times 10^7$ W/m$^3$

2.21  (a)  $\dot{S}_{m,F}|_{peak} = 0.65\dfrac{Mu_o^2}{\tau}(1 - \frac{t}{\tau})$ (each of the front brakes)

          $\dot{S}_{m,F}|_{peak} = 0.35\dfrac{Mu_o^2}{\tau}(1 - \frac{t}{\tau})$ (each of the rear brakes)

     (b)  $\dot{S}_{m,F}|_{peak} = 120.3$ kW

2.22  (d)  $T_s = 40.49°$C

2.23  (a)  $\dot{n}_{r,CH_4} = -3.671$ kg/m$^3$-s,

     (b)  $\dot{n}_{r,CH_4} = -0.5219$ kg/m$^3$-s,

     (c)  $\dot{n}_{r,CH_4} = -0.3150$ kg/m$^3$-s

2.24  (a)  $\dot{M}_{lg} = 0.8593$ g/s, (b) $\dot{M}_{r,CH_4} = 0.2683$ g/s, (c) $\dot{S}_{e,J} = 1.811 \times 10^5$ W

2.25  (a)  $\dot{s}_{e,J} = 5.62 \times 10^8$ W/m$^3$, (b) $\Delta\varphi = 2.18$ V, (c) $J_e = 1.8$ A

     (d)  $\dot{s}_{e,J} = 2.81 \times 10^8$ W/m$^3$, $\Delta\varphi = 3.08$ V, $J_e = 1.3$ A

2.26  (a)  $q_c = 49{,}808$ W/m$^2$, (b) $q_h = 56{,}848$ W/m$^2$

2.27  (c)  $T_{max} = 45.03°$C   at   $t = 594$   s,   (d)   $T(t) = T_\infty + [a/(1/\tau - 1/\tau_r)][\exp(-t/\tau_r) - \exp(-t/\tau)]$, $a = \dot{S}_{r,o}/\rho c_p V$, $\tau = \rho c_p V R_t$

2.30  (a)  wet alumina: $\dot{s}_{e,m} = 3.338$ W/m$^3$, (b) dry alumina: $\dot{s}_{e,m} = 16.69$ W/m$^3$

     (c)  dry sandy soil: $\dot{s}_{e,m} = 1{,}446$ W/m$^3$

2.33  (b)  $q_x(x = x_2) = 1.378 \times 10^6$ W/m$^2$

2.36  (b)  $\dot{S}_{e,\alpha}/A = 637.0$ W/m$^2$, $\dot{S}_{e,\epsilon}/A = -157.8$ W/m$^2$, (c) $\dot{S}/A = 479.2$ W/m$^2$

2.37  (b)  $\dot{m}_{lg} = 5.074 \times 10^{-2}$ kg/m$^2$-s, (c) $D(t) = D(t = 0) - \dfrac{2\dot{m}_{lg}}{\rho_l}t$, (d) $t = 36.57$ s

2.38  (b)  $q_{k,e} = -8 \times 10^4$ W/m$^2$

2.39  (b)  $q_{k,t} = -140$ W/m$^2$

2.40  (b)  $q_{k,s} = 8.989 \times 10^9$ W/m$^2$

2.41  (a)  $R_{t,1-\infty} = 0.1889°$C/W, $\Delta(\dot{S}_{r,c})_1 = 45.37$ W, (b) $\dot{N}_{r,ATP} = 6.283 \times 10^{-18}$ kg ATP/cell-s,

     (c)  $D_{ATP} = 0.2484\,\mu$m, (d) $D_{ATP} = 1.009\,\mu$m

2.42      $\alpha_{S,p} = 260.0\,\mu$V/K

2.43 (a) $\lambda = 1.12\,\mu\text{m}$ (near-infrared range), (c) $\Delta E_{e,F} > 3.11$ eV, (d) $\dot{N}_{blue} = 2.01 \times 10^{20}$ photon/s, $\dot{N}_{red} = 3.52 \times 10^{20}$ photon/s

2.44 (a) plot $E_e^* = (P/E_e^*)\sin E_e^* + \cos E_e^*$, $E_e^* = \beta l_m$, $P = 6$, (b) $-1 \leq (P/E_e^*)\sin E_e^* + \cos E_e^* \leq 1$

2.45 (a) 358.2 m/s, (b) 13.11 1/m, (c) 7,666 W/m$^3$, (d) $\sigma_{ac} = 18.86$ 1/m

2.46 (a) $\sigma_{ac} = 15.79$ 1/m $= 1.372$ dB/cm, (b) experiment: $\sigma_{ac} = 3.1$ dB/cm $= 35.69$ 1/m

**3.1** (a) for $T = 300$ K, $k_{pr} = 424.7$ W/m-K, (b) for $T = 300$ K, $\Delta k(\%) = 6\%$

3.2 (a) $k = 0.02205$ W/m-K, (b) $\Delta k(\%) = 9.68$ %

3.3 (c) $(\rho c_p k)^{1/2}$ (argon) $= 4.210$, (air) $= 5.620$, (helium) $= 11.22$, (hydrogen) $= 15.42$ W-s$^{1/2}$/m$^2$-K

3.9 $L = 0.6$ nm: $k = 0.5914$ W/m-K, $L = 6$ nm: $k = 1.218$ W/m-K

3.12 (a) $Q_{k,2\text{-}1} = -100$ W, (b) $Q_{k,2\text{-}1} = -83.3$ W, (c) $Q_{k,2\text{-}1} = -82.3$ W

3.13 (a) $A_k R_{k,1\text{-}2} = 2.5 \times 10^{-5}\,^\circ\text{C}/(\text{W/m}^2)$, (b) $A_k R_{k,1\text{-}2} = 7.4 \times 10^{-1}\,^\circ\text{C}/(\text{W/m}^2)$,

   (c) $T_1 = 60.02^\circ\text{C}$, (d) $R_k$-value (copper) $= 1.4 \times 10^{-4}\,^\circ\text{F}/(\text{Btu/hr})$, $R_k$-value(silica aerogel) $= 4.2^\circ\text{F}/(\text{Btu/hr})$

3.14 $\Delta Q\% = 63.3$ %

3.15 (b) parallel: $\langle k \rangle = 14.4$ W/m-K, series: $\langle k \rangle = 0.044$ W/m-K, random: $\langle k \rangle = 0.19$ W/m-K

3.16 (b) (i) $Q_{1\text{-}2} = 4.703$ W, (ii) $Q_{1\text{-}2} = 4.492$ W

3.17 (b) $Q_{k,2\text{-}1} = 8.408 \times 10^2$ W, (c) $\dot{M}_{lg} = 3.960$ g/s, (d) $T_{2'} = -10.45^\circ\text{C}$

3.18 (b) (i) $T_s = 43.35^\circ\text{C}$, (ii) $T_s = 76.92^\circ\text{C}$

3.19 (b) $\langle k \rangle = 0.373$ W/m-K, (c) $T_g = 1,643$ K, (d) $Q_{g,1}/\dot{S}_{r,c} = 0.854$, $Q_{g,2}/\dot{S}_{r,c} = 0.146$

3.20 (b) $\dot{M}_{lg} = 0.051$ kg/s, (c) $\Delta t = 0.01$ s

3.23 (c) $Q_{k,1\text{-}2} = -7.423 \times 10^{-2}$ W, (d) $Q_{k,1\text{-}2} = -7.635 \times 10^{-2}$ W

3.24 for $p = 10^5$ Pa, $\Delta T_c = 4.9^\circ\text{C}$, for $p = 10^6$ Pa, $\Delta T_c = 2.7^\circ\text{C}$

3.25 (b) for $A_k R_{k,c} = 10^{-4}$ K/(W/m$^2$), $T_h = 105^\circ\text{C}$, for $A_k R_{k,c} = 4 \times 10^{-2}$ K/(W/m$^2$), $T_h = 1,469^\circ\text{C}$

3.26 (i) $Q_{k,2\text{-}1}$(no blanket) $= 1.006 \times 10^3$ W, (ii) $Q_{k,2\text{-}1}$(with blanket) $= 6.662 \times 10^2$ W

3.27 (b) $L = 5.984$ mm

3.29 (b) $A_p = 2.59 \times 10^{-5}$ m$^2$, (c) $T_{max} = 41.8^\circ\text{C}$, (d) $q_c = -5,009$ W/m$^2$, $q_h = 17,069$ W/m$^2$

3.30 (a) $Q_c(2.11\text{ A}) = -0.079$ W, (b) $Q_c(1.06\text{ A}) = -0.047$ W,

   (c) $Q_c(4.22\text{ A}) = 0.048$ W

3.31 (a) $T_c = 231$ K, (b) $J_e = 2.245$ A, (c) $T_c = 250.2$ K

3.33 (c) $T(t \to \infty) = -18.40^\circ\text{C}$

3.34 (c) $T_1 = 71.36^\circ\text{C}$

3.35 (b) $R_{k,1\text{-}2} = 106.1^\circ\text{C/W}$, (c) $T_1 = 60.34^\circ\text{C}$

3.36 (b) $e_{e,o} = 160$ V/m

3.40 (b) $T_c = [3R_eJ_c^2 + 2(T_h/R_k) + \frac{1}{2}R_eJ_e^2R_k\alpha_S]/[4\alpha_SJ_e + (2/R_k) + \alpha_S^2R_kJ_e^2]$,

$T_c = 209.4$ K,

$J_e = 2.3$ A, (c) $\Delta\varphi = 0.36$ V.

3.41 (b) $J_e = 1.210 \times 10^{-2}$ A, (c) $T_{c,min} = 223.3$ K, (d) $Q_{c,max} = -5.731 \times 10^{-4}$ W

3.42 (i) (b) $R_{k,1\text{-}2} = 0.4584$ K/W, (c) $Q_{k,1\text{-}2} = 21.81$ W,

(ii) (b) $R_{k,1\text{-}2} = 0.6575$ K/W, (c) $Q_{k,1\text{-}2} = 15.21$ W

3.44 (b) $T_1 = -19.63°$C, $Q_{k,1\text{-}2} = 61.69$ W, (c) $\dot{M}_{ls} = 665.7$ g/hr

3.45 (b) $L_1(t = \Delta t) = 6.127$ $\mu$m, $R_{k,1\text{-}2} = 145.6$ kW

3.46 (a) $t = 0.8378$ s, (b) $t = 20.35$ s

3.47 $\delta_\alpha/[2(\alpha\tau)^{1/2}] = 0.4310$

3.48 (a) (i) $T(x = 0, t = 10^{-6}$ s$) = 1.594 \times 10^5°$C, (ii) $T(x = 0, t = 10^{-6}$ s$) = 1.594 \times 10^4°$C

(b) (i) $\delta_\alpha(t = 10^{-6}$ s$) = 6.609$ $\mu$m, (ii) $\delta_\alpha(t = 10^{-4}$ s$) = 66.09$ $\mu$m

3.49 (a) $t = 3.872$ s, (b) $t = 31.53$ s

3.50 (i) $t = 2,970$ s, (ii) $t = 1,458$ s, (iii) $t = 972.1$ s

3.51 (a) $t = 5.6$ min, (b) $t = 35$ min

3.52 $t = 7.7$ min

3.53 $t = 24$ s

3.54 (b) first-degree burn: $x = 8.6 \pm 0.4$ mm; second-degree burn: $x = 5.8 \pm 0.4$ mm; third-degree burn: $x = 4.75 \pm 0.4$ mm

3.55 (a) $T(x = 1$ mm$) = 215.8$ K, (b) $T(x = 3$ mm$) = 291.9$ K, (c) $q_{\rho ck} = 20,627$ W

3.56 (b) $T_s(x = L, t = 1.5$ s$) = 71°$C

3.57 $T_{12} \simeq T_1(t = 0)$

3.58 $t = 7.8$ $\mu$s

3.59 (a) $T(x = 4$ mm, $t = 600$ s$) = 42.15°$C

(b) $T(x = 4$ mm, $t = 600$ s$) = 64.30°$C

3.60 $t = 7.87$ min

3.61 (b) $t = 5.127$ ms

3.62 $u_b = 51.5$ cm/min

3.63 (b) $t = 177.8$ s, (c) $t = 675.7$ s, (d) $T_1(t \to \infty) = 230.2°$C

3.64 (b) $\dot{S}_{e,\sigma} = 26.93$ W, $t = 1.402$ $\mu$s

3.65 (b) $T(x = 0, t = 2$ s$) = 91.41°$C

3.66 $T_1(t) - T_2 = [T_1(t = 0) - T_2]e^{-t/\tau_1} + a_1\tau_1(1 - e^{-t/\tau_1})$

3.67 (b) (i) $T_1(t) = 2.6°$C, (ii) $T_1(t) = 66.9°$C

3.68 (b) $t = 6.236$ s, $L = 3.120$ m, (c) $Fo_R = 150.3$, (d) $N_{k,1} = 1.803 \times 10^{-3}$

3.70 (c) $L = 8.427$ m

3.71 $T_e(t \to \infty) = 1,672$ K

3.72 (b) (i) $T_1(t = 4$ s$) = 66.97°$C, (ii) $T_1(t = 4$ s$) = 254.8°$C

3.73 (a) $Q_k = 109.5$ W, $\langle k \rangle_{yy} = 0.6255$ W/m-K, (b) $\langle k \rangle_{yy} = 0.8375$ W/m-K

3.74      $T^*(x^* = 0.125, y^* = 0.125) = 0.03044$ for $N = 21$

3.75   (b)   $T_1 = 4,804°C$,   $T_2 = 409.8°C$,   $T_3 = 142.9°C$,   $T_4 = 121.3°C$,   $T_5 = 114.4°C$

3.76   (b)   $Q_{k,h-c} = 4.283$ W, (c) $\langle k \rangle = 42.83$ W/m-K

3.77      $\delta_\alpha = 47$ $\mu$m

3.79   (a)   $t = 74$ s,   $\int q_k dt = 3.341 \times 10^5$ J/m$^2$,   (b)   $t = 21$ s,   $\int q_k dt = 3.341 \times 10^5$ J/m$^2$

3.80      $t_1 = 286.6$ s, $t_2 = 40.31$ s

3.81   (b)   $t = 31.8$ s

3.82   (b)   $\tau_{rr}(r = 0) = \tau_{\theta\theta}(r = 0) = 1.725 \times 10^8$ Pa

3.83   (b)   $T_R = 550.6°C$

3.84   (b)   $Q_c - Q_{k,h-c} = 0.5R_e J_e^2 - \alpha_S J_e T_c$, $Q_{k,h-c} + Q_{K,h-\infty} = 0.5R_e J_e^2 + \alpha_S J_e T_h$,     $T_q - T_c = Q_c R_{k,c}$, $T_h - T_p = Q_{k,h-\infty} R_{k,c}$,     $T_h - T_c = Q_{k,h-c} R_{k,h-c}$, $T_p - T_\infty = Q_{k,h-\infty} R_{k,s}$.

3.85   (a)   $d^2/dy^2 = -\omega^2 A e^{i(kjl_m - \omega t)}$, $\omega(k) = 2(F/ml_m)^{1/2} \mid \sin kl_m/2 \mid$, (c) for Pb $u_g = (Fl_m/m)^{1/2} = 1,157$ m/s

3.89   (b)   (i) $\dot{S}_{r,c} = 743.7$ W, (ii) $\dot{S}_{r,c} = 743.7$ W

3.90   (c)   $t = 3.658$ ms

3.91   (b)   $x_1 = 4.68$ mm, $x_2 = 5.63$ mm, $x_3 = 8.59$ mm

3.92   (a)   $T_1(t = t_l) = 1.172 \times 10^{4°}C$, (b) $t_{f,1} = 65.07$ ps, (c) $t_{f,2} = 6.990$ ms

**4.1**   (a)   $E_b = 201,584$ W/m$^2$, (b) $Q_{r,\epsilon} = 82,649$ W
      (c)   $F_{0.39T - 0.77T} = 0.13\%$, $F_{0.77T - 25T} = 99.55\%$, $F_{25T - 1000T} = 0.32\%$

4.2   (b)   aluminum: $(Q_{r,\rho})_1 = 18,437$ W, $(Q_{r,\alpha})_1 = 1,823$ W, nickel: $(Q_{r,\rho})_2 = 13,270$ W, $(Q_{r,\alpha})_2 = 6,990$ W, paper: $(Q_{r,\rho})_3 = 1,103$ W, $(Q_{r,\alpha})_3 = 19,247$ W
      (c)   aluminum: $(Q_{r,\epsilon})_1 = 98.94$ W, $(Q_{r,o})_1 = 18,536$ W, nickel: $(Q_{r,\epsilon})_2 = 379.3$ W, $(Q_{r,o})_2 = 13,649$ W, paper: $(Q_{r,\epsilon})_3 = 1,044$ W, $(Q_{r,o})_3 = 2,057$ W
      (d)   aluminum: $Q_{r,1} = -1,724$ W, nickel: $Q_{r,2} = -6,611$ W, paper: $Q_{r,3} = -18,203$ W

4.3   (a)   $(\epsilon_r)_1 = 0.09$, $(\epsilon_r)_2 = 0.29$, $(\epsilon_r)_3 = 0.65$

4.4   (a)   $q_{r,\epsilon}(\text{visible}) = 3.350 \times 10^5$ W/m$^2$, (b) $q_{r,\epsilon}(\text{near infrared}) = 2.864 \times 10^6$ W/m$^2$,
      (c)   $q_{r,\epsilon}(\text{remaining}) = 8,874$ W/m$^2$

4.7   (a)   $T = 477.7$ K, (b) $\lambda_{max} = 6.067$ $\mu$m, (c) $T_{lg} = 373.2$ K, $E_{b,\lambda}(T = 373.2$ K, $\lambda = 1.5$ $\mu$m$) = 3.680 \times 10^{-4}$ W/m$^2$-$\mu$m

4.8   (a)   $\epsilon_{r,\lambda}(\text{SiC}) = 0.8301$, $\epsilon_{r,\lambda}(\text{Al}) = 0.008324$

4.9      (i) white potassium zirconium silicate, (ii) black-oxidized copper, (iii) aluminum foil

4.10   (a)   $F_{1-2} = 1/9$ (b) $F_{1-1} = 0.7067$, $F_{2-3} = 0.12$
      (c)   $F_{1-2} = 0.08$, $F_{2-1} = 0.32$, (d) $F_{1-3} = 0.085$, $F_{2-1} = 0.415$
      (e)   $F_{1-2} = 0.003861$, $F_{2-3} = 0.7529$

4.11 (a) $F_{1\text{-}2} = 0.2, l^* = w^* = 1, w^* = a^* = 1, 1/R_1^* = 1/R_2^* = 1.80$,

(b) $l = 1.02a$ for the discs, $l = a$ for the plates

4.12 (a) $\epsilon_{r,1}' = 1/\{[D/(4L + D) \times (1 - \epsilon_{r,1})/\epsilon_{r,1}] + 1\}$

4.13 (b) (i) oxygen: $q_{r,1\text{-}2} = 1.17$ W/m$^2$, (i) hydrogen: $q_{r,1\text{-}2} = 0.14$ W/m$^2$

(ii) oxygen: $q_{r,1\text{-}2} = 0.596$ W/m$^2$, (ii) hydrogen: $q_{r,1\text{-}2} = 0.0730$ W/m$^2$

4.14 (a) $Q_{r,1}/\dot{S}_{lg} = 4.7\%$, (b) $\Delta Q_{r,1} = 4.5\%$

4.15 (b) $Q_{r,1} = 82$ W

4.16 $R_{r,Sigma} = 2/(A_r \epsilon_r) - 1/A_r$

4.17 (b) $\dot{S}_{e,J} = 1,280$ W

4.18 (b) $T_1 = 808.6°$C, (c) $T_1 = 806.6°$C

4.20 (b) $q_{r,1\text{-}2} = (E_{b,1} - E_{b,2})/(2[\dfrac{1 - \epsilon_r}{\epsilon_r} + \dfrac{1}{2\epsilon + \epsilon_r(1 - \epsilon)}])$.

4.21 (b) (i) $q_{r,2\text{-}1} = -245.7$ W/m$^2$, (ii) $q_{r,2\text{-}1} = -210.6$ W/m$^2$,

(iii) $q_{r,2\text{-}1} = -179.8$ W/m$^2$, (iv) $q_{r,2\text{-}1} = -134.0$ W/m$^2$

4.22 (b) $Q_{r,1\text{-}2} = 1180$ W, (c) $Q_{r,1\text{-}2} = 182.8$ W

4.23 (b) $Q_{r,1-2} = 10,765$ W, (c) $Q_{r,2} = 19,300$ W

4.24 (b) $F_{1\text{-}2} = 0.125, F_{1\text{-}3} = 0.875, F_{2\text{-}3} = 0.8958$; (c) $T_1 = 1,520.9$ K

4.25 (a) $Q_{r,1\text{-}2} = -54.13$ W, (c) $Q_{r,1\text{-}2} = 114.1$ W, (d) $T_3 = 400$ K

4.26 $Q_{r,2} = -201,322$ W

4.27 (b) $Q_{r,1-2} = 3.860$ kW, (c) $Q_{r,1-2} = 0.7018$ kW,

(d) $Q_{r,1-2} = (A_{r,1}/2)(E_{b,1} - E_{b,2})$ for $F_{1-2} \to 0$

4.28 (a) $t = 208.3$ s, (b) $T(x = 0, t) = 4,741°$C

4.29 (b) $\dot{M}_l = 0.5992$ g/s

4.30 (b) $\Delta t = 15.36$ ns

4.31 (c) $T_2 = \dfrac{1}{\sigma_{SB}}\{\dfrac{\alpha_{r,2}}{2\epsilon_{r,2}}[(q_{r,o})_a + (q_{r,o})_b]\}^{1/4}$

4.32 (b) $(q_{r,i})_f = 9.52 \times 10^4$ W/m$^2$, (c) $\dot{S}_{e,\sigma} = 2.989 \times 10^5$ W, (d) $\Delta t = 71.14$ s

4.33 (a) $(q_{r,i})_f = 1.826 \times 10^4$ W/m$^2$, (b) $q_{r,i} = 319.6$ W/m$^2$

4.34 (a) $q_{ku,3} = 6,576$ W/m$^2$, (b) $Q_{r,1}(IR + visible) = 2,835$ W,

$Q_{r,1}(UV) = 148,960$ W, (c) $T_{3,max} = 529.9$ K

4.35 (a) (i) $\langle Q_u \rangle_{L\text{-}0} = 500$ W, (ii) $\langle Q_u \rangle_{L\text{-}0} = 362$ W, (b) (i) $\eta = 31.22\%$, (ii) $\eta = 22.60\%$

4.36 (a) $\dot{S}_{e,\sigma}/A = 738$ W/m$^2$, (b) $d\langle T \rangle_L/dt = 40$ K/day

4.37 (c) (i) $Q_{r,1,t}/A_1 = -1,242$ W/m$^2$, (ii) $Q_{r,1,b}/A = -209$ W/m$^2$

(d) (i) $(dV_1/dt)/A_1 = -0.478$ $\mu$m/s, (ii) $(dV_1/dt)/A_1 = -0.0805$ $\mu$m/s

4.38 (a) $\langle Q_u \rangle_{L\text{-}0} = 466.4$ W, (b) $\eta = 19.42\%$

4.39 (a) $\langle Q_u \rangle_{L\text{-}0} = 1,376$ W, (b) $Q_1 = -34.15$ W

4.40 (b) $Q_1 = 827.7$ W

4.41 (a) $Q_{r,1\text{-}2} = -3.008$ W

4.42 (b) $T_1 = 860.5$ K

4.43 (b) $Q_{2\text{-}1} = 3,235$ W, (c) $Q_{2\text{-}1} = 25.65$ W

4.44 (b) $R_k + R_{r,\Sigma} = \dfrac{(1 - \epsilon)(l_1 + l_2)}{A_r k_s} + \dfrac{(2 - \epsilon_r)\epsilon(l_1 + l_2)}{4A_r \epsilon_r \sigma_{SB} T^3 l_2}$

4.45 $\epsilon = 0.8028, \langle k_r \rangle = 0.0004399$ W/m-K at $T = 1,000$ K

4.46  (b)  $t(T_1 = 600\text{ K}) = 162$ s

4.47  (a)  $R_{r,\Sigma} = \dfrac{l_1}{A_r(1-\epsilon)^{2/3}k_s} + \dfrac{1}{4A_r(1-\epsilon)^{2/3}\sigma_{SB}T^3}\dfrac{2-\epsilon_r}{\epsilon_r}$

4.48  (a)  (i) $\sigma_{ex} = 83.78$ 1/m, (ii) $\sigma_{ex} = 8.259 \times 10^5$ 1/m

    (b)  (i) $\sigma_{ex} = 185.6$ 1/m, (ii) $\sigma_{ex} = 1.531 \times 10^6$ 1/m

4.49  (a)  $u_p = 0.48$ m/s, (b) $N_{r,1} = 1.2 \times 10^{-6} < 0.1$

4.50  (i)  $q_1 = 2.305$ W/m$^2$, (ii) $q_1 = 2.305$ W/m$^2$

4.51  (i)  $Q_{1-2} = 127.8$ W, (ii) $Q_{1-2} = 71.93$ W

4.52  (b)  $t(T_1 = 500°\text{C}) = 5$ ms

4.53  (a)  (i) $Q_{k,1-2}/L = -25.7$ W/m, (ii) $Q_{r,1-2}/L = -1.743$ W, (b) $R_2 = 2.042 \times 10^8$ m

4.54  (b)  $t = 181.1$ s

4.57  (b)  $\dot{m}_{ls} = 8.293$ g/m$^2$-s

4.59  (b)  Ge: $dT_1/dt = 36.54$ K/s, Si: $dT_1/dt = 1{,}708$ K/s,

    (c)  $\epsilon_r A_{r,1}\sigma_{sb}(T_1^4 - T_\infty^4) + (T_1 - T_\infty)/[\ln(4L_1/D_1)/2\pi kL_1] = -(\rho c_p V)_1 dT_1/dt + (\dot{S}_{e,\sigma})_1$,

    (d)  Ge: $T_1(t \to \infty) = 298.5$ K, Si: $T_1(t \to \infty) = 304.5$ K

4.60  (b)  $\langle\alpha\rangle = 3.087 \times 10^{-7}$ m$^2$/s

4.61  (b)  $\dot{M}_{lg} = 89.42$ μg/s, (d) $\dot{M}_{lg} = 27.99$ μg/s

**5.1**  (b)  $Q(x = 0) = 767.3$ W

5.2  (a)  $R_{k,u}/R_{uL} = 0.2586$, (b) $R_{k,u}/R_{uL} = 6.535 \times 10^{-3}$

5.3  (b)  $T_{f,2} = 367.9°\text{C}$, (c) $T_{f,2} = 2520°\text{C}$

5.5      $(Q_{k,u})_{1-2} = 140.3$ W

5.7  (b)  (i) $(Q_{k,u})_{1-2} = -3.214$ W, $(q_{k,u})_{1-2} = -4.092 \times 10^6$ W/m$^2$

    (ii) $(Q_{k,u})_{1-2} = -7.855 \times 10^{-2}$ W, $(q_{k,u})_{1-2} = -1.000 \times 10^5$ W/m$^2$

5.8  (b)  $\dot{M}_l = 1.160 \times 10^{-7}$ kg/s, $\dot{m}_l = 0.1477$ kg/s-m$^2$

5.9  (b)  $\dot{M}_{lg} = 2.220$ g/s, $T_{f,2} = 186.7°\text{C}$

5.10  (b)  $T_{f,2} = 23.18°\text{C}$

5.11      $\dot{M}_l = 0.5476$ kg/s

5.12  (a)  $\langle k \rangle = 0.63$ W/m-K, (b) $\langle k_r \rangle = 0.19$ W/m-K, (c) $u_{f,1} = 1.30$ m/s

5.13      $Tu = 0.2162$

5.15  (a)  $Ze = 8.684$, (b) $u_f = 1.037$ m/s

5.16  (a)  $T_{f,2} = 2{,}944°\text{C}$, (b) $u_{f,1} = 3.744$ m/s

5.17  (b)  $T_s = T_{f,2} = 1{,}476$ K, $Q_{r,2-p} = 40{,}259$ W, (c) $\eta = 37.62\%$

5.18  (b)  $T_{f,2} = 197.0°\text{C}$, (c) $T_{f,3} = 2{,}747°\text{C}$, (d) $T_{f,4} = 2{,}564°\text{C}$, (e) $\Delta T_{excess} = 177°\text{C}$

5.19      $T_{f,2} = 2{,}039$ K (for $q_{loss} = 10^5$ W/m$^2$)

5.21  (b)  $T_s = 1{,}040$ K, (c) $\eta = 60.90\%$

5.22  (b)  $\dot{M}_f = 1.800$ g/s

5.23  (b)  $T_{f,2} = 2{,}472°\text{C}$

5.24  (b)  $T_{f,2} = 3{,}134°\text{C}$

5.25  (b)  $T_{f,2} = 203.7°\text{C}$

**6.2**  (a)  (i) $q_{ku,L} = -50{,}703$ W/m$^2$, (ii) $q_{ku,L} = -2{,}269$ W/m$^2$, (iii) $q_{ku,L} = -39.59$ W/m$^2$

  (b)  (i) $\delta_{\alpha,L} = 5.801$ mm, (ii) $\delta_{\alpha,L} = 3.680$ mm, (iii) $\delta_{\alpha,L} = 22.39$ mm

  (c)  $(q_{ku})_{(i)}, \mathrm{Pr}\to 0 = -46{,}115$ W/m$^2$, $\delta_{\alpha,L,\mathrm{Pr}\to 0} = 7.805$ mm.

6.3  (a)  (i) $\langle Q_{ku}\rangle_L = 1.502$ W, (ii) $\langle Q_{ku}\rangle_L = 65.23$ W, (b) (i) $\delta_\alpha = 14.16$ mm, (ii) $\delta_\alpha = 5.336$ mm

6.7  (a)  (i) $\langle \mathrm{Nu}\rangle_L = 119.8$, (ii) $\langle \mathrm{Nu}\rangle_L = 378.8$, (iii) $\langle \mathrm{Nu}\rangle_L = 2{,}335$

  (b)  (i) $A_{ku}\langle R_{ku}\rangle_L = 3.326 \times 10^{-1}\,{}^\circ$C/(W/m$^2$),
       (ii) $A_{ku}\langle R_{ku}\rangle_L = 1.052 \times 10^{-1}\,{}^\circ$C/(W/m$^2$),
       (iii) $A_{ku}\langle R_{ku}\rangle_L = 1.711 \times 10^{-2}\,{}^\circ$C/(W/m$^2$)

6.7  (c)  (i) $\langle Q_{ku}\rangle_L = 150.3$ W, (ii) $\langle Q_{ku}\rangle_L = 475.4$ W, (iii) $\langle Q_{ku}\rangle_L = 2{,}930$ W

6.8  (b)  $u_{f,\infty} = 3.78$ m/s

6.9  (b)  $T_{f,\infty} = 277.20$ K, (c) $\langle Q_{ku}\rangle_L = -653.5$ W, $Q_{r,1} = 100.8$ W, ice would melt

6.10  (a)  single nozzle: $\langle \mathrm{Nu}\rangle_L = 46.43$, $A_{ku}\langle R_{ku}\rangle_L = 8.413 \times 10^{-2}\,{}^\circ$C/(W/m$^2$), $\langle Q_{ku}\rangle_L = 406.5$ W

  (b)  multiple nozzles: $\langle \mathrm{Nu}\rangle_L = 28.35$, $A_{ku}\langle R_{ku}\rangle_L = 4.593 \times 10^{-2}\,{}^\circ$C/(W/m$^2$), $\langle Q_{ku}\rangle_L = 744.6$ W

6.11  (c)  $\dot{S}_{m,F}/A_{ku} = 75{,}391$ W/m$^2$, $L_n = 0.5425$ cm

6.12  (b)  $t = 20.31$ s

6.13  (b)  (i) parallel flow: $t = 2.465$ s (ii) perpendicular flow: $t = 1.123$ s

6.14  (a)  vertical: $\langle \mathrm{Nu}\rangle_L = 43.08$, $A_{ku}\langle R_{ku}\rangle_L = 2.232 \times 10^{-1}\,{}^\circ$C/(W/m$^2$), $\langle Q_{ku}\rangle_L = -7.390$ W

6.14  (b)  horizontal: $\langle \mathrm{Nu}\rangle_D = 18.55$, $A_{ku}\langle R_{ku}\rangle_D = 2.073 \times 10^{-1}\,{}^\circ$C/(W/m$^2$), $\langle Q_{ku}\rangle_D = -7.956$ W

6.15  (b)  $\langle Q_{ku}\rangle_L = 411.9$ W, (c) $Q_{r,s\text{-}w} = 692.0$ W, (d) $\eta = 5.66\%$

6.18  (b)  (i) $\langle Q_{ku}\rangle_L = 938.5$ W, (ii) $\langle Q_{ku}\rangle_L = 1{,}163$ W

6.19  (a)  $\dot{S}_{e,J} = 2{,}045$ W, (b) $Q_{ku,CHF} = 2{,}160$ W, (c) $T_s = 108.9^\circ$C

  (d)  $A_{ku}\langle R_{ku}\rangle_D = 8.575 \times 10^{-6}\,{}^\circ$C/(W/m$^2$), $\langle \mathrm{Nu}\rangle_D = 8{,}587$

6.20      $T_{s,1} = 103.5^\circ$C

6.21  (b)  $\langle Q_{ku}\rangle_L = -9.058 \times 10^3$ W, (c) $\dot{M}_{lg} = -4.013$ g/s

6.22  (b)  $\langle Q_{ku}\rangle_L = 595.2$ W, (c) $\Delta\varphi = 109.1$ V, $J_e = 5.455$ A

6.23  (b)  $\langle Q_{ku}\rangle_L = 8.718 \times 10^6$ W

6.24  (a)  (i) $\langle q_{ku}\rangle_L = 1.099 \times 10^6$ W/m$^2$, (ii) $\langle q_{ku}\rangle_L = 2 \times 10^6$ W/m$^2$

  (b)  (i) $\langle q_{ku}\rangle_L = 1.896 \times 10^4$ W/m$^2$, (ii) $\langle q_{ku}\rangle_L = 9.384 \times 10^5$ W/m$^2$

6.25  (a)  $\langle Q_{ku}\rangle_{D,s} = 195.9$ W, (b) $\langle Q_{ku}\rangle_{D,c} = 630.5$ W, (c) $L = 1.08$ cm, (d) $T_2 = 0.97^\circ$C

6.26  (b)  $T_2 = 817.3$ K.

6.27  (b)  $T_s = 1{,}094$ K

6.28      $T_{f,\infty} - T_{s,L} = 8.71^\circ$C

6.29  (b)  $t = 17.77$ s $< 20$ s

6.30  (b)  (i) $\langle Q_{ku}\rangle_D = 378.6$ W, (ii) $\langle Q_{ku}\rangle_L = 91.61$ W, (c) $\delta_v/D = 0.6438 < 1.0$

6.31  (b)  $r_{tr} = 21.70$ cm, (c) $\langle Q_{ku}\rangle_L/(T_s - T_{f,\infty}) = 23.80$ W/$^\circ$C, (d) $T_s = 860.3^\circ$C

6.32 (a) $t = 71.4$ min, (b) $t = 6.7$ min
6.33 (a) $T_s = 1{,}146$ K, (b) $T_s = 722.6$ K
6.34 (d) $T_s^*(r = 0, t) = 0.2779$
6.35 (b) $\mathrm{Bi}_D = 4.121 \times 10^{-4} < 0.1$, (c) $t = 2.602$ ms
6.36 (a) $\langle Q_{ku} \rangle_w = 47.62$ W, (b) $\langle Q_{ku} \rangle_w = 357.1$ W
6.37 (a) $T_1(t = 4\ \mathrm{s}) = 346.2$ K, (b) $t = 17.0$ min, (c) $\mathrm{Bi}_l = 3.89 \times 10^{-3}$
6.38 (b) $R_{k,sl-b} = 1.25^\circ$C/W, $R_{k,sl-s} = 6.25^\circ$C/W, $\langle R_{ku} \rangle_D = 2.54^\circ$C/W
     (c) $Q_{k,sl-b} = 5.76$ W, (d) $Q_{k,sl-\infty} = 4.23$ W, (e) $\dot{S}_{sl} = 9.99$ W
6.39 (b) $\langle \mathrm{Nu} \rangle_D = 413.5$, $T_{r,max} = 318$ K
6.40 (b) $T_p = 511^\circ$C, (c) $T_p = 64.5^\circ$C
6.41 (c) $\mathrm{Bi}_l = 0.05$, (d) $T_1 = 297.7$ K, (e) $T_1 = 275.1$ K
6.42 (b) $\mathrm{Bi}_D = 3.042 \times 10^{-3}$, (c) $t = 6.685 \times 10^{-3}$ s
6.43 (b) $T_s = 352.9^\circ$C, $l/k_s < 1.36 \times 10^{-3}\,^\circ$C/(W/m$^2$)
6.44 (b) $\eta_f = 0.9426$, (c) $T_s = 82.57^\circ$C, (d) $\Gamma_f = 8.287$
6.45 (b) $T_1(t = t_0 = 1\ \mathrm{hr}) = 18.41^\circ$C
6.46 (b) $R_c = 5.2$ mm, (c) $R_{k,1\text{-}2} = 0.3369^\circ$C/W, $(R_{ku})_{D,2} = 1.319^\circ$C/W,
     (d) $R_c = 5.241$ mm
6.48 (b) $T_1 = 54.45^\circ$C, (c) $T_1 = 50.24^\circ$C
6.49 (b) $\dot{S}_{r,c} = 9{,}024$ W, (c) $\dot{M}_{O_2} = -0.6010$ g/s, (d) $\dot{M}_{O_2} = -0.002997$ g/s
6.50 (d) $T_s = 282.9$ K, (e) $\langle Q_{ku} \rangle_L = -1{,}369$ W, $Q_{k,u} = -67.39$ W
6.51 (b) $t$(droplet vanishes) $= 382$ s, (c) $L = 38.2$ m
6.52 (b) $\dot{M}_{lg} = 3.120 \times 10^{-3}$ g/s, $T_s = 285.6$ K, (c) $\Delta t = 578.2$ s
6.53 (b) $\langle Q_{ku} \rangle_L = 2.186$ W, (c) $\dot{M}_{lg} = 8.808 \times 10^{-7}$ kg/s, $\dot{S}_{lg} = -2.034$ W
     (d) $dT_c/dt = -6.029 \times 10^{-3}\,^\circ$C/s
6.56 (a) $\langle \mathrm{Nu} \rangle_L = [0.664 \mathrm{Re}_{L,t}^{1/2} + 0.037(\mathrm{Re}_L^{4/5} - \mathrm{Re}_{L,t}^{4/5})]\mathrm{Pr}^{1/3}$
6.58 (b) $T_s = 474.2$ K
6.59 (b) (i) $\langle Q_{ku} \rangle_w = 3.335$ W, (ii) $\langle Q_{ku} \rangle_w = 1.956$ W
6.60 (c) $T_h = 304.51$ K $= 31.51^\circ$C

**7.1** (b) $\langle \mathrm{Nu} \rangle_{D,H} = 38.22$, (c) $NTU = 0.6901$, (d) $\epsilon_{he} = 0.4985$
     (e) $\langle R_u \rangle_L = 0.1080^\circ$C/W, (f) $\langle Q_{ku} \rangle_{L\text{-}0} = 1{,}019$ W, (g) $\langle T_f \rangle_L = 74.84^\circ$C
7.2 (b) laminar flow: $\langle u_f \rangle = 0.061$ m/s, turbulent flow: $\langle u_f \rangle = 1.42$ m/s
7.3 (a) $\lim_{NTU \to 0} \langle T_f \rangle_L = \langle T_f \rangle_0 - NTU(\langle T_f \rangle_0 - T_s)$,
     (b) $NTU = 4L \langle \mathrm{Nu} \rangle_D k_k / D^2 \dot{m}_f c_{p,f}$,
     (c) $NTU$ decreases as $L$ decreases or $D$ increases
7.4 (a) (i) $NTU = 111.5$, (ii) $NTU = 2.957 \times 10^4$
7.6 (b) $\langle T_f \rangle_L = -43.73^\circ$C, (c) $\langle Q_u \rangle_{L\text{-}0} = 224.4$ W
7.7 (b) $NTU = 13.12$, $\langle Q_u \rangle_{L\text{-}0} = 1.276$ W, (c) $NTU = 18.89$, $\langle Q_u \rangle_{L\text{-}0} = 1.276$ W
7.8 (b) $T_s = 92.74^\circ$C $< T_{lg}$, (c) $T_s = 399.0^\circ$C $> T_{lg}$
7.9 (b) $\langle Q_u \rangle_{L\text{-}0} = 926.5$ W, (c) $\langle T_f \rangle_L = 129.5^\circ$C
7.10 (b) $\langle Q_{ku} \rangle_{D,h} = -61.80$ W, (c) $\dot{M}_{lg} = 4.459 \times 10^{-4}$ kg/s, $x_L = 0.05410$
7.11 (b) $\langle Q_{ku} \rangle_D = 28.93$ W, (c) $\dot{M}_{lg} = 2.298 \times 10^{-4}$ kg/s, $x_L = 0.6298$
7.12 (b) $T_s = 560.8^\circ$C

7.13 (b) $NTU = 21.57$, $\langle T_f \rangle_L = 30°C$ for $N = 400$
$NTU = 32.36$, $\langle T_f \rangle_L = 30°C$ for $N = 600$

7.14 (b) $\langle Q_{ku} \rangle_L = 3{,}022$ W, $\langle Q_u \rangle_{L-0} = 3{,}158$ W

7.15 (b) $\langle \mathrm{Nu} \rangle_{D,p} = 53.10$, (c) $NTU = 2.770$, (d) $T_s = 20(°C) + 1.049\dot{S}_{e,J}$
(e) $\langle T_f \rangle_L = 20(°C) + 0.9832\dot{S}_{e,J}$, (f) $\dot{S}_{e,J} < 864.6$ W, $\langle T_f \rangle_L < 870.3°C$

7.16 (b) $\langle k \rangle = 0.6521$ W/m-K, (c) $NTU = 3.471$, (d) $\mathrm{Bi}_L = 1.025 \times 10^4$, (e) $\tau_s = 1.159$ hr,
(f) $\int_0^{4\tau_s} \langle Q_u \rangle_{L-0}(t)dt = 2.562 \times 10^9$ J

7.17 (b) $T_s = 484.6°C$, (c) $Q_{r,s-\infty} = 822.0$ W

7.18 (b) $\dot{S}_{e,J} = 10.80$ W

7.19 (b) $Q_c/A_c = 310.2$ W/m$^2$

7.20 (b) $\langle \mathrm{Nu} \rangle_D = 85.39$, (c) $\langle R_{ku} \rangle_D = 2.527 \times 10^{-5}$ K/W, (d) $R_{k,s-c} = 2.036 \times 10^{-3}$ K/W
(e) $\langle Q_u \rangle_{L-0} = -1.209 \times 10^4$ W, $\dot{M}_{sl} = 36.24$ g/s

7.21 (a) (i) $\langle \mathrm{Nu} \rangle_D = 14{,}778$ and $\langle q_{ku} \rangle_D = 9{,}946$ kW/m$^2$
(ii) $\langle \mathrm{Nu} \rangle_D = 2{,}246$ and $\langle q_{ku} \rangle_D = -237.9$ kW/m$^2$

7.22 (b) (i) $Q_{1-2} = 681.7$ W, (ii) $Q_{1-2} = 545.2$ W, (iii) $Q_{1-2} = 567.1$ W, (c) $l_a = 2$ cm

7.24 (b) $\langle T_{f,c} \rangle_0 = 623.6°C$, (c) $N = 5.81 \simeq 6$

7.25 (b) $\langle T_{f,h} \rangle_L = 1{,}767$ K, (c) $\langle T_{f,h} \rangle_0 = 450.0$ K and $\langle T_{f,c} \rangle_L = 343.7$ K
(d) $\langle Q_u \rangle_{L-0} = 1{,}161$ W and $\eta = 89.36\%$

7.26 (b) $\langle T_{f,c} \rangle_L = 27.35°C$, $\langle T_{f,h} \rangle_L = 27.12°C$, (c) $\langle Q_u \rangle_{L-0} = 171.7$ W

7.27 (b) $R_{ku,c} = 0.0250$ K/W, $R_{ku,h} = 0.2250$ K/W, (c) $\langle T_{f,c} \rangle_L = 31.43°C$

7.28 (b) $\langle Q_u \rangle_{L-0} = 6{,}623$ W, (c) $\eta = 90.73\%$

7.29 (b) $NTU = 1.106$, (c) $C_r = 0.07212$, (d) $\epsilon_{he} = 0.655$, (e) $\langle T_{f,c} \rangle_L = 34.13°C$,
(f) $\langle T_{f,h} \rangle_L = 47.8°C$, (g) $\langle Q_u \rangle_{L-0} = 908.4$ W

7.30 (b) $NTU = 0.07679$, (c) $\epsilon_{eh} = 0.07391$, (d) $\langle T_{f,c} \rangle_L = 307.4$ K, (e) $\langle Q_u \rangle_{L-0} = 32.09$ kW

7.31 (b) $\eta = 95.0\%$

7.32 (i) (b) $NTU = 0.8501$, (c) $\langle Q_u \rangle_{L-0} = 23{,}970$ W, (d) $\dot{M}_{lg} = 0.01090$ kg/s
(ii) (b) $NTU = 1.422$, (c) $\langle Q_u \rangle_{L-0} = 31{,}823$ W, (d) $\dot{M}_{lg} = 0.01446$ kg/s

7.33 (b) $L = 12.45$ m

7.34 (a) $\langle Q_{u,c} \rangle_{L-0} = 33.01$ W, (b) $L = 2.410$ m, (c) $[\Delta \langle T_{f,c} \rangle]_{max} = 66.10°C$

7.38 (a) $(T_i - T_{i-1})/(R_k)_{i \, \mathrm{to} \, i-1} + (T_i - T_{i+1})/(R_k)_{i \, \mathrm{to} \, i+1} +$
$(\dot{n}c_p)_f \Delta V_i (T_i - \langle T_f \rangle_0) = -(\rho c_p \Delta V)_i dT_i/dt + 2\sigma_{ac} I_{ac}(x_i) \Delta V_i,$
(c) $T_{max} \cong 47.69°C$

7.39 (b) $\langle Q_u \rangle_{L-0} = 187.5$ W, $T_1 - \langle T_f \rangle_0 = 0.3592°C$, (c) $\dot{N}_{r,\mathrm{ATP}} = 0.002148$ kg ATP/s.

7.40 (b) $R_{k,i} = 8.919 \times 10^{-6}$ K/W, $R_{k,c} = 2.875 \times 10^{-5}$ K/W, $\langle R_{ku} \rangle_D = 4.591 \times 10^{-7}$ K/W,
$R_\Sigma = 3.813 \times 10^{-5}$ K/W, (c) $\langle Q_u \rangle_{l-0} = 1.078 \times 10^6$ W, (d) $\dot{M}_{ls} = 3.232$ kg/s,
(e) $t = 241.4$ s, (f) $\langle T_f \rangle_l = 234.0$ K

7.41 (b) $R_k = 20.8$ K/W, (c) $L_{gap} = 3.3$ mm, $T_{gc} = 103.3°C$, $T_g = 58.8°C$

7.42 (b) first stream $\langle T_{f,c} \rangle_0 = 258.4$ K, second stream $\langle T_{f,c} \rangle_0 = 196.8$ K

7.43 (b) $T_s(t = t_f) = 1,112$ K

**8.1** (b) $J_e^2 R_{e,o} = 22.18$ W

8.2 (b) $\langle T_f \rangle_0 = 3.559°$C, (c) $\dot{S}_{e,J} = 5.683$ W, (d) $T_s = 69.75°$C

8.3 (c) $\langle Q_u \rangle_{L\text{-}0} = -6.47$ W, $J_e = 0.8$ A, $T_c = 290.5$ K, $T_h = 333.3$ K, $\langle T_f \rangle_L = 293.0$ K, $\langle R_{ku} \rangle_w = 1.515$ K/W, $R_{k,h\text{-}c} = 1.235$ K/W, $\langle R_u \rangle_L = 0.4183$ K/W, $Q_{k,h\text{-}c} = 36.65$ W, $\dot{S}_{e,P} = -46.02$ W, $(\dot{S}_{e,J})_c = 5.04$ W

8.4 (b) $T_{s,1}(t) = \langle T_f \rangle_1(t) = 400°$C for $t > 10$ s

8.5 (b) $N_{te} = 30$

# Some Thermodynamic Relations*

Thermodynamics is the only physical theory of a general nature of which I
am convinced that it will never be overthrown.

– A. Einstein

This appendix is found on the Web site www.cambridge.org/kaviany. In this
Appendix some of the relations among independent thermodynamic parameters
are reviewed. These are used in the derivation of the thermal energy conservation
equation in Appendix B. Also reviewed in this Appendix are the heat of phase
change and the heat of reaction, which are included in the specific enthalpy.

**A.1** Simple Compressible Substance*

**A.2** Phase Change and Heat of Phase Change*

**A.3** Chemical Reaction and Heat of Reaction*

**A.4** References*

A.1 Simple, Compressible Substance*

A.1.1 Internal Energy and Specific Heat Capacity at Constant Volume*

A.1.2 Specific Enthalpy and Specific Heat Capacity at Constant Pressure*

A.2 Phase Change and Heat of Phase Change*

A.3 Chemical Reaction and Heat of Reaction*

A.4 References*

# APPENDIX B

# Derivation of Differential-Volume Energy Equation*

Research is to see what everybody else sees, and to think what nobody else has thought.

– A. Szent-Gyorgyi

This appendix is found on the Web site www.cambridge.org/kaviany. In this Appendix the differential-volume thermal energy equation is derived. This is the most general energy conservation equation for an infinitesimal differential volume. It states that the divergence of the heat flux vector is balanced with the energy storage, divergence of the kinetic energy flux, pressure and viscous stress work, the volumetric gravitational and electromagnetic-force work, and conversion of energy. For convenience, the kinetic energy term is replaced by a proper substitution using the dot product of the momentum conservation equation with the velocity vector. Since the momentum equation is a statement of the Newton second law and since the product of force and velocity is generally referred to as the mechanical energy, the result is the so-called mechanical energy equation. We then write the mass and momentum conservation equations, beginning with the mass conservation equation.

**B.1** Total Energy Equation*
**B.2** Mechanical Energy Equation*
**B.3** Thermal Energy Equation*
**B.4** Thermal Energy Equation: Enthalpy Formulation*
**B.5** Thermal Energy Equation: Temperature Formulation*
**B.6** Conservation Equations in Cartesian and Cylindrical Coordinates*
**B.7** Bounding-Surface Energy Equation with Phase Change*
**B.8** References*

# Tables of Thermochemical and Thermophysical Properties

I should like to write "through measuring to knowing," as a motto above every physics laboratory.

– H. K. Onnes

## C.1 Tables

### Unit Conversion, Universal Constants, Dimensionless Numbers, Energy Conversion Relations, and Geometrical Relations

### Periodic Table and Phase Transitions

### Atmospheric Thermophysical Properties*

### Electrical and Acoustic Properties

* This table is found on the Web site www.cambridge.org/kaviany.

## Thermal Conductivity

## Thermophysical Properties of Solids

## Surface-Radiation Properties

## Mass Transfer and Thermochemical Properties*

## Thermophysical Properties of Fluids

## Liquid-Gas Surface Tension*

## Saturated Liquid-Vapor Properties*

# C.2 References**

* This table is found on the Web site www.cambridge.org/kaviany.

** References are found on the Web site www.cambridge.org/kaviany.

## Unit Conversion, Universal Constants, Dimensionless Numbers, Energy Conversion Relations, and Geometrical Relations

Table C.1(a). *Conversion between SI and English units (adapted from [9])*

| Conversion | Multiplier |
|---|---|
| atm to Pa | $1.01325 \times 10^5$ |
| bar to Pa | $10^5$ |
| Btu to J | $1.055 \times 10^3$ |
| Btu/hr to W | 0.2931 |
| Btu/hr-ft-°F to W/m-K | 1.7306 |
| cal to J | 4.1868 |
| dyne/cm$^2$ to Pa | 0.1 |
| eV to J | $1.6022 \times 10^{-19}$ |
| ft$^2$/s to m$^2$/s | 0.09290 |
| gal to liter | 3.785 |
| gal to m$^3$ | $3.785 \times 10^{-3}$ |
| hp to W | 745.7 |
| in to m | 0.0254 |
| J to Btu | $9.4782 \times 10^{-4}$ |
| J/kg to Btu/lbm | $4.299 \times 10^{-4}$ |
| J/kg-K to Btu/lbm-°F | $2.3885 \times 10^{-4}$ |
| liter to m$^3$ | $10^{-3}$ |
| mm Hg to Pa | 133.32 |
| mile to km | 1.6093 |
| poise to Pa-s | 0.1 |
| psi to Pa | $6.8964 \times 10^3$ |
| lbf/ft$^2$ to Pa | 47.880 |
| lbf-s/ft$^2$ to Pa-s | 47.880 |
| lbm to kg | 0.45359 |
| lbm/ft$^3$ to kg/m$^3$ | 16.0185 |
| W/m-K to Btu/hr-ft-°F | 0.57782 |

Table C.1(b). *Universal constants (adapted from [8])*

| Constant | | Magnitude |
|---|---|---|
| Avogadro number | $N_A$ | $6.0221 \times 10^{26}$ molecule/kmole |
| Boltzmann constant | $k_B$ | $1.3807 \times 10^{-23}$ J/K $= 8.618 \times 10^{-5}$ eV/K |
| electron charge | $e_c$ | $1.6022 \times 10^{-19}$ C |
| electron mass | $m_e$ | $9.109 \times 10^{-31}$ kg |
| free space, electrical permittivity | $\epsilon_o$ | $8.8542 \times 10^{-12}$ A$^2$-s$^2$/N-m$^2$ |
| free space, magnetic permeability | $\mu_o$ | $1.2566 \times 10^{-6}$ N/A$^2$ |
| Newton (gravitational) constant | $G_N$ | $6.67259 \times 10^{-11}$ m$^3$/kg-s$^2$ |
| Planck constant | $h_P$ | $6.626075 \times 10^{-34}$ J-s $= 4.316 \times 10^{-15}$ eV-s |
| speed of light in vacuum | $c_o$ | $2.99792 \times 10^8$ m/s |
| standard gravitational acceleration | $g_o$ | $9.8067$ m/s$^2$ |
| Stefan-Boltzmann constant | $\sigma_{SB} = \frac{2\pi^5}{15} \frac{k_B^4}{h_P^3 c_o^2}$ | $5.6705 \times 10^{-8}$ W/m$^2$-K$^4$ |
| universal gas constant | $R_g = k_B N_A$ | $8.3145 \times 10^3$ J/kmole-K |

Table C.1(c). *Dimensionless numbers*

| Name | | Relation |
|---|---|---|
| Archimedes or Galileo number | $\mathrm{Ar}_L$ or $\mathrm{Ga}_L$ | $F_{\Delta\rho}/F_\mu = (g\Delta h_{lg}\rho L^3)/\rho_l v_l^2$ |
| Biot number | $\mathrm{Bi}_L$ (or $N_{ku,s}$) | $R_{k,s}/\langle R_{ku}\rangle_L = \langle\mathrm{Nu}\rangle_L k_f/k_s$ |
| boiling or Kutateladze number | $\mathrm{Ku}$ | $q_{ku}/(\dot{m}\Delta h_{lg})$ |
| Bond number | $\mathrm{Bo}_L$ | $F_{\Delta\rho}/F_\sigma = g\Delta\rho_{lg}L^2/\sigma$ |
| conduction-radiation number | $N_r$ | $R_{k,s}/(R_{r,\Sigma}/4\sigma_{\mathrm{SB}}T_m^3) = 4\sigma_{\mathrm{SB}}T^3L/k_s$ |
| dimensionless critical heat flux | $q_{ku,CHF}^*$ | $q_{ku,CHF}(\rho_g^2/\sigma g\Delta\rho_{lg})^{1/2}/\rho_g\Delta h_{lg}$ |
| Fourier number | $\mathrm{Fo}_L$ | $t/t_\alpha = t\alpha/L^2$ |
| Froude number | $\mathrm{Fr}_D$ | $F_u/F_g = [(\rho u^2)/(gD)]^{1/2}$ |
| Grashof number | $\mathrm{Gr}_L$ | $F_{\Delta T}/F_\mu = g\beta_f\Delta TL^3/v_f^2$ |
| internal-external conduction number | $N_{k,i}$ | $R_{k,i}/R_{k,i-j}$ |
| Jakob number | $\mathrm{Ja}_l$ | $c_{p,l}(T_{ls}-T_{l,o})/\Delta h_{lg}$ or $c_{p,l}(T_s-T_{lg})/\Delta h_{lg}$ |
| Knudsen number | $\mathrm{Kn}_L$ | $\lambda_m(\text{mean-free path})/L(\text{linear dimension})$ |
| Lewis number | $\mathrm{Le}$ | $D_m/\alpha_f$ |
| Nusselt number | $\langle\mathrm{Nu}\rangle_L = N_{ku,f}$ | $R_{k,f}/\langle R_{ku}\rangle_L$ |
| number of transfer units | $NTU$ | $R_{u,f}/\langle R_{ku}\rangle_D = 1/(\dot{M}c_p)_f\langle R_{ku}\rangle_D$ |
| Péclet number | $\mathrm{Pe}_L$ (or $N_u$) | $R_{k,f}/R_{u,f} = u_fL/\alpha_f$ |
| Prandtl number | $\mathrm{Pr}$ | $v_f/\alpha_f = (\mu c_p)_f/k_f$ |
| ratio of stream heat capacitances | $C_r$ | $(\dot{M}c_p)_{min}/(\dot{M}c_p)_{max}$ |
| Rayleigh number | $\mathrm{Ra}_L$ | $\mathrm{Gr}_L\mathrm{Pr} = g\beta_f\Delta TL^3/\alpha_f v_f$ |
| Reynolds number | $\mathrm{Re}_L$ | $F_u/F_\mu = u_fL/v_f$ |
| Stefan number | $\mathrm{Ste}_l$ | $c_{p,l}(T_{l,o}-T_{sl})/\Delta h_{sl}$ |
| Weber number | $\mathrm{We}_D$ | $F_u/F_\sigma = \rho_l u_d^2D/\sigma$ |
| Zel'dovich number | $\mathrm{Ze}$ | $\Delta E_a\Delta T/R_gT^2$ |

Table C.1(d). *Summary of volumetric $\dot{S}_i/V$ (W/m$^3$) and surface $\dot{S}_i/A$(W/m$^2$) thermal energy conversion mechanisms and their properties*

| Mechanism of Energy Conversion | Relation |
|---|---|
| chemical reaction: exo- or endothermic reaction | $\dot{S}_{r,c}/V = -\Delta h_{r,F}a_re^{-\Delta E_a/R_gT}$ |
| nuclear reaction: fission reaction heating | $\dot{S}_{r,fi}/V = -n_nu_n(A_{nu}/V)\Delta h_{r,fi}$ |
| nuclear reaction: fusion reaction heating | $\dot{S}_{r,fu}/V = -n_1n_2(Au)_{12}\Delta h_{r,fu}$ |
| nuclear reaction: radioactive decay heating | $\dot{S}_{r,\tau}/V = \dot{s}_{r,o}\exp[-t/\tau - 1/2(r-r_o)^2\sigma_{ex}^2]$ |
| physical bond: phase-change cooling/heating | $\dot{S}_{ij}/V = -\dot{n}_{ij}\Delta h_{ij}$ or $\dot{S}_{ij}/A = -\dot{m}_{ij}\Delta h_{ij}$ |
| electromagnetic: dielectric (microwave) heating | $\dot{S}_{e,m}/V = 2\pi f\epsilon_o\epsilon_{ec}e_e^2$ |
| electromagnetic: Joule heating | $\dot{S}_{e,J}/V = \rho_ej_e^2 = j_e^2/\sigma_e$ |
| electromagnetic: Peltier cooling/heating | $\dot{S}_{e,P}/A = \pm\alpha_{\mathrm{P}}j_e = \pm\alpha_{\mathrm{S}}Tj_e$ |
| electromagnetic: Thomson heating/cooling | $\dot{S}_{e,T}/A = \alpha_{\mathrm{T}}j_e(T_2-T_1)$ |
| electromagnetic: surface absorption heating | $\dot{S}_{e,\alpha}/A = (1-\rho_r)q_{r,i} = \alpha_rq_{r,i}$ |
| electromagnetic: volumetric absorption heating | $\dot{S}_{e,\sigma}/V = (1-\rho_r)q_{r,i}\sigma_{ex}e^{-\sigma_{ex}x}$ |
| electromagnetic: surface emission | $\dot{S}_{e,\epsilon}/A = -\epsilon_r\sigma_{\mathrm{SB}}T^4$ |
| mechanical: pressure-compressibility cooling/heating | $\dot{S}_{m,p}/V = \beta T(\partial/\partial t + \mathbf{u}\cdot\nabla)p$ |
| mechanical: viscous heating | $\dot{S}_{m,\mu}/V = \mathbf{S}_\mu : \nabla\mathbf{u}$ |
| mechanical: surface-friction heating | $\dot{S}_{m,F}/A = \mu_Fp_c\Delta u_i$ |
| mechanical: volumetric ultrasonic heating | $\dot{S}_{m,ac}/V = 2\sigma_{ac}I_{ac}$ |

Table C.1(e). *Geometrical relations for several objects*

| Object Geometry | Parameters | Area, $A$ | Volume, $V$ |
|---|---|---|---|
| cylinder (circular cross section) | length $L$ radius $R$ | $2\pi RL + 2\pi R^2$ | $\pi R^2 L$ |
| (rectangular cross section) | length $L$ width $w$ height $H$ | $2Lw + 2LH + 2wH$ | $LwH$ |
| cube | length $L$ | $6L^2$ | $L^3$ |
| sphere | radius $R$ | $4\pi R^2$ | $\dfrac{4}{3}\pi R^3$ |
| cone (right circular) | length $L$ base radius $R$ | $\pi R(R^2 + L^2)^{1/2} + \pi R^2$ | $\dfrac{1}{3}\pi R^2 L$ |
| frustum (right cone) | length $L$ upper radius $R_2$ lower radius $R_1$ | $\pi\{R_1^2 + R_2^2 + (R_1 + R_2)$ $\times [(R_1 - R_2)^2 + L^2]^{1/2}\}$ | $\dfrac{\pi}{3}L(R_1^2 + R_2^2 + R_1 R_2)$ |
| tetrahedron | each side length $L$ | $3^{1/2}L^2$ | $\dfrac{2^{1/2}}{12}L^3$ |
| octahedron | each side length $L$ | $3^{1/2}2L^2$ | $\dfrac{2^{1/2}}{3}L^3$ |
| ellipsoid | axes $R_1, R_2, R_3$ | | $\dfrac{4}{3}\pi R_1 R_2 R_3$ |

Table C.1(f).  *Unit prefixes*

| Factor | Prefix | Symbol | Factor | Prefix | Symbol |
|---|---|---|---|---|---|
| $10^{24}$ | yotta | Y | $10^{-24}$ | yocto | y |
| $10^{21}$ | zetta | Z | $10^{-21}$ | zepto | z |
| $10^{18}$ | exa | E | $10^{-18}$ | atto | a |
| $10^{15}$ | peta | P | $10^{-15}$ | femto | f |
| $10^{12}$ | tera | T | $10^{-12}$ | pico | p |
| $10^{9}$ | giga | G | $10^{-9}$ | nano | n |
| $10^{6}$ | mega | M | $10^{-6}$ | micro | $\mu$ |
| $10^{3}$ | kilo | k | $10^{-3}$ | milli | m |
| $10^{2}$ | hecto | h | $10^{-2}$ | centi | c |
| $10^{1}$ | deka | da | $10^{-1}$ | deci | d |

One angstrom is $1\ \text{Å} = 10^{-10}$ m.

## Periodic Table and Phase Transitions

Table C.2. *Periodic table with thermophysical properties at thermodynamic STP (adapted from [4, 8, 9])*

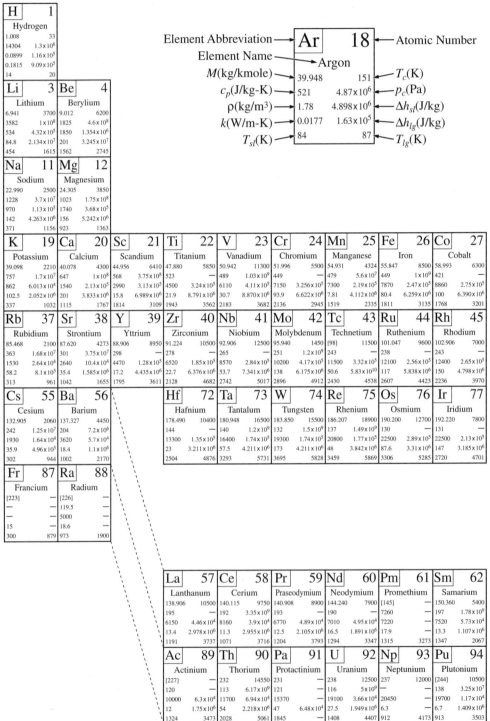

## Table C.2 (continued)

**He 2 — Helium**
| 4.002 | 5 |
|---|---|
| 5197 | 2.27×10⁵ |
| 0.179 | 2.1×10⁴ |
| 0.152 | 2.1×10⁵ |
| 3 | 4 |

**B 5 — Boron**
| 10.811 | — |
|---|---|
| 1026 | — |
| 2340 | 1.933×10⁶ |
| 27.4 | — |
| 2365 | 4275 |

**C 6 — Carbon**
| 12.011 | — |
|---|---|
| 569 | — |
| 3510 | — |
| 2300 | — |
| 4100 | 4473 |

**N 7 — Nitrogen**
| 14.007 | 126 |
|---|---|
| 1039 | 3.39×10⁶ |
| 1.24 | — |
| 0.026 | 3.99×10⁵ |
| 63 | 77 |

**O 8 — Oxygen**
| 15.999 | 154 |
|---|---|
| 919 | 5.04×10⁶ |
| 1.43 | — |
| 0.274 | 4.26×10⁵ |
| 54 | 90 |

**F 9 — Fluorine**
| 18.998 | 144 |
|---|---|
| 828 | 5.22×10⁶ |
| 1.7 | — |
| 0.0279 | 3.44×10⁵ |
| 54 | 95 |

**Ne 10 — Neon**
| 20.180 | 44 |
|---|---|
| 1030 | 2.76×10⁶ |
| 0.9 | 1.6×10⁵ |
| 0.0493 | 8.89×10⁴ |
| 24 | 27 |

**Al 13 — Aluminum**
| 26.982 | 8650 |
|---|---|
| 897 | — |
| 2700 | 3.97×10⁵ |
| 237 | 1.0875×10⁷ |
| 933 | 2793 |

**Si 14 — Silicon**
| 28.086 | 5159 |
|---|---|
| 678 | — |
| 2330 | 1.802×10⁶ |
| 149 | 1.4×10⁷ |
| 1687 | 3540 |

**P 15 — Phosphorus**
| 30.974 | 994 |
|---|---|
| 791 | — |
| 2690 | 8.48×10⁴ |
| 12.1 | 1.799×10⁶ |
| 317 | 553 |

**S 16 — Sulfur**
| 32.066 | 1314 |
|---|---|
| 732 | 2.07×10⁷ |
| 2070 | 5.34×10⁴ |
| 0.205 | 2.87×10⁵ |
| 388 | 718 |

**Cl 17 — Chlorine**
| 35.453 | 416 |
|---|---|
| 479 | 7.977×10⁶ |
| 3.21 | 1.8×10⁵ |
| 0.0089 | 5.76×10⁵ |
| 172 | 239 |

**Ar 18 — Argon**
| 39.948 | 151 |
|---|---|
| 521 | 4.87×10⁶ |
| 1.78 | 4.898×10³ |
| 0.0177 | 1.63×10⁵ |
| 84 | 87 |

**Ni 28 — Nickel**
| 58.693 | 6294 |
|---|---|
| 444 | — |
| 8900 | 2.91×10⁵ |
| 90.9 | 6.309×10⁶ |
| 1728 | 3187 |

**Cu 29 — Copper**
| 63.546 | 8500 |
|---|---|
| 385 | — |
| 8960 | 2.07×10⁵ |
| 401 | 4.726×10⁶ |
| 1358 | 2836 |

**Zn 30 — Zinc**
| 65.390 | 2169 |
|---|---|
| 388 | — |
| 7140 | 1.12×10⁵ |
| 116 | 1.767×10⁶ |
| 693 | 1180 |

**Ga 31 — Gallium**
| 69.723 | 7620 |
|---|---|
| 371 | — |
| 5910 | 7.97×10⁴ |
| 40.6 | 3.688×10⁶ |
| 303 | 2477 |

**Ge 32 — Germanium**
| 72.610 | 5642 |
|---|---|
| 310 | — |
| 5323 | 5.09×10⁵ |
| 60.2 | 4.557×10⁶ |
| 1211 | 3107 |

**As 33 — Arsenic**
| 74.992 | 1673 |
|---|---|
| 343 | — |
| 5727 | — |
| 50.2 | 1.7×10⁶ |
| 1090 | 888 |

**Se 34 — Selenium**
| 78.960 | 1766 |
|---|---|
| 339 | 2.72×10⁷ |
| 4810 | 6.62×10⁴ |
| 2.04 | 1.2×10⁶ |
| 494 | 958 |

**Br 35 — Bromine**
| 79.904 | 588 |
|---|---|
| 226 | 1.03×10⁷ |
| 3120 | 6.6×10⁴ |
| 0.122 | 1.9×10⁵ |
| 266 | 332 |

**Kr 36 — Krypton**
| 83.800 | 210 |
|---|---|
| 248 | 5.5×10⁶ |
| 3.73 | 1.9×10⁵ |
| 0.0095 | 1.08×10⁵ |
| 116 | 120 |

**Pd 46 — Palladium**
| 106.420 | — |
|---|---|
| 251 | — |
| 12020 | 4.17×10⁵ |
| 71.8 | 5.567×10⁶ |
| 2896 | 4912 |

**Ag 47 — Silver**
| 107.868 | 7460 |
|---|---|
| 235 | — |
| 10500 | 1.02×10⁵ |
| 429 | 2.323×10⁶ |
| 1235 | 2436 |

**Cd 48 — Cadmium**
| 112.411 | 1903 |
|---|---|
| 232 | — |
| 8690 | 5.39×10⁴ |
| 96.9 | 8.86×10⁵ |
| 594 | 1040 |

**In 49 — Indium**
| 114.820 | 4377 |
|---|---|
| 233 | — |
| 7310 | 2.86×10⁴ |
| 81.8 | 2.019×10⁶ |
| 430 | 2346 |

**Sn 50 — Tin**
| 118.710 | 8000 |
|---|---|
| 228 | — |
| 7280 | 6.07×10⁴ |
| 66.8 | 2.495×10⁶ |
| 505 | 2876 |

**Sb 51 — Antimony**
| 121.757 | 2989 |
|---|---|
| 207 | — |
| 6680 | 1.66×10⁵ |
| 24.4 | — |
| 904 | 1860 |

**Te 52 — Tellurium**
| 127.600 | 2329 |
|---|---|
| 197 | — |
| 6240 | 1.4×10⁶ |
| 2.35 | 8.95×10⁶ |
| 723 | 1261 |

**I 53 — Iodine**
| 126.904 | 819 |
|---|---|
| 145 | — |
| 4953 | — |
| 0.449 | 3.29×10⁵ |
| 387 | 457 |

**Xe 54 — Xenon**
| 131.290 | 289 |
|---|---|
| 158 | 5.84×10⁶ |
| 5.89 | 1.7×10⁶ |
| 0.0057 | 9.63×10⁴ |
| 161 | 165 |

**Pt 78 — Platinum**
| 195.080 | 8280 |
|---|---|
| 133 | — |
| 21500 | 1.15×10⁵ |
| 71.6 | 2.612×10⁶ |
| 2042 | 4100 |

**Au 79 — Gold**
| 196.967 | 9500 |
|---|---|
| 129 | — |
| 19300 | 6.43×10⁴ |
| 318 | 1.701×10⁶ |
| 1337 | 3130 |

**Hg 80 — Mercury**
| 200.590 | 1750 |
|---|---|
| 140 | 1.5×10⁸ |
| 13534 | 1.14×10⁴ |
| 8.3 | 2.96×10⁶ |
| 234 | 630 |

**Tl 81 — Thallium**
| 204.383 | 3219 |
|---|---|
| 129 | — |
| 11800 | 2.06×10⁴ |
| 46.1 | 8.06×10⁵ |
| 577 | 1746 |

**Pb 82 — Lead**
| 207.200 | 5400 |
|---|---|
| 129 | — |
| 11300 | 2.32×10⁴ |
| 35.3 | 8.58×10⁵ |
| 600 | 2022 |

**Bi 83 — Bismuth**
| 208.980 | 4620 |
|---|---|
| 122 | — |
| 9790 | 5.33×10⁴ |
| 7.92 | 7.3×10⁵ |
| 544 | 1837 |

**Po 84 — Polonium**
| [209] | 2281 |
|---|---|
| 125 | — |
| 9320 | 5.98×10⁴ |
| 20 | — |
| 527 | 1033 |

**At 85 — Astatine**
| [210] | — |
|---|---|
| 140 | — |
| Ñ | — |
| 1.7 | — |
| 575 | 650 |

**Rn 86 — Radon**
| [222] | 377 |
|---|---|
| 155 | — |
| 9.73 | 1.2×10⁴ |
| 0.0036 | 7.52×10⁴ |
| 202 | 211 |

**Eu 63 — Europium**
| 151.965 | 4600 |
|---|---|
| 182 | — |
| 5240 | 6.06×10⁴ |
| 13.9 | 9.44×10⁵ |
| 1095 | 1869 |

**Gd 64 — Gadolinium**
| 157.250 | 8670 |
|---|---|
| 236 | — |
| 7900 | 6.36×10⁴ |
| 10.5 | 2.286×10⁶ |
| 1585 | 3546 |

**Tb 65 — Terbium**
| 158.925 | — |
|---|---|
| 173 | — |
| 8230 | 6.79×10⁴ |
| 11.1 | 1.42×10⁶ |
| 1629 | 3503 |

**Dy 66 — Dysprosium**
| 162.500 | — |
|---|---|
| 165 | — |
| 8550 | 6.79×10⁴ |
| 10.7 | 2.083×10⁶ |
| 1685 | 3141 |

**Ho 67 — Holmium**
| 164.930 | — |
|---|---|
| 165 | — |
| 8800 | 1.03×10⁵ |
| 16.2 | 1.461×10⁶ |
| 1747 | 2973 |

**Er 68 — Erbium**
| 167.260 | 7250 |
|---|---|
| 168 | — |
| 9070 | 1.19×10⁵ |
| 14.5 | 1.563×10⁶ |
| 1802 | 3141 |

**Tm 69 — Thulium**
| 168.934 | 6430 |
|---|---|
| 160 | — |
| 9320 | 9.9×10⁴ |
| 16.9* | 1.130×10⁶ |
| 1818 | 2223 |

**Yb 70 — Ytterbium**
| 173.040 | 8950 |
|---|---|
| 155 | — |
| 6900 | 1.43×10⁴ |
| 34.9* | 7.45×10⁵ |
| 1092 | 1469 |

**Lu 71 — Lutetium**
| 174.967 | — |
|---|---|
| 154 | — |
| 9840 | 1.26×10⁵ |
| 16.4* | 2.034×10⁶ |
| 1936 | 3675 |

**Am 95 — Americium**
| [243] | — |
|---|---|
| — | — |
| 13670 | — |
| 10 | — |
| 1449 | 2284 |

**Cm 96 — Curium**
| [247] | — |
|---|---|
| — | — |
| 13300 | — |
| 10 | — |
| 1618 | — |

**Bk 97 — Berkelium**
| [247] | — |
|---|---|
| — | — |
| 14790 | — |
| 10 | — |
| 1323 | — |

**Cf 98 — Californium**
| [251] | — |
|---|---|
| — | — |
| — | — |
| 10 | — |
| 1173 | — |

**Es 99 — Einsteinium**
| [252] | — |
|---|---|
| — | — |
| — | — |
| 10 | — |
| 1133 | — |

**Fm 100 — Fermium**
| [257] | — |
|---|---|
| — | — |
| — | — |
| 10 | — |
| 1800 | — |

**Md 101 — Mendelevium**
| [258] | — |
|---|---|
| — | — |
| — | — |
| 10 | — |
| 1100 | — |

**No 102 — Nobelium**
| [259] | — |
|---|---|
| — | — |
| — | — |
| 10 | — |
| 1100 | — |

**Lr 103 — Lawrencium**
| [260] | — |
|---|---|
| — | — |
| — | — |
| 10 | — |
| 1900 | — |

*Note:* A white elemental letter with a black background designates a semiconductor

Table C.3. *Designation and physical properties for some refrigerants in order of increasing boiling temperature; $T_{lg}$ is for one atm pressure (adapted from [1])\**

Table C.4. *Phase transitions for some fluid (low melting-temperature) substances (at one atm pressure) (adapted from [9])\**

Table C.5(a). *Heat of melting for some organic and inorganic substances with melting temperatures between 260 and 530 K (at one atm pressure), arranged in order of increasing melting temperature $T_{sl}$ (adapted from [7]).*

| Substance | Formula | $T_{sl}$, K | $\Delta h_{sl}$, J/kg |
|---|---|---|---|
| nitrogen tetroxide | $N_2O_4$ | 263.9 | $1.59 \times 10^5$ |
| water | $H_2O$ | 273.2 | $3.34 \times 10^5$ |
| phosphorus oxychloride | $POCl_3$ | 274.2 | $8.54 \times 10^4$ |
| hydrazine | $H_4N_2$ | 274.6 | $3.93 \times 10^5$ |
| nitrobenzene | $C_6H_5NO_2$ | 278.9 | $9.41 \times 10^4$ |
| sulphur trioxide | $SO_3$ | 290 | $1.07 \times 10^5$ |
| molybdenum fluoride | $MoF_6$ | 290.7 | $2.06 \times 10^4$ |
| vanadium fluoride | $VF_5$ | 292.7 | $3.42 \times 10^5$ |
| phenylhydrazine | $C_6H_8N_2$ | 292.8 | $1.52 \times 10^5$ |
| iodine chloride | $ICl$ | 300.6 | $7.15 \times 10^4$ |
| tin bromide | $SnBr_4$ | 304.2 | $2.74 \times 10^4$ |
| arsenic bromide | $AsBr_3$ | 304.3 | $3.72 \times 10^4$ |
| paraffin (eicosane)[†] | $C_{20}H_{42}$ | 310.0 | $2.17 \times 10^5$ |
| titanium bromide | $TiBr_4$ | 312.2 | $3.51 \times 10^4$ |
| sodium acetate trihydrate | $CH_3COONa,3H_2O$ | 331.2 | $1.86 \times 10^5$ |
| gallium chloride | $GaCl_3$ | 351.1 | $6.32 \times 10^4$ |
| aluminum tribromide | $AlBr_3$ | 370.7 | $4.22 \times 10^4$ |
| gallium bromide | $GaBr_3$ | 394.7 | $3.91 \times 10^4$ |
| silver nitrate | $AgNO_3$ | 485.2 | $6.77 \times 10^4$ |
| bismuth trichloride | $BiCl_3$ | 503.2 | $3.46 \times 10^4$ |
| tin | $Sn$ | 505.0 | $6.07 \times 10^4$ |
| lithium nitrate | $LiNO_3$ | 526.2 | $3.61 \times 10^5$ |

[†]A group of latent-heat parrafins/waxes, called Rubitherm, have varying saturated hydrocarbon chain lengths and distinct $T_{sl}$ from 282.2 K to 384.8 K. The heat of fusion $\Delta h_{sl}$ is about $2 \times 10^5$ J/kg.

Table C.5(b). *Heat of adsorption for some gas-porous solid pairs [18].*

| Porous Solid (Adsorbent) | $\Delta h_{ad}$, J/kg-gas Gas (Adsorbate) | | | |
|---|---|---|---|---|
| | $H_2O$ | $N_2$ | $CO_2$ | $CH_4$ |
| 13X zeolite | $2.8 \times 10^6$ | $6.0 \times 10^5$ | $7.8 \times 10^5$ | $8.6 \times 10^5$ |
| 4A zeolite | $4.2 \times 10^6$ | | | |
| alumina | $1.8 \times 10^6$ | $3.0 \times 10^5$ | $7.6 \times 10^5$ | $6.5 \times 10^5$ |
| activated carbon | $9.3 \times 10^5$ | $5.1 \times 10^5$ | $3.3 \times 10^5$ | $1.2 \times 10^6$ |

\* This table is found on the Web site www.cambridge.org/kaviany.

Table C.6. *Maximum Jakob number $Ja_l = c_{p,l}(T_{lg} - T_{sl})/\Delta h_{lg}$ and Stefan number $Ste_l = c_{p,l}(T_{lg} - T_{sl})/\Delta h_{sl}$ for some substances (at one atm pressure) (adapted from [9])\**

## Atmospheric Thermophysical Properties

Table C.7. *Thermophysical properties of the earth atmosphere (adapted from [9])\**

## Electrical and Acoustic Properties

Table C.8. *Electrical resistivity for some pure metals, $\rho_e(ohm\text{-}m)$ (adapted from [19])*

| $T$, K | aluminum | cesium | chromium | copper | gold |
|---|---|---|---|---|---|
| 1 | $0.000100 \times 10^{-8}$ | $0.0026 \times 10^{-8}$ | | $0.00200 \times 10^{-8}$ | $0.0220 \times 10^{-8}$ |
| 100 | $0.442 \times 10^{-8}$ | $5.28 \times 10^{-8}$ | $1.6 \times 10^{-8}$ | $0.348 \times 10^{-8}$ | $0.650 \times 10^{-8}$ |
| 200 | $1.587 \times 10^{-8}$ | $12.2 \times 10^{-8}$ | $7.7 \times 10^{-8}$ | $1.046 \times 10^{-8}$ | $1.462 \times 10^{-8}$ |
| 300 | $2.733 \times 10^{-8}$ | $21.0 \times 10^{-8}$ | $12.7 \times 10^{-8}$ | $1.725 \times 10^{-8}$ | $2.271 \times 10^{-8}$ |
| 500 | $4.99 \times 10^{-8}$ | | $20.1 \times 10^{-8}$ | $3.090 \times 10^{-8}$ | $3.97 \times 10^{-8}$ |
| 700 | $7.35 \times 10^{-8}$ | | $29.5 \times 10^{-8}$ | $4.514 \times 10^{-8}$ | $5.82 \times 10^{-8}$ |
| 900 | $10.18 \times 10^{-8}$ | | $39.9 \times 10^{-8}$ | $6.041 \times 10^{-8}$ | $7.86 \times 10^{-8}$ |

| $T$, K | iron | lead | magnesium | molybdenum | nickel |
|---|---|---|---|---|---|
| 1 | $0.0225 \times 10^{-8}$ | | $0.0062 \times 10^{-8}$ | $0.00070 \times 10^{-8}$ | $0.0032 \times 10^{-8}$ |
| 100 | $1.28 \times 10^{-8}$ | $6.4 \times 10^{-8}$ | $0.91 \times 10^{-8}$ | $0.858 \times 10^{-8}$ | $0.96 \times 10^{-8}$ |
| 200 | $5.20 \times 10^{-8}$ | $13.6 \times 10^{-8}$ | $2.75 \times 10^{-8}$ | $3.13 \times 10^{-8}$ | $3.67 \times 10^{-8}$ |
| 300 | $9.98 \times 10^{-8}$ | $21.3 \times 10^{-8}$ | $4.51 \times 10^{-8}$ | $5.52 \times 10^{-8}$ | $7.20 \times 10^{-8}$ |
| 500 | $23.7 \times 10^{-8}$ | $38.3 \times 10^{-8}$ | $7.86 \times 10^{-8}$ | $10.6 \times 10^{-8}$ | $17.7 \times 10^{-8}$ |
| 700 | $44.0 \times 10^{-8}$ | | $11.2 \times 10^{-8}$ | $15.8 \times 10^{-8}$ | $32.1 \times 10^{-8}$ |
| 900 | | | $14.4 \times 10^{-8}$ | $21.2 \times 10^{-8}$ | $38.6 \times 10^{-8}$ |

| $T$, K | palladium | platinum | silver | tantalum | tungsten |
|---|---|---|---|---|---|
| 1 | $0.0200 \times 10^{-8}$ | $0.002 \times 10^{-8}$ | $0.00100 \times 10^{-8}$ | $0.10 \times 10^{-8}$ | $0.000016 \times 10^{-8}$ |
| 100 | $2.62 \times 10^{-8}$ | $2.755 \times 10^{-8}$ | $0.418 \times 10^{-8}$ | $3.64 \times 10^{-8}$ | $1.02 \times 10^{-8}$ |
| 200 | $6.88 \times 10^{-8}$ | $6.77 \times 10^{-8}$ | $1.029 \times 10^{-8}$ | $8.66 \times 10^{-8}$ | $3.18 \times 10^{-8}$ |
| 300 | $10.80 \times 10^{-8}$ | $10.8 \times 10^{-8}$ | $1.629 \times 10^{-8}$ | $13.5 \times 10^{-8}$ | $5.44 \times 10^{-8}$ |
| 500 | $17.94 \times 10^{-8}$ | $18.3 \times 10^{-8}$ | $2.87 \times 10^{-8}$ | $22.9 \times 10^{-8}$ | $10.3 \times 10^{-8}$ |
| 700 | $24.2 \times 10^{-8}$ | $25.4 \times 10^{-8}$ | $4.21 \times 10^{-8}$ | $31.8 \times 10^{-8}$ | $15.7 \times 10^{-8}$ |
| 900 | $29.4 \times 10^{-8}$ | $32.0 \times 10^{-8}$ | $5.64 \times 10^{-8}$ | $40.1 \times 10^{-8}$ | $21.5 \times 10^{-8}$ |

| $T$, K | vanadium | zinc | zirconium |
|---|---|---|---|
| 1 | | $0.0100 \times 10^{-8}$ | $0.250x \times 10^{-8}$ |
| 100 | $4.01 \times 10^{-8}$ | $1.60 \times 10^{-8}$ | $9.79 \times 10^{-8}$ |
| 200 | $12.4 \times 10^{-8}$ | $3.83 \times 10^{-8}$ | $26.3 \times 10^{-8}$ |
| 300 | $20.2 \times 10^{-8}$ | $6.06 \times 10^{-8}$ | $43.3 \times 10^{-8}$ |
| 500 | $34.8 \times 10^{-8}$ | $10.82 \times 10^{-8}$ | $76.5 \times 10^{-8}$ |
| 700 | $47.2 \times 10^{-8}$ | | $104.2 \times 10^{-8}$ |
| 900 | $58.7 \times 10^{-8}$ | | $123.1 \times 10^{-8}$ |

\* This table is found on the Web site www.cambridge.org/kaviany.

Table C.9(a). *Thermoelectric properties for some materials, $T = 300\ K$ (adapted from [15])*

| Material | $\alpha_{S,i}$, $\mu$V/K | $\rho_e$, ohm-m | $k$, W/m-K |
|---|---|---|---|
| $p$- and $n$-type germanium (Ge) | 420, −528 | $3.3 \times 10^{-6}, 6.8 \times 10^{-7}$ | 64.4, 64.0 |
| $n$-type silicon (Si) | −450 | $4.0 \times 10^{-5}$ | 144.6 |
| $n$-type nickel (Ni) | −19 | $7.2 \times 10^{-8}$ | 91 |
| $n$-type antimony (Sb) | +40 | $4.17 \times 10^{-8}$ | 24.3 |
| $p$-type bismuth (Bi) | −60 | $1.00 \times 10^{-6}$ | 7.9 |
| $p$- and $n$-type bismuth telluride ($Bi_2Te_3$) | 230, −210 | $1.00 \times 10^{-5}$ | 1.70, 1.45 |
| $n$-type germanium ($n$-Ge) | −138 | $1.89 \times 10^{-6}$ | 60 |
| $p$-type silicon germanium alloy ($p$-$Si_{0.7}Ge_{0.3}$) | 121 | $1.14 \times 10^{-5}$ | 4.90 |
| $n$-type silicon germanium alloy ($n$-$Si_{0.7}Ge_{0.3}$) | −108 | $0.85 \times 10^{-5}$ | 4.59 |

$\mu$V/°C $= 10^{-6}$ V/°C.

Table C.9(b). *Thermoelectric properties for p- and n-type silicon-germanium alloy ($Si_{0.7}Ge_{0.3}$) at different temperatures. The optimized junction figure of merit is also shown (adapted from [15])*

| $T$, K | $\rho_{e,p}$ $\mu$ohm-m | $\rho_{e,n}$ | $\alpha_{S,p}$ $\mu$V/K | $-\alpha_{S,n}$ | $k_p$ W/m-°C | $k_n$ | $c_{p,p} = c_{p,n}$ J/kg-°C | $(Z_e)_{pn}$, 1/K | $(Z_eT)_{pn}$ |
|---|---|---|---|---|---|---|---|---|---|
| 300 | 11.4 | 8.5 | 121 | 108 | 4.90 | 4.59 | 645 | 0.000279 | 0.084 |
| 400 | 13.2 | 10.1 | 144 | 136 | 4.80 | 4.45 | 647 | 0.000365 | 0.146 |
| 500 | 15.1 | 11.7 | 161 | 165 | 4.70 | 4.31 | 649 | 0.000441 | 0.220 |
| 600 | 17.1 | 13.6 | 176 | 190 | 4.57 | 4.19 | 658 | 0.000499 | 0.299 |
| 700 | 19.3 | 16.1 | 189 | 213 | 4.41 | 4.08 | 672 | 0.000538 | 0.377 |
| 800 | 21.6 | 18.6 | 203 | 233 | 4.25 | 3.99 | 687 | 0.000574 | 0.459 |
| 900 | 24.1 | 20.6 | 216 | 247 | 4.12 | 3.93 | 703 | 0.000596 | 0.537 |
| 1000 | 26.8 | 21.1 | 228 | 253 | 4.07 | 3.93 | 719 | 0.000605 | 0.605 |
| 1100 | 29.6 | 19.7 | 237 | 250 | 4.14 | 4.01 | 736 | 0.000595 | 0.655 |
| 1200 | 31.9 | 17.8 | 240 | 242 | 4.38 | 4.20 | 754 | 0.000555 | 0.666 |
| 1300 | 32.6 | 19.1 | 233 | 243 | 4.84 | 4.53 | 772 | 0.000474 | 0.616 |

$\mu$ohm-m $= 10^{-6}$ ohm-m.

Table C.10.  *Dielectric loss factor for some substances (adapted from [10])*

| Substance | $T$, K | $f = 10^7$ Hz $\epsilon_{ec}$ or $\langle\epsilon_{ec}\rangle$ | $f = 10^9$ Hz $\epsilon_{ec}$ or $\langle\epsilon_{ec}\rangle$ | $f = 3 \times 10^9$ Hz $\epsilon_{ec}$ or $\langle\epsilon_{ec}\rangle$ |
|---|---|---|---|---|
| *Food* | | | | |
| beef steak, | | | | |
|    bottom round | 298 | 1300 | 39 | 12 |
|    frozen lean | 273 | | 0.72 | 0.3 |
| butter | 273 | | | 0.39 |
| potato, raw | 298 | 47.8 | 19.6 | 15.7 |
| turkey, cooked | 298 | | 68.0 | 14.0 |
| *Forest products* | | | | |
| Douglas fir | 293 | | 0.04 | 0.02 |
| paper, royal grey | 355 | | 0.216 | 0.235 |
| *Minerals and ceramics* | | | | |
| barium titanate | 298 | 8.55 | 55.0 | 180.0 |
| fused silica | 298 | 0.0001 | 0.0002 | 0.0002 |
| marble, dry | 298 | 0.33 | | 0.22 |
| porcelain | 298 | 0.0018 | 0.008 | 0.01 |
| ruby mica, muscovite | 298 | 0.0016 | | 0.0016 |
| sandy soil, dry | 298 | 0.04 | 0.026 | 0.016 |
| *Oils and waxes* | | | | |
| cable oil | 298 | 0.009 | 0.009 | 0.0043 |
| wax | 298 | 0.0006 | 0.00083 | 0.0011 |
| *Plastics* | | | | |
| cellulose acetate | | 0.07 | 0.072 | 0.094 |
| laminated fiberglass | 297 | 0.17 | 0.108 | 0.128 |
| Plexiglas, perspex | 300 | 0.027 | 0.017 | 0.015 |
| polyamide | 298 | 0.07 | 0.043 | 0.036 |
| polyethylene | 296 | | 0.0024 | 0.0026 |
| polythene | 297 | 0.0004 | | 0.0007 |
| *Rubber* | | | | |
| cyclized, pliobond | 297 | | | 0.28 |
| natural crepe | 298 | | | 0.0065 |
| silicon | 25 | | 0.245 | 0.48 |
| *Water* | | | | |
| distilled | 298 | | 1.2 | 12.00 |
| distilled | 368 | | 0.364 | 2.44 |
| ice, pure | 261 | 0.07 | | 0.003 |
| with 0.5 molal NaCl | 298 | | 269.0 | 41.87 |

Table C.11. *Acoustic absorption coefficient, speed of sound, and density, for some biological tissues, gases, liquids and solids under STP conditions. (f = 1 MHz)*

| Material | $\sigma_{ac}$, 1/m | $a_s$, m/s | $\rho$, kg/m$^3$ |
|---|---|---|---|
| *Tissue* | | | |
| blood | 1.4 | 1566 | 1040 |
| blood vessel | 20 | 153.0 | 1080 |
| bone | 161 | 3445 | 1820 |
| cartilage | 58 | 1665 | |
| fat | 7 | 1478 | 920 |
| muscle | 14 | 1552 | 1040 |
| skin | 31 | 1519 | |
| *Gases (STP)* | | | |
| air | 13.7 | 343.0 | 1.233 |
| nitrogen | 16.4 | 356.0 | 1.250 |
| oxygen | 19.2 | 328.2 | 1.363 |
| argon | 18.7 | 320.3 | 1.780 |
| neon | 35 | 435 | 0.9 |
| water ($T = 373$ K) | 10.8 | 404.8 | 0.5831 |
| *Liquids* | | | |
| water | 0.025 | 1481 | 1000 |
| sulphuric acid (0.1 M) | 0.028 | 1510 | 1002 |
| Aluminum chloride (0.1 M) | 0.026 | 1534 | 1004 |
| mercury | 0.005 | 1450 | 13600 |
| benzene | 0.825 | 1324 | 876.0 |
| trichloromethane | 0.364 | 920.0 | 1481 |
| glycerine | 3 | 1980 | 1260 |
| *Metals* | | | |
| copper(Grain size, 0.16 mm) | 0.255 | 3560 | 8933 |
| iron(Grain size, 0.25 mm) | 0.230 | 5130 | 7870 |
| magnesium(Grain size, 0.028 mm) | 0.0000032 | 4470 | 1740 |
| *Porous materials* | | | |
| Water-filled Berea sandstone (fast wave) | 0.28 | 4834 | 2640 |
| Water-filled Berea sandstone (slow wave) | 182.3 | 658.4 | 2640 |
| silt clay | 11.5 | 1650 | 1270 |
| *Plastics* | | | |
| nylon (6/6, black) | 7.368 | 2770 | 1120 |
| PVC (polyvinyl chlorine) | 5.158 | 2380 | 1380 |
| PVC (polysulfone) | 12.23 | 2240 | 1240 |
| acrylic | 2.947 | 2750 | 1190 |

## Thermal Conductivity

Table C.12.  *Thermal conductivity for some gases as a function of temperature, k(W/m-K) (adapted from [8])*

| Substance | \( T \), K | | | | | | | | |
| --- | --- | --- | --- | --- | --- | --- | --- | --- | --- |
| | 100 | 200 | 250 | 300 | 400 | 500 | 600 | 800 | 1000 |
| acetone | | | 0.0080 | 0.0115 | 0.0201 | 0.0310 | | | |
| acetylene | | 0.0117 | 0.0162 | 0.0213 | 0.0332 | 0.0452 | 0.054 | | |
| air | 0.0092 | 0.0181 | 0.0223 | 0.0267 | 0.0331 | 0.0395 | 0.0456 | 0.0569 | 0.0672 |
| ammonia | | 0.0153 | 0.0197 | 0.0246 | 0.0364 | 0.0506 | 0.0656 | 0.0977 | |
| argon | 0.0065 | 0.0124 | 0.0152 | 0.0177 | 0.0223 | 0.0264 | 0.0301 | 0.0369 | 0.0427 |
| benzene | | | 0.0077 | 0.0104 | 0.0195 | 0.0335 | 0.0524 | | |
| bromine | | | 0.0038 | 0.0048 | 0.0067 | | | | |
| *n*-butane | | | 0.0117 | 0.0160 | 0.0264 | 0.0337 | | | |
| isobutane | | | 0.0163 | 0.0272 | 0.0385 | | | | |
| carbon dioxide | | 0.0095 | 0.0129 | 0.0166 | 0.0244 | 0.0323 | 0.0403 | 0.0560 | 0.0680 |
| carbon monoxide | | 0.0175 | 0.0214 | 0.0252 | 0.0323 | 0.0386 | | | 0.0644 |
| carbon tetrachloride | | | | 0.0053 | 0.0067 | 0.0099 | 0.0126 | | |
| chlorine | | 0.0054 | 0.0071 | 0.0089 | 0.0124 | 0.0156 | 0.0190 | | |
| deuterium | 0.058 | 0.101 | 0.122 | 0.141 | 0.176 | | | | |
| ethane | | 0.0102 | 0.0156 | 0.0218 | 0.0360 | 0.0516 | 0.0685 | 0.107 | 0.164 |
| ethanol | | | 0.0092 | 0.0147 | 0.0245 | 0.0327 | | | |
| ethyl alcohol | | | | | 0.0245 | 0.0327 | | | |
| ethylene | | 0.0105 | 0.0152 | 0.0204 | 0.0342 | 0.0491 | 0.0653 | 0.0944 | 0.138 |
| fluorine | 0.0090 | 0.0184 | 0.0231 | 0.0279 | 0.0371 | 0.0455 | 0.0527 | 0.0618 | |
| helium | 0.0073 | 0.1151 | 0.1338 | 0.150 | 0.180 | 0.211 | 0.247 | 0.307 | 0.363 |
| *n*-heptane | | | 0.0082 | 0.0120 | 0.0214 | 0.0325 | 0.0447 | 0.0709 | 0.097 |
| *n*-hexane | | | 0.0090 | 0.0128 | 0.0232 | 0.0355 | 0.0499 | 0.0870 | 0.136 |
| hydrogen | 0.068 | 0.128 | 0.156 | 0.182 | 0.221 | 0.256 | 0.291 | 0.360 | 0.428 |
| hydrogen sulfide | | | 0.0114 | 0.0147 | 0.0212 | | | | |
| methane | 0.0106 | 0.0218 | 0.0277 | 0.0343 | 0.0484 | 0.0671 | 0.0948 | 0.124 | 0.169 |
| methyl alcohol | | | | | 0.0249 | 0.0351 | | | |
| nitrogen | 0.0094 | 0.0183 | 0.0222 | 0.0260 | 0.0325 | 0.0386 | 0.0441 | 0.0541 | 0.0631 |
| nitric oxide | 0.0090 | 0.0178 | 0.0219 | 0.0259 | 0.0331 | 0.0396 | 0.0462 | 0.0595 | 0.723 |
| *n*-octane | | | | | 0.0226 | 0.0334 | | | |
| oxygen | 0.0091 | 0.0182 | 0.0225 | 0.0267 | 0.0343 | 0.0412 | 0.0480 | 0.0603 | 0.717 |
| *n*-pentane | | | 0.0107 | 0.0152 | 0.0250 | 0.0362 | | | |
| propane | | 0.0077 | 0.0129 | 0.0183 | 0.0295 | 0.0417 | | | |
| toluene | | | 0.0116 | 0.0146 | 0.0240 | 0.0349 | 0.0461 | | |
| water | | | | 0.0181 | 0.0264 | 0.0357 | 0.0464 | 0.0710 | 0.0981 |

Table C.13. *Thermal conductivity for some nonmetallic liquids as a function of temperature, k(W/m-K) (adapted from [8])*

| Substance | $T$, K | | | | | | | | | | |
|---|---|---|---|---|---|---|---|---|---|---|---|
| | 180 | 220 | 240 | 260 | 280 | 300 | 320 | 340 | 360 | 380 | 400 |
| acetone | 0.204 | 0.189 | 0.180 | 0.171 | 0.162 | 0.153 | 0.145 | 0.137 | 0.128 | 0.119 | 0.111 |
| ammonia | | 0.620 | 0.581 | 0.542 | 0.504 | 0.467 | 0.425 | 0.383 | 0.337 | 0.286 | 0.230 |
| benzene | | | | | 0.153 | 0.143 | 0.134 | 0.131 | 0.125 | 0.119 | 0.114 |
| bromine | | | | | 0.127 | 0.122 | 0.118 | 0.114 | 0.109 | 0.104 | 0.099 |
| butane | 0.161 | 0.141 | 0.131 | 0.121 | | | | | | | |
| carbon dioxide | | 0.181 | 0.153 | 0.127 | 0.100 | 0.085 | | | | | |
| chlorine | 0.189 | 0.172 | 0.163 | 0.154 | 0.144 | 0.134 | 0.124 | 0.113 | 0.101 | 0.088 | 0.072 |
| ethane | 0.153 | 0.126 | 0.112 | 0.099 | 0.085 | 0.067 | | | | | |
| ethylene | 0.176 | 0.133 | 0.112 | 0.091 | 0.066 | | | | | | |
| heptane | | | | 0.138 | 0.132 | 0.127 | 0.121 | 0.116 | 0.111 | 0.108 | 0.104 |
| hexane | | 0.147 | 0.142 | 0.137 | 0.133 | 0.127 | 0.122 | 0.115 | 0.110 | 0.105 | 0.100 |
| isobutane | 0.143 | 0.126 | 0.117 | | | | | | | | |
| methanol | 0.242 | 0.227 | 0.220 | 0.213 | 0.207 | 0.200 | 0.193 | 0.187 | 0.183 | 0.179 | 0.175 |
| nonane | | | 0.145 | 0.140 | 0.134 | 0.129 | 0.123 | 0.118 | 0.112 | 0.107 | 0.101 |
| octane | 0.124 | 0.116 | 0.111 | 0.107 | 0.102 | 0.097 | 0.092 | 0.088 | 0.083 | | |
| pentane | 0.156 | 0.143 | 0.137 | 0.129 | 0.123 | 0.116 | 0.109 | 0.102 | 0.096 | 0.089 | 0.080 |
| propane | 0.165 | 0.139 | 0.123 | 0.115 | 0.106 | 0.096 | 0.087 | 0.077 | 0.068 | | |
| propylene | 0.177 | 0.152 | 0.141 | 0.130 | 0.121 | 0.111 | | | | | |
| R-11 | 0.120 | 0.109 | 0.104 | 0.098 | 0.092 | 0.087 | 0.082 | 0.076 | 0.070 | 0.064 | 0.059 |
| R-12 | 0.112 | 0.097 | 0.091 | 0.083 | 0.076 | 0.068 | 0.061 | 0.054 | 0.047 | 0.039 | |
| R-13 | 0.098 | 0.076 | 0.067 | 0.057 | 0.046 | 0.036 | | | | | |
| R-21 | 0.147 | 0.132 | 0.124 | 0.117 | 0.109 | 0.102 | 0.094 | 0.086 | 0.080 | 0.071 | 0.063 |
| R-22 | 0.146 | 0.126 | 0.117 | 0.107 | 0.097 | 0.087 | 0.076 | 0.067 | | | |
| R-114 | 0.095 | 0.085 | 0.080 | 0.074 | 0.069 | 0.064 | 0.059 | 0.054 | 0.048 | | |
| R-C318 | | | 0.085 | 0.078 | 0.071 | 0.065 | 0.058 | 0.051 | 0.044 | | |
| R-502 | 0.111 | 0.095 | 0.087 | 0.079 | 0.072 | 0.064 | 0.056 | 0.040 | | | |
| sulfur dioxide | | 0.245 | 0.232 | 0.220 | 0.207 | 0.195 | 0.182 | 0.156 | | | |
| toluene | 0.160 | 0.153 | 0.148 | 0.143 | 0.138 | 0.133 | 0.128 | 0.123 | 0.118 | 0.113 | 0.108 |
| water | | | | | 0.582 | 0.606 | 0.640 | 0.660 | 0.674 | 0.683 | 0.688 |
| xenon | 0.067 | 0.051 | 0.044 | 0.037 | 0.030 | | | | | | |

Table C.14. *Thermal conductivity for some solids as a function of temperature, k(W/m-K) (adapted from [8])*

| | T, K | | | | | | | | | | |
|---|---|---|---|---|---|---|---|---|---|---|---|
| Substance | 100 | 200 | 250 | 300 | 400 | 500 | 600 | 800 | 1000 | 1200 | 1400 |
| alumina | 520 | 80 | 60 | 36 | 27 | 20 | 16 | 10 | 7.6 | 6.3 | 5.7 |
| aluminum | 300 | 237 | 235 | 237 | 240 | 236 | 231 | 218 | | | |
| antimony | 46 | 30 | 26.7 | 24.3 | 21.3 | 19.5 | 18.3 | 16.8 | | | |
| arsenic | | 69 | 58 | 50 | 41 | 35 | | | | | |
| barium | | 19.4 | 18.6 | 18.2 | | | | | | | |
| barium titanate | 13 | 5.9 | 4.8 | 4.1 | 3.5 | 3.0 | | | | | |
| beryllia | 1490 | 480 | 376 | 272 | 198 | 153 | 114 | 60 | 33 | 22 | 17 |
| beryllium | 890 | 300 | 235 | 200 | 160 | 139 | 126 | 106 | 91 | 79 | 69 |
| bismuth | 19 | 9.7 | 8.6 | 7.9 | 7.0 | 6.6 | | | | | |
| boron | 190 | 55 | | | | | | | | | |
| brass | 46 | 74 | 92 | 114 | 134 | 143 | 146 | 150 | | | |
| cadmium | 103 | 99 | 98 | 97 | 95 | 92 | | | | | |
| calcium | | 220 | 210 | 201 | 189 | 182 | 178 | 153 | 116 | | |
| carbon | | | | | | | | | | | |
| (amorphous) | 0.67 | 1.18 | 1.40 | 1.60 | 1.87 | 2.06 | 2.19 | 2.37 | 2.5 | 2.8 | 3.2 |
| cerium | 6.0 | 9.0 | 10.3 | 11.4 | 13.3 | 15.0 | 16.5 | 19.3 | 21.8 | | |
| cesium | 40 | 37 | 36 | 36 | | | | | | | |
| chromium | 160 | 110 | 100 | 94 | 91 | 86 | 81 | 71 | 65 | 62 | 59 |
| cobalt | 165 | 120 | 110 | 100 | 85 | 75 | 67 | 58 | 52 | 49 | 42 |
| constantan | 19 | 21 | 22 | 23 | 27 | | | | | | |
| copper | 480 | 413 | 406 | 401 | 393 | 386 | 379 | 366 | 352 | 339 | |
| duralumin | 101 | 138 | 151 | 174 | 187 | 188 | | | | | |
| dysprosium | 10 | 10 | 10 | 11 | 11 | 12 | 12 | 14 | 15 | 17 | 18 |
| erbium | 12 | 15 | 15 | 14 | 14 | 14 | 14 | 15 | 16 | 17 | 19 |
| eureka | | 20 | 21 | 22 | 24 | 25 | | | | | |
| gadolinium | 15 | 12 | 11 | 10 | | | | | | | |
| German silver | 17 | 21 | 23 | 24 | 25 | 27 | | | | | |
| germanium | 230 | 97 | 75 | 60 | 43 | 34 | 27 | 20 | 17 | 17 | |
| gold | 327 | 323 | 321 | 317 | 311 | 304 | 298 | 284 | 270 | 255 | |
| hafnium | 26 | 24 | 24 | 23 | 23 | 22 | 21 | 21 | 21 | 21 | 21 |
| holmium | 14 | 15 | 16 | 16 | 17 | 18 | | | | | |
| indium | 98 | 90 | 86 | 82 | 75 | | | | | | |
| iridium | 172 | 153 | 150 | 147 | 144 | 141 | 138 | 132 | 126 | 120 | 114 |
| iron | 134 | 94 | 87 | 80 | 70 | 61 | 55 | 43 | 32 | 28 | 31 |
| lanthanum | 10 | 12 | 13 | 14 | 15 | 16 | 18 | 21 | 23 | | |
| lead | 39.7 | 36.7 | 36.0 | 35.3 | 34.0 | 32.8 | 31.4 | | | | |
| lithium | 105 | 90 | 87 | 85 | 80 | | | | | | |
| lutetium | 28 | 25 | 24 | 23 | 22 | 22 | | | | | |
| magnesium | 169 | 159 | 157 | 156 | 153 | 151 | 149 | 146 | | | |
| manganese | 5.8 | 7.2 | 7.5 | 7.8 | 7.9 | 8.0 | | | | | |
| manganin | 11 | 14 | 18 | 21 | 23 | 27 | | | | | |
| mercury | 32 | 29 | | | | | | | | | |
| molybdenum | 180 | 143 | 140 | 138 | 134 | 130 | 126 | 118 | 112 | 105 | 100 |
| monel | 16 | 20 | 22 | 23 | 25 | 27 | | | | | |
| nickel | 165 | 105 | 98 | 91 | 80 | 72 | 66 | 68 | 72 | 76 | 80 |
| niobium | 55 | 53 | 53 | 54 | 55 | 57 | 58 | 61 | 64 | 68 | 71 |

*(continued)*

Table C.14  (continued)

| Substance | T, K | | | | | | | | | | |
|---|---|---|---|---|---|---|---|---|---|---|---|
| | 100 | 200 | 250 | 300 | 400 | 500 | 600 | 800 | 1000 | 1200 | 1400 |
| osmium | 115 | 91 | 89 | 88 | 87 | 87 | | | | | |
| palladium | 77 | 72 | 72 | 72 | 74 | 76 | 80 | 87 | 94 | 102 | 107 |
| phosphorus | 31 | 18 | 14 | 12 | | | | | | | |
| platinum | 78 | 73 | 72 | 72 | 72 | 72 | 73 | 86 | 79 | 83 | 87 |
| plutonium | 3.3 | 4.8 | 5.7 | 6.7 | 9.6 | | | | | | |
| potassium | 107 | 104 | 103 | 102 | | | | | | | |
| potassium chloride | 25 | 11 | 8 | 7 | 6 | 5 | | | | | |
| praseodymium | 8 | 10 | 12 | 13 | 14 | 15 | 16 | 18 | 22 | | |
| pyrex | 0.6 | 0.9 | 1.0 | | | | | | | | |
| rhenium | 59 | 51 | 49 | 48 | 46 | 45 | 44 | 44 | 45 | 46 | 47 |
| rhodium | 190 | 154 | 152 | 150 | 146 | 141 | 136 | 127 | 121 | 116 | 112 |
| rubber | 0.1 | 0.2 | 0.2 | | | | | | | | |
| rubidium | 60 | 59 | 59 | 58 | | | | | | | |
| ruthenium | 155 | 118 | 117 | 116 | 114 | 111 | 108 | 102 | 98 | 94 | 91 |
| salt | 24 | 11 | 9 | 7 | 4 | 3 | | | | | |
| sapphire | 480 | 82 | 58 | 46 | 32 | 24 | | | | | |
| silicon | 880 | 260 | 190 | 149 | 99 | 76 | 62 | 42 | 31 | 26 | 24 |
| silver | 450 | 430 | 429 | 429 | 425 | 419 | 412 | 396 | 379 | 361 | |
| sodium | 136 | 142 | 143 | 141 | | | | | | | |
| solder | 39 | 16 | 13 | 10 | 8 | 6 | | | | | |
| sulfur | 0.6 | 0.4 | 0.3 | 0.3 | | | | | | | |
| tantalum | 59 | 58 | 57 | 58 | 58 | 59 | 59 | 59 | 60 | 61 | 62 |
| teflon | 0.23 | 0.25 | 0.25 | 0.26 | | | | | | | |
| thallium | 56 | 49 | 48 | 46 | 44 | 42 | | | | | |
| thoria | | | 16 | 13 | 10 | 8 | 6 | 5 | 4 | 3 | 3 |
| thorium | 60 | 55 | 54 | 54 | 55 | 55 | 56 | 57 | 58 | 59 | 60 |
| thulium | 14 | 16 | 17 | 17 | | | | | | | |
| tin | 85 | 73 | 70 | 67 | 62 | 60 | | | | | |
| titanium | 31 | 25 | 23 | 21 | 20 | 20 | 19 | 19 | 21 | 22 | 24 |
| titanium oxide | 20 | 9 | 9 | 8 | 6 | 5 | 4 | 3 | 3 | 3 | 3 |
| tungsten | 208 | 185 | 180 | 173 | 159 | 146 | 137 | 125 | 118 | 112 | 108 |
| uranium | 22 | 25 | 26 | 28 | 30 | 33 | 34 | 39 | 44 | 49 | 56 |
| vanadium | 36 | 31 | 31 | 31 | 31 | 32 | 33 | 36 | 38 | 41 | 43 |
| zinc | 117 | 118 | 118 | 116 | 111 | 107 | 103 | | | | |
| zirconium | 33 | 25 | 24 | 23 | 22 | 21 | 21 | 21 | 23 | 26 | 28 |

Table C.15.  *Thermal conductivity and density for some building and insulation materials (mostly porous) (adapted from [1, 7, 11])*

| Substance | $\langle \rho \rangle$, kg/m$^3$ | $T$, K | $\langle k \rangle$, W/m-K |
|---|---|---|---|
| aerogel, silica, opacified | 140 | 393 | 0.023 |
|  |  | 563 | 0.045 |
| asbestos-cement boards | 1920 | 293 | 0.74 |
| asbestos sheets | 889 | 324 | 0.17 |
| asbestos slate | 1790 | 273 | 0.15 |
|  | 1790 | 333 | 0.20 |
| asbestos | 469 | 73 | 0.074 |
|  | 469 | 273 | 0.16 |
|  | 580 | 273 | 0.15 |
|  | 580 | 373 | 0.19 |
|  | 580 | 473 | 0.21 |
|  | 580 | 673 | 0.22 |
|  | 697 | 73 | 0.16 |
|  | 697 | 273 | 0.23 |
| aluminum foil (7 air spaces per 10 cm) | 3 | 311 | 0.043 |
|  |  | 450 | 0.066 |
| ashes, wood |  | 273–373 | 0.071 |
| asphalt | 2110 | 293 | 0.74 |
| *blankets and felts:* |  |  |  |
| aluminosilicate fibers | 96–128 | 297 | 0.036 |
| *bricks:* |  |  |  |
| alumina (92–99 Al$_2$O$_3$) fused |  | 700 | 3.1 |
| alumina (64–65 Al$_2$O$_3$) |  | 1588 | 4.7 |
|  | 1840 | 1073 | 1.1 |
|  | 1840 | 1373 | 1.1 |
| building brick work |  | 293 | 0.69 |
| carbon | 1550 |  | 5.2 |
| chrome brick (32 Cr$_2$O$_3$) | 3200 | 473 | 1.2 |
|  | 3200 | 923 | 1.5 |
|  | 3200 | 1588 | 1.7 |
| diatomaceous earth, natural, across strata | 444 | 477 | 0.088 |
|  | 444 | 1144 | 0.13 |
| diatomaceous earth, natural, parallel to strata | 444 | 477 | 0.14 |
|  | 444 | 1144 | 0.18 |
| diatomaceous earth, molded and fired | 610 | 477 | 0.24 |
|  | 610 | 1144 | 0.31 |
| diatomaceous earth and clay, molded and fired | 678 | 477 | 0.24 |
|  | 678 | 1144 | 0.33 |
| diatomaceous earth, high burn, large pores | 590 | 473 | 0.23 |
|  | 590 | 1273 | 0.59 |
| fire clay (Missouri) |  | 473 | 1.0 |
|  |  | 873 | 1.5 |
|  |  | 1273 | 1.6 |
|  |  | 1673 | 1.8 |

*(continued)*

Table C.15  *(continued)*

| Substance | $\langle \rho \rangle$, kg/m$^3$ | $T$, K | $\langle k \rangle$, W/m-K |
|---|---|---|---|
| kaolin insulating brick | 430 | 773 | 0.26 |
|  | 430 | 1423 | 0.45 |
| kaolin insulating fire brick | 300 | 473 | 0.087 |
|  | 300 | 1033 | 0.196 |
| magnesite (86.8 MgO, 6.3 Fe$_2$O$_3$, 3 CaO, 2.6 SiO$_2$) | 2530 | 477 | 3.8 |
|  | 2530 | 923 | 2.8 |
|  | 2530 | 1473 | 1.9 |
| silicon carbide brick, recrystallized | 2070 | 873 | 18.5 |
|  | 2070 | 1073 | 16 |
|  | 2070 | 1273 | 14 |
|  | 2070 | 1473 | 12 |
|  | 2070 | 1673 | 11 |
| calcium carbonate, natural | 2590 | 303 | 2.3 |
| white marble |  |  | 2.9 |
| chalk | 1500 |  | 0.69 |
| calcium sulfate (4H$_2$O), artificial | 1360 | 313 | 0.38 |
| plaster (artificial) | 2110 | 348 | 0.74 |
| (building) | 1250 | 298 | 0.43 |
| cambric (varnished) |  | 308 | 0.16 |
| carbon, gas |  | 273–373 | 3.5 |
| carbon stock | 1500 | 89 | 0.95 |
|  |  | 273 | 6.2 |
| cardboard, corrugated |  |  | 0.064 |
| celluloid | 1400 | 303 | 0.21 |
| cements (insulating) | 380–480 | 297 | 0.071 |
| charcoal flakes | 191 | 353 | 0.074 |
|  | 240 | 353 | 0.088 |
| clinker (granular) |  | 273–973 | 0.47 |
| coke, petroleum |  | 373 | 5.9 |
|  |  | 773 | 5.0 |
| coke, petroleum (20–100 mesh) | 990 | 673 | 0.95 |
| coke (powdered) |  | 273–373 | 0.19 |
| concrete (cinder) |  | 303 | 0.35 |
| (stone) |  | 303 | 0.93 |
| (1:4 dry) |  | 303 | 0.76 |
| cotton wool | 80 | 303 | 0.042 |
| cork board | 160 | 311 | 0.043 |
| cork (regranulated) | 130 | 1144 | 0.045 |
| (ground) | 150 | 477 | 0.043 |

Table C.15  *(continued)*

| Substance | $\langle \rho \rangle$, kg/m$^3$ | $T$, K | $\langle k \rangle$, W/m-K |
|---|---|---|---|
| diatomaceous earth powder, coarse | 320 | 1144 | 0.062 |
| | 320 | 477 | 0.14 |
| fine | 275 | 1144 | 0.069 |
| | 444 | | 0.13 |
| molded pipe covering | 416 | | 0.088 |
| | 990 | 1144 | 0.40 |
| dolomite | 2680 | 323 | 1.7 |
| ebonite | | | 0.17 |
| enamel, silicate | 610 | | 0.87–1.3 |
| felt, wool | 330 | 303 | 0.052 |
| fiber insulating board | 237 | 294 | 0.048 |
| fiber, red | 1290 | 298 | 0.47 |
| (with binder, baked) | | 293–370 | 0.17 |
| fiberglass, paper-faced | 40 | 300 | 0.035 |
| gas carbon | | 273–373 | 3.5 |
| glass | | | 0.35-1.3 |
| borosilicate type | 2230 | 303–348 | 1.1 |
| window glass | | | 0.52–1.1 |
| soda glass | | | 0.52–0.76 |
| granite (74% w $SiO_2$, 14% w $Al_2O_3$) | 2,600 | 293 | 2.1–2.4 |
| graphite, longitudinal | | 293 | 160 |
| powdered, through 100 mesh | 480 | 313 | 0.18 |
| gypsum (molded and dry) | 1200 | 293 | 0.43 |
| hair felt (perpendicular to fibers) | 270 | 303 | 0.036 |
| ice | 921 | 273 | 2.3 |
| infusorial earth, see diatomaceous earth | | | |
| kapok | 14 | 293 | 0.035 |
| lampblack | 160 | 313 | 0.066 |
| lava | | | 0.85 |
| leather, sole | 999 | | 0.16 |
| limestone (15.3 vol. % $H_2O$) | 1000 | 297 | 0.93 |
| linen | | 303 | 0.087 |
| magnesia (powdered) | 796 | 320 | 0.61 |
| magnesia (light carbonate) | 210 | 294 | 0.059 |
| magnesium oxide (compressed) | 799 | 293 | 0.55 |
| marble (95% w carbonate) | 2,700 | 293 | 2.2–2.3 |
| mica (perpendicular to planes) | | 323 | 0.43 |
| mill shavings | | | 0.057–0.087 |
| mineral wool | 150 | 303 | 0.039 |
| | 316 | 303 | 0.042 |

*(continued)*

Table C.15  *(continued)*

| Substance | $\langle \rho \rangle$, kg/m$^3$ | $T$, K | $\langle k \rangle$, W/m-K |
|---|---|---|---|
| paper | | | 0.13 |
| paraffin wax | | 273 | 0.24 |
| petroleum coke | | 373 | 5.9 |
| | | 773 | 5.0 |
| pipe insulation (slag or glass) | 48–64 | 297 | 0.033 |
| | 160–240 | 310 | 0.048 |
| polyestyrene, rigid | 30–60 | 300 | 0.028 |
| polyurethane, foam (formed in place) | 70 | 300 | 0.026 |
| porcelain (kaolin, quartz, and feldspar) | | 473 | 1.5 |
| portland cement | | 363 | 0.29 |
| pumice stone | | 294–339 | 0.24 |
| pyroxylin plastics | | | 0.13 |
| rubber (hard) | 1200 | 273 | 0.15 |
| (para) | | 294 | 0.19 |
| (soft) | | 294 | 0.13–0.16 |
| sand (dry) | 1520 | 293 | 0.33 |
| sandstone | 2200 | 313 | 1.83 |
| sawdust | 190 | 294 | 0.05 |
| silk | 100 | | 0.045 |
| varnished | | 311 | 0.17 |
| slag, blast furnace | | 297–400 | 0.11 |
| slag wool | 190 | 303 | 0.038 |
| slate | | 367 | 1.5 |
| snow | 556 | 273 | 0.47 |
| soil (earth) | 1500 | 300 | 0.52 |
| sulfur (monoclinic) | | 373 | 0.16–0.17 |
| (rhombic) | | 294 | 0.28 |
| wall board, insulating type | 237 | 294 | 0.048 |
| wall board, stiff paste board | 690 | 303 | 0.07 |
| wood shavings | 140 | 303 | 0.059 |
| wood (across grain): | | | |
| balsa | 11–0–130 | 303 | 0.043–0.052 |
| oak | 825 | 288 | 0.21 |
| maple | 716 | 323 | 0.19 |
| pine, white | 545 | 288 | 0.15 |
| teak | 641 | 288 | 0.17 |
| white fir | 450 | 333 | 0.11 |
| wood (parallel to grain): | | | |
| pine | 551 | 294 | 0.35 |
| wool, animal | 110 | 303 | 0.036 |

## Thermophysical Properties of Solids

Table C.16. *Thermophysical properties for some metallic solids (at $T = 300$ K) (adapted from [9])*

| Substance | $T_{sl}$, K | $\rho$, kg/m³ | $c_p$, J/kg-K | $k$, W/m-K | $\alpha$, m²/s | $\beta_s$, 1/K |
|---|---|---|---|---|---|---|
| aluminum | | | | | | $2.25 \times 10^{-5}$ |
|   pure | 933 | 2702 | 903 | 237 | $97.1 \times 10^{-6}$ | |
|   Duralumin | 775 | 2770 | 875 | 174 | $71.8 \times 10^{-6}$ | |
|     (4.4 Cu, 1.0 Mg, 0.75 Mn, 0.4 Si) | | | | | | |
|   alloy 195, cast (4.5 Cu) | | 2790 | 883 | 168 | $68.1 \times 10^{-6}$ | |
| beryllium | 1550 | 1850 | 1825 | 200 | $59.2 \times 10^{-6}$ | $1.13 \times 10^{-5}$ |
| bismuth | 545 | 9780 | 122 | 7.9 | $6.59 \times 10^{-6}$ | $1.34 \times 10^{-5}$ |
| cadmium | 594 | 8650 | 231 | 97 | $48.4 \times 10^{-6}$ | $3.08 \times 10^{-5}$ |
| copper | | | | | | |
|   pure | 1358 | 8933 | 385 | 401 | $117 \times 10^{-6}$ | $1.67 \times 10^{-5}$ |
|   brass (30 Zn) | 1188 | 8530 | 380 | 111 | $34.2 \times 10^{-6}$ | $1.80 \times 10^{-5}$ |
|   commercial bronze (10 Al) | 1293 | 8800 | 420 | 52 | $14.1 \times 10^{-6}$ | |
|   constantan (40 Ni) | | 8920 | 420 | 22.7 | $6.06 \times 10^{-6}$ | $1.69 \times 10^{-5}$ |
|   constantan (45 Ni) | | 8860 | | 23 | | |
|   electrolytic tough pitch | | 8950 | 385 | 386 | $112 \times 10^{-6}$ | |
|     (Cu + Ag 99.90 minimum) | | | | | | |
|   German silver (15 Ni, 22 Zn) | | 8618 | 410 | 116 | $32.8 \times 10^{-6}$ | |
| gold | 1336 | 19300 | 129 | 317 | $127 \times 10^{-6}$ | $1.42 \times 10^{-5}$ |
| iron | | | | | | $1.18 \times 10^{-5}$ |
| pure | 1810 | 7870 | 447 | 80.2 | $22.8 \times 10^{-6}$ | |
|   Armco (99.75 pure) | | 7870 | 447 | 72.7 | $20.7 \times 10^{-6}$ | |
|   cast (4 C) | | 7272 | 420 | 51 | $16.7 \times 10^{-6}$ | |
| carbon steel | | | | | | |
|   AISI 1010 | | 7830 | 434 | 64 | $18.8 \times 10^{-6}$ | $1.15 \times 10^{-5}$ |
|     (0.1 C, 0.4 Mn) | | | | | | |
|   AISI 1042, annealed | | 7840 | 460 | 50 | $13.9 \times 10^{-6}$ | |
|     (0.42 C, 0.64 Mn, 0.063 Ni, 0.13 Cu) | | | | | | |
|   AISI 4130, hardened | | 7840 | 460 | 43 | $11.9 \times 10^{-6}$ | |
|     and tempered | | | | | | |
|     (0.3 C, 0.5 Mn, 0.3 Si, 0.95 Cr, 0.5 Mo) | | | | | | |
| stainless steel | | | | | | |
|   AISI 302 | | 8055 | 480 | 15 | $3.88 \times 10^{-6}$ | |
|     (0.15 C, 2 Mn, 1 Si, 16–18 Cr, 6–8 Ni) | | | | | | |
|   AISI 304 | 1670 | 7900 | 477 | 15 | $3.98 \times 10^{-6}$ | |
|     (0.08 C, 2 Mn, 1 Si, | | | | | | |
|       18–20 Cr, 8–10 Ni) | | | | | | |

*(continued)*

Table C.16  *(continued)*

| Substance | $T_{sl}$, K | $\rho$, kg/m$^3$ | $c_p$, J/kg-K | $k$, W/m-K | $\alpha$, m$^2$/s | $\beta_s$, 1/K |
|---|---|---|---|---|---|---|
| stainless steel (Continued) | | | | | | $1.71 \times 10^{-5}$ |
| AISI 316 | | 8238 | 468 | 13 | $3.37 \times 10^{-6}$ | |
| (0.08 C, 2 Mn, 1 Si, 16-18 Cr, | | | | | | |
| 10-14 Ni, 2-3 Mo) | | | | | | |
| AISI 410 | | 7770 | 460 | 25 | $7.00 \times 10^{-6}$ | |
| (0.15 C, 1 Mn, 1 Si, 11.5-13 Cr) | | | | | | |
| lead | 601 | 11340 | 129 | 35.3 | $24.1 \times 10^{-6}$ | $2.93 \times 10^{-5}$ |
| magnesium | 929 | 1740 | 1024 | 156 | $87.6 \times 10^{-6}$ | $8.2 \times 10^{-6}$ |
| manganese | 300 | 7740 | 448 | 7.8 | $2.25 \times 10^{-6}$ | $2.28 \times 10^{-5}$ |
| molybdenum | 2894 | 10240 | 251 | 138 | $53.6 \times 10^{-6}$ | $4.8 \times 10^{-6}$ |
| nickel | | | | | | |
| pure | 1728 | 8900 | 444 | 91 | $23.0 \times 10^{-6}$ | $1.19 \times 10^{-5}$ |
| Inconel-X-750 | 1665 | 8510 | 439 | 11.7 | $3.13 \times 10^{-6}$ | |
| (15.5 Cr, 1 Nb, 2.5 Ti, 0.7 Al, 7 Fe) | | | | | | |
| nichrome (20 Cr) | 1672 | 8314 | 460 | 13 | $3.40 \times 10^{-6}$ | |
| nimonic (20 Cr, 0.4 Ti) | | 8370 | 461 | 11.7 | $3.03 \times 10^{-6}$ | |
| Hasteloy B (38 Mo, 5 Fe) | | 9240 | 381 | 12.2 | $3.47 \times 10^{-6}$ | |
| cupro-nickel (50 Cu) | | 8800 | 421 | 19.5 | $5.26 \times 10^{-6}$ | $1.31 \times 10^{-5}$ |
| Chromel-P (10 Cu) | | 8730 | | 17 | | |
| Alumel (2 Mn, 2 Al) | | 8600 | | 48 | | |
| palladium | 1827 | 12020 | 244 | 71.8 | $24.5 \times 10^{-6}$ | $11.8 \times 10^{-5}$ |
| platinum | | | | | | |
| pure | 2045 | 21450 | 133 | 71.6 | $25.1 \times 10^{-6}$ | $8.8 \times 10^{-6}$ |
| 60 Pt-40 Rh (40 Rh) | 1800 | 16630 | 162 | 47 | $174 \times 10^{-6}$ | |
| silicon | 1685 | 2230 | 712 | 148 | $89.2 \times 10^{-6}$ | $2.6 \times 10^{-6}$ |
| silver | 1235 | 10500 | 235 | 429 | $174 \times 10^{-6}$ | $1.97 \times 10^{-5}$ |
| tantalum | 3293 | 16600 | 140 | 57.5 | $24.7 \times 10^{-6}$ | $6.3 \times 10^{-6}$ |
| tin | 505 | 7310 | 227 | 66.6 | $40.1 \times 10^{-6}$ | $2.34 \times 10^{-5}$ |
| titanium | | | | | | |
| pure | 1993 | 4500 | 522 | 21.9 | $9.32 \times 10^{-6}$ | $8.0 \times 10^{-5}$ |
| Ti-6Al-4V | | 4420 | 610 | 5.8 | $2.15 \times 10^{-6}$ | |
| Ti-2Al-2Mn | | 4510 | 466 | 8.4 | $4.0 \times 10^{-6}$ | |
| tungsten | 3660 | 19300 | 132 | 174 | $68.3 \times 10^{-6}$ | $4.5 \times 10^{-6}$ |
| zinc | 693 | 7140 | 389 | 116 | $41.8 \times 10^{-6}$ | $3.24 \times 10^{-5}$ |
| zirconium | | | | | | |
| pure | 2125 | 6570 | 278 | 22.7 | $12.4 \times 10^{-6}$ | $5.7 \times 10^{-6}$ |
| zircaloy-4 | | 6560 | 285 | 14.2 | $7.60 \times 10^{-6}$ | |
| (1.2–1.75 Sn, 0.18-0.24 Fe, | | | | | | |
| 0.07-0.13 Cr) | | | | | | |

Table C.17. *Thermophysical properties for some nonmetallic solids (adapted from [9])*

| Substance | $T$, K | $\rho$, kg/m³ | $c_p$, J/kg-K | $k$, W/m-K | $\alpha$, m²/s | $\beta_s$, 1/K |
|---|---|---|---|---|---|---|
| acrylic (general purpose) | 293 | 1180 | 1500 | 0.2 | $0.1130 \times 10^{-6}$ | $7.45 \times 10^{-5}$ |
| aerogel, silica | 393 | 140 | | 0.023 | | $6.5 \times 10^{-6}$ |
| alumina (aluminum oxide, polycrystalline) | 293 | 3975 | 765 | 36.0 | $11.9 \times 10^{-6}$ | $4.6 \times 10^{-6}$ |
| asbestos | 73 | 469 | | 0.074 | | |
| | 273 | 469 | | 0.16 | | |
| | 273 | 577 | 816 | 0.15 | | |
| | 373 | 577 | 816 | 0.19 | | |
| | 473 | 577 | | 0.21 | | |
| | 673 | 577 | | 0.22 | | |
| | 73 | 697 | | 0.16 | | |
| | 273 | 697 | | 0.23 | | |
| asphalt | 293 | 2110 | 920 | 0.06 | $0.03 \times 10^{-6}$ | |
| bakelite | 293 | 1270 | 1590 | 0.23 | $0.11 \times 10^{-6}$ | $2.2 \times 10^{-5}$ |
| beryllia | 293 | 3000 | 1030 | 270 | $88 \times 10^{-6}$ | |
| brick | 293 | 1925 | 835 | 0.72 | $0.45 \times 10^{-6}$ | |
| brick, dry | 293 | 1760–1810 | 837 | 0.38–0.52 | $0.28$–$0.34 \times 10^{-6}$ | |
| brick, fireclay | 293 | 2640 | 960 | 1.0 | $0.39 \times 10^{-6}$ | |
| carbon, amorphous | 293 | 1950 | 724 | 1.6 | $1.13 \times 10^{-6}$ | |
| cardboard, corrugated | | | | 0.064 | | |
| clay | 293 | 1460 | 879 | 1.28 | $1.0 \times 10^{-6}$ | |
| coal, anthracite | 293 | 1200–1510 | 1260 | 0.26 | $0.13$–$0.15 \times 10^{-6}$ | |
| coal, powdered | 303 | 737 | 1300 | 0.12 | $0.13 \times 10^{-6}$ | |
| concrete | 293 | 1910–2310 | 879 | 0.81–1.40 | $0.49$–$0.70 \times 10^{-6}$ | $1.44 \times 10^{-5}$ |
| cork, board | 303 | 160 | | 0.043 | | |
| cork, expanded scrap | 293 | 44.9–119 | 1880 | 0.036 | $0.15$–$0.44 \times 10^{-6}$ | |
| cork, ground | 303 | 151 | | 0.043 | | |
| cotton | 293 | 80.1 | 1300 | 0.059 | $0.58 \times 10^{-6}$ | |
| diamond (a crystalline carbon) | 293 | 3500 | 509 | 2300 | $1290 \times 10^{-6}$ | |
| diatomaceous earth | 311 | 320 | | 0.062 | | |
| | 1044 | 320 | | 0.14 | | |
| earth, coarse gravelly | 293 | 2050 | 1840 | 0.52 | $1.4 \times 10^{-6}$ | |
| felt, wool | 303 | 330 | | 0.052 | | |
| fiber, insulating board | 294 | 237 | | 0.048 | | |
| fiber, red | 293 | 1290 | | 0.47 | | |
| glass, borosilicate ($SiO_2$, $B_2O_3$, $CaO$, ...) | 303 | 2230 | 835 | 1.1 | $0.5907 \times 10^{-6}$ | $7.0 \times 10^{-6}$ |
| glass, soda-lime ($SiO_2$, $NaO$, $K_2O$, $CaO$, ...) | 300 | 2500 | 750 | 1.4 | $0.7467 \times 10^{-6}$ | |
| glass, wool | 293 | 200 | 670 | 0.040 | $0.28 \times 10^{-6}$ | |
| glass plate | 293 | 2710 | 837 | 0.76 | $0.34 \times 10^{-6}$ | $7.0 \times 10^{-6}$ |
| granite | | | | 1.7–4.0 | | |
| graphite (pyrolytic) | | | | | | |
|    perpendicular to layers | 300 | 2210 | 1122 | 5.70 | $2.299 \times 10^{-6}$ | |
|    parallel to layers | 300 | 2210 | 1122 | 19.50 | $784.4 \times 10^{-6}$ | |
|    hardboard | 293 | 1000 | 1380 | 0.15 | $0.11 \times 10^{-6}$ | |
| ice | 273 | 913 | 1930 | 2.2 | $1.2 \times 10^{-6}$ | |

*(continued)*

Table C.17  *(continued)*

| Substance | $T$, K | $\rho$, kg/m$^3$ | $c_p$, J/kg-K | $k$, W/m-K | $\alpha$, m$^2$/s | $\beta_s$, 1/K |
|---|---|---|---|---|---|---|
| magnesite | 293 | 3025 | 1130 | 4.0 | $1.2 \times 10^{-6}$ | |
| magnesia | 293 | 3635 | 943 | 48 | $14 \times 10^{-6}$ | $1.35 \times 10^{-5}$ |
| marble | 293 | 2500–2710 | 808 | 2.8 | $1.4 \times 10^{-6}$ | |
| nylon 6/6 | 293 | 1140 | 1700 | 0.25 | $0.129 \times 10^{-6}$ | $7.65 \times 10^{-5}$ |
| paper | 293 | 930 | 1340 | 0.011 | $0.01 \times 10^{-6}$ | |
| plaster board | 293 | 800 | | 0.17 | | |
| Plexiglas | 300 | 1190 | 1465 | 0.19 | $0.109 \times 10^{-6}$ | $7.4 \times 10^{-5}$ |
| plywood | 293 | 540 | 1220 | 0.12 | $0.18 \times 10^{-6}$ | |
| polystyrene | 293 | 1050 | 1800 | 0.14 | $0.07407 \times 10^{-6}$ | $7.98 \times 10^{-5}$ |
| polyester (thermoset) | 293 | 1650 | 1800 | 0.17 | $0.05724 \times 10^{-6}$ | |
| polyurethane | 293 | 1060 | 1600 | 0.2 | $0.1179 \times 10^{-6}$ | $1.10 \times 10^{-4}$ |
| polyurethane foam (formed in place) | 293 | 70 | 1050 | 0.026 | $0.36 \times 10^{-6}$ | |
| Pyrex | 293 | 2250 | 835 | 1.4 | $0.74 \times 10^{-6}$ | |
| quartz (see silica crystalline) | | | | | | |
| rubber (elastomer), foam | 293 | 70 | | 0.03 | | |
| rubber (elastomer), hard | 273 | 1200 | 2010 | 0.15 | $0.06219 \times 10^{-6}$ | |
| rubber (elastomer), soft | 300 | 1100 | 2010 | 0.13 | $0.05880 \times 10^{-6}$ | $7.2 \times 10^{-5}$ |
| tissue (human) | | | | | | |
| skin | 300 | | | 0.37 | | |
| fat | 300 | 920 | | 0.20 | | |
| muscle | 300 | 1040 | | 0.41 | | |
| salt | 293 | | 854 | 7.1 | | |
| sandstone | 293 | 2160–2310 | 712 | 1.6–2.1 | $1.1$–$1.3 \times 10^{-6}$ | |
| sapphire (mostly crystalline alumina) | 293 | 3975 | 765 | 46 | $15 \times 10^{-6}$ | |
| silica (crystalline, quartz) perpendicular to crystalline axis | 300 | 2650 | 745 | 6.21 | $3.145 \times 10^{-6}$ | $7.5 \times 10^{-6}$ |
| parallel to crystalline axis | 300 | 2650 | 745 | 10.4 | $5.268 \times 10^{-6}$ | |
| silica (amorphous) | 293 | 2220 | 745 | 1.38 | $0.843 \times 10^{-6}$ | $4 \times 10^{-7}$ |
| silicon carbide (high purity) | 293 | 3160 | 675 | 490 | $230 \times 10^{-6}$ | |
| silicon carbide (sintered) | 293 | 3160 | 715 | 114 | $50.46 \times 10^{-6}$ | $1.1 \times 10^{-6}$ |
| silicon carbide (sintered $\alpha$) | 473 | 3160 | 675 | 102.6 | $48.1 \times 10^{-6}$ | |
| silicon carbide (CVD) | 273 | 3160 | 675 | 40 | $18.8 \times 10^{-6}$ | $2.2 \times 10^{-6}$ |
| silk | 293 | 57.7 | 1380 | 0.036 | $0.44 \times 10^{-6}$ | |
| soil | 293 | 2050 | 1840 | 0.52 | $0.14 \times 10^{-6}$ | |
| Teflon | 293 | 2200 | 350 | 0.26 | $0.34 \times 10^{-6}$ | |
| thoria | 293 | 4160 | 710 | 14 | $4.7 \times 10^{-6}$ | |
| vermiculite | 293 | 120 | 840 | 0.06 | $0.60 \times 10^{-6}$ | |
| wood, fir (20% moisture) radial | 293 | 416–421 | 2720 | 0.14 | $0.12 \times 10^{-6}$ | $5.4 \times 10^{-5}$ |
| wood, oak radial | 293 | 609-801 | 2390 | 0.17–0.21 | $0.11$–$0.12 \times 10^{-6}$ | $4.9 \times 10^{-6}$ |
| wood, pine | 293 | 525 | 2750 | 0.12 | $0.54 \times 10^{-6}$ | $5.4 \times 10^{-4}$ |
| yttria (tablet) | 300 | 5046 | 545 | 4.64 | $1.295 \times 10^{-6}$ | $8.5 \times 10^{-6}$ |
| yttria (crystalline) | 300 | 5061 | 545 | 15 | $5.438 \times 10^{-6}$ | |
| zirconia (plasma sprayed, also stabilized zirconia) | 373 | 5680 | 610 | 1.675 | $0.4834 \times 10^{-6}$ | $7.0 \times 10^{-6}$ |
| zirconia (pure) | 373 | 5680 | 610 | 6.15 | $1.775 \times 10^{-6}$ | |

## Surface-Radiation Properties

Table C.18. *Total, hemispherical emissivity for some surfaces (adapted from [7])*

| Substance | $T$, K | $\epsilon_r$ |
|---|---|---|
| *Metals and their oxides* | | |
| aluminum, highly polished | 488–848 | 0.039–0.057 |
| aluminum, commercial sheet | 373 | 0.09 |
| aluminum, rough polish | 373 | 0.18 |
| aluminum, oxidized at 600°C | 473–873 | 0.11–0.19 |
| alumina (aluminum oxide) | 400–600 | 0.78–0.69 |
| | 600–1000 | 0.69–0.54 |
| | 1000–1400 | .54–.42 |
| aluminum alloys | 298 | 0.09 |
| antimony, polished | 308–533 | 0.28–0.31 |
| bismuth, bright | 348 | 0.34 |
| brass, highly polished | 518–628 | 0.028–0.037 |
| brass, oxidized at 600°C | 473–873 | 0.61–0.59 |
| chromium, polished | 313–1373 | 0.08–0.36 |
| copper, polished | 388 | 0.023 |
| cuprous oxide | 1073–1373 | 0.66–0.54 |
| molten copper | 1348–1548 | 0.16–0.13 |
| gold, polished | 498–898 | 0.018–0.035 |
| Inconel | 503–1273 | 0.32–0.78 |
| iron, polished | 698–1298 | 0.14–0.38 |
| iron, oxidized surfaces | 293–1473 | 0.61–0.89 |
| iron, molten | 1573–1673 | 0.29 |
| lead | 398–498 | 0.057–0.075 |
| lead, oxidized at 323 K | 473 | 0.63 |
| magnesium, polished | 308–533 | 0.07–0.13 |
| magnesium oxide | 548–1098 | 0.55–0.20 |
| mercury | 273–373 | 0.09–0.12 |
| molybdenum, polished | 308–3023 | 0.05–0.29 |
| monel | 298–1148 | 0.14–0.53 |
| nickel, polished | 297 | 0.045 |
| nickel oxide | 923–1528 | 0.59–0.86 |
| nickel-copper, polished | 373 | 0.059 |
| nichrome wire, bright | 323–1273 | 0.65–0.79 |
| nichrome wire, oxidized | 323–773 | 0.95–0.98 |
| platinum, polished | 498–898 | 0.054–0.104 |
| silver, polished | 498–898 | 0.020–0.032 |
| stainless steel, polished | 373 | 0.074 |
| tantalum | 1613–3273 | 0.19–0.31 |
| thorium oxide | 548–773 | 0.58–0.36 |
| | 773–1098 | 0.36–0.21 |
| tin, polished | 298 | 0.043, 0.064 |
| tungsten, polished | 373 | 0.066 |
| zinc, polished | 498–598 | 0.045–0.053 |
| zinc, oxidized at 673 K | 673 | 0.11 |

*(continued)*

Table C.18  *(continued)*

| Substance | $T, \mathrm{K}$ | $\epsilon_r$ |
|---|---|---|
| *Refractories, building materials, paints, and miscellaneous* | | |
| asbestos | 296–643 | 0.93–0.96 |
| brick, building | 1273 | 0.45 |
| brick, fireclay | 1273 | 0.75 |
| brick, white refractory | 1373 | 0.29 |
| carbon, filament | 1313–1678 | 0.526 |
| carborundum (87% SiC; density 2.3 g/cm$^3$) | 1283–1673 | 0.92–0.81 |
| concrete tiles | 1273 | 0.63 |
| concrete, rough | 311 | 0.94 |
| enamel, white fused | 293 | 0.90 |
| glass, smooth | 293 | 0.94 |
| glass, Pyrex, lead, and soda | 533–813 | 0.95–0.85 |
| gypsum, 5 mm thick on smooth or blackened plate | 293 | 0.903 |
| ice, smooth | 273 | 0.966 |
| ice, rough crystals | 273 | 0.985 |
| magnesite refractory brick | 1273 | 0.38 |
| marble, light gray, polished | 293 | 0.93 |
| paints, lacquers, and varnishes | 293–418 | 0.80–0.98 |
| paints and lacquers, aluminum | 296–588 | 0.27–0.67 |
| paper | 293–308 | 0.91–0.95 |
| plaster, rough lime | 283–361 | 0.91 |
| porcelain, glazed | 293 | 0.92 |
| quartz, rough, fused | 293 | 0.93 |
| quartz, glass, 1.98 mm thick | 553–1113 | 0.90–0.41 |
| quartz, glass, 6.88 mm thick | 553–1112 | 0.93–0.47 |
| quartz, opaque | 553–1113 | 092–0.68 |
| rubber, hard, glossy plate | 296 | 0.94 |
| rubber, soft, gray, rough (reclaimed) | 298 | 0.86 |
| sandstone | 308–533 | 0.83–0.90 |
| serpentine, polished | 296 | 0.90 |
| silica (98% SiO$_2$; Fe-free); 10 $\mu$m grain | 1283–1838 | 0.42–0.33 |
| silicon carbide | 423–923 | 0.83–0.96 |
| slate | 308 | 0.67–0.80 |
| soot, candle | 363–533 | 0.95 |
| Teflon | 300 | 0.85 |
| water | 273–373 | 0.95–0.963 |
| wood, sawdust | 308 | 0.75 |
| wood, oak, planed | 293 | 0.90 |
| wood, beech | 343 | 0.94 |
| zirconium silicate | 513–1103 | 0.92–0.52 |

Table C.19. *Total, hemispherical emissivity and solar absorptivity for some surfaces (adapted from [7])*

| Substance | $T$, K | $\epsilon_r$ | $\alpha_r$ |
|---|---|---|---|
| aluminum foil | 293 | 0.025 | 0.10 |
| aluminum alloy, on a DC-6 aircraft | 338 | 0.16 | 0.54 |
| aluminum, Reflectal | 248 | 0.79 | 0.23 |
| beryllium | 423 | 0.18 | 0.77 |
| black paint | 248 | 0.95 | 0.975 |
| chromium plate | 308 | 0.15 | 0.78 |
| copper, electroplated | 293 | 0.03 | 0.47 |
| copper, black-oxidized in Ebonol C | 308 | 0.16 | 0.91 |
| glass | 248 | 0.83 | 0.13 |
| gold | 293 | 0.025 | 0.21 |
| nickel, electroplated | 293 | 0.03 | 0.22 |
| nickel, Tabor solar absorber | 308 | 0.11 | 0.85 |
| silica, Corning Glass | 308 | 0.84 | 0.08 |
| silicon solar cell | 308 | 0.32 | 0.94 |
| silver Chromatone paint | 293 | 0.24 | 0.20 |
| stainless steel, type 410 | 308 | 0.13 | 0.76 |
| Teflon | 300 | 0.85 | 0.12 |
| titanium, oxidized at 450°C | 308 | 0.21 | 0.80 |
| titanium, anodized | 248 | 0.73 | 0.51 |
| white epoxy paint | 248 | 0.88 | 0.25 |
| white potassium zirconium silicate inorganic spacecraft coating | 293 | 0.89 | 0.13 |
| zinc, blackened | 308 | 0.12 | 0.89 |

## Mass Transfer and Thermochemical Properties of Gaseous Fuels

Table C.20(a). *Mass diffusivity for some gaseous species at one atm pressure (adapted from [9,14])**

Table C.20(b). *Temperature dependence of mass diffusion coefficient for some gaseous species, in air at one atm pressure (adapted from [11])**

Table C.21(a). *Thermochemical properties for some gaseous fuels mixed and reacting with air at thermodynamic STP (adapted from [2])**

Table C.21(b). *Thermochemical properties of some industrial gaseous fuels at $T = 289$ K and one atm pressure (adapted from [16])**

* This table is found on the Web site www.cambridge.org/kaviany.

## Thermophysical Properties of Fluids

Table C.22.  *Thermophysical properties for some gases at one atm pressure (adapted from [11])*

| $T$, K | $k$, W/m-K | $\rho$, kg/m$^3$ | $c_p$, J/kg-K | $\nu$, m$^2$/s | $\alpha$, m$^2$/s | Pr |
|---|---|---|---|---|---|---|
| | | | air | | | |
| 150 | 0.0158 | 2.355 | 1017 | $4.52 \times 10^{-6}$ | $6.600 \times 10^{-6}$ | 0.69 |
| 200 | 0.0197 | 1.767 | 1009 | $7.69 \times 10^{-6}$ | $11.05 \times 10^{-6}$ | 0.69 |
| 250 | 0.0235 | 1.413 | 1009 | $11.42 \times 10^{-6}$ | $16.48 \times 10^{-6}$ | 0.69 |
| 260 | 0.0242 | 1.360 | 1009 | $12.23 \times 10^{-6}$ | $17.64 \times 10^{-6}$ | 0.69 |
| 270 | 0.0249 | 1.311 | 1009 | $13.06 \times 10^{-6}$ | $18.82 \times 10^{-6}$ | 0.69 |
| 280 | 0.0255 | 1.265 | 1008 | $13.91 \times 10^{-6}$ | $20.00 \times 10^{-6}$ | 0.69 |
| 290 | 0.0261 | 1.220 | 1007 | $14.77 \times 10^{-6}$ | $21.24 \times 10^{-6}$ | 0.69 |
| 300 | 0.0267 | 1.177 | 1005 | $15.66 \times 10^{-6}$ | $22.57 \times 10^{-6}$ | 0.69 |
| 310 | 0.0274 | 1.141 | 1005 | $16.54 \times 10^{-6}$ | $23.89 \times 10^{-6}$ | 0.69 |
| 320 | 0.0281 | 1.106 | 1006 | $17.44 \times 10^{-6}$ | $25.26 \times 10^{-6}$ | 0.69 |
| 330 | 0.0287 | 1.073 | 1006 | $18.37 \times 10^{-6}$ | $26.59 \times 10^{-6}$ | 0.69 |
| 340 | 0.0294 | 1.042 | 1007 | $19.32 \times 10^{-6}$ | $28.02 \times 10^{-6}$ | 0.69 |
| 350 | 0.0300 | 1.012 | 1007 | $20.30 \times 10^{-6}$ | $29.44 \times 10^{-6}$ | 0.69 |
| 360 | 0.0306 | 0.983 | 1007 | $21.30 \times 10^{-6}$ | $30.91 \times 10^{-6}$ | 0.69 |
| 370 | 0.0313 | 0.956 | 1008 | $22.32 \times 10^{-6}$ | $32.48 \times 10^{-6}$ | 0.69 |
| 380 | 0.0319 | 0.931 | 1008 | $23.36 \times 10^{-6}$ | $33.99 \times 10^{-6}$ | 0.69 |
| 390 | 0.0325 | 0.906 | 1009 | $24.42 \times 10^{-6}$ | $35.55 \times 10^{-6}$ | 0.69 |
| 400 | 0.0331 | 0.883 | 1009 | $25.50 \times 10^{-6}$ | $37.15 \times 10^{-6}$ | 0.69 |
| 500 | 0.0395 | 0.706 | 1017 | $37.30 \times 10^{-6}$ | $54.18 \times 10^{-6}$ | 0.69 |
| 600 | 0.0456 | 0.589 | 1038 | $50.50 \times 10^{-6}$ | $73.11 \times 10^{-6}$ | 0.69 |
| 700 | 0.0513 | 0.507 | 1065 | $65.15 \times 10^{-6}$ | $93.16 \times 10^{-6}$ | 0.70 |
| 800 | 0.0564 | 0.442 | 1089 | $81.20 \times 10^{-6}$ | $116.13 \times 10^{-6}$ | 0.70 |
| 900 | 0.0625 | 0.392 | 1111 | $98.60 \times 10^{-6}$ | $141.44 \times 10^{-6}$ | 0.70 |
| 1000 | 0.0672 | 0.354 | 1130 | $117.3 \times 10^{-6}$ | $167.99 \times 10^{-6}$ | 0.70 |
| 1500 | 0.0870 | 0.235 | 1202 | $229.0 \times 10^{-6}$ | $327.82 \times 10^{-6}$ | 0.70 |
| 2000 | 0.1032 | 0.176 | 1244 | $368.0 \times 10^{-6}$ | $524.79 \times 10^{-6}$ | 0.70 |
| 3000 | 0.4860 | 0.114 | 2726 | $841.0 \times 10^{-6}$ | $1564.89 \times 10^{-6}$ | 0.54 |
| | | | ammonia (NH$_3$) | | | |
| 250 | 0.0198 | 0.842 | 2200 | $9.70 \times 10^{-6}$ | $10.69 \times 10^{-6}$ | 0.91 |
| 300 | 0.0246 | 0.703 | 2200 | $14.30 \times 10^{-6}$ | $15.91 \times 10^{-6}$ | 0.90 |
| 400 | 0.0364 | 0.520 | 2270 | $26.60 \times 10^{-6}$ | $30.84 \times 10^{-6}$ | 0.86 |
| 500 | 0.0511 | 0.413 | 2420 | $42.50 \times 10^{-6}$ | $51.13 \times 10^{-6}$ | 0.83 |
| | | | argon (Ar) | | | |
| 150 | 0.0096 | 3.28 | 527 | $3.80 \times 10^{-6}$ | $5.550 \times 10^{-6}$ | 0.68 |
| 250 | 0.0151 | 1.95 | 523 | $10.11 \times 10^{-6}$ | $14.81 \times 10^{-6}$ | 0.68 |
| 300 | 0.0176 | 1.622 | 521 | $14.1 \times 10^{-6}$ | $17.47 \times 10^{-6}$ | 0.68 |
| 400 | 0.0223 | 1.217 | 520 | $23.5 \times 10^{-6}$ | $35.24 \times 10^{-6}$ | 0.67 |
| 500 | 0.0265 | 0.973 | 520 | $34.6 \times 10^{-6}$ | $52.38 \times 10^{-6}$ | 0.66 |
| 600 | 0.0302 | 0.811 | 520 | $47.3 \times 10^{-6}$ | $71.61 \times 10^{-6}$ | 0.66 |
| 800 | 0.0369 | 0.608 | 520 | $76.6 \times 10^{-6}$ | $116.71 \times 10^{-6}$ | 0.66 |
| 1000 | 0.0427 | 0.487 | 520 | $111.2 \times 10^{-6}$ | $168.61 \times 10^{-6}$ | 0.66 |
| 1500 | 0.0551 | 0.324 | 520 | $218.0 \times 10^{-6}$ | $327.04 \times 10^{-6}$ | 0.67 |

Table C.22  *(continued)*

| $T$, K | $k$, W/m-K | $\rho$, kg/m$^3$ | $c_p$, J/kg-K | $\nu$, m$^2$/s | $\alpha$, m$^2$/s | Pr |
|---|---|---|---|---|---|---|
| | | | carbon dioxide (CO$_2$) | | | |
| 250 | 0.01435 | 2.15 | 782 | $5.97 \times 10^{-6}$ | $8.540 \times 10^{-6}$ | 0.70 |
| 300 | 0.01810 | 1.788 | 844 | $8.50 \times 10^{-6}$ | $11.99 \times 10^{-6}$ | 0.71 |
| 400 | 0.0259 | 1.341 | 937 | $14.6 \times 10^{-6}$ | $20.61 \times 10^{-6}$ | 0.71 |
| 500 | 0.0333 | 1.073 | 1011 | $21.9 \times 10^{-6}$ | $30.70 \times 10^{-6}$ | 0.71 |
| 600 | 0.0407 | 0.894 | 1074 | $30.3 \times 10^{-6}$ | $42.39 \times 10^{-6}$ | 0.71 |
| 800 | 0.0544 | 0.671 | 1168 | $49.8 \times 10^{-6}$ | $69.41 \times 10^{-6}$ | 0.72 |
| 1000 | 0.0665 | 0.537 | 1232 | $72.3 \times 10^{-6}$ | $100.52 \times 10^{-6}$ | 0.72 |
| 1500 | 0.0945 | 0.358 | 1329 | $143.8 \times 10^{-6}$ | $198.62 \times 10^{-6}$ | 0.72 |
| 2000 | 0.1176 | 0.268 | 1371 | $231.0 \times 10^{-6}$ | $320.06 \times 10^{-6}$ | 0.72 |
| | | | helium (He) | | | |
| 50 | 0.046 | 0.974 | 5200 | $6.63 \times 10^{-6}$ | $9.080 \times 10^{-6}$ | 0.73 |
| 150 | 0.096 | 0.325 | 5200 | $40.0 \times 10^{-6}$ | $56.80 \times 10^{-6}$ | 0.70 |
| 250 | 0.133 | 0.195 | 5200 | $92.0 \times 10^{-6}$ | $131.06 \times 10^{-6}$ | 0.70 |
| 300 | 0.149 | 0.1624 | 5200 | $124.0 \times 10^{-6}$ | $176.44 \times 10^{-6}$ | 0.70 |
| 400 | 0.178 | 0.1218 | 5200 | $200.0 \times 10^{-6}$ | $281.04 \times 10^{-6}$ | 0.71 |
| 500 | 0.205 | 0.0974 | 5200 | $290.0 \times 10^{-6}$ | $404.75 \times 10^{-6}$ | 0.72 |
| 600 | 0.229 | 0.0812 | 5200 | $390.0 \times 10^{-6}$ | $542.35 \times 10^{-6}$ | 0.72 |
| 800 | 0.273 | 0.0609 | 5200 | $620.0 \times 10^{-6}$ | $862.07 \times 10^{-6}$ | 0.72 |
| 1000 | 0.313 | 0.0487 | 5200 | $890.0 \times 10^{-6}$ | $1235.98 \times 10^{-6}$ | 0.72 |
| | | | hydrogen (H$_2$) | | | |
| 20 | 0.0158 | 1.219 | 10400 | $0.893 \times 10^{-6}$ | $1.250 \times 10^{-6}$ | 0.72 |
| 40 | 0.0302 | 0.6094 | 10300 | $3.38 \times 10^{-6}$ | $4.81 \times 10^{-6}$ | 0.70 |
| 60 | 0.0451 | 0.4062 | 10660 | $7.06 \times 10^{-6}$ | $10.42 \times 10^{-6}$ | 0.68 |
| 80 | 0.0621 | 0.3047 | 11790 | $11.7 \times 10^{-6}$ | $17.29 \times 10^{-6}$ | 0.68 |
| 100 | 0.0805 | 0.2437 | 13320 | $17.3 \times 10^{-6}$ | $24.80 \times 10^{-6}$ | 0.70 |
| 150 | 0.125 | 0.1625 | 16170 | $34.4 \times 10^{-6}$ | $47.57 \times 10^{-6}$ | 0.73 |
| 200 | 0.158 | 0.1219 | 15910 | $55.8 \times 10^{-6}$ | $81.47 \times 10^{-6}$ | 0.68 |
| 250 | 0.181 | 0.0975 | 15250 | $81.1 \times 10^{-6}$ | $121.73 \times 10^{-6}$ | 0.67 |
| 300 | 0.198 | 0.0812 | 14780 | $109.9 \times 10^{-6}$ | $164.98 \times 10^{-6}$ | 0.67 |
| 400 | 0.227 | 0.0609 | 14400 | $177.6 \times 10^{-6}$ | $258.85 \times 10^{-6}$ | 0.69 |
| 500 | 0.259 | 0.0487 | 14530 | $258.1 \times 10^{-6}$ | $366.02 \times 10^{-6}$ | 0.70 |
| 600 | 0.299 | 0.0406 | 14400 | $350.9 \times 10^{-6}$ | $511.43 \times 10^{-6}$ | 0.69 |
| 800 | 0.365 | 0.0305 | 14530 | $572.5 \times 10^{-6}$ | $868.75 \times 10^{-6}$ | 0.66 |
| 1000 | 0.423 | 0.0244 | 14760 | $841.2 \times 10^{-6}$ | $1174.53 \times 10^{-6}$ | 0.72 |
| 1500 | 0.587 | 0.0164 | 16000 | $1560 \times 10^{-6}$ | $2237.04 \times 10^{-6}$ | 0.70 |
| 2000 | 0.751 | 0.0123 | 17050 | $2510 \times 10^{-6}$ | $3581.05 \times 10^{-6}$ | 0.70 |

*(continued)*

Table C.22  *(continued)*

| $T$, K | $k$, W/m-K | $\rho$, kg/m$^3$ | $c_p$, J/kg-K | $\nu$, m$^2$/s | $\alpha$, m$^2$/s | Pr |
|---|---|---|---|---|---|---|
| | | | nitrogen (N$_2$) | | | |
| 150 | 0.0157 | 2.276 | 1050 | $4.53 \times 10^{-6}$ | $6.570 \times 10^{-6}$ | 0.69 |
| 250 | 0.0234 | 1.366 | 1044 | $11.3 \times 10^{-6}$ | $16.41 \times 10^{-6}$ | 0.69 |
| 300 | 0.0267 | 1.138 | 1043 | $15.5 \times 10^{-6}$ | $22.49 \times 10^{-6}$ | 0.69 |
| 400 | 0.0326 | 0.854 | 1047 | $25.2 \times 10^{-6}$ | $36.46 \times 10^{-6}$ | 0.69 |
| 500 | 0.0383 | 0.683 | 1057 | $36.7 \times 10^{-6}$ | $53.05 \times 10^{-6}$ | 0.69 |
| 600 | 0.044 | 0.569 | 1075 | $49.7 \times 10^{-6}$ | $71.93 \times 10^{-6}$ | 0.69 |
| 800 | 0.055 | 0.427 | 1123 | $80.0 \times 10^{-6}$ | $114.70 \times 10^{-6}$ | 0.70 |
| 1000 | 0.066 | 0.341 | 1167 | $115.6 \times 10^{-6}$ | $165.85 \times 10^{-6}$ | 0.70 |
| 1500 | 0.091 | 0.228 | 1244 | $226.0 \times 10^{-6}$ | $320.84 \times 10^{-6}$ | 0.70 |
| 2000 | 0.114 | 0.171 | 1287 | $61.9 \times 10^{-6}$ | $518.00 \times 10^{-6}$ | 0.70 |
| | | | oxygen (O$_2$) | | | |
| 150 | 0.0148 | 2.60 | 890 | $4.39 \times 10^{-6}$ | $6.400 \times 10^{-6}$ | 0.69 |
| 250 | 0.0234 | 1.559 | 910 | $11.4 \times 10^{-6}$ | $16.49 \times 10^{-6}$ | 0.69 |
| 300 | 0.0274 | 1.299 | 920 | $15.8 \times 10^{-6}$ | $22.93 \times 10^{-6}$ | 0.69 |
| 400 | 0.0348 | 0.975 | 945 | $26.1 \times 10^{-6}$ | $37.77 \times 10^{-6}$ | 0.69 |
| 500 | 0.042 | 0.780 | 970 | $38.3 \times 10^{-6}$ | $55.51 \times 10^{-6}$ | 0.69 |
| 600 | 0.049 | 0.650 | 1000 | $52.5 \times 10^{-6}$ | $75.38 \times 10^{-6}$ | 0.69 |
| 800 | 0.062 | 0.487 | 1050 | $84.5 \times 10^{-6}$ | $121.25 \times 10^{-6}$ | 0.70 |
| 1000 | 0.074 | 0.390 | 1085 | $122.0 \times 10^{-6}$ | $174.88 \times 10^{-6}$ | 0.70 |
| 1500 | 0.101 | 0.260 | 1140 | $239 \times 10^{-6}$ | $340.76 \times 10^{-6}$ | 0.70 |
| 2000 | 0.126 | 0.195 | 1180 | $384 \times 10^{-6}$ | $547.59 \times 10^{-6}$ | 0.70 |
| | | | water (vapor) (H$_2$O) | | | |
| 400 | 0.0277 | 0.555 | 1900 | $25.2 \times 10^{-6}$ | $26.27 \times 10^{-6}$ | 0.96 |
| 500 | 0.0365 | 0.441 | 1947 | $40.1 \times 10^{-6}$ | $42.51 \times 10^{-6}$ | 0.94 |
| 600 | 0.046 | 0.366 | 2003 | $58.5 \times 10^{-6}$ | $65.75 \times 10^{-6}$ | 0.93 |
| 800 | 0.066 | 0.275 | 2130 | $102.3 \times 10^{-6}$ | $112.68 \times 10^{-6}$ | 0.91 |
| 1000 | 0.088 | 0.220 | 2267 | $155.8 \times 10^{-6}$ | $176.44 \times 10^{-6}$ | 0.88 |
| 1500 | 0.148 | 0.146 | 2594 | $336.0 \times 10^{-6}$ | $390.79 \times 10^{-6}$ | 0.86 |
| 2000 | 0.206 | 0.109 | 2832 | $575.0 \times 10^{-6}$ | $667.34 \times 10^{-6}$ | 0.86 |

Table C.23. *Thermophysical properties for some liquids (adapted from [8])*

| $T$, K | $k$, W/m-K | $\rho$, kg/m$^3$ | $c_p$, J/kg-K | $\nu$, m$^2$/s | $\alpha$, m$^2$/s | Pr | $\beta$, K$^{-1}$ |
|---|---|---|---|---|---|---|---|
| | | | ammonia (NH$_3$) | | | | |
| 220 | 0.547 | 707.2 | 4462 | $0.444 \times 10^{-6}$ | $0.173 \times 10^{-6}$ | 2.70 | |
| 230 | 0.547 | 695.2 | 4466 | $0.415 \times 10^{-6}$ | $0.177 \times 10^{-6}$ | 2.38 | |
| 240 | 0.548 | 683.0 | 4473 | $0.393 \times 10^{-6}$ | $0.179 \times 10^{-6}$ | 2.17 | |
| 250 | 0.548 | 670.4 | 4499 | $0.383 \times 10^{-6}$ | $0.181 \times 10^{-6}$ | 2.11 | |
| 260 | 0.545 | 657.4 | 4547 | $0.379 \times 10^{-6}$ | $0.182 \times 10^{-6}$ | 2.08 | |
| 270 | 0.541 | 644.1 | 4613 | $0.375 \times 10^{-6}$ | $0.182 \times 10^{-6}$ | 2.06 | |
| 280 | 0.530 | 630.3 | 4690 | $0.370 \times 10^{-6}$ | $0.181 \times 10^{-6}$ | 2.04 | |
| 290 | 0.523 | 616.0 | 4773 | $0.362 \times 10^{-6}$ | $0.179 \times 10^{-6}$ | 2.03 | 0.00245 |
| 300 | 0.511 | 600.9 | 4862 | $0.359 \times 10^{-6}$ | $0.175 \times 10^{-6}$ | 2.01 | |
| 310 | 0.496 | 585.5 | 4966 | $0.346 \times 10^{-6}$ | $0.171 \times 10^{-6}$ | 2.00 | |
| 320 | 0.481 | 569.3 | 5081 | $0.333 \times 10^{-6}$ | $0.167 \times 10^{-6}$ | 1.99 | |
| | | | eutectic calcium chloride solution (29.9% CaCl$_2$) | | | | |
| 220 | 0.398 | 1322 | 2600 | $39.76 \times 10^{-6}$ | $0.116 \times 10^{-6}$ | 343 | |
| 230 | 0.411 | 1317 | 2630 | $28.38 \times 10^{-6}$ | $0.119 \times 10^{-6}$ | 239 | |
| 240 | 0.425 | 1312 | 2650 | $19.52 \times 10^{-6}$ | $0.122 \times 10^{-6}$ | 160 | |
| 250 | 0.440 | 1307 | 2680 | $12.88 \times 10^{-6}$ | $0.126 \times 10^{-6}$ | 109 | |
| 260 | 0.455 | 1302 | 2700 | $8.18 \times 10^{-6}$ | $0.129 \times 10^{-6}$ | 63.7 | |
| 270 | 0.469 | 1298 | 2730 | $5.16 \times 10^{-6}$ | $0.132 \times 10^{-6}$ | 39.2 | |
| 280 | 0.481 | 1293 | 2750 | $3.66 \times 10^{-6}$ | $0.135 \times 10^{-6}$ | 27.1 | |
| 290 | 0.495 | 1288 | 2780 | $2.91 \times 10^{-6}$ | $0.138 \times 10^{-6}$ | 21.7 | |
| 300 | 0.507 | 1283 | 2800 | $2.41 \times 10^{-6}$ | $0.141 \times 10^{-6}$ | 17.1 | |
| 310 | 0.519 | 1279 | 2830 | $2.03 \times 10^{-6}$ | $0.144 \times 10^{-6}$ | 14.1 | |
| 320 | 0.531 | 1274 | 2860 | $1.73 \times 10^{-6}$ | $0.146 \times 10^{-6}$ | 11.9 | |
| | | | carbon dioxide (CO$_2$) | | | | |
| 220 | 0.0809 | 1173 | 1800 | $0.119 \times 10^{-6}$ | $37.84 \times 10^{-9}$ | 3.11 | |
| 230 | 0.0964 | 1118 | 1900 | $0.118 \times 10^{-6}$ | $45.73 \times 10^{-9}$ | 2.61 | |
| 240 | 0.109 | 1084 | 2000 | $0.117 \times 10^{-6}$ | $51.33 \times 10^{-9}$ | 2.29 | |
| 250 | 0.114 | 1046 | 2100 | $0.115 \times 10^{-6}$ | $53.93 \times 10^{-9}$ | 2.15 | |
| 260 | 0.112 | 997.8 | 2200 | $0.113 \times 10^{-6}$ | $52.27 \times 10^{-9}$ | 2.18 | |
| 270 | 0.107 | 944.5 | 2400 | $0.110 \times 10^{-6}$ | $47.45 \times 10^{-9}$ | 2.33 | |
| 280 | 0.0995 | 880.4 | 2900 | $0.103 \times 10^{-6}$ | $38.99 \times 10^{-9}$ | 2.67 | |
| 290 | 0.0902 | 798.8 | 4100 | $0.0940 \times 10^{-6}$ | $26.36 \times 10^{-9}$ | 3,71 | 0.014 |
| 300 | 0.0754 | 650.2 | 27000 | $0.0841 \times 10^{-6}$ | $8.610 \times 10^{-9}$ | 21.3 | |

*(continued)*

Table C.23  *(continued)*

| $T$, K | $k$, W/m-K | $\rho$, kg/m$^3$ | $c_p$, J/kg-K | $\nu$, m$^2$/s | $\alpha$, m$^2$/s | Pr | $\beta$, K$^{-1}$ |
|---|---|---|---|---|---|---|---|
| | | | | engine oil (unused) | | | |
| 270 | 0.147 | 900.6 | 1790 | $4.79 \times 10^{-3}$ | $91.7 \times 10^{-9}$ | 52600 | |
| 290 | 0.145 | 889.7 | 1870 | $1.47 \times 10^{-3}$ | $87.8 \times 10^{-9}$ | 15900 | 0.0007 |
| 310 | 0.145 | 877.8 | 1950 | $417 \times 10^{-6}$ | $87.0 \times 10^{-9}$ | 4000 | |
| 330 | 0.141 | 865.8 | 2040 | $108 \times 10^{-6}$ | $80.5 \times 10^{-9}$ | 1320 | |
| 350 | 0.138 | 8553.7 | 2120 | $44.5 \times 10^{-6}$ | $77.4 \times 10^{-9}$ | 574 | |
| 370 | 0.137 | 841.7 | 2210 | $22.9 \times 10^{-6}$ | $74.3 \times 10^{-9}$ | 308 | |
| 390 | 0.135 | 830.6 | 2300 | $13.6 \times 10^{-6}$ | $71.6 \times 10^{-9}$ | 190 | |
| 410 | 0.133 | 818.7 | 2380 | $8.65 \times 10^{-6}$ | $69.0 \times 10^{-9}$ | 125 | |
| 430 | 0.132 | 807.5 | 2470 | $5.93 \times 10^{-6}$ | $66.6 \times 10^{-9}$ | 89 | |
| | | | | ethylene glycol [$C_2H_4(OH)_2$] | | | |
| 270 | 0.241 | 1133 | 2280 | $63.27 \times 10^{-6}$ | $93.3 \times 10^{-9}$ | 677 | |
| 290 | 0.248 | 1119 | 2370 | $25.92 \times 10^{-6}$ | $93.8 \times 10^{-9}$ | 276 | 0.000626 |
| 310 | 0.255 | 1103 | 2460 | $10.26 \times 10^{-6}$ | $93.9 \times 10^{-9}$ | 110 | |
| 330 | 0.259 | 1090 | 2550 | $5.34 \times 10^{-6}$ | $93.3 \times 10^{-9}$ | 57.3 | |
| 350 | 0.261 | 1079 | 2640 | $3.25 \times 10^{-6}$ | $92.3 \times 10^{-9}$ | 35.2 | |
| 370 | 0.263 | 1061 | 2730 | $2.17 \times 10^{-6}$ | $91.0 \times 10^{-9}$ | 23.9 | |
| | | | | glycerine [$C_3H_5(OH)_3$] | | | |
| 270 | 0.281 | 1278 | 2240 | $9.90 \times 10^{-3}$ | $98.8 \times 10^{-9}$ | $101 \times 10^3$ | |
| 280 | 0.283 | 1272 | 2300 | $4.59 \times 10^{-3}$ | $97.0 \times 10^{-9}$ | $47.1 \times 10^3$ | |
| 290 | 0.285 | 1266 | 2370 | $1.73 \times 10^{-3}$ | $95.2 \times 10^{-9}$ | $18.0 \times 10^3$ | 0.0005 |
| 300 | 0.286 | 1260 | 2425 | $0.705 \times 10^{-3}$ | $93.4 \times 10^{-9}$ | $7.52 \times 10^3$ | |
| 310 | 0.286 | 1254 | 2490 | $0.307 \times 10^{-3}$ | $91.9 \times 10^{-9}$ | $3.33 \times 10^3$ | |
| 320 | 0.287 | 1247 | 2560 | $0.171 \times 10^{-3}$ | $89.9 \times 10^{-9}$ | $1.88 \times 10^3$ | |
| | | | | mercury (Hg) | | | |
| 270 | 8.13 | 13634 | 140 | $126 \times 10^{-9}$ | $4.25 \times 10^{-6}$ | 0.0294 | |
| 290 | 8.62 | 13585 | 139 | $116 \times 10^{-9}$ | $4.56 \times 10^{-6}$ | 0.0255 | 0.000181 |
| 320 | 9.33 | 13512 | 139 | $105 \times 10^{-9}$ | $4.98 \times 10^{-6}$ | 0.0211 | |
| 370 | 10.4 | 13390 | 137 | $93.5 \times 10^{-9}$ | $5.68 \times 10^{-6}$ | 0.0165 | |
| 420 | 11.4 | 13270 | 136 | $85.8 \times 10^{-9}$ | $6.31 \times 10^{-6}$ | 0.0136 | |
| 470 | 12.3 | 13151 | 156 | $80.5 \times 10^{-9}$ | $6.88 \times 10^{-6}$ | 0.0117 | |
| 520 | 13.1 | 13031 | 137 | $76.7 \times 10^{-9}$ | $7.38 \times 10^{-6}$ | 0.0104 | |
| 580 | 14.0 | 12810 | 134 | $67.7 \times 10^{-9}$ | $8.12 \times 10^{-6}$ | 0.0084 | |

Table C.23 *(continued)*

| $T$, K | $k$, W/m-K | $\rho$, kg/m$^3$ | $c_p$, J/kg-K | $\nu$, m$^2$/s | $\alpha$, m$^2$/s | Pr | $\beta$, K$^{-1}$ |
|---|---|---|---|---|---|---|---|
| | | | methyl chloride (CH$_3$Cl) | | | | |
| 220 | 0.217 | 1058 | 1474 | $321 \times 10^{-9}$ | $138 \times 10^{-9}$ | 2.31 | |
| 230 | 0.211 | 1039 | 1480 | $319 \times 10^{-9}$ | $138 \times 10^{-9}$ | 2.32 | |
| 240 | 0.205 | 1021 | 1489 | $315 \times 10^{-9}$ | $135 \times 10^{-9}$ | 2.34 | |
| 250 | 0.198 | 1004 | 1499 | $310 \times 10^{-9}$ | $131 \times 10^{-9}$ | 2.37 | |
| 260 | 0.190 | 986.8 | 1514 | $307 \times 10^{-9}$ | $127 \times 10^{-9}$ | 2.41 | |
| 270 | 0.181 | 968.0 | 1532 | $303 \times 10^{-9}$ | $123 \times 10^{-9}$ | 2.47 | |
| 280 | 0.173 | 948.3 | 1553 | $299 \times 10^{-9}$ | $118 \times 10^{-9}$ | 2.53 | |
| 290 | 0.166 | 928.9 | 1578 | $294 \times 10^{-9}$ | $113 \times 10^{-9}$ | 2.61 | |
| 300 | 0.157 | 914.2 | 1607 | $290 \times 10^{-9}$ | $108 \times 10^{-9}$ | 2.69 | 0.000126 |
| 310 | 0.147 | 891.2 | 1640 | $283 \times 10^{-9}$ | $102 \times 10^{-9}$ | 2.80 | |
| 320 | 0.136 | 867.7 | 1677 | $276 \times 10^{-9}$ | $94.4 \times 10^{-9}$ | 2.93 | |
| | | | Refrigerant-12 | | | | |
| 220 | 0.068 | 1555 | 872.1 | $319 \times 10^{-9}$ | $49.7 \times 10^{-9}$ | 6.4 | 0.0828 |
| 230 | 0.069 | 1527 | 881.7 | $288 \times 10^{-9}$ | $51.0 \times 10^{-9}$ | 5.6 | |
| 240 | 0.069 | 1498 | 892.2 | $261 \times 10^{-9}$ | $52.2 \times 10^{-9}$ | 5.0 | |
| 250 | 0.070 | 1469 | 903.7 | $240 \times 10^{-9}$ | $53.5 \times 10^{-9}$ | 4.5 | |
| 260 | 0.072 | 1438 | 916.3 | $225 \times 10^{-9}$ | $54.7 \times 10^{-9}$ | 4.1 | |
| 270 | 0.073 | 1407 | 930.1 | $216 \times 10^{-9}$ | $55.5 \times 10^{-9}$ | 3.9 | |
| 280 | 0.073 | 1374 | 945.0 | $206 \times 10^{-9}$ | $55.9 \times 10^{-9}$ | 3.7 | |
| 290 | 0.073 | 1340 | 960.9 | $200 \times 10^{-9}$ | $56.0 \times 10^{-9}$ | 3.5 | |
| 300 | 0.072 | 1306 | 946.6 | $195 \times 10^{-9}$ | $56.0 \times 10^{-9}$ | 3.5 | |
| 310 | 0.070 | 1268 | 982.9 | $192 \times 10^{-9}$ | $55.7 \times 10^{-9}$ | 3.5 | |
| 320 | 0.068 | 1228 | 1016 | $190 \times 10^{-9}$ | $54.7 \times 10^{-9}$ | 3.5 | |
| | | | sulfur dioxide (SO$_2$) | | | | |
| 220 | 0.244 | 1568 | 1358 | $502 \times 10^{-9}$ | $114 \times 10^{-9}$ | 4.39 | |
| 230 | 0.237 | 1544 | 1360 | $442 \times 10^{-9}$ | $113 \times 10^{-9}$ | 3.89 | |
| 240 | 0.232 | 1520 | 1361 | $387 \times 10^{-9}$ | $112 \times 10^{-9}$ | 3.44 | |
| 250 | 0.227 | 1495 | 1362 | $338 \times 10^{-9}$ | $111 \times 10^{-9}$ | 3.04 | |
| 260 | 0.220 | 1471 | 1363 | $299 \times 10^{-9}$ | $110 \times 10^{-9}$ | 2.71 | |
| 270 | 0.213 | 1446 | 1364 | $266 \times 10^{-9}$ | $109 \times 10^{-9}$ | 2.45 | |
| 280 | 0.206 | 1420 | 1364 | $240 \times 10^{-9}$ | $107 \times 10^{-9}$ | 2.24 | |
| 290 | 0.201 | 1394 | 1365 | $217 \times 10^{-9}$ | $106 \times 10^{-9}$ | 2.05 | 0.00194 |
| 300 | 0.194 | 1367 | 1366 | $196 \times 10^{-9}$ | $104 \times 10^{-9}$ | 1.88 | |
| 310 | 0.187 | 1338 | 1367 | $178 \times 10^{-9}$ | $102 \times 10^{-9}$ | 1.74 | |
| 320 | 0.179 | 1308 | 1368 | $165 \times 10^{-9}$ | $101 \times 10^{-9}$ | 1.64 | |

*(continued)*

Table C.23  *(continued)*

| $T$, K | $k$, W/m-K | $\rho$, kg/m$^3$ | $c_p$, J/kg-K | $\nu$, m$^2$/s | $\alpha$, m$^2$/s | Pr | $\beta$, K$^{-1}$ |
|---|---|---|---|---|---|---|---|
| | | | water ($H_2O$) | | | | |
| 270 | 0.545 | 1002 | 4222 | $191 \times 10^{-8}$ | $129 \times 10^{-9}$ | 14.7 | |
| 290 | 0.590 | 1000 | 4186 | $113 \times 10^{-8}$ | $141 \times 10^{-9}$ | 8.02 | 0.000203 |
| 310 | 0.623 | 995.3 | 4178 | $711 \times 10^{-9}$ | $150 \times 10^{-9}$ | 4.74 | 0.000273 |
| 330 | 0.648 | 986.8 | 4183 | $505 \times 10^{-9}$ | $154 \times 10^{-9}$ | 3.22 | |
| 350 | 0.665 | 975.7 | 4194 | $381 \times 10^{-9}$ | $163 \times 10^{-9}$ | 2.34 | |
| 370 | 0.678 | 962.5 | 4213 | $305 \times 10^{-9}$ | $167 \times 10^{-9}$ | 1.81 | |
| 390 | 0.684 | 947.5 | 4244 | $254 \times 10^{-9}$ | $171 \times 10^{-9}$ | 1.490 | |
| 410 | 0.684 | 930.8 | 4278 | $219 \times 10^{-9}$ | $172 \times 10^{-9}$ | 1.245 | |
| 430 | 0.681 | 912.4 | 4332 | $194 \times 10^{-9}$ | $173 \times 10^{-9}$ | 1.120 | |
| 450 | 0.676 | 892.0 | 4406 | $176 \times 10^{-9}$ | $172 \times 10^{-9}$ | 1.018 | |
| 470 | 0.667 | 870.0 | 4492 | $162 \times 10^{-9}$ | $171 \times 10^{-9}$ | 0.947 | |
| 490 | 0.655 | 846.0 | 4593 | $152 \times 10^{-9}$ | $168 \times 10^{-9}$ | 0.898 | |
| 510 | 0.638 | 819.6 | 4734 | $144 \times 10^{-9}$ | $165 \times 10^{-9}$ | 0.874 | |
| 530 | 0.615 | 790.3 | 4919 | $138 \times 10^{-9}$ | $159 \times 10^{-9}$ | 0.874 | |
| 550 | 0.585 | 757.5 | 5169 | $135 \times 10^{-9}$ | $150 \times 10^{-9}$ | 0.905 | |
| 570 | 0.546 | 719.9 | 5649 | $135 \times 10^{-9}$ | $134 \times 10^{-9}$ | 1.003 | |

Table C.24. *Thermophysical properties for some liquid metals (adapted from [7, 11])*

| Substance | $T$, K | $\rho_l$, kg/m$^3$ | $c_{p,l}$, kJ/kg-K | $\nu_l$, m$^2$/s | $k_l$, W/m-K | $\alpha_l$, m$^2$/s | Pr$_l$ |
|---|---|---|---|---|---|---|---|
| Bi | 589 | 10011 | 0.1444 | $1.617 \times 10^{-7}$ | 16.4 | $0.138 \times 10^{-5}$ | 0.0142 |
| $T_{sl} = 544$ K | 811 | 9739 | 0.1545 | $1.133 \times 10^{-7}$ | 15.6 | $1.035 \times 10^{-5}$ | 0.0110 |
| | 1033 | 9467 | 0.1645 | $0.8343 \times 10^{-7}$ | 15.6 | $1.001 \times 10^{-5}$ | 0.0083 |
| Hg | 270 | 13634 | 0.140 | $126 \times 10^{-9}$ | 8.13 | $4.25 \times 10^{-6}$ | 0.0294 |
| $T_{sl} = 234$ K | 420 | 13270 | 0.130 | $85.8 \times 10^{-9}$ | 11.4 | $6.61 \times 10^{-6}$ | 0.0136 |
| | 580 | 12820 | 0.134 | $67.7 \times 10^{-9}$ | 14.0 | $8.12 \times 10^{-6}$ | 0.0084 |
| Pb | 644 | 10540 | 0.159 | $2.276 \times 10^{-7}$ | 16.1 | $1.084 \times 10^{-5}$ | 0.024 |
| $T_{sl} = 600$ K | 755 | 10412 | 0.155 | $1.849 \times 10^{-7}$ | 15.6 | $1.223 \times 10^{-5}$ | 0.017 |
| | 977 | 10140 | | $1.347 \times 10^{-7}$ | 14.9 | | |
| K | 422 | 807.3 | 0.80 | $4.608 \times 10^{-7}$ | 45.0 | $6.99 \times 10^{-5}$ | 0.0066 |
| $T_{sl} = 337$ K | 700 | 741.7 | 0.75 | $2.397 \times 10^{-7}$ | 39.5 | $7.07 \times 10^{-5}$ | 0.0034 |
| | 977 | 674.4 | 0.75 | $1.905 \times 10^{-7}$ | 33.1 | $6.55 \times 10^{-5}$ | 0.0029 |
| Li | 500 | 514 | 4.340 | $1.033 \times 10^{-6}$ | 43.7 | $1.949 \times 10^{-5}$ | 0.053 |
| $T_{sl} = 453$ K | 700 | 493 | 4.190 | $0.726 \times 10^{-6}$ | 48.4 | $2.342 \times 10^{-5}$ | 0.031 |
| | 900 | 473 | 4.160 | $0.522 \times 10^{-6}$ | 55.9 | $2.900 \times 10^{-5}$ | 0.018 |
| Na | 366 | 929.1 | 1.38 | $7.516 \times 10^{-7}$ | 86.2 | $6.71 \times 10^{-5}$ | 0.011 |
| $T_{sl} = 371$ K | 644 | 860.2 | 1.30 | $3.270 \times 10^{-7}$ | 72.3 | $6.48 \times 10^{-5}$ | 0.0051 |
| | 977 | 778.5 | 1.26 | $2.285 \times 10^{-7}$ | 59.7 | $6.12 \times 10^{-5}$ | 0.0037 |
| NaK | 366 | 887.4 | 1.130 | $6.522 \times 10^{-7}$ | 25.6 | $2.552 \times 10^{-5}$ | 0.026 |
| (45/55) | 644 | 821.7 | 1.055 | $2.871 \times 10^{-7}$ | 27.5 | $3.17 \times 10^{-5}$ | 0.0091 |
| $T_{sl} = 292$ K | 977 | 740.1 | 1.043 | $2.174 \times 10^{-7}$ | 28.9 | $3.74 \times 10^{-5}$ | 0.0058 |
| NaK | 366 | 849.0 | 0.946 | $5.797 \times 10^{-7}$ | 24.4 | $3.05 \times 10^{-5}$ | 0.019 |
| (22/78) | 672 | 775.3 | 0.879 | $2.666 \times 10^{-7}$ | 26.7 | $3.92 \times 10^{-5}$ | 0.0068 |
| $T_{sl} = 262$ K | 1033 | 690.4 | 0.883 | $2.118 \times 10^{-7}$ | | | |
| PbBi | 422 | 10524 | 0.147 | | 9.05 | $0.586 \times 10^{-5}$ | |
| (44.5/55.5) | 644 | 10236 | 0.147 | $1.496 \times 10^{-7}$ | 11.86 | $0.790 \times 10^{-5}$ | 0.0189 |
| $T_{sl} = 398$ K | 922 | 9835 | | $1.171 \times 10^{-7}$ | | | |

## Liquid-Gas Surface Tension

Table C.25. *Surface tension for some organic (and water as a reference) liquids with their vapors at various temperatures, $\sigma(N/m)$ (adapted from [8])\**

## Saturated Liquid-Vapor Properties

Table C.26. *Thermophysical properties for some saturated fluids (at one atm pressure) (adapted from [3])\**

Table C.27. *Thermophysical properties for saturated water\**

Table C.28. *Thermophysical properties for saturated Refrigerant 134a\**

Table C.29. *Thermophysical properties for saturated sodium\**

## C.2 References**

* This table is found on the Web site www.cambridge.org/kaviany.
** References are found on the Web site www.cambridge.org/kaviany.

# Subject Index

A man travels the world in search of what he needs and returns home to find it.

– G. Moore